Galactic Dynamics

James Binney and Scott Tremaine

GALACTIC DYNAMICS

PRINCETON UNIVERSITY PRESS
Princeton, New Jersey

Copyright © 1987 by Princeton University Press

Published by Princeton University Press, 41 William Street,
Princeton, New Jersey 08540
In the United Kingdom: Princeton University Press,
Guildford, Surrey

All Rights Reserved

Library of Congress Cataloging in Publication Data will be
found on the last printed page of this book

ISBN 0-691-08444-0 (cloth)
　　　 0-691-08445-9 (pbk.)

This book has been composed in Times Roman

Clothbound editions of Princeton University Press books
are printed on acid-free paper, and binding materials are
chosen for strength and durability. Paperbacks, although
satisfactory for personal collections, are not usually
suitable for library rebinding

Printed in the United States of America by
Princeton University Press,
Princeton, New Jersey

Contents

3 The Orbits of Stars   103

4 Equilibria of Collisionless Systems   187

Contents

Preface

Always majestic, usually spectacularly beautiful, galaxies are the fundamental building blocks of the Universe. The inquiring mind cannot help asking how they formed, how they function, and what will become of them in the distant future. The principal tool used in answering these questions is stellar dynamics, the study of the motion of large numbers of point masses orbiting under the influence of their mutual self-gravity. The main aim of this book is to provide the reader with a background in stellar dynamics at the level required to carry out research in the field of galactic structure.

Although galaxies are important in their own right, they also offer the prospect of providing clues to new laws of physics—for example, the formation of galaxies is intimately related to the properties of the early Universe; the dark matter which we believe exists in galaxies may be composed of some unknown species of elementary particle; and galaxies have frequently been used as enormous laboratories to study the laws of physics in extreme conditions. In addition, the study of galaxies offers the physicist an opportunity to apply many of the powerful tools of theoretical physics which have been developed in other fields: classical, celestial, and modern Hamiltonian mechanics, fluid mechanics, statistical mechanics, and plasma physics provide the most relevant backgrounds, and although there is little need for quantum mechanics, the mathematical techniques developed in an introductory quantum mechanics course are in constant use.

Despite the breadth of the background that is drawn upon, the study of galactic dynamics will carry the physicist to the frontiers of knowledge faster than almost any other branch of theoretical physics, in part because the fundamental issues in the subject are easy to understand for anyone with an undergraduate training in physics, in part because theorists are scrambling to keep pace with the flood of new observations—a flood that will dramatically increase when the Hubble Space Telescope is launched and the next generation of large Earth-based telescopes becomes available—and, what is perhaps most important, because the theoretical effort that has been devoted to the study of galactic structure over the last few decades has been much smaller than that given to many other fields of comparable interest and importance.

The view adopted in this book is that galactic dynamics is a branch of theoretical physics. Accordingly, the book has been designed for readers with a standard undergraduate preparation in physics. By contrast, we have assumed no background in astronomy (although the reader who has had an introductory astronomy course should certainly benefit from it). The relevant observational facts about galaxies are summarized in the introductory chapter.

One of the pleasures of writing a textbook in a rapidly developing field is the opportunity to advertise one's own views on the proper organization and relative importance of various topics and areas. This is especially true in a field like galactic structure, where only a handful of influential textbooks have been written in the last half-century. The classic texts, Chandrasekhar's *Principles of Stellar Dynamics* (1942) and Ogorodnikov's *Dynamics of Stellar Systems* (1965), are now so out of date as to be principally of historical interest. The next major contribution was Dimitri Mihalas's undergraduate textbook *Galactic Astronomy* (1968), which is now out of print. The present book grew out of efforts made by J. P. Ostriker around 1975 to encourage the revision and expansion of Mihalas's book into a two-volume graduate text, with the first volume devoted to observations and the second to theory. The first (observational) volume was again titled *Galactic Astronomy*, and was published by Mihalas and Binney in 1981. The present volume is devoted to theory rather than observation, and is intended to serve as a companion and complement to *Galactic Astronomy*.

All of the chapters except the Introduction contain sets of problems at the end. Many of the problems are intended to elucidate topics that are not fully covered in the main text. The degree of difficulty is indicated by a number in square brackets at the start of each problem, ranging from [1] (easy) to [3] (difficult). Six chapters are accompanied by one or more appendixes. These will be found at the back of the book; Appendix 5.B, for example, is the second appendix related to the material of Chapter 5.

One vexing issue in astrophysical notation is how to indicate approximate equality. We use "$=$" to denote equality to several significant digits, "$\approx$" for equality to order of magnitude, and "$\simeq$" for everything in between. The ends of proofs are indicated by a sideways triangle, "$\triangleleft$".

Although we have tried to keep jargon to a minimum, we were unable to resist the economy of a few abbreviations: "distribution function" is written throughout the book as "DF", "initial mass function" as "IMF", and "root mean square" as "RMS".

We are deeply indebted to many colleagues for their patient and enthusiastic support during the preparation of this book: John Bahcall, Mordehai Milgrom, J. P. Ostriker, Martin Rees, and Hugo van Woerden arranged extended visits to the Institute for Advanced Study, Princeton, to the Weizmann Institute, Rehovot, to Princeton University Observatory, to the Institute of Astronomy, Cambridge, and to the Kapteyn Laboratory, Groningen, during which much of this book was written; Sandra Faber, Ken Freeman, Ortwin Gerhard, Douglas Heggie, Steve Kent, Blane Little, Dimitri Mihalas, Lyman Spitzer, and Simon White provided many helpful comments on one or more draft chapters; Alar Toomre gave us detailed and constructive criticism of the entire manuscript, as did the participants in the Institute for Advanced Study "shotgun" seminar organized by Nick Kylafis and Stefano Casertano. Other colleagues have made special efforts to provide us with data and original prints and drawings for our illustrations: David Malin, Allan Sandage, and François Schweizer filled our shopping list for more than two dozen photographs of galaxies, clusters, and other objects; Alar Toomre contributed several exquisitely crafted illustrations; Tjeerd van Albada replotted an N-body simulation for our benefit; and Sverre Aarseth furnished a special compact version of one of his N-body codes for Appendix 4.B. We owe a special debt to Dimitri Mihalas for his generous contributions of time and advice in the early stages of this project.

Finally, we thank J. P. Ostriker for providing the initial impetus for this project, for helping to determine the organization and structure of the book, for clarifying and improving many of the arguments, for serving as our champion at Princeton University Press, and, most important, for teaching us much—perhaps most—of what we know of stellar dynamics.

Galactic Dynamics

1 Introduction

A **stellar system** is a gravitationally bound assemblage of stars. Stellar systems vary over some fourteen orders of magnitude in size and mass, from binary stars, to star clusters containing 10^2 to 10^6 stars, through galaxies containing 10^{10} to 10^{12} stars, to vast clusters containing thousands of galaxies.

The behavior of these systems is determined by Newton's laws of motion and Newton's law of gravity, and the study of the dynamics of stellar systems is the branch of theoretical physics called **stellar dynamics**. (As yet, there is no direct evidence for stellar systems in which special or general relativistic effects are important, although such systems may well be present at the centers of some galaxies.) Stellar dynamics is directly related to at least three other areas of theoretical physics. Superficially, it is closest to celestial mechanics, since both involve the study of orbits in a gravitational potential; however, much of the mathematical formalism of celestial mechanics is of little use in stellar dynamics, since it is based on perturbation expansions in powers of mass, eccentricity, and inclination, which do not converge when applied to most stellar systems. The most fundamental connections of stellar dynamics are with classical statistical mechanics, since the number of stars in a star cluster or galaxy is so large that a statistical treatment of the dynamics is necessary. Finally, many of the mathematical tools that have been developed to study stellar systems are borrowed from plasma physics, which also involves the study of large assemblies of particles interacting via long-range forces.

For an initial orientation to the physics of stellar systems, it is useful to summarize a few orders of magnitude. Our Sun is located in a stellar system called the **Milky Way** or simply **the Galaxy**. The Galaxy contains about 10^{11} visible stars, as well as 10^{10} solar masses of gas (written $10^{10}\,\mathrm{M_\odot}$; $1\,\mathrm{M_\odot} = 1.99 \times 10^{33}$ g),[1] distributed in tens of thousands of gas clouds with a wide range of masses and sizes. Since the gas contains only a small fraction of the total mass, it has little direct influence on the dynamics; however, the dense gas clouds are the sites of new star formation and thereby play an important role in the long-term evolution of the Galaxy.

---

[1] See Appendix 1.A for a tabulation of useful constants.

3

Most of the stars in the Galaxy travel on nearly circular orbits in a thin disk whose radius is of order 10 kiloparsecs (1 kiloparsec $\equiv$ 1 kpc $\equiv$ 3.086×10^{21} cm), and thickness of order 1 kpc. The typical circular speed of a star in the disk is about $200\,\mathrm{km\,s^{-1}}$, so that the time required to complete one orbit at 10 kpc is $3 \times 10^8\,\mathrm{yr}$. The dispersion in the velocities of stars at a given position is about $40\,\mathrm{km\,s^{-1}}$. Since the age of the Galaxy is about $10^{10}\,\mathrm{yr} \equiv 10\,\mathrm{Gyr}$, a typical disk star has completed over thirty revolutions, and hence it is reasonable to assume that the Galaxy is approximately in a steady state at the present time. The steady-state approximation is important because it permits us to decouple the questions of the present-day *equilibrium* and *structure* of the Galaxy from the far thornier issue of the *formation* of the Galaxy; at present there is no single widely accepted theory of galaxy formation, and so this complex but important issue will be only touched on in the present book.

Since the orbital period of stars near the Sun is a factor 10^5 or 10^6 times longer than the interval over which we have recorded observations, we are forced to base our entire understanding of galactic structure on what amounts to an instantaneous snapshot of the system. To a limited extent, the snapshot can be supplemented by measurements of the angular velocities (or **proper motions**) of stars that are close enough so that their position on the sky has changed noticeably over the last hundred years or so; and by **line-of-sight velocities** of stars measured from Doppler shifts in their spectra. Thus the *positions* and *velocities* of some stars can be determined, but their *accelerations* are completely inaccessible to current observational techniques.

Using these rough estimates for the dimensions of the Galaxy, we can calculate the typical mean free path of a star before it collides with another star. For an assembly of particles moving on straight-line orbits, the mean free path is $\lambda = 1/(n\sigma)$, where n is the number density and σ is the cross-section. Let us make the crude assumption that all stars are like the Sun so that the cross-section for collision is $\sigma = \pi(2\,\mathrm{R_\odot})^2$, where $\mathrm{R_\odot} = 6.96 \times 10^{10}$ cm is the solar radius.[2] If we spread 10^{11} stars uniformly over a disk of radius 10 kpc and thickness 1 kpc, then the number density of stars in the disk is $0.3\,\mathrm{pc^{-3}}$ (1 parsec $\equiv$ 1 pc $\equiv 3.086 \times 10^{18}$ cm). The mean free path is then $\lambda = 1.5 \times 10^{33}\,\mathrm{cm} = 5 \times 10^{14}\,\mathrm{pc}$. The interval between collisions is approximately λ/v, where v is the random velocity of stars at a given location. For $v = 40\,\mathrm{km\,s^{-1}}$ we find the collision interval to be about $10^{19}\,\mathrm{yr}$, a factor of 10^9 longer than the

---

[2] This calculation neglects the enhancement in collision cross-section due to the gravitational attraction of the passing stars, but this only enhances the collision rate by a factor of 100 or so, and hence will not affect our conclusion. See §8.4.5 for more detail.

age of the Galaxy. Evidently, collisions between stars are so rare that they are of absolutely no importance to the dynamics of the Galaxy, and for most purposes the stars can be approximated as mass points.[3]

Since direct collisions are rare, each star's motion is determined solely by the gravitational attraction of all the other stars in the galaxy. A useful first approximation for the gravitational field in the galaxy is obtained by imagining that the mass of the galaxy is continuously distributed, rather than concentrated into discrete mass points. Thus we begin in Chapter 2 with a description of Newtonian potential theory, developing the tools to describe the smoothed gravitational fields of stellar systems. In Chapter 3 we investigate the orbits of stars in given gravitational potentials, providing both quantitative and qualitative descriptions of the behavior of stellar orbits in a variety of force fields. In Chapter 4 we consider how to close the loop by finding pairs of potentials and orbit distributions that are **self-consistent** in the sense that the combined mass of the stars provides the density distribution that generates the required potential.

The models constructed in Chapter 4 are **stationary** as well as self-consistent, that is, the density at each point is constant in time because the arrivals and departures of stars exactly balance in every volume element. Clearly this is the kind of model that is needed to describe a galaxy that is many revolutions old and hence presumably in a steady state. However, simply satisfying the condition of stationarity is inadequate, since many stationary systems are unstable; in such a system the smallest deviation from the mathematical model considered will cause the system to evolve away from the model to some quite different configuration. Hence Chapter 5 is devoted to a discussion of the theory of stability of stellar systems.

In Chapter 6 we describe some of the fascinating phenomena that are peculiar to galactic disks. These include the beautiful spiral patterns that are usually seen in disk galaxies, especially in the young, bright stars and gas; the curious barlike structures seen at the centers of about half of all disk galaxies; and the warps that are present in many spiral galaxy disks (including the disk of our own Galaxy).

Chapter 7 is devoted to the interplay between stellar systems. We describe the physics of collisions between galaxies or clusters, and the influence of the surrounding galaxy on the evolution of a small star cluster orbiting within it, through such processes as tidal stripping, dynamical friction, and shock heating. We also study the effect of irregularities

---

[3] The rarity of encounters is of course very fortunate, since the passage of a star within even 10^{14} cm of the Sun would cause major perturbations to the Earth's orbit and hence have disastrous consequences for life on Earth.

in the galactic gravitational field—generated, for example, by molecular clouds or spiral arms—on the orbits of disk stars.

Even though stellar collisions are extremely rare, the gravitational tugs of stars passing nearby will continually randomize the orbit of any star by a series of small kicks. This process, which is somewhat similar to Brownian motion, is the mechanism—analogous to collisions of molecules in a gas—by which the members of a stellar system gradually approach equipartition of energy. It operates so slowly that it can be neglected in galaxies. However, the cumulative effects of many of these random gravitational kicks have largely determined the evolution and present form of many star clusters. Chapter 8 describes the kinetic theory of stellar systems, that is, the analysis of the evolution of stellar systems toward thermodynamic equilibrium as a result of stellar encounters. The results described here can be directly applied to observations of globular clusters in our Galaxy, and also have implications for the evolution of clusters of galaxies and galactic nuclei. We also investigate the interesting behavior of hypothetical stellar systems that are so dense that collisions and near-misses between stars occur frequently.

Chapter 9 lies somewhat outside the mainstream of the book, since it is concerned with chemical rather than dynamical evolution of galaxies. It has been included because the interplay between chemical and dynamical evolution is observationally important for studies of our Galaxy and of galaxies at large cosmological redshifts. In our own Galaxy, the inexorable pollution of the interstellar gas by the debris of exploding stars enables us to estimate the ages of stars, and thus to gain information about the evolution of their orbits. Similarly, chemical evolution of distant galaxies changes their luminosities and colors, and it is important to understand these changes if we wish to test theories of the dynamical evolution of galaxies.

The final chapter is devoted to what is perhaps the most important unresolved puzzle in extragalactic astronomy—the nature, origin, and distribution of matter in the Universe. We shall show that the techniques of stellar dynamics developed in previous chapters lead to the surprising conclusion that most of the mass in galaxies is in some invisible form. Dark matter may comprise over 90% of the matter in the Universe. The composition of dark matter is completely unknown; neutrinos, supermassive black holes, and low mass stars are just three of the wide variety of candidates that have been proposed. There is no known way to detect this material except through its gravitational effects, and in this sense galactic dynamics is so far the only tool with which astronomers can study the properties of this major constituent of the Universe.

The reader is assumed to be familiar with classical mechanics at the level of Goldstein (1980) or Landau and Lifshitz (1976), and with mathematical physics at the level of Arfken (1970), Margenau and Murphy (1956), or Mathews and Walker (1970). In Chapters 5 and 6 we will also draw on the basic equations of fluid dynamics. Brief summaries of the required background material are given in Appendixes 1.B (mathematics), 1.D (mechanics), and 1.E (fluid mechanics), while Appendix 1.A contains a list of constants, and Appendix 1.C summarizes the properties of special functions that will be used in this book. No previous exposure to astronomy is necessary, but the context of the book will be clearer to the reader who has grasped the basic observational data on galaxies as described in Mihalas and Binney (1981). A very brief summary of the most relevant data is given in the following section. For a general introduction to astronomy, consult Unsöld and Baschek (1983) or the superb introductory textbook by Shu (1982).

It has been said that "galaxies are to astronomy what atoms are to physics" (Sandage 1961). The analogy is correct in that galaxies are relatively isolated systems that generally maintain a unique identity throughout their lives, except for occasional collisions and mergers with other galaxies. The analogy is also appropriate because a galaxy forms a dynamical and chemical *unit*: the atoms in your body, for example, have come from all over the Galaxy, but few or none of them have come from other galaxies. However, galaxies are unlike atoms in that their structure does not follow in a simple way from basic physical laws. Rather, the present configuration of the Galaxy is the result of a myriad of chemical, dynamical, and other processes that may or may not act in the same way in other galaxies. Furthermore, the Galaxy is continually evolving, so that the present state is simply a transient form during a slow evolutionary sequence that has yet to be completed. Perhaps a more appropriate analogy is that galaxies are to astronomy as ecosystems are to biology: this reflects more accurately their complexity, their relative isolation, and their ongoing evolution.

1.1 An Overview of the Observations

In this section we review the observational data about stars, galaxies, and clusters that we will use later in this book. A far more comprehensive discussion is given in the book *Galactic Astronomy* (Mihalas & Binney 1981), which is a companion to the present book and will be referred to simply as MB.

1 Stars

The luminosity of the Sun is $L_\odot = 3.83 \times 10^{33}$ erg s^{-1}. More precisely, this is the **bolometric** luminosity, the total rate of energy output integrated over all wavelengths.[4] The bolometric luminosity is difficult to measure accurately, in part because the Earth's atmosphere is opaque at most wavelengths. Hence the luminosity of stars is usually measured in one or more specified wavelength bands, in particular the **visual** or V band, centered on $\lambda = 0.55\mu$ ($1\mu = 1$ micron$= 10^{-4}$ cm $= 10^4$ Å), the **blue** or B band, centered on $\lambda = 0.44\mu$, and the **ultraviolet** or U band centered on $\lambda = 0.365\mu$, all with width $\Delta\lambda/\lambda \simeq 0.2$ (see MB, §2-6). Thus, for example, the luminosity of Sirius (the brightest star in the sky) in the B and V bands is written

$$L_B = 42.5\,L_{\odot B} \quad ; \quad L_V = 23.3\,L_{\odot V}. \tag{1-1}$$

Similarly, the nearest star is Proxima Centauri, which has

$$L_B = 1.7 \times 10^{-5}\,L_{\odot B} \quad ; \quad L_V = 5.6 \times 10^{-5}\,L_{\odot V}. \tag{1-2}$$

This notation is usually simplified by dropping the subscript B or V from $L_\odot$, when it is clear from the context that the luminosity refers to the B or V band.

Luminosities are often expressed in a logarithmic scale, by defining the **absolute magnitude** as

$$M \equiv -2.5\log_{10} L + \text{constant}, \tag{1-3}$$

where the constant is chosen separately for each wavelength band; in B and V the constants are chosen so that the solar absolute magnitude is

$$M_{\odot B} = 5.48, \quad M_{\odot V} = 4.83. \tag{1-4}$$

(The original definition of the zero of the magnitude scale is of only historical interest.) Thus, Sirius has absolute magnitude $M_V = 1.41$, $M_B = 1.41$, and Proxima Centauri has $M_V = 15.45$, $M_B = 17.42$. The **flux** from a star of luminosity L at distance d is $f = L/(4\pi d^2)$, and a logarithmic measure of the flux from a star is provided by the **apparent magnitude**

$$m \equiv -2.5\log_{10}\left[\frac{L(10\,\text{pc})^2}{d^2}\right] + \text{constant} = M + 5\log_{10}(d/10\,\text{pc}); \tag{1-5}$$

---

[4] See Allen (1973) for an extensive tabulation of the quantities and relations discussed in this subsection.

thus, the absolute magnitude is the apparent magnitude that the star would have at a distance of 10 parsecs. Sirius is at a distance of 2.7 pc and has apparent magnitude $m_V = -1.45$, while Proxima Centauri is at 1.3 pc and has apparent magnitude $m_V = 11.05$. For comparison, the faintest stars visible to the naked eye have $m_V \approx 6$. Note that *faint* stars have *large* magnitudes. For brevity the apparent magnitudes m_B and m_V are often abbreviated simply as B and V.

The **distance modulus** $m - M = 5 \log_{10}(d/10 \, \text{pc})$ is often used as a measure of distance.

The **color** of a star is measured by the ratio of the luminosity in two wavelength bands, for example by L_V/L_B or equivalently by $M_B - M_V = m_B - m_V = B - V$. Sirius has color $B - V = 0.00$ and Proxima Centauri has $B - V = 1.97$. Stellar spectra are approximately black-body and hence the color is a measure of the temperature at the surface of the star.

A more precise measure of the surface temperature is the **effective temperature** T_{eff}, defined as the temperature of the black body with the same radius and luminosity as the star in question. If the stellar radius is R, then the Stefan-Boltzmann law implies that the bolometric luminosity is

$$L = 4\pi R^2 \sigma T_{\text{eff}}^4, \tag{1-6}$$

where $\sigma = 5.67 \times 10^{-5} \, \text{erg s}^{-1} \, \text{cm}^{-2} \, \text{K}^{-4}$. The relation between $B - V$ color and effective temperature is given in Allen (1973) and MB §3-5.

An alternative measure of the surface temperature of a star is its **spectral class**, which is assigned on the basis of the relative prominence of various absorption lines in the stellar spectrum. In order of decreasing temperature, the spectral classes are labeled O, B, A, F, G, K, and M, and each class is divided into ten subclasses by the numbers $0, 1, \ldots, 9$. Thus a B0 star is slightly cooler than an O9 star. Using this scheme, experienced observers can determine the effective temperature of a star to within about 10% from a quick examination of the stellar spectrum. For example, Sirius has spectral class A1 and effective temperature 10,000 K, while Proxima Centauri has spectral class M5 and $T_{\text{eff}} = 3000$ K.

The Sun is a quite ordinary G2 star, with $T_{\text{eff}} = 5770$ K.

The **Hertzsprung-Russell**, **HR**, or **color-magnitude** diagram is a plot of absolute magnitude as a function of color (spectral type or effective temperature sometimes replaces color, since all three quantities are closely related). Figure 1-1 shows the HR diagram for about 1000 nearby stars with accurate distances. The most striking feature is that most of the stars lie in a well-defined band stretching from about $M_V = 3$, $B - V = 0.5$, to $M_V = 15$, $B - V = 2.0$. This band, known as the **main**

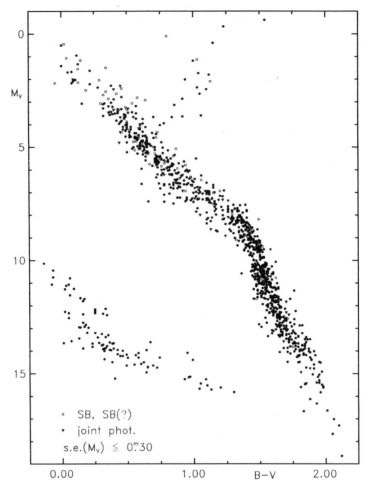

Figure 1-1. The color-magnitude diagram for 1094 stars within 25 parsecs. The sample is restricted to stars for which the uncertainty in the absolute magnitude is less than 0.3. Updated version of a diagram published in Gliese et al. (1986).

sequence, consists of stars that are burning hydrogen in their cores. In this stage of a star's life, the mass and chemical composition of the star uniquely determine both the surface temperature and the total luminosity. Thus the main sequence is a mass sequence, with more massive stars at the upper left (high luminosity, high temperature) and less massive stars at the lower right (low luminosity and temperature). The very faint main-sequence stars with $M_V \approx 16$ have a mass of about $0.1\,\mathrm{M_\odot}$; the apparent lack of still fainter stars arises because such stars never ignite hydrogen in their centers and hence are generally invisible. At the other end of the main sequence, there is also a lack of main-sequence stars brighter than $M_V = 3$; this is simply a consequence of the rarity of these stars.

The radius of a star can be determined from its luminosity and effective temperature by equation (1-6). Stars that are bright and cool (i.e., M_V large and negative, $B - V$ large and positive, upper right of the diagram) must be large, while stars that are faint and hot (lower left of the diagram) must be very small. Since the main sequence crosses from upper left to lower right, this argument suggests that radius is not a strong function of luminosity along the main sequence. Detailed analysis bears this out: between $M_V = 4.8$ ($1\,M_\odot$) and $M_V = 15$ ($0.1\,M_\odot$), the radius varies by only a factor of ten, from $1\,R_\odot$ to $0.1\,R_\odot$, despite a change of a factor of 10^4 in luminosity. Main sequence stars are sometimes called **dwarf** stars, to distinguish them from the larger giant stars that we discuss below.

The HR diagram also contains a number of faint blue stars around $M_V = 14$, $B - V = 0.3$. These are **white dwarfs**, stars that have exhausted their nuclear fuel and are gradually cooling to invisibility. As their location in the diagram suggests, white dwarf stars are very small, typically with radii of order $10^{-2}\,R_\odot$. In fact, white dwarfs are so dense that the electron gas in the interior of the star is degenerate; in other words, gravitational contraction is resisted, not by thermal pressure as in main-sequence stars, but rather by the Fermi energy of the star's cold, degenerate electron gas.

Figure 1-1 also contains a few stars that lie above the main sequence. These are **red giant** stars, stars that have exhausted hydrogen in their cores and are now burning hydrogen in a thick shell. The red giants are rare compared to main-sequence stars because the red-giant phase in a star's life is much shorter than its main-sequence phase. On the other hand, red giants are so luminous that they dominate the total luminosity of many stellar systems, such as elliptical galaxies. Another consequence of the high luminosity of red giants is that a far larger fraction of red giants is found in flux-limited samples than in volume-limited samples. For example, roughly half of the 100 brightest stars ($m_V < 2.6$) are giants, but *none* of the 100 nearest stars are giants.

Figure 1-2 is the HR diagram of a globular cluster. The advantage of analyzing a cluster is that all the stars are at the same distance, so that individual distance determinations are unnecessary. Moreover, all the stars are presumably the same age and have the same chemical composition; thus age and composition differences and distance errors do not blur the HR diagram. This cluster is at a distance $d = 4.2\,\text{kpc}$, so

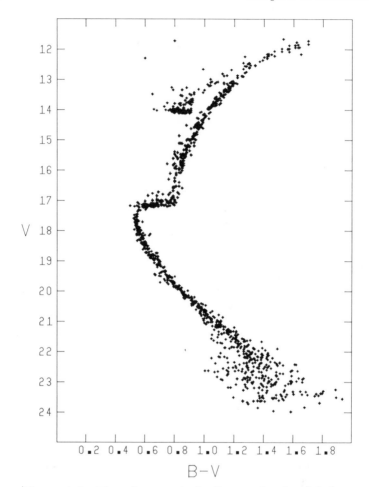

Figure 1-2. The color-magnitude diagram for the globular cluster 47 Tucanae. Here there are more giants relative to main-sequence stars than in Figure 1-1, but this enhancement is artificial because the bright star sample was collected over a wider area to improve statistics. (From Harris & Hesser 1987.)

that $V = 21$ corresponds to $M_V = 7.9$. The main sequence fainter than $M_V \approx 9$ ($V \approx 22$) is masked by contamination from foreground stars. The giants lie on a well-defined locus called the **giant branch**, which extends from $V = 17$ to $V = 12$ ($M_V = 4$ to $M_V = -1$); the brightest giants have radii of about $30\,R_\odot$. As stars evolve, they climb the giant branch, until core helium burning is initiated at the tip of the giant branch ($V = 12$, $M_V = -1$, $B - V = 1.6$). The stars then cross to the **horizontal branch** (the clump of stars at $V = 14$, $M_V = 1$, $B - V = 0.8$),

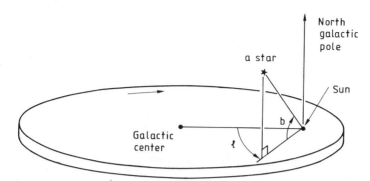

Figure 1-3. A schematic picture of the Sun's location in the Galaxy, illustrating the galactic coordinate system. An arrow points in the direction of galactic rotation.

where they remain until the helium in the core is exhausted (see MB, §§3-5 to 3-7 for more detail). Fits of theoretical models of stellar evolution to these observations yield an age of 15 to 18 Gyr (Harris et al. 1983, VandenBerg 1983).

2 The Galaxy

Most of the visible stars in the Galaxy lie in a flattened, roughly ax-isymmetric, disklike structure known as the **galactic disk**. On clear, dark nights the cumulative light from the myriad of faint disk stars is visible as a luminous band stretching across the sky, called the **Milky Way**. The midplane of this disk is called the **galactic plane** and serves as the equator of the **galactic coordinates** (ℓ, b), where ℓ is the **galactic longitude** and b is the **galactic latitude**. The **galactic poles** are in the directions normal to the galactic plane at $b = \pm 90°$. The galactic co-ordinate system is a heliocentric system in which the zero of longitude points to the galactic center (see Figure 1-3).

The Sun is located at a distance R_0 from the center of the Galaxy, for which the best current estimate is $R_0 = 8.5 \pm 1$ kpc. Observations of other galaxies of similar form suggest that the surface brightness of the disk is an exponential function of radius,

$$I(R) = I_0 \exp(-R/R_d), \tag{1-7}$$

where the **disk scale length** $R_d = 3.5 \pm 0.5$ kpc (de Vaucouleurs & Pence 1978). Thus the Sun lies farther from the galactic center than about

70% of the total disk light. This concentration of luminosity toward the galactic center is not apparent to the naked eye, since interstellar dust absorbs the light from distant disk stars (the optical depth in the V band is unity at a distance of only about 0.6 kpc; see Spitzer 1978). By contrast, the Galaxy is nearly transparent in the direction of the galactic poles.

The stars of the disk travel in nearly circular orbits around the galactic center. The speed of a hypothetical star in a circular orbit of radius R in the galactic equator is denoted $v_c(R)$, and a plot of $v_c(R)$ versus R is called the **circular-speed curve**. The circular speed at the solar radius R_0 is

$$v_0 \equiv v_c(R_0) = 220 \pm 15 \, \text{km s}^{-1}. \tag{1-8}$$

Local measurements of stellar motions also yield the slope of the circular-speed curve through the two **Oort constants**:

$$A \equiv \tfrac{1}{2} \left(\frac{v_c}{R} - \frac{dv_c}{dR} \right)_{R_0} \quad ; \quad B \equiv -\tfrac{1}{2} \left(\frac{v_c}{R} + \frac{dv_c}{dR} \right)_{R_0}. \tag{1-9}$$

The best current estimates are $A = 14.5 \pm 1.5 \, \text{km s}^{-1} \, \text{kpc}^{-1}$, $B = -12 \pm 3 \, \text{km s}^{-1} \, \text{kpc}^{-1}$. These values imply $(v_c/R)_{R_0} = A - B = 26.5 \pm 4 \, \text{km s}^{-1} \, \text{kpc}^{-1}$, consistent with the value $25.9 \, \text{km s}^{-1} \, \text{kpc}^{-1}$ obtained from the standard values $v_c = 220 \, \text{km s}^{-1}$, $R_0 = 8.5 \, \text{kpc}$.

The difference between the Sun's actual velocity and the mean velocity of the stars in its vicinity is known as the **solar motion**. The exact value of the solar motion depends on the spectral class of the stars used to determine the mean velocity, since different spectral classes have slightly different mean velocities [this effect is known as the **asymmetric drift** and is discussed in detail in §4.2.1(a)]. The most useful procedure for our purposes is to use the velocity of a hypothetical set of stars in precisely circular orbits. The velocity of these stars at the solar position is called the **local standard of rest**. The solar motion relative to the local standard of rest is (MB, §6-4)

$$16.5 \, \text{km s}^{-1} \text{ in the direction } \ell = 53°, \; b = 25°. \tag{1-10}$$

Most of our detailed knowledge of the properties of stars comes from observations of stars within a few hundred parsecs of the Sun. This distance is much smaller than the disk scale length, and hence it is reasonable to assume that the overall properties of the distribution of stars (number density of stars versus luminosity or mass, chemical composition, kinematics, etc.) are constant within this region, even though there may be large-scale gradients in these properties across the galactic disk. To formalize this assumption, we define the **solar neighborhood** to be a volume centered on the Sun that is much smaller than the overall size of the Galaxy and yet large enough to contain a statistically useful sample of stars. Obviously the definition is quite loose, but the concept of the solar neighborhood is nevertheless extremely useful. The appropriate size of the volume depends on which stars we wish to investigate: for white dwarfs, which are both common and faint, the "solar neighborhood" may consist of a sphere of radius 10 pc centered on the Sun, while for the bright but rare O and B stars, the solar neighborhood may be considered to extend as far as 1 kpc from the Sun. The appropriate shape of the volume also depends on what we want to investigate: if we were discussing properties that depend on distance from the midplane of the disk, such as the mass density in stars, the solar neighborhood would be a thin slice centered on the midplane, while for properties that are integrated over the disk thickness, such as surface density, the appropriate volume would be a cylinder.

The thickness of the galactic disk is different for different classes of stars. In particular, older stars are found at greater distances from the galactic plane. Let us define the **characteristic thickness** of the disk to be the ratio of the disk's surface density to its volume density at the galactic plane. Then the O and B stars, which are massive and young (typical ages $\lesssim 10^7$ yr) have a characteristic thickness of only about 200 pc, while G stars like the Sun (typical ages 5×10^9 yr), have a characteristic thickness of 700 pc (Bahcall & Soneira 1980). This correlation is believed to arise because the irregular gravitational fields of spiral arms and/or molecular clouds gradually increase the random velocities of disk stars (see §7.5).

A great deal of survey work over the past fifty years has been devoted to cataloging the contents of the solar neighborhood. The current inventory is summarized in Table 1-1. The visible stars include all main-sequence and giant stars; for a detailed breakdown, see Table 4-5 of MB. The category "stellar remnants" includes white dwarfs and neutron stars, while "gas" includes atomic and molecular hydrogen, ionized gas, and a small contribution from interstellar dust. "Other matter" denotes matter of unknown composition whose presence is detected only through its gravitational effects [see §4.2.1(b)].

In addition to the disk, the Galaxy also contains a **spheroid**, a roughly spherical distribution of stars whose chemical composition, kinematics, and evolutionary history are quite different from stars in the disk. The spheroid stars are generally older than the disk stars and are believed to have formed at the time of formation of the Galaxy, in contrast

Table 1-1. Inventory of the solar neighborhood

Component	Volume density ($M_\odot\,pc^{-3}$)	Surface density ($M_\odot\,pc^{-2}$)	Luminosity density ($L_\odot\,pc^{-3}$)	Surface brightness ($L_\odot\,pc^{-2}$)
Visible stars	0.044	26.4	0.067	15
Stellar remnants	0.028	18.2	0	0
Gas	0.042	5.3	0	0
Other matter	0.07	25	0	0
Total	0.18	75	0.067	15

NOTES: Volume densities are taken from §4.2.1(b). Surface densities are computed from volume densities following Bahcall and Soneira (1980). Luminosity density is measured in the V band and is taken from Table 4-5 of MB. Surface brightness is taken from Bahcall and Soneira (1980) and de Vaucouleurs and Pence (1978). Volume and luminosity densities are measured in the galactic plane.

to disk stars, which have formed at a steady rate throughout the history of the Galaxy. The spheroid stars contain fewer heavy elements than the disk stars, presumably because the interstellar gas from which the disk stars form has been polluted by the metal-rich debris of exploding supernovae. In contrast to disk stars, spheroid stars exhibit little or no net rotation but large random velocities. Any stellar system with these properties (low heavy element abundances, no young stars, little or no rotation) is called a **Population II** system or a **halo population**, while a system with properties similar to the disk is called a **Population I** system or a **disk population**. The galactic spheroid is relatively small compared to the disk, comprising only about 15%–30% of the total luminosity of the Galaxy.

A summary of properties of the Galaxy is provided in Table 1-2; for more detail, see §2.7.

3 Other Galaxies

The nearest galaxy to the Milky Way is the **Large Magellanic Cloud**, a relatively modest galaxy only 60 kpc from the Sun that is visible to the naked eye in the southern hemisphere as a faint patch of light (see Figure 1-4). The nearest large disk galaxy similar to our own is called **M31** or the **Andromeda galaxy**, and is located at a distance of 700 kpc (Figure 1-5). Only the central parts of this galaxy are visible to the naked eye, but on long-exposure photographs the galaxy is seen to extend over almost three degrees on the sky.

Table 1-2. Properties of the Galaxy

Global properties:	
Disk scale length R_d	$3.5 \pm 0.5\,\mathrm{kpc}$
Disk luminosity (V band)	$1.2 \times 10^{10}\,\mathrm{L_\odot}$
Spheroid luminosity (V band)	$2 \times 10^9\,\mathrm{L_\odot}$
Total luminosity (V band)	$1.4 \times 10^{10}\,\mathrm{L_\odot}$
Disk mass-to-light ratio Υ_V	$5\Upsilon_\odot$
Disk mass	$6 \times 10^{10}\,\mathrm{M_\odot}$
Hubble type	Sbc
Solar neighborhood properties:	
Solar radius R_0	$8.5 \pm 1\,\mathrm{kpc}$
Circular speed v_0	$220 \pm 15\,\mathrm{km\,s^{-1}}$
Angular speed $\Omega_0 = v_0/R_0$	$25.9 \pm 4\,\mathrm{km\,s^{-1}\,kpc^{-1}}$
	$= 8.4 \pm 1 \times 10^{-16}\,\mathrm{s^{-1}}$
Rotation period $2\pi/\Omega_0$	$2.4 \times 10^8\,\mathrm{yr}$
Oort's A constant	$14.5 \pm 1.5\,\mathrm{km\,s^{-1}\,kpc^{-1}}$
Oort's B constant	$-12 \pm 3\,\mathrm{km\,s^{-1}\,kpc^{-1}}$
Epicycle frequency $\kappa_0 = \sqrt{-4B(A-B)}$	$36 \pm 10\,\mathrm{km\,s^{-1}\,kpc^{-1}}$
Vertical frequency $\nu_0 = \sqrt{4\pi G\rho_0}$	$(3.2 \pm 0.5) \times 10^{-15}\,\mathrm{s^{-1}}$
Vertical period $2\pi/\nu_0$	$6.2 \times 10^7\,\mathrm{yr}$
Radial dispersion at $z = 0$	$30\,\mathrm{km\,s^{-1}}$
Radial dispersion (z-averaged)	$45\,\mathrm{km\,s^{-1}}$
Metallicity Z	$\simeq Z_\odot = 0.02$

NOTES: See §2.7 for more detail on global properties. Solar vertical frequency has been computed using total density ρ_0 from Table 1-1 and §4.2.1(b). Radial velocity dispersion is a mass-weighted average, taken from Wielen (1977).

Our Galaxy is just one member of a vast sea of over 10^9 galaxies stretching to a distance of at least several thousand Mpc (1 megaparsec$\equiv$ 1 Mpc $= 3.086 \times 10^{24}\,\mathrm{cm}$). The detailed local information that we can deduce about our own Galaxy is complemented by the less detailed but global picture that we have of other galaxies.

Before we describe the properties of galaxies, it is useful to review how distances to galaxies are measured. For galaxies within a few Mpc the distance can be determined by measuring the flux from the brightest stars in the galaxy, which have nearly the same luminosity in all galaxies of a given type, or by measuring the flux from Cepheid variable stars, which have a known period-luminosity relation (MB, §3-8). At larger distances, where individual stars are no longer visible, we use other less accurate calibrators, such as the diameters of **HII regions**, spheres of ionized gas surrounding O and B stars, or the flux from the brightest

Figure 1-4. The Large Magellanic Cloud, a barred irregular galaxy (Hubble class SBm) that is the nearest neighbor galaxy to our own (distance 50 kpc). The length of the main barlike structure is about 180', or six times the diameter of the full moon. Courtesy of D. F. Malin and the Royal Observatory, Edinburgh.

globular clusters in the galaxy, or correlations between luminosity and rotation speed or morphology. At still larger distances, $\gtrsim 10\,\mathrm{Mpc}$, the line-of-sight velocity of a galaxy relative to the galactic center, v_p, is correlated with its distance l. For $l \lesssim 20\,\mathrm{Mpc}$, this correlation is complicated by the gravitational effects of the nearby Virgo cluster of galaxies (§10.3.2), but at sufficiently large distances, $l \gtrsim 25\,\mathrm{Mpc}$, the velocity satisfies **Hubble's law**

$$v_p = H_0 l, \tag{1-11}$$

where H_0 is the **Hubble constant**. The value of the Hubble constant is difficult to determine, and at the present time is probably uncertain by a factor of about two. To keep this uncertainty explicit we write

$$H_0 \equiv 100 h \,\mathrm{km\,s^{-1}\,Mpc^{-1}}, \tag{1-12}$$

Figure 1-5. The Andromeda galaxy, M31, the nearest large disk galaxy similar to our own. It is at a distance of about 700 kpc and its Hubble class is Sb. The Andromeda galaxy is accompanied by two dwarf elliptical galaxies, NGC 205 (E5) and M32 (E2). The distance between the two dwarfs is about 1°. Courtesy of A. R. Sandage.

where the dimensionless parameter h is probably between 0.5 and 1.0. A typical galaxy satisfies Hubble's law to within a few hundred $\mathrm{km\,s^{-1}}$; thus, for example, a galaxy with a velocity of 5,000 $\mathrm{km\,s^{-1}}$ is known to be at a distance of $50h^{-1}\,\mathrm{Mpc}$ to within about 5%. The uncertainty in

Figure 1-6. M87, a large elliptical galaxy located in the core of the Virgo cluster of galaxies at a distance of about 15 Mpc. In Hubble's system this galaxy is classified as E1. Most of the faint starlike objects are globular clusters belonging to M87. Courtesy of A. R. Sandage.

Hubble's constant affects the whole distance scale of the Universe and hence is reflected as an uncertainty in many of the average properties of galaxies; for example, the mean density of galaxies scales as h^3 and the mean luminosity of a galaxy of a given type scales as h^{-2}.

Galaxies are found in a wide range of shapes, sizes, and masses but can usefully be divided into four main types according to **Hubble's classification system** (see Sandage 1961 or MB, §5-1, for a more complete description of Hubble's system).

(a) Elliptical galaxies   These are smooth, featureless Population II systems containing little or no gas or dust (see Figure 1-6). The fraction of bright galaxies that are elliptical is a function of the local density, ranging from about 10% in low-density regions to 40% in dense clusters of galaxies (Dressler 1980). The **isophotes** (contours of constant surface brightness) are approximately concentric ellipses, with axis ratio b/a ranging from 1 to about 0.3. Elliptical galaxies are denoted by the symbols E0, E1, etc., where the brightest isophotes of a galaxy of type En have axis ratio $b/a = 1 - n/10$. The **ellipticity** is $\epsilon = 1 - b/a$. Thus the most elongated elliptical galaxies are of type E7. Since we see only the projected brightness distribution, it is impossible to determine directly whether elliptical galaxies are axisymmetric or triaxial.

The surface brightness of an elliptical galaxy falls off smoothly with radius. Often the outermost parts of a galaxy are undetectable against the background night-sky brightness. The surface-brightness profiles of most elliptical galaxies can be fit by the $R^{1/4}$ or **de Vaucouleurs** (1948) law,

$$I(R) = I(0) \exp(-kR^{0.25}) \equiv I_e \exp\{-7.67[(R/R_e)^{0.25} - 1]\}, \quad (1\text{-}13)$$

where the **effective radius** R_e is the radius of the isophote containing half of the total luminosity and I_e is the surface brightness at R_e. The effective radius is typically $3h^{-1}$ kpc for bright ellipticals and is smaller for fainter galaxies (Kormendy 1977). Another successful fitting formula is the **Hubble-Reynolds law** (Reynolds 1913; Hubble 1930),

$$I_H(R) = \frac{I_0 R_H^2}{(R + R_H)^2}, \quad (1\text{-}14)$$

where the radius R_H is typically $0.1R_e$.

Because galaxies do not generally have a sharp outer edge, it is conventional to specify their total spatial extent by the **Holmberg radius**, the radius of the isophote corresponding to surface brightness 26.5 mag $(\text{arcsec})^{-2}$ (in the B band). This is roughly 1%–2% of sky brightness and is the lowest surface brightness that can in general be reliably measured.

The luminosities of elliptical galaxies range over a factor of 10^7. The **luminosity function** $\phi(L)$ describes the relative numbers of galaxies of different luminosities, and is defined so that $\phi(L)dL$ is the number of galaxies in the luminosity interval $L \to L + dL$ in a representative unit volume of the Universe. A convenient analytic approximation to $\phi(L)$ is **Schechter's law** (MB, §5-3),

$$\phi(L)dL = n_\star \left(\frac{L}{L_\star}\right)^\alpha \exp(-L/L_\star)\frac{dL}{L_\star}, \quad (1\text{-}15)$$

where $n_\star = 1.2 \times 10^{-2} h^3 \, \mathrm{Mpc}^{-3}$, $\alpha = -1.25$, and $L_\star = 1.0 \times 10^{10} h^{-2} \, L_\odot$ in the visual band (Kirshner et al. 1983). Note that according to Schechter's law, $\int_L^\infty \phi(L) dL$ diverges as $L \to 0$; of course, the total number of galaxies per unit volume must be finite so that Schechter's law must fail at sufficiently low luminosities, but the divergence is an accurate reflection of the large number of faint galaxies found at the limits of detectability.

Most giant elliptical galaxies exhibit little or no rotation, even those with highly elongated isophotes. Their stars have random velocities along the line of sight whose root mean square (hereafter RMS) dispersion σ_p can be measured from the Doppler broadening of spectral lines. The velocity dispersion in the inner few kiloparsecs is correlated with luminosity according to the **Faber-Jackson law**,

$$\sigma_p \simeq 220 (L/L_\star)^{0.25} \, \mathrm{km \, s}^{-1}. \tag{1-16}$$

(b) Lenticular galaxies   These are galaxies with a prominent disk that contains no gas, dust, bright young stars, or spiral arms. Lenticular disks are smooth and featureless, like elliptical galaxies, but obey the exponential surface-brightness law (1-7) characteristic of spiral galaxies. Lenticulars are labeled by the notation S0 in Hubble's classification scheme. They are very rare in low-density regions, comprising less than 10% of all bright galaxies, but up to half of all galaxies in high-density regions are S0's.

The lenticulars form a transition class between ellipticals and spirals. The transition is smooth and continuous, so that there are S0 galaxies that might well be classified as E7, and others that have sometimes been classified as spirals.

The strong dependence of the fractional abundance of S0 galaxies on the local density is obviously an important—but still controversial—clue to the mechanism of galaxy formation.

(c) Spiral galaxies   These are galaxies, like our own and M31, which contain a prominent disk composed of Population I stars, gas, and dust. In all these systems the disk contains **spiral arms**, filaments of bright O and B stars, gas, and dust, in which large numbers of stars are currently forming. The spiral arms vary greatly in their length and prominence from one spiral galaxy to another but are almost always present (see Figure 1-5 and §6.1).

In low-density regions of the Universe, almost 80% of all bright galaxies are spirals, but the fraction drops to 10% in dense regions such as cluster cores.

The distribution of surface brightness in spiral galaxy disks obeys the exponential law (1-7) (Freeman 1970). The typical disk scale length is $R_d \simeq 3h^{-1}$ kpc, and the central surface brightness is remarkably constant at $I_0 \simeq 140 L_\odot \, pc^{-2}$ (independent of the Hubble constant).

The circular-speed curves of most spiral galaxies are nearly flat, $v_c(R)$ independent of R, except near the center, where the circular speed drops to zero (see Figure 10-1). Typical circular speeds are between 200 and $300 \, km \, s^{-1}$. It is a remarkable fact that the circular speed curves still remain flat even at radii well beyond the outer edge of the visible galaxy, thus implying the presence of invisible or dark mass in the outer parts of the galaxy (see §10.1.6).

Spiral galaxies also contain a spheroid of Population II stars. The luminosity of the spheroid relative to the disk correlates well with a number of other properties of the galaxy, in particular the fraction of the disk mass in gas, the color of the disk, and how tightly the spiral arms are wound. This correlation is the basis of Hubble's classification of spiral galaxies. Hubble divided spiral galaxies into a sequence of four classes or types, called Sa, Sb, Sc, Sd. Along the sequence Sa→Sd the relative luminosity of the spheroid decreases, the relative mass of gas increases, and the spiral arms become more loosely wound. The spiral arms also become more clumpy, so that individual patches of young stars and HII regions become visible. This sequence is illustrated by comparing M31 (Figure 1-5), an Sb galaxy, and M51 (Figure 6-1), an Sc galaxy. Note that the arms are much more prominent and open in M51, while the spheroid is much less prominent. Our own Galaxy appears to be intermediate between Sb and Sc, so its Hubble type is written as Sbc.

(d) Irregular galaxies Any classification scheme has to contain an attic—a class into which objects that conform to no particular pattern can be placed. Since the time of Hubble, nonconformist galaxies have been dumped into the irregular class (denoted Irr). A minority of Irr galaxies are spiral or elliptical galaxies that have been violently distorted by a recent encounter with a neighbor. The galaxy NGC5195 shown in Figure 6-1 is probably of this type. However, the majority of Irr galaxies are simply low-luminosity gas-rich systems such as the Magellanic Clouds (see Figure 1-4). These galaxies are designated Sm or Im to indicate their kinship to the Magellanic Clouds.

Magellanic irregulars are extremely common—more than a third of our Galaxy's near neighbors are of this type—but they do not feature prominently in the usual galaxy catalogs, because any flux-limited catalog is strongly biased against intrinsically faint systems.

Much of the luminosity of Im galaxies is emitted by massive young stars and large HII regions. These systems are gas-rich—the interstellar medium of such a galaxy often contains more than 30% of the system's mass—and like spiral galaxies they are probably quite flattened. However, Im galaxies, unlike spirals, are not dominated by massive disks of stars on nearly circular orbits. This is in part because Im galaxies are often barred, but mainly because turbulent velocities in the interstellar gas are not much smaller than the circular speeds of these systems: the circular speed in an Im galaxy is rarely greater than $50 \,\mathrm{km\,s^{-1}}$, while interstellar gas has a characteristic random velocity $v_{\mathrm{rand}} \simeq 10 \,\mathrm{km\,s^{-1}}$. Hence stars formed with random velocities comparable to those of the clouds in which they are born have significant orbital eccentricities from birth. The irregular appearance of Im galaxies arises in part from the large random velocities, and in part from the dominance of the optical emission by a small number of bright young stars and HII regions (see Gallagher & Hunter 1984 for a review).

All of the galaxy types we have described, except for the Irr attic, form a single sequence: E→S0→Sa→Sb→Sc→Sd→Im. This sequence is known as the **Hubble sequence**. Galaxies near the start of the sequence (**early-type** galaxies) have little or no cool gas and dust, and consist mostly of old Population II stars; galaxies near the end (**late-type** galaxies) are rich in gas, dust, and young stars.

4 Open and Globular Clusters

A typical galaxy contains myriads of small stellar systems containing between 10^2 and 10^6 stars. These systems are called **star clusters** and can be divided into two main types.

Galactic or open clusters are Population I systems found in the galactic disk (see Figure 1-7). A typical open cluster contains 10^2 to 10^3 stars and has a radius of order 1 to 10 pc (see also Table 1-3 and MB, §3-5). New open clusters are formed continuously, and most of the ones we see are younger than a few times 10^8 yr. It appears that older clusters have been disrupted, probably by gravitational perturbations from passing interstellar gas clouds (see §7.2). There are believed to be about 10^5 open clusters in the Galaxy.

Figure 1-7. (Left) The open cluster NGC 2477; (right) the globular cluster 47 Tucanae, at a distance of 4.6 kpc. The latter cluster has a luminosity of about $6 \times 10^5 \, L_\odot$. Only the giant stars are visible in this photograph. Courtesy of D. F. Malin and the Anglo-Australian Telescope Board.

Globular clusters are Population II systems containing between 10^4 and 10^6 stars. Our Galaxy contains about 200 globular clusters, forming a roughly spherical distribution concentrated toward the galactic center. Like other Population II systems, globular clusters contain no gas, dust, or young stars. The clusters themselves are almost spherical and appear to be dynamically stable and very long lived (Figure 1-7). Unlike open clusters, globular clusters are generally very old and are believed to be relics of the formation of the Galaxy itself. The stellar density in the center of a globular cluster is much higher than in most galaxies; a typical value is $10^4 \, M_\odot \, pc^{-3}$, compared with $0.05 \, M_\odot \, pc^{-3}$ in the solar neighborhood. Since globular clusters are very inhomogeneous, a single measure of their radius is insufficient, and three separate radii are often quoted: the **core radius**, where the surface brightness has fallen to half its central value; the **median radius**, the radius of a sphere that contains half of all the light; and the **limiting** or **tidal radius**, the outer limit of the cluster where the density drops to zero. Typical values of these and other cluster parameters are given in Table 1-3.

Table 1-3. Parameters of globular and open clusters

	Globular	Open
Central density ρ_0	$8 \times 10^3\,M_\odot\,pc^{-3}$	$100\,M_\odot\,pc^{-3}$
Core radius r_c	$1.5\,pc$	$1\,pc$
Median radius r_h	$10\,pc$	$2\,pc$
Tidal radius r_t	$50\,pc$	$10\,pc$
Central velocity dispersion σ_0 (line-of-sight)	$7\,km\,s^{-1}$	$1\,km\,s^{-1}$
Mass-to-light ratio Υ	$2\Upsilon_\odot$	$1\Upsilon_\odot$
Mass M	$6 \times 10^5\,M_\odot$	$250\,M_\odot$
Lifetime	$10^{10}\,yr$	$2 \times 10^8\,yr$

NOTES: Values for globular clusters are medians from the compilation of Peterson and King (1975). Values for open clusters are typical values from the literature. The central densities are especially uncertain: individual clusters may have central densities that differ by a factor of 100 from the values quoted.

The brightest globular clusters are as luminous as the faintest dwarf elliptical galaxies. However, a typical low-luminosity galaxy is a very low surface-brightness object with a radius of several kpc, while a bright globular cluster has a much smaller radius ($\lesssim 100\,pc$) and a correspondingly higher surface brightness.

5 Groups and Clusters of Galaxies

Galaxies are not distributed uniformly across the Universe. Most galaxies belong to a rich and complex hierarchy of structure that includes binary galaxies, small groups of a few galaxies in close proximity, enormous voids in which the number density of galaxies is greatly depleted, and giant clusters containing thousands of galaxies.

Associations that contain only a handful of galaxies are called **groups of galaxies** while bigger associations are called **clusters of galaxies** (see Table 1-4). The exact dividing line between groups and clusters is arbitrary, but for many purposes it is useful to define a cluster to be an association that has more than fifty members whose luminosities lie within two magnitudes (a factor of about six) of the third brightest galaxy in the association. Figure 1-8 shows examples of a group and the central region of a cluster. The line-of-sight velocity dispersion in a group is generally about $150\,km\,s^{-1}$, and in a large cluster it is about $1000\,km\,s^{-1}$.

Table 1-4. Parameters of groups and clusters of galaxies

	Groups	Clusters
Core radius r_c	$250h^{-1}\,\mathrm{kpc}$	$250h^{-1}\,\mathrm{kpc}$
Median radius r_h	$0.7h^{-1}\,\mathrm{Mpc}$	$3h^{-1}\,\mathrm{Mpc}$
Velocity dispersion σ (line-of-sight)	$150\,\mathrm{km\,s^{-1}}$	$800\,\mathrm{km\,s^{-1}}$
Mass-to-light ratio Υ	$250h\Upsilon_\odot$	$250h\Upsilon_\odot$
Mass M	$2 \times 10^{13}h^{-1}\,\mathrm{M_\odot}$	$10^{15}h^{-1}\,\mathrm{M_\odot}$

NOTES: Values for groups taken from Huchra and Geller (1982). Values for clusters are typical estimates from the literature. Median radii and total masses are especially uncertain, since the outer boundary of a group or cluster is difficult to determine.

The relative proportions of different galaxy types vary from the center to the outer regions of dense clusters: the ellipticals are more concentrated to the center than the spirals, with the S0's having an intermediate degree of concentration.

Like galaxies, groups and clusters may be regarded for many purposes as assemblies of point masses orbiting under their mutual gravitational attraction. However, groups and clusters are dynamically younger than galaxies. Even in the central parts of a rich cluster, galaxies have completed only a few orbits since the cluster formed, and in many groups and clusters, especially in the outer parts, galaxies are still falling toward the cluster center for the first time. For example, the Galaxy and M31 are the two dominant members of a group called the **Local Group**, which is still collapsing (§10.2.1); it is quite likely that at the end of this collapse in $\approx 10^{10}\,\mathrm{yr}$ the two galaxies will merge together into a single giant stellar system.

Another important difference between galaxies and clusters of galaxies is that the fractional volume of a cluster that is occupied by galaxies ($\approx 10^{-3}$) is much larger than the fractional volume of a galaxy that is occupied by stars ($\approx 10^{-22}$). Thus collisions and encounters between galaxies in a cluster are much more frequent, and play a more important role than do collisions between stars in a galaxy.

The radial distribution of galaxies in a cluster is similar to the radial distribution of stars in an elliptical galaxy and can be described by the $R^{1/4}$ law [eq. (1-13)] with effective radius $R_e \simeq 1\text{–}2h^{-1}\,\mathrm{Mpc}$. The profiles of clusters are difficult to determine accurately, both because the measurements involve a collection of discrete galaxies rather than a smooth surface brightness distribution, and because the outermost parts have not yet reached dynamical equilibrium. In some cases better infor-

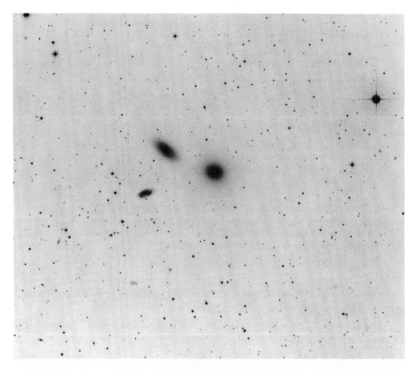

Figure 1-8. (Top) The compact group of galaxies containing the galaxy NGC 3379, at a distance of about $8h^{-1}$ Mpc. (Bottom) The central core of the cluster of galaxies CA0340-538, a rich cluster at a distance of $180h^{-1}$ Mpc (see Lucey et al. 1983). Courtesy of D. F. Malin and the Royal Observatory, Edinburgh (top) and the Anglo-Australian Telescope Board (bottom).

Table 1-5. Parameters of a typical galactic nucleus

Central density ρ_0	$5 \times 10^6 \, \text{M}_\odot \, \text{pc}^{-3}$
Core radius r_c	$1 \, \text{pc}$
Central velocity dispersion σ_0 (line-of-sight)	$150 \, \text{km} \, \text{s}^{-1}$
Mass-to-light ratio Υ	$10 \Upsilon_\odot$
Mass M	$10^8 \, \text{M}_\odot$

mation on the shape and extent of the cluster is obtained by measuring the distribution of X-ray surface brightness in the cluster. The X-rays are emitted by hot gas that fills the cluster potential well.

6 Galactic Nuclei

The centers of many galaxies appear to contain small, dense stellar systems called **galactic nuclei**. These systems are very difficult to study because they are generally too small to be resolved by ground-based optical telescopes. They are of great interest because a substantial fraction of them, the **active galactic nuclei**, are strong sources of optical, radio, and X-ray emission, possibly arising from the accretion of matter onto a massive (10^6 to $10^9 \, \text{M}_\odot$) black hole at their centers. The theoretical exploration of these systems is still in its infancy, but one exciting possibility is that interactions among these stars in a dense stellar system lead inevitably to the formation of a black hole, perhaps through stellar coalescence as a result of collisions (see §8.4.5 and MB §9-4).

Table 1-5 contains approximate estimates of the main dynamical properties of a typical nucleus.

2 Potential Theory

Much of the mass of a galaxy resides in stars. To compute the potential of a large collection of stars, we should in principle simply add the point-mass potentials of all the stars together. Of course, this is not practicable for the $\approx 10^{11}$ stars in a typical galaxy, and for most purposes it is sufficient to model the potential as arising from a smoothed-out density that is everywhere proportional to the local star density. In particular, in §4.1 we shall show that we obtain an excellent approximation to the orbit of a single star through a galaxy like our own by treating the star as a test particle that moves in a smooth potential of this kind. In this chapter we show how the force field of such an idealized galaxy can be calculated.

The chapter is divided into eight sections. We start by deriving some general results, and then in §2.1 we specialize to discuss the simplest potentials, those of spherical bodies. In §2.2 we describe some non-spherical density distributions that also have simple potentials. These special systems give us useful insight into the potentials of real galaxies and provide useful prototypes, but they do not form a good basis for detailed modeling. Therefore in §2.3 and §2.4 we show how to calculate the potential of any moderately flattened or elongated body, and in §2.6 we treat the potentials of perfectly thin disks. In §2.7 we apply these results to our own Galaxy, and finally in §2.8 we describe how the results derived in this chapter are used in computer simulations of galaxies.

Sections 2.3 to 2.6 are rather mathematical, and readers who do not enjoy that kind of thing, and are willing to take a few results on trust, may prefer to move straight from §2.2 to §2.7.

1 General results

Our goal is to calculate the force $\mathbf{F}(\mathbf{x})$ on a unit mass at position $\mathbf{x}$ that is generated by the gravitational attraction of a distribution of mass $\rho(\mathbf{x})$. According to Isaac Newton's inverse-square law of gravitation, the force $\mathbf{F}(\mathbf{x})$ may be obtained by summing the small contributions

$$\delta\mathbf{F}(\mathbf{x}) = G\frac{\mathbf{x}' - \mathbf{x}}{|\mathbf{x}' - \mathbf{x}|^3}\delta m(\mathbf{x}') = G\frac{\mathbf{x}' - \mathbf{x}}{|\mathbf{x}' - \mathbf{x}|^3}\rho(\mathbf{x}')\delta^3\mathbf{x}' \qquad (2\text{-}1)$$

to the overall force from each small element of volume $\delta^3 \mathbf{x}'$ located at $\mathbf{x}'$. Thus

$$\mathbf{F}(\mathbf{x}) = G \int \frac{\mathbf{x}' - \mathbf{x}}{|\mathbf{x}' - \mathbf{x}|^3} \rho(\mathbf{x}') d^3 \mathbf{x}'. \tag{2-2}$$

If we define the **gravitational potential** $\Phi(\mathbf{x})$ by

$$\Phi(\mathbf{x}) = -G \int \frac{\rho(\mathbf{x}')}{|\mathbf{x}' - \mathbf{x}|} d^3 \mathbf{x}', \tag{2-3}$$

and notice that

$$\nabla_\mathbf{x} \left(\frac{1}{|\mathbf{x}' - \mathbf{x}|} \right) = \frac{\mathbf{x}' - \mathbf{x}}{|\mathbf{x}' - \mathbf{x}|^3}, \tag{2-4}$$

we find that we may write $\mathbf{F}$ as

$$\begin{aligned} \mathbf{F}(\mathbf{x}) &= \nabla_\mathbf{x} \int \frac{G\rho(\mathbf{x}')}{|\mathbf{x}' - \mathbf{x}|} d^3 \mathbf{x}' \\ &= -\nabla\Phi, \end{aligned} \tag{2-5}$$

where for brevity we have dropped the subscript $\mathbf{x}$ on the gradient operator ∇. Since the force is determined by the gradient of a potential, the gravitational force is conservative (cf. Appendix 1.D.1).

The potential is useful because, being a scalar field, it is easier to visualize than the vector force field. Also, in many situations the best way to obtain $\mathbf{F}$ is first to calculate the potential and then to take its gradient.

If we take the divergence of equation (2-2), we find

$$\nabla \cdot \mathbf{F}(\mathbf{x}) = G \int \nabla_\mathbf{x} \cdot \left(\frac{\mathbf{x}' - \mathbf{x}}{|\mathbf{x}' - \mathbf{x}|^3} \right) \rho(\mathbf{x}') d^3 \mathbf{x}'. \tag{2-6}$$

Now

$$\nabla_\mathbf{x} \cdot \left(\frac{\mathbf{x}' - \mathbf{x}}{|\mathbf{x}' - \mathbf{x}|^3} \right) = -\frac{3}{|\mathbf{x}' - \mathbf{x}|} + \frac{3(\mathbf{x}' - \mathbf{x}) \cdot (\mathbf{x}' - \mathbf{x})}{|\mathbf{x}' - \mathbf{x}|^5}. \tag{2-7}$$

When $\mathbf{x}' - \mathbf{x} \neq 0$ we may cancel the factor $|\mathbf{x}' - \mathbf{x}|^2$ from top and bottom of the last term in this equation to conclude that

$$\nabla_\mathbf{x} \cdot \left(\frac{\mathbf{x}' - \mathbf{x}}{|\mathbf{x}' - \mathbf{x}|^3} \right) = 0 \qquad (\mathbf{x}' \neq \mathbf{x}). \tag{2-8}$$

Therefore, any contribution to the integral of equation (2-6) must come from the point $\mathbf{x}' = \mathbf{x}$, and we may restrict the volume of integration to a small sphere of radius h centered on this point. Since, for sufficiently small h, the density will be almost constant through this volume, we can take $\rho(\mathbf{x}')$ out of the integral. The remaining terms of the integrand may then be arranged as follows:

$$
\begin{aligned}
\boldsymbol{\nabla} \cdot \mathbf{F}(\mathbf{x}) &= G\rho(\mathbf{x}) \int_{|\mathbf{x}'-\mathbf{x}|\leq h} \boldsymbol{\nabla}_{\mathbf{x}} \cdot \left(\frac{\mathbf{x}'-\mathbf{x}}{|\mathbf{x}'-\mathbf{x}|^3} \right) d^3\mathbf{x}' \\
&= -G\rho(\mathbf{x}) \int_{|\mathbf{x}'-\mathbf{x}|\leq h} \boldsymbol{\nabla}'_{\mathbf{x}} \cdot \left(\frac{\mathbf{x}'-\mathbf{x}}{|\mathbf{x}'-\mathbf{x}|^3} \right) d^3\mathbf{x}' \qquad \text{(2-9a)} \\
&= -G\rho(\mathbf{x}) \int_{|\mathbf{x}'-\mathbf{x}|= h} \frac{(\mathbf{x}'-\mathbf{x}) \cdot d^2\mathbf{S}'}{|\mathbf{x}'-\mathbf{x}|^3}.
\end{aligned}
$$

The last step in this sequence uses the divergence theorem to convert the volume integral into a surface integral [eq. (1B-42)]. Now on the sphere $|\mathbf{x}'-\mathbf{x}| = h$ we have $d^2\mathbf{S}' = (\mathbf{x}'-\mathbf{x})hd^2\Omega$, where $d^2\Omega$ is a small element of solid angle. Hence equation (2-9a) becomes

$$
\boldsymbol{\nabla} \cdot \mathbf{F}(\mathbf{x}) = -G\rho(\mathbf{x}) \int d^2\Omega = -4\pi G\rho(\mathbf{x}). \qquad \text{(2-9b)}
$$

If we substitute from equation (2-5) for $\boldsymbol{\nabla} \cdot \mathbf{F}$, we obtain **Poisson's equation** relating the potential Φ to the density ρ;

$$
\nabla^2\Phi = 4\pi G\rho. \qquad \text{(2-10)}
$$

Equation (2-10) provides a route to Φ, and then to $\mathbf{F}$ that is often more convenient than equation (2-2) or equation (2-3). In the special case $\rho = 0$ we have **Laplace's equation**,

$$
\nabla^2\Phi = 0. \qquad \text{(2-11)}
$$

We may use Poisson's equation to derive a useful generalization of equation (2-8). A unit point mass at $\mathbf{x}'$ has density $\rho(\mathbf{x}) = \delta(\mathbf{x}-\mathbf{x}')$, where δ is the Dirac delta function [eq. (1C-1)], and potential $-G|\mathbf{x}-\mathbf{x}'|^{-1}$. Hence equation (2-10) yields

$$
\nabla^2_{\mathbf{x}} \left(\frac{1}{|\mathbf{x}-\mathbf{x}'|} \right) = -4\pi\delta(\mathbf{x}-\mathbf{x}') \quad \text{or} \quad \boldsymbol{\nabla}_{\mathbf{x}} \cdot \left(\frac{\mathbf{x}'-\mathbf{x}}{|\mathbf{x}-\mathbf{x}'|^3} \right) = -4\pi\delta(\mathbf{x}-\mathbf{x}').
$$
$$
\text{(2-12)}
$$

If we integrate both sides of equation (2-10) over an arbitrary volume containing total mass M, and then apply the divergence theorem, we obtain

$$
4\pi G \int \rho d^3\mathbf{x} = 4\pi GM = \int \nabla^2\Phi d^3\mathbf{x} = \int \boldsymbol{\nabla}\Phi \cdot d^2\mathbf{S}. \qquad \text{(2-13)}
$$

This result is **Gauss's theorem**, which may be stated in words as *the integral of the normal component of $\nabla\Phi$ over any closed surface equals $4\pi G$ times the mass contained within that surface.*

We have seen that the gravitational force is conservative, that is, that the work done against gravitational forces in moving two stars from infinity to a given configuration is independent of the path along which they are moved, and is defined to be the potential energy of the configuration. Similarly, the work done against gravitational forces in assembling an arbitrary continuous distribution of mass $\rho(\mathbf{x})$ is independent of the details of how the mass was assembled, and is defined to be equal to the potential energy of the mass distribution or simply the **potential energy**. An expression for the potential energy can be obtained by the following argument.

Suppose that some of the mass is already in place so that the density and potential are $\rho(\mathbf{x})$ and $\Phi(\mathbf{x})$. If we now bring in a small mass δm from infinity to position $\mathbf{x}$, the work done is $\delta m\Phi(\mathbf{x})$. Thus, if we add a small increment of density $\delta\rho(\mathbf{x})$, the change in potential energy is

$$\delta W = \int \delta\rho(\mathbf{x})\Phi(\mathbf{x})d^3\mathbf{x}. \qquad (2\text{-}14)$$

According to Poisson's equation the resulting change in potential $\delta\Phi(\mathbf{x})$ satisfies $\nabla^2(\delta\Phi) = 4\pi G(\delta\rho)$, and hence

$$\delta W = \frac{1}{4\pi G} \int \Phi\nabla^2(\delta\Phi)\, d^3\mathbf{x}. \qquad (2\text{-}15)$$

Using the divergence theorem in the form (1B-43), we may write this as

$$\delta W = \frac{1}{4\pi G} \int \Phi\nabla(\delta\Phi)\cdot d^2\mathbf{S} - \frac{1}{4\pi G} \int \nabla\Phi\cdot\nabla(\delta\Phi)d^3\mathbf{x}, \qquad (2\text{-}16)$$

where the surface integral vanishes because $\Phi \propto r^{-1}$, $|\nabla\delta\Phi| \propto r^{-2}$ as $r \to \infty$, so the integrand $\propto r^{-3}$ while the total surface area $\propto r^2$. But $\nabla\Phi\cdot\nabla(\delta\Phi) = \frac{1}{2}\delta|(\nabla\Phi)|^2$. Hence

$$\delta W = -\frac{1}{8\pi G}\delta\left(\int |\nabla\Phi|^2 d^3\mathbf{x}\right). \qquad (2\text{-}17)$$

If we now sum up all of the contributions δW, we have a simple expression for the potential energy,

$$W = -\frac{1}{8\pi G} \int |\nabla\Phi|^2 d^3\mathbf{x}. \qquad (2\text{-}18)$$

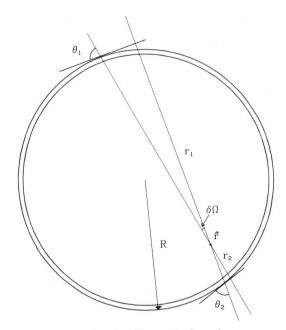

Figure 2-1. Proof of Newton's first theorem.

To obtain an alternative expression for W, we may again apply the divergence theorem and replace $\nabla^2\Phi$ by $4\pi G\rho$ to obtain

$$W = \tfrac{1}{2} \int \rho(\mathbf{x})\Phi(\mathbf{x})d^3\mathbf{x}. \qquad (2\text{-}19)$$

2.1 Spherical Systems

1 Newton's Theorems

Newton proved two results that enable us to calculate the gravitational potential of any spherically symmetric distribution of matter very easily. We may state these as:

Newton's First Theorem *A body that is inside a spherical shell of matter experiences no net gravitational force from that shell.*

Newton's Second Theorem *The gravitational force on a body that lies outside a closed spherical shell of matter is the same as it would be if all the shell's matter were concentrated into a point at its center.*

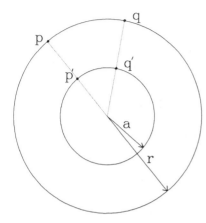

Figure 2-2. Proof of Newton's second theorem.

Figure 2-1 illustrates the proof of Newton's first theorem. Consider the cone associated with an elementary solid angle $d\Omega$ centered on the point $\mathbf{r}$. This cone intersects the spherical shell of matter at two points, at distances r_1 and r_2 from $\mathbf{r}$. Elementary geometrical considerations assure us that the angles θ_1 and θ_2 are equal, and therefore that the masses δm_1 and δm_2 contained within $\delta\Omega$ where it intersects the shell are in the ratio $\delta m_1/\delta m_2 = (r_1/r_2)^2$. Hence $\delta m_2/r_2^2 = \delta m_1/r_1^2$ and a particle placed at $\mathbf{r}$ is attracted equally in opposite directions. Summing over all elementary cones centered on $\mathbf{r}$, one concludes that the body at $\mathbf{r}$ experiences no net force from the shell.◁

An important corollary of Newton's first theorem is that the gravitational potential inside an empty spherical shell is constant because $\nabla\Phi = -\mathbf{F} = 0$. Thus we may evaluate the potential $\Phi(\mathbf{r})$ inside the shell by calculating the integral expression (2-3) for $\mathbf{r}$ located at any interior point. The most convenient place for $\mathbf{r}$ is the center of the shell, for then all points on the shell are at the same distance R, and one immediately has

$$\Phi = -\frac{GM}{R}. \tag{2-20}$$

The proof of Newton's second theorem eluded Newton for more than ten years. Yet with hindsight it is easy. The trick is to compare the potential at a point $\mathbf{p}$ located a distance r from the center of a spherical shell of mass M and radius a $(r > a)$, with the potential at a point $\mathbf{p}'$ located a distance a from the center of a shell of mass M and radius r. Figure 2-2 illustrates the proof. Consider the contribution $\delta\Phi$ to the potential at $\mathbf{p}$ from the portion of the given sphere with solid angle $\delta\Omega$ located at $\mathbf{q}'$. Evidently

$$\delta\Phi = -\frac{GM}{|\mathbf{p} - \mathbf{q}'|}\frac{\delta\Omega}{4\pi}. \qquad (2\text{-}21a)$$

But the contribution $\delta\Phi'$ of the matter near $\mathbf{q}$ to the potential at $\mathbf{p}'$ is clearly

$$\delta\Phi' = -\frac{GM}{|\mathbf{p}' - \mathbf{q}|}\frac{\delta\Omega}{4\pi}. \qquad (2\text{-}21b)$$

Finally, as $|\mathbf{p} - \mathbf{q}'| = |\mathbf{p}' - \mathbf{q}|$ by symmetry, it follows that $\delta\Phi = \delta\Phi'$, and then by summation over all points $\mathbf{q}$ and $\mathbf{q}'$ that $\Phi = \Phi'$. But we already know that $\Phi' = -GM/r$, therefore $\Phi = -GM/r$, which is exactly the potential that would be generated by concentrating the entire mass of the given sphere at its center.◁

The Newtonian gravitational potentials of different spherical shells add linearly, so we may calculate the gravitational potential at $\mathbf{r}$ generated by an arbitrary spherically symmetric density distribution $\rho(\mathbf{r}')$ in two parts by adding the contributions to the potential produced by shells (i) with $r' < r$, and (ii) with $r' > r$. In this way we obtain

$$\Phi(r) = -4\pi G \left[\frac{1}{r}\int_0^r \rho(r')r'^2\,dr' + \int_r^\infty \rho(r')r'\,dr'\right]. \qquad (2\text{-}22)$$

From Newton's second theorem or from equation (2-22) it follows that the gravitational attraction of the density distribution $\rho(r')$ on a unit test mass at radius r is entirely determined by the mass interior to r:

$$\mathbf{F}(r) = -\frac{d\Phi}{dr}\hat{\mathbf{e}}_r = -\frac{GM(r)}{r^2}\hat{\mathbf{e}}_r, \qquad (2\text{-}23a)$$

where

$$M(r) = 4\pi \int_0^r \rho(r')r'^2\,dr'. \qquad (2\text{-}23b)$$

An important property of a spherical matter distribution is its **circular speed** $v_c(r)$, defined to be the speed of a test particle in a circular orbit at radius r. Once we have $\Phi(r)$ or $\mathbf{F}(r)$, we may readily evaluate v_c from

$$v_c^2 = r\frac{d\Phi}{dr} = r|\mathbf{F}| = \frac{GM(r)}{r}. \qquad (2\text{-}24)$$

The circular speed measures the mass interior to r. A second important quantity is the **escape speed** v_e defined by

$$v_e(r) = \sqrt{2|\Phi(r)|}. \qquad (2\text{-}25)$$

A star at r can escape from the gravitational force field represented by Φ only if it has a speed at least as great as $v_e(r)$, for only then does its (positive) kinetic energy $\frac{1}{2}v^2$ exceed the absolute value of its (negative) potential energy Φ.

2 Potentials of Some Simple Systems

It is instructive to discuss the potentials generated by several simple density distributions:

(a) Point mass   In this case

$$\Phi(r) = -\frac{GM}{r} \quad ; \quad v_c(r) = \sqrt{\frac{GM}{r}} \quad ; \quad v_e(r) = \sqrt{\frac{2GM}{r}}. \quad (2\text{-}26)$$

Any circular speed that declines with increasing radius like $r^{-1/2}$ is frequently referred to as **Keplerian** because Kepler first understood that $v_c \propto r^{-1/2}$ in the solar system.

(b) Homogeneous sphere   If the density is some constant ρ, we have $M(r) = \frac{4}{3}\pi r^3 \rho$ and

$$v_c = \sqrt{\frac{4\pi G\rho}{3}} r. \quad (2\text{-}27)$$

Thus in this case the circular velocity rises linearly with radius, and the orbital period of a mass on a circular orbit is

$$T = \frac{2\pi r}{v_c} = \sqrt{\frac{3\pi}{G\rho}}, \quad (2\text{-}28)$$

independent of the radius of its orbit.

If a test mass is released from rest at radius r in the gravitational field of a homogeneous body, its equation of motion is

$$\frac{d^2 r}{dt^2} = -\frac{GM(r)}{r^2} = -\frac{4\pi G\rho}{3} r, \quad (2\text{-}29)$$

which is the equation of motion of a harmonic oscillator of angular frequency $2\pi/T$. Therefore no matter what is the initial value of r, the test mass will reach $r = 0$ in a quarter of a period, or in a time

$$t_{\text{dyn}} = \frac{T}{4} = \sqrt{\frac{3\pi}{16G\rho}}. \quad (2\text{-}30)$$

Although this result is only correct for a homogeneous sphere, we shall define the **dynamical time** of a system with mean density ρ by equation

(2-30).[1] The dynamical time is approximately equal to the time required for an orbiting star to travel halfway across a system of this mean density.

From equation (2-22) it follows that if the density vanishes for $r > a$, the gravitational potential is

$$\Phi(r) = \begin{cases} -2\pi G\rho(a^2 - \frac{1}{3}r^2), & r < a \\ -\dfrac{4\pi G\rho a^3}{3r}, & r > a, \end{cases} \qquad (2\text{-}31)$$

which can be used to compute the escape speed.

(c) Isochrone potential   We might expect that a spherical galaxy has roughly constant density near its center, while the density falls to zero at sufficiently large radii. The potential of a galaxy of this type would be proportional to $r^2 - constant$ at small radii and to r^{-1} at large radii. A simple potential with these properties is the **isochrone potential**

$$\Phi(r) = -\frac{GM}{b + \sqrt{b^2 + r^2}}. \qquad (2\text{-}32)$$

The linear scale of the system that generates this potential is set by the constant b, while M is the system's total mass. The potential (2-32) owes its name to a property of its orbits that we shall derive in §3.1. By equation (2-24) we have for the circular speed at radius r

$$v_c^2(r) = \frac{GMr^2}{(b + a)^2 a}, \qquad (2\text{-}33a)$$

where

$$a \equiv \sqrt{b^2 + r^2}. \qquad (2\text{-}33b)$$

When r is very large $v_c \simeq \sqrt{GM/r}$, as required for a system of finite mass and extent. By Poisson's equation (2-10) and equation (1B-51) for ∇^2 in spherical coordinates, the density associated with the isochrone potential is

$$\rho(r) = \frac{1}{4\pi G}\frac{1}{r^2}\frac{d}{dr}\left(r^2\frac{d\Phi}{dr}\right) = M\left[\frac{3(b+a)a^2 - r^2(b+3a)}{4\pi(b+a)^3 a^3}\right]. \qquad (2\text{-}34)$$

Thus the central density is

$$\rho(0) = \frac{3M}{16\pi G b^3}, \qquad (2\text{-}35)$$

---

[1] A homogeneous sphere of pressureless material of density ρ released from rest at $t = 0$ will collapse to a point at $t = t_{\rm dyn}/\sqrt{2}$ (Problem 3-4). Consequently $t_{\rm dyn}/\sqrt{2}$ is often referred to as the **free-fall time** of a system of density ρ.

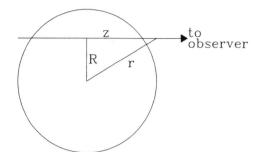

Figure 2-3. Projection of a spherical body along the line of sight.

and at large radii the density tends to

$$\rho(r) \simeq \frac{bM}{2\pi r^4} \quad (r \gg b). \tag{2-36}$$

(d) Modified Hubble profile   The surface brightnesses of many elliptical galaxies may be approximated over a large range of radii by the Hubble-Reynolds law $I_H(R)$ [eq. (1-14)]. It is possible to solve for the spherical luminosity density $j(r)$ that generates a given circularly symmetric brightness distribution $I(R)$ (see Problem 2-10). However, the resulting formulae for the luminosity distribution of a galaxy that obeys the Hubble-Reynolds law are cumbersome (Hubble 1930). Fortunately, the simple luminosity density

$$j_h(r) = j_0 \left[1 + \left(\frac{r}{a} \right)^2 \right]^{-\frac{3}{2}}, \tag{2-37}$$

where a is the **core radius**, gives rise to a surface brightness distribution that is similar to I_H (Rood et al. 1972). In fact, in the notation of Figure 2-3 we have that

$$I_h(R) = 2 \int_0^\infty j_h(r) \, dz = 2j_0 \int_0^\infty \left[1 + \left(\frac{R}{a} \right)^2 + \left(\frac{z}{a} \right)^2 \right]^{-\frac{3}{2}} dz. \tag{2-38}$$

Using the substitution $y \equiv z/\sqrt{a^2 + R^2}$, we obtain the **modified Hubble profile**

$$I_h(R) = \frac{2j_0 a}{1 + (R/a)^2} \int_0^\infty \frac{dy}{(1 + y^2)^{\frac{3}{2}}} = \frac{2j_0 a}{1 + (R/a)^2}. \tag{2-39}$$

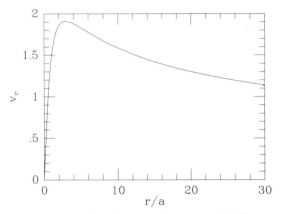

Figure 2-4. Circular speed versus radius for a body whose projected density follows the modified Hubble profile (2-39). The circular speed v_c is plotted in units of $\sqrt{Gj_0\Upsilon a^2}$.

Thus $I_h(R) \propto R^{-2}$ at large R and $I_h(R) \to constant$ as $R \to 0$, just as in the Hubble-Reynolds law (1-14).

Since the brightness distributions of many elliptical galaxies are fairly well fitted by the Hubble law, we conclude that the three-dimensional luminosity densities of elliptical galaxies cannot be very different from that specified by equation (2-37). Thus using equation (2-22) we can calculate the potential that would be generated by such a galaxy if its mass were distributed in the same way as its light. If $\rho(r) = \Upsilon j(r)$, where Υ is the constant **mass-to-light ratio** in the galaxy, one has

$$M_h(r) = 4\pi a^3 \Upsilon j_0 \left\{ \ln\left[\frac{r}{a} + \sqrt{\frac{r^2}{a^2} + 1} \right] - \frac{r}{a}\left(\frac{r^2}{a^2} + 1 \right)^{-\frac{1}{2}} \right\}, \quad (2\text{-}40)$$

$$\Phi_h = -\frac{GM_h(r)}{r} - \frac{4\pi G \Upsilon j_0 a^2}{\sqrt{1 + (r/a)^2}}. \quad (2\text{-}41)$$

One serious difficulty with both the Hubble-Reynolds and modified Hubble profiles is that the mass diverges logarithmically at large r; from equation (2-40) $M_h \simeq 4\pi a^3 \Upsilon j_0 \left[\ln(2r/a) - 1 \right]$ for $r \gg a$. In practice, a galaxy must have a finite mass, so $j(r)$ must fall below $j_h(r)$ at sufficiently large r. Nevertheless, the potential Φ_h is finite, and in fact rather nearly equal to $-GM_h(r)/r$ whenever $r \gg a$. This behavior indicates that from the gravitational point of view the density distribution of equation (2-37) is highly concentrated toward the core and behaves much like a point mass at large r. The circular speed is shown in Figure 2-4. It peaks at $r = 2.9a$ and then falls nearly as steeply as in the Keplerian case.

(e) Power-law density profile   Many galaxies have luminosity profiles that approximate a power law over a large range in radius. Consider the structure of a system whose mass density drops off as some power of the radius:

$$\rho(r) = \rho_0 \left(\frac{r_0}{r}\right)^\alpha. \tag{2-42}$$

The surface density of this system is

$$\Sigma(R) = \frac{\rho_0 r_0^\alpha}{R^{\alpha-1}} \frac{(-\frac{1}{2})!(\frac{\alpha-3}{2})!}{(\frac{\alpha-2}{2})!}. \tag{2-43}$$

We assume that $\alpha < 3$, since only in this case is the mass interior to r finite, namely

$$M(r) = \frac{4\pi\rho_0 r_0^\alpha}{3 - \alpha} r^{(3-\alpha)}. \tag{2-44}$$

From equations (2-44) and (2-24) the circular speed is

$$v_c^2(r) = \frac{4\pi G\rho_0 r_0^\alpha}{3 - \alpha} r^{(2-\alpha)}. \tag{2-45}$$

In Chapter 8 of MB we saw that the circular-speed curves of many galaxies are remarkably flat. Equation (2-45) suggests that the mass density in these galaxies may be proportional to r^{-2}. In Chapter 4 we shall find that this is the density profile characteristic of a self-consistent stellar-dynamical model called the **singular isothermal sphere**.

Equation (2-44) shows that $M(r)$ diverges at large r for all $\alpha < 3$. However, when $\alpha > 2$, the potential difference in these models between radius r and infinity, is finite. Thus the escape speed $v_e(r)$ from radius r is given by

$$v_e^2(r) = 2 \int_r^\infty \frac{GM(r')}{r'^2} dr' = \frac{8\pi G\rho_0 r_0^\alpha}{(3 - \alpha)(\alpha - 2)} r^{(2-\alpha)}$$
$$= 2\frac{v_c^2(r)}{\alpha - 2} \quad (\alpha > 2). \tag{2-46}$$

Over the range $3 > \alpha > 2$, $(v_e/v_c)^2$ rises from the value 2 that is characteristic of a point mass, toward infinity. Since the light distributions of elliptical galaxies suggest $\alpha \simeq 3$ at large r [see eq. (2-37)], while the circular speeds in spiral galaxies suggest $\alpha \simeq 2$ (since their rotation curves are flat), it is clear that the escape speeds of galaxies will be very uncertain. Furthermore, since $M(r)$ diverges as $r \to \infty$ for α in the range $2 < \alpha < 3$, it is also clear that we shall only be able to determine the total mass and luminosity of any galaxy if we can push observations out to radii where the rotation speed becomes Keplerian ($v_c \propto r^{-1/2}$) and the luminosity density falls more steeply than $j(r) \propto r^{-3}$.

2.2 Potential-Density Pairs for Flattened Systems

Later in this chapter we will show how to obtain the gravitational potentials of disks and spheroids of every sort. However, we shall find that the calculation of the gravitational potential and force field that are generated by a given distribution of matter is generally an arduous task that all too often leads to cumbersome formulae involving special functions, or numerical calculations. Fortunately, for many purposes it suffices to represent a galaxy by a simple model that has the same gross structure as the galaxy. In this section we describe families of potentials that are generated by fairly simple and realistic density distributions. These potentials help us to understand how the gravitational potential of an initially spherical body is affected by flattening the body, and in later chapters we shall use several of these potentials to illustrate features of dynamics in axisymmetric galaxies.

1 Plummer-Kuzmin Models

Consider the spherical potential

$$\Phi_P = -\frac{GM}{\sqrt{r^2 + b^2}}. \tag{2-47a}$$

By direct differentiation we find

$$\nabla^2 \Phi_P = \frac{1}{r^2} \frac{d}{dr} \left(r^2 \frac{d\Phi_P}{dr} \right) = \frac{3GMb^2}{(r^2 + b^2)^{5/2}}. \tag{2-48}$$

Thus from Poisson's equation we have that the density corresponding to the potential (2-47a) is

$$\rho_P(r) = \left(\frac{3M}{4\pi b^3} \right) \left(1 + \frac{r^2}{b^2} \right)^{-\frac{5}{2}}. \tag{2-47b}$$

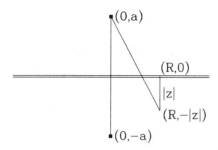

Figure 2-5. At the point $(R, -|z|)$ below Kuzmin's disk, the potential is identical with that of a point mass located distance a above the disk's center.

In 1911 H. C. Plummer used the potential-density pair that is described by equations (2-47) to fit observations of globular clusters. It is therefore known as **Plummer's model**. We shall encounter it again in §4.4.3(a) as a member of the family of stellar systems known as polytropes.

Now consider the axisymmetric potential

$$\Phi_K(R, z) = -\frac{GM}{\sqrt{R^2 + (a + |z|)^2}}. \tag{2-49a}$$

As Figure 2-5 indicates, at points with $z < 0$, Φ_K is identical with the potential of a point mass M located at the point $(R, z) = (0, a)$; and when $z > 0$, Φ_K coincides with the potential generated by a point mass at $(0, -a)$. Hence $\nabla^2 \Phi_K$ must vanish everywhere except on the plane $z = 0$. By applying Gauss's theorem to a flat volume that contains a small portion of the plane $z = 0$, we conclude that Φ_K is generated by the surface density

$$\Sigma_K(R) = \frac{aM}{2\pi(R^2 + a^2)^{\frac{3}{2}}}. \tag{2-49b}$$

The potential-density pair of equations (2-49) was introduced by Kuzmin (1956), but it is often referred to as "Toomre's model 1" because it became widely known in the West only after Toomre (1962) unknowingly rederived it.

Consider next the potential

$$\Phi_M(R, z) = -\frac{GM}{\sqrt{R^2 + (a + \sqrt{z^2 + b^2})^2}}. \tag{2-50a}$$

When $a = 0$, Φ_M reduces to Plummer's spherical potential (2-47a), and when $b = 0$, Φ_M reduces to Kuzmin's very flattened potential (2-49a). Thus, depending on the choice of the two parameters a and b, Φ_M can represent the potential of anything from an infinitesimally thin disk to a spherical system. If we calculate $\nabla^2 \Phi_M$, we find that the mass distribution with which it is associated is

$$\rho_M(R, z) = \left(\frac{b^2 M}{4\pi}\right) \frac{aR^2 + (a + 3\sqrt{z^2 + b^2})(a + \sqrt{z^2 + b^2})^2}{\left[R^2 + (a + \sqrt{z^2 + b^2})^2\right]^{5/2}(z^2 + b^2)^{3/2}}. \quad (2\text{-}50b)$$

The potential-density pair of equations (2-50) was introduced by Miyamoto and Nagai (1975). In Figure 2-6 we show contour plots of $\rho_M(R, z)$ for various values of b/a. When $b/a \simeq 0.2$, these are qualitatively similar to the light distributions of disk galaxies, although there are quantitative differences. For example, from equation (2-50b) we have that $\rho(R, 0) \propto R^{-3}$ when R is large, whereas the brightness profiles of disks fall off at least as fast as $\exp(-R/R_d)$ [eq. (1-7)].

Since Poisson's equation is linear in Φ and ρ, the difference between any two potential-density pairs is itself a potential-density pair. Therefore, if we differentiate equations (2-49) with respect to a^2, we obtain a new potential-density pair. For example, if we differentiate equation (2-49a) $(n-1)$ times with respect to a^2 and set $a^2 = 2nR_T^2$, we obtain the potential of the disk whose surface density is

$$\Sigma_K^n(R) = \Sigma_0 \left(1 + \frac{R^2}{2nR_T^2}\right)^{-(n+\frac{1}{2})}, \quad (2\text{-}51)$$

where Σ_0 is a suitable constant. The nth of these models is known as **Toomre's model n**.

In the limit $n \to \infty$, Σ_K^n tends to the Gaussian distribution

$$\Sigma_K^\infty(R) = \Sigma_0 \exp\left(-\frac{R^2}{2R_T^2}\right). \quad (2\text{-}52)$$

Similarly, if we differentiate equations (2-50) n times with respect to b^2, we obtain a new potential-density pair (Satoh 1980). As $n \to \infty$ the pair obtained takes on the relatively simple form

$$\Phi_M^\infty(R, z) = -\frac{GM}{S}, \quad (2\text{-}53a)$$

$$\rho_M^\infty(R, z) = \frac{ab^2 M}{4\pi S^3(z^2 + b^2)} \left[\frac{1}{\sqrt{z^2 + b^2}} + \frac{3}{a}\left(1 - \frac{R^2 + z^2}{S^2}\right)\right], \quad (2\text{-}53b)$$

where

$$S^2 \equiv R^2 + z^2 + a\left(a + 2\sqrt{z^2 + b^2}\right). \quad (2\text{-}53c)$$

Figure 2-7 shows that at large b/a the isodensity surfaces Φ_M^∞ are more nearly elliptical than those of Φ_M.

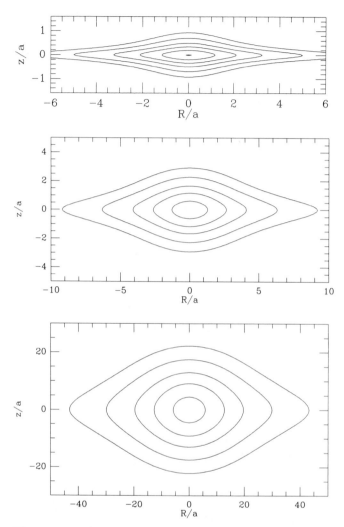

Figure 2-6. Contours of equal density in the (R, z) plane for the Miyamoto-Nagai density distribution (2-50b) when: $b/a = 0.2$ (top); $b/a = 1$ (middle); $b/a = 10$ (bottom). Contour levels are $f \times (1, 0.3, 0.1, 0.03, \ldots)$, where: $f = M/a^3$ (top); $f = 0.1M/a^3$ (middle); $f = 0.0001M/a^3$ (bottom).

2 Logarithmic Potentials

Since the Plummer-Kuzmin models have finite mass, the circular speed associated with these potentials falls off in Keplerian fashion $v_c \propto R^{-1/2}$ at large R. However, in §8-4 of MB it was shown that the rotation curves of spiral galaxies tend to be flat or rising at large radii. If at large R, $v_c \propto$

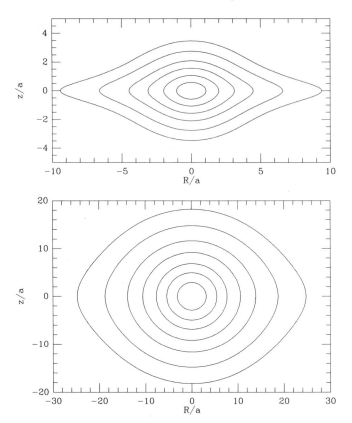

Figure 2-7. Contours of equal density in the (R,z) plane for Satoh's density distribution (2-53b) when: $b/a = 1$ (top); $b/a = 10$ (bottom). Contour levels are $f \times (1, 0.3, 0.1, 0.03, \ldots)$, where: $f = 0.1M/a^3$ (top); $f = 0.001M/a^3$ (bottom).

v_0, a constant, then $d\Phi/dR \propto R^{-1}$, and hence $\Phi \propto v_0^2 \ln R + constant$ in this region. Therefore, consider the potential

$$\Phi_L = \tfrac{1}{2}v_0^2 \ln \left(R_c^2 + R^2 + \frac{z^2}{q_\Phi^2} \right) + \text{constant}, \qquad (2\text{-}54a)$$

where R_c and v_0 are constants, and $q_\Phi \leq 1$. The density distribution to which Φ_L corresponds is

$$\rho_L(R,z) = \left(\frac{v_0^2}{4\pi G q_\Phi^2} \right) \frac{(2q_\Phi^2 + 1)R_c^2 + R^2 + 2(1 - \tfrac{1}{2}q_\Phi^{-2})z^2}{(R_c^2 + R^2 + z^2 q_\Phi^{-2})^2}. \qquad (2\text{-}54b)$$

At small R and z, ρ_L tends to the value $\rho_L(0,0) = (4\pi G)^{-1}(2 + q_\Phi^{-2})(v_0/R_c)^2$, and when R or $|z|$ is large, ρ_L falls off as R^{-2} or z^{-2}.

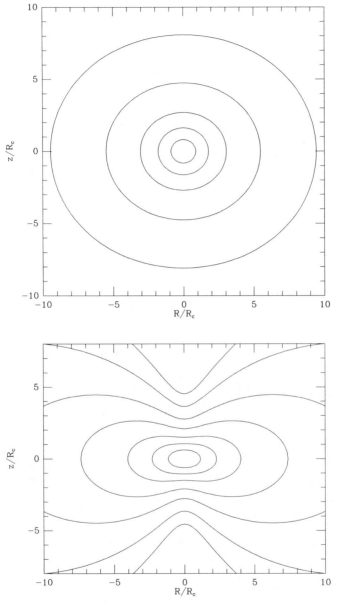

Figure 2-8. Contours of equal density in the (R, z) plane for ρ_L [eq. (2-54b)] when: $q_\Phi = 0.95$ (top); $q_\Phi = 0.7$ (bottom). In each case the contour levels are $0.1v_0^2/(GR_c^2) \times (1, 0.3, 0.1, \ldots)$. When $q_\Phi = 0.7$ the density is negative near the z-axis for $|z| \gtrsim 7R_c$.

The equipotential surfaces of Φ_L are ellipses of axial ratio q_Φ, but Figure 2-8 shows that the equidensity surfaces are rather flatter. In fact, if we define the axial ratio q_ρ of the isodensity surfaces by the ratio z_m/R_m of the distances down the z and R axes at which a given isodensity surface cuts the z axis and the x or y axis, we find

$$q_\rho^2 = \frac{1 + 4q_\Phi^2}{2 + 3/q_\Phi^2} \quad (r \ll R_c) \tag{2-55a}$$

or

$$q_\rho^2 = q_\Phi^4 \left(2 - \frac{1}{q_\Phi^2} \right) \quad (r \gg R_c). \tag{2-55b}$$

In either case the potential is only about a third as flattened as the density distribution. ρ_L becomes negative on the z-axis when $q_\Phi < 1/\sqrt{2}$.

The circular speed at radius R in the equatorial plane of Φ_L is

$$v_c = \frac{v_0 R}{\sqrt{R_c^2 + R^2}}. \tag{2-54c}$$

3 Poisson's Equation in Very Flattened Systems

In any axisymmetric system with density $\rho(R, z)$, Poisson's equation can be written [eq. (1B-50)]

$$\frac{\partial^2 \Phi}{\partial z^2} = 4\pi G \rho(R, z) + \frac{1}{R} \frac{\partial}{\partial R} (R F_R), \tag{2-56}$$

where $F_R = -(\partial \Phi/\partial R)$ is the radial force. Now consider, for example, the Miyamoto-Nagai potential-density pair given by equations (2-50). As the parameter $b \to 0$, the density distribution becomes more and more flattened, and the density in the $z = 0$ plane at fixed R becomes larger and larger as b^{-1}. However, the radial force F_R remains well behaved as $b \to 0$; indeed, in the limit $b = 0$, $F_R = -(\partial \Phi_K/\partial R)$, where $\Phi_K(R, z)$ is simply the Kuzmin potential (2-49). Thus, near $z = 0$ the first term on the right side of equation (2-56) becomes very large compared to the second, and Poisson's equation simplifies to the form

$$\frac{\partial^2 \Phi(R, z)}{\partial z^2} = 4\pi G \rho(R, z). \tag{2-57}$$

This simple form of Poisson's equation applies to almost any thin disk system. It implies that the vertical variation of the potential at a given radius R depends only on the density distribution at that radius. Effectively, this means that the solution of Poisson's equation in a thin disk can be decomposed into two steps: (i) Approximate the thin disk as a surface density layer of zero thickness and determine the potential in the plane of the disk $\Phi(R, 0)$ using the models of this section or the more general techniques of §2.6. (ii) At each radius R solve equation (2-57) to find the vertical variation of $\Phi(R, z)$.

2.3 Ellipsoidal Systems*

Bodies whose isodensity surfaces are similar concentric ellipsoids are interesting primarily because the isodensity surfaces of elliptical galaxies are very nearly elliptical (see MB, §5.2). Furthermore, Newton's theorems for spherical bodies can be generalized to include ellipsoidal bodies. Consequently, models with ellipsoidal isodensity surfaces are easy to construct. Also, the general results that we obtain for these systems enable us to develop a feel for the relationship between any body and the potential it generates.

The problem of calculating the gravitational potential of an ellipsoidal distribution of matter challenged some of the best minds of the eighteenth century. Accounts of this work, which relied on geometrical arguments to a degree that is uncommon today, are given by Kellogg (1953) and Chandrasekhar (1969). In this section we obtain only the potentials of oblate figures of rotation. Our analysis, which employs a very different approach from that adopted by Maclaurin and Ivory in the eighteenth century, is easily generalized to prolate figures of rotation, and can, at rather greater cost, be extended to ellipsoidal bodies that have three distinct principal axes. However, in the interests of brevity we shall merely cite the corresponding results for prolate and triaxial bodies.

1 Axisymmetric Systems

Spheroidal bodies call for spheroidal coordinates, so we consider the form of Laplace's equation, $\nabla^2 \Phi = 0$, in **oblate spheroidal coordinates**. These coordinates employ the usual azimuthal angle ϕ of cylindrical

---

* This section contains more advanced material and can be skipped on a first reading.

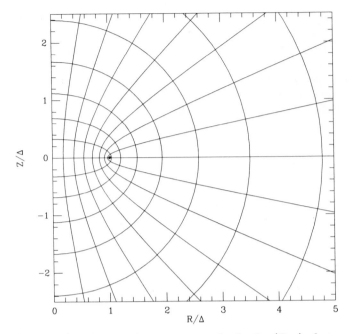

Figure 2-9. Curves of constant u and v in the (R, z) plane. Semi-ellipses are curves of constant u, and hyperbolae are curves of constant v. The common focus of all curves is marked by a dot.

coordinates, but replace the coordinates (R, z) with new coordinates (u, v) that are defined by

$$R = \Delta \cosh u \sin v \quad ; \quad z = \Delta \sinh u \cos v, \qquad (2\text{-}58)$$

where Δ is a constant. Figure 2-9 shows the curves of constant u and v in the (R, z) plane. The curves $u = constant$ are confocal ellipses with foci at $(R, z) = (\pm\Delta, 0)$, and the curves $v = constant$ are the hyperbolae formed by the normals to these ellipses. In order to ensure that each point has a unique v-coordinate, we exclude the disk $z = 0$, $R \leq \Delta$ from the space to be considered.

If we vary in turn u, v, and ϕ by small amounts while holding the other two coordinates constant, the point (u, v, ϕ) moves parallel to the three orthogonal unit vectors $\hat{\mathbf{e}}_u$, $\hat{\mathbf{e}}_v$, $\hat{\mathbf{e}}_\phi$ by the distances $h_u \delta u$, $h_v \delta v$ and $h_\phi \delta \phi$, where the scale factors are

$$
\begin{aligned}
h_u &= \Delta\sqrt{\sinh^2 u + \cos^2 v}, \\
h_v &= \Delta\sqrt{\sinh^2 u + \cos^2 v}, \\
h_\phi &= \Delta \cosh u \sin v.
\end{aligned}
\qquad (2\text{-}59)
$$

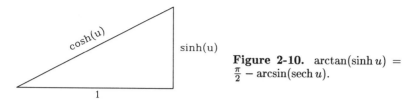

Figure 2-10. $\arctan(\sinh u) = \frac{\pi}{2} - \arcsin(\operatorname{sech} u)$.

Hence the gradient of a potential Φ may be expressed in these coordinates as [see eq. (1B-39)]

$$\nabla\Phi = \frac{1}{\Delta\sqrt{\sinh^2 u + \cos^2 v}}\left[\frac{\partial\Phi}{\partial u}\hat{\mathbf{e}}_u + \frac{\partial\Phi}{\partial v}\hat{\mathbf{e}}_v\right] + \frac{1}{\Delta\cosh u \sin v}\frac{\partial\Phi}{\partial\phi}\hat{\mathbf{e}}_\phi.$$

(2-60)

We further have [see eq. (1B-52)]

$$\nabla^2\Phi = \frac{1}{\Delta^2(\sinh^2 u + \cos^2 v)}\left[\frac{1}{\cosh u}\frac{\partial}{\partial u}\left(\cosh u\frac{\partial\Phi}{\partial u}\right)\right.$$
$$\left.+\frac{1}{\sin v}\frac{\partial}{\partial v}\left(\sin v\frac{\partial\Phi}{\partial v}\right)\right] + \frac{1}{\Delta^2\cosh^2 u \sin^2 v}\frac{\partial^2\Phi}{\partial\phi^2}.$$

(2-61)

In §2.6.4 we shall employ equation (2-61) in all its glory. However, for the moment we concentrate on potentials that are functions $\Phi(u)$ of only the "radial" coordinate u. For potentials of this class, $\nabla^2\Phi = 0$ reduces to

$$\frac{d}{du}\left(\cosh u\frac{d\Phi}{du}\right) = 0.$$

(2-62)

Hence either

$$\Phi = \Phi_0 \qquad \text{(a constant)}, \tag{2-63a}$$

or $(d\Phi/du) = A\operatorname{sech} u$, where A is a constant. Integrating this last equation we find

$$\Phi = A\left[\arctan(\sinh u) + \psi_0\right] = A\left[\frac{\pi}{2} - \arcsin(\operatorname{sech} u) + \psi_0\right], \tag{2-63b}$$

where we have used Figure 2-10 to rearrange the contents of the square parentheses.

For u large, $\operatorname{sech} u \to 0$, so a potential of the form (2-63b) varies as

$$\Phi \simeq A\left(\frac{\pi}{2} + \psi_0 - \operatorname{sech} u\right) \simeq A\left(\frac{\pi}{2} + \psi_0 - \frac{\Delta}{r}\right), \tag{2-64}$$

where r is the usual spherical radius. Hence, if we set $\psi_0 = -\frac{1}{2}\pi$ and $A = GM/\Delta$, the Φ given by equation (2-63b) tends to zero at infinity

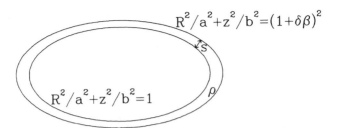

Figure 2-11. A homoeoid of density ρ is bounded by the surfaces $R^2/a^2 + z^2/b^2 = 1$ and $R^2/a^2 + z^2/b^2 = (1 + \delta\beta)^2$. The perpendicular distance s between the bounding surfaces varies with position around the homoeoid.

like the gravitational potential of a body of mass M. Thus we are led to consider the potential defined by

$$\Phi = -\frac{GM}{\Delta} \times \begin{cases} \arcsin(\operatorname{sech} u_0) & (u < u_0), \\ \arcsin(\operatorname{sech} u) & (u \geq u_0). \end{cases} \qquad (2\text{-}65)$$

This potential is everywhere continuous and solves $\nabla^2 \Phi = 0$ everywhere except on the spheroid $u = u_0$. Hence it is the gravitational potential of a shell of material on the surface $u = u_0$. This shell has principal semi-axes of lengths $a \equiv \Delta \cosh u_0$ and $b \equiv \Delta \sinh u_0$. Hence the shell's **eccentricity**

$$e \equiv \sqrt{1 - \frac{b^2}{a^2}} = \operatorname{sech} u_0, \qquad (2\text{-}66)$$

and we may rewrite equations (2-65) as

$$\Phi = -\frac{GM}{ae} \times \begin{cases} \arcsin(e) & (u < u_0), \\ \arcsin(\operatorname{sech} u) & (u \geq u_0). \end{cases} \qquad (2\text{-}67)$$

We can find the surface density of the shell $u = u_0$ by applying Gauss's theorem (2-13) to the potential (2-67). Since $\nabla\Phi = 0$ inside the shell, by equation (2-60) the surface density of the shell is

$$\begin{aligned} \Sigma(u, v) = \frac{\hat{\mathbf{e}}_u \cdot \nabla\Phi}{4\pi G} &= \frac{1}{4\pi G \Delta \sqrt{\sinh^2 u_0 + \cos^2 v}} \left(\frac{d\Phi}{du}\right)_{u=u_0+} \\ &= \frac{M}{4\pi a \sqrt{b^2 + a^2 e^2 \cos^2 v}}, \end{aligned} \qquad (2\text{-}68)$$

where evaluation at u_0+ denotes the limiting value as $u \to u_0$ from above. We now show that (2-68) has a simple physical interpretation:

$\Sigma(u, v)$ is the surface density of the thin shell of uniform density that is bounded by the two surfaces $\beta = 1$ and $\beta = 1 + \delta\beta$ of the set of similar spheroids,

$$\left(\frac{R^2}{a^2} + \frac{z^2}{b^2}\right) = \beta^2 = \text{constant}. \tag{2-69}$$

Figure 2-11 illustrates the proof. The small perpendicular **s** in the figure runs between the shell's inner and outer skins. Thus $\mathbf{s} = s\nabla\beta/|\nabla\beta|$, and at any point on the surface we have $\delta\beta = (\mathbf{s} \cdot \nabla\beta) = s|\nabla\beta|$. Hence $s = \delta\beta/|\nabla\beta|$ and the surface density of the body is

$$\tilde{\Sigma} = \rho s = \frac{\rho\delta\beta}{|\nabla\beta|} = \left(\frac{R^2}{a^4} + \frac{z^2}{b^4}\right)^{-\frac{1}{2}} \rho\beta\delta\beta. \tag{2-70}$$

Finally, writing $R = a\sin v$, $z = b\cos v$, and $e = \sqrt{1 - b^2/a^2}$, we find

$$\tilde{\Sigma} = \frac{ab\rho\beta\delta\beta}{\sqrt{b^2 + a^2 e^2 \cos^2 v}}. \tag{2-71}$$

Comparing (2-71) with (2-68) we see that the potential (2-67) is indeed generated by a thin shell of uniform density that is bounded by similar spheroids. We call such a shell a **thin homoeoid** and have:

Homoeoid Theorem *The exterior isopotential surfaces of a homoeoidal shell of negligible thickness are the spheroids that are confocal with the shell itself. Inside the shell the potential is constant.*

This theorem helps us to understand qualitatively the potential of an inhomogeneous spheroidal body. Each shell of the body makes a contribution to the potential that is constant on surfaces that have the same shape as the shell near its surface, and become progressively rounder as one goes farther out. In the next section we shall interpret this tendency of the isopotentials to become spherical at large radii in terms of the rapid decay of the higher **multipole components** of the gravitational field. The shape of the isopotential surface at a distance r from the center of an inhomogeneous spheroidal body represents a compromise between the rather round contributions of the central shells, and the more aspherical contributions of the shells just interior to r; shells outside r make no contribution. Thus, if the body is very centrally concentrated, the isopotentials near its edge will be nearly round, while a more homogeneous spheroidal body will have more flattened isopotentials.

The homoeoid theorem applies only to a *thin* homoeoid. But equation (2-67) yields a remarkable property of any homoeoid—that is, of any shell, no matter how thick, whose inner and outer surfaces are similar (*not* confocal) ellipsoids:

Newton's Third Theorem *A mass that is inside a homoeoid experiences no net gravitational force from the homoeoid.*

Proof. Break the given homoeoid into a series of thin homoeoids. The interior of the thick homoeoid lies in the interior of each of its component thin homoeoids, and the interior potential of each thin homoeoid is constant. Hence the aggregate interior potential is constant and generates no gravitational force.◁

We now use equation (2-67) to calculate the gravitational potential of a body whose isodensity surfaces are the similar spheroids

$$\text{constant} = m^2 \equiv R^2 + \frac{z^2}{1 - e^2}, \tag{2-72}$$

i.e., a body in which $\rho = \rho(m^2)$. The volume inside an oblate spheroidal shell with semi-axis lengths a and b is

$$V = \tfrac{4}{3}\pi a^2 b = \tfrac{4}{3}\pi a^3 \sqrt{1 - e^2}, \tag{2-73}$$

so the mass of the thin homoeoid that is bounded by the surfaces m and $m + \delta m$ is

$$\delta M = \tfrac{4}{3}\pi \rho(m^2)\sqrt{1 - e^2}\,\delta(m^3). \tag{2-74}$$

There is a unique family of confocal spheroids such that one member of the family coincides with the homoeoid m. Let $u_m(R_0, z_0)$ be the label of the member of this family that passes through the point (R_0, z_0) at which the potential is required. Then if (R_0, z_0) lies inside the homoeoid m, we have on setting $a = m$ in equation (2-67) and substituting for δM from equation (2-74) that the contribution of m to the potential at (R_0, z_0) is

$$\delta\Phi_{\text{int}} \equiv \delta\Phi(R_0, z_0) = -2\pi G\frac{\sqrt{1 - e^2}}{e}\,\arcsin(e)\rho(m^2)\delta(m^2). \tag{2-75a}$$

Similarly, if (R_0, z_0) lies outside the homoeoid,

$$\delta\Phi_{\text{ext}} \equiv \delta\Phi(R_0, z_0) = -2\pi G\frac{\sqrt{1 - e^2}}{e}\,\arcsin(\operatorname{sech} u_m)\rho(m^2)\delta(m^2). \tag{2-75b}$$

The potential of the entire body is the sum of contributions (2-75) from all the homoeoids that make up the body. If we define

$$\psi(m) \equiv \int_0^{m^2} \rho(m^2)dm^2, \tag{2-76}$$

the sum of the $\delta\Phi_{\text{int}}$ is

$$\sum_{m>m_0} \delta\Phi_{\text{int}} = -2\pi G \frac{\sqrt{1-e^2}}{e} \arcsin(e) \left[\psi(\infty) - \psi(m_0)\right], \qquad (2\text{-}77)$$

where m_0 is the label of the homoeoid that passes through (R_0, z_0):

$$m_0^2 \equiv R_0^2 + \frac{z_0^2}{1-e^2}. \qquad (2\text{-}78)$$

Similarly,

$$\sum_{m<m_0} \delta\Phi_{\text{ext}} = -2\pi G \frac{\sqrt{1-e^2}}{e} \int_0^{m_0^2} \arcsin(\text{sech}\, u_m)\rho(m^2)dm^2. \quad (2\text{-}79)$$

Integrating equation (2-79) by parts,

$$\sum_{m<m_0} \delta\Phi_{\text{ext}} = -2\pi G \frac{\sqrt{1-e^2}}{e}$$

$$\times \left\{ [\psi(m)\arcsin(\text{sech}\, u_m)]_{m=0}^{m_0} - \int_{m=0}^{m_0} \frac{\psi(m)d\,\text{sech}\, u_m}{\sqrt{1-\text{sech}^2 u_m}} \right\}.$$
$$(2\text{-}80)$$

The quantity u_m appearing in equation (2-80) is a function $u_m(R_0, z_0)$ by virtue of the condition that u_m label the spheroid through (R_0, z_0) which is confocal with the homoeoid $m = constant$. Let the Δ parameter of the family of spheroids containing m be Δ_m, and let $u_\star$ be the label of the homoeoid m within this family. Then $m = \Delta_m \cosh u_\star$ and $\sqrt{1-e^2}\, m = \Delta_m \sinh u_\star$, so $\Delta_m = me$, and we have from equation (2-58) that

$$\frac{R_0^2}{\Delta_m^2 \cosh^2 u_m} + \frac{z_0^2}{\Delta_m^2 \sinh^2 u_m} = 1$$
$$\text{implies} \quad \frac{R_0^2}{1+\sinh^2 u_m} + \frac{z_0^2}{\sinh^2 u_m} = m^2 e^2, \qquad (2\text{-}81)$$

which is the required equation for u_m. Thus, in particular, $m = 0$ implies $\sinh u_m = \infty$, and $m = m_0$ implies $\sinh u_m = \sqrt{(1-e^2)}/e^2$. Inserting these limits into equation (2-80), and adding the result to equation (2-77), we find

$$\Phi(R_0, z_0) = -2\pi G \frac{\sqrt{1-e^2}}{e}$$

$$\times \left[\psi(\infty)\arcsin e - \int_{\sinh u_m = \sqrt{(1-e^2)}/e}^{\infty} \psi(m) \frac{d\sinh u_m}{1+\sinh^2 u_m} \right]. \qquad (2\text{-}82)$$

We can cast this equation into a slightly more aesthetic form if we define a new variable of integration

$$\tau \equiv a_0^2 e^2 \left[\sinh^2 u_m - \left(\frac{1}{e^2} - 1 \right) \right],$$
(2-83)

where a_0 is any constant. Then equation (2-81) becomes

$$\frac{R_0^2}{\tau + a_0^2} + \frac{z_0^2}{\tau + b_0^2} = \frac{m^2}{a_0^2} \quad (b_0 \equiv \sqrt{1 - e^2} a_0),$$
(2-84a)

and equation (2-82) becomes

$$\Phi(R_0, z_0) = -2\pi G \frac{\sqrt{1 - e^2}}{e}$$
$$\times \left[\psi(\infty) \arcsin e - \frac{a_0 e}{2} \int_0^\infty \frac{\psi(m) d\tau}{(\tau + a_0^2)\sqrt{\tau + b_0^2}} \right].$$
(2-84b)

It is instructive to apply equations (2-84) to the determination of the interior potential of a homogeneous spheroid of density ρ_0 and eccentricity e that has semi-axes of lengths a_1 and $a_3 = \sqrt{1 - e^2} a_1$. For this case, equation (2-76) yields

$$\psi(m) = \rho_0 \times \begin{cases} m^2, & m^2 < a_1^2, \\ a_1^2, & (m^2 \geq a_1^2). \end{cases}$$
(2-85)

Equation (2-84b) takes on a particularly simple form if we set the arbitrary constant a_0 equal to a_1. If (R_0, z_0) lies inside the spheroid, $m(\tau)$ is always smaller than a_0. Hence, we may substitute from equation (2-84a) and (2-85) into (2-84b), to obtain

$$\Phi(R_0, z_0) = -2\pi G \rho_0 a_1^2 \frac{\sqrt{1 - e^2}}{e} \left[\arcsin e \right.$$
$$\left. - \frac{a_1 e}{2} \int_0^\infty \left(\frac{R_0^2}{\tau + a_1^2} + \frac{z_0^2}{\tau + a_3^2} \right) \frac{d\tau}{(\tau + a_1^2)\sqrt{\tau + a_3^2}} \right].$$
(2-86)

This potential is quadratic in the coordinates and may be written

$$\Phi(\mathbf{x}) = -\pi G \rho_0 (I a_1^2 - A_1 R^2 - A_3 z^2),$$
(2-87)

where the dimensionless coefficients I and A_i are given in Table 2-1. An expression for the exterior potential of the homogeneous spheroid is given in Table 2-2.

Table 2-1. Formulae for the dimensionless quantities $I \equiv a_2 a_3 a_1^{-1} \int_0^\infty \Delta^{-1} d\tau$ and $A_i \equiv a_1 a_2 a_3 \int_0^\infty \Delta^{-1}(a_i^2+\tau)^{-1} d\tau$ that occur in equations (2-87) and Table 2-2. $\Delta^2 \equiv \prod_{i=1}^3 (a_i^2 + \tau)$.

	$a_1 = a_2 > a_3$ (oblate)	$a_1 = a_2 < a_3$ (prolate)	$a_1 > a_2 > a_3$ (triaxial)
	$e \equiv \sqrt{1 - a_3^2/a_1^2}$	$e \equiv \sqrt{1 - a_1^2/a_3^2}$	$k \equiv \sqrt{\dfrac{a_1^2 - a_2^2}{a_1^2 - a_3^2}}$; $k'^2 = 1 - k^2$; $\theta \equiv \arccos\left(\dfrac{a_3}{a_1}\right)$
I	$\dfrac{2\sqrt{1-e^2}}{e} \arcsin e$	$\dfrac{\sqrt{1-e^2}}{e} \ln\left(\dfrac{1+e}{1-e}\right)$	$2\dfrac{a_2 a_3}{a_1^2} \dfrac{F(\theta, k)}{\sin\theta}$
A_1	$\dfrac{\sqrt{1-e^2}}{e^2}\left[\dfrac{\arcsin e}{e} - \sqrt{1-e^2}\right]$	$\dfrac{1-e^2}{e^2}\left[\dfrac{1}{1-e^2} - \dfrac{1}{2e}\ln\left(\dfrac{1+e}{1-e}\right)\right]$	$2\dfrac{a_2 a_3}{a_1^2} \dfrac{F(\theta, k) - E(\theta, k)}{k^2 \sin^3\theta}$
A_2	$= A_1$	$= A_1$	$2\dfrac{a_2 a_3}{a_1^2} \dfrac{E(\theta,k) - k'^2 F(\theta,k) - (a_3/a_2) k^2 \sin\theta}{k^2 k'^2 \sin^3\theta}$
A_3	$2\dfrac{\sqrt{1-e^2}}{e^2}\left[\dfrac{1}{\sqrt{1-e^2}} - \dfrac{\arcsin e}{e}\right]$	$2\dfrac{1-e^2}{e^2}\left[\dfrac{1}{2e}\ln\left(\dfrac{1+e}{1-e}\right) - 1\right]$	$2\dfrac{a_2 a_3}{a_1^2} \dfrac{(a_2/a_3)\sin\theta - E(\theta,k)}{k'^2 \sin^3\theta}$

Table 2-2. Potentials and potential-energy tensors of ellipsoidal bodies.

Thin shell	$\Phi(\mathbf{x}_{\text{int}}) = -\dfrac{Ga_1}{2a_2 a_3} I(\mathbf{a}) M_{\text{shell}}$	$\Phi(\mathbf{x}_{\text{ext}}) = -\dfrac{Ga_1'}{2a_2' a_3'} I(\mathbf{a}') M_{\text{shell}}$
Homogeneous	$\Phi(\mathbf{x}_{\text{int}}) = -\pi G\rho[I(\mathbf{a})a_1^2 - \sum_{i=1}^3 A_i(\mathbf{a})x_i^2]$	$W_{ij} = -\dfrac{8}{15}\pi^2 G\rho^2 a_1 a_2 a_3 A_i a_i^2 \delta_{ij}$
	$\Phi(\mathbf{x}_{\text{ext}}) = -\pi G\rho \dfrac{a_1 a_2 a_3}{a_1' a_2' a_3'}[I(\mathbf{a}')a_1'^2 - \sum_{i=1}^3 A_i(\mathbf{a}')x_i^2]$	$W = -\dfrac{8}{15}\pi^2 G\rho^2 a_1 a_2 a_3 I a_1^2$
Inhomogeneous	$\Phi(\mathbf{x}) = -\pi G\dfrac{a_2 a_3}{a_1}\displaystyle\int_0^\infty \{\psi(\infty) - \psi[m(\tau, \mathbf{x})]\}\dfrac{d\tau}{\Delta}$	$W_{ij} = -2\pi^2 G\dfrac{a_2 a_3}{a_1^4}\mathbf{S} A_i a_i^2 \delta_{ij}$ $\qquad W = -2\pi^2 G\dfrac{a_2 a_3}{a_1^2}\mathbf{S} I \delta_{ij}$

NOTES: I and A_i as in Table 2-1. If $\sum_{i=1}^3 x_i^2/[a_i^2 + \lambda(\mathbf{x})] = 1$, then $a_i'^2 \equiv a_i^2 + \lambda(\mathbf{x})$; $\Delta^2(\tau) \equiv \prod_{i=1}^3 (a_i^2 + \tau)$; $m^2(\tau, \mathbf{x}) \equiv a_1^2 \sum_{i=1}^3 x_i^2/(a_i^2 + \tau)$; $\psi(m) \equiv \int_0^{m^2} \rho(\mathbf{x}) dm^2$; $\mathbf{S} \equiv \int_0^\infty dm^2 \rho(m^2) \int_0^{m^2} \rho(m'^2) m'^2 dm'^2$.

With the help of equations (2-76) and (2-84) we obtain the gravitational force generated by a spheroidal system as

$$\mathbf{F} = -\nabla\Phi = -\pi G\sqrt{1-e^2}\, a_0 \int_0^\infty \frac{\rho(m^2)\nabla m^2 d\tau}{(\tau + a_0^2)\sqrt{\tau + b_0^2}}. \tag{2-88}$$

We may use equation (2-88) to find the circular speed $v_c(R)$ in the equatorial plane of an oblate spheroidal galaxy. The radial component of the force (2-88) is

$$F_R = -2\pi G\sqrt{1-e^2}\, a_0^3 R \int_0^\infty \frac{\rho(m^2)d\tau}{(\tau + a_0^2)^2\sqrt{\tau + b_0^2}}. \tag{2-89}$$

In the equatorial plane $z = 0$, equation (2-84a) yields

$$m = \frac{a_0 R}{\sqrt{\tau + a_0^2}}. \tag{2-90a}$$

Hence

$$\frac{d\tau}{(\tau + a_0^2)^2} = -\frac{2m}{R^2 a_0^2}dm, \tag{2-90b}$$

and equation (2-89) yields

$$v_c^2(R) = 4\pi G\sqrt{1-e^2} \int_0^R \frac{\rho(m^2)m^2 dm}{\sqrt{R^2 - m^2 e^2}}. \tag{2-91}$$

Let us see how these formulae work out in a specific case. Consider the oblate spheroidal density distribution

$$\rho(m^2) = \rho_0 \left[1 + \left(\frac{m}{a_0}\right)^2\right]^{-\frac{3}{2}}, \tag{2-92}$$

where a_0 is the core radius and e is the eccentricity of the system. In the limit $e \to 0$ this reduces to the modified Hubble profile (2-37). We substitute for ρ into equation (2-91) to obtain

$$v_c^2(R) = 4\pi G\rho_0 a_0^3 \frac{\sqrt{1-e^2}}{e} \int_0^R \frac{m^2 dm}{(a_0^2 + m^2)^{\frac{3}{2}}\sqrt{R^2/e^2 - m^2}}. \tag{2-93}$$

By making the substitution

$$m = \frac{R\sin\theta}{e\sqrt{1 + (R/ea_0)^2\cos^2\theta}} \tag{2-94}$$

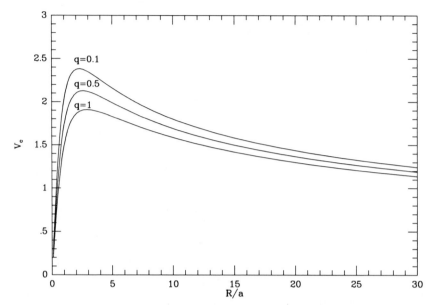

Figure 2-12. Circular speed versus radius for three bodies with the same face-on projected density profile (the modified Hubble profile) but different axis ratios $q = b/a$. Though all three bodies have the same total mass inside a spheroid of given semi-major axis, v_c increases with flattening $1 - q$.

one may show that the integral of equation (2-93) equals

$$\frac{ek}{R} \left[F(\theta_m, k) - E(\theta_m, k) \right], \tag{2-95}$$

where F and E are incomplete elliptic integrals (see Appendix 1.C.4),

$$k \equiv \left[\left(\frac{a_0 e}{R} \right)^2 + 1 \right]^{-\frac{1}{2}}, \quad \text{and} \quad \theta_m \equiv \arcsin \sqrt{\frac{e^2 a_0^2 + R^2}{a_0^2 + R^2}}. \tag{2-96a}$$

Hence

$$v_c^2(R) = 4\pi G \rho_0 a_0^3 \frac{\sqrt{1 - e^2}}{R} k \left[F(\theta_m, k) - E(\theta_m, k) \right]. \tag{2-96b}$$

We may use this result to investigate how strongly a galaxy's circular speed is affected by its shape. In Figure 2-12 we plot the circular-speed curves of three galaxies whose density profiles are given by equation (2-92) for axis ratio $q = \sqrt{1 - e^2} = 1$ (spherical system), $q = 0.5$ (E5 galaxy), and $q = 0.1$ (the flatness characteristic of spiral galaxies). The

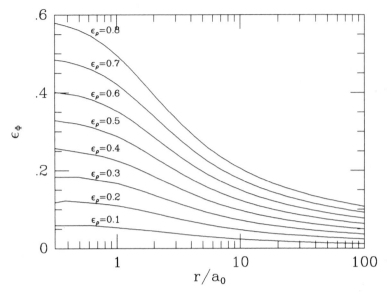

Figure 2-13. The ellipticity ϵ_Φ of an equipotential surface versus the surface's semi-major axis length r. Each curve is labeled by the ellipticity $\epsilon_\rho = 1 - q$ of the body with density (2-92) that generates the corresponding potential. Notice the rapidity with which the equipotential surfaces become spherical at large r/a_0.

central density has in each case been adjusted so as to hold constant the mass $M(a)$ interior to the spheroid of semi-major axis a. Notice that the peak circular speed of the $q = 0.1$ model is about 20% higher than that of the spherical system, as one might have expected from the greater compactness of the flattened model.

For the density distribution defined by equation (2-92), Figure 2-13 shows the ellipticity ϵ_Φ of the isopotential surfaces for several values of the ellipticity $\epsilon_\rho \equiv 1 - q$ of the density distribution. One sees that in the core $r < a_0$, $\epsilon_\Phi \gtrsim \frac{1}{2}\epsilon_\rho$, while at a few core radii $\epsilon_\Phi \approx \frac{1}{3}\epsilon_\rho$. Thus the potential is generally flattened only about a third as much as the density.

2 Triaxial Systems

The results of this section can all be generalized to yield the potentials of bodies whose isodensity surfaces are the triaxial ellipsoids on which the Cartesian coordinates (x_1, x_2, x_3) satisfy the condition that

$$m^2 \equiv a_1^2 \sum_{i=1}^{3} \frac{x_i^2}{a_i^2} \qquad (2\text{-}97)$$

be constant (Ferrers 1877; Chandrasekhar 1969). In particular, a thin shell of uniform material whose inner and outer skins are the surfaces m and $m + \delta m$, generates an exterior potential that is constant on the ellipsoidal surfaces

$$\text{constant} \equiv m^2 = a_1^2 \sum_{i=1}^{3} \frac{x_i^2}{a_i^2 + \tau}, \qquad (2\text{-}98)$$

where $\tau \geq 0$ labels the isopotential surfaces. There is no gravitational field inside such a shell.

We may find the gravitational potential of any body in which $\rho = \rho(m^2)$ by breaking the body down into thin triaxial homoeoids: the triaxial analog of equation (2-84b) is

$$\Phi(\mathbf{x}) = -\pi G \left(\frac{a_2 a_3}{a_1} \right) \int_0^\infty \frac{[\psi(\infty) - \psi(m)] d\tau}{\sqrt{(\tau + a_1^2)(\tau + a_2^2)(\tau + a_3^2)}}, \qquad (2\text{-}99)$$

where $\psi(m)$ is again defined by (2-76) and $m = m(\mathbf{x}, \tau)$ through equation (2-98).

(a) Ferrers potentials   A particularly simple application of equation (2-99) is to the case in which

$$\rho(m^2) = \rho_0 \left(1 - m^2/a_1^2 \right)^n \quad (m \leq a_1) \; ; \quad \rho(m^2) = 0 \quad (m > a_1), \quad (2\text{-}100)$$

where $m = m(\mathbf{x})$ through equation (2-97). By (2-76) we now have

$$\psi(\infty) - \psi(m) = \frac{\rho_0 a_1^2}{n+1} \left(1 - \frac{m^2}{a_1^2} \right)^{(n+1)} \quad (m \leq a_1). \qquad (2\text{-}101)$$

Hence the internal potential of a body whose density is of the form (2-100) is

$$\Phi(\mathbf{x}) = -\frac{\pi G \rho_0 a_1 a_2 a_3}{n+1} \int_0^\infty \left(1 - \sum_{i=1}^{3} \frac{x_i^2}{\tau + a_i^2} \right)^{(n+1)}$$
$$\times \frac{d\tau}{\sqrt{(\tau + a_1^2)(\tau + a_2^2)(\tau + a_3^2)}}. \qquad (2\text{-}102)$$

If n is an integer, the bracket involving $\mathbf{x}$ in equation (2-102) can be multiplied out, and the potential at any point obtained as a sum of terms of the form $A_{pqr} x_1^p x_2^q x_3^r$, where the coefficients A_{pqr} are independent of $\mathbf{x}$. Potentials of this simple form are ideally suited to numerical studies of orbits in triaxial galaxies, such as we shall describe in §3.3 and §3.4. We shall refer to these potentials as **Ferrers potentials**.

The $n = 0$ Ferrers potential arises from a homogeneous ellipsoid with semi-axes a_1, a_2, a_3. Expressions for the interior and exterior potential of such bodies can be derived from equations (2-99) and (2-102) and are given in Tables 2-1 and 2-2.

2.4 Multipole Expansion

In the last section we found the potentials generated by bodies whose isodensity surfaces are similar concentric ellipsoids. We now show how to calculate the potential of an arbitrary distribution of matter.

Our first step is to obtain the potential of a thin spherical shell of variable surface density. If the shell is of negligible thickness, the task of solving Poisson's equation $\nabla^2 \Phi = 4\pi G \rho$ reduces to that of solving Laplace's equation $\nabla^2 \Phi = 0$ inside and outside the shell, subject to suitable boundary conditions at infinity, at the origin, and on the shell. Now in spherical coordinates Laplace's equation is [eq. (1B-51)]

$$\frac{1}{r^2} \frac{\partial}{\partial r} \left(r^2 \frac{\partial \Phi}{\partial r} \right) + \frac{1}{r^2 \sin \theta} \frac{\partial}{\partial \theta} \left(\sin \theta \frac{\partial \Phi}{\partial \theta} \right) + \frac{1}{r^2 \sin^2 \theta} \frac{\partial^2 \Phi}{\partial \phi^2} = 0. \quad (2\text{-}103)$$

This may be solved by the method of **separation of variables**. We seek special solutions that are the product of functions of one variable only:

$$\Phi(r, \theta, \phi) = R(r) P(\theta) Q(\phi). \quad (2\text{-}104a)$$

Substituting equation (2-104a) into (2-103) and rearranging, we obtain

$$\frac{\sin^2 \theta}{R} \frac{d}{dr} \left(r^2 \frac{dR}{dr} \right) + \frac{\sin \theta}{P} \frac{d}{d\theta} \left(\sin \theta \frac{dP}{d\theta} \right) = -\frac{1}{Q} \frac{d^2 Q}{d\phi^2}. \quad (2\text{-}104b)$$

The left side of this equation does not depend on ϕ, and the right side does not depend on r or θ. It follows that both sides are equal to some constant, say m^2. Hence

$$-\frac{1}{Q} \frac{d^2 Q}{d\phi^2} = m^2, \quad (2\text{-}105a)$$

$$\frac{\sin^2\theta}{R}\frac{d}{dr}\left(r^2\frac{dR}{dr}\right) + \frac{\sin\theta}{P}\frac{d}{d\theta}\left(\sin\theta\frac{dP}{d\theta}\right) = m^2. \qquad \text{(2-105b)}$$

Equation (2-105a) may be immediately integrated to

$$Q(\phi) = Q_m^+ e^{im\phi} + Q_m^- e^{-im\phi}. \qquad \text{(2-106a)}$$

We require Φ to be a periodic function of ϕ with period 2π, so m can take only integer values. Since equations (2-105) depend only on m^2, we could restrict our attention to non-negative values of m without loss of generality. However, if we allow m to take both positive and negative values, then the second exponential in equation (2-106a) becomes redundant, and we may write simply

$$Q = Q_m e^{im\phi} \quad (m = \ldots, -1, 0, 1, \ldots). \qquad \text{(2-106b)}$$

Equation (2-105b) can be written

$$\frac{1}{R}\frac{d}{dr}\left(r^2\frac{dR}{dr}\right) = \frac{m^2}{\sin^2\theta} - \frac{1}{P\sin\theta}\frac{d}{d\theta}\left(\sin\theta\frac{dP}{d\theta}\right). \qquad \text{(2-107)}$$

Since the left side of this equation does not depend on θ and the right side does not depend on r, both sides must equal some constant, which we write as $l(l+1)$. Thus equation (2-107) implies

$$\frac{1}{R}\frac{d}{dr}\left(r^2\frac{dR}{dr}\right) - l(l+1)R = 0, \qquad \text{(2-108a)}$$

and in terms of $x \equiv \cos\theta$,

$$\frac{d}{dx}\left[(1-x^2)\frac{dP}{dx}\right] - \frac{m^2}{1-x^2}P + l(l+1)P = 0. \qquad \text{(2-108b)}$$

Two linearly independent solutions of equation (2-108a) are

$$R(r) = Ar^l \quad \text{and} \quad R(r) = Br^{-(l+1)}. \qquad \text{(2-109)}$$

The solutions of equation (2-108b) are associated Legendre functions $P_l^{|m|}(x)$ (see Appendix 1.C.5). Physically acceptable solutions exist only when l is a non-negative integer and $|m| \leq l$. When $m = 0$ the solutions are simply polynomials in x, called Legendre polynomials $P_l(x)$.

Rather than write out the product $P_l^{|m|}(\cos\theta)e^{im\phi}$ again and again, it is helpful to define the spherical harmonic $Y_l^m(\theta, \phi)$, which is equal to $P_l^{|m|}(\cos\theta)e^{im\phi}$ times a constant chosen so that the Y_l^m satisfy the orthogonality relation [see eq. (1C-29)]

$$\int_0^\pi \sin\theta d\theta \int_0^{2\pi} d\phi\, Y_l^{m*}(\theta,\phi)\, Y_{l'}^{m'}(\theta,\phi) = \delta_{ll'}\delta_{mm'}, \qquad (2\text{-}110)$$

where δ_{ij} is unity if $i=j$ and zero otherwise.

Putting all these results together, we have from equations (2-104a), (2-107b), and (2-109) that

$$\Phi_{lm}(r,\theta,\phi) = \left[A_{lm}r^l + B_{lm}r^{-(l+1)}\right]Y_l^m(\theta,\phi) \qquad (2\text{-}111)$$

is a solution of $\nabla^2\Phi = 0$ for all non-negative integers l and integer m in the range $-l \le m \le l$.

Now let us apply the special solutions defined by equation (2-111) to the problem of determining the potential of a thin shell of radius a and surface density $\sigma(\theta,\phi)$. We write the potential as

$$\Phi_{\text{int}}(r,\theta,\phi) = \sum_{l=0}^{\infty}\sum_{m=-l}^{l}\left[A_{lm}r^l + B_{lm}r^{-(l+1)}\right]Y_l^m(\theta,\phi) \quad (r \le a),$$
$$(2\text{-}112a)$$

and

$$\Phi_{\text{ext}}(r,\theta,\phi) = \sum_{l=0}^{\infty}\sum_{m=-l}^{l}\left[C_{lm}r^l + D_{lm}r^{-(l+1)}\right]Y_l^m(\theta,\phi) \quad (r \ge a).$$
$$(2\text{-}112b)$$

The potential at the center must be finite, so $B_{lm} = 0$ for all l,m. Similarly, the potential at infinity must be finite, so $C_{lm} = 0$ for $l > 0$, and we may choose the zero of the potential so that $C_{00} = 0$. Now let us expand the surface density as

$$\sigma(\theta,\phi) = \sum_{l=0}^{\infty}\sum_{m=-l}^{l}\sigma_{lm}Y_l^m(\theta,\phi), \qquad (2\text{-}113)$$

where the σ_{lm} are numbers yet to be determined. To obtain the coefficient $\sigma_{l'm'}$, we multiply both sides of equation (2-113) by $Y_{l'}^{m'*}(\theta,\phi)\sin\theta$ and integrate over all angles. With equation (2-110) we find

$$\int_0^\pi \sin\theta d\theta \int_0^{2\pi} d\phi\, Y_{l'}^{m'*}(\theta,\phi)\sigma(\theta,\phi) = \sigma_{l'm'}. \qquad (2\text{-}114)$$

Notice that $\sigma_{00} = M/(2a^2\sqrt{\pi})$, where M is the mass of the shell.

Gauss's theorem [eq. (2-13)] applied to a small piece of the shell tells us that

$$\left(\frac{\partial \Phi_{\text{ext}}}{\partial r}\right)_{r=a} - \left(\frac{\partial \Phi_{\text{int}}}{\partial r}\right)_{r=a} = 4\pi G \sigma(\theta, \phi), \qquad (2\text{-}115)$$

so inserting equations (2-112) into equation (2-115), we obtain

$$-\sum_{l=0}^{\infty} \sum_{m=-l}^{l} \left[(l+1)D_{lm}a^{-(l+2)} + lA_{lm}a^{(l-1)}\right] Y_l^m(\theta, \phi) =$$
$$4\pi G \sum_{l=0}^{\infty} \sum_{m=-l}^{l} \sigma_{lm} Y_l^m(\theta, \phi). \qquad (2\text{-}116)$$

Furthermore, $\Phi_{\text{ext}}(a, \theta, \phi)$ must equal $\Phi_{\text{int}}(a, \theta, \phi)$ because zero work can be done in passing through an infinitesimally thin shell, so from equations (2-112) we have

$$\sum_{l=0}^{\infty} \sum_{m=-l}^{l} A_{lm}a^l Y_l^m(\theta, \phi) = \sum_{l=0}^{\infty} \sum_{m=-l}^{l} D_{lm}a^{-(l+1)} Y_l^m(\theta, \phi). \qquad (2\text{-}117)$$

The coefficients $A_{lm}a^l$ etc. of each spherical harmonic Y_l^m on each side of equation (2-117) must be equal, as can be shown by multiplying both sides of the equation by $Y_{l'}^{m'*}(\theta, \phi)$, integrating over angles, and using the orthogonality relation (2-110). Therefore, from equation (2-117) we have

$$D_{lm} = A_{lm}a^{(2l+1)}. \qquad (2\text{-}118)$$

By a similar argument we obtain from equation (2-116) that

$$A_{lm} = -4\pi G a^{-(l-1)} \frac{\sigma_{lm}}{2l+1} \quad ; \quad D_{lm} = -4\pi G a^{(l+2)} \frac{\sigma_{lm}}{2l+1}. \qquad (2\text{-}119)$$

Collecting these results together, we have from equations (2-112) that

$$\Phi_{\text{int}}(r, \theta, \phi) = -4\pi G a \sum_{l=0}^{\infty} \left(\frac{r}{a}\right)^l \sum_{m=-l}^{l} \frac{\sigma_{lm}}{2l+1} Y_l^m(\theta, \phi)$$
$$\Phi_{\text{ext}}(r, \theta, \phi) = -4\pi G a \sum_{l=0}^{\infty} \left(\frac{a}{r}\right)^{(l+1)} \sum_{m=-l}^{l} \frac{\sigma_{lm}}{2l+1} Y_l^m(\theta, \phi), \qquad (2\text{-}120)$$

where the σ_{lm} are given by equation (2-114).

We may now evaluate the potential of a solid body by breaking it down into a series of spherical shells. We let $\delta\sigma_{lm}(a)$ be the σ-coefficient of the shell lying between a and $a + \delta a$, and $\delta\Phi(r, \theta, \phi; a)$ be the corresponding potential at r. Then we have by equation (2-114)

$$\delta\sigma_{lm}(a) = \int_0^\pi \sin\theta d\theta \int_0^{2\pi} d\phi\, Y_l^{m*}(\theta, \phi)\rho(a, \theta, \phi)\delta a \equiv \rho_{lm}(a)\delta a.$$
(2-121)

Substituting these values of σ_{lm} into equations (2-120) and integrating over all a, we obtain the potential at r generated by the entire collection of shells:

$$\Phi(r, \theta, \phi) = \sum_{a=0}^r \delta\Phi_{\text{ext}} + \sum_{r=a}^\infty \delta\Phi_{\text{int}}$$

$$= -4\pi G \sum_{l,m} \frac{Y_l^m(\theta, \phi)}{2l+1}\left[\frac{1}{r^{(l+1)}}\int_0^r \rho_{lm}(a)a^{(l+2)}da\right.$$
(2-122)

$$\left. + r^l \int_r^\infty \rho_{lm}(a)\frac{da}{a^{(l-1)}}\right].$$

This equation gives the potential generated by the body as an expansion in **multipoles**: the terms associated with $l = m = 0$ are the **monopole** terms, those associated with $l = 1$ are **dipole** terms, those with $l = 2$ are **quadrupole** terms, and so forth. Similar expansions will be familiar from electrostatics (e.g., Jackson 1975). The monopole terms are the same as the two terms in equation (2-22). Since there is no gravitational analog of negative charge, the monopole terms are always at least as important as any of the terms arising from higher-order multipoles. In particular, if one places the origin of coordinates at the center of mass of the system, the dipole term (which is often dominant in the electrostatic case) vanishes identically outside any matter distribution. While the monopole terms generate a circular speed curve $v_c(r) = \sqrt{GM(r)/r}$ that never declines with increasing r more steeply than in the Keplerian case ($v_c \propto r^{-1/2}$), the higher-order multipoles may cause the circular speed to fall more steeply with increasing radius. However, this behavior never extends over a large range in r.

As an illustration of the effectiveness of the multipole expansion, we show in Figure 2-14 the contours of Satoh's potential $\Phi_M^\infty(R, z)$ [eq. (2-53a)], together with the approximations to this potential that one obtains from equation (2-122) if one includes only terms with $l \le 2$ or 8. The flexibility of the multipole expansion makes it an excellent tool for numerical work, and it is used in the N-body models that will be described in §2.8 and §4.7.

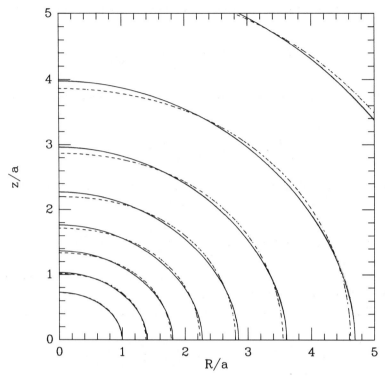

Figure 2-14. Equipotentials of Satoh's density distribution (2-53b) with $b/a = 1$. Full curves show the exact equipotentials computed from equation (2-53a), and dashed curves show the estimate provided by equation (2-122) with the sum over l extending to $l = 2$. Contours based on the sum to $l = 8$ are also plotted (dotted contours) but almost overlie the full curves.

2.5 Potential Energy Tensors

In §4.3 we shall encounter the tensor **W** that is defined by

$$W_{jk} = -\int \rho(\mathbf{x}) x_j \frac{\partial \Phi}{\partial x_k} d^3\mathbf{x}, \qquad (2\text{-}123)$$

where ρ and Φ are the density and potential of some body, and the integral is to be taken over all space. In this section we deduce some useful properties of **W**, which is known as the **Chandrasekhar potential energy tensor**.

If we substitute for Φ from equation (2-3), **W** becomes

$$W_{jk} = G \int \rho(\mathbf{x}) x_j \frac{\partial}{\partial x_k} \int \frac{\rho(\mathbf{x}')}{|\mathbf{x}' - \mathbf{x}|} d^3\mathbf{x}' d^3\mathbf{x}. \qquad (2\text{-}124)$$

Since the range of the integration over $\mathbf{x}'$ does not depend on $\mathbf{x}$, we may carry the differentiation inside the integral to find

$$W_{jk} = G \iint \rho(\mathbf{x})\rho(\mathbf{x}') \frac{x_j(x_k' - x_k)}{|\mathbf{x}' - \mathbf{x}|^3} d^3\mathbf{x}' d^3\mathbf{x}. \qquad (2\text{-}125a)$$

Furthermore, since $\mathbf{x}$ and $\mathbf{x}'$ are dummy variables of integration, we may relabel them and write

$$W_{jk} = G \iint \rho(\mathbf{x}')\rho(\mathbf{x}) \frac{x_j'(x_k - x_k')}{|\mathbf{x} - \mathbf{x}'|^3} d^3\mathbf{x} d^3\mathbf{x}'. \qquad (2\text{-}125b)$$

Finally, on interchanging the order of integration in equation (2-125b) and adding the result to equation (2-125a), we obtain

$$W_{jk} = -\tfrac{1}{2}G \iint \rho(\mathbf{x})\rho(\mathbf{x}') \frac{(x_j' - x_j)(x_k' - x_k)}{|\mathbf{x}' - \mathbf{x}|^3} d^3\mathbf{x}' d^3\mathbf{x}. \qquad (2\text{-}126)$$

From this expression we draw the important inference that the tensor **W** is **symmetric**, that is, that $W_{jk} = W_{kj}$.

When we take the **trace** of both sides of equation (2-126), we find

$$\text{trace}(\mathbf{W}) \equiv \sum_{j=1}^{3} W_{jj} = -\tfrac{1}{2}G \int \rho(\mathbf{x}) \int \frac{\rho(\mathbf{x}')}{|\mathbf{x}' - \mathbf{x}|} d^3\mathbf{x}' d^3\mathbf{x}$$
$$= \tfrac{1}{2} \int \rho(\mathbf{x})\Phi(\mathbf{x}) d^3\mathbf{x}. \qquad (2\text{-}127)$$

Comparing this with equation (2-19) we see that $\text{trace}(\mathbf{W})$ is simply the total gravitational potential energy W.

For the important special case of spherical bodies, some useful expressions for W can be derived directly from equation (2-123). Taking its trace we have

$$W = -\int \rho\mathbf{x} \cdot \nabla\Phi d^3\mathbf{x}. \qquad (2\text{-}128)$$

But if the body is spherical,

$$\nabla\Phi = \frac{d\Phi}{dr}\hat{\mathbf{e}}_r = \frac{GM(r)}{r^2}\hat{\mathbf{e}}_r, \qquad (2\text{-}129)$$

where $M(r)$ is the total mass interior to r. Substituting this expression into equation (2-128) and integrating over all directions of $\mathbf{r}$, we obtain

$$W = -4\pi G \int_0^\infty \rho M r\, dr. \qquad (2\text{-}130)$$

As an example of the use of this formula, consider the potential energy of a homogeneous sphere of radius a and density ρ. We have $M(r) = \frac{4}{3}\pi\rho r^3$, and therefore

$$W = -\frac{16\pi^2}{3}G\rho^2 \int_0^a r^4 dr = -\frac{16}{15}\pi^2 G\rho^2 a^5 = -\frac{3}{5}\frac{GM^2}{a}. \qquad (2\text{-}131)$$

Sometimes it is useful to characterize the size of a system that lacks a sharp boundary by quoting a radius r_g, called the **gravitational radius**, that is simply related to the system's mass and potential energy:

$$r_g \equiv \frac{GM^2}{|W|}. \qquad (2\text{-}132)$$

The potential energy tensor of a spherical body is **diagonal**, that is, $W_{jk} = 0$ for $j \neq k$ (see Problem 2-4). Furthermore, since spherical symmetry requires that the three diagonal components of $\mathbf{W}$ be equal to one another, $\mathbf{W}$ may be written

$$W_{jk} = \tfrac{1}{3}W\delta_{jk}. \qquad (2\text{-}133)$$

Such tensors are said to be **isotropic**.

If the body is flattened along some axis, say the x_3 axis, W_{33} will be smaller than the other components because for most pairs of matter elements, $(x_3 - x_3')^2 < (x_1 - x_1')^2$.

P. H. Roberts (1962) evaluated the expression on the right side of equation (2-126) for density distributions that are constant on similar concentric ellipsoids. Writing the density $\rho(m^2)$, where m is defined by equation (2-97), and defining $\psi(m)$ by equation (2-76), he obtained the result

$$W_{jk} = -\pi^2 G \frac{a_2 a_3}{a_1^2} \left(\frac{a_j}{a_1}\right)^2 A_j \delta_{jk} \int_0^\infty [\psi(\infty) - \psi(m)]^2 \, dm, \qquad (2\text{-}134)$$

where the A_j are the quantitities given in Table 2-1. Notice that the right side of equation (2-134) comprises a constant times the product of two factors: (i) a factor $(a_j/a_1)^2 A_j \delta_{jk}$ that depends only on the axial ratios (a_2/a_1) etc.; and (ii) a factor $\int [\psi(\infty) - \psi(m)]^2 \, dm$ that is independent of the body's ellipticity and is the same for all components of the tensor; this integral can be evaluated from a knowledge of the radial density structure alone. In particular, ratios of potential energy terms, for example (W_{11}/W_{33}), depend only on the body's ellipticity, and *are entirely independent of the radial density structure*. We shall exploit this useful result in §4.3. In Table 2-2 we give expressions for the potential energy tensors of homogeneous ellipsoids.

2.6 Potentials of Disks*

Most of the light emitted by a typical spiral galaxy comes from a thin disk. Whether the mass of such a galaxy is equally concentrated into a disk is a question that can only be answered by a study of the system's dynamics. But we may anticipate that a substantial fraction of the galaxy's mass is concentrated in the disk, and it is therefore useful to be able to calculate the gravitational field generated by an idealized disk of zero thickness. We shall be mainly concerned with axisymmetric disks.

There are three different techniques for calculating the potential of a thin disk. In the first approach, we treat the disk as a very flat ellipsoid and use the formulae of §2.3 to obtain the potential. This approach helps us to gain an intuitive understanding of the relationship between a disk's radial density distribution and its potential, but it is not well suited to detailed calculations of disk potentials. Therefore later in this section we develop two other techniques that are better suited to the calculation of specific disk potentials.

1 Disks as Flattened Spheroids

Consider first how we can represent a non-uniform disk as a collection of highly flattened homoeoids. A homogeneous spheroid of density ρ, semi-axes a and c, and axial ratio $q = c/a$ has mass $M(a) = \frac{4}{3}\pi\rho q a^3$ [eq. (2-73)] and surface density (see Figure 2-15)

$$\Sigma(a, R) = 2\rho q\sqrt{a^2 - R^2}, \qquad (2\text{-}135)$$

where R is the usual cylindrical radius. Differentiating these expressions with respect to a, we obtain the mass $\delta M(a)$ and the surface density $\delta\Sigma(a, R)$ of the thin homoeoid of density ρ, semi-major axis a, thickness δa, and axial ratio q,

$$\delta M(a) = 4\pi\rho q a^2 \delta a \quad ; \quad \delta\Sigma(a, R) = \frac{2\rho q a}{\sqrt{a^2 - R^2}}\delta a. \qquad (2\text{-}136)$$

If we now let q tend to zero while holding $2\rho q a \equiv \Sigma_0$ constant, we obtain the mass and surface density of an infinitely flattened homoeoid:

$$\delta M(a) = 2\pi\Sigma_0 a\delta a \quad ; \quad \delta\Sigma(a, R) = \frac{\Sigma_0\delta a}{\sqrt{a^2 - R^2}}. \qquad (2\text{-}137)$$

---

* This section contains more advanced material and can be skipped on a first reading.

to observer

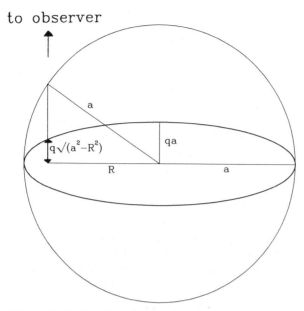

Figure 2-15. A spheroid of axis ratio q and semi-major axis a is viewed along a line of sight that cuts the spheroid's equatorial plane perpendicularly at radius R. This line of sight cuts through the spheroid for a distance $2q\sqrt{a^2 - R^2}$.

We may construct a razor-thin disk of known surface density $\Sigma(R)$ by finding the family of homoeoids whose combined surface density equals $\Sigma(R)$ at all R. In mathematical language, we have to find the function $\Sigma_0(a)$ that satisfies the integral equation

$$\Sigma(R) = \sum_{a \geq R} \delta\Sigma(a, R) = \int_R^\infty \frac{\Sigma_0(a)da}{\sqrt{a^2 - R^2}}. \qquad (2\text{-}138a)$$

This is an Abel integral equation. Its solution is [eq. (1B-59)]

$$\Sigma_0(a) = -\frac{2}{\pi}\frac{d}{da}\int_a^\infty \frac{\Sigma(R)RdR}{\sqrt{R^2 - a^2}}. \qquad (2\text{-}138b)$$

In equation (2-138a), note that some of the mass at radius R comes from homoeoids having $a > R$. When we calculate the attraction of the disk at R, this portion of matter will not contribute, because a point in the disk at radius R is an interior point of all homoeoids with $a > R$. Just how much of the matter that is physically interior to R

gets assigned to homoeoids having $a > R$ depends on the surface density exterior to R. Thus two disks can have identical density distributions for $R < r_0$ and yet have very different force fields at r_0. In this respect disks differ from spherical distributions of mass, for which the force at r_0 depends only on the density at $r < r_0$. In fact, the surface density of a disk at $R > r_0$ affects the attraction at r_0 because the annulus of material exterior to r_0 actually pulls a star placed at radius r_0 outward, thus partially compensating the inward attraction of the interior matter. At points in a disk where little matter is pulling outward, for example on the perimeter of a sharp-edged disk, the circular speed can be much higher than at the edge of the spherical body with the same total mass and radius.

The potential of a disk of known $\Sigma(R)$ can be found by inserting the distribution of thin homoeoid densities that is given by equation (2-138b) into the formulae of §2.3. This program has been carried through by Kuzmin (1952) and Brandt (1960), and employed in some studies of galactic rotation curves (for references see Brandt & Belton 1962), but one can immediately see that it will lead to a cumbersome double integral over values of R and a. Instead, we shall calculate the potential of a disk of arbitrary surface density $\Sigma(R)$ by two alternative routes.

2 Disk Potentials via Elliptic Integrals

If $\mathbf{x}' = (R', \phi', 0)$ is a given point in a thin disk and $\mathbf{x} = (R, \phi, z)$ is any other point, then $\Phi(\mathbf{x})$ is given by the integral

$$\Phi(\mathbf{x}) = -G \int_0^\infty \Sigma(R') R' dR' \int_0^{2\pi} \frac{d\phi'}{|\mathbf{x} - \mathbf{x}'|}. \tag{2-139}$$

Since the potential of an axisymmetric disk cannot depend on ϕ, we may evaluate Φ for $\phi = 0$. In this case

$$\begin{aligned}
|\mathbf{x} - \mathbf{x}'|^2 &= R^2 + R'^2 - 2RR' \cos\phi' + z^2 \\
&= [(R + R')^2 + z^2][1 - k^2 \cos^2(\tfrac{1}{2}\phi')],
\end{aligned} \tag{2-140a}$$

where

$$k^2 \equiv \frac{4RR'}{(R + R')^2 + z^2}, \tag{2-140b}$$

and we have used the relation $1 + \cos\phi' = 2\cos^2(\tfrac{1}{2}\phi')$. It is straight-forward to show that $k^2 \leq 1$ for all possible values of R, R', and z.

Substituting from equation (2-140a) into equation (2-139) and making the substitution $x = \cos(\frac{1}{2}\phi')$, we find

$$\Phi(R, z) = -4G \int_0^\infty \frac{\Sigma(R')R'dR'}{\sqrt{(R + R')^2 + z^2}} \int_0^1 \frac{dx}{\sqrt{(1 - x^2)(1 - k^2 x^2)}}.$$
(2-141)

By equation (1C-15), the integral over x in equation (2-141) equals $F(\frac{\pi}{2}, k) \equiv K(k)$. K is a complete elliptic integral and is tabulated by, for example, Abramowitz and Stegun (1965). Thus we have finally

$$\Phi(R, z) = -4G \int_0^\infty K\left(\sqrt{\frac{4RR'}{(R + R')^2 + z^2}}\right) \frac{\Sigma(R')R'dR'}{\sqrt{(R + R')^2 + z^2}}.$$
(2-142a)

A more compact form of equation (2-142a) is

$$\Phi(R, z) = -\frac{2G}{\sqrt{R}} \int_0^\infty K(k)k\Sigma(R')\sqrt{R'}dR'.$$
(2-142b)

If we differentiate this expression with respect to R, we obtain

$$\frac{\partial \Phi}{\partial R}(R, z) = \frac{G}{R^{\frac{3}{2}}} \int_0^\infty \left[kK - 2R\frac{d(kK)}{dk}\frac{\partial k}{\partial R}\right] \Sigma\sqrt{R'}\,dR'.$$
(2-143)

The derivative of K that occurs in equation (2-143) may be eliminated with the help of the identity [eq. (1C-16)]

$$\frac{d(kK)}{dk} = \frac{E(k)}{1 - k^2},$$
(2-144)

where $E(k) \equiv E(\frac{\pi}{2}, k)$ is a complete elliptic integral. Furthermore, from equation (2-140b) we have that

$$2R\frac{\partial k}{\partial R} = \frac{k^3}{4}\left(\frac{R'}{R} - \frac{R}{R'} + \frac{z^2}{RR'}\right).$$
(2-145)

With equations (2-144) and (2-145), equation (2-143) becomes

$$R\frac{\partial \Phi}{\partial R}(R, z) = \frac{G}{\sqrt{R}} \int_0^\infty \left[K(k) - \frac{1}{4}\left(\frac{k^2}{1 - k^2}\right)\right.$$
$$\left. \times \left(\frac{R'}{R} - \frac{R}{R'} + \frac{z^2}{RR'}\right)E(k)\right] k\Sigma(R')\sqrt{R'}dR'.$$
(2-146)

If we were to set $z = 0$ in equation (2-146) we would obtain a formal expression for the square of the circular speed at R. Unfortunately, when $z = 0$, k approaches unity as the variable of integration R' passes through the value R, and near $k = 1$ both K and $(1 - k^2)^{-1}$ diverge. This behavior is inconvenient from the point of view of numerical work; however, the singularity is integrable and equation (2-146) still offers one of the best routes to the circular speed of a thin disk. Fortunately $(\partial\Phi/\partial R)$ is always a slowly varying function of z near $z = 0$, so one may estimate its value at $z = 0$ by evaluating the integral of equation (2-146) for z small but non-zero. The integral in equations (2-142) may also be evaluated for some small but finite value of z.

3 Disk Potentials via Bessel Functions

An alternative expression for $\Phi(R, z)$ was given by Toomre (1962). In Toomre's method we solve Laplace's equation $\nabla^2 \Phi = 0$ subject to appropriate boundary conditions on the disk and at infinity. In cylindrical coordinates Laplace's equation is

$$\frac{1}{R}\frac{\partial}{\partial R}\left(R\frac{\partial \Phi}{\partial R}\right) + \frac{\partial^2 \Phi}{\partial z^2} = 0. \tag{2-147}$$

Writing

$$\Phi(R, z) = J(R)Z(z), \tag{2-148}$$

we obtain by the method of separation of variables (see §2.4)

$$\frac{1}{J(R)R}\frac{d}{dR}\left(R\frac{dJ}{dR}\right) = \frac{-1}{Z(z)}\frac{d^2 Z}{dz^2} = -k^2, \tag{2-149}$$

where k is an arbitrary real or complex number. Thus

$$\frac{d^2 Z}{dz^2} - k^2 Z = 0, \tag{2-150}$$

$$\frac{1}{R}\frac{d}{dR}\left(R\frac{dJ}{dR}\right) + k^2 J(R) = 0. \tag{2-151a}$$

Equation (2-150) may be immediately integrated to

$$Z(z) = S\exp(\pm kz), \tag{2-152}$$

where S is a constant. Equation (2-151a) is simplified if we make the substitution $u = kR$;

$$\frac{1}{u}\frac{d}{du}\left(u\frac{dJ}{du}\right) + J(u) = 0. \tag{2-151b}$$

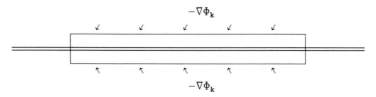

Figure 2-16. The disk mass within the box shown in cross-section equals $-(4\pi G)^{-1}$ times the integral of the normal component of $\nabla\Phi_k$ over the surface of the box. The horizontal component of $\nabla\Phi_k$ is due to the gravitational attraction from the rest of the galaxy.

The solution to equation (2-151b) that is of interest, is the one that remains finite at $u = 0$ ($R = 0$). This is conventionally written $J_0(u) = J_0(kR)$ and is called the cylindrical Bessel function of order zero (Appendix 1.C.7). Summarizing these results, we have that the functions

$$\Phi_\pm(R, z) = \exp(\pm kz)J_0(kR) \qquad (2\text{-}153)$$

are solutions of $\nabla^2\Phi = 0$.

Now consider the function

$$\Phi_k(R, z) = \exp(-k|z|)J_0(kR), \qquad (2\text{-}154a)$$

where k is real and positive. $\Phi_k \to 0$ when $|z| \to \infty$, and furthermore $\Phi_k \to 0$ as $R \to \infty$ since $J_0(u) \to 0$ as $u \to \infty$. Thus Φ_k satisfies all the conditions required for it to be the potential generated by an isolated density distribution. Furthermore, for $z > 0$, Φ_k coincides with Φ_-, and for $z < 0$, Φ_k coincides with Φ_+. Therefore, Φ_k solves $\nabla^2\Phi = 0$ everywhere except in the plane $z = 0$. At $z = 0$, Φ_k does not satisfy Laplace's equation because its gradient suffers a discontinuity. Figure 2-16 illustrates how we may use Gauss's theorem [eq. (2-13)] to evaluate the surface density $\Phi_k(R)$ of the sheet that generates this discontinuity. We have that

$$\lim_{z\to0+}\left(\frac{\partial\Phi_k}{\partial z}\right) = -kJ_0(kR) \quad \text{and} \quad \lim_{z\to0-}\left(\frac{\partial\Phi_k}{\partial z}\right) = +kJ_0(kR).$$
$$(2\text{-}155)$$

The integral of $\nabla\Phi_k$ over the closed unit surface that is shown in the figure must equal $4\pi G\Sigma_k$ from which it follows that [2]

$$\Sigma_k(R) = -\frac{k}{2\pi G}J_0(kR). \qquad (2\text{-}154b)$$

---

[2] Equations (2-154) give an especially simple potential-density pair. A general procedure for generating potential-density pairs is described by Clutton-Brock (1972).

We now use equations (2-154) to find the potential generated by a disk of arbitrary surface density $\Sigma(R)$. If we can find a function $S(k)$ such that

$$\Sigma(R) = \int_0^\infty S(k)\Sigma_k(R)dk = -\frac{1}{2\pi G}\int_0^\infty S(k)J_0(kR)kdk, \quad (2\text{-}156)$$

then we will have

$$\Phi(R, z) = \int_0^\infty S(k)\Phi_k(R, z)dk = \int_0^\infty S(k)J_0(kR)e^{-k|z|}dk. \quad (2\text{-}157)$$

Equation (2-156) states that $S(k)$ is the Hankel transform of $(-2\pi G\Sigma)$ [eq. (1C-47b)]. Hankel transforms have properties that are very similar to those of the familiar Fourier transforms. In particular, they may be inverted by use of equation (1C-47a). We find

$$S(k) = -2\pi G \int_0^\infty J_0(kR)\Sigma(R)RdR. \quad (2\text{-}158)$$

If we eliminate $S(k)$ between this equation and equation (2-157), we obtain finally

$$\Phi(R, z) = -2\pi G \int_0^\infty dk e^{-k|z|} J_0(kR) \int_0^\infty \Sigma(R')J_0(kR')R'dR'. \quad (2\text{-}159)$$

A quantity of particular interest is the circular speed $v_c(R)$. Setting $z = 0$ in equation (2-157) and differentiating both sides with the help of the identity $dJ_0(x)/dx = -J_1(x)$ [eq. (1C-42)], we obtain

$$v_c^2(R) = R\left(\frac{\partial\Phi}{\partial R}\right)_{z=0} = -R\int_0^\infty S(k)J_1(kR)kdk. \quad (2\text{-}160)$$

We illustrate the utility of these formulae with three examples.

(a) Rotation curve of Mestel's disk   Mestel (1963) considered the properties of a disk whose surface density is given by

$$\Sigma(R) = \frac{\Sigma_0 R_0}{R}, \quad (2\text{-}161)$$

that is, a disk in which surface density is inversely proportional to radius. According to equation (2-158),[3] $S(k)$ for this disk is given by

$$S(k) = -2\pi G\Sigma_0 R_0 \int_0^\infty J_0(kR)dR. \quad (2\text{-}162)$$

---

[3] Strictly, we have not justified equation (2-158) in the case where Σ diverges as $R \to 0$. However, the circular speed will barely change at radii of interest if we set $\Sigma(R) = \Sigma_{\max}$ for all sufficiently small R.

Using equation (1C-43) we find

$$S(k) = -\frac{2\pi G\Sigma_0 R_0}{k}. \tag{2-163}$$

Substituting this result for $S(k)$ into equation (2-160) we then obtain

$$v_c^2(R) = 2\pi G\Sigma_0 R_0 R \int_0^\infty J_1(kR)dk = 2\pi G\Sigma_0 R_0. \tag{2-164}$$

Thus the circular speed of Mestel's disk is constant. Furthermore, for Mestel's particular surface-density law (2-161), $v_c(R)$ is given by the simple formula

$$v_c^2(R) = \frac{GM(R)}{R}, \tag{2-165a}$$

where

$$M(R) = 2\pi \int_0^R \Sigma(R')R'dR' \tag{2-165b}$$

is the mass interior to R. This is precisely analogous to equation (2-24) for a spherical system. Although we have argued that in general the circular speed is affected by the mass exterior to R, for the particular case of Mestel's disk the simple formula (2-165b) happens to give the correct answer.

(b) Rotation curve of an exponential disk   Setting $\Sigma(R) = \Sigma_0 e^{-R/R_d}$ we obtain from equation (2-158), with the help of formula 6.623.2 of Gradshteyn and Ryzhik (1965),

$$S(k) = -\frac{2\pi G\Sigma_0 R_d^2}{[1 + (kR_d)^2]^{\frac{3}{2}}} \tag{2-166}$$

for the $S(k)$ of the exponential disk. Then from equation (2-157) we obtain for the potential

$$\Phi(R, z) = -2\pi G\Sigma_0 R_d^2 \int_0^\infty \frac{J_0(kR)e^{-k|z|}}{[1 + (kR_d)^2]^{\frac{3}{2}}} dk. \tag{2-167}$$

If we set $z = 0$, formula (6.552.1) of Gradshteyn and Ryzhik (1965) and formula 9.6.27 of Abramowitz and Stegun (1965) enable us to evaluate this integral. There results

$$\Phi(R, 0) = -\pi G\Sigma_0 R[I_0(y)K_1(y) - I_1(y)K_0(y)], \quad \text{where} \quad y \equiv \frac{R}{2R_d}. \tag{2-168}$$

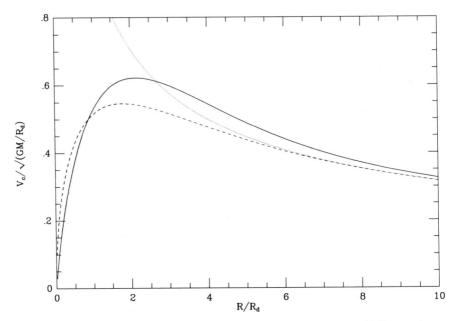

Figure 2-17. The circular-speed curves of: an exponential disk (full curve); a point with the same total mass (dotted curve); the spherical body for which $M(r)$ is given by equation (2-170) (dashed curve).

In equation (2-168) I_n and K_n are modified Bessel functions of the first and second kinds (see Appendix 1.C.7).

If we differentiate equation (2-168) with respect to R we obtain the circular speed of the exponential disk (Freeman 1970):

$$v_c^2(R) = R\frac{\partial\Phi}{\partial R} = 4\pi G\Sigma_0 R_d y^2 \left[I_0(y)K_0(y) - I_1(y)K_1(y)\right]. \quad (2\text{-}169)$$

In Figure 2-17 we show this circular speed together with the circular speed of the spherical body that has as much mass $M_s(r)$ interior to $r = R$ as the exponential disk, that is,

$$M_s(R) = M_d(R) = 2\pi \int_0^R \Sigma_0 \exp(-R'/R_d)R'dR'$$
$$= 2\pi\Sigma_0 R_d^2\left[1 - \exp(-R/R_d)\left(1 + \frac{R}{R_d}\right)\right]. \quad (2\text{-}170)$$

Notice that the disk achieves a peak circular speed that is about 15% higher than that of the equivalent spherical distribution. The dashed

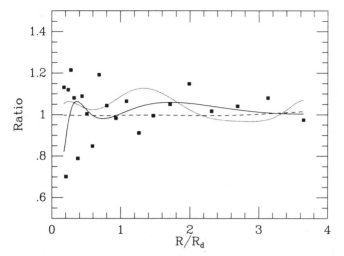

Figure 2-18. The surface density Σ of a disk cannot be reliably recovered from noisy measurements of the circular speed v_c. The squares show the ratio $v_c(\text{data})/v_c(\text{true})$ of simulated measurements of a perfectly exponential disk. When a spline is least-squares fitted to these points (full curve), equation (2-174) yields the surface densities shown by the dotted curve as the ratio $\Sigma(\text{estimate})/\Sigma(\text{true})$. The dashed curve shows the same ratio when noise-free data are run through the same program.

line in Figure 2-17 also gives the Keplerian circular speed of a system in which the entire mass of the disk is concentrated at the center. Notice that the disk's circular speed approaches the Keplerian speed only slowly and from above, whereas the circular speed of the equivalent spherical distribution tends rapidly to the Keplerian speed from below.

The potential energy of an exponential disk is

$$W = 2\pi \int_0^\infty \Phi(R,0)\Sigma(R)R\,dR$$

$$= -16\pi^2 G\Sigma_0^2 R_d^3 \int_0^\infty e^{-2y}\left[I_0(y)K_1(y) - I_1(y)K_0(y)\right]y^2\,dy \qquad (2\text{-}171)$$

$$\simeq -11.6 G\Sigma_0^2 R_d^3.$$

(c) Deducing $\Sigma(R)$ given $v_c(R)$ Applying the usual inversion formula for Hankel transforms [eq. (1C-47)] to equation (2-160), we find

$$S(k) = -\int_0^\infty v_c^2(R')J_1(kR')\,dR', \qquad (2\text{-}172)$$

and thus from equation (2-156),

$$\Sigma(R) = -\frac{1}{2\pi G} \int_0^\infty J_0(kR)dk \int_0^\infty v_c^2(R') \frac{\partial J_0(kR')}{\partial R'} dR'. \qquad (2\text{-}173)$$

Here we have again exploited the relation $J_0' = -J_1$. Integrating the inner integral of equation (2-173) by parts subject to the conditions $v_c(0) = v_c(\infty) = 0$, and then inverting the order of integrations, we obtain

$$\Sigma(R) = \frac{1}{2\pi G} \int_0^\infty \frac{dv_c^2}{dR'} dR' \int_0^\infty J_0(kR)J_0(kR')dk$$

$$= \frac{1}{\pi^2 G} \left[\frac{1}{R} \int_0^R \frac{dv_c^2}{dR'} K\left(\frac{R'}{R}\right) dR' + \int_R^\infty \frac{dv_c^2}{dR'} K\left(\frac{R}{R'}\right) \frac{dR'}{R'} \right],$$
$$(2\text{-}174)$$

where we have evaluated explicitly the standard integral occurring in this equation in terms of the complete elliptic integral K [eq. (1C-45)]. K becomes very large when $R \simeq R'$. Thus from equation (2-174) we conclude that the surface density at R is in the main determined by the value of the gradient of v_c^2 at radii R' that lie close to R. We should not be surprised by this conclusion; $\Sigma(R)$ must depend on the difference between the actual variation of v_c with R and the Keplerian variation $v_c \propto R'^{-1/2}$, which would imply $\Sigma(R) = 0$ for all R. From the observational point of view, however, formula (2-174) is of relatively little use since (dv_c^2/dR) is usually subject to significant observational error, and if we simply insert our best estimate of (dv_c^2/dR) into the right side of equation (2-174), we will derive surface densities $\Sigma(R)$ that vary with radius in an erratic and unphysical way. Figure 2-18 illustrates this point. In practice we must instead use equation (2-160) to fit a model of Σ with a few adjustable parameters to the observations of v_c. Unless very high quality observations of v_c are available covering a wide range of radii, there is inevitably considerable uncertainty in the exact form of $\Sigma(R)$ (Bahcall et al. 1982; Kalnajs 1983).

4 Potentials of Non-Axisymmetric Disks

The potentials of non-axisymmetric disks may be obtained by a straightforward extension of the method of separation of variables that led to equation (2-159) (see Problem 2-9). Here we discuss two other ways in which to determine the potential of a non-axisymmetric disk.

(a) Disk potentials using oblate spheroidal coordinates   In §5.3 we shall require expressions for disk potentials that are obtained by using the method of separation of variables to solve Laplace's equation in the coordinates defined by equations (2-58). If we substitute $\Phi = U(u)V(v)e^{im\phi}$ into equation (2-61) and separate variables in the usual way, we find that U and V satisfy

$$\frac{m^2}{\sin^2 v} - \frac{1}{V \sin v}\frac{d}{dv}\left(\sin v \frac{dV}{dv}\right) = l(l+1), \qquad (2\text{-}175a)$$

$$\frac{m^2}{\cosh^2 u} + \frac{1}{U \cosh u}\frac{d}{du}\left(\cosh u \frac{dU}{du}\right) = l(l+1), \qquad (2\text{-}175b)$$

where $l(l+1)$ is the separation constant. The left side of equation (2-175a) is the same as the right side of equation (2-107) with v substituted for θ. Furthermore, the boundary conditions $(dV/dv) = 0$ at $v = 0$ or π that must be satisfied by a physically acceptable function V are the same as the conditions we imposed on the function $P(\theta)$ that occurs in equation (2-107). Hence

$$V(v, \phi) = V_{lm}\, Y_l^m(v, \phi), \qquad (2\text{-}176)$$

where V_{lm} is a constant and Y_l^m is the spherical harmonic defined by equation (1C-27). Since the potential must be symmetrical about the plane of the disk $v = \frac{\pi}{2}$, we must restrict ourselves to values of l and m for which $Y_l^m(v, \phi)$ is an even function of $\cos v$. Hence we require $l - m$ to be even.

  If we change the independent variable in equation (2-175b) to $\varsigma = i \sinh u$, the equation becomes the associated Legendre equation (2-108b) with x replaced by the complex variable ς. If the potential is to vanish at infinity (large u), it must be proportional to the fundamental solution of equation (2-108b) that vanishes at large x. This is written $Q_l^m(x)$ (see Appendix 1.C.5). Hence the functions

$$\Phi_{lm}(u, v, \phi) = \left[\frac{V_{lm}}{Q_l^m(0)}\right] Q_l^m(i \sinh u)\, Y_l^m(v, \phi), \qquad (l-m) \text{ even},$$
$$(2\text{-}177a)$$

satisfy Laplace's equation everywhere outside the excluded disk $u = 0$ ($z = 0$, $R \le \Delta$) and vanish at infinity. However, there is a discontinuity in the gradient of Φ_{lm} on the excluded disk. By Gauss's theorem, this discontinuity is generated by a surface density $\Sigma_{lm}(v, \phi)$; thus from equation (2-60) we have

$$4\pi G\Sigma_{lm}(v, \phi) = 2\left(\hat{\mathbf{e}}_u \cdot \boldsymbol{\nabla}\Phi_{lm}\right)_{u\to 0+} = \frac{2}{\Delta|\cos v|} \lim_{u\to 0+} \left(\frac{\partial \Phi_{lm}}{\partial u}\right)$$

$$= 2V_{lm}i \lim_{\varsigma \to 0} \left[\frac{d\ln Q_l^m(\varsigma)}{d\varsigma}\right] \frac{Y_l^m(v, \phi)}{\Delta|\cos v|}.$$

$$(2\text{-}178)$$

Using equations (1C-18) and (1C-7) to evaluate the limit in equation (2-178), we then find (Hunter 1963)

$$\Sigma_{lm} = -\left(\frac{2V_{lm}}{\pi^2 G\Delta g_{lm}}\right)\frac{Y_l^m(v, \phi)}{|\cos v|}, \qquad (2\text{-}179a)$$

where

$$g_{lm} \equiv \frac{(l+m)!(l-m)!}{2^{(2l-1)}\left[\left(\frac{l+m}{2}\right)!\left(\frac{l-m}{2}\right)!\right]^2}. \qquad (2\text{-}179b)$$

A general disk potential, which is a sum over l and m of potentials of the form (2-177a), is generated by the surface density $\Sigma(v, \phi)$ that is the sum of surface densities $\Sigma_{lm}(v, \phi)$. According to equation (2-179a), $-2V_{lm}/(\pi^2 G\Delta g_{lm})$ is the coefficient of $Y_l^m(v, \phi)$ in the expansion of $|\cos v|\Sigma(v, \phi)$ in spherical harmonics. Thus with the orthogonality relation (1C-23b) we have

$$\frac{2V_{lm}}{\pi^2 G\Delta g_{lm}} = -\int_0^{2\pi} d\phi \int_0^\pi \Sigma(v, \phi)Y_l^{m*}(v, \phi)|\cos v|\sin v\, dv. \qquad (2\text{-}180)$$

The integrand in equation (2-180) is symmetrical about $v = \frac{\pi}{2}$, so we may restrict the v integration to the range $(0, \frac{\pi}{2})$ and double the result. Hence

$$V_{lm} = -\frac{\pi^2 G g_{lm}}{\Delta}\int_0^{2\pi} d\phi \int_0^\Delta \Sigma(R, \phi)Y_l^{m*}\left(\arcsin\left(\frac{R}{\Delta}\right), \phi\right)R\, dR. \qquad (2\text{-}177b)$$

Equations (2-177) are useful only for disks that have clearly defined outer edges. A different technique that was introduced by Kalnajs is more flexible:

(b) Disk potentials via logarithmic spirals   By equation (2-139) the potential $\Phi(R, \phi)$ at any point in the plane of a non-axisymmetric disk is

$$\Phi(R, \phi) = -G\int_0^\infty R'dR'\int_0^{2\pi} \frac{\Sigma(R', \phi')d\phi'}{\sqrt{R'^2 + R^2 - 2RR'\cos(\phi' - \phi)}}. \qquad (2\text{-}181)$$

The integral in this expression can be simplified if we define a new radial coordinate,

$$u \equiv \ln R, \qquad (2\text{-}182)$$

and introduce the **reduced potential** V and the **reduced surface density** S by

$$\sqrt{R}\,\Phi \equiv V(u, \phi) = e^{u/2}\Phi\,[R(u), \phi]$$

$$R^{\frac{3}{2}}\Sigma \equiv S(u, \phi) = e^{3u/2}\Sigma\,[R(u), \phi]. \qquad (2\text{-}183)$$

With these substitutions (2-181) becomes

$$V(u, \phi) = -G \int_{-\infty}^{\infty} du' \int_{0}^{2\pi} K(u - u', \phi - \phi')S(u', \phi')d\phi', \qquad (2\text{-}184a)$$

where

$$K(u - u', \phi - \phi') \equiv \frac{1}{\sqrt{2\,[\cosh(u - u') - \cos(\phi - \phi')]}}. \qquad (2\text{-}184b)$$

Now consider the reduced potential $V_{lm}(u, \phi)$ that is generated by the particular reduced surface density

$$S_{\alpha m}(u, \phi) = e^{i(\alpha u + m\phi)}, \qquad (2\text{-}185)$$

where α is a real number and m is an integer. We have

$$V_{\alpha m}(u, \phi) = -G \int_{-\infty}^{\infty} du' \int_{0}^{2\pi} K(u - u', \phi - \phi')e^{i(\alpha u' + m\phi')}d\phi'$$

$$= -Ge^{i(\alpha u + m\phi)} \int_{-\infty}^{\infty} du' \int_{0}^{2\pi} K(u - u', \phi - \phi')e^{i[\alpha(u'-u)+m(\phi'-\phi)]}d\phi'.$$

$$(2\text{-}186)$$

If we change to new variables of integration $u'' \equiv u - u'$ and $\phi'' \equiv \phi - \phi'$, equation (2-186) becomes

$$V_{\alpha m} = -GN(\alpha, m)e^{i(\alpha u + m\phi)}, \qquad (2\text{-}187)$$

where

$$N(\alpha, m) \equiv \int_{-\infty}^{\infty} du'' \int_{0}^{2\pi} K(u'', \phi'')e^{-i(\alpha u'' + m\phi'')}d\phi''$$

$$= \pi \frac{[\frac{1}{2}(m - \frac{1}{2} + i\alpha)]![\frac{1}{2}(m - \frac{1}{2} - i\alpha)]!}{[\frac{1}{2}(m + \frac{1}{2} + i\alpha)]![\frac{1}{2}(m + \frac{1}{2} - i\alpha)]!}. \qquad (2\text{-}188)$$

The kernel $N(\alpha, m)$ is real and even[4] in both α and m. Hence the potential generated by an arbitrary linear combination

$$S(u, \phi) \equiv \sum_{m=-\infty}^{\infty} \int_{-\infty}^{\infty} A_m(\alpha) e^{i(\alpha u + m\phi)} d\alpha \qquad (2\text{-}189a)$$

of surface densities of the form (2-185) is given by (Kalnajs 1971)

$$V(u, \phi) = G \sum_{m=-\infty}^{\infty} \int_{-\infty}^{\infty} N(\alpha, m) A_m(\alpha) e^{i(\alpha u + m\phi)} d\alpha. \qquad (2\text{-}189b)$$

Furthermore, from equation (2-189a) we see that $A_m(\alpha)$ is nothing but the Fourier transform of the reduced surface density $S(u, \phi)$:

$$A_m(\alpha) = \frac{1}{(2\pi)^2} \int_{-\infty}^{\infty} du \int_{0}^{2\pi} S(u, \phi) e^{-i(\alpha u + m\phi)} d\phi. \qquad (2\text{-}189c)$$

So we may use equation (2-189b) to obtain the potential in the plane $z = 0$ that is generated by any given distribution of surface density. Since $\alpha u + m\phi = \alpha \ln R + m\phi$ is constant on **logarithmic spirals**, equations (2-189) determine the potential of a disk by decomposing the density into spirals. Note that this derivation does not produce a simple expression for the value of the potential away from the plane $z = 0$ (Problem 2-11 remedies this defect).

2.7 The Potential of Our Galaxy

J. H. Oort was the first astronomer to succeed in synthesizing observations of the motions of stars in the solar neighborhood into the currently accepted picture of our Galaxy. Oort's main aim was to explain the kinematics of the solar neighborhood, but he also initiated the process of interpreting the kinematics in dynamical terms by constructing the first modern mass model of our Galaxy (Oort 1932). Since Oort's pioneering work, and especially since observations of the 21-cm line of neutral hydrogen (§8-3 of MB) first yielded reliable estimates of the circular speed at points interior to the solar radius, many models of our Galaxy have been proposed. For several years after the publication by Kwee et al. (1954) of the first good 21-cm rotation curve, the standard mass model was that of Schmidt (1956), who modeled the Galaxy as a

---

[4] This statement follows because $(z!)^* = (z^*)!$.

superposition of four highly flattened spheroids of the type discussed in §2.3.

More recent models of our Galaxy assume that it is a normal Sbc system and break it down into disk and spheroidal components of the type that we find in external galaxies (see §5-2 of MB). In particular, the surface brightness of the disk is assumed to be exponential, while the spheroid is assumed to follow either the $R^{1/4}$ law (1-13) or the Hubble-Reynolds law (1-14).

The scale lengths and densities of these components can be estimated in two different ways. In one approach, which we shall call the **photometric method**, one first estimates the luminosity densities and scale lengths from star counts or from measurements of the integrated light from unresolved stars. Then one chooses a position-independent mass-to-light ratio Υ for each component and multiplies the luminosity density of each component by Υ to determine the corresponding mass density. In the second method, which we may call the **dynamical method**, one assumes that each component has an unknown but position-independent Υ and then chooses the associated scale lengths and mass densities to match known dynamical quantities, such as the circular speed at a number of radii inside and outside the solar circle and the escape velocity from the solar neighborhood. Thus the photometric method relies heavily on our ability to observe directly some of the Galaxy's components, while the dynamical method determines the characteristics of the galactic mass model without considering the light put out by stars. Reviews of galactic mass models of both types have been given by Bahcall (1986) and Caldwell and Ostriker (1981), respectively.

(a) Photometric method   De Vaucouleurs and Pence (1978) and Bahcall and Soneira (1980) have used the photometric method to construct detailed galactic models, in which the galactic spheroid is assumed to follow the $R^{1/4}$ law (1-13), with the same effective radius R_e as the galactic globular cluster system. De Vaucouleurs and Pence find $R_e \simeq \frac{1}{3}R_0$, where R_0 is the distance of the Sun from the galactic center. (These authors assume $R_0 = 8\,\text{kpc}$, slightly smaller than our preferred value of 8.5 kpc.) Once R_e has been chosen, the parameter I_e in the $R^{1/4}$ law can be determined in three independent ways: (i) from the density of high-velocity stars in the solar neighborhood; (ii) from counts of faint stars at high galactic latitude; and (iii) from measurements of the sky brightness at different points along the galactic meridian $l = 0$. All these methods are subject to substantial uncertainties.

(i) The most direct method is based on the local density of high-velocity stars (Bahcall et al. 1983). The principle that underlies this method is that any model of the galactic spheroid predicts a definite number of stars whose proper motions are larger than some specified quantity. Since all stars brighter than a given flux level with large proper motions can be detected in a suitable survey, and the proportion of these that are high velocity stars can be ascertained from distances and radial velocities, one can, in principle, match the model to the observations. Unfortunately the only survey of this type, that of Schmidt (1975), contains only 18 high-velocity stars, so the normalization that is obtained in this way is subject to substantial error.

(ii) If we know the luminosity function of the spheroid stars (see §4-2 of MB), any model of the spheroid luminosity density predicts the number of stars in each magnitude interval that should be observed in each direction. Of course disk stars will also be seen down any line of sight, and these must be subtracted from the observed counts before the spheroid's density can be estimated. Bahcall and Soneira describe how star counts at different galactic latitudes and in different colors can be used to effect this separation and to check the resulting model for internal consistency.

(iii) De Vaucouleurs and Pence determined the spheroid's density by measuring the variation with galactic latitude b of the sky brightness along galactic longitude $\ell = 0$. This measurement is extremely difficult because the contribution of the spheroid to the sky brightness is considerably smaller than the contributions from air-glow, zodiacal light (see §5-2 of MB), and unresolved disk stars. Hence the spheroid's luminosity is obtained as the small difference between large numbers, which is necessarily a hazardous proceeding.

De Vaucouleurs and Pence conclude in this way that the total visual luminosity of the galactic spheroid[5] $L_V(\text{spheroid}) = 6 \times 10^9 \, (R_0/8\,\text{kpc})^2 \, L_\odot$. Bahcall and Soneira infer a substantially smaller luminosity, $L_V(\text{spheroid}) = 2 \times 10^9 \, (R_0/8\,\text{kpc})^3 \, L_\odot$, but with an uncertainty that does not exclude the larger luminosity of de Vaucouleurs and Pence. In Table 1-2 we adopt the Bahcall-Soneira value.

The surface brightness of the disk near the Sun is easily determined from star counts. De Vaucouleurs and Pence find $I_V(R_0) = 13\,L_\odot\,\text{pc}^{-2}$; for comparison Bahcall and Soneira find $I_V(R_0) = 16\,L_\odot\,\text{pc}^{-2}$, and we take $15\,L_\odot\,\text{pc}^{-2}$ in Table 1-1. The disk scale length R_d is less certain. De Vaucouleurs and Pence estimate $R_d = 0.44R_0$ by assuming that the

---

[5] As seen face-on from outside, the spheroid would appear fainter by about 10% due to absorption in the disk.

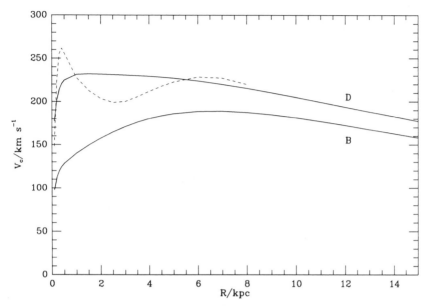

Figure 2-19. Circular speed versus radius for two photometric models of our Galaxy: de Vaucouleurs and Pence (D); Bahcall and Soneira (B). The dashed curve is the rotation curve interior to R_0 inferred by Burton and Gordon (1978) from CO and HI observations.

central surface brightness of the galactic disk is $21.67 \, \text{mag arcsec}^{-2}$ by analogy with the disks of other galaxies (see §5-2 of MB), and Bahcall and Soneira adopt the same value. We use the slightly smaller value $R_d = 3.5 \, \text{kpc} = 0.41 R_0$ in this book (Table 1-2). With $R_d = 0.44 R_0$, the total luminosity of the disk is found to be $L_V(\text{disk}) = 1.0 \times 10^{10} \, (R_0/8 \, \text{kpc})^2 \, \text{L}_\odot$ (de Vaucouleurs & Pence 1978) or $L_V(\text{disk}) = 1.2 \times 10^{10} \, (R_0/8 \, \text{kpc})^2 \, \text{L}_\odot$ (Bahcall & Soneira 1980).

In §4.2 we shall find that the mass-to-light ratio of the galactic disk is $\Upsilon_V \simeq 5 \Upsilon_\odot$, while spheroidal systems are found to have mass-to-light ratios $\Upsilon_V \approx 12h\Upsilon_\odot$, where h is the ratio of the Hubble constant to $100 \, \text{km s}^{-1} \, \text{Mpc}^{-1}$ [see eq. (1-12)]. If we set $R_0 = 8 \, \text{kpc}$ for consistency with Bahcall and Soneira, set $h = 0.75$, and use the resulting mass-to-light ratios to convert the luminosity models of de Vaucouleurs and Pence and of Bahcall and Soneira into mass models, we obtain from equations (2-24) and (2-169) the circular speeds in the galactic plane that are given by the full curves in Figure 2-19. The dashed curve in that figure shows the rotation curve obtained by Burton and Gordon (1978) from observations of neutral hydrogen and carbon monoxide when one assumes $R_0 = 8 \, \text{kpc}$ and $v_c(R_0) = 220 \, \text{km s}^{-1}$ (see §8-3 of MB).

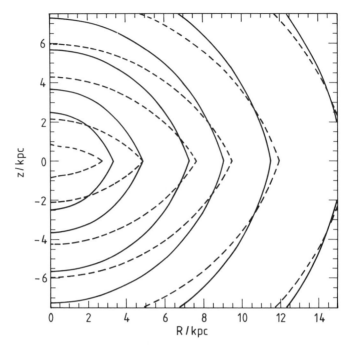

Figure 2-20. Isopotential contours for the de Vaucouleurs and Pence (1978) model of our Galaxy (full curves). The dashed curves are isopotential curves for the disk's contribution to the overall potential.

With the mass-to-light ratios we have chosen, the photometrically predicted circular speed at R_0 of $190-215\,\mathrm{km\,s^{-1}}$ does not differ greatly from the values $v_c \simeq 220\,\mathrm{km\,s^{-1}}$ that have been derived in several kinematic studies of the galaxy (e.g., Gunn et al. 1979). The full curves D and B in Figure 2-19 imply Oort constants ($A = 15.9\,\mathrm{km\,s^{-1}\,kpc^{-1}}$, $B = -10.9\,\mathrm{km\,s^{-1}\,kpc^{-1}}$ for curve D, and $A = 12.7\,\mathrm{km\,s^{-1}\,kpc^{-1}}$, $B = -10.7\,\mathrm{km\,s^{-1}\,kpc^{-1}}$ for curve B) that are in satisfactory agreement with other estimates (see §8-2 of MB and Table 1-2).

Figure 2-20 shows isopotential contours for the de Vaucouleurs and Pence model.

While the general shape and amplitude of the curves D and B in Figure 2-19 are similar to the dashed observed rotation curve, the agreement is not good. This disagreement, together with indications from observations of external galaxies that these systems are surrounded by large amounts of dark matter (see Chapter 10), have caused Bahcall and Soneira to add to their model an unseen halo component in which the density[6]

$$\rho(r) = \frac{\rho_0}{1 + (r/r_c)^2}. \tag{2-190}$$

Hence the final Bahcall and Soneira model is not based on the pure photometric method, but derives from a hybrid approach.

The unseen halo component cannot obey equation (2-190) out to arbitrarily large radius, since the total mass in the halo would diverge. Unfortunately, most dynamical measurements are quite insensitive to the mass distribution in the outer parts of the halo. This is a consequence of Newton's first and third theorems, which state that the halo mass outside any radius r exerts no net gravitational force on a particle inside r so long as the halo is spherical or ellipsoidal. There is only one dynamical constraint on the extent of the halo that can be obtained from local measurements: namely, that the escape speed from the Galaxy must exceed the largest speed of any star in the solar neighborhood.[7] A number of high-velocity stars near the Sun are known to have velocities relative to an inertial frame of about $500 \, \mathrm{km \, s^{-1}}$ (Carney & Latham 1987). To see what this tells us about the extent of the dark halo, we consider a simple model in which the Galaxy is spherical, with a circular speed that is constant out to some maximum radius $r_\star$. The spherical model is much simpler than flattened models, and flattening generally has only a small effect on the escape speed, while the assumption of a constant circular speed represents a reasonable extrapolation of the observed galactic rotation curve out to $R \simeq 2R_0$ and is consistent with the rotation curves measured for other galaxies to much greater distances. Thus the mass contained within radius r is

$$M(r) = \begin{cases} \dfrac{v_c^2 r}{G}, & \text{at } r < r_\star \\ \dfrac{v_c^2 r_\star}{G} \equiv M_\star, & \text{at } r > r_\star, \end{cases} \tag{2-191}$$

and using equations (2-23) and (2-25) we find that the escape speed v_e satisfies

$$v_e^2 = \begin{cases} 2v_c^2 \left[1 + \ln(r_\star/r)\right] & \text{at } r < r_\star \\ \dfrac{2v_c^2 r_\star}{r} & \text{at } r > r_\star. \end{cases} \tag{2-192}$$

Adopting $v_c = 220 \, \mathrm{km \, s^{-1}}$ and $v_e = 500 \, \mathrm{km \, s^{-1}}$ at $r = R_0 = 8.5 \, \mathrm{kpc}$, we find the outer limit of the halo to be $r_\star \simeq 4.9R_0 \simeq 41 \, \mathrm{kpc}$, and the

---

[6] Bahcall and Soneira estimate $r_c \approx 2 \, \mathrm{kpc}$, and $\rho(R_0) \approx 0.01 \, \mathrm{M_\odot \, pc^{-3}}$.

[7] There are two small loopholes in this argument: (i) some stars may only be bound to the Local Group of galaxies, not the Galaxy itself; (ii) some stars may have recently achieved escape speed as a result of disruption of a very tight binary system.

total mass of the Galaxy to be $M_\star \simeq 4.6 \times 10^{11} \, M_\odot$. These are lower
limits to $r_\star$ and $M_\star$ since the escape speed may exceed $500 \, \mathrm{km \, s^{-1}}$ by
a substantial amount. For comparison, the total mass of the galactic
disk is about $6 \times 10^{10} \, M_\odot$ (based on $L_V(\mathrm{disk}) = 1.2 \times 10^{10} \, L_\odot$ and
$\Upsilon_V = 5\Upsilon_\odot$), and the spheroid mass is at most comparable to the disk
mass and may be much smaller. Thus the unseen halo contains *at least*
a factor of six more mass than the visible components of the Galaxy,
and possibly much more.

(b) Dynamical method   By contrast, the model of Caldwell
and Ostriker (1981) is based on the pure dynamical approach. This
model optimizes the fit of suitably parametrized mass models to thir-
teen adopted constraints: (1) the distance to the galactic center ($R_0 \simeq$
$8.8 \, \mathrm{kpc}$); (2) Oort's constant $A \simeq 12.6 \, \mathrm{km \, s^{-1} \, kpc^{-1}}$; (3) Oort's constant $B \simeq -13.1 \, \mathrm{km \, s^{-1} \, kpc^{-1}}$; (4) the total column density of material
at the solar radius $\Sigma_0 \simeq 101 \, M_\odot \, \mathrm{pc^{-2}}$; (5) Hubble-Reynolds scale length
$R_H \simeq 99 \, \mathrm{pc}$ [eq. (1-14)] of the spheroid; (6–10) the tangent-point velocities (see §8-1 of MB) of HI and CO line emission at five longitudes
$0 < l < 90°$; (11–12) the circular speed at $R \simeq 1.6 R_0$ and at $R \simeq 7 R_0$;
and (13) the escape speed $v_e(R_0) \simeq 640 \, \mathrm{km \, s^{-1}}$.

The final model comprises a Hubble-Reynolds spheroid with $R_H =$
$103 \, \mathrm{pc}$, a dark halo of the form (2-190) and a disk with surface density

$$\Sigma(R) = \Sigma(0)(e^{-R/R_1} - e^{-R/R_2}) \quad R_1 > R_2. \tag{2-193}$$

Thus the surface density of the Caldwell-Ostriker disk vanishes at the
center and falls exponentially at $R \gg R_1 - R_2$. Caldwell and Ostriker
adopted $\sqrt{R_1^2 - R_2^2} = 3.58 \, \mathrm{kpc}$ and $\sqrt{R_1^2 - R_2^2} = 0.103 \, \mathrm{kpc}$. Their dark
halo has scale length $r_c = 7.82 \, \mathrm{kpc}$ and is sharply truncated at an outer
radius $r_\star = 150 \, \mathrm{kpc}$.

2.8 Numerical Methods

Much of our present understanding of galactic dynamics has emerged
from numerical models of galaxies. The simplest type of model is one
in which a number of particles move under the influence of their mutual gravitational attraction. The principal limitation on what can be
achieved with these **N–body models** is the computational expense involved in calculating the forces acting on their particles, so many authors
have investigated efficient algorithms for calculating these forces.

Somewhat surprisingly, the optimum strategy is not to use point masses. The force that two point masses exert on one another becomes large at small separations. Consequently, the velocities of point masses may change rapidly, and extremely small time steps must be used in the numerical integration of the equations of motion if the motion is to be followed accurately. Using small time steps is undesirable since the calculation then consumes a great deal of computer time.[8] However, large forces and accelerations are avoided if the particles are extended. For example, a commonly used model for the force exerted by the jth particle on the ith particle is

$$\mathbf{F}_{ij} = \frac{Gm^2(\mathbf{x}_j - \mathbf{x}_i)}{(\epsilon^2 + \mid \mathbf{x}_i - \mathbf{x}_j \mid^2)^{3/2}}, \qquad (2\text{-}194)$$

where both particles are assumed to be of mass m. The maximum force, which occurs when $\mid \mathbf{x}_i - \mathbf{x}_j \mid^2 = \frac{1}{2}\epsilon^2$, has magnitude $2Gm^2/(3^{3/2}\epsilon^2)$. The parameter ϵ is often called the **softening parameter**, and the corresponding potential is called a **softened point-mass potential**. In these calculations the particles should be thought of not as stars, but as bunches of stars that are located in the same small volume of space and have very similar velocities. In this section we shall discuss efficient methods for calculating the forces acting in a system of such softened particles. These methods are well suited to galaxy simulations, but are less well suited to studies of the dynamics of small stellar systems, such as star clusters, in which "collisional" processes are important (see §4.1 and Chapter 8).

At each time step we have to calculate the sum over $i = 1, \ldots, N$ of the forces (2-194) to which the jth particle is subject. There are several ways of doing this:

(a) Direct summation   The forces may be calculated directly from equation (2-194). However, unless N is fairly small, these summations are costly because we have to evaluate the $\frac{1}{2}N(N - 1)$ radicals $(\epsilon^2 + \mid \mathbf{x}_i - \mathbf{x}_j \mid^2)^{3/2}$ that form the denominators in equation (2-194). By employing various expedients it is possible to calculate the forces in this way for values of N up to about 5000 (Aarseth 1972, 1985; Ahmad & Cohen 1973), but if one wishes to use more than a few thousand particles, less direct methods must be employed. A direct-summation program is reproduced in Appendix 4.B.

---

[8] This problem is particularly severe because pairs of points may occasionally form tightly bound binaries with very short orbital periods.

(b) Fourier method on a Cartesian grid   If we divide the space occupied by the model into a large number J of cells, we may obtain an estimate of the potential at the center of any cell by assuming that all the material in each cell, lies at the center of the cell and then evaluating the sum

$$\Phi_\alpha = \sum_{\beta=1}^{J} G_{\alpha\beta} M_\beta, \qquad (2\text{-}195)$$

where M_β is the mass of the particles in the βth cell and $G_{\alpha\beta}$ is defined to be the potential at the center of the αth cell that is generated by unit mass at the center of the βth cell. If the position vectors of the centers of the two cells are $\mathbf{x}_\alpha$ and $\mathbf{x}_\beta$, we have

$$G_{\alpha\beta} = -\frac{G}{\sqrt{\epsilon^2 + |\,\mathbf{x}_\alpha - \mathbf{x}_\beta\,|^2}}. \qquad (2\text{-}196)$$

$G_{\alpha\beta} M_\beta$ will be a good estimate of the potential generated at $\mathbf{x}_\alpha$ by the matter in the βth cell unless the two cells are nearly adjacent to one another. Hence if the number J of cells becomes very large, $G_{\alpha\beta} M_\beta$ will be a good approximation to the contribution of the βth cell to $\Phi(\mathbf{x}_\alpha)$ for the overwhelming majority of values of α.

Since the locations $\mathbf{x}_\alpha$ of the centers of the cells are fixed, we may use equation (2-196) to evaluate $G_{\alpha\beta}$ and store the results at the start of the calculation. Then at each time step $\Phi(\mathbf{x}_\alpha)$ may be estimated from equation (2-195) without extracting a large number of radicals, as is required if we use equation (2-194) directly. Furthermore, since $G_{\alpha\beta}$ depends only on the difference $(\mathbf{x}_\alpha - \mathbf{x}_\beta)$, there is a highly economical method, based on **discrete Fourier transforms**, for the evaluation of the sum over cells that appears in equation (2-195). By way of preparation for describing this method, we now derive two key results concerning discrete Fourier transforms (see, e.g., Press et al. 1986 for more detail).

If $\{x_k\}$ $(k = -K, \ldots, K-1)$ is a set of $2K$ numbers, the discrete Fourier transform of this set is defined to be the set of numbers

$$\hat{x}_p \equiv \frac{1}{\sqrt{2K}} \sum_{k=-K}^{K-1} x_k e^{-ikp\pi/K} \quad (p = -K, \ldots, K-1). \qquad (2\text{-}197a)$$

We have:

Discrete Fourier Transform Theorem *If the $2K$ numbers $\hat{x}_p$ are defined by equation (2-197a), then*

$$x_{k'} = \frac{1}{\sqrt{2K}} \sum_{p=-K}^{K-1} \hat{x}_p e^{ik'p\pi/K}. \qquad (2\text{-}197b)$$

Proof: We multiply both sides of equation (2-197a) by $(1/\sqrt{2K})e^{ik'p\pi/K}$ and sum over p, to obtain

$$\frac{1}{\sqrt{2K}}\sum_{p=-K}^{K-1}\hat{x}_p e^{ik'p\pi/K} = \frac{1}{2K}\sum_{p=-K}^{K-1}\sum_{k=-K}^{K-1} x_k e^{i(k'-k)p\pi/K}$$
$$= \frac{1}{2K}\sum_{k=-K}^{K-1} x_k \sum_{p=-K}^{K-1} e^{i(k'-k)p\pi/K}. \qquad (2\text{-}198)$$

The inner sum on the right side of equation (2-198) is a geometric progression, with sum $2K$ if $k' = k$, or (since k and k' are integers)

$$e^{-i(k'-k)\pi}\frac{1 - e^{(k'-k)2\pi i}}{1 - e^{i(k'-k)\pi/K}} = 0 \quad \text{if} \ \ k' \neq k.$$

Thus the only contributing term in the outer sum on the right of equation (2-198) is that for which $k = k'$.◁

In 1965 Cooley and Tukey published an algorithm, known as the **fast Fourier transform**, by which the discrete transform $\hat{x}_p$ of $2K$ numbers x_k can be evaluated in only $4K\log_2(2K)$ multiplications and additions.[9] For large K this is many fewer operations than the $(2K)^2$ operations required by a direct evaluation of the sums of equation (2-197a).

Discrete Fourier transforms share many of the properties of continuous Fourier transforms.[10] To demonstrate these properties, it is necessary to define the quantities x_k for k outside the range $[-K, K-1]$ by the rule

$$x_k = x_{k+2mK} \qquad \text{for all integer } m, \qquad (2\text{-}199)$$

i.e., to assume that x_k is periodic with period $2K$. Note that $\hat{x}_p$ is already periodic with period $2K$ [see eq. (2-197a)]. We may now prove:

Discrete Fourier Convolution Theorem If the three sets of $2K$ numbers $\{x_k\}, \{y_k\}, \{z_k\}$ $(k = -K, \ldots, K-1)$ are related by

$$z_k = \frac{1}{\sqrt{2K}}\sum_{k'=-K}^{K-1} y_{(k-k')}x_{k'}, \qquad (2\text{-}200a)$$

then

$$\hat{z}_p = \hat{y}_p\hat{x}_p. \qquad (2\text{-}200b)$$

[9] The Cooley-Tukey algorithm requires that $K = p^m$, where m is an integer and p is a prime number; usually $p = 2$.

[10] However, note that they are logically entirely distinct, and disaster will attend any attempt to invert a continuous Fourier transform as if it were a discrete transform, or vice versa.

Proof: We take the discrete Fourier transform of both sides of equation (2-200a) and then rearrange the resulting double sum. We have

$$\hat{z}_p = \frac{1}{2K} \sum_{k=-K}^{K-1} e^{-ikp\pi/K} \sum_{k'=-K}^{K-1} y_{(k-k')} x_{k'}$$
$$= \frac{1}{2K} \sum_{k'=-K}^{K-1} x_{k'} e^{-ipk'\pi/K} \sum_{k=-K}^{K-1} y_{(k-k')} e^{-ip(k-k')\pi/K}. \tag{2-201}$$

If we now define $k'' \equiv (k-k')$, the inner sum in equation (2-201) becomes $\sqrt{2K}\hat{y}_p$. This is independent of k', so it may be taken out of the outer sum, which then yields $\sqrt{2K}\hat{x}_p$.◁

Fast Fourier transforms provide a very economical way of evaluating convolutions such as (2-200a): Direct evaluation of the z_k would require $(2K)^2$ additions and multiplications, to be compared with $2K[6\log_2(2K) + 1]$ operations required to evaluate the two transforms $\hat{x}_p$ and $\hat{y}_p$, to form $\hat{z}_p$ from them and then to recover z_k from $\hat{z}_p$.

For simplicity we describe the use of discrete Fourier transforms in finding the gravitational potentials of N-body models for the case of a planar mass distribution such as a stellar disk; the generalization to three dimensions is straightforward. We assume that the cells are square and use units in which each cell is one unit of length on a side and the gravitational constant is unity. The cell centers may then be at $x_{lm} = (l, m)$, where $l, m = 0, \ldots, K - 1$. Hence the number of cells $J = K^2$ and equations (2-195) and (2-196) become

$$\Phi_{lm} = \sum_{l'=0}^{K-1} \sum_{m'=0}^{K-1} G(l - l', m - m') M_{l'm'}, \tag{2-202a}$$

where

$$G(l - l', m - m') \equiv \frac{-1}{\sqrt{\epsilon^2 + (l - l')^2 + (m - m')^2}}, \tag{2-202b}$$

and we have introduced a new notation for G to stress that G depends only on the differences $l - l'$ and $m - m'$.

The sums in equation (2-202a) are almost of the same form as those appearing in equation (2-200a). The only differences are: (i) in equation (2-200a) the index is summed from $-K$, whereas in equation (2-202a) the indices start at 0; and (ii) the functions $M_{l'm'}$ and $G(p, q)$ are not periodic. However, we may extend the range of the summation in equation (2-202a) by defining $M_{l'm'} = 0$ whenever either l' or m' is in the range $-K$ to -1, without affecting the value of any of the Φ_{lm}. Furthermore, we can make $M_{l'm'}$ periodic since the sum (2-202a) does not involve any values of l' and m' outside the first period. We are also free to make G periodic by defining it to be the periodic function that agrees with equation (2-202b) for $l - l'$ and $m - m'$ in the range $[-K, K - 1]$. (We can do this for the following reason: the only values of the indices l and m that are of interest are in the range $[0, K - 1]$ since we only need the potential where we have particles. Thus the only values of $l - l'$ and $m - m'$ that are of interest and for which the coefficient of G is non-zero lie in the range $[-K, K - 1]$.)

We may now apply the discrete Fourier convolution theorem to show

$$\hat{\Phi}_{pq} = 2K\hat{G}_{pq}\hat{M}_{pq}. \qquad (2\text{-}203)$$

Thus a fast alternative to the direct evaluation of the expression (2-202a) for Φ_{lm} is to take the Fourier transform of $M_{l'm'}$, multiply each element of the resulting matrix $\hat{M}_{pq}$ by $\hat{G}_{pq}$ to form $\hat{\Phi}_{pq}$, and then use (2-197) to recover the Φ_{lm}. The matrix $\hat{G}_{pq}$ need not be recalculated after the first time step, so the work to be performed at each time step consists of the evaluation of two two-dimensional transforms, which can be accomplished with $2[4K\log_2(2K)]^2$ multiplications and additions, and $(2K)^2$ further multiplications for the formation of $\hat{\Phi}_{pq}$. Thus if we use Fourier transforms we can obtain all the Φ_{lm} with about $(2K)^2[1 + 4\log_2(2K)]^2$ arithmetic operations. If we were to evaluate the double sum (2-202a) directly, we could obtain the Φ_{lm} only after K^4 additions and multiplications. For $K \gtrsim 16$ the transforms provide a more economical route to the Φ_{lm}.

Many studies of the dynamics of two- and three-dimensional model galaxies have employed Fourier transforms on Cartesian grids to find $\Phi(\mathbf{x})$, and hence $\nabla\Phi(\mathbf{x})$ (Hockney & Brownrigg 1974; Hohl & Zang 1979; Miller 1978; Efstathiou et al. 1982). These calculations typically distribute matter over $(32)^3$ cells and can involve enormous numbers of particles—100,000 particles in a simulation is not unusual. This has the advantage that the statistical uncertainty in the results of these calculations that arises from random fluctuations in the numbers of particles in different volumes is very small.

Many valuable results have been obtained using fast Fourier transforms on Cartesian grids, but two examples will show that there are some interesting problems that require alternative approaches:

(i) Consider the use of this method to simulate the collision of two galaxies. Initially one would like the galaxies to be well separated, but then most of the space inside the smallest rectangular box that contains both galaxies would be empty. Consequently, the majority of the cells would be empty, and the largest part of the work performed in calculating the Φ_{lm} would be wasted on obtaining the potential at points where we do not need it. Thus the calculation will be prohibitively expensive unless the galaxies are close to one another before we start the calculation, and this could lead to misleading results. By contrast, if we evaluate the forces by direct summation over all particles, no difficulty arises when the galaxies are far from one another, but of course we then are severely restricted in the number of particles comprising each galaxy.

(ii) When the potential is calculated on a Cartesian grid, it is difficult to follow systems such as elliptical galaxies in which there are strong density contrasts. This is because the quantity Φ_α that is defined by equation (2-196) is a good approximation to $\Phi(\mathbf{x}_\alpha)$ only if the particle density is nearly constant in all cells that are located near $\mathbf{x}_\alpha$. If all cells are of the same size, it may be impossible to ensure that the cells are sufficiently small that in the densest regions the density is nearly constant from cell to cell. It is possible to overcome this problem by augmenting the force field that is obtained by the Fourier technique with a separate estimate of the force field that is generated by the particles that lie in the cells nearest to the point of observation. Such an estimate can be obtained by direct summation over particles (Efstathiou & Eastwood 1981). A correction of this type has been used in simulations of the expanding Universe (e.g., Efstathiou et al. 1982), but it increases the computational expense considerably and is rarely worthwhile in simulations of galaxy dynamics.

Both of these difficulties can be remedied in alternative algorithms that we now describe.

(c) Fourier method in polar coordinates   In §2.6 we saw that in terms of the modified cylindrical coordinates $u = \ln R$ and ϕ, the reduced potential $V(u, \phi) = e^{u/2}\Phi[R(u), \phi]$ of a flat disk is related to the reduced surface density $S(u, \phi) = e^{3u/2}\Sigma[R(u), \phi]$ by

$$V(u, \phi) = -\frac{G}{\sqrt{2}} \int_{-\infty}^{\infty} du' \int_0^{2\pi} \frac{S(u', \phi')d\phi'}{\sqrt{\cosh(u - u') - \cos(\phi - \phi')}}, \quad (2\text{-}204)$$

If we break the disk up into cells across which u and ϕ change by Δu and $\Delta\phi$, respectively, the integral on the right side of this expression becomes a double sum over all cells

$$V_{lm} \simeq \sum_{l'm'} K(l' - l, m' - m)M_{l'm'}, \quad (2\text{-}205a)$$

where

$$M_{lm} \equiv \int \int_{\text{cell}(l,m)} S(u, \phi) du \, d\phi \qquad \text{(2-205b)}$$

and

$$K(l - l', m - m') \equiv -\frac{2^{-\frac{1}{2}}G}{\sqrt{\cosh(u_l - u_{l'}) - \cos(\phi_m - \phi_{m'})}}. \qquad \text{(2-205c)}$$

Since in equation (2-204a) K depends only on the differences $| l - l' |$ and $| m - m' |$, the double sum in this equation can be evaluated by the same Fourier technique that we described for the case of a Cartesian grid.

Note that when $(l, m) = (l', m')$, K is not defined. However, by taking $S(u', \phi') = constant$ and approximating $\cosh(u - u') - \cos(\phi - \phi')$ by $\frac{1}{2}(u - u')^2 + \frac{1}{2}(\phi - \phi')^2$, it can be shown that the contribution of the material in the $(l, m)^{\text{th}}$ cell to V_{lm} will be roughly M_{lm} times

$$K(0,0) = -2G \left[\frac{1}{\Delta\phi} \text{arcsinh} \left(\frac{\Delta\phi}{\Delta u} \right) + \frac{1}{\Delta u} \text{arcsinh} \left(\frac{\Delta u}{\Delta\phi} \right) \right]. \qquad \text{(2-206)}$$

A rectangular grid in (u, ϕ) space generates cells in (R, ϕ) space that become smaller and smaller as $R = 0$ is approached. This makes a potential-finding scheme based on equations (2-205) well suited to numerical simulations of centrally concentrated disks.

(d) Potential from multipole expansion Consider the potential generated by a particle whose spherical coordinates are $(r_\alpha, \theta_\alpha, \phi_\alpha)$. This can be regarded as the potential of a spherical shell whose total mass m_α is the same as that of the particle, but whose surface density is zero everywhere except at the particle's location. From equations (2-120) and (2-114) the potential of such a shell is,

$$\Phi(r < r_\alpha, \theta, \phi) = -4\pi G \sum_{l=0}^{\infty} \frac{r^l}{r_\alpha^{l-1}} \sum_{m=-l}^{l} \tilde{\sigma}_{lm}(\alpha) \, Y_l^m(\theta, \phi) \qquad \text{(2-207a)}$$

$$\Phi(r > r_\alpha, \theta, \phi) = -4\pi G \sum_{l=0}^{\infty} \frac{r_\alpha^{l+2}}{r^{l+1}} \sum_{m=-l}^{l} \tilde{\sigma}_{lm}(\alpha) \, Y_l^m(\theta, \phi), \qquad \text{(2-207b)}$$

where

$$\tilde{\sigma}_{lm}(\alpha) = \frac{m_\alpha}{4\pi(2l + 1)r_\alpha^2} Y_l^{m*}(\theta_\alpha, \phi_\alpha). \qquad \text{(2-207c)}$$

Now imagine that at each time step in an N-body simulation, we carry out the following steps for some small predetermined integer L.

(i) We make a crude estimate of the location of the deepest point in the simulation's potential well. This will usually lie at the point of highest particle density.

(ii) We establish a system of spherical coordinates around the point selected in the previous step, and for all $l \leq L$ we obtain from equation (2-207c) the $(L+1)^2$ coefficients $\tilde{\sigma}_{lm}$ of each particle $\alpha = 1, \ldots, N$.

(iii) We then estimate the overall potential at (r, θ, ϕ) as

$$\tilde{\Phi}(r, \theta, \phi) = -4\pi G \sum_{l=0}^{L} \sum_{m=-l}^{l} \left[r^l \sum_{r_\alpha < r} \frac{\tilde{\sigma}_{lm}(\alpha)}{r_\alpha^{l-1}} \right.$$
$$\left. + \frac{1}{r^{l+1}} \sum_{r_\alpha > r} r_\alpha^{l+2} \tilde{\sigma}_{lm}(\alpha) \right] Y_l^m(\theta, \phi).$$
(2-208)

In the limit $L \to \infty$, $\tilde{\Phi}$ converges to the potential generated by our system of point particles. For finite L, $\tilde{\Phi}$ is the potential generated by a series of spherical shells of radius r_α on which the density is strongly peaked toward the point with angular coordinates $(\theta_\alpha, \phi_\alpha)$. Hence one can use $\tilde{\Phi}$ as an approximation to the true potential Φ. If the system is only moderately aspherical, $\tilde{\Phi}$ can be a good estimate of Φ even with L as small as 4.

The computational expense of calculating the $\tilde{\sigma}_{lm}$ is proportional to $(L+1)^2 N$. If the particles are sorted by increasing radius, intelligent programming of the evaluation of the sums in equation (2-208) enables one to obtain $\tilde{\Phi}$ at the position of every particle at an expense that is also proportional to N. To compute $\tilde{\Phi}$ at subsequent time steps, the particles must be resorted in radius, but this can be done cheaply since the ordering will be almost the same if the time step is short enough. Thus the computational expense of a simulation in which the forces are calculated from $\tilde{\Phi}$ is proportional to the total number of particles in the simulation. Consequently, the cost of this technique compares favorably with that of the Fourier technique for obtaining the potential generated by large numbers of particles, and it is very much less expensive than calculating the forces by direct summation over all particles.

An important advantage of using the multipole expansion rather than the Fourier method to obtain Φ is that the contribution to $\nabla\Phi$ from any particle α is inaccurate only at points that lie closer to the particle than some fraction ϵ of the radial coordinate r_α (ϵ is inversely proportional to L). Thus the spatial resolution of the force field that one obtains from $\tilde{\Phi}$ is very large near the center of the system, and this makes the multipole expansion ideal for simulations of highly centrally concentrated systems such as elliptical galaxies. It can also be adapted to handle colliding galaxies.

Van Albada (1982), Villumsen (1982), and McGlynn (1982, 1984) have all performed N-body simulations using multipole expansions. McGlynn (1982) discusses some refinements that help to reduce two-body relaxation in simulations of this type and minimize problems that can arise from an unfortunate choice of center. This approach to finding the forces acting in N-body simulations can be extended to schemes using expansions in other coordinate systems such as cylindrical and bispherical coordinates (Villumsen 1984; Piran & Villumsen 1987).

Problems

2-1. [1] Astronauts orbiting an unexplored planet find that (i) the surface of the planet is precisely spherical; and (ii) the potential exterior to the planetary surface is $\Phi = -GM/r$ exactly, that is, there are no non-zero multipole moments of higher order than the monopole. Can they conclude from these observations that the mass distribution in the interior of the planet is spherically symmetric? If not, give a simple example of a nonspherical mass distribution that would reproduce the observations.

2-2. [1] Show that the gravitational potential energy of a spherical system can be written

$$W = -\frac{G}{2} \int_0^\infty \frac{M^2(r) dr}{r^2}, \qquad (2\text{P-1})$$

where $M(r)$ is the mass interior to radius r.

2-3. [1] Show that the potential generated by the spherical density distribution (Jaffe 1983)

$$\rho(r) = \left(\frac{M}{4\pi r_J^3}\right) \frac{r_J^4}{r^2(r+r_J)^2} \quad \text{is} \quad \Phi(r) = \frac{GM}{r_J} \ln\left(\frac{r}{r+r_J}\right), \qquad (2\text{P-2})$$

where M and r_J are constants. Verify that the total mass of the system is M. Show that the circular speed is approximately constant at $r \ll r_J$, and falls off as $v_c \propto r^{-\frac{1}{2}}$ at $r \gg r_J$.

2-4. [1] Prove that the Chandrasekhar potential energy tensor for any spherical body has the form $W_{jk} = \frac{1}{3}W\delta_{jk}$, where W is the potential energy.

2-5. [1] Defining **prolate spheroidal coordinates** (u, v) by $R = a \sinh u \sin v$, $z = a \cosh u \cos v$, where a is a constant, show that $R^2 + (a + |z|)^2 = a(\cosh u + |\cos v|)^2$. Hence show that the potential (2-49a) of Kuzmin's disk can be written

$$\Phi_K(u, v) = -\left(\frac{GM}{a}\right) \frac{\cosh u - |\cos v|}{\sinh^2 u + \sin^2 v}. \qquad (2P\text{-}3)$$

In §3.5 we show that this potential is an example of a Stäckel potential, in which orbits admit an extra isolating integral.

2-6. [2] Consider an axisymmetric body whose density distribution is $\rho(R, z)$ and total mass is $M = \int \rho(R, z) d^3\mathbf{r}$. Assume that the body has finite extent $[\rho(R, z) = 0$ for $r^2 = R^2 + z^2 > r_{\max}^2]$ and is symmetric about its equator, that is, $\rho(R, -z) = \rho(R, z)$.

(a) Show that at distances large compared to $r_{\max}$, the potential arising from this body can be written in the form

$$\Phi(R, z) \simeq -\frac{GM}{r} - \frac{G}{4} \frac{(R^2 - 2z^2)}{r^5} \int \rho(R', z')(R'^2 - 2z'^2) d^3\mathbf{r}', \qquad (2P\text{-}4)$$

where the fractional error is of order $(r_{\max}/r)^2$ smaller than the second term.

(b) Show that at large distances from an exponential disk with surface density $\Sigma(R) = \Sigma_0 \exp(-R/R_d)$, the potential has the form

$$\Phi(R, z) \simeq -\frac{GM}{r}\left[1 + \frac{3R_d^2(R^2 - 2z^2)}{2r^4}\right], \qquad (2P\text{-}5)$$

where M is the mass of the disk.

2-7. [2] Prove that the external potentials and force fields of any two confocal spheroids of uniform density and equal mass are everywhere the same.

2-8. [2] Use equation (2-99) to show that a prolate body with density $\rho = \rho_0(1 + R^2/a_1^2 + z^2/a_3^2)^{-2}$, where $a_3 > a_1$, generates the potential

$$\Phi(u, v) = -\pi G a_1^2 a_3 \rho_0 \int_0^\infty \frac{\sqrt{a_3^2 + \tau}\, d\tau}{(\tau + a_3^2 + \lambda)(\tau + a_3^2 + \mu)}, \qquad (2P\text{-}6)$$

where (u, v) are oblate spheroidal coordinates defined by equation (2-58) with $\Delta^2 = a_3^2 - a_1^2$, and we have written $\lambda \equiv \Delta^2 \sinh^2 u$, $\mu \equiv -\Delta^2 \cos^2 v$. Decompose the integral in (2P-6) into partial fractions to show (without evaluating the integrals) that Φ is of the special Stäckel form discussed in §3.5. Finally, show that

$$\Phi(u, v) = -\left(\frac{2\pi G a_1^2 a_3 \rho_0}{\Delta^2}\right) \frac{f(\Delta \sinh u) - f(i\Delta \cos v)}{\sinh^2 u + \cos^2 v}, \qquad (2P\text{-}7a)$$

where

$$f(z) \equiv z \arctan(z/a_3). \qquad (2P\text{-}7b)$$

[Hint: To ensure convergence of the integrals, you may wish to add $(\tau + a_3^2)^{-\frac{1}{2}}$ to one of the integrands and subtract it from the other.] De Zeeuw (1985) calls the body with this potential the **perfect prolate spheroid**, because it is the only prolate axisymmetric density distribution of constant ellipticity whose potential is of the Stäckel form.

2-9. [2] Show that the analog to equation (2-159) that relates the potential $\Phi(R, \phi, z)$ to the surface density $\Sigma(R, \phi)$ for a non-axisymmetric disk is

$$\Phi(R, \phi, z) = -G \sum_{m=-\infty}^{\infty} e^{im\phi} \int_0^{\infty} dk e^{-k|z|} J_m(kR) \int_0^{\infty} dR' R' J_m(kR')$$

$$\times \int_0^{2\pi} \Sigma(R', \phi') e^{-im\phi'} d\phi',$$

(2P-8)

where $J_m(u)$ is the cylindrical Bessel function of order m.

2-10. [2] The purpose of this problem is to reproduce an elegant method due to Schwarzschild (1954) of evaluating the potential energy W of a finite spherical system that has a constant mass-to-light ratio Υ.

(a) Show that the surface brightness $I(R)$ and luminosity density $j(r)$ are related by the formula

$$I(R) = 2 \int_r^{\infty} \frac{j(r) r \, dr}{\sqrt{r^2 - R^2}}.$$

(2P-9)

(b) Invert equation (2P-9) using Abel's formula [see eq. (1B-59)] to obtain

$$j(r) = -\frac{1}{\pi} \int_r^{\infty} \frac{dI(R)}{dR} \frac{dR}{\sqrt{R^2 - r^2}}.$$

(2P-10)

(c) The **strip brightness** $S(x)$ is defined so that $S(x)dx$ is the total luminosity in a strip of width dx that passes a distance x from the projected center of the system. Show that

$$S(x) = 2 \int_x^{\infty} \frac{I(R) R \, dR}{\sqrt{R^2 - x^2}}.$$

(2P-11)

(d) Show that the strip brightness and luminosity density are related by

$$j(x) = -\frac{1}{2\pi x} \frac{dS(x)}{dx}.$$

(2P-12)

(e) Prove that the mass interior to radius r is given by

$$M(r) = -2\Upsilon \int_0^r \frac{dS}{dx} x \, dx.$$

(2P-13)

(f) Using equation (2-130) for W show that

$$W = -2G\Upsilon^2 \int_0^{\infty} S^2(x) \, dx.$$

(2P-14)

This result was used by Schwarzschild and Bernstein (1955) in one of the first measurements of the mass-to-light ratio of a globular cluster. It is less popular nowadays because the computation of W can be carried out numerically, and because strip brightnesses are difficult to measure accurately in the presence of background stars.

2-11. [3] We have derived relations between the potential and surface density of non-axisymmetric disks by solving Laplace's equation in oblate spheroidal coordinates [see §2.6.4(a)] and cylindrical coordinates (Problem 2-9). Derive a relation of this kind by solving Laplace's equation in spherical coordinates, and show that the result is identical with the formula Kalnajs derived using logarithmic spirals [eq. (2-189)]. [Hint: You may need associated Legendre functions $P_\lambda^m(x)$, where λ is a complex number. Equations (1C-18) and (1C-7) may also be helpful.]

2-12. [3] Show that the circular speed $v_c(R)$ in a thin axisymmetric disk of surface density $\Sigma(R)$ may be written in the form (Mestel 1963)

$$v_c^2(R) = \frac{GM(R)}{R} + 2\pi \sum_{k=1}^{\infty} \alpha_{2k}\left[(2k+1)R^{-(2k+1)}\int_0^R \Sigma(R')R'^{2k+1}dR'\right.$$
$$\left. - 2kR^{2k}\int_R^\infty \Sigma(R')R'^{-2k}dR'\right],$$
$$\tag{2P-15}$$

where

$$\alpha_k = \pi\left[\frac{(2k)!}{2^k(k!)^2}\right]^2. \tag{2P-16}$$

[Hint: Start with equation (2-139) and expand $|\mathbf{x}-\mathbf{x}'|^{-1}$ in Legendre polynomials using equation (1C-23).]

2-13. [3] (Suggested by H. Dejonghe) Prove that the surface density $\Sigma(x,y)$ and potential $\Phi(x,y)$ in a disk occupying the $z=0$ plane are related by

$$\Sigma(x',y') = \frac{1}{4\pi^2 G}\iint \frac{dx\,dy}{|\mathbf{x}-\mathbf{x}'|}\left(\frac{\partial^2\Phi}{\partial x^2}+\frac{\partial^2\Phi}{\partial y^2}\right). \tag{2P-17}$$

(Hint: You may wish to use the results of §5.3.1.)

3 The Orbits of Stars

In this chapter we focus on the orbits of individual stars in given potentials. Thus we ask the questions, "What kinds of orbits are possible in a spherically symmetric, or an axially symmetric potential? How are these orbits modified if we distort the potential into a barlike form?" We shall obtain a number of useful analytic results for the simpler potentials, and use these results to develop an intuitive understanding of how stars move in more complex potentials.

All of the work in this chapter is based on a fundamental approximation: although galaxies are composed of stars, we shall neglect the forces from individual stars and consider only the large-scale forces from the overall mass distribution, which is made up of thousands of millions of stars. In other words, we assume that the force fields of galaxies are *smooth*, neglecting small-scale irregularities due to individual stars or larger objects like globular clusters or molecular clouds. As we shall see in Chapters 4 and 7, the force fields of galaxies *are* sufficiently smooth that these irregularities can affect the orbits of stars only after many tens of dynamical times.

We begin our study with the simplest case: that of a star moving in a static, spherically symmetric potential. This potential is the appropriate one for globular clusters, which are usually nearly spherical, but, more important, the results we obtain provide a valuable point of reference for the subsequent discussions of axisymmetric and barlike potentials.

3.1 Orbits in Static Spherical Potentials

The motion of a star in a **centrally directed field of force** is greatly simplified by the familiar law of conservation of angular momentum (see Appendix 1.D). Thus if

$$\mathbf{r} = r\hat{\mathbf{e}}_r \tag{3-1}$$

denotes the position vector of the star with respect to the center, and the radial force per unit mass is

$$\mathbf{F} = F(r)\hat{\mathbf{e}}_r, \tag{3-2}$$

103

the equation of motion of the star is

$$\frac{d^2 \mathbf{r}}{dt^2} = F(r)\hat{\mathbf{e}}_r. \tag{3-3}$$

If we remember that the cross product of any vector with itself is zero, we have

$$\frac{d}{dt}\left(\mathbf{r}\times\frac{d\mathbf{r}}{dt}\right) = \frac{d\mathbf{r}}{dt}\times\frac{d\mathbf{r}}{dt} + \mathbf{r}\times\frac{d^2\mathbf{r}}{dt^2} = F(r)\mathbf{r}\times\hat{\mathbf{e}}_r = 0. \tag{3-4}$$

Equation (3-4) says that $(\mathbf{r}\times\dot{\mathbf{r}})$ is some constant vector, say $\mathbf{L}$:

$$\mathbf{r}\times\frac{d\mathbf{r}}{dt} = \mathbf{L}. \tag{3-5}$$

Of course, $\mathbf{L}$ is simply the angular momentum per unit mass; geometrically it is a vector perpendicular to the plane defined by the star's instantaneous position and velocity vectors. Since this vector is constant, we conclude that the star moves in a plane, the **orbital plane**. This greatly simplifies the determination of the star's orbit, for now that we have established that the star moves in a plane, we may simply use plane polar coordinates (r, ψ) in which the center of attraction is at $r = 0$ and ψ is the azimuthal angle in the orbital plane. Writing $\mathbf{r} = r\hat{\mathbf{e}}_r$ and using equations (1B-21) for the derivatives of unit vectors in equation (3-3), we find

$$\ddot{r} - r\dot{\psi}^2 = F(r) \tag{3-6a}$$

$$2\dot{r}\dot{\psi} + r\ddot{\psi} = 0. \tag{3-6b}$$

After multiplication by r, equation (3-6b) may be immediately integrated to

$$r^2\dot{\psi} = \text{constant} \equiv L. \tag{3-7}$$

It is not hard to show that L is actually the length of the vector $\mathbf{r}\times\dot{\mathbf{r}}$, and hence that (3-7) is just a restatement of the conservation of angular momentum. Geometrically, L is equal to twice the rate at which the radius vector sweeps out area.

To proceed further we use equation (3-7) to replace time t by angle ψ as the independent variable in equation (3-6a). Since (3-7) implies

$$\frac{d}{dt} = \frac{L}{r^2}\frac{d}{d\psi}, \tag{3-8}$$

equation (3-6a) becomes

$$\frac{L^2}{r^2}\frac{d}{d\psi}\left(\frac{1}{r^2}\frac{dr}{d\psi}\right) - \frac{L^2}{r^3} = F(r). \tag{3-9}$$

This equation can be simplified by the substitution

$$u \equiv \frac{1}{r}, \tag{3-10a}$$

which puts (3-9) into the form

$$\frac{d^2u}{d\psi^2} + u = -\frac{F(1/u)}{L^2u^2}. \tag{3-10b}$$

The solutions of this equation are of two types: along **unbound** orbits $r \to \infty$ and hence $u \to 0$, while on **bound** orbits r and u oscillate between finite limits. Thus each bound orbit is associated with a periodic solution of this equation. We give three analytic examples later in this section, but in general the solutions of equation (3-10b) must be obtained numerically. Some additional insight is gained by deriving a "radial energy" equation from equation (3-10b) in much the same way as we derive the conservation of kinetic plus potential energy in Appendix 1.D; we multiply (3-10b) by $du/d\psi$ and integrate over ψ to obtain

$$\left(\frac{du}{d\psi}\right)^2 + \frac{2\Phi}{L^2} + u^2 = \text{constant} \equiv \frac{2E}{L^2}, \tag{3-11}$$

where $\Phi(r)$ is a potential derived from $F(r)$ using the relation $F(r) = -(d\Phi/dr) = +u^2(d\Phi/du)$. When we rewrite equation (3-11) in the form

$$E = \tfrac{1}{2}\left(\frac{dr}{dt}\right)^2 + \tfrac{1}{2}\left(r\frac{d\psi}{dt}\right)^2 + \Phi(r), \tag{3-12}$$

it becomes apparent that the constant E is just the energy per unit mass of the star. For bound orbits the equation $(du/d\psi) = 0$ or

$$u^2 + \frac{2[\Phi(1/u) - E]}{L^2} = 0 \tag{3-13}$$

will have roots u_1 and u_2 between which the star oscillates radially as it revolves in ψ. Thus the orbit is contained between an inner radius $r_1 = u_1^{-1}$, known as the **pericenter** distance, and an outer radius $r_2 = u_2^{-1}$, called the **apocenter** distance. The **radial period** T_r is the time required for the star to travel from apocenter to pericenter and back. To

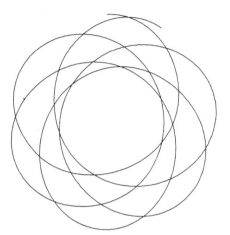

Figure 3-1. A typical orbit in a
spherical potential forms a rosette.

determine T_r we use equation (3-7) to eliminate $\dot{\psi}$ from equation (3-12).
We find

$$\left(\frac{dr}{dt}\right)^2 = 2(E - \Phi) - \frac{L^2}{r^2}. \tag{3-14}$$

Equation (3-14) may be rewritten

$$\frac{dr}{dt} = \pm\sqrt{2[E - \Phi(r)] - \frac{L^2}{r^2}}. \tag{3-15}$$

The two possible signs arise because the star moves alternately in and
out. Comparing (3-15) with (3-13) we see that $\dot{r} = 0$ at the pericenter
and apocenter distances r_1 and r_2, as of course it must. From equation
(3-15) it follows that the radial period is

$$T_r = 2\int_{r_1}^{r_2} \frac{dr}{\sqrt{2[E - \Phi(r)] - L^2/r^2}}. \tag{3-16}$$

In traveling from pericenter to apocenter and back, the azimuthal
angle ψ increases by an amount

$$\Delta\psi = 2\int_{r_1}^{r_2} \frac{d\psi}{dr}dr = 2\int_{r_1}^{r_2} \frac{L}{r^2}\frac{dt}{dr}dr. \tag{3-17a}$$

Substituting for dt/dr from (3-15) this becomes

$$\Delta\psi = 2L\int_{r_1}^{r_2} \frac{dr}{r^2\sqrt{2[E - \Phi(r)] - L^2/r^2}}. \tag{3-17b}$$

The **azimuthal period** is

$$T_\psi = \frac{2\pi}{\Delta\psi}T_r; \qquad (3\text{-}18)$$

the mean angular speed of the particle is $2\pi/T_\psi$. In general $\Delta\psi/2\pi$ will not be a rational number. Hence the orbit will not be closed: a typical orbit resembles a rosette and eventually passes through every point in the annulus between the circles of radii r_1 and r_2 (see Figure 3-1). There are, however, two important special potentials in which all bound orbits are closed.

(a) Spherical harmonic oscillator   We call a potential of the form $\Phi(r) = \frac{1}{2}\Omega^2 r^2 + constant$ a spherical harmonic oscillator potential. As we saw in §2.1.2(b), this potential is generated by a homogeneous sphere. Equation (3-10b) can be solved analytically in this case, but it is actually simpler to use Cartesian coordinates (x, y) defined by $x = r\cos\psi$, $y = r\sin\psi$. In these coordinates, the equations of motion are simply

$$\ddot{x} = -\Omega^2 x \quad ; \quad \ddot{y} = -\Omega^2 y, \qquad (3\text{-}19a)$$

with solutions

$$x = X\cos(\Omega t + \epsilon_x) \quad ; \quad y = Y\cos(\Omega t + \epsilon_y), \qquad (3\text{-}19b)$$

where X, Y, ϵ_x, and ϵ_y are arbitrary constants. Every orbit is closed since the periods of the oscillations in x and y are identical. The orbits form ellipses centered on the center of attraction.

(b) Kepler potential   When the star is acted on by an inverse-square law of force $F(r) = -GM/r^2$ due to a point mass M, $F = -GMu^2$, and equation (3-10b) becomes

$$\frac{d^2u}{d\psi^2} + u = \frac{GM}{L^2}, \qquad (3\text{-}20)$$

the general solution of which is

$$u(\psi) = C\cos(\psi - \psi_0) + \frac{GM}{L^2}, \qquad (3\text{-}21)$$

where $C > 0$ and ψ_0 are arbitrary constants. Defining the orbit's **eccentricity** by

$$e = \frac{CL^2}{GM} \qquad (3\text{-}22a)$$

and its **semi-major axis** by

$$a \equiv \frac{L^2}{GM(1 - e^2)},$$
(3-22b)

equation (3-21) may be rewritten

$$r(\psi) = \frac{a(1 - e^2)}{1 + e\cos(\psi - \psi_0)}.$$
(3-23)

An orbit for which $e \geq 1$ is unbound, since $r \to \infty$ as $(\psi - \psi_0) \to$ arccos$(1/e)$; the orbit forms a hyperbola if $e > 1$ and a parabola if $e = 1$. The star's asymptotic speed v_∞ as $r \to \infty$ is related to e and L by (see Problem 3-1)

$$e^2 = 1 + \left(\frac{Lv_\infty}{GM}\right)^2.$$
(3-24)

Orbits for which $e < 1$ are bound, for then r is finite for all values of ψ. Furthermore, r is now a periodic function of ψ with period 2π, so the star returns to its original radial coordinate after exactly one revolution in ψ. Thus these orbits are closed, and one may readily show from equation (3-23) that they form ellipses with the attracting center at one focus. The pericenter and apocenter distances are

$$r_1 = a(1 - e) \quad \text{and} \quad r_2 = a(1 + e),$$
(3-25)

and the radial and azimuthal periods are

$$T_r = T_\psi = 2\pi\sqrt{\frac{a^3}{GM}}.$$
(3-26)

The angle $\psi - \psi_0$ is known as the **true anomaly**. A useful parametric representation of the orbit is

$$r = a(1 - e\cos\eta) \quad ; \quad t = (T_r/2\pi)(\eta - e\sin\eta),$$
(3-27)

where η is the **eccentric anomaly**.

From (3-11) the energy per unit mass of any Kepler orbit is

$$E = -\frac{GM}{2a}.$$
(3-28)

From these results [or from eq. (3-17b)] we see that a star on a Kepler orbit completes a radial oscillation in the time required for ψ to increase by $\Delta\psi = 2\pi$, whereas a star that orbits in a harmonic potential has already completed a radial oscillation by the time ψ has increased by $\Delta\psi = \pi$. Since galaxies are more extended than point masses, and less extended than homogeneous spheres, a typical star in a spherical galaxy completes a radial oscillation after its angular coordinate has increased by an amount that lies somewhere in between these two extremes; $\pi < \Delta\psi < 2\pi$. Thus, we expect a star to oscillate from its apocenter through its pericenter and back in a shorter time than is required for one complete azimuthal cycle about the galactic center.

(c) Isochrone potential An explicit demonstration of this behavior is furnished by orbits in the isochrone potential of equation (2-32) (Hénon 1959). It is convenient to define an auxiliary variable s by

$$s \equiv -\frac{GM}{b\Phi} = 1 + \sqrt{\frac{r^2}{b^2} + 1}. \tag{3-29}$$

Solving equation (3-29) for r, we find that

$$\frac{r^2}{b^2} = s^2 \left(1 - \frac{2}{s}\right), \qquad s \geq 2. \tag{3-30}$$

Given this one-to-one relationship between s and r, we may employ s as a radial coordinate in place of r. The integrals (3-16) and (3-17b) for T_r and $\Delta\psi$ both involve the infinitesimal quantity

$$dI \equiv \frac{dr}{\sqrt{2[E - \Phi] - L^2/r^2}}. \tag{3-31}$$

When we use equation (3-30) to eliminate r from this expression, we find

$$dI = \frac{b(s-1)ds}{\sqrt{2Es^2 - 2(2E - GM/b)s - (4GM/b) - (L^2/b^2)}}. \tag{3-32}$$

As the star moves from pericenter r_1 to apocenter r_2, s varies from the smaller root s_1 of the quadratic expression in the denominator of equation (3-32) to the larger root s_2. Thus, combining equations (3-16) and (3-32), we find

$$T_r = \frac{2b}{\sqrt{-2E}} \int_{s_1}^{s_2} \frac{(s-1)ds}{\sqrt{(s_2 - s)(s - s_1)}} = \frac{2\pi b}{\sqrt{-2E}} \left[1 - \tfrac{1}{2}(s_1 + s_2)\right], \tag{3-33}$$

where we have assumed $E < 0$ since we are dealing with bound orbits. But from the denominator of equation (3-32) it follows that the roots s_1 and s_2 obey

$$s_1 + s_2 = 2\left(1 - \frac{GM}{2Eb}\right), \tag{3-34a}$$

and so the radial period

$$T_r = \frac{2\pi GM}{(-2E)^{\frac{3}{2}}}, \tag{3-34b}$$

exactly as in the Keplerian case. Note that T_r depends on the energy E but not on the angular momentum L—it is this property that gives the isochrone its name.

Equation (3-17b), for the increment $\Delta\psi$ in azimuthal angle per cycle in the radial direction, yields

$$\Delta\psi = 2L\int_{s_1}^{s_2} \frac{dI}{r^2} = \frac{2Lb^2}{\sqrt{-2E}}\int_{s_1}^{s_2} \frac{(s-2)(s-1)s\,ds}{\sqrt{(s_2-s)(s-s_1)}}$$
$$= \pi\left[1 + \frac{L}{\sqrt{L^2 + 4GMb}}\right]. \tag{3-35}$$

From this expression we see that $\pi < \Delta\psi < 2\pi$. The only orbits for which $\Delta\psi$ approaches the value 2π characteristic of Keplerian motion are those with $L^2 \gg 4GMb$. Such orbits never approach the core $r < b$ of the potential, and hence always move in a near-Keplerian force field.

It is sometimes useful to consider that an orbit in a non-Keplerian force field forms an approximate ellipse, though one that **precesses** by $\psi_p = \Delta\psi - 2\pi$ in the time needed for one radial oscillation. For the orbit shown in Figure 3-1, this precession is in the sense opposite to the rotation of the star itself. The angular velocity Ω_p of the rotating frame in which the ellipse appears closed is

$$\Omega_p = \frac{\psi_p}{T_r} = \frac{\Delta\psi - 2\pi}{T_r}. \tag{3-36}$$

Hence we say that Ω_p is the **precession rate** of the ellipse. We shall return to these ideas in §6.2.1(a).

1 Constants and Integrals of the Motion

A **constant of motion** in a given force field is any function $C(\mathbf{x}, \mathbf{v}; t)$ of the coordinates, velocities, and time that is constant along any stellar orbit; that is, if the position and velocity along an orbit are given by $\mathbf{x}(t)$ and $\mathbf{v}(t) = d\mathbf{x}/dt$,

$$C[\mathbf{x}(t_1), \mathbf{v}(t_1); t_1] = C[\mathbf{x}(t_2), \mathbf{v}(t_2); t_2] \qquad (3\text{-}37)$$

for any t_1 and t_2.

An **integral of motion** $I(\mathbf{x}, \mathbf{v})$ is any function of the phase-space coordinates $(\mathbf{x}, \mathbf{v})$ alone that is constant along any orbit:

$$I[\mathbf{x}(t_1), \mathbf{v}(t_1)] = I[\mathbf{x}(t_2), \mathbf{v}(t_2)]. \qquad (3\text{-}38)$$

While every integral is a constant of the motion, the converse is not true.[1] For example, on a circular orbit in a spherical potential the azimuthal coordinate ψ satisfies $\psi = \Omega t + \psi_0$, where Ω is the star's constant angular speed and ψ_0 is its azimuth at $t = 0$. Hence $C \equiv t - \psi/\Omega$ is a constant of the motion, but it is not an integral because it depends on time as well as the phase-space coordinates.

Any orbit in any force field always has six independent constants of motion. Indeed, since the initial phase-space coordinates $(\mathbf{x}_0, \mathbf{v}_0) \equiv [\mathbf{x}(0), \mathbf{v}(0)]$ can always be determined from $[\mathbf{x}(t), \mathbf{v}(t)]$ by integrating the equations of motion backward, $(\mathbf{x}_0, \mathbf{v}_0)$ can be regarded as six constants of motion.

In certain important cases, a few integrals can be written down easily: In any static potential $\Phi(\mathbf{x})$, the energy $E(\mathbf{x}, \mathbf{v}) = \frac{1}{2}v^2 + \Phi$ is an integral of motion. If a potential $\Phi(R, z, t)$ is axisymmetric about the z-axis, the z-component of the angular momentum is an integral, and in a spherical potential $\Phi(r, t)$ the three components of the angular momentum vector $\mathbf{L} = \mathbf{r} \times \mathbf{v}$ constitute three integrals of motion. However we shall find that it is usually impossible to obtain simple expressions for all the integrals of motion.

These concepts and their significance for the geometry of orbits in phase space are nicely illustrated by the example of motion in a spherically symmetric potential. In this case the energy E and the three components of the angular momentum per unit mass $\mathbf{L} = \mathbf{x} \times \mathbf{v}$ constitute four integrals. However, we shall find it more convenient to use $L = |\mathbf{L}|$ and the two independent components of the unit vector $\hat{\mathbf{n}} = \mathbf{L}/L$ as integrals in place of $\mathbf{L}$. We have seen that $\hat{\mathbf{n}}$ defines the orbital plane within which the position vector $\mathbf{r}$ and the velocity vector $\mathbf{v}$ must lie, and since the four coordinates r, ψ, $v_r = \dot{r}$, and $v_\psi = r\dot{\psi}$ suffice to specify the positions of $\mathbf{r}$ and $\mathbf{v}$ within this plane, we conclude that the two independent

[1] In some books constants of motion are called "integrals of motion" and integrals of motion are called "conservative integrals of motion".

components of $\hat{\mathbf{n}}$ restrict the star's phase point to a four-dimensional region of phase space. Furthermore, L and the energy E restrict the phase
point to that two-dimensional surface in this four-dimensional region on
which $v_r = \pm\sqrt{2[E - \Phi(r)] - L^2/r^2}$ and $v_\psi = L/r$. In §3.5 we shall see
that this surface is a torus and that the sign ambiguity in v_r is analogous
to the sign ambiguity in the z-coordinate of a point on the sphere $r^2 = 1$
when one specifies the point through its x and y coordinates. Thus for
practical purposes, given E, L, and $\hat{\mathbf{n}}$, the position and velocity of the
star can be specified by two quantities, for example r and ψ.

Is there a fifth integral of motion in a spherical potential? To study
this question, we examine motion in the potential

$$\Phi(r) = -GM\left(\frac{1}{r} + \frac{a}{r^2}\right). \tag{3-39}$$

For this potential, equation (3-10b) becomes

$$\frac{d^2u}{d\psi^2} + \left(1 - \frac{2GMa}{L^2}\right)u = \frac{GM}{L^2}, \tag{3-40}$$

the general solution of which is

$$u = C\cos\left(\frac{\psi - \psi_0}{K}\right) + \frac{GMK^2}{L^2}, \tag{3-41a}$$

where

$$K \equiv \left(1 - \frac{2GMa}{L^2}\right)^{-\frac{1}{2}}. \tag{3-41b}$$

Hence

$$\psi_0 = \psi - K\,\text{Arccos}\left[\frac{1}{C}\left(\frac{1}{r} - \frac{GMK^2}{L^2}\right)\right], \tag{3-42}$$

where $t = \text{Arccos}\,x$ is the general solution of $x = \cos t$, and C can be
expressed in terms of E and L by

$$E = \frac{1}{2}\frac{C^2L^2}{K^2} - \frac{1}{2}\left(\frac{GMK}{L}\right)^2. \tag{3-43}$$

Notice that by equation (3-42), the quantity ψ_0 depends only on the
parameters of the potential and the phase-space coordinates (through
L, r, and C). The function $\text{Arccos}\,x$ is multiple-valued, but by the proper
choice of solution we can ensure that ψ_0 is constant as the particle moves
along its orbit. Hence ψ_0 is an integral of motion.

Now suppose that we know E, L, ψ_0, and the radial coordinate r. Since we have four numbers—three integrals and one coordinate—it is natural to ask how we might use these numbers to determine the azimuthal coordinate ψ. We rewrite equation (3-42) in the form

$$\psi = \psi_0 \pm K \arccos \left[\frac{1}{C} \left(\frac{1}{r} - \frac{GMK^2}{L^2} \right) \right] + 2nK\pi, \qquad (3\text{-}44)$$

where $\arccos(x)$ is defined to be the value of $\text{Arccos}\,(x)$ that lies between 0 and π, and n is an arbitrary integer. If K is irrational—as nearly all real numbers are—then by a suitable choice of the integer n, we can make ψ modulo 2π approximate any given number as closely as we please. Thus for any values of E, L and r, an orbit that is known to have a given value of the integral ψ_0 can have an azimuthal angle as close as we please to any number between 0 and 2π.

On the other hand, if K is rational these problems do not arise. The simplest and most important case is that of the Kepler potential, when $a = 0$ and $K = 1$. Equation (3-44) now becomes

$$\psi = \psi_0 \pm \arccos \left[\frac{1}{C} \left(\frac{1}{r} - \frac{GM}{L^2} \right) \right] + 2n\pi, \qquad (3\text{-}45)$$

which yields only two values of ψ modulo 2π for given E, L and r.

Geometrically, the fifth integral $\psi_0(\mathbf{x}, \mathbf{v})$ defines a curve in phase space by the intersection of the five-dimensional surface $\psi_0(\mathbf{x}, \mathbf{v}) = constant$ with the two-dimensional surface on which $E = constant$ and $\mathbf{L} = constant$. Figure 3-1 can be regarded as a projection of this curve. In the Keplerian case $K = 1$, the curve closes on itself, and hence does not cover the surface $E = constant$, $\mathbf{L} = constant$. But when K is irrational, the curve is endless and densely covers the surface of constant E and $\mathbf{L}$.

Integrals like ψ_0 for irrational K that fail to confine orbits are called **non-isolating integrals**. All other integrals, such as E, $\mathbf{L}$, or the function ψ_0 when $K = 1$, are called **isolating integrals**. A more precise definition is the following: Consider the orbit that passes through a given point $(\mathbf{x}_0, \mathbf{v}_0)$. This orbit lies in the n-dimensional region of phase space S_n defined by the conditions $I_1 = constant$, $I_2 = constant, \ldots, I_{6-n} = constant$, where the I_i are any integrals that are already known to be isolating. The integral $I(\mathbf{x}, \mathbf{v})$ is isolating with respect to the orbit if there is some n-dimensional region of S_n in which there is no point that lies arbitrarily close to the surface $I(\mathbf{x}, \mathbf{v}) = I(\mathbf{x}_0, \mathbf{v}_0)$. Otherwise I is non-isolating. Isolating integrals are of great practical and theoretical importance, whereas non-isolating integrals are of essentially no value for galactic dynamics.

3.2 Orbits in Axisymmetric Potentials

Few galaxies are even approximately spherical, but many approximate figures of revolution. Thus in this section we begin to explore the types of orbits that are possible in many real galaxies. As in Chapter 2, we shall usually employ a cylindrical coordinate system (R, ϕ, z) centered on the galactic nucleus, and shall align the z-axis with the galaxy's symmetry axis.

Stars whose motions are confined to the equatorial plane of an axisymmetric galaxy have no way of perceiving that the potential in which they move is not spherically symmetric. Therefore their orbits will be identical with those we discussed in the last section; the radial coordinate R of a star on such an orbit oscillates around some mean value as the star revolves around the center, and the orbit again forms a rosette figure.

1 Motion in the Meridional Plane

The situation is much more complex and interesting for stars whose motions carry them out of the equatorial plane of the system. The study of such general orbits in axisymmetric galaxies can be reduced to a two-dimensional problem by exploiting the conservation of the z-component of angular momentum of any star. Let the potential, which we assume to be symmetric about the plane $z = 0$, be $\Phi(R, z)$. Then the general equation of motion of the star is

$$\frac{d^2\mathbf{r}}{dt^2} = -\boldsymbol{\nabla}\Phi(R, z). \tag{3-46}$$

Writing $\mathbf{r}$ and $\boldsymbol{\nabla}\Phi$ in terms of their components parallel to the unit vectors $\hat{\mathbf{e}}_R$, $\hat{\mathbf{e}}_\phi$, and $\hat{\mathbf{e}}_z$, we have

$$\mathbf{r} = R\hat{\mathbf{e}}_R + z\hat{\mathbf{e}}_z \tag{3-47a}$$

and

$$\boldsymbol{\nabla}\Phi = \frac{\partial\Phi}{\partial R}\hat{\mathbf{e}}_R + \frac{\partial\Phi}{\partial z}\hat{\mathbf{e}}_z. \tag{3-47b}$$

We now use equation (1B-24) for the acceleration in cylindrical coordinates to obtain

$$\ddot{R} - R\dot{\phi}^2 = -\frac{\partial\Phi}{\partial R}, \tag{3-48a}$$

$$\frac{d}{dt}\left(R^2\dot{\phi}\right) = 0, \tag{3-48b}$$

$$\ddot{z} = -\frac{\partial\Phi}{\partial z}. \tag{3-48c}$$

Equation (3-48b) expresses conservation of the component of angular momentum about the z-axis, $L_z = R^2\dot\phi$, while equations (3-48a) and (3-48c) describe the coupled oscillations of the star in the R and z-directions. We may eliminate $\dot\phi$ from equation (3-48a) to obtain equations for the evolution of R and z:

$$\ddot R = -\frac{\partial\Phi_{\rm eff}}{\partial R} \quad ; \quad \ddot z = -\frac{\partial\Phi_{\rm eff}}{\partial z}, \tag{3-49a}$$

where

$$\Phi_{\rm eff} \equiv \Phi(R, z) + \frac{L_z^2}{2R^2}. \tag{3-49b}$$

Thus the three-dimensional motion of a star in an axisymmetric potential $\Phi(R, z)$ can be reduced to the motion of the star in a plane. This (non-uniformly) rotating plane with Cartesian coordinates (R, z) is often called the **meridional plane**, and $\Phi_{\rm eff}(R, z)$ is called the **effective potential**.

Figure 3-2 shows contour plots of the effective potential

$$\Phi_{\rm eff} = \tfrac12 v_0^2 \ln\left(R^2 + \frac{z^2}{q^2}\right) + \frac{L_z^2}{2R^2}, \tag{3-50}$$

for $L_z = 0.2$ and axial ratios $q = 0.9$ and 0.5. This potential resembles the potential experienced by a star in an oblate spheroidal galaxy that has a constant circular speed v_0 (§2.2.2). Notice that $\Phi_{\rm eff}$ rises very steeply near the z-axis—it is as if the axis of symmetry were protected by a centrifugal barrier.

The minimum in $\Phi_{\rm eff}$ has a simple physical significance. The minimum occurs where

$$0 = \frac{\partial\Phi_{\rm eff}}{\partial R} = \frac{\partial\Phi}{\partial R} - \frac{L_z^2}{R^3}$$
$$0 = \frac{\partial\Phi_{\rm eff}}{\partial z}. \tag{3-51}$$

The second of these conditions is satisfied anywhere in the equatorial plane $z = 0$, and the first is satisfied at the radius R_g where

$$\left(\frac{\partial\Phi}{\partial R}\right)_{(R_g,0)} = \frac{L_z^2}{R_g^3} = R_g\dot\phi^2. \tag{3-52}$$

This is simply the condition for a circular orbit with angular speed $\dot\phi$. Thus the minimum of $\Phi_{\rm eff}$ occurs at the radius at which a circular orbit has angular momentum L_z, and the value of $\Phi_{\rm eff}$ at the minimum is the energy of this circular orbit.

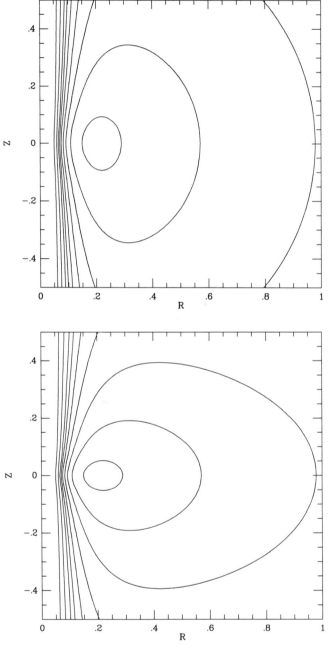

Figure 3-2. Level contours of the effective potential
of equation (3-50) when $L_z = 0.2$ and: $q = 0.9$
(top); $q = 0.5$ (bottom). Contours are shown for
$\Phi_{\text{eff}} = -1, -0.5, 0, 0.5, 1, 1.5, 2, 3, 5$, assuming $v_0 = 1$.

The energy E of a star moving in an arbitrary axisymmetric potential Φ is given by

$$E = \tfrac{1}{2}[\dot{R}^2 + (R\dot{\phi})^2 + \dot{z}^2] + \Phi = \tfrac{1}{2}(\dot{R}^2 + \dot{z}^2) + \left(\Phi + \frac{L_z^2}{2R^2}\right)$$
$$= \tfrac{1}{2}(\dot{R}^2 + \dot{z}^2) + \Phi_{\text{eff}}. \tag{3-53}$$

Thus the effective potential Φ_{eff} is the sum of the gravitational potential energy of the orbiting star and the kinetic energy associated with its motion in the ϕ-direction. Any difference between Φ_{eff} and E is simply kinetic energy of the motion in the (R, z) plane. Since the kinetic energy is non-negative, the orbit is restricted to the area in the meridional plane satisfying the inequality $E \geq \Phi_{\text{eff}}$. The curve bounding this area is called the **zero-velocity curve**, since the orbit can only reach this curve if its velocity is instantaneously zero.

Unless the gravitational potential Φ is of some special form, equations (3-49a) cannot be solved analytically. However, we may follow the evolution of $R(t)$ and $z(t)$ by integrating the equations of motion numerically, starting from a variety of initial conditions. Figure 3-3 shows the result of two such integrations for the potential (3-50) with $q = 0.9$ (see Richstone 1982). The orbits shown are of stars of the same energy and angular momentum, yet they look quite different in real space, and hence the phase points of stars on these orbits must move through different regions of phase space. Is this because the equations of motion (3-49a) admit a second isolating integral $I(R, z, \dot{R}, \dot{z})$ in addition to E, and I takes different values along these two orbits?

2 Surfaces of Section

The phase space associated with the motion we are considering has four dimensions, R, z, $\dot{R}$, and $\dot{z}$, and the four-dimensional motion of the phase-space point of an individual star is much too complicated to visualize. Nonetheless, we can determine whether orbits in the (R, z) plane admit an additional isolating integral by use of a simple graphical device. Since the energy $E(R, z, \dot{R}, \dot{z})$ is conserved, we could plot the motion of the representative point in a three-dimensional space, say $(R, z, \dot{R})$, and then $\dot{z}$ would be determined (to within a sign) by the known value of E. However, even three-dimensional spaces are difficult to draw on paper, so we simply show the points where the representative point of the star crosses some plane in the phase space, say the $z = 0$ plane. To remove the sign ambiguity in $\dot{z}$, we plot the $(R, \dot{R})$ coordinates only when $\dot{z} > 0$.

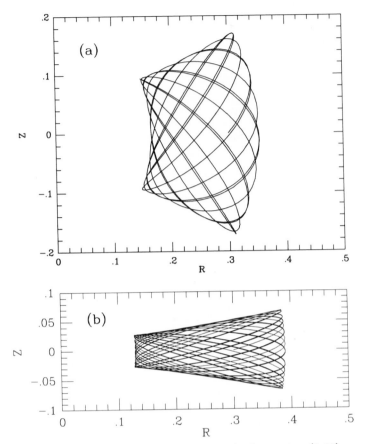

Figure 3-3. Two orbits in the potential of equation (3-50) with $q = 0.9$. Both orbits are at energy $E = -0.8$ and angular momentum $L_z = 0.2$, and we assume $v_0 = 1$.

In other words, we plot the values of R and $\dot{R}$ every time the star crosses the equator going upward. Such plots were first used by Poincaré and are called **surfaces of section**. Note that the orbit is restricted to an area in the surface of section defined by the constraint $E \geq \frac{1}{2}\dot{R}^2 + \Phi_{\text{eff}}$; the curve bounding this area is often called the zero-velocity curve of the surface of section since it can only be reached by an orbit with $\dot{z} = 0$.

Now suppose the star's motion in the (R, z) plane respects an additional isolating integral I. Then the trajectory of its phase point in the three-dimensional space with coordinates $(R, z, \dot{R})$ is restricted by the condition $I = constant$, to a two-dimensional region, which must cut the plane $z = 0$ through $(R, z, \dot{R})$ space in a curve, called the orbit's **invariant curve**. Thus, if I exists for any orbit, the points generated by

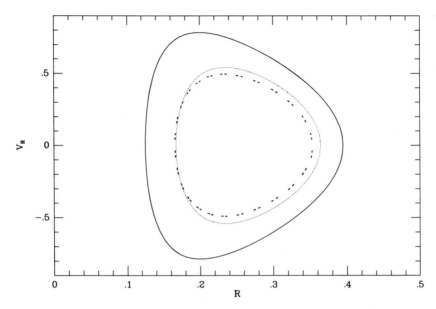

Figure 3-4. Points generated by the orbit of Figure 3-3a in the $(R, \dot{R})$ surface of section. If the total angular momentum L of the orbit were conserved, the points would fall on the dotted curve. The full curve is the zero-velocity curve at the energy of this orbit.

that orbit in the surface of section must lie on a curve (or perhaps at a finite number of discrete points in degenerate cases). Conversely, if E is the only isolating integral admitted by motion in the $(R, \dot{R})$ plane, the points generated by each orbit will eventually fill one or more areas in the surface of section.

Figure 3-4 shows the points generated by the orbit of Figure 3-3a in the $(R, \dot{R})$ surface of section. It is apparent that these points lie on a smooth curve, and thus that some isolating integral I seems to be respected by this orbit. It is often (but not invariably) found that for realistic galactic potentials, orbits in the (R, z) plane do admit an integral of this type. When this is the case, I is called the **third integral** since it is in addition to the two classical integrals E and L_z.

We may form an intuitive picture of the nature of the third integral by considering two special cases. If the potential Φ is spherical, we know that the total angular momentum L is an integral. This suggests that for a nearly spherical potential, the third integral may be approximated by L. The dotted curve in Figure 3-4 shows the curve on which the points generated by the orbit of Figure 3-3a would lie if the third integral were L, and Figure 3-5 shows the actual time evolution of L along that or-

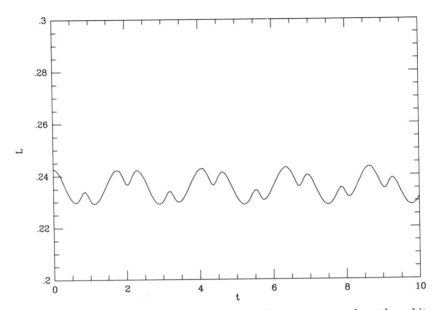

Figure 3-5. The total angular momentum is almost constant along the orbit shown in Figure 3-4.

bit. From these two figures we see that L is an approximately conserved quantity even for orbits in potentials that are significantly flattened. We may think of these orbits as approximately planar and with more or less fixed eccentricity. The approximate orbital planes have a fixed inclination to the z-axis but precess about this axis, at a rate that gradually tends to zero as the potential becomes more and more spherical.

In §3.6 we shall show that when the potential is that of a disk galaxy, we may think of the star as conducting almost independent oscillations in the z and R-directions. The frequency characteristic of the z-oscillations then proves to be sufficiently greater than that associated with the R-oscillations, that we can reproduce a typical surface of section by treating the z-oscillations as motion in a slowly varying one-dimensional potential.

3 Nearly Circular Orbits: Epicycles and the Velocity Ellipsoid

In disk galaxies many stars are on nearly circular orbits, so it is useful to derive approximate solutions to equations (3-49a) that are valid for such orbits. We define x by

$$x \equiv R - R_g, \tag{3-54}$$

where R_g is the solution of equation (3-52). Thus $(x, z) = (0, 0)$ are the coordinates in the meridional plane of the minimum in Φ_{eff}. When we expand Φ_{eff} in a Taylor series about this point, we obtain

$$\Phi_{\text{eff}} = \frac{1}{2} \left(\frac{\partial^2 \Phi_{\text{eff}}}{\partial R^2} \right)_{(R_g, 0)} x^2 + \frac{1}{2} \left(\frac{\partial^2 \Phi_{\text{eff}}}{\partial z^2} \right)_{(R_g, 0)} z^2 + O(xz^2) + \text{constant.}$$
(3-55)

Note that the term that is proportional to xz vanishes because Φ_{eff} is symmetric about $z = 0$. The equations of motion (3-49a) become very simple in the **epicycle approximation** in which we neglect all terms in Φ_{eff} of order xz^2 or higher powers of x and z. We define two new quantities by

$$\kappa^2 \equiv \left(\frac{\partial^2 \Phi_{\text{eff}}}{\partial R^2} \right)_{(R_g, 0)} \quad ; \quad \nu^2 \equiv \left(\frac{\partial^2 \Phi_{\text{eff}}}{\partial z^2} \right)_{(R_g, 0)}, \quad (3\text{-}56)$$

for then equations (3-49a) become

$$\ddot{x} = -\kappa^2 x, \quad (3\text{-}57\text{a})$$

$$\ddot{z} = -\nu^2 z. \quad (3\text{-}57\text{b})$$

According to these equations, x and z evolve like the displacements of two harmonic oscillators, with frequencies κ and ν, respectively. The two frequencies κ and ν are called the **epicycle frequency** and the **vertical frequency**. If we substitute from equation (3-49b) for Φ_{eff} we obtain

$$\kappa^2 = \left(\frac{\partial^2 \Phi}{\partial R^2} \right)_{(R_g, 0)} + \frac{3L_z^2}{R_g^4}, \quad (3\text{-}58\text{a})$$

$$\nu^2 = \left(\frac{\partial^2 \Phi}{\partial z^2} \right)_{(R_g, 0)}. \quad (3\text{-}58\text{b})$$

Since the circular frequency is given by

$$\Omega^2(R) = \frac{1}{R} \left(\frac{\partial \Phi}{\partial R} \right)_{(R, 0)} = \frac{L_z^2}{R^4}, \quad (3\text{-}58\text{c})$$

equation (3-58a) may be written

$$\kappa^2 = \left(R \frac{d\Omega^2}{dR} + 4\Omega^2 \right)_{R_g}. \quad (3\text{-}59)$$

In practice the circular frequency Ω in a galaxy rarely increases as a function of radius. Very near the center of a galaxy, where the circular speed rises approximately linearly with radius, Ω is nearly constant and $\kappa \simeq 2\Omega$. Elsewhere Ω declines with radius, though rarely faster than the Keplerian falloff, $\Omega \propto R^{-3/2}$, which yields $\kappa = \Omega$. Thus, in general,

$$\Omega \lesssim \kappa \lesssim 2\Omega. \tag{3-60}$$

The largest values of κ/Ω are associated with the most homogeneous mass distributions.

In §1.1.2 we introduced Oort's constant A by the relation [eq. (1-9)]

$$A \equiv \tfrac{1}{2}\left(\frac{v_c}{R} - \frac{dv_c}{dR}\right)_{R_0} = -\tfrac{1}{2}\left(R\frac{d\Omega}{dR}\right)_{R_0}, \tag{3-61a}$$

where $v_c(R) = \Omega R$ is the circular speed at radius R in the disk of our Galaxy, and R_0 is the distance of the Sun from the galactic center. Oort's constant B is such that the circular frequency at the radius of the Sun is given by

$$\Omega_0 = A - B. \tag{3-61b}$$

Thus

$$B = -\left(\tfrac{1}{2}R\frac{d\Omega}{dR} + \Omega\right)_{R_0}. \tag{3-61c}$$

Equations (3-59) and (3-61) give κ at the Sun as

$$\kappa_0^2 = -4B(A - B) = -4B\Omega_0. \tag{3-62}$$

Using values of A and B from Table 1-2 we find that the epicycle frequency at the Sun is $\kappa_0 = 36 \pm 10\,\mathrm{km\,s^{-1}\,kpc^{-1}}$, and that the ratio κ_0/Ω_0 at the Sun is

$$\frac{\kappa_0}{\Omega_0} = 2\sqrt{\frac{-B}{A-B}} = 1.3 \pm 0.2. \tag{3-63}$$

Consequently the Sun makes about 1.3 oscillations in the radial direction in the time it takes to complete an orbit around the galactic center. Hence its orbit does not close on itself in an inertial frame, but forms a rosette figure like those discussed above for stars in spherically symmetric potentials.

The equations of motion (3-57) lead to two integrals, namely, the energies

$$E_R \equiv \tfrac{1}{2}(\dot{x}^2 + \kappa^2 x^2) \quad ; \quad E_z \equiv \tfrac{1}{2}(\dot{z}^2 + \nu^2 z^2) \tag{3-64}$$

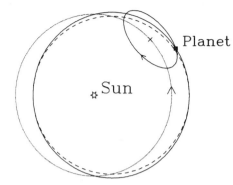

Planet

$\ast$ Sun

Figure 3-6. An elliptical Kepler orbit (dashed curve) is well approximated by the superposition of retrograde motion at angular frequency κ around a small ellipse with axis ratio $\frac{1}{2}$, and prograde motion of the ellipse's center at angular frequency Ω around a circle (dotted curve).

of the two oscillators. Thus if the star's orbit is sufficiently nearly circular that our truncation of the series for Φ_{eff} [eq. (3-55)] is justified, then the orbit admits three integrals of motion: E_R, E_z, and L_z. These are all isolating integrals.

From equations (3-53), (3-55), (3-56), and (3-64) we see that the energy of such a star is made up of three parts:

$$E = E_R + E_z + \Phi_{\mathrm{eff}}(R_g, z). \qquad (3\text{-}65)$$

Thus the three integrals of motion can equally be chosen as (E_R, E_z, L_z) or (E, E_z, L_z).

From §2.2.3 it is clear that in a disk galaxy such as our own, we expect $\Phi_{\mathrm{eff}} \propto z^2$ only for values of z for which $\rho_{\mathrm{disk}}(z) \simeq constant$, i.e., for $z \lesssim 300\,\mathrm{pc}$. For stars that do not rise above this height, equation (3-57b) yields

$$z = Z \cos(\nu t + \varsigma), \qquad (3\text{-}66)$$

where Z and ς are arbitrary constants. However, the orbits of the majority of disk stars carry these stars further above the plane than $300\,\mathrm{pc}$. Therefore the epicycle approximation does not form a reliable guide to the z-motion of the majority of disk stars. The great value of this approximation lies rather in its ability to describe the motions in the equatorial plane of any star on a nearly circular orbit. So far we have described only the radial component of this motion, so we now turn to the azimuthal motion.

Equation (3-57a), which governs the radial motion, has the general solution

$$x(t) = X \cos(\kappa t + \psi), \qquad (3\text{-}67)$$

where X and ψ are arbitrary constants. Now let $\Omega_g = L_z/R_g^2$ be the angular speed of the circular orbit with angular momentum L_z. Since L_z is conserved, we have

$$\dot{\phi} = \frac{L_z}{R^2} = \frac{L_z}{R_g^2}\left(1 + \frac{x}{R_g}\right)^{-2}$$
$$\simeq \Omega_g\left(1 - \frac{2x}{R_g}\right). \tag{3-68}$$

Substituting for x from (3-67) and integrating, we obtain

$$\phi = \Omega_g t + \phi_0 - \frac{2\Omega_g X}{\kappa R_g}\sin(\kappa t + \psi). \tag{3-69}$$

The nature of the motion described by these equations can be clarified by erecting Cartesian axes (x, y, z) with origin at the **guiding center** (or **epicenter**) $R = R_g$, $\phi = \Omega_g t + \phi_0$. The x and z coordinates have already been defined, and the y coordinate is perpendicular to both; to first order we have

$$y = -\frac{2\Omega_g}{\kappa}X\sin(\kappa t + \psi)$$
$$\equiv -Y\sin(\kappa t + \psi). \tag{3-70}$$

Equations (3-67) and (3-70) are the complete solution for the orbit in the epicycle approximation. The motion in the z-direction is independent of the motion in x and y. In the (x, y) plane the star moves on an ellipse called the **epicycle** around the guiding center (see Figure 3-6). The lengths of the semi-axes of the epicycle are in the ratio

$$\frac{X}{Y} = \frac{\kappa}{2\Omega_g}. \tag{3-71}$$

For a harmonic oscillator potential $X/Y = 1$ and for a Keplerian potential $X/Y = \frac{1}{2}$; in general, the inequality (3-60) shows that $Y > X$, and the epicycle is elongated in the tangential direction.[2] From equation (3-63), $X/Y \simeq 0.7$ in the solar neighborhood. The star travels around the epicycle in a retrograde direction, with period $2\pi/\kappa$.

Consider the motion of a star that moves on an epicyclic orbit, as viewed by an astronomer who sits at the guiding center of the star's

---

[2] Epicycles were invented by Hipparchus to describe the motion of the planets about the Sun. Hipparchus's work was not very successful, largely because he used circular epicycles with $X/Y = 1$. If only he had used epicycles with the proper axis ratio $X/Y = \frac{1}{2}$!

orbit. At different times in the orbit the astronomer's proper motion measurements yield the maximum values κX and κY of the velocities $\dot{x}$ and $\dot{y}$ that are given by equations (3-67) and (3-70). Since by equation (3-71), $X/Y = \kappa/(2\Omega_g)$, these measurements yield important information about the galactic potential. However, the epicycle period is much longer than an astronomer's lifetime, so he might try to find the result more quickly by averaging the results from many stars whose orbits differ only in their epicycle phases ψ to obtain $\overline{\dot{x}^2}$ and $\overline{\dot{y}^2}$. He will find

$$\frac{\overline{\dot{y}^2}}{\overline{\dot{x}^2}} = \frac{\overline{[v_\phi - v_c(R_g)]^2}}{\overline{v_R^2}} = \frac{4\Omega_g^2}{\kappa^2} \simeq \frac{4\Omega_0^2}{\kappa_0^2} = \frac{A - B}{-B}. \tag{3-72}$$

However, this is not a practical procedure since in general we do not know the location of the guiding center of any given star. Instead, we can only measure v_R and $v_\phi(R_0) - v_c(R_0)$ for a group of stars, each of which has its own epicentric radius R_g, as they pass near the Sun at radius R_0. Let us therefore calculate this latter quantity:

$$v_\phi(R_0) - v_c(R_0) = R_0(\dot{\phi} - \Omega_0) = R_0(\dot{\phi} - \Omega_g + \Omega_g - \Omega_0)$$
$$\simeq R_0\left[(\dot{\phi} - \Omega_g) - \left(\frac{d\Omega}{dR}\right)_{R_0} x\right]. \tag{3-73a}$$

With equations (3-70) and (3-68) this becomes

$$v_\phi - v_c(R_0) \simeq -R_0 x \left[\frac{2\Omega_g}{R_g} + \left(\frac{d\Omega}{dR}\right)_{R_0}\right]. \tag{3-73b}$$

In the coefficient of the small quantity x we can approximate Ω_g/R_g by Ω_0/R_0. Hence

$$v_\phi - v_c(R_0) \simeq -x\left[2\Omega_0 + R_0\left(\frac{d\Omega}{dR}\right)_{R_0}\right]. \tag{3-73c}$$

Finally using equations (3-61) to introduce Oort's constants we obtain

$$v_\phi - v_c(R_0) \simeq 2Bx. \tag{3-74}$$

Similarly, we may neglect the dependence of κ on R_g to obtain with equation (3-62)

$$\overline{v_R^2} = \overline{\kappa^2 x^2} \simeq -4B(A - B)\overline{x^2}, \tag{3-75}$$

so, forming an average over many different stars, we find

$$\frac{\overline{[v_\phi - v_c(R_0)]^2}}{\overline{v_R^2}} \simeq \frac{-B}{A - B} = \frac{\kappa_0^2}{4\Omega_0^2} \simeq 0.45. \tag{3-76}$$

In Chapter 8 of MB this important result was used to estimate Oort's constant B. In §4.2.1(c) we shall rederive it from a rather different point of view.

Note that the axis ratio in equation (3-76) is the inverse of the ratio of velocities given by (3-72). Individual epicycles are elongated in the azimuthal direction, but the velocity ellipsoid is elongated in the radial direction.

3.3 Orbits in Planar Non-Axisymmetric Potentials

Many, possibly most, galaxies have non-axisymmetric structures. These are most evident near the centers of disk galaxies where one often finds a bright stellar bar. This bar is frequently associated with other non-axisymmetric components (Kormendy 1982). Furthermore, we saw in Chapter 5 of MB that elliptical galaxies may not be axisymmetric either. Evidently we need to understand how stars orbit in a non-axisymmetric potential if we are to model these systems successfully.

We start with the simplest possible problem, namely, planar motion in a nonrotating potential. Toward the end of this section we generalize the discussion to two-dimensional motion in potentials whose figures rotate steadily, and in the next section we show how an understanding of two-dimensional motion can be used in problems involving three-dimensional potentials.

1 Two-Dimensional Nonrotating Potential

Consider the logarithmic potential (cf. §2.2.2)

$$\Phi_L(x,y) = \tfrac{1}{2}v_0^2 \ln\left(R_c^2 + x^2 + \frac{y^2}{q^2}\right) \quad (q \le 1). \tag{3-77}$$

This potential has the following properties that make it useful for orbit calculations:

(i) The equipotentials have constant axial ratio q, so that the influence of the non-axisymmetry is similar at all radii.

(ii) For $R = \sqrt{x^2 + y^2} \ll R_c$, we may expand Φ_L in powers of R/R_c and find

$$\Phi_L(x,y) \simeq \frac{v_0^2}{2R_c^2}\left(x^2 + \frac{y^2}{q^2}\right) + \text{constant} \quad (R \ll R_c), \tag{3-78}$$

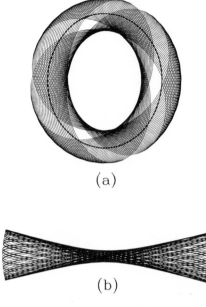

(a)

Figure 3-7. Two orbits of a common energy, $E = -0.337$, in the potential Φ_L of equation (3-77) when $v_0 = 1$, $q = 0.9$ and $R_c = 0.14$: (a) a loop orbit; (b) a box orbit. The closed parent of the loop orbits is superposed on the orbit of (a).

(b)

which is just the potential of the two-dimensional harmonic oscillator. In §2.3 we saw that gravitational potentials of this form are generated by homogeneous ellipsoids. Thus for $R < R_c$, Φ_L approximates the potential of a homogeneous density distribution.

(iii) For $R \gg R_c$ and $q = 1$, $\Phi_L \simeq v_0^2 \ln R$, which yields a circular speed $v_c \simeq v_0$ that is nearly constant. Thus the radial component of the force generated by Φ_L with $q = 1$ is consistent with the flat rotation curves of many disk galaxies.

The simplest orbits in Φ_L are those that are confined to $R \ll R_c$; when Φ_L is of the form (3-78), the orbit is the sum of independent harmonic motions parallel to the x and y axes. The frequencies of these motions are $\omega_x = v_0/R_c$ and $\omega_y = v_0/qR_c$, and unless the frequencies are commensurable (i.e., unless $\omega_x/\omega_y = n/m$ for some integers n and m), the star eventually passes close to every point inside a rectangular box. These orbits are therefore known as **box orbits**.[3] Such orbits have no particular sense of circulation about the center and respect two integrals of the motion, which we may take to be the energies E_x and E_y of the independent oscillations parallel to the coordinate axes.

To investigate orbits at larger radii $R \gtrsim R_c$, we must use numerical integrations. One example is shown in Figure 3-7a. The star now rotates in a fixed sense about the center of the potential while oscillating

---

[3] The curve traced by a box orbit is sometimes called a **Lissajous figure** and is easily displayed on an oscilloscope.

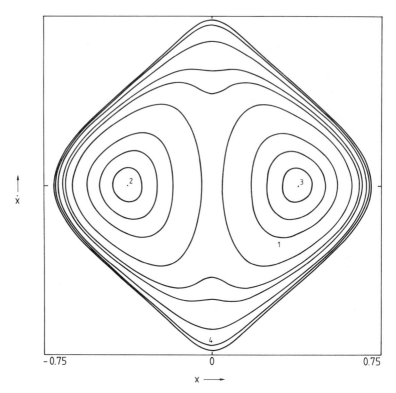

Figure 3-8. The $(x, \dot{x})$ surface of section formed by orbits in Φ_L of the same energy as the orbits depicted in Figure 3-7. The isopotential surface of this energy cuts the long axis at $x = 0.7$.

in radius. Orbits of this type are called **loop orbits**. Any star launched from $R \gg R_c$ in the tangential direction with a speed of order v_0 will follow a loop orbit. The initial tangential velocity of the star determines the width of the elliptical annulus to which the star is confined, just as the width of the annulus filled by an orbit in an axisymmetric potential varies as a function of angular momentum. This suggests that stars on loop orbits in Φ_L may respect an integral that is some sort of generalization of the angular momentum L_z.

We may test this hypothesis by generating a surface of section. Figure 3-8 is the surface of section $y = 0$, $\dot{y} > 0$ generated by orbits in Φ_L of the same energy as the orbits shown in Figure 3-7. The boundary curve in this figure arises from the energy constraint

$$\tfrac{1}{2}\dot{x}^2 + \Phi_L(x, 0) \leq \tfrac{1}{2}(\dot{x}^2 + \dot{y}^2) + \Phi_L(x, 0) = E_{y=0} \qquad (3\text{-}79)$$

Each closed curve in this figure corresponds to a different orbit. All these orbits clearly respect an integral I_2 in addition to the energy.

There are two types of closed curve in Figure 3-8 corresponding to two basic types of orbit. Figure 3-7a shows the spatial form of the loop orbit that generates the curve marked 1 in Figure 3-8. At a given energy there is a whole family of such orbits that differ in the width of the elliptical annuli within which they are confined. The unique orbit of this family that circulates in an anti-clockwise sense and closes on itself after one revolution is the **closed loop orbit**, which is also shown in Figure 3-7a. In the surface of section this orbit generates the single point 3. Orbits with non-zero annular widths generate the curves that loop around the point 3. Naturally, there are loop orbits that circulate in a clockwise sense in addition to the anti-clockwise orbits; in the surface of section their representative curves loop around the point 2.

Figure 3-7b shows the spatial form of an orbit of the second type. This orbit generates the curve numbered 4 in the surface of section of Figure 3-8. It is called a **box orbit** because it can be thought of as a distorted form of a box orbit of the two-dimensional harmonic oscillator. All the curves in the surface of section that are symmetric about the origin, rather than centered on one of the points 2 or 3, correspond to box orbits. These orbits differ from loop orbits in two ways: (i) in the course of time a star on any of them passes arbitrarily close to the center of the potential, and (ii) stars on these orbits have no unique sense of rotation about the center. In the surface of section 3-8 the outermost curve corresponds to the orbit on which $y = \dot{y} = 0$; on this orbit the star simply oscillates back and forth along the x-axis. We call this the **closed long-axis orbit**. The curves interior to this bounding curve that also center on the origin correspond to less and less elongated box orbits. Figure 3-9 shows one of the least elongated box orbits and one of the most eccentric loop orbits. Spatially, these two orbits resemble one another, but the loop orbit has a fixed sense of rotation about the center, which the box orbit does not.

It is interesting to compare the curves of Figure 3-8 with the curves generated by the integrals that we encountered earlier in this chapter. For example, if the angular momentum L_z were an integral, the curves on the surface of section $y = 0$, $\dot{y} > 0$ would be given by the relation

$$(L_z)_{y=0} = x\dot{y} = x\sqrt{2E - v_0^2 \ln\left(x^2 + R_c^2\right) - \dot{x}^2}. \qquad (3\text{-}80)$$

These curves are shown as dashed curves in Figure 3-10. They resemble the curves in Figure 3-8 only near the closed loop orbits 2 and 3, and cannot reproduce the curves generated by box orbits at all. Similarly,

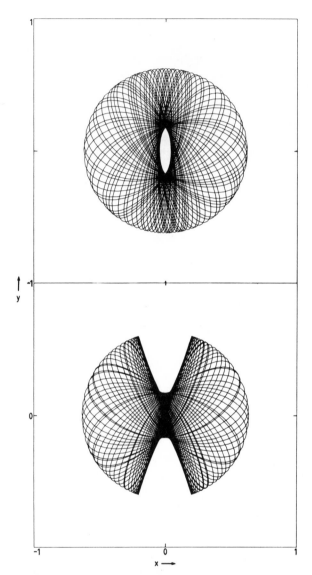

Figure 3-9. One of the most eccentric loop orbits and one of the least elongated box orbits in the potential $\Phi_L(q = 0.9, R_c = 0.14)$ at the energy of Figures 3-7 and 3-8.

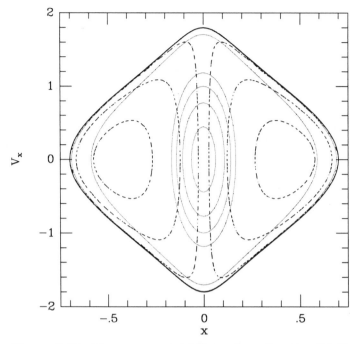

Figure 3-10. The appearance of the surface of section 3-8 if orbits were (a) at constant angular momentum [eq. (3-80); dashed curves], or (b) at constant E_x [eq. (3-81a); inner dotted curves], or (c) at constant E'_x [eq. (3-81b); outer dotted curves].

if the extra integral were the energy E_x of the x-component of motion in the harmonic potential (3-78), the curves in Figure 3-8 would be contours of constant

$$E_x = \tfrac{1}{2}\dot{x}^2 + \frac{v_0^2}{2R_c^2}x^2. \qquad (3\text{-}81a)$$

The dotted ellipses near the center of Figure 3-10 are curves of constant E_x. They resemble the curves in Figure 3-8 that are generated by the box orbits only in that they are symmetrical about the x-axis. Figure 3-10 shows that a better approximation to the invariant curves of box orbits is provided by the quantity

$$E'_x \equiv \tfrac{1}{2}\dot{x}^2 + \Phi(x,0), \qquad (3\text{-}81b)$$

which may be thought of as the energy invested in motion parallel to the potential's long axis. In a sense the integrals respected by box and loop orbits are analogous to E'_x and L_z, respectively.

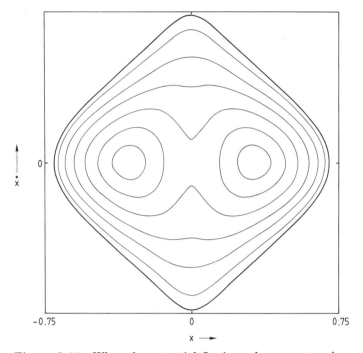

Figure 3-11. When the potential Φ_L is made more strongly barred by diminishing q, the proportion of orbits that are boxes grows at the expense of the loops: the figure shows the same surface of section as Figure 3-8 but for $q = 0.8$ rather than $q = 0.9$.

Figures 3-7 and 3-8 suggest an intimate connection between **closed orbits** and **families of nonclosed orbits**. We say that the clockwise closed loop orbit is the **parent** of the family of clockwise loop orbits. Similarly, the closed long-axis orbit $y = 0$ is the parent of the box orbits.

The closed orbits that are the parents of families are all **stable**, since members of their families that are initially close to them remain close at all times. In fact, we may think of any member of the family as engaged in stable oscillations about the parent closed orbit. A simple example of this state of affairs is provided by orbits in an axisymmetric potential. In a two-dimensional axisymmetric potential there are at each energy only two stable closed orbits—the clockwise and the anti-clockwise circular orbits.[4] All other orbits, having finite eccentricity, belong to families whose parents are these two orbits.

---

[4] Clearly special potentials such as the Kepler potential, in which all orbits are closed, must be excepted from this statement.

The correspondence between closed orbits and families of closed orbits enables us to trace the evolution of the orbital structure of a potential as the energy of the orbits or the shape of the potential is altered, simply by tracing the evolution of the stable closed orbits. For example, consider how the orbital structure supported by Φ_L [eq. (3-77)] evolves as we pass from the axisymmetric potential that is obtained when $q = 1$ to the barred potentials that are obtained when $q < 1$. When $q = 1$, L_z is an integral and so the surface of section is similar to the dashed curves in Figure 3-10. The only stable closed orbits are circular, and all orbits are loop orbits. When we make q slightly smaller than one, the long-axis orbit becomes stable and parents a family of elongated box orbits that oscillate about the axial orbit. As q is diminished more and more below one, a larger and larger portion of phase space comes to be occupied by box rather than loop orbits. Comparison of Figures 3-8 and 3-11 shows that this evolution manifests itself in the surface of section by the growth of the band of box orbits that runs around the outside of Figure 3-11 at the expense of the two bull's-eyes in that figure that are associated with the loop orbits. In configuration space the closed loop orbits become more and more elongated, with the result that less and less epicyclic motion needs to be added to one of these closed orbits to fill in the hole at its center and thus terminate the sequence of loop orbits. The erosion of the bull's-eyes in the surface of section is associated with this process.

The appearance of the surface of section also depends on the energy of its orbits. Figure 3-12 shows a surface of section for motion in Φ_L ($q = 0.9, R_c = 0.14$) at a lower energy than that of Figure 3-8. The changes in the surface of section are closely related to changes in the size and shape of the box and loop orbits. Box orbits that reach radii much greater than the core radius R_c have rather narrow waists (see Figure 3-9), and closed loop orbits of the same energy are nearly circular. If we consider box orbits and closed loop orbits of progressively smaller dimensions, the waists of the box orbits become steadily less narrow, and the closed orbits become progressively more eccentric as the dimensions of the orbits approach R_c. Eventually, at an energy E_c, the closed loop orbit degenerates into a line parallel to the short axis of the potential. Loop orbits do not exist at energies less than E_c. At $E < E_c$, all orbits are box orbits. The absence of loop orbits at $E < E_c$ is not unexpected since we saw above [eq. (3-78)] that when $x^2 + y^2 \ll R_c^2$, the potential is essentially that of the two-dimensional harmonic oscillator, none of whose orbits are loops. At these energies the only closed orbits are the short- and the long-axis closed orbits, and we expect both of these orbits to be stable. In fact, the short-axis orbit becomes unstable at the energy

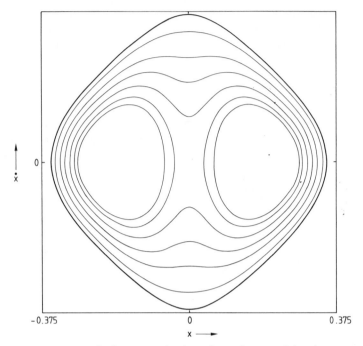

Figure 3-12. At low energies in a barred potential a large fraction of all orbits are boxes: the figure shows the same surface of section as 3-8 but for the energy whose isopotential surface cuts the x-axis at $x = 0.35$ rather than at $x = 0.7$ as in Figure 3-8.

E_c at which the loop orbits first appear. One says that the stable short-axis orbit of the low-energy regime **bifurcates** into the stable clockwise and anti-clockwise loop orbits. Stable closed orbits often appear in pairs like this.

Many two-dimensional barred potentials have orbital structures that resemble that of Φ_L. In particular:

 (i) Most orbits in these potentials respect a second integral in addition to energy.

 (ii) The majority of orbits in these potentials can be classified as either loop orbits or box orbits. The loop orbits have a fixed sense of rotation and never carry the star near the center, while the box orbits have no fixed sense of rotation and allow the star to pass arbitrarily close to the center.

(iii) When the axial ratio of the isopotential curves is close to unity, most of the phase space is filled with loop orbits, but as the axial ratio changes away from unity, box orbits become more common. However, we shall see in §3.5 that certain barred potentials have considerably more complex orbital structures.

2 Two-Dimensional Rotating Potential

It is likely that the figures of non-axisymmetric galaxies rotate with respect to inertial space, so that it is necessary to study orbits in rotating potentials. Suppose, therefore, that the frame of reference (x, y, z) in which the potential Φ_L is static rotates steadily at angular velocity $\mathbf{\Omega}_b = \Omega_b \hat{\mathbf{e}}_z$, where $\Omega_b > 0$. The equations of motion in this rotating coordinate system are [see eq. (1D-42)]

$$\ddot{\mathbf{r}} = -\nabla\Phi - 2(\mathbf{\Omega}_b \times \dot{\mathbf{r}}) - \mathbf{\Omega}_b \times (\mathbf{\Omega}_b \times \mathbf{r}), \qquad (3\text{-}82)$$

where the terms $-2(\mathbf{\Omega}_b \times \dot{\mathbf{r}})$ and $-\mathbf{\Omega}_b \times (\mathbf{\Omega}_b \times \mathbf{r})$ represent the Coriolis and centrifugal forces.

If we take the dot product of equation (3-82) with $\dot{\mathbf{r}}$, we may rearrange the resulting equation to read

$$\frac{dE_J}{dt} = 0, \qquad (3\text{-}83a)$$

where

$$E_J \equiv \tfrac{1}{2}\dot{\mathbf{r}}^2 + \Phi - \tfrac{1}{2}|\mathbf{\Omega}_b \times \mathbf{r}|^2. \qquad (3\text{-}83b)$$

E_J is known as **Jacobi's integral**. This new integral can be related to the energy $E = \tfrac{1}{2}v_{\text{in}}^2 + \Phi$, where v_{in} is the velocity in an inertial coordinate system. Since [see eq. (1D-38)]

$$\dot{\mathbf{r}} = \mathbf{v}_{\text{in}} - \mathbf{\Omega}_b \times \mathbf{r}, \qquad (3\text{-}84)$$

we have

$$E = \tfrac{1}{2}\left(\dot{\mathbf{r}} + \mathbf{\Omega}_b \times \mathbf{r}\right)^2 + \Phi = E_J + \dot{\mathbf{r}} \cdot (\mathbf{\Omega}_b \times \mathbf{r}) + |\mathbf{\Omega}_b \times \mathbf{r}|^2. \qquad (3\text{-}85)$$

But the angular momentum

$$\mathbf{L} = \mathbf{r} \times \mathbf{v}_{\text{in}} = \mathbf{r} \times \dot{\mathbf{r}} + \mathbf{r} \times (\mathbf{\Omega}_b \times \mathbf{r}), \qquad (3\text{-}86)$$

hence

$$\mathbf{\Omega}_b \cdot \mathbf{L} = \mathbf{\Omega}_b \cdot (\mathbf{r} \times \dot{\mathbf{r}}) + \mathbf{\Omega}_b \cdot [\mathbf{r} \times (\mathbf{\Omega}_b \times \mathbf{r})], \qquad (3\text{-}87)$$

and using the vector identity (1B-8), we have

$$E_J = E - \mathbf{\Omega}_b \cdot \mathbf{L}. \tag{3-88}$$

In other words, in a rotating non-axisymmetric potential, neither E nor $\mathbf{L}$ is conserved, but the combination $E - \mathbf{\Omega}_b \cdot \mathbf{L}$ *is* conserved. Notice that Jacobi's integral is the sum of $\frac{1}{2}\dot{\mathbf{r}}^2 + \Phi$, which would be the energy if the frame were not rotating, and the quantity $-\frac{1}{2}|\mathbf{\Omega}_b \times \mathbf{r}|^2 = -\frac{1}{2}\Omega^2 R^2$, which can be thought of as the "potential energy" to which the centrifugal "force" gives rise.

If we define an effective potential

$$\Phi_{\text{eff}} = \Phi - \tfrac{1}{2}\Omega_b^2 R^2, \tag{3-89a}$$

the equation of motion (3-82) becomes

$$\ddot{\mathbf{r}} = -\boldsymbol{\nabla}\Phi_{\text{eff}} - 2(\mathbf{\Omega}_b \times \dot{\mathbf{r}}), \tag{3-89b}$$

and Jacobi's integral is

$$E_J = \tfrac{1}{2}|\dot{\mathbf{r}}|^2 + \Phi_{\text{eff}}. \tag{3-90}$$

The surface $\Phi_{\text{eff}} = E_J$ is often called the **zero-velocity surface**. All regions in which $\Phi_{\text{eff}} > E_J$ are forbidden to the star. Thus, although the solution of the differential equations for the orbit in a rotating potential may be difficult, we can at least define forbidden regions into which the star cannot penetrate.

Figure 3-13 shows contours of Φ_{eff} for the potential Φ_L of equation (3-77). Φ_{eff} is characterized by five stationary points, marked L_1 to L_5, at which both $(\partial\Phi_{\text{eff}}/\partial x)$ and $(\partial\Phi_{\text{eff}}/\partial y)$ vanish. These points are sometimes called **Lagrange points** after similar points that play an important role in the restricted three-body problem discussed by Lagrange (Szebehely 1967). The central stationary point L_3 in Figure 3-13 is, of course, a minimum of the potential and is surrounded by a region in which the centrifugal potential $-\frac{1}{2}\Omega_b^2 R^2$ makes only a small contribution to Φ_{eff}. The four points L_1, L_2, L_4, and L_5 are the four points at which it is possible for a star to travel on a circular orbit while appearing to be stationary in the rotating frame. On this orbit the gravitational and centrifugal forces precisely balance. We shall refer to the annulus bounded by circles through L_1, L_2 and L_4, L_5 as the **region of corotation**. The stationary points L_1 and L_2 on the x-axis are saddle points, while the stationary points L_4 and L_5 along the y-axis are maxima of the effective potential. Stars with values of E_J smaller than

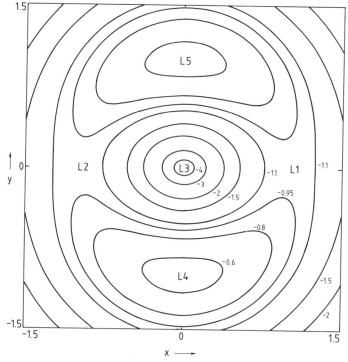

Figure 3-13. Contours of constant effective potential Φ_{eff} when the potential is given by equation (3-77) with $v_0 = 1$, $q = 0.8$, $R_c = 0.1$, and $\Omega_b = 1$. The point marked L_3 is a minimum of Φ_{eff}, while those marked L_4 and L_5 are maxima. Φ_{eff} has saddle points at L_1 and L_2.

the value Φ_c taken by Φ_{eff} at L_1 and L_2 cannot move from the center of the potential to infinity. By contrast, a star for which E_J exceeds Φ_c, or any star that is initially outside the contour through L_1 and L_2, can *in principle* escape to infinity. However, it cannot be assumed that a star of the latter class will *necessarily* escape, because the Coriolis force prevents stars from accelerating steadily in the direction of $-\nabla\Phi_{\text{eff}}$.

Given the equations of motion (3-82), it is a straightforward matter to calculate numerically orbits in a given rotating potential, and we shall do so later in this section. However, it is helpful first to consider motion near each of the Lagrange points L_1 to L_5. Since these are stationary points of Φ_{eff}, it follows that if we expand Φ_{eff} around one of these points

$\mathbf{r}_L = (x_L, y_L)$ in powers of $(x - x_L)$ and $(y - y_L)$, we have

$$
\Phi_{\text{eff}}(x, y) = \Phi_{\text{eff}}(x_L, y_L) + \frac{1}{2}\left(\frac{\partial^2 \Phi_{\text{eff}}}{\partial x^2}\right)_{\mathbf{r}_L} (x - x_L)^2
$$
$$
+ \left(\frac{\partial^2 \Phi_{\text{eff}}}{\partial x \partial y}\right)_{\mathbf{r}_L} (x - x_L)(y - y_L) + \frac{1}{2}\left(\frac{\partial^2 \Phi_{\text{eff}}}{\partial y^2}\right)_{\mathbf{r}_L} (y - y_L)^2 + \ldots
$$
$$(3\text{-}91)$$

Furthermore, for any barlike potential whose principal axes lie along the coordinate axes, $\left(\frac{\partial^2 \Phi_{\text{eff}}}{\partial x \partial y}\right) = 0$ at $\mathbf{r}_L$ by symmetry. Hence, if we retain only low-order terms in equation (3-91) and define

$$
\xi \equiv x - x_L \quad ; \quad \eta \equiv y - y_L, \tag{3-92}
$$

and

$$
\Phi_{xx} \equiv \left(\frac{\partial^2 \Phi_{\text{eff}}}{\partial x^2}\right)_{\mathbf{r}_L} \quad ; \quad \Phi_{yy} \equiv \left(\frac{\partial^2 \Phi_{\text{eff}}}{\partial y^2}\right)_{\mathbf{r}_L}, \tag{3-93}
$$

the equations of motion (3-89b) become for a star near $\mathbf{r}_L$,

$$
\ddot{\xi} = 2\Omega_b \dot{\eta} - \Phi_{xx}\xi \quad ; \quad \ddot{\eta} = -2\Omega_b \dot{\xi} - \Phi_{yy}\eta. \tag{3-94}
$$

This is a pair of linear differential equations with constant coefficients. The general solution can be found by substituting $\xi = X \exp(\lambda t)$, $\eta = Y \exp(\lambda t)$, where X, Y, and λ are complex constants. With these substitutions, equations (3-94) become

$$
(\lambda^2 + \Phi_{xx})X - 2\lambda\Omega_b Y = 0 \quad ; \quad 2\lambda\Omega_b X + (\lambda^2 + \Phi_{yy})Y = 0. \tag{3-95}
$$

These simultaneous equations have a nontrivial solution for X and Y only if the determinant of the matrix

$$
\begin{pmatrix} \lambda^2 + \Phi_{xx} & -2\lambda\Omega_b \\ 2\lambda\Omega_b & \lambda^2 + \Phi_{yy} \end{pmatrix} \tag{3-96}
$$

vanishes. Thus we require

$$
\lambda^4 + \lambda^2\left(\Phi_{xx} + \Phi_{yy} + 4\Omega_b^2\right) + \Phi_{xx}\Phi_{yy} = 0. \tag{3-97}
$$

This is the **characteristic equation** for λ. It has four roots, which may be either real or complex. If λ is a root, $-\lambda$ is also a root, so if there is any root that has non-zero real part $\text{Re}(\lambda) = \gamma$, the general solution to equations (3-94) will contain terms that cause $|\xi|$ and $|\eta|$ to grow exponentially in time; $|\xi| \propto \exp(|\gamma|t)$ and $|\eta| \propto \exp(|\gamma|t)$. Under these circumstances essentially all orbits rapidly carry the star far from the

Lagrange point, and the approximation on which equations (3-94) rest breaks down. In this case the Lagrange point is said to be **unstable**.

When the roots of equation (3-97) are pure imaginary, say $\lambda = \pm i\alpha$ or $\pm i\beta$, with $0 \leq \alpha \leq \beta$ real, the general solution to equations (3-94) is

$$
\begin{aligned}
\xi &= X_1 \cos(\alpha t + \phi_1) + X_2 \cos(\beta t + \phi_2), \\
\eta &= Y_1 \sin(\alpha t + \phi_1) + Y_2 \sin(\beta t + \phi_2).
\end{aligned}
\tag{3-98}
$$

Substituting these equations into the differential equations (3-94), we find that X_1 and Y_1 and X_2 and Y_2 are related by

$$
Y_1 = \frac{\Phi_{xx} - \alpha^2}{2\Omega_b \alpha} X_1 = \frac{2\Omega_b \alpha}{\Phi_{yy} - \alpha^2} X_1,
\tag{3-99a}
$$

and

$$
Y_2 = \frac{\Phi_{xx} - \beta^2}{2\Omega_b \beta} X_2 = \frac{2\Omega_b \beta}{\Phi_{yy} - \beta^2} X_2.
\tag{3-99b}
$$

The conditions that both roots λ^2 of the quadratic equation (3-97) in λ^2 be real and negative, and hence that the Lagrange point be stable, are

$$
\Phi_{xx} \Phi_{yy} > 0,
\tag{3-100a}
$$

and

$$
\Phi_{xx} + \Phi_{yy} + 4\Omega_b^2 > 2\sqrt{\Phi_{xx}\Phi_{yy}}.
\tag{3-100b}
$$

At saddle points of Φ_{eff} such as L_1 and L_2, Φ_{xx} and Φ_{yy} have opposite signs, so these Lagrange points are always unstable. At a minimum of Φ_{eff}, such as L_3, Φ_{xx} and Φ_{yy} are both positive, so that the first of the two inequalities (3-100) is obviously satisfied. The second is also satisfied since, as both sides of the inequality are positive, it may be squared and reformed into the condition

$$
(\Phi_{xx} - \Phi_{yy})^2 + 8\Omega_b^2 (\Phi_{xx} + \Phi_{yy}) + 16\Omega_b^2 > 0,
\tag{3-101}
$$

which is clearly satisfied whenever both Φ_{xx} and Φ_{yy} are positive. Hence L_3 is stable.

When Φ_{xx} and Φ_{yy} are positive, one may show from equation (3-97) that $\alpha < \beta$ satisfy

$$
\alpha^2 < \Phi_{xx} < \beta^2.
\tag{3-102}
$$

Also, when $\Omega_b^2/\Phi_{xx} \to 0$, α^2 tends to Φ_{xx}, which we assume to be smaller than Φ_{yy}, and β^2 tends to Φ_{yy}.

The stability of the Lagrange points L_4 and L_5 depends on the details of the potential. For the potential Φ_L of equation (3-77) we have

$$\Phi_{\text{eff}} = \tfrac{1}{2}v_0^2 \ln\left(R_c^2 + x^2 + \frac{y^2}{q^2}\right) - \tfrac{1}{2}\Omega_b^2(x^2 + y^2), \qquad (3\text{-}103)$$

so that L_4 and L_5 occur at $(0, \pm y_L)$, where

$$y_L \equiv \sqrt{\frac{v_0^2}{\Omega_b^2} - q^2 R_c^2}, \qquad (3\text{-}104)$$

and we see that L_4, L_5 are present only if $\Omega_b < v_0/(qR_c)$. Differentiating the potential again we find

$$\Phi_{xx}(0, y_L) = -\Omega_b^2(1 - q^2)$$
$$\Phi_{yy}(0, y_L) = -2\Omega_b^2\left[1 - q^2\left(\frac{\Omega_b R_c}{v_0}\right)^2\right]. \qquad (3\text{-}105)$$

Hence $\Phi_{xx}\Phi_{yy}$ is positive if the Lagrange points exist, and

$$\Phi_{xx} + \Phi_{yy} + 4\Omega_b^2 = \Omega_b^2\left[1 + q^2 + 2q^2\left(\frac{\Omega_b R_c}{v_0}\right)^2\right] \qquad (3\text{-}106)$$

is always positive. It follows, as for a minimum in Φ_{eff}, that in potentials of the form (3-77) L_4 and L_5 are always stable, despite being maxima of the effective potential.

For future use we note that in the limit where the core radius is small, $(\Omega_b R_c/v_0) \to 0$, we obtain, to leading order in the ellipticity $e = \sqrt{1 - q^2}$ of the potential, $\alpha^2 = e^2\Omega_b^2 = -\Phi_{xx}$ and $\beta^2 = 2\Omega_b^2$.

Equations (3-98) describing the motion about a stable Lagrange point show that each orbit is a superposition of motion at frequencies α and β around two ellipses. The shapes of these ellipses and the sense of the star's motion on them are determined by equations (3-99). For example, in the case of small R_c and e, the α-ellipse around the point L_4 is highly elongated in the ξ-direction (the tangential direction), while the β-ellipse has $Y_2 = -X_2/\sqrt{2}$. The star therefore moves around the β-ellipse in the sense opposite to that of the rotation of the potential. The β-ellipse is simply the familiar epicycle from §3.2.3, while the α-ellipse represents a slow wallowing in the weak non-axisymmetric component of Φ_L.

Now consider motion about the central Lagrange point L_3. From equations (3-99) and the inequality (3-102), it follows that $Y_1/X_1 > 0$.

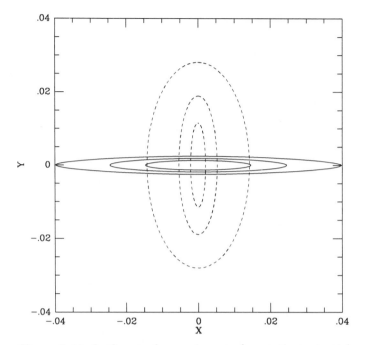

Figure 3-14. In the near-harmonic core of a rotating potential, the closed orbits are elongated ellipses. Stars on the orbits shown as full curves circulate about the center in the same sense as the potential's figure rotates. On the dashed orbits, stars circulate in the opposite sense.

Thus the star's motion around the α-ellipse has the same sense as the rotation of the potential; such an orbit is said to be **prograde** or **direct**. When $\Omega_b^2 \ll |\Phi_{xx}|$, it is straightforward to show from equations (3-97) and (3-99) that $X_1 \gg Y_1$ and hence that this prograde motion runs almost parallel to the long axis of the potential—this is the long-axis orbit familiar to us from our study of nonrotating bars. Conversely the star moves around the β-ellipse in the sense opposite to that of the rotation of the potential (the motion is **retrograde**), and $|X_2| < |Y_2|$. When Ω_b^2/Φ_{xx} is small, the β-ellipse goes over into the short-axis orbit of a nonrotating potential. A general prograde orbit around L_3 is made up of motion on the β-ellipse around a guiding center that moves around the α-ellipse, and conversely for retrograde orbits.

We now turn to a study of orbits that are not confined to the vicinity of a Lagrange point. We adopt the potential (3-77) with $q = 0.8$, $R_c = 0.03$, $v_0 = 1$, and $\Omega_b = 1$. This choice places the corotation annulus near $R_{CR} = 30R_c$. Jacobi's integral [eq. (3-83b)] now plays the role

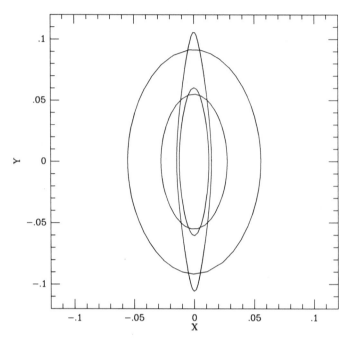

Figure 3-15. Closed orbits at two energies higher than those shown in 3-14. Just outside the potential's near-harmonic core there are at each energy two prograde closed orbits aligned parallel to the potential's short axis. One of these orbits (the less elongated) is stable, while the other is unstable.

that energy played in our similar investigation of orbits in nonrotating potentials, and by a slight abuse of language we shall refer to E_J as the "energy". At radii $R \lesssim R_c$ the two important sequences of stable closed orbits in the nonrotating case are the long- and the short-axis orbits. Figure 3-14 confirms the prediction of our analytic treatment that in the presence of rotation these become oval in shape. Orbits of both sequences are stable and therefore parent families of nonclosed orbits.

Consider now the evolution of the orbital structure as we leave the core region. At an energy E_1, similar to that at which loop orbits first appeared in the nonrotating case, pairs of prograde orbits like those shown in Figure 3-15 appear. Only one member of the pair is stable. When it first appears, the stable orbit is highly elongated parallel to the short axis, but as the energy is increased it becomes more round. Eventually the decrease in the elongation of this orbit with increasing energy is reversed, the orbit again becomes highly elongated parallel to the short axis and finally disappears along with its unstable companion orbit at

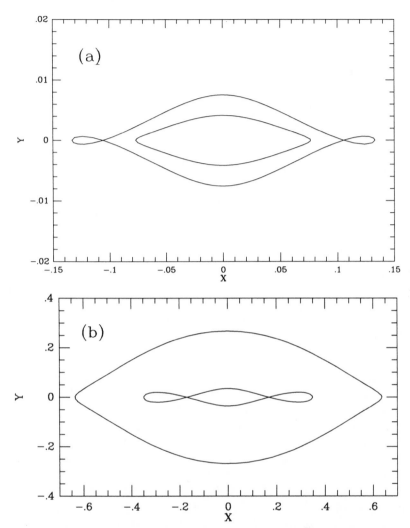

Figure 3-16. Near the energy at which the orbit pairs shown in Figure 3-15 appear, the closed long-axis orbits develop loops (a). At higher energies these expand, until the orbits open out into roughly elliptical figures (b). Notice that in (a) the x- and y-scales are different. In each diagram, two orbits at two different energies are shown.

an energy E_2.[5] In the notation of Contopoulos and Papayannopoulos (1980) these stable orbits are said to belong to the **sequence x$_2$**, while their unstable companions are of the **sequence x$_3$**.

---

[5] In the theory of weak bars, the energies E_1 and E_2 at which these prograde orbits appear and disappear are associated with the first and second inner Lindblad radii, respectively [see eq. (3-122)].

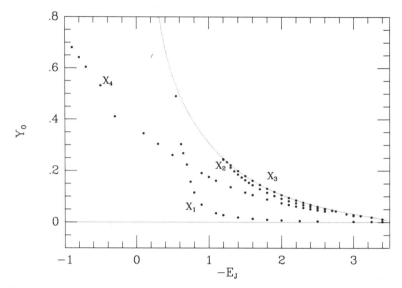

Figure 3-17. A plot of the Jacobi integral E_J of closed orbits in $\Phi_L(q = 0.8, R_c = 0.03, \Omega_b = 1)$ against the value of y at which the orbit cuts the potential's short axis. The dotted curve shows the relation $\Phi_{\text{eff}}(0, y) = E_J$. The families of orbits x_1–x_4 are marked.

The sequence of long-axis orbits (often called the **sequence $\mathbf{x_1}$**) suffers a very significant transition near E_2. Figure 3-16 shows the form of these orbits on either side of the transition. On the low-energy side of the transition (Figure 3-16a) the long-axis orbits cross themselves. On the high-energy side the orbits follow an ellipse-like curve that is elongated in the same sense as the potential and does not cross itself (Figure 3-16b). The orbits first become rounder and then adopt progressively more complex shapes as they approach the corotation region in which the Lagrange points L_1, L_2, L_4, and L_5 are located.

In the vicinity of the corotation annulus, there are important sequences of closed orbits on which stars move around one of the Lagrange points L_4 or L_5, rather than about the center.

Essentially all closed orbits that carry stars well outside the corotation region are nearly circular. In fact, the potential's figure spins much more rapidly than these stars circulate on their orbits, so the non-axisymmetric forces on such stars tend to be averaged out. One finds that at large radii prograde orbits tend to align with the bar, while retrograde orbits align perpendicular to the bar.

These results are summarized in Figure 3-17. In this figure we plot against the value of E_J for each closed orbit the distance y at which it

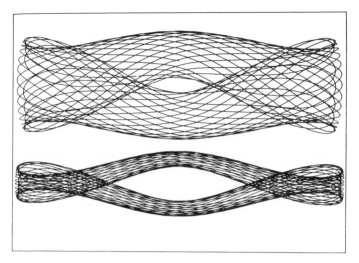

Figure 3-18. Two nonclosed orbits of a common energy in the rotating potential Φ_L.

crosses the short axis of the potential. Each sequence of closed orbits generates a continuous curve in this diagram known as the **characteristic curve** of that sequence.

The stable closed orbits we have described are all associated with substantial families of nonclosed orbits. Figure 3-18 shows some of these. As in the nonrotating case, a star on one of these nonclosed orbits may be considered to be executing stable oscillations about one of the fundamental closed orbits. In potentials of the form (3-77) essentially all orbits belong to one of these families. This is not always true, however, as we explain in §3.5.

It is important to distinguish between orbits that enhance the elongation of the potential and those that oppose it. The overall mass distribution of a galaxy must be elongated in the same sense as the potential, which suggests that most stars are on orbits on which they spend the majority of their time nearer to the potential's long axis than to its short axis. Interior to the corotation radius, the only orbits that satisfy this criterion are orbits of the family parented by the long-axis orbits, which therefore must be the most heavily populated orbits in any self-consistent bar. The shapes of these orbits range from butterfly-like at radii comparable to the core radius R_c, to nearly rectangular between R_c and the **inner Lindblad radius** (see below), to oval between this radius and corotation.

To an observer in an inertial frame of reference, stars on orbits belonging to the long-axis family circulate about the center of the potential in the same sense as the potential rotates. One part of the circulation seen by such an observer is due to the rotation of the frame of reference in which the potential is static. A second component of circulation is due to the mean streaming motion of such stars when referred to the rotating frame of the potential. Both components of circulation diminish toward zero if the angular velocity of the potential is reduced to zero. Near corotation the dominant component arises from the rotation of the frame of reference of the potential, while at small radii the more important component is the mean streaming motion of the stars through the rotating frame of reference. The closed long-axis orbits give rise to the strongest streaming motion.

3 Weak Bars

Before we leave the subject of orbits in planar non-axisymmetric potentials, we derive an analytic description of loop orbits in weak bars.

(a) Lindblad resonances　　　We assume that the figure of the potential rotates at some steady pattern speed Ω_b, and we seek to represent a general loop orbit as a superposition of the circular motion of a guiding center and small oscillations around this guiding center. Hence our treatment of orbits in weak non-axisymmetric potentials will be closely related to the epicycle theory of nearly circular orbits in an axisymmetric potential.

Let (R, φ) be polar coordinates in the frame that rotates with the potential, such that the line $\varphi = 0$ coincides with the long axis of the potential. By equations (1B-24), (1B-37), and (3-82), the equations of motion in this frame are

$$\ddot{R} - R\dot{\varphi}^2 = -\frac{\partial \Phi}{\partial R} + 2R\dot{\varphi}\Omega_b + \Omega_b^2 R \qquad (3\text{-}107a)$$

$$R\ddot{\varphi} + 2\dot{R}\dot{\varphi} = -\frac{1}{R}\frac{\partial \Phi}{\partial \varphi} - 2\dot{R}\Omega_b. \qquad (3\text{-}107b)$$

Since we assume that the bar is weak, we may write

$$\Phi(R, \varphi) = \Phi_0(R) + \Phi_1(R, \varphi), \qquad (3\text{-}108)$$

where $|\Phi_1/\Phi_0| \ll 1$. We divide R and φ into zeroth- and first-order parts

$$R(t) = R_0 + R_1(t) \quad ; \quad \varphi(t) = \varphi_0(t) + \varphi_1(t) \qquad (3\text{-}109)$$

by substituting these expressions into equation (3-107a) and requiring that the zeroth-order terms should separately sum to zero. Thus

$$R_0\dot{\varphi}_0^2 = \left(\frac{d\Phi_0}{dR}\right)_{R_0} - 2R_0\dot{\varphi}_0\Omega_b - \Omega_b^2 R_0, \qquad (3\text{-}110a)$$

which yields

$$R_0\left(\dot{\varphi}_0 + \Omega_b\right)^2 = \left(\frac{d\Phi_0}{dR}\right)_{R_0}. \qquad (3\text{-}110b)$$

This is the usual equation for centrifugal equilibrium at R_0. If we define $\Omega_0 \equiv \Omega(R_0)$, where

$$\Omega(R) \equiv \pm\sqrt{R\frac{d\Phi_0}{dR}} \qquad (3\text{-}111)$$

is the circular frequency at R in the potential Φ_0, equation (3-110b) for the angular speed of the guiding center (R_0, φ_0) becomes

$$\dot{\varphi}_0 = \Omega_0 - \Omega_b. \qquad (3\text{-}112)$$

We assume that $\Omega_b > 0$, so that $\Omega_0 > 0$ for prograde orbits and $\Omega_0 < 0$ for retrograde orbits. We choose $t = 0$ such that

$$\varphi_0 = (\Omega_0 - \Omega_b)t. \qquad (3\text{-}113)$$

The first order terms in the equations of motion (3-107) now yield

$$\ddot{R}_1 + \left(\frac{d^2\Phi_0}{dR^2} - \Omega^2\right)_{R_0} R_1 - 2R_0\Omega_0\dot{\varphi}_1 = -\left.\frac{\partial\Phi_1}{\partial R}\right|_{R_0}, \qquad (3\text{-}114a)$$

$$\ddot{\varphi}_1 + 2\Omega_0\frac{\dot{R}_1}{R_0} = -\frac{1}{R_0^2}\left.\frac{\partial\Phi_1}{\partial\varphi}\right|_{R_0}. \qquad (3\text{-}114b)$$

To proceed further we must choose a specific form of Φ_1; we set

$$\Phi_1(R, \varphi) = \Phi_b(R)\cos(m\varphi), \qquad (3\text{-}115)$$

where m is a positive integer. In practice we are mostly concerned with the case $m = 2$ since the potential is then barred. If $\varphi = 0$ is to coincide with the long axis of the potential, we must have $\Phi_b < 0$.

So far we have assumed only that the angular velocity $\dot{\varphi}_1$ is small, not that φ_1 is itself small. This distinction will be important when we consider what happens at resonances, but for the moment we assume that $\varphi_1 \ll 1$ and hence that $\varphi(t)$ always remains close to $(\Omega_0 - \Omega_b)t$.

With this assumption we may replace φ by φ_0 in the expressions for $\partial\Phi_1/\partial R$ and $\partial\Phi_1/\partial\varphi$ to yield

$$\ddot{R}_1 + \left(\frac{d^2\Phi_0}{dR^2} - \Omega^2\right)_{R_0} R_1 - 2R_0\Omega_0\dot{\varphi}_1 = -\left(\frac{d\Phi_b}{dR}\right)_{R_0} \cos\left[m(\Omega_0 - \Omega_b)t\right],$$

(3-116a)

$$\ddot{\varphi}_1 + 2\Omega_0\frac{\dot{R}_1}{R_0} = \frac{m\Phi_b(R_0)}{R_0^2} \sin\left[m(\Omega_0 - \Omega_b)t\right].$$

(3-116b)

Integrating the second of these equations, we obtain

$$\dot{\varphi}_1 = -2\Omega_0\frac{R_1}{R_0} - \frac{\Phi_b(R_0)}{R_0^2(\Omega_0 - \Omega_b)} \cos\left[m(\Omega_0 - \Omega_b)t\right] + \text{constant.} \quad (3\text{-}117)$$

We now eliminate $\dot{\varphi}_1$ from equation (3-116a) to find

$$\ddot{R}_1 + \kappa_0^2 R_1 = -\left[\frac{d\Phi_b}{dR} + \frac{2\Omega\Phi_b}{R(\Omega - \Omega_b)}\right]_{R_0} \cos\left[m(\Omega_0 - \Omega_b)t\right] + \text{constant,}$$

(3-118a)

where

$$\kappa_0^2 \equiv \left(\frac{d^2\Phi_0}{dR^2} + 3\Omega^2\right)_{R_0} = \left(R\frac{d\Omega^2}{dR} + 4\Omega^2\right)_{R_0} \quad (3\text{-}118b)$$

is the usual epicycle frequency [eq. (3-59)]. The constant in equation (3-118a) is unimportant since it can be absorbed by a shift $R_1 \to R_1 + constant$.

Equation (3-118a) is the equation of motion of a harmonic oscillator of natural frequency κ_0 that is driven at frequency $m(\Omega_0 - \Omega_b)$. The general solution to this equation is

$$R_1(t) = C_1 \cos(\kappa_0 t + \psi) - \left[\frac{d\Phi_b}{dR} + \frac{2\Omega\Phi_b}{R(\Omega - \Omega_b)}\right]_{R_0} \frac{\cos\left[m(\Omega_0 - \Omega_b)t\right]}{\Delta},$$

(3-119a)

where C_1 and ψ are arbitrary constants, and

$$\Delta \equiv \kappa_0^2 - m^2(\Omega_0 - \Omega_b)^2. \quad (3\text{-}119b)$$

If we use equation (3-113) to eliminate t from equation (3-119a), we find

$$R_1(\varphi_0) = C_1 \cos\left(\frac{\kappa_0\varphi_0}{\Omega_0 - \Omega_b} + \psi\right) + C_2 \cos(m\varphi_0), \quad (3\text{-}120a)$$

where

$$C_2 \equiv -\frac{1}{\Delta}\left[\frac{d\Phi_b}{dR} + \frac{2\Omega\Phi_b}{R(\Omega - \Omega_b)}\right]_{R_0}. \quad (3\text{-}120b)$$

If $C_1 = 0$, $R_1(\varphi_0)$ becomes periodic in φ_0 with period $2\pi/m$, and thus the orbit that corresponds to $C_1 = 0$ is a closed loop orbit. The orbits that are described by equations (3-120) with $C_1 \neq 0$ are the nonclosed loop orbits. In the following we set $C_1 = 0$ so that we may study the closed loop orbits.

The right side of equation (3-120a) for R_1 becomes singular at a number of values of R_0:

(i) **Corotation resonance.** When

$$\Omega_0 = \Omega_b, \tag{3-121}$$

$\dot{\varphi}_0 = 0$, and the guiding center corotates with the potential.

(ii) **Lindblad resonances.** When

$$m(\Omega_0 - \Omega_b) = \pm\kappa_0, \tag{3-122}$$

the star encounters successive crests of the potential at a frequency that coincides with the frequency of its natural radial oscillations. Radii at which such resonances occur are called **Lindblad radii** after the Swedish astronomer Bertil Lindblad (1895–1965). The plus sign in equation (3-122) corresponds to the case in which the star overtakes the potential; this is called an **inner Lindblad resonance**. If $(\Omega_0 - \Omega_b) = -\kappa_0/m$, the crests of the potential sweep by the more slowly rotating star at the resonant frequency κ_0, and R_0 is said to be the radius of the **outer Lindblad resonance**.

There is a simple connection between the two types of resonance. A circular orbit has two natural frequencies. If the star is displaced radially, it oscillates at the epicycle frequency κ_0. On the other hand, if the star is displaced azimuthally in such a way that it is still on a ciruclar orbit, then it will continue on a circular orbit displaced from the original one. Thus the star is neutrally stable to displacements of this form; in other words, its natural azimuthal frequency is zero. The two types of resonance arise between the forcing frequency seen by the star, $m(\Omega_0 - \Omega_b)$, and the two natural frequencies κ_0 and 0.

Figure 6-10 shows plots of Ω, $\Omega + \frac{1}{2}\kappa$ and $\Omega - \frac{1}{2}\kappa$ for two circular-speed curves typical of galaxies. A galaxy may have 0, 1, 2, or more Lindblad resonances. The Lindblad and corotation resonances play a central role in the study of bars and spiral structure, and we shall encounter them again in this chapter and in Chapter 6.

From equation (3-120a) it follows that for $m = 2$ the closed loop orbit is aligned with the bar whenever $C_2 > 0$, and is aligned perpendicular to the bar when $C_2 < 0$. When R_0 passes through a Lindblad

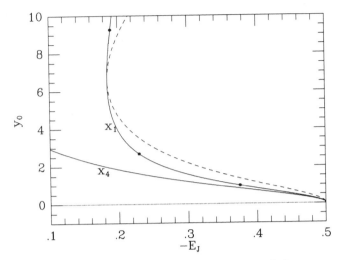

Figure 3-19. The full curves are the characteristic curves of the prograde (upper) and retrograde (lower) circular orbits in the isochrone potential (2-32) when a rotating frame of reference is employed. The dashed curve shows the relation $\Phi_{\rm eff}(0,y) = E_J$, and the dots mark the positions of the Lindblad resonances when a small non-axisymmetric component is added to the potential.

resonance, the sign of C_2, and therefore the orientation of the closed loop orbits, changes. The orientation of the closed loop orbits also changes near the corotation resonance.

It is interesting to relate the results of our analytic treatment of orbits in weak bars to the orbital structure of a strong bar that we obtained numerically in the last subsection. Figure 3-19 is helpful in this connection. The full curves in this figure show, for a particular rotating frame of reference, the relationship between E_J and the radii of prograde and retrograde circular orbits in the isochrone potential (2-32). As in Figure 3-17, the dashed curve marks the relation $\Phi_{\rm eff}(0,y) = E_J$. This curve touches the curve of the prograde circular orbits at the corotation resonance. If in the given frame we were to add a small non-axisymmetric component to the potential, the orbits marked by large dots would lie at the Lindblad resonances (from right to left, the first and second inner Lindblad resonances and the outer Lindblad resonance). We call the radius of the first inner Lindblad resonance[6] $R_{\rm IL1}$, and similarly $R_{\rm IL2}$, $R_{\rm OL}$, and $R_{\rm CR}$ for the radii of the other Lindblad resonances and of corotation. At energies at which the orbit lies within $R_{\rm IL1}$ or moves through

---

[6] Also called the **inner inner Lindblad resonance**.

the region $R_{\mathrm{IL2}} < R < R_{\mathrm{CR}}$, equations (3-120) with $C_1 = 0$ describe orbits of the sequences x_1 and x_4, while at energies such that the orbit moves through the region $R_{\mathrm{IL1}} < R < R_{\mathrm{IL2}}$, these equations describe orbits of the family x_2. Thus equations (3-120) describe only the families of orbits in a barred potential that are parented by a nearly circular orbit. However, when the non-axisymmetric component of the potential is very weak, most of phase space is occupied by such orbits. As the non-axisymmetry of the potential becomes stronger, families of orbits that are not described by equations (3-120) become more important.

(b) Orbits trapped at resonance   When R_0 approaches the radius of either a Lindblad resonance or the corotation resonance, the value of R_1 that is predicted by equations (3-120) becomes large, and our linearized treatment of the equations of motion breaks down. However, one can modify the analysis to cope with these resonances. We now discuss the necessary modifications for the case of the corotation resonance. The case of the Lindblad resonances is described in Goldreich and Tremaine (1981).

The appropriate modification is suggested by our investigation of orbits near the Lagrange points L_4 and L_5 in the potential Φ_L [eq. (3-77)], when the core radius R_c and ellipticity e approach zero. In this limit the non-axisymmetric part of the potential is proportional to e^2, and hence we have an example of a weak bar when $e \to 0$. We found that a star's orbit was a superposition of motion at frequencies α and β around two ellipses. In the limit $e \to 0$, the β-ellipse represents the familiar epicyclic motion and will not be considered further. The α-ellipse is highly elongated in the azimuthal direction, with axis ratio $|Y_1/X_1| = e$, and its frequency is small, $\alpha = e\Omega_b$.

These results suggest we consider the approximation in which R_1, $\dot{R}_1$, and $\dot{\phi}_1$ are small but ϕ_1 is not. Specifically, if the bar strength Φ_1 is proportional to some small parameter that we may call e^2, we assume that ϕ_1 is of order unity, R_1 is of order e, and the time derivative of any quantity is smaller than that quantity by of order e. Let us place the guiding center at L_5 $[\Omega(R_0) = \Omega_b;\ \varphi_0 = \frac{1}{2}\pi]$ and use equation (3-118b) to write the equations of motion (3-114) as

$$\ddot{R}_1 + \left(\kappa_0^2 - 4\Omega_0^2\right) R_1 - 2R_0\Omega_0\dot{\varphi}_1 = -\frac{\partial \Phi_1}{\partial R}, \qquad (3\text{-}123a)$$

$$\ddot{\varphi}_1 + 2\Omega_0\frac{\dot{R}_1}{R_0} = -\frac{1}{R_0^2}\frac{\partial \Phi_1}{\partial \varphi}. \qquad (3\text{-}123b)$$

According to our ordering, the terms on the left side of the first line are of order e^3, e, and e, respectively, while the term on the right side is of order e^2. All the terms on the second line are of order e^2. Hence we may simplify the first line by keeping only the terms of order e:

$$\left(\kappa_0^2 - 4\Omega_0^2\right) R_1 - 2R_0\Omega_0\dot{\varphi}_1 = 0. \tag{3-124}$$

Substituting equation (3-124) into equation (3-123b) to eliminate R_1, we find:

$$\ddot{\varphi}_1 \left(\frac{\kappa_0^2}{\kappa_0^2 - 4\Omega_0^2}\right) = -\frac{1}{R_0^2}\frac{\partial\Phi_1}{\partial\varphi}\bigg|_{(R_0,\varphi_0+\varphi_1)}. \tag{3-125}$$

Substituting from equation (3-115) for Φ_1 we obtain with $m = 2$

$$\ddot{\varphi}_1 = -\frac{2\Phi_b}{R_0^2}\left(\frac{4\Omega_0^2 - \kappa_0^2}{\kappa_0^2}\right)\sin\left[2(\varphi_0 + \varphi_1)\right]. \tag{3-126}$$

By inequality (3-60) we have that $4\Omega_0^2 > \kappa_0^2$. Also we have $\Phi_b < 0$ and $\varphi_0 = \frac{\pi}{2}$, and so equation (3-126) becomes

$$\frac{d^2\psi}{dt^2} = -p^2\sin\psi, \tag{3-127a}$$

where

$$\psi \equiv 2\varphi_1 \quad \text{and} \quad p^2 \equiv \frac{4}{R_0^2}\,|\Phi_b(R_0)|\,\frac{4\Omega_0^2 - \kappa_0^2}{\kappa_0^2}. \tag{3-127b}$$

Equation (3-127a) is simply the equation of a pendulum. Notice that the singularity in R_1 that appeared at corotation in equation (3-120) has disappeared in this more careful analysis. Notice also the interesting fact that the stable equilibrium point of the pendulum, $\varphi_1 = 0$, is at the *maximum*, not the minimum, of the potential Φ_1. If the integral of motion

$$E_p = \tfrac{1}{2}\dot{\psi}^2 - p^2\cos\psi \tag{3-128}$$

is less than p^2, the star oscillates slowly or **librates** about the Lagrange point, whereas if $E_p > p^2$, the star is not trapped by the bar but **circulates** about the center of the galaxy. For small-amplitude librations, the libration frequency is p, consistent with our assumption that the oscillation frequency is of order e when Φ_b is of order e^2.

We may obtain the shape of the orbit from equation (3-124) by using equation (3-128) to eliminate $\dot{\varphi}_1 = \tfrac{1}{2}\dot{\psi}$:

$$R_1 = -\frac{2R_0\Omega_0\dot{\varphi}_1}{4\Omega_0^2 - \kappa_0^2} = \pm\left(\frac{R_0\Omega_0}{4\Omega_0^2 - \kappa_0^2}\right)\sqrt{2\left[E_p + p^2\cos(2\varphi_1)\right]}. \tag{3-129}$$

We leave as an exercise the demonstration that when $E_p \gg p^2$, equation (3-129) describes the same orbits as are obtained from (3-120a) with $C_1 = 0$ and $\Omega \neq \Omega_b$.

Notice that the analysis of this subsection complements the analysis of motion near the Lagrange points in §3.3.2. The earlier analysis is valid for small oscillations around a Lagrange point of an arbitrary two-dimensional rotating potential, while the present analysis is valid for excursions of any amplitude in azimuth around the points L_4 and L_5 of a nearly axisymmetric potential.

3.4 Orbits in Three-Dimensional Triaxial Potentials

In §3.3 we have studied orbits in two-dimensional potentials. We now use the knowledge we gained there to assemble a qualitative picture of the orbital structure in a three-dimensional triaxial potential. Our picture is based on the idea developed in the last section, that a generic orbit may be represented as an oscillation about an underlying closed orbit. By mapping the principal sequences of closed orbits, we can develop a classification scheme for the nonclosed orbits. Furthermore, from a knowledge of the structure of the underlying closed orbit, it is possible to estimate many of the properties of orbits that are of interest from the point of view of galactic structure.

1 Nonrotating Potentials

If a star is initially in one of the symmetry planes of the potential and has a velocity vector that lies within that same plane, it is clear that the entire orbit of the star will lie within the plane. Therefore from our study of orbits in planar potentials we may infer the existence of a number of closed orbits at each energy; within the core of the potential we have the three axial orbits, and outside the core we have these three orbits plus three loop orbits, one about each axis. Let us consider which of these orbits may be stable and therefore parent families of nonclosed orbits.

Well inside the core of the potential, the motion of a star becomes practically equivalent to the motion of a three-dimensional harmonic oscillator. Hence in this region all orbits are members of the family of box orbits. It is convenient to consider the long-axis orbit to be the parent of this family, though both the short- and the middle-axis orbits are stable also.

Outside the core, where closed loop orbits in all three planes are possible, it is immediately apparent that both the short- and the middle-axis orbits must be unstable: the middle axis is the short axis of the plane that contains the long and the middle axes, and therefore if loop orbits are possible in this plane we know from our study of motion in two-dimensional potentials that motion up this axis must be unstable. Similarly, the short axis of the three-dimensional potential is the short axis of the plane that contains the long and the short axes, so the short-axis orbit must also be unstable. Hence the only axial orbit of the full potential that can be stable is the long-axis orbit. We shall see that this orbit in fact parents a large family of nonclosed orbits.

Consider next the closed loop orbits. Well outside the core of the potential there is at each energy one of these about each axis. Two of them parent important families of nonclosed orbits—the **long-** and the **short-axis tube orbits**—but both numerical experiments (Heisler et al. 1982) and analytical arguments (Binney 1981, 1982) show that the middle-axis loop orbit is unstable and hence parents no family of nonclosed orbits.

The long-axis orbit of a nonrotating triaxial potential parents a family of orbits that are natural three-dimensional generalizations of the two-dimensional box orbits. Figure 3-20a shows a typical orbit of this family. If the potential is centrally concentrated, these orbits are narrow near the center, and their eight corners touch the zero-velocity surface $\Phi(\mathbf{x}) = E$.

Members of the family of orbits whose parent is the closed long-axis loop orbit vary in shape from elliptical annuli that lie in the plane containing the short and the intermediate axes of the potential (Figure 3-20d), to long flaring tubes centered on the long axis (Figure 3-20c). Stars on these orbits circulate about the long axis while executing oscillations parallel to the long axis. They never touch the zero-velocity surface.

The closed short-axis loop orbit parents a family of nonclosed orbits that resemble the annular orbits that were discussed in §3.2 in connection with axisymmetric potentials (Figure 3-20b).

2 Rotating Potentials

The full orbital structure of a rotating triaxial potential is extremely complex. Deep in the core of the potential, the stable closed orbits comprise (i) two axial orbits in the equatorial plane, equivalent to the

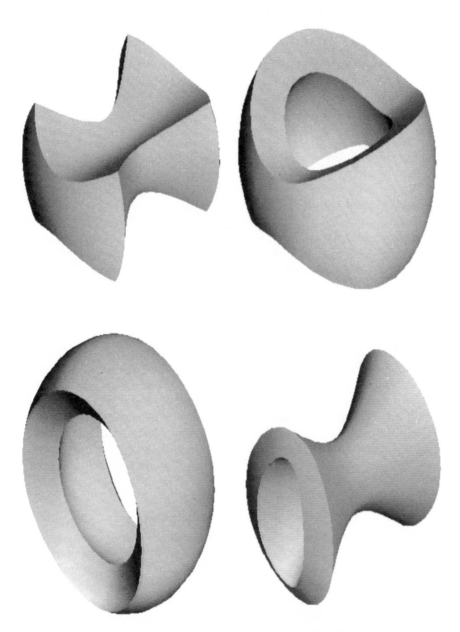

Figure 3-20. Orbits in a nonrotating triaxial potential. Clockwise from top left: (a) box orbit; (b) short-axis tube orbit; (c) inner long-axis tube orbit; (d) outer long-axis tube orbit. [Courtesy of T. Statler; see Statler (1986).]

prograde and retrograde orbits around the central Lagrange point L_3 of a planar potential (see §3.3.2); (ii) an orbit along the rotation axis of the potential.

Beyond the core region the orbital structure is more complex. We have seen that outside the core of a nonrotating potential there are three major families of orbits—one of box orbits and two of loop orbits. Rotation of the figure of the potential modifies the orbits as follows. The box orbits acquire a prograde sense of rotation but retain the same general shape so long as the rotation rate is small. The loop orbits about the rotation axis of the potential now fall into two distinct classes, prograde and retrograde loops. At certain radii these loops become unstable to perturbations perpendicular to the equatorial plane of the galaxy and join onto other families of closed loop orbits. The loop orbits about the long axis of the potential likewise break into two distinct classes, whose orbits differ in their sense of rotation about the long axis. These orbits do not lie in the plane perpendicular to the long axis, but are mirror images of each other on reflection in this plane. They are interesting because they demonstrate that closed orbits such as those on which one would expect to find any gas that is spiraling into the center of a galaxy need not be confined to principal planes of the galaxy (van Albada et al. 1981).

The discussion in this section is only intended to give a rough qualitative guide to the taxonomy of orbits in triaxial potentials. The interested reader can find more information in Athanassoula et al. (1983), Heisler et al. (1982), and Pfenniger (1984a), or, better still, by writing an orbit integration program and investigating the orbital structure of model potentials at first-hand. For a summary of numerical methods for orbit integration, see Press et al. (1986).

3.5 The Phase-Space Structure of Orbits*

In §3.2 and §3.3 we found that orbits in two-dimensional potentials usually admit two isolating integrals, while the fully three-dimensional orbits discussed in the last section ususally admit three isolating integrals. Orbits such as those discussed so far, which have at least as many isolating integrals as spatial dimensions, we call **regular**. In this section we discuss the general properties of regular orbits and show that they have an especially simple structure in phase space. In §3.6 and §4.4 we

---

* This section contains more advanced material and can be skipped on a first reading.

shall exploit special properties of this phase-space structure to simplify greatly several aspects of galactic dynamics. At the end of this section we briefly discuss orbits that have fewer isolating integrals than spatial dimensions, that is, **irregular** orbits. Such orbits are much harder to work with than regular orbits, and we shall not have much to say about them in this book, even though they are likely to play an important role in the structure of barred galaxies.

1 Orbits in Stäckel Potentials

We start our study of regular orbits by displaying the especially simple examples provided by orbits in a class of potentials called Stäckel potentials. These orbits exhibit all the important features of the numerically-integrated orbits in planar and three-dimensional bars discussed in §3.3 and §3.4, and yet all these features can be deduced analytically. For simplicity we develop the theory only for planar orbits, merely stating the corresponding results for three-dimensional orbits.

In §3.3.1 we remarked that box orbits in a planar nonrotating bar potential resemble Lissajous figures generated by two-dimensional harmonic motion, while loop orbits have many features in common with orbits in circularly symmetric potentials. Let us examine these parallels more closely, and see what hints they offer toward the construction of more realistic analytic models of orbits in typical galactic potentials.

The orbits of a two-dimensional harmonic oscillator admit two isolating integrals, $E_x \equiv \frac{1}{2}(v_x^2 + \omega_x^2 x^2)$ and $E_y \equiv E - E_x = \frac{1}{2}(v_y^2 + \omega_y^2 y^2)$. At each point in the portion of the (x, y) plane visited by the orbit, the particle can have one of four velocity vectors. These velocities arise from the ambiguity in the signs of v_x and v_y when we are given only E_x and E_y: $v_x(x) = \pm\sqrt{2E_x - \omega_x^2 x^2}$; $v_y(y) = \pm\sqrt{2E_y - \omega_y^2 y^2}$. The boundaries of the orbit are the lines on which $v_x = 0$ or $v_y = 0$.

Consider now planar orbits in a circularly symmetric potential $\Phi(r)$. These orbits fill annuli. At each point in the allowed annulus two velocity vectors are possible: $v_r(r) = \pm\sqrt{2(E - \Phi) - L_z^2/r^2}$, $v_\phi(r) = L_z/r$. The boundaries of the orbit are the curves $v_r = 0$.

These examples have a number of important points in common.

(i) The boundaries of orbits are found by equating to zero one component of velocity in a coordinate system that reflects the symmetry of the potential.

(ii) The momenta in this privileged coordinate system can be written as functions of one variable only: $v_x(x)$ and $v_y(y)$ in the case of the

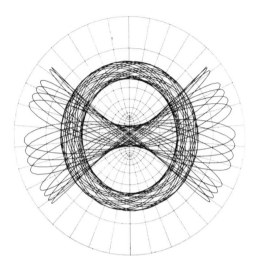

Figure 3-21. The boundaries of loop and
box orbits in barred potentials approximately
coincide with the coordinate curves of a
system of spheroidal coordinates. The figure
shows two orbits in the potential Φ_L of
equation (3-77), and a number of curves on
which the coordinates u and v defined by
equations (3-130) are constant.

harmonic oscillator; and $p_r = v_r(r)$ and $p_\phi = L_z$ (which depends on
neither coordinate) in the case of motion in a circularly symmetric
potential.

(iii) These expressions arise by splitting up a function of the coordinates
times the Hamiltonian function into two parts, each of which is
a function of only one coordinate and its conjugate momentum:
$H \equiv \frac{1}{2}|\mathbf{v}|^2 + \frac{1}{2}(\omega_x^2 x^2 + \omega_y^2 y^2) = E_x(x, v_x) + E_y(y, v_y)$ in the case
of the harmonic oscillator, and $r^2 H = r^2\left[\frac{1}{2}v_r^2 + \Phi(r)\right] + \frac{1}{2}p_\phi^2$ in the
case of motion in a circularly symmetric potential.

The first of these observations suggests that we look for a coordinate system whose coordinate curves run parallel to the edges of box
and/or loop orbits. One attractive coordinate system is suggested by
the meridional plane of the spheroidal coordinates introduced in §2.3.1.
Thus Figure 3-21 suggests that if we set

$$x = \Delta \sinh u \cos v \quad ; \quad y = \Delta \cosh u \sin v \qquad (3\text{-}130)$$

with Δ a constant, we might be able to find a barred potential in which
box orbits are bounded top and bottom by hyperbolae of constant v,

and right and left by segments of the ellipses of constant u. Similarly, each loop orbit could fill the annulus between two ellipses of constant u.

Now that we have chosen a coordinate system, item (iii) above suggests that we next write the Hamiltonian function in terms of u, v, and their conjugate momenta. We first express the Lagrangian $\mathcal{L} \equiv \frac{1}{2}|\dot{\mathbf{r}}|^2 - \Phi$ in terms of $(u, v, \dot{u}, \dot{v})$. With the substitutions $x \rightarrow z$ and $y \rightarrow R$, our coordinate system becomes identical with that defined for the meridional plane by equations (2-58). Hence [see eq. (1B-15)]

$$|\dot{\mathbf{r}}|^2 = \left|h_u \dot{u}\hat{\mathbf{e}}_u + h_v \dot{v}\hat{\mathbf{e}}_v\right|^2, \tag{3-131a}$$

where the scale factors h_u and h_v are given by equations (2-59). Thus

$$|\dot{\mathbf{r}}|^2 = \Delta^2 \left(\sinh^2 u + \cos^2 v\right)\left(\dot{u}^2 + \dot{v}^2\right), \tag{3-131b}$$

so

$$\mathcal{L} = \tfrac{1}{2}\Delta^2 \left(\sinh^2 u + \cos^2 v\right)\left(\dot{u}^2 + \dot{v}^2\right) - \Phi, \tag{3-132}$$

and the momenta are [eq. (1D-44)]

$$\begin{aligned} p_u &\equiv \frac{\partial \mathcal{L}}{\partial \dot{u}} = \Delta^2 \left(\sinh^2 u + \cos^2 v\right)\dot{u} \\ p_v &\equiv \frac{\partial \mathcal{L}}{\partial \dot{v}} = \Delta^2 \left(\sinh^2 u + \cos^2 v\right)\dot{v}. \end{aligned} \tag{3-133}$$

From equation (1D-45) it now follows that the Hamiltonian is

$$\begin{aligned} H(u, v, p_u, p_v) = p_u\dot{u} + p_v\dot{v} - \mathcal{L} &= \tfrac{1}{2}\Delta^2 \left(\sinh^2 u + \cos^2 v\right)\left(\dot{u}^2 + \dot{v}^2\right) + \Phi \\ &= \frac{p_u^2 + p_v^2}{2\Delta^2(\sinh^2 u + \cos^2 v)} + \Phi. \end{aligned} \tag{3-134}$$

Since H has no explicit dependence on time, it is a constant of the motion, the numerical value of which is the particle's energy E.

The examples of motion in harmonic and circular potentials suggest that we seek a form of $\Phi(u, v)$ that will enable us to split a multiple of the equation $H(u, v, p_u, p_v) = E$ into a part involving only u and p_u and a part that involves only v and p_v. Evidently we require that $(\sinh^2 u + \cos^2 v)\Phi$ be of the form $U(u) - V(v)$, i.e., that[7]

$$\Phi(u, v) = \frac{U(u) - V(v)}{\sinh^2 u + \cos^2 v}, \tag{3-135}$$

---

[7] The denominator of equation (3-135) vanishes when $u = 0$, $v = \frac{\pi}{2}$. However, we may avoid an unphysical singularity in Φ by choosing U and V such that $U(0) = V(\frac{\pi}{2})$.

for then we may rewrite $H = E$ as

$$E \sinh^2 u - \frac{p_u^2}{2\Delta^2} - U(u) = \frac{p_v^2}{2\Delta^2} - V(v) - E \cos^2 v. \qquad (3\text{-}136)$$

It can be shown that potentials of the form (3-135) are generated by bodies resembling real galaxies (see Problems 2-5 and 2-8), so there are interesting physical systems for which equation (3-136) is valid.

If the analogy with the harmonic oscillator holds, p_u will be a function of u only, and similarly for p_v. Under these circumstances, the left side of equation (3-136) does not depend on v, and the right side does not depend on u, so both sides must equal some constant, say I_2. Hence we would then have

$$p_u = \pm \Delta \sqrt{2 \left[E \sinh^2 u - I_2 - U(u) \right]} \qquad (3\text{-}137a)$$

$$p_v = \pm \Delta \sqrt{2 \left[E \cos^2 v + I_2 + V(v) \right]}. \qquad (3\text{-}137b)$$

It is a straightforward exercise to show that the analogy with the harmonic oscillator *does* hold, and thus that the quantity I_2 defined by equations (3-137) *is* an integral, by direct time differentiation of both sides of either of equations (3-137), followed by elimination of $\dot{u}$ and $\dot{p}_u$ with Hamilton's equations (1D-49) (Problem 3-14).

Eliminating E between equations (3-137), we obtain the integral I_2 as a function of the phase-space coordinates:

$$I_2(u, v, p_u, p_v) = \frac{\sinh^2 u \left(\dfrac{p_v^2}{2\Delta^2} - V \right) - \cos^2 v \left(\dfrac{p_u^2}{2\Delta^2} + U \right)}{\sinh^2 u + \cos^2 v}. \qquad (3\text{-}138)$$

Potentials of the form (3-135) are called **Stäckel** potentials after the German mathematician P. Stäckel, who showed that the coordinates (u, v) are actually the *only*[8] coordinates that allow the Hamiltonian $H = \frac{1}{2}|\mathbf{v}|^2 + \Phi$ to be separated in the manner of equation (3-136) (Goldstein 1980, §10-4). One may show (Problem 3-15) that in the limit $\Delta \to 0$, the potential becomes axisymmetric and $\Delta^2 I_2 \to \frac{1}{2} L_z^2$. Thus, I_2 plays the role of a generalized angular momentum.

As an example of the use of equations (3-137) we investigate the shapes they predict for orbits in the potential obtained by choosing in equation (3-135)

$$U(u) = -W \sinh u \arctan \left(\frac{\Delta \sinh u}{a_3} \right)$$

$$V(v) = W \cos v \operatorname{arctanh} \left(\frac{\Delta \cos v}{a_3} \right), \qquad (3\text{-}139)$$

---

[8] Aside from coordinates (f, g) that are obtained from (u, v) by taking functions $f(u)$ and $g(v)$ of u and v alone.

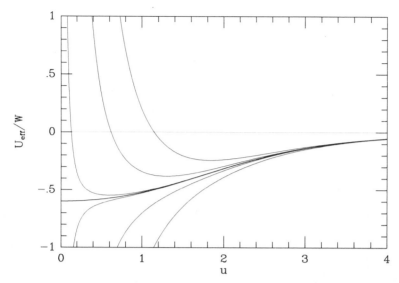

Figure 3-22. A plot of the effective potential U_{eff} defined by equation (3-140a). The function U is given by equation (3-139) with $\Delta = 0.6$ and $a_3 = 1$. From top to bottom the curves correspond to $I_2 = 1, 0.25, 0.01, 0, -0.01, -0.25, -1$.

where W, Δ, and a_3 are constants.[9]

An orbit of specified E and I_2 can explore all values of u and v for which equations (3-137) predict positive p_u^2 and p_v^2. In Figure 3-22 we plot the "effective potential" for u,

$$U_{\text{eff}}(u) \equiv \frac{I_2 + U(u)}{\sinh^2 u}, \qquad (3\text{-}140\text{a})$$

for several values of I_2. By equation (3-137a), the only permitted values of u are those for which U_{eff} is smaller than the star's energy E. If $I_2 \leq 0$, all values of $|u|$ smaller than some maximum value $u_{\max}$ are permitted, but if $I_2 > 0$, u must exceed a minimum value $u_{\min}$. The permitted values of v are similarly governed by the requirement that the effective potential

$$V_{\text{eff}}(v) \equiv -\frac{I_2 + V(v)}{\cos^2 v} \qquad (3\text{-}140\text{b})$$

not exceed E. From equations (3-139) it can be seen that as $\cos v \to 0$, the ratio $V(v)/\cos^2 v$ remains finite. Hence in this limit, $V_{\text{eff}} \to +\infty$

---

[9] With these choices for U and V, the potential (3-135) becomes the potential in the meridional plane of the **perfect prolate spheroid** introduced in Problem 2-8. Hence the planar orbits described here are equivalent to orbits in the potential of this spheroid that have zero angular momentum about the spheroid's symmetry axis.

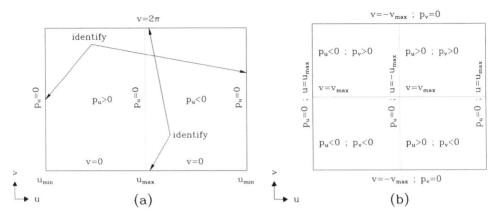

Figure 3-23. Sketch maps of the orbital tori of (a) a loop orbit, and (b) a box orbit.

when $I_2 < 0$, so permitted values of v cannot then exceed a maximum value $v_{max} < \frac{\pi}{2}$, while when $I_2 \geq 0$, all values of v are permitted. Thus orbits with $I_2 > 0$ are confined only by the ellipses $u = u_{min}$ and $u = u_{max}$, while orbits with $I_2 < 0$ are confined top and bottom by portions of the hyperbolae $v = \pm v_{max}$, and right and left by portions of the ellipses $u = u_{max}$. We recognize the former orbits as loops and the latter as boxes. Thus we have reached a remarkable conclusion: *the Stäckel potential provides an analytic model with all the qualitative features of the numerical results described in §3.3.* [For a much more detailed discussion of analytic models of box and loop orbits, see de Zeeuw (1985).]

Hitherto, we have assumed that u and v vary over the ranges $0 \leq u < \infty$ and $0 \leq v < 2\pi$. This convention works well for loop orbits, along which u and v then vary continuously (except at $v = 2\pi$). But a slightly different convention is needed to ensure that u and v vary continuously along box orbits; henceforth we shall employ the conventions

$$0 < u_{min} \leq u \leq u_{max} \quad \text{and} \quad 0 \leq v < 2\pi \quad \text{(loop orbit)}$$
$$0 > -u_{max} \leq u \leq u_{max} \quad \text{and} \quad -\tfrac{1}{2}\pi \leq v \leq \tfrac{1}{2}\pi \quad \text{(box orbit)}.$$
$$(3\text{-}141)$$

Equations (3-137) determine not only the structure of orbits in two-dimensional configuration space, but also their structure in four-dimensional phase space. Since these orbits admit two isolating integrals H and I_2, each orbit is confined to a two-dimensional surface in phase space. By varying the parameters u and v in equations (3-137), we may roam at will over these surfaces. Let us map one of these surfaces, say the surface of a loop orbit. We know that in the case of a loop orbit, u and v satisfy $u_{min} \leq u \leq u_{max}$ and $0 \leq v \leq 2\pi$. Furthermore, for any values of u and v, *two* values of p_u are possible, corresponding to the two signs in equation (3-137a). Hence our map of this orbital surface will be of the form shown in Figure 3-23a. Points on the left boundary of the figure correspond to the same phase-space points $(u_{\min}, v, p_u = 0, p_v)$ as points on the right edge of the figure. Similarly, the phase-space points $(u, v = 0, p_u, p_v(0))$ are represented at both top and bottom of the figure. Hence we obtain a more complete representation of the orbital surface by cutting out the marked rectangle and taping together first the top and bottom edges, and then the left and right edges. In this way we form a doughnut-shaped surface. In mathematical language, the orbital surface is said to be topologically equivalent to the two-torus in four-space.

Figure 3-23b shows the map of the phase-space surface of a typical box orbit. Since both p_u and p_v are now subject to sign uncertainties, each pair of numbers (u, v) corresponds to four phase-space points. As in the case of the loop orbits, points on the left and right edges of the rectangle represent the same phase-space points twice over, so a more realistic representation of the orbital surface is obtained by taping these edges together. Similarly, the top and bottom edges should be joined up. Hence the orbital surfaces of box orbits are also topologically equivalent to the two-torus in four-space.

Few people can visualize four-dimensional phase space. Fortunately the portion of phase space that is associated with orbits of any given energy E is three-dimensional just like ordinary space, and it is interesting to imagine how this three-space is filled with orbital two-tori. The tori of orbits of energy E form a one-parameter family, within which individual tori may be labeled by the value of I_2 on the corresponding orbit. The structure of these tori may be laid bare by slicing the three-space $H = E$ along some plane, say the plane $v = 0$. Each torus cuts this plane in a smooth closed curve, so the nested sequence of tori shows up in the cross-section $\{v = 0\}$ as a nested sequence of **invariant curves** like tree rings. Equation (3-137a) is the equation of these curves, which are shown in Figure 3-24. The orbits for which u crosses zero are box orbits, while those for which the sign of u is fixed are loops. The resemblance of Figure 3-24 to many of the surfaces of section of §3.3 is obvious. Clearly the numerically integrated orbits discussed in §3.3 also form nested sequences of tori.

(a) Actions We have seen that the basic structures in phase space are orbital tori. How should we label these tori? The obvious labels are E and I_2, but it turns out that better labels are provided by the areas inside the invariant curves in which the tori puncture the surfaces

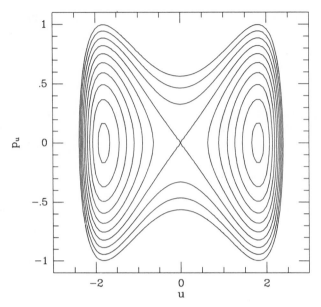

Figure 3-24. The (u, p_u), $v = 0$ surface of section for
motion at $E = -0.25$ in the Stäckel potential defined by
equations (3-135) and (3-139) with $\Delta = 0.6$ and $a_3 = 1$.
Each curve is a contour of constant I_2 [eq. (3-137a)]. The
invariant curves of box orbits run round the outside of
the figure, while the bull's eyes at right are the invariant
curves of prograde loop orbits. Temporarily suspending
the convention that loops always have $u > 0$, we show
the invariant curves of retrograde loops as the bull's eyes
at left.

of section $v = constant$ and $u = constant$. The area inside one of the
invariant curves shown in Figure 3-24 is

$$2\pi J_r \equiv \int_{u_{\min}}^{u_{\max}} du \int_{-|p_u(u)|}^{|p_u(u)|} dp_u, \qquad (3\text{-}142)$$

where $|p_u(u)|$ is the value of p_u on the upper portion of the invariant
curve [the positive root of eq. (3-137a)]. Performing the inner integral
in equation (3-142) and rearranging the outer integral in the usual way,
we have

$$J_r = \frac{1}{\pi} \int_{u_{\min}}^{u_{\max}} |p_u(u)|\, du = \frac{1}{2\pi} \oint p_u(u)\, du, \qquad (3\text{-}143a)$$

where $p_u(u)$ is given by equation (3-137a), and the circuit is to be taken
in the *clockwise* sense in order to ensure that $J_r \geq 0$. J_r is called the

radial action. Since v does not occur in equation (3-137a), the radial action is the same for every cut $v = constant$.

By analogy with equation (3-143a), we define the **azimuthal action** as

$$J_a = \frac{1}{2\pi} \oint p_v(v)dv. \tag{3-143b}$$

Once again, J_a is the same for every cut $u = constant$ through a given torus. For future reference, note that $2\pi J_a$ can be displayed as an area in the surface of section $\{u = u_0\}$, where the constant u_0 is chosen such that $u = u_0$ is the equation of the closed loop orbit of the given energy.

Why do we state that actions constitute better labels for the tori than the integrals H and I_2, of which they are rather horrible functions? One reason is that the actions alone enjoy the property of **adiabatic invariance** that will be discussed in the next section. A more important reason is that in addition to labels of the tori, we need coordinates to tell us where we are on any given torus, and the actions, unlike H and I_2, generate beautiful coordinates for the tori, the so-called **angle variables**.

(b) Angle variables   Equations (3-143) and (3-137) give J_r and J_a as functions of E and I_2, and therefore of (u, v, p_u, p_v). We should like to use J_r and J_a as new phase-space coordinates, and seek to this end two new phase-space functions, θ_r and θ_a, such that the four variables $(J_r, J_a, \theta_r, \theta_a)$ form a set of canonical coordinates for the whole phase space, the actions constituting the momenta of the set and the angles the coordinates.[10] Since the actions are well-defined functions of the phase-space coordinates, the Poisson bracket [see eq. (1D-60)] $[f, J_r]$ of J_r with any phase-space function f is also well defined, and we may define the new coordinate curves as follows. On each torus—defined by a pair of values (J_r, J_a)—we assign to some phase-space point w_0 the angle coordinates $\theta_r = \theta_a = 0$. Then we mark out the θ_r-axis by integrating from this point the coupled ordinary differential equations

$$\frac{dw_\alpha}{d\theta_r} = [w_\alpha, J_r] \quad (\alpha = 1, \ldots, 4), \tag{3-144a}$$

where w_α is any of the phase-space coordinates (u, v, p_u, p_v). Then from every point $w(\theta_r, \theta_a = 0)$ on the solution to these equations, we integrate the differential equations

$$\frac{dw_\alpha}{d\theta_a} = [w_\alpha, J_a] \quad (\alpha = 1, \ldots, 4). \tag{3-144b}$$

---

[10] Since the momenta of a canonical set always "commute" with one another, that is, have vanishing mutual Poisson bracket, we shall be able to construct such a canonical set only if $[J_r, J_a] = 0$. For a proof that this condition is satisfied, see Problem 3-17.

One can now show the remarkable result [see Arnold (1978) §49 for details] that the paths we generate in this way all close on themselves; in fact, $\mathbf{w}(\theta_r + 2\pi, \theta_a) = \mathbf{w}(\theta_r, \theta_a)$, and $\mathbf{w}(\theta_r, \theta_a + 2\pi) = \mathbf{w}(\theta_r, \theta_a)$ for any θ_r and θ_a. Hence our paths form a neat grid on the surface of the torus, and the numbers (θ_r, θ_a) can be used as coordinates on the torus.[11]

For many purposes, it is expedient to use integrals as phase-space coordinates, since then at most two of a star's four coordinates evolve in time. But we also wish to employ canonical coordinates, in order that Hamilton's equations apply and phase-space volumes are easily expressed in terms of differentials of the coordinates [see eq. (1D-71)]. The angles and actions just described provide a set of coordinates that satisfy these criteria. Furthermore, one may show that actions are the *only* integrals that can be complemented with conjugate coordinates to make up a canonical set, and it is from this fact that their importance for dynamics derives.

Unfortunately, it is often difficult to carry through the construction we have described to obtain explicit expressions $\mathbf{w}(\mathbf{J}, \boldsymbol{\theta})$ for the ordinary phase-space coordinates in terms of the angles and actions. However, we shall see in the next paragraph that the mere knowledge that such a construction is *in principle* possible can be extremely valuable. In the rare cases in which we can obtain expressions of the form $\mathbf{w}(\mathbf{J}, \boldsymbol{\theta})$, equation (3-150b) enables us to solve immediately for the temporal development of any orbit.

(c) Action space   In Chapter 4 we shall develop the idea that galaxies are made up of orbits, and we shall find it helpful to think of whole orbits as single points in an abstract space. Any isolating integrals can serve as coordinates for such a representation, but the most advantageous coordinates are the actions. Thus we define **action space** to be the imaginary space whose Cartesian coordinates are the actions. A typical example is shown in Figure 3-25. Points on the axes represent orbits for which only one of the integrals (3-143) is non-zero. These are the closed orbits; points on the radial-action axis represent closed long-axis orbits, while points on the azimuthal-action axis represent closed loop orbits. The origin represents the orbit of a star that just sits at the center of the potential. Above and to the left of a curve running diagonally through the space, the orbits are loop orbits; the boxes lie below and to the right of this line. Orbits of a common energy lie on

---

[11] If we try to repeat this construction with, say, H and I_2 replacing the J_i in equations (3-144), the curves we would generate would *not* close on themselves, and we would not obtain a coordinate system valid everywhere on the torus.

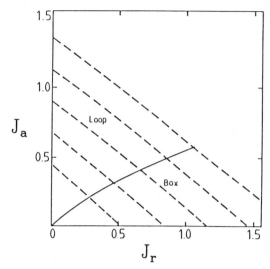

Figure 3-25. The action space of orbits in the
potential Φ_L of equation (3-77) when $q = 0.9$
and $R_c = 0.14$. Energy is constant on dashed
lines. (From Binney & Spergel 1984.)

curves that slope diagonally from upper left to lower right. Every point
in the positive quadrant, all the way to infinity, represents a bound orbit.

A small region R in action space represents a group of orbits. The
volume of four-dimensional phase space occupied by these orbits is

$$V = \int_D d^2\mathbf{x}\,d^2\mathbf{v}, \qquad (3\text{-}145)$$

where D is the region of phase space visited by stars on the orbits of
R. Since the coordinate set $(\mathbf{J}, \boldsymbol{\theta})$ is canonical, $d^2\mathbf{x}\,d^2\mathbf{v} = d^2\mathbf{J}\,d^2\boldsymbol{\theta}$ [see
eq. (1D-71)] and thus

$$V = \int_D d^2\mathbf{J}\,d^2\boldsymbol{\theta}. \qquad (3\text{-}146)$$

But for any orbit the angle variables cover the range $(0, 2\pi)$, so we may
immediately integrate over the angles to find

$$V = (2\pi)^2 \int_R d^2\mathbf{J} = (2\pi)^2 A, \qquad (3\text{-}147)$$

where A is the area of the region R. Thus the area of a piece of action
space is directly proportional to the volume of phase space occupied by
its orbits.

(d) Generalization to three dimensions   All the results we have
described can be generalized to motion in three-dimensional Stäckel potentials. These potentials are most conveniently expressed as functions
of **confocal ellipsoidal coordinates** (λ, μ, ν). These coordinates are characterized by two lengths, $\Delta_1 \leq \Delta_2$. In the (x, y) plane the coordinates
reduce to $\lambda = \Delta_1^2 \sinh^2 u$, $\mu = -\Delta_1^2 \cos^2 v$. The coordinate ν is approximately $-\Delta_2^2 \cos^2 \theta$, where θ is the usual colatitude of spherical polar coordinates. In close analogy with our derivation of equations (3-137), the
Hamiltonian for motion in three-dimensional Stäckel potentials yields
expressions for the momenta p_λ, p_μ, and p_ν as functions of only their re-
spective coordinates, E and two additional isolating integrals I_2 and I_3.
These expressions demonstrate that each orbital surface is a three-torus
embedded in six-dimensional phase space. There are now three actions,
$J_r \equiv (2\pi)^{-1} \oint p_\lambda d\lambda$, $J_a \equiv (2\pi)^{-1} \oint p_\mu d\mu$, and the **latitudinal action**
$J_l \equiv (2\pi)^{-1} \oint p_\nu d\nu$, and three angle variables θ_r, θ_a, and θ_l conjugate
to the actions. Action space is three-dimensional, the phase-space vol-
ume occupied by a set of orbits being $(2\pi)^3$ times the volume filled by
these orbits in action space.

In the limit $\Delta_1 \to 0$, the potentials become oblate figures of revolu-
tion and $J_a \to |L_z|$, where $\mathbf{L}$ is the angular momentum vector. If Δ_2 is
now allowed to tend to zero, the potentials become spherically symmet-
ric, and $J_l \to L - |L_z|$. In the limit $\Delta_1 \to \infty$, $\Delta_2 \to \infty$, the potentials
become those of three-dimensional harmonic oscillators, with actions
$J_r = E_x/\omega_x$, $J_a = E_y/\omega_y$, and $J_l = E_z/\omega_z$, where the E_i and ω_i are
the energies and angular frequencies of the oscillators. De Zeeuw (1985)
gives a full account of orbits in three-dimensional Stäckel potentials.

2 Structure of Regular Orbits in General Potentials

A two-dimensional Stäckel potential (3-135) is specified by two arbitrary functions of one variable. Similarly, a three-dimensional Stäckel
potential involves three arbitrary functions of one variable. Since the
densities, and therefore the potentials, of galaxies are specified by arbitrary functions of three variables, few, if any, galactic potentials can be
of Stäckel's form.

Why then have we devoted so much space to Stäckel potentials?
Because orbits in these potentials turn out to have the same structure
as the much more general class of regular orbits, that is, orbits that
admit as many isolating integrals as they have spatial dimensions. In
particular, one may show [see Arnold (1978) for details] that any three-
dimensional orbit that admits three isolating integrals, say H, I_2, and

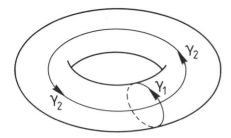

Figure 3-26. Two closed paths on a torus that cannot be deformed into one another, nor contracted to single points.

I_3, with vanishing Poisson bracket $[I_2, I_3]$, forms a three-torus in phase space.

The actions J_r, J_a and J_l of this torus are defined by integrals of the form

$$J_i = \frac{1}{2\pi} \oint_{\gamma_i} \mathbf{v} \cdot d\mathbf{x} \quad (i = r, a, l), \tag{3-148}$$

where γ_i is a closed path around the torus. The nature of the paths employed in equation (3-148) is most easily grasped by considering closed paths on a two-torus. Figure 3-26 shows two closed paths on a two-torus. The path γ_1 loops around the torus by the narrow way, while γ_2 goes around by the longer path. If rubber bands were laid along these paths, we should find that by sliding the bands over the surface of the torus, we could neither contract them to a single point, nor bring them alongside one another. This property of having sets of irreducible, irreconcilable closed paths is characteristic of a torus. One may show that the integral (3-148) is the same for any two paths around the orbital three-torus that *can* be distorted into one another by sliding them over the torus [see Arnold (1978) §49]. Furthermore, because of this invariance of (3-148) under deformation of the defining path, the value of this integral around *any* closed path γ can be expressed as a linear combination of its values around any three paths that cannot be deformed into one another. Actions J_i may be defined by choosing any three such paths γ_i. However, for most purposes it is convenient to choose paths γ_r and γ_a that are natural generalizations of the paths $\{v = const, p_v = const\}$, $\{u = const, p_u = const\}$, which we used to define the radial and azimuthal actions (3-143), and a third path γ_l that generalizes the path in the phase space of a three-dimensional Stäckel potential along which only ν and p_ν vary.

Coordinates for the orbital tori are again provided by angle variables conjugate to the actions. On incrementing any angle variable by 2π one again returns to the same point in phase space. Thus each regular orbit enjoys all the structure of an orbit in a Stäckel potential, and action-

angle variables enjoy all the beautiful properties that we derived for
Stäckel potentials.

(a) Quasi-periodic motion   With the passage of time, the phase
point of a star on any regular orbit moves over a torus. What does this
motion look like when referred to the angle coordinates? Since the $(\mathbf{J}, \boldsymbol{\theta})$
system is canonical, Hamilton's equations hold. Hence

$$\dot{\theta}_i = \frac{\partial H}{\partial J_i} \quad (i = r, a, l). \tag{3-149}$$

Now H is constant on any torus, so both H and $(\partial H/\partial J_i)$ are functions
of the J_k, and not the θ_k. Furthermore, the actions do not change in
time since the phase point remains always on the same torus. Therefore,
the orbit is characterized by three constant frequencies

$$\omega_i \equiv \frac{\partial H}{\partial J_i} \quad (i = r, a, l), \tag{3-150a}$$

and integrating equation (3-149) we find that the angle coordinates of
an orbiting star increase linearly in time:

$$\theta_i(t) = \omega_i t + \theta_i(0) \quad (i = r, a, l). \tag{3-150b}$$

Since nothing is changed physically when we increase any of the
angle variables by 2π, the Cartesian phase-space coordinates $(\mathbf{x}, \mathbf{v})$ must
be periodic functions of the θ_i with period 2π. Any such function can
be expressed as a Fourier series. Hence

$$\mathbf{x}(\boldsymbol{\theta}, \mathbf{J}) = \sum_{l,m,n=-\infty}^{\infty} \mathbf{X}_{lmn}(\mathbf{J}) \exp\left[i(l\theta_r + m\theta_a + n\theta_l)\right]. \tag{3-151}$$

Substituting from equation (3-150b) for the time dependence of the angle
coordinates along an orbit, we find that the time evolution of the position
vector of any star can be written

$$\mathbf{x}(t) = \sum_{l,m,n=-\infty}^{\infty} \widetilde{\mathbf{X}}_{lmn} \exp\left[i(l\omega_r + m\omega_a + n\omega_l)t\right], \tag{3-152a}$$

where

$$\widetilde{\mathbf{X}}_{lmn} \equiv \mathbf{X}_{lmn} \exp\left\{i[l\theta_r(0) + m\theta_a(0) + n\theta_l(0)]\right\}. \tag{3-152b}$$

Thus the Fourier transform of the time evolution of any stellar coordinate involves only integer combinations of the three fundamental frequencies. Functions of the form (3-152a) are said to be **conditionally
periodic** or **quasi-periodic** functions of time.

We are now in a position to prove a result that plays a crucial role
in the theory of time-independent stellar systems.

Time Averages Theorem *If the frequencies* (3-150a) *are incommensurable, the average time that the phase point of a star on a regular orbit spends in any region D of its torus is proportional to the integral $V(D) = \int_D d^3\boldsymbol{\theta}$ through D.*

Proof: Let f_D be the function such that $f_D(\boldsymbol{\theta}) = 1$ when the point $\boldsymbol{\theta}$ lies in D, and is zero otherwise. We may expand f_D in a Fourier series

$$f_D = \sum_{l,m,n=-\infty}^{\infty} F_{lmn} \exp\left[i(l\theta_r + m\theta_a + n\theta_l)\right] = \sum_{l=-\infty}^{\infty} F_{\mathbf{l}} \exp(i\mathbf{l} \cdot \boldsymbol{\theta}),$$
(3-153)

where we have introduced a compact notation $\mathbf{l} \equiv (l, m, n)$. Now

$$\int_{\text{torus}} f_D(\boldsymbol{\theta}) d^3\boldsymbol{\theta} = \int_D d^3\boldsymbol{\theta} = V(D).$$
(3-154a)

With equation (3-153) we therefore have

$$V(D) = \int_{\text{torus}} f_D(\boldsymbol{\theta}) d^3\boldsymbol{\theta} = \sum_{\mathbf{l}=-\infty}^{\infty} F_{\mathbf{l}} \prod_{k=1}^{3} \int_0^{2\pi} \exp(il_k \psi) \, d\psi = (2\pi)^3 F_0.$$
(3-154b)

On the other hand, the fraction of the interval $(0, T)$ during which the star's phase point lies in D is

$$\tau_T(D) = \frac{1}{T} \int_0^T f_D[\boldsymbol{\theta}(t)] dt,$$
(3-155)

where $\boldsymbol{\theta}(t)$ is the position of the star's phase point at time t. With equations (3-150b) and (3-153), equation (3-155) becomes

$$\tau_T(D) = \frac{1}{T} \sum_{\mathbf{l}} e^{i\mathbf{l}\cdot\boldsymbol{\theta}(0)} \int_0^T F_{\mathbf{l}} e^{i(\mathbf{l}\cdot\boldsymbol{\omega})t} \, dt$$

$$= F_0 + \frac{1}{T} \sum_{\mathbf{l}\neq 0} e^{i\mathbf{l}\cdot\boldsymbol{\theta}(0)} F_{\mathbf{l}} \left[\frac{e^{i(\mathbf{l}\cdot\boldsymbol{\omega})T} - 1}{i\mathbf{l} \cdot \boldsymbol{\omega}}\right].$$
(3-156)

Thus

$$\lim_{T\to\infty} \tau_T(D) = F_0 = \frac{V(D)}{(2\pi)^3},$$
(3-157)

which completes the proof.◁

Note that if the frequencies are commensurable, $(\mathbf{l} \cdot \boldsymbol{\omega})$ vanishes for some $\mathbf{l} \neq 0$ and the second equality in equation (3-156) becomes invalid. In fact, if $\omega_i : \omega_j = m : n$, say, then by equations (3-150b)

$I_4 \equiv (n\theta_i - m\theta_j)$ becomes an isolating integral that confines the phase point of the star to a spiral on the torus. Motion in a spherical potential provides a familiar example of this phenomenon. Only two independent frequencies, Ω and κ, characterize this motion, so there are four isolating integrals—for example E and the three components of the angular momentum **L**. If there is more than one commensurability between the three frequencies ω_i, then yet a fifth isolating integral arises. For example, in Keplerian motion $\Omega = \kappa$, and there are five isolating integrals (§3.1).

(b) Action spaces of general potentials   The important difference between the orbital structures of Stäckel potentials and those of more general potentials concerns the way in which the orbits are packed together in phase space. Stäckel potentials support only the major orbit families introduced in §3.3 and §3.4. A general potential, by contrast, usually supports not only the major families, but also a number of minor families of regular orbits. The portion of phase space occupied by each minor orbit family is filled by an array of nested tori on which one can define angles and actions just as in the case of the major families. But these minor families, unlike the major families, do not permit their tori to be linked together into a single global array. Consequently, the phase space associated with a general potential does not admit a global set of action-angle coordinates, and this changes the structure of action space: whereas every point in the action space of a Stäckel potential represents one, and only one, orbit, in the action space of a general potential, some points do not correspond to any orbits at all, while others are claimed by more than one orbit. For more information about these phenomena, see Binney and Spergel (1984).

3 Irregular Orbits

Galactic dynamics would be a lot easier than it is if all orbits were regular. Unfortunately, for realistic galactic potentials the concept of regularity is at best an idealization, and at worst a delusion. To see how thin is the ice over which we skate when we presume (as we often shall) that most orbits are regular, consider the orbital structure of a potential that is very closely related to the potential (3-77) that generated the splendidly regular surfaces of section shown in Figures 3-8 and 3-11. In polar coordinates the potential (3-77) that gives rise to Figure 3-8 is

$$\Phi_L(R,\phi) = \tfrac{1}{2}v_0^2 \ln\left[\tfrac{1}{2}R^2(q^{-2}+1) - \tfrac{1}{2}R^2(q^{-2}-1)\cos 2\phi + R_c^2\right].$$
$$(3\text{-}158)$$

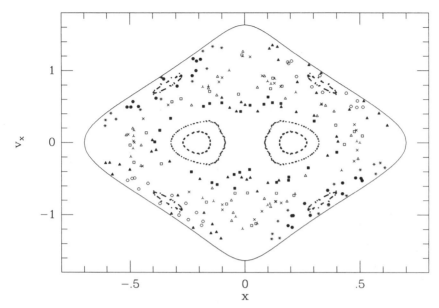

Figure 3-27. The $(x, \dot{x})$ $y = 0$ surface of section for motion in the potential Φ_N of equation (3-161). Each irregular orbit is denoted by a different symbol.

Now consider motion in the potential

$$\Phi_N(R, \phi) = \tfrac{1}{2}v_0^2 \ln \left[\tfrac{1}{2}R^2(q^{-2} + 1) - \tfrac{1}{2}R^2(q^{-2} - 1) \cos 2\phi \right.$$
$$\left. - \left(\frac{R^3}{R_e} \right) \cos 2\phi + R_c^2 \right], \tag{3-159}$$

where R_e is a constant. The only difference between Φ_L and Φ_N is the presence of the term $(R^3/R_e) \cos 2\phi$ in the logarithm. Let us set $R_e = 1.5$, $R_c = 0.14$, and $q = 0.9$, and study the surface of section generated by orbits in Φ_N that is most nearly equivalent to the surface of section for Φ_L shown in Figure 3-8. Figure 3-27 shows the surface of section associated with Φ_N in which the zero-velocity curve cuts the x-axis at the same coordinate, $x = 0.7$, as in Figure 3-8. Near the center, the points generated by individual orbits lie on smooth curves, indicating that these orbits respect a second isolating integral in addition to energy. But outside this region, the points generated by individual orbits seem to occur at random. What causes this erratic behavior?

Figure 3-28, which is a surface of section generated by orbits in Φ_L when $R_c = 0.14$ and $q = 0.6$, will help us to understand the origin of the disorder in Figure 3-28. Much of Figure 3-28 is taken up with a

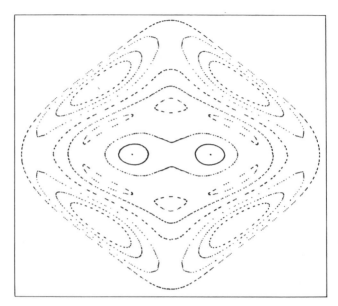

Figure 3-28. A surface of section for motion in the potential Φ_L when $q = 0.6$. The energy of this surface of section is similar to that of Figure 3-8 for motion in Φ_L when $q = 0.9$. Notice that strengthening the non-axisymmetry of the potential causes a significant proportion of the surface of section to be taken up with the "islands" of resonant orbit families.

series of islands. At the center of each island sits the point in which the phase-space trajectory of a closed orbit strikes the surface of section.[12] There are sometimes several islands in a chain because the trajectory of the closed orbit associated with this chain cuts the surface of section several times before it closes on itself. The trajectory of this closed orbit is enclosed by the tori of neighboring nonclosed orbits; the contours of the islands in the surface of section are formed by cross-sections through these tori. It is not hard to see that the central closed orbit of a chain of islands is always stable, and that any stable closed orbit gives rise to a chain of islands.

By equation (3-150b) a regular orbit is closed whenever its characteristic frequencies ω_k are commensurable. Furthermore, in the portion of phase space that is occupied by any single family of orbits, the frequencies are by equation (3-150a) continuous functions of the actions. Therefore, any family of regular orbits in which the frequencies are not

---

[12] Any orbit that punctures the surface of section at only a finite number of points is necessarily closed.

entirely independent of the actions must contain an infinite number of closed orbits corresponding to the infinite number of rational numbers that lie in any finite range of the real line. In general, an infinite subset of these closed orbits sire small orbit families. The special feature of a Stäckel potential is that there are only a finite number of orbit families because only a finite number of these closed orbits are fully stable rather than marginally stable, and hence entrap neighboring nonclosed orbits. A study of the stability of closed orbits will reveal how remarkable is this state of affairs.

(a) Floquet analysis   We may determine the stability of a closed orbit by means of first-order perturbation theory. If $[\mathbf{X}(t), \mathbf{V}(t)]$ are the phase-space coordinates of a closed orbit in a two-dimensional potential, then we may write the coordinates of a neighboring orbit in the form $[\mathbf{X}(t) + \mathbf{x}(t), \mathbf{V}(t) + \mathbf{v}(t)]$, where (x/X) and (v/V) are small initially. The equations of motion for this orbit read

$$\dot{\mathbf{X}} + \dot{\mathbf{x}} = \mathbf{V} + \mathbf{v},$$

$$\dot{\mathbf{V}} + \dot{\mathbf{v}} = -(\boldsymbol{\nabla}\Phi)_{(\mathbf{X}+\mathbf{x})} \simeq -[\boldsymbol{\nabla}\Phi + (\mathbf{x}\cdot\boldsymbol{\nabla})(\boldsymbol{\nabla}\Phi)]_{\mathbf{X}} \qquad (3\text{-}160)$$

Canceling the zero-order terms and writing the remaining terms out in matrix form, we obtain

$$\frac{d}{dt}\begin{pmatrix} x \\ y \\ v_x \\ v_y \end{pmatrix} = -\begin{pmatrix} 0 & 0 & -1 & 0 \\ 0 & 0 & 0 & -1 \\ \Phi_{xx} & \Phi_{xy} & 0 & 0 \\ \Phi_{xy} & \Phi_{yy} & 0 & 0 \end{pmatrix}\begin{pmatrix} x \\ y \\ v_x \\ v_y \end{pmatrix}, \qquad (3\text{-}161)$$

where $\Phi_{xx} \equiv (\partial^2\Phi/\partial x^2)$ etc. The partial derivatives in the matrix of equation (3-161) are to be evaluated along the closed path $\mathbf{X}(t)$ and are therefore periodic functions of time, with the same period T as the underlying closed orbit. A theorem of Floquet (see Margenau & Murphy 1956) states that any solution of a system of periodic differential equations like (3-161) can be written as a linear combination of functions of the form

$$\begin{pmatrix} x(t) \\ y(t) \\ v_x(t) \\ v_y(t) \end{pmatrix} = \begin{pmatrix} x_0 \\ y_0 \\ v_{x0} \\ v_{y0} \end{pmatrix} e^{\mu t} P_\mu(t), \qquad (3\text{-}162)$$

where $P_\mu(t)$ is periodic with the same period T as the equations themselves. In general, there are four possible values of the quantity μ and its associated eigenvector $(x_0, y_0, v_{x0}, v_{y0})$. In our case, two of these values of μ are equal to zero—these represent motions in which the orbit under study remains on the closed orbit about which we have perturbed the equations of motion, but either lags or leads the point $(\mathbf{X}, \mathbf{V})$. The two other values of μ are either pure real or pure imaginary: either $\mu_3 = +\alpha$ and $\mu_4 = -\alpha$, or $\mu_3 = +i\omega$ and $\mu_4 = -i\omega$, where α and ω are real. If $\mu_3 = \alpha$, the closed orbit is unstable because almost every orbit that starts off close to the closed orbit deviates from it at an exponential rate. If $\mu_3 = i\omega$, the orbit is stable. In a potential that supports only a finite number of orbit families, $\mu_3 = \mu_4 = 0$ for all but a finite number of closed orbits. Hence such a potential must satisfy some very special conditions and we cannot expect an arbitrarily chosen potential to be of this type.[13]   We shall call such potentials **integrable**. The Stäckel potentials are especially simple integrable potentials.

One possible approach to the study of motion in a general potential Φ is to find an integrable potential Φ_I that is similar to Φ, and then to treat orbits in Φ as orbits in Φ_I that have been perturbed by the small quantity $\delta\Phi = (\Phi - \Phi_I)$. An elegant introduction to this approach will be found in the review by Berry (1978), while Gerhard (1985) offers an example of its application to Stäckel potentials. The perturbation $\delta\Phi$ changes the quantities μ that are associated with the closed orbits in Φ_I, so that we no longer have $\mu = 0$ for all but a finite number of closed orbits.

Suppose the perturbation stabilizes the orbit in Φ_I for which the fundamental frequencies ω_1 and ω_2 are in the ratio $\omega_1 : \omega_2 = m : n$. The period of this orbit is $T = m(2\pi/\omega_1) = n(2\pi/\omega_2)$, and it generates of order $m \approx n$ points in the surface of section. Hence it gives rise to a chain of order m islands in the surface of section, and it will be necessary to follow nonclosed orbits in its neighborhood for many multiples of T before the structure of these islands shows up clearly in a computed surface of section. If m and n are large, this will be impracticable, and the surface of section that we obtain from the computer will not show the structures of the islands but will at best give an idea of how the islands are arranged. If the islands are small and are arranged on a smooth curve, we shall (correctly) conclude that the orbit respects two isolating integrals. However, the islands may rather be arranged on a very complex curve or on no discernible curve at all: in this case our numerically obtained surface of section will show very little structure, and we would (erroneously) conclude that energy is the only isolating integral. The

---

[13] These considerations are readily generalized to orbits in three-dimensional potentials. The matrix that occurs in the analog of equation (3-161) is then a 6×6 matrix. Of the six possible values of μ, two are still zero and the remaining four fall into pairs that differ only in a sign. The underlying closed orbit is unstable if either of these pairs is real.

chaotic region in Figure 3-28 may arise in this way. Clearly, the practical definition of an isolating integral is a function of the timescale over which the orbit is followed and the resolution with which we view the orbit in phase space.

The approach to orbits in non-integrable potentials that we have just described has been fully developed by Kolmogorov, Arnold, and Moser, and is generally called KAM theory. We have implicitly assumed that the phase-space trajectory of any orbit in a general potential is quasi-periodic and moves on a torus. Actually, this is an oversimplification. In most potentials, part of phase space is occupied by orbits that are not confined to tori, the so-called **irregular orbits**. The central result of KAM theory, the so-called KAM theorem, is the demonstration that quasi-periodic orbits always occupy a non-zero volume of phase space when the potential is sufficiently close to integrable (Moser 1973). However, this general result is not powerful enough to be directly applicable to galactic dynamics: we have seen that in practice many orbits that are actually quasi-periodic appear to be irregular over galactic timescales. Conversely, many irregular orbits are so tightly hemmed in by the tori of neighboring regular orbits that for long periods they *appear* to be regular. An interesting, and at present very open, question is the degree to which a long timescale is introduced into galactic dynamics by the large number of dynamical times that may be required for an irregular orbit to slip through a crack between confining tori[14]—a process known as **Arnold diffusion**.

3.6 Slowly Varying Potentials*

So far in this chapter we have been concerned with motion in potentials that are time-independent in either an inertial or a rotating frame. It is sometimes necessary to consider how stars move in potentials that are time-dependent. The nature of the problem posed by a time-varying potential depends on the speed with which the potential evolves. In this section we shall confine ourselves to potentials that evolve slowly. Examples of important processes that can be handled in this way include:

(i) Interactions between the individual stars at the core of a dense stellar system (such as a globular cluster or galaxy nucleus) cause the core to contract (§8.4). The timescale on which this contraction

---

[14] In much the same way, an alpha particle in an unstable atomic nucleus may require a prodigious number of characteristic times to tunnel out to freedom.

* This section contains more advanced material and can be skipped on a first reading.

occurs proves to be very long compared with the orbital times of individual stars.

(ii) In Chapter 9 we shall see that stars of galaxies and globular clusters lose substantial quantities of mass as they gradually evolve and shed their envelopes into interstellar, and then perhaps intergalactic, space (Richstone & Potter 1982).

(iii) A number of lines of evidence suggest that the disks of spiral and S0 galaxies may have formed by gas settling into the equatorial planes of the preexisting spheroidal components of these galaxies (§9.2). In this case the orbits of the stars of the spheroidal component will undergo a slow evolution as the gravitational potential of the disk gains in strength.

Potential variations that are slow compared to a typical orbital frequency are called **adiabatic**. We now show that for stars on regular orbits, the action integrals that were introduced in §3.5 are constant during such adiabatic changes of the potential. For this reason these integrals are often called **adiabatic invariants**.

(a) Adiabatic invariance of actions   Suppose we have a sequence of potentials $\Phi_\lambda(\mathbf{x})$ that depend continuously on the parameter λ. For each fixed λ assume that the orbits supported by Φ_λ are regular and thus that phase space is filled by arrays of nested tori on which the phase points of individual stars move. We consider what happens when λ is changed from its initial value, say $\lambda = \lambda_0$, to a new value λ_1. After this change has occurred, each star's phase point will start to move on a torus of the set that belongs to Φ_{λ_1}. In general, two stellar phase points that started out on the same torus of Φ_{λ_0} will move onto two different tori of Φ_{λ_1}. But if λ is changed very slowly compared to all the characteristic times $2\pi/\omega_k$ associated with motion on each torus, all phase points that are initially on a given torus of Φ_{λ_0} will be equally affected by the variation of λ. This statement follows from the time averages theorem of §3.5.2(a), which shows that all stars spend the same fraction of their time in each portion of the torus. Thus all phase points that start on the same torus of Φ_{λ_0} will end on a single torus of Φ_{λ_1}. Said in other language, any two stars that are initially on a common orbit (but at different phases) will still be on a common orbit after the variation of λ is complete.

Suppose the variation of λ starts at time $t = 0$ and is complete by time t_1, and let $\mathbf{H}_t$ be the time-evolution operator defined in equation (1D-50). Then we have just seen that $\mathbf{H}_{t_1}$, which is a canonical map [see Appendix 1.D.4(c)], maps tori of Φ_{λ_0} onto tori of Φ_{λ_1}. These facts guarantee that actions are adiabatically invariant, for the following reason. Choose three closed curves γ_i, $(i = r, a, l)$ on any torus M of Φ_{λ_0} that through the integrals (3-148) generate the actions J_i of this torus. Then, since $\mathbf{H}_{t_1}$ is the endpoint of a continuous deformation of phase space into itself, the images $\mathbf{H}_{t_1}(\gamma_i)$ of these curves are suitable curves along which to evaluate the actions J_i' of $\mathbf{H}_{t_1}(M)$, the torus to which M is mapped by $\mathbf{H}_{t_1}$. But by a corollary to the Poincaré invariant theorem [Appendix 1.D.4(b)], we have that if γ is any closed curve and $\mathbf{H}_{t_1}(\gamma)$ is its image under the canonical map $\mathbf{H}_{t_1}$, then

$$\oint_{\mathbf{H}_{t_1}(\gamma)} \mathbf{v} \cdot d\mathbf{x} = \oint_\gamma \mathbf{v} \cdot d\mathbf{x}. \tag{3-163}$$

Hence $J_i' = J_i$, and the actions of stars do not change if the potential evolves sufficiently slowly.

(b) Applications   We illustrate these ideas with a number of simple examples. More sophisticated applications of adiabatic invariants will be found in Young (1980), Goodman and Binney (1984), and Binney and May (1986).

We first consider the harmonic oscillator whose potential is

$$\Phi = \tfrac{1}{2}\omega^2 x^2. \tag{3-164}$$

The action integral is (see Problem 3-16)

$$J = \frac{1}{2\omega}\left[v^2 + (\omega x)^2\right] = \frac{E}{\omega}. \tag{3-165}$$

In terms of the amplitude of oscillation X this reads

$$J = \tfrac{1}{2}\omega X^2. \tag{3-166}$$

Now suppose that the spring is slowly stiffened by a factor $s^2 > 1$. After this operation the natural frequency is

$$\omega' = s\omega, \tag{3-167}$$

and by the adiabatic invariance of J, the new amplitude X' satisfies

$$\tfrac{1}{2}\omega' X'^2 = J = \tfrac{1}{2}\omega X^2. \tag{3-168}$$

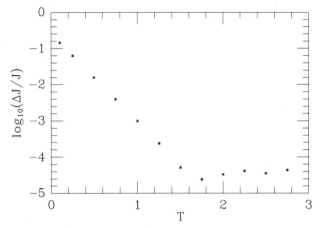

Figure 3-29. Checking the invariance of the action integral (3-165) when the natural frequency of a harmonic oscillator is varied according to equation (3-171). For $T \lesssim 1.6$ the fractional change in J declines approximately as $\Delta J/J \propto \exp(-7T)$. ΔJ is the RMS change in the action on integrating the oscillator's equation of motion from $t = -20T$ to $t = 20T$, using eight equally spaced phases. Note that $\dot{\omega} \neq 0$ at each end of the calculation, and in these circumstances the true action differs slightly from (3-165). For $T \gtrsim 1.6$ this error in equation (3-165) dominates the calculated value of ΔJ (see Lichtenberg & Lieberman (1983) eq. (2.3.27)].

Thus the amplitude is diminished to

$$X' = \frac{X}{\sqrt{s}}. \tag{3-169}$$

On the other hand, the energy $E = \omega J$, has increased to[15]

$$E' = \omega' J = s\omega J = sE. \tag{3-170}$$

It is interesting to investigate how rapidly we may change the frequency ω without destroying the invariance of J. Let ω vary with time according to

$$\omega(t) = \pi\sqrt{3 + \frac{2}{\pi}\arctan\left(\frac{2t}{T}\right)}. \tag{3-171}$$

---

[15] The simplest proof of this result uses quantum mechanics. The energy of a harmonic oscillator is $E = (n + \frac{1}{2})\hbar\omega$ where n is an integer. When ω is slowly varied, n cannot change discontinuously and hence must remain constant. Therefore $E/\omega = E'/\omega'$. Of course, for galaxies n is rather large.

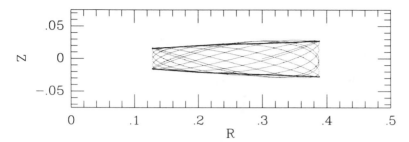

Figure 3-30. The envelope of an orbit in the effective potential (3-50) with $q = 0.5$ (dotted curve) is well modeled by equation (3-173) (full curves).

Thus the frequency changes from $\omega = \sqrt{2}\pi$ at $t \ll -T$ to $\omega = 2\pi$ at $t \gg T$ over a characteristic time T. In Figure 3-29 we show the results of numerically integrating the oscillator's equation of motion with ω given by equation (3-171). We plot the RMS difference ΔJ between the initial and final values of J for various values of T and eight different phases of the oscillator at $t = -20$. Evidently for $T \gtrsim 1.5$, J is well conserved. Since the initial and final periods of oscillation are 2 and 1, we conclude that the potential does not have to change very slowly for J to be well conserved. In fact, one can show that the fractional change in J is in general less than $\exp(-\omega T)$ for $\omega T \gg 1$.

As a second example of the use of adiabatic invariants, consider the shapes shown in Figure 3-3 of the orbits in the meridional plane of an axisymmetric galaxy. In §3.2.2 we remarked that disk stars in the solar neighborhood oscillate perpendicular to the galactic plane considerably more rapidly than they oscillate in the radial direction. Therefore, if we consider the radial coordinate $R(t)$ of a disk star to be a known function of time, we may consider the equation of motion (3-48c) of the z-coordinate to describe motion in a slowly varying potential. If the amplitude of the z-oscillations is small, we may expand $(\partial\Phi/\partial z)$ about $z = 0$ to find

$$\ddot{z} \simeq -\omega^2 z \quad \text{where} \quad \omega(t) \equiv \left(\frac{\partial^2 \Phi}{\partial z^2}\right)^{\frac{1}{2}}_{[R(t),0]} \equiv \sqrt{\Phi_{zz}[R(t),0]}. \quad (3\text{-}172)$$

If the action integral of this harmonic oscillator is conserved, we expect the amplitude $Z(R)$ to satisfy [see eq. (3-166)]

$$Z(R) = Z(R_0) \left[\frac{\Phi_{zz}(R_0,0)}{\Phi_{zz}(R,0)}\right]^{\frac{1}{4}}. \quad (3\text{-}173)$$

Figure 3-30 compares the prediction of equation (3-173) with the true shape of an orbit in the effective potential (3-50). Evidently the gross structure of these orbits can be understood in terms of adiabatic invariants.

As another example of the use of adiabatic invariants, consider the motion of a star on a loop orbit in a slowly varying planar potential $\Phi(R, \phi)$. The azimuthal action is

$$J_a = \frac{1}{2\pi} \int_0^{2\pi} p_\phi d\phi, \tag{3-174}$$

where $p_\phi = R^2 \dot{\phi}$. We now conduct the following experiment. Initially the potential Φ is axisymmetric. Then $p_\phi = L_z$ is an integral, and we can perform the integral in (3-174) to obtain $J_a = L_z$. We now slowly distort the potential in some arbitrary fashion into a new axisymmetric configuration. At the end of this operation, the azimuthal action still has value J_a and is again equal to the angular momentum L_z. Thus the star will finish the experiment with the same angular momentum with which it started,[16] even though its instantaneous angular momentum, $R^2 \dot{\phi}$, will have been changing during most of the experiment. Of course, if the potential remains axisymmetric throughout, p_ϕ remains an integral at all times and is exactly conserved no matter how rapidly the potential is varied.

The radial action of any loop orbit in a nonrotating potential can be obtained from the area inside the orbit's invariant curve in the $\phi = 0$ surface of section. Thus

$$J_r = \frac{1}{\pi} \int_{R_{\min}}^{R_{\max}} \dot{R} dR. \tag{3-175}$$

If the potential is axisymmetric, we have by conservation of energy $\dot{R} = \pm\sqrt{2(E - \Phi) - L_z^2/R^2}$. Hence

$$J_r = \frac{1}{\pi} \int_{R_{\min}}^{R_{\max}} \sqrt{2(E - \Phi) - \frac{L_z^2}{R^2}} \, dR. \tag{3-176}$$

For a Kepler potential $\Phi = -GM/R$, this yields

$$J_r = \frac{GM}{\sqrt{2|E|}} - L = L\left[\frac{1}{\sqrt{1 - e^2}} - 1\right] = \sqrt{GMa}\left(1 - \sqrt{1 - e^2}\right), \tag{3-177}$$

---

[16] This statement does not apply for stars that switch from loop to box orbits and back again as the potential is varied (Binney & Spergel 1984). These stars will generally be on highly eccentric orbits initially.

where e is the eccentricity of the orbit [eq. (3-22)], and we have replaced L_z by L. In the isochrone potential (2-32) one obtains (see Problem 3-19)

$$J_r = \frac{GM}{\sqrt{2|E|}} - \tfrac{1}{2}L\left(1 + \sqrt{1 + \frac{4bGM}{L^2}}\right). \tag{3-178}$$

Finally, if the star is on a nearly circular orbit, which can be described by epicycle theory, equation (3-175) becomes with equations (3-54) and (3-67)

$$J_r = \frac{1}{2\pi}\int_0^{\frac{2\pi}{\kappa}} (X\kappa \sin \kappa t)^2\, dt = \tfrac{1}{2}\kappa X^2. \tag{3-179}$$

This result can also be obtained by recognizing that in the epicycle approximation the radial motion is harmonic with frequency κ and amplitude X. Hence equation (3-166) for the adiabatic invariant in a harmonic oscillator yields equation (3-179) directly. Notice also that equation (3-179) is consistent with the exact expression (3-177) for the Keplerian case, using the relations $\Omega = \kappa = \sqrt{GM/a^3}$ and $X \simeq ae$.

If the ratio of the frequencies Ω and κ associated with a nearly circular orbit is unchanged by a slow variation of the potential, equation (3-179) implies that the eccentricity X/R_g of the orbit is adiabatically invariant:

$$\text{constant} = J_r = \tfrac{1}{2}X^2\Omega\left(\frac{\kappa}{\Omega}\right) = \tfrac{1}{2}L^2\left(\frac{X}{R_g}\right)^2\left(\frac{\kappa}{\Omega}\right). \tag{3-180}$$

Problems

3-1. [1] From the relationship $\dot{r} = \dot{\psi}(dr/d\psi)$ and equation (3-21), show for a Kepler orbit that $\sin(\psi - \psi_0) = (L\dot{r}/eGM)$. Derive equation (3-24) by considering this expression in the limit $r \to \infty$.

3-2. [1] Show that the energy of a circular orbit in the isochrone potential (2-32) is $E_c = -GM/(2a)$, where $a = \sqrt{b^2 + r^2}$. Let the angular momentum of this orbit be $L_c(E)$. Show that

$$L_c = \sqrt{GMb}\left(x^{-\frac{1}{2}} - x^{\frac{1}{2}}\right), \qquad \text{where} \qquad x \equiv -\frac{2Eb}{GM}. \tag{3P-1}$$

3-3. [2] $\Delta\psi$ denotes the increment in azimuthal angle during one complete radial cycle of an orbit. For the isochrone potential, $\Delta\psi$ is given by equation (3-35).

(a) What is $\Delta\psi$ along an orbit in the potential (3-39)?

(b) Prove in the epicycle approximation that along orbits in a potential with circular frequency $\Omega(R)$,

$$\Delta\psi = 2\pi \left(4 + \frac{d\ln\Omega}{d\ln R}\right)^{-\frac{1}{2}}. \tag{3P-2}$$

(c) Show that the exact expressions for $\Delta\psi$ in the isochrone potential and the potential of equation (3-39) reduce for orbits of small eccentricity to the epicycle result just derived.

(d) What is $\Delta\psi$ in the epicycle approximation for orbits in the potential $\Phi \propto r^{-\alpha}$ when $\alpha < 2$?

3-4. [1] Prove that if a homogeneous sphere of a pressureless fluid with density ρ is released from rest, it will collapse to a point in time $t_{ff} = \frac{1}{4}\sqrt{3\pi/(2G\rho)}$. The time t_{ff} is called the **free-fall time** of a system of density ρ.

3-5. [1] A star orbiting in a spherical potential suffers an arbitrary instantaneous velocity change while it is at pericenter. Show that the pericenter distance of the ensuing orbit cannot be larger than the initial pericenter distance.

3-6. [1] In a spherically symmetric system, the apocenter and pericenter distances are given by the roots of equation (3-13). Show that if $E < 0$ and the potential $\Phi(r)$ is generated by a non-negative density distribution, this equation has either zero or two roots. [Hint: Take the second derivative of (3-13) with respect to u and use Poisson's equation (Contopoulos 1954).]

3-7. [1] Prove that circular orbits in a given potential are unstable if the angular momentum per unit mass on a circular orbit decreases outward.

3-8. [1] For what spherically symmetric potential is a possible trajectory $r = ae^{b\psi}$?

3-9. [1] Prove that at any point in an axisymmetric system at which the local density is negligible, the epicycle, vertical, and circular frequencies κ, ν, and Ω [eqs. (3-58)] are related by $\kappa^2 + \nu^2 = 2\Omega^2$.

3-10. [3] Plot a $(y, \dot{y})$, $(x = 0, \dot{x} > 0)$ surface of section for motion in the potential Φ_L of equation (3-77) when $q = 0.9$. Qualitatively relate the structure of this surface of section to the structure of the $(x, \dot{x})$ surface of section shown in Figure 3-8.

3-11. [3] Sketch the structure of the $(x = 0, \dot{x} > 0)$ surface of section for motion at energy E in a Kepler potential when (a) the (x, y) coordinates are inertial, and (b) the coordinates rotate at 0.75 times the circular frequency Ω at the energy E. (Hint: See Binney et al. 1985.)

3-12. [3] Consider two point masses M_1 and $M_2 > M_1$, which travel in a circular orbit about their center of mass under their mutual attraction. (a) Show that the Lagrange point L_4 of this system forms an equilateral triangle with the two masses. (b) Show that motion near L_4 is stable if $M_1/(M_1 + M_2) < 0.03852$ (see Szebehely 1967).

3-13. [2] Show that for prolate spheroidal coordinates ($R = \Delta \sinh u \sin v$, $z = \Delta \cosh u \cos v$) we can obtain the momenta in the form $p_u(u)$ and $p_v(v)$, provided that the potential Φ is of the form

$$\Phi(u,v) = \frac{U(u) - V(v)}{\sinh^2 u + \sin^2 v}. \tag{3P-3}$$

3-14. [3] For motion in a potential of the form (3-135), obtain

$$\dot{p}_u = \frac{2H \sinh u \cosh u - (dU/du)}{\sinh^2 u + \cos^2 v}, \tag{3P-4}$$

where (u,v) are the prolate spheroidal coordinates defined by equations (3-130), by (a) differentiating equation (3-137a) with respect to t and then using $\dot{u} = (\partial H/\partial p_u)$, and (b) from $\dot{p}_u = -(\partial H/\partial u)$.

3-15. [2] For the coordinates defined by equation (3-130), show that in the limit $\Delta \to 0$ we have $\Delta \sinh u \to \Delta \cosh u \to r$ and $v \to \phi$, the usual polar angle. Hence show that in this limit $\Delta^2 I_2 \to \frac{1}{2} L_z^2$, where I_2 is the integral defined by equation (3-138).

3-16. [2] (a) Show that the action of the harmonic oscillator with Hamiltonian $H = \frac{1}{2}(v^2 + \omega^2 x^2)$ is H/ω. (b) From the defining equation (3-144), show that the angle variable may be written

$$\theta(x,v) = \arctan\left(\frac{v}{\omega x}\right). \tag{3P-5}$$

[Hint: Solve for $x(\theta)$ and $v(\theta)$, and then eliminate unwanted constants of integration by inserting your expressions into the Hamiltonian.]

3-17. [2] Show that the radial and angular actions defined by equations (3-143) have vanishing mutual Poisson bracket: $[J_r, J_a] = 0$. {Hint: If $f(a,b)$ is any differentiable function of two variables, and A and B are any differentiable functions of the phase-space variables, then $[A, f] = [A, a](\partial f/\partial a) + [A, b](\partial f/\partial b)$.}

3-18. [2] Show that adiabatic invariance of actions implies that closed orbits remain closed when the potential is adiabatically deformed. An initially circular orbit in a spherical potential Φ does not remain closed when Φ is squashed along any line that is not parallel to the orbit's original angular-momentum vector. Why does this statement remain true no matter how slowly Φ is squashed?

3-19. [2] From equations (3-34b) and (3-150a), show that the radial action J_r of an orbit in the isochrone potential (2-32) is related to the energy E and angular momentum L of this orbit by

$$J_r = \sqrt{GMb}\left[x^{-\frac{1}{2}} - f(L)\right],\qquad\text{(3P-6)}$$

where $x \equiv -2Eb/(GM)$ and f is some function. Use equation (3P-1) to show that $f(L) = (\sqrt{l^2+1} - l)^{-1} = \sqrt{l^2+1} + l$, where $l \equiv L/(2\sqrt{GMb})$, and hence show that the Hamiltonian of a particle orbiting in the isochrone potential may be written

$$H(J_r, L) = -\frac{GM}{2b}\left(\frac{J_r}{\sqrt{GMb}} + l + \sqrt{l^2+1}\right)^{-2}.\qquad\text{(3P-7)}$$

4 Equilibria of Collisionless Systems

There is a fundamental difference between galaxies and the systems that are normally dealt with in statistical mechanics, such as molecules in a box. This difference lies in the nature of the forces that act between the constituent particles. The force between two gas molecules is small unless the molecules are very close to each other, when they repel each other strongly. Consequently gas molecules are subject to violent and short-lived accelerations as they collide with one another, interspersed with longer periods when they move at nearly constant velocity. The gravitational force that acts between the stars of a galaxy is of an entirely different nature. To see this, consider the force with which the stars in the cone shown in Figure 4-1 attract the star that is at the cone's apex. The force toward any star falls off with distance r as r^{-2}, but if the density of stars is uniform, the number of attracting stars per unit length of the cone increases as r^2. Therefore, if the density of stars is constant throughout the cone, equal lengths of the cone will attract the star at the apex with equal force. Of course, if the density of attracting stars is exactly homogeneous, the star at the apex will experience no net force because it will be pulled equally in all directions. But if the density of attracting stars falls off in one direction more slowly than in the opposing direction, the star at the apex will be subject to a net force. This simple argument indicates that the net gravitational force acting on a star in a galaxy is determined by the gross structure of the galaxy rather than by whether the star happens to lie close to some other star. Consequently the force on any star will not vary rapidly, and each star may be supposed to accelerate smoothly through the force field that is generated by the galaxy as a whole. This chapter is concerned with models that are based on this approximation, and we start by investigating more quantitatively under what circumstances it is valid.

1 The Relaxation Time

Consider a galaxy of N identical stars of mass m and focus on the motion of an individual star across this system. We seek an order-of-magnitude estimate of the difference between the true velocity of this star after it

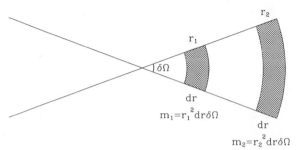

Figure 4-1. If the density of stars were everywhere the same, the stars in each of the shaded segments would make equal contributions to the net force on a star at the cone's apex. Thus the acceleration of a star at the apex is determined by the large-scale gradient in the density of stars within the galaxy.

has crossed the system and the velocity that it would have had at that time if the mass of the other stars had been smoothly distributed through the system rather than concentrated into individual stars. Suppose our star passes within distance b of another star (Figure 4-2). Clearly the amount $\delta\mathbf{v}$ by which the encounter deflects the velocity $\mathbf{v}$ of our star will depend on the masses and speeds of the two stars as well as on b. In §7.1 we calculate $\delta\mathbf{v}$ exactly [see eqs. (7-10)]. However, for our present purposes it is more instructive to obtain a crude estimate of $\delta\mathbf{v}$ for those encounters in which $|\delta\mathbf{v}|/v \ll 1$ and the perturbing star is almost stationary during the encounter. In this case we may calculate the component $\delta\mathbf{v}_\perp$ of $\delta\mathbf{v}$ that is perpendicular to $\mathbf{v}$ by assuming that our star passes the perturber on a straight-line trajectory, and integrating the perpendicular force $\mathbf{F}_\perp$ that gives rise to $\delta\mathbf{v}_\perp$ along this trajectory. We place the origin of time at the instant of closest approach of the two stars, and find in the notation of Figure 4-2

$$F_\perp = \frac{Gm^2}{b^2 + x^2}\cos\theta = \frac{Gm^2 b}{(b^2 + x^2)^{\frac{3}{2}}} \simeq \frac{Gm^2}{b^2}\left[1 + \left(\frac{vt}{b}\right)^2\right]^{-\frac{3}{2}}. \qquad (4\text{-}1)$$

But by Newton's laws

$$m\dot{\mathbf{v}}_\perp = \mathbf{F}_\perp, \qquad (4\text{-}2)$$

so substituting equation (4-2) into equation (4-1) and integrating with respect to time we have

$$|\delta\mathbf{v}_\perp| \simeq \frac{Gm}{bv}\int_{-\infty}^{\infty}\left(1 + s^2\right)^{-\frac{3}{2}}ds = \frac{2Gm}{bv}. \qquad (4\text{-}3)$$

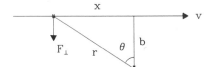

Figure 4-2. A field star approaches the test star at speed v and impact parameter b. We estimate the resulting impulse to the test star by approximating the field star's trajectory as a straight line.

Thus $|\delta\mathbf{v}_\perp|$ is roughly equal to the force at closest approach, Gm/b^2, times the duration of this force b/v. Now the surface density of stars in our hypothetical galaxy is of order $N/\pi R^2$, where R is the galaxy's characteristic radius, so that in crossing the galaxy once our star suffers

$$\delta n = \frac{N}{\pi R^2} 2\pi b\,db = \frac{2N}{R^2} b\,db \qquad (4\text{-}4)$$

encounters with impact parameters in the range b to $b + db$. Each such encounter produces a perturbation $\delta\mathbf{v}_\perp$ to the star's velocity, but because these small perturbations are randomly oriented around $\mathbf{v}$, the mean value $\overline{\delta\mathbf{v}_\perp}$ of these vectors is zero. But by summing their squares, we conclude that $v_\perp^2$ changes by an amount

$$\delta v_\perp^2 \simeq \left(\frac{2Gm}{bv}\right)^2 \frac{2N}{R^2} b\,db. \qquad (4\text{-}5)$$

Notice that our assumption of a straight-line trajectory breaks down, and equation (4-5) becomes invalid, when $|\,\delta\mathbf{v}_\perp|\simeq v$; from equation (4-3) this occurs if the impact parameter $b \lesssim b_{\min} \equiv Gm/v^2$. Integrating equation (4-5) over all values of b between $b_{\min}$ and the largest possible impact parameter R, we find

$$\Delta v_\perp^2 \equiv \int_{b_{\min}}^{R} \delta v_\perp^2 \simeq 8N\left(\frac{Gm}{Rv}\right)^2 \ln\Lambda, \qquad (4\text{-}6a)$$

where

$$\ln\Lambda \equiv \ln\left(\frac{R}{b_{\min}}\right). \qquad (4\text{-}6b)$$

In §4.2 we shall show that the typical speed v of a star in a galaxy is related to the mass Nm and radius R of the galaxy by

$$v^2 \approx \frac{GNm}{R}. \qquad (4\text{-}7)$$

If we eliminate R from equation (4-6a) using equation (4-7), we have

$$\frac{\Delta v_\perp^2}{v^2} = \frac{8\ln\Lambda}{N}. \qquad (4\text{-}8)$$

If the star makes many crossings of the galaxy, $v_\perp^2$ will change by of order $\Delta v_\perp^2$ at each crossing, so that the number of crossings n_{relax} that are required for its velocity to change by of order itself is given by

$$n_{\text{relax}} = \frac{N}{8 \ln \Lambda}. \qquad (4\text{-}9)$$

The **relaxation time** may be defined as $t_{\text{relax}} = n_{\text{relax}} \times t_{\text{cross}}$, where $t_{\text{cross}} = R/v$ is the **crossing time**. The appropriate value of Λ is $\Lambda = R/b_{\text{min}} \approx Rv^2/(Gm) \approx N$ by equation (4-7). Hence we conclude that individual stellar encounters will perturb a star from the course it would take if the other matter of the system were perfectly smoothly distributed only over of order $0.1N/\ln N$ crossing times. Consequently, even if N is as small as 50, each star is deflected from its mean trajectory only after several crossing times, and it is possible to obtain some understanding of the dynamics of even small systems by investigating the orbits of the stars in a suitable mean potential. This chapter is devoted to this level of analysis, which effectively constitutes a complete analysis for systems that are less than $0.1N/\ln N$ crossing times old.

Galaxies typically have $N \approx 10^{11}$ stars and are a few hundred crossing times old, so for these systems stellar encounters are entirely unimportant. In a globular cluster, on the other hand, $N \approx 10^5$ and the crossing time $t_{\text{cross}} \approx 10^5 \, \text{yr}$, so that stellar encounters may be important over the cluster lifetime of $10^{10} \, \text{yr}$. Indeed, in the core of a globular cluster, where t_{cross} is very short and $N \approx 10^4$, encounters play a key role. A rich cluster of galaxies has $N \approx 10^3$ and $t_{\text{cross}} \approx 10^9 \, \text{yr}$, so that here again encounters may or may not be important, depending on the detailed structure of the overall system. But in the case of a cluster of galaxies, or of a globular cluster, as in the case of a galaxy, the fundamental dynamics is that of a **collisionless system** in which the constituent particles (galaxies or stars) move under the influence of the *mean* potential generated by all the other particles. In this chapter we study this problem, deferring until Chapter 8 discussion of the longer-term evolution that is driven by encounters.

4.1 The Collisionless Boltzmann Equation

Imagine a large number of stars moving under the influence of a smooth potential $\Phi(\mathbf{x}, t)$. At any time t, a full description of the state of any collisionless system is given by specifying the number of stars $f(\mathbf{x}, \mathbf{v}, t)d^3\mathbf{x}\,d^3\mathbf{v}$ having positions in the small volume $d^3\mathbf{x}$ centered on $\mathbf{x}$ and velocities in the small range $d^3\mathbf{v}$ centered on $\mathbf{v}$. The quantity

$f(\mathbf{x}, \mathbf{v}, t)$ is called the **distribution function** or **phase-space density** of the system. Throughout this book we abbreviate "distribution function" to "DF". Clearly $f \geq 0$ everywhere in phase space.

If we know the initial coordinates and velocities of every star, Newton's laws enable us to evaluate their positions and velocities at any later time. Thus, given $f(\mathbf{x}, \mathbf{v}, t_0)$, it should be possible to calculate $f(\mathbf{x}, \mathbf{v}, t)$ for any t using only the information that is contained in $f(\mathbf{x}, \mathbf{v}, t_0)$. To this end, consider the flow of points in phase space that arises as stars move along their orbits. The coordinates in phase space are

$$(\mathbf{x}, \mathbf{v}) \equiv \mathbf{w} \equiv (w_1, \ldots, w_6) \qquad (4\text{-}10a)$$

say, so the velocity of this flow may be written

$$\dot{\mathbf{w}} = (\dot{\mathbf{x}}, \dot{\mathbf{v}}) = (\mathbf{v}, -\boldsymbol{\nabla}\Phi) : \qquad (4\text{-}10b)$$

$\dot{\mathbf{w}}$ is a six-dimensional vector that bears the same relationship to the six-dimensional vector $\mathbf{w}$ as the three-dimensional fluid flow velocity $\mathbf{u} = \dot{\mathbf{x}}$ does to the position vector $\mathbf{x}$ in an ordinary fluid flow.

A characteristic of the flow described by $\dot{\mathbf{w}}$ is that it conserves stars; in the absence of encounters stars do not jump from one point in phase space to another, but rather drift smoothly through the space. Therefore the density of stars $f(\mathbf{w}, t)$ satisfies a continuity equation analogous to that satisfied by the density $\rho(\mathbf{x}, t)$ of an ordinary fluid flow [see eq. (1E-3)];

$$\frac{\partial f}{\partial t} + \sum_{\alpha=1}^{6} \frac{\partial(f\dot{w}_\alpha)}{\partial w_\alpha} = 0. \qquad (4\text{-}11)$$

The physical content of this equation can be seen by integrating it over some volume of phase space. The first term then describes the rate at which the stock of stars inside this volume is increasing, while an application of the divergence theorem shows that the second term describes the rate at which stars flow out of this volume.

The flow described by $\dot{\mathbf{w}}$ is very special, for it has the property that

$$\sum_{\alpha=1}^{6} \frac{\partial \dot{w}_\alpha}{\partial w_\alpha} = \sum_{i=1}^{3} \left(\frac{\partial v_i}{\partial x_i} + \frac{\partial \dot{v}_i}{\partial v_i} \right) = \sum_{i=1}^{3} -\frac{\partial}{\partial v_i} \left(\frac{\partial \Phi}{\partial x_i} \right) = 0. \qquad (4\text{-}12)$$

Here $(\partial v_i / \partial x_i) = 0$ because v_i and x_i are independent coordinates of phase space, and the last step in equation (4-12) follows because $\boldsymbol{\nabla}\Phi$

does not depend on the velocities. If we use equation (4-12) to simplify equation (4-11), we obtain the **collisionless Boltzmann equation:**[1]

$$\frac{\partial f}{\partial t} + \sum_{\alpha=1}^{6} \dot{w}_\alpha \frac{\partial f}{\partial w_\alpha} = 0, \qquad (4\text{-}13\text{a})$$

i.e.,

$$\frac{\partial f}{\partial t} + \sum_{i=1}^{3} \left(v_i \frac{\partial f}{\partial x_i} - \frac{\partial \Phi}{\partial x_i} \frac{\partial f}{\partial v_i} \right) = 0, \qquad (4\text{-}13\text{b})$$

or, in vector notation,

$$\frac{\partial f}{\partial t} + \mathbf{v} \cdot \nabla f - \nabla \Phi \cdot \frac{\partial f}{\partial \mathbf{v}} = 0. \qquad (4\text{-}13\text{c})$$

Equation (4-13) is the fundamental equation of stellar dynamics. It is a special case of Liouville's theorem (§8.1.2).

The meaning of the collisionless Boltzmann equation can be clarified by extending to six dimensions the concept of the convective or Lagrangian derivative [see eq. (1E-7)]. We define

$$\frac{df}{dt} \equiv \frac{\partial f}{\partial t} + \sum_{\alpha=1}^{6} \dot{w}_\alpha \frac{\partial f}{\partial w_\alpha}. \qquad (4\text{-}14)$$

(df/dt) represents the rate of change of the density of phase points as seen by an observer who moves through phase space with a star at velocity $\dot{\mathbf{w}}$. The collisionless Boltzmann equation (4-13a) is then simply

$$\frac{df}{dt} = 0. \qquad (4\text{-}13\text{d})$$

In words, the flow of stellar phase points through phase space is incompressible; the phase-space density f around the phase point of a given star always remains the same.[2]

Notice that encounters between stars cause $\dot{\mathbf{v}}$ to differ from the smoothed gravitational force $-\nabla \Phi$ and therefore invalidate equation

---

[1] Often also called the Vlasov equation, although it is a simplified version of an equation derived by L. Boltzmann in 1872. See Hénon (1983).

[2] A simple example of an incompressible flow in phase space is provided by an idealized marathon race in which all runners travel at constant speeds: at the start of the course, the density of runners is large but they travel at a wide variety of speeds; at the finish, the density is low, but at any given time all runners passing the post have nearly the same speed.

(4-12). Encounters can be thought of as leading to an additional collision term appearing on the right side of equations (4-13). We shall return to this concept in §8.3.1.

The quantity f appearing in equations (4-13) was defined to be the number of stars per unit volume of phase space. However, there are many possible sets of phase-space coordinates. Let the six-dimensional vector $\mathbf{W}$ represent some arbitrary set of phase-space coordinates, and $\mathbf{w} = (\mathbf{x}, \mathbf{v})$ the usual Cartesian coordinates. Then the number of stars in the interval $d^6\mathbf{W}$ may be written $N(\mathbf{W})d^6\mathbf{W} = f(\mathbf{w})d^6\mathbf{w}$. In arbitrary coordinates, the volume elements $d^6\mathbf{W}$ and $d^6\mathbf{w}$ are not necessarily the same, and hence the phase-space density f and the coordinate density N may differ. However, if the coordinates $\mathbf{W}$ are canonical, we have by equation (1D-71) that the volume element $d^6\mathbf{W} = d^6\mathbf{w}$, and therefore that $N = f$; in words, *in any canonical coordinate system, the coordinate density and the phase-space density are identical.*

Equation (4-11) is equally true when f represents the mass density, or the luminosity density in phase space, and therefore equations (4-13) hold for these interpretations of f also. Note also that if several species of stars are present in a system, for example giants and dwarfs, then the density of each species must separately satisfy the collisionless Boltzmann equation.

(a) Collisionless Boltzmann equation in arbitrary coordinates
It is sometimes convenient to express the collisionless Boltzmann equation in coordinates other than the usual Cartesian variables $(\mathbf{x}, \mathbf{v})$. In principle, we may obtain the form of equation (4-13b) in any coordinate system by a straightforward application of the chain rule for changing partial derivatives. However, this procedure is tedious, and there is a far simpler procedure. The collisionless Boltzmann equation in the form (4-13d) is just the statement that f is constant along the trajectories of stellar phase points. This is true in any coordinates, whether canonical or not. Hence equation (4-14) implies that equation (4-13a) is valid in any coordinates.

For example, in cylindrical coordinates we have

$$\frac{df}{dt} = \frac{\partial f}{\partial t} + \dot{R}\frac{\partial f}{\partial R} + \dot{\phi}\frac{\partial f}{\partial \phi} + \dot{z}\frac{\partial f}{\partial z} + \dot{v}_R\frac{\partial f}{\partial v_R} + \dot{v}_\phi\frac{\partial f}{\partial v_\phi} + \dot{v}_z\frac{\partial f}{\partial v_z} = 0. \quad (4\text{-}15)$$

Furthermore, $\dot{R} = v_R$, $\dot{\phi} = (v_\phi/R)$, $\dot{z} = v_z$, and we have that [cf. eqs. (3-48)]

$$\dot{v}_R = -\frac{\partial \Phi}{\partial R} + \frac{v_\phi^2}{R} \quad ; \quad \dot{v}_\phi = -\frac{1}{R}\frac{\partial \Phi}{\partial \phi} - \frac{v_R v_\phi}{R} \quad ; \quad \dot{v}_z = -\frac{\partial \Phi}{\partial z}. \quad (4\text{-}16)$$

Thus the collisionless Boltzmann equation in cylindrical coordinates reads

$$\frac{\partial f}{\partial t} + v_R \frac{\partial f}{\partial R} + \frac{v_\phi}{R} \frac{\partial f}{\partial \phi} + v_z \frac{\partial f}{\partial z} + \left(\frac{v_\phi^2}{R} - \frac{\partial \Phi}{\partial R} \right) \frac{\partial f}{\partial v_R}$$
$$- \frac{1}{R} \left(v_R v_\phi + \frac{\partial \Phi}{\partial \phi} \right) \frac{\partial f}{\partial v_\phi} - \frac{\partial \Phi}{\partial z} \frac{\partial f}{\partial v_z} = 0. \tag{4-17}$$

Conversion to spherical and rotating coordinates is discussed in Problems 4-3 and 4-2.

1 The Coarse-Grained Distribution Function

We have defined f to be the phase-space density of stars of some particular type. Clearly, we can determine the density of discrete objects such as stars only by counting them in volumes large enough to contain many stars. Hence, with this naive definition of f it is difficult to assign meaning to a DF that varies radically within a volume expected to contain only one or two stars. Unfortunately, we shall see in §4.7 that there are physically interesting circumstances in which the collisionless Boltzmann equation generates just such rapidly fluctuating DFs out of slowly varying DFs. What are we to make of this situation?

This problem is eliminated if we interpret f as a *probability density*: if at time $t = 0$ there is a certain probability that some star lies in a region of phase space D_0, then at time t this same probability attaches to the phase-space volume D_t to which Newton's laws have by then moved the phase points of D_0. Hence the probability density obeys equation (4-11), and thus its corollaries, equations (4-13). Furthermore, there is no difficulty in interpreting rapid fluctuations in a probability density. With this interpretation, f plays the role in stellar dynamics that the wave function plays in quantum mechanics; it is not itself measurable, but we may use it to obtain the answer to any observational question by evaluating the expectation value of some phase-space function $Q(\mathbf{x}, \mathbf{v})$. An example will illustrate this point.

Given the DF f_M of M dwarfs in the solar neighborhood, it would be interesting to calculate the probability P that an M dwarf lies within 1 pc of the Sun. Interpreting f_M as a probability density, we have

$$P = \int d^3\mathbf{v} \int_{|\mathbf{x}-\mathbf{x}_\odot|<1\,\mathrm{pc}} f_M(\mathbf{x}, \mathbf{v}) d^3\mathbf{x}$$
$$= \int d^3\mathbf{v} \int Q_1(\mathbf{x}, \mathbf{v}) f_M(\mathbf{x}, \mathbf{v}) d^3\mathbf{x} = \langle Q_1 \rangle, \tag{4-18a}$$

where $Q_1 = 1$ if $|\mathbf{x} - \mathbf{x}_\odot| < 1\,\mathrm{pc}$ and is zero otherwise. Notice that this equation would have no meaning if f_M were interpreted as a phase-space density. Similarly, the expectation value of the z-velocity dispersion of these stars is

$$\sigma_z^2 = P^{-1} \int d^3\mathbf{v} \int v_z^2 Q_1 f_M d^3\mathbf{x} = P^{-1} \langle v_z^2 Q_1 \rangle. \qquad (4\text{-}18b)$$

It is sometimes useful to employ a **coarse-grained distribution function** $\overline{f}$, whose value at any phase-space point $(\mathbf{x}, \mathbf{v})$ is the average value of f in some specified small volume centered on $(\mathbf{x}, \mathbf{v})$. While $\overline{f}(\mathbf{x}, \mathbf{v})$ is an interesting quantity, the function $\overline{f}$ is of limited practical value because it does *not* satisfy the collisionless Boltzmann, or any other simple evolutionary equation (see §4.7). For contrast with $\overline{f}$, f is sometimes called the **fine-grained distribution function.**

4.2 The Jeans Equations

The DF f is a function of seven variables, and thus the complete solution of the collisionless Boltzmann equation is usually very difficult. However, we can often gain valuable insights by taking moments of the collisionless Boltzmann equation. For example, if we simply integrate equation (4-13b) over all possible velocities, we obtain

$$\int \frac{\partial f}{\partial t} d^3\mathbf{v} + \int v_i \frac{\partial f}{\partial x_i} d^3\mathbf{v} - \frac{\partial \Phi}{\partial x_i} \int \frac{\partial f}{\partial v_i} d^3\mathbf{v} = 0, \qquad (4\text{-}19)$$

where we have employed the summation convention (see Appendix 1.B.1). The range of velocities over which we are integrating does not depend on time, so the partial derivative $\partial/\partial t$ in the first term of this equation may be taken outside the integral. Similarly, since v_i does not depend on x_i, the partial derivative $\partial/\partial x_i$ in the second term of the equation may be taken outside the integral sign. Furthermore, the last term on the left side of the equation vanishes on application of the divergence theorem and use of the fact that $f(\mathbf{x}, \mathbf{v}, t) = 0$ for sufficiently large $v \equiv |\mathbf{v}|$, i.e., there are no stars that move infinitely fast. Therefore, if we define the spatial density of stars $\nu(\mathbf{x})$ and the mean stellar velocity $\overline{\mathbf{v}}(\mathbf{x})$ by

$$\nu \equiv \int f d^3\mathbf{v} \quad ; \quad \overline{v}_i \equiv \frac{1}{\nu} \int f v_i d^3\mathbf{v}, \qquad (4\text{-}20)$$

we have that

$$\frac{\partial \nu}{\partial t} + \frac{\partial(\nu \overline{v}_i)}{\partial x_i} = 0. \qquad (4\text{-}21)$$

Clearly equation (4-21) is a continuity equation like equation (1E-3). If we now multiply equation (4-13b) by v_j and integrate over all velocities, we obtain

$$\frac{\partial}{\partial t}\int fv_j d^3\mathbf{v} + \int v_i v_j \frac{\partial f}{\partial x_i}d^3\mathbf{v} - \frac{\partial \Phi}{\partial x_i}\int v_j \frac{\partial f}{\partial v_i}d^3\mathbf{v} = 0. \qquad (4\text{-}22)$$

The last term on the right side can be transformed by applying the divergence theorem and using the fact that f vanishes for large v:

$$\int v_j \frac{\partial f}{\partial v_i}d^3\mathbf{v} = -\int \frac{\partial v_j}{\partial v_i}fd^3\mathbf{v} = -\int \delta_{ij}fd^3\mathbf{v} = -\delta_{ij}\nu. \qquad (4\text{-}23)$$

Thus equation (4-22) may be rewritten

$$\frac{\partial(\nu\overline{v_j})}{\partial t} + \frac{\partial(\nu\overline{v_i v_j})}{\partial x_i} + \nu\frac{\partial \Phi}{\partial x_j} = 0, \qquad (4\text{-}24a)$$

where

$$\overline{v_i v_j} \equiv \frac{1}{\nu}\int v_i v_j fd^3\mathbf{v}. \qquad (4\text{-}24b)$$

Equation (4-24a) can be put into a more familiar form by subtracting from it v_j times the equation of continuity (4-21) to yield

$$\nu\frac{\partial\overline{v_j}}{\partial t} - \overline{v_j}\frac{\partial(\nu\overline{v_i})}{\partial x_i} + \frac{\partial(\nu\overline{v_i v_j})}{\partial x_i} = -\nu\frac{\partial \Phi}{\partial x_j}, \qquad (4\text{-}25)$$

and then noting that the mean value of $v_i v_j$ may be broken into a part $\overline{v_i}\,\overline{v_j}$ that is due to streaming motion and a part

$$\sigma_{ij}^2 \equiv \overline{(v_i - \overline{v_i})(v_j - \overline{v_j})} = \overline{v_i v_j} - \overline{v_i}\,\overline{v_j} \qquad (4\text{-}26)$$

that arises because the stars near any given point $\mathbf{x}$ do not all have the same velocity. Then using equation (4-26) in equation (4-25), we obtain the analog of Euler's equation (1E-8) of fluid flow;

$$\nu\frac{\partial\overline{v_j}}{\partial t} + \nu\overline{v_i}\frac{\partial\overline{v_j}}{\partial x_i} = -\nu\frac{\partial \Phi}{\partial x_j} - \frac{\partial(\nu\sigma_{ij}^2)}{\partial x_i}. \qquad (4\text{-}27)$$

The left side and the first term on the right side of equation (4-27) take exactly the same form as in the ordinary Euler equation. The last term on the right side of equation (4-27) represents something akin to the pressure force $-\nabla p$. More exactly, $-\nu\sigma_{ij}^2$ is a **stress tensor** that describes an anisotropic pressure. Since equations (4-21), (4-24), and

(4-27) were first applied to stellar dynamics by Sir James Jeans (1919), we call them the **Jeans equations**.[3]

The tensor $\boldsymbol{\sigma}^2$ is manifestly symmetric [eq. (4-26)], and therefore we know from matrix algebra that at any point $\mathbf{x}$ we may choose a set of orthogonal axes $\hat{\mathbf{e}}_i(\mathbf{x})$ in which $\boldsymbol{\sigma}^2$ is diagonal, that is, $\sigma_{ij}^2 = \sigma_{ii}^2 \delta_{ij}$ (no summation over i). The ellipsoid that has the diagonalizing coordinate axes $\hat{\mathbf{e}}_i(\mathbf{x})$ for its principal axes and σ_{11}, σ_{22} and σ_{33} for its semi-axis lengths is called the **velocity ellipsoid** at $\mathbf{x}$.

We shall see below that equation (4-27) is valuable for its ability to relate observationally accessible quantities, like the streaming velocity, velocity dispersion, and so forth. But its fundamental defect must be recognized: we have no analog of the equation of state of a fluid system to relate the six independent components of the tensor $\boldsymbol{\sigma}^2$ to the density ν. The reader may object that if we multiply the collisionless Boltzmann equation (4-13b) through by $v_i v_k$ and integrate over all velocities, we obtain a new set of differential equations for $\boldsymbol{\sigma}^2$ which might supply the missing link between $\boldsymbol{\sigma}^2$ and ν. Unfortunately, these equations involve quantities like $\overline{v_i v_j v_k}$ for which we would require still further equations. Thus these additional equations are of no use unless we can in some way truncate or *close* this regression to ever higher moments of the velocity distribution. In practice, one resolves the difficulty by making some assumption about the form of the tensor $\boldsymbol{\sigma}^2$. The value of this approach depends on whether this assumption is physically well founded.

We may obtain the Jeans equations in cylindrical coordinates by taking moments of equation (4-17). For simplicity we assume that the system under study is axisymmetric and hence that all derivatives with respect to ϕ vanish. Integrating equation (4-17) over all velocities, we obtain in close analogy with our derivation of equation (4-21)

$$\frac{\partial \nu}{\partial t} + \frac{1}{R}\frac{\partial (R\nu \bar{v}_R)}{\partial R} + \frac{\partial (\nu \bar{v}_z)}{\partial z} = 0. \tag{4-28}$$

When we multiply equation (4-17) by v_R or v_z and then integrate over all velocities, we obtain, respectively,

$$\frac{\partial (\nu \bar{v}_R)}{\partial t} + \frac{\partial (\nu \overline{v_R^2})}{\partial R} + \frac{\partial (\nu \overline{v_R v_z})}{\partial z} + \nu \left(\frac{\overline{v_R^2} - \overline{v_\phi^2}}{R} + \frac{\partial \Phi}{\partial R} \right) = 0, \tag{4-29a}$$

$$\frac{\partial (\nu \bar{v}_\phi)}{\partial t} + \frac{\partial (\nu \overline{v_R v_\phi})}{\partial R} + \frac{\partial (\nu \overline{v_\phi v_z})}{\partial z} + \frac{2\nu}{R}\overline{v_\phi v_R} = 0, \tag{4-29b}$$

---

[3] They were originally derived by Maxwell, but he already has a set of equations named after him.

and

$$\frac{\partial(\nu\bar{v}_z)}{\partial t} + \frac{\partial(\nu\overline{v_R v_z})}{\partial R} + \frac{\partial(\nu\overline{v_z^2})}{\partial z} + \frac{\nu\overline{v_R v_z}}{R} + \nu\frac{\partial\Phi}{\partial z} = 0. \quad (4\text{-}29c)$$

By an analogous sequence of steps one may use equation (4P-2) to derive the Jeans equations for a spherically symmetric stellar system. The case of greatest practical importance is when the system is in a steady state and $\bar{v}_r = \bar{v}_\theta = 0$. We find on integrating $\bar{v}_r$ times equation (4P-2) that

$$\frac{d(\nu\overline{v_r^2})}{dr} + \frac{\nu}{r}\left[2\overline{v_r^2} - \left(\overline{v_\theta^2} + \overline{v_\phi^2}\right)\right] = -\nu\frac{d\Phi}{dr}. \quad (4\text{-}30)$$

1 Applications of the Jeans Equations

(a) Asymmetric drift   In §6-4 of MB it is shown that subsystems of our Galaxy for which the quantity $\langle\Pi^2\rangle = \overline{v_R^2}$ is large rotate about the galactic center more slowly than the local standard of rest (that is, the local circular speed). If we define the asymmetric drift v_a of a stellar population to be the difference between the local standard of rest and the mean rotation velocity of this population, the discussion of §6-4 of MB may be summarized by the empirical relationship $v_a \equiv v_c - \bar{v}_\phi \simeq \overline{v_R^2}/D$, where $D \simeq 120\,\mathrm{km\,s^{-1}}$. We can now show that this relationship is a consequence of equation (4-29a).

We suppose the disk is in a steady state and is symmetric about its equator. Then, since the Sun lies close to the galactic equator, we may evaluate equation at $z = 0$, and assume that $(\partial\nu/\partial z) = 0$ by symmetry, to find

$$\frac{R}{\nu}\frac{\partial(\nu\overline{v_R^2})}{\partial R} + R\frac{\partial(\overline{v_R v_z})}{\partial z} + \overline{v_R^2} - \overline{v_\phi^2} + R\frac{\partial\Phi}{\partial R} = 0 \quad (z = 0). \quad (4\text{-}31)$$

Now we define the azimuthal velocity dispersion σ_ϕ^2 by

$$\sigma_\phi^2 = \overline{(v_\phi - \bar{v}_\phi)^2} = \overline{v_\phi^2} - \bar{v}_\phi^2, \quad (4\text{-}32)$$

and substitute $R(\partial\Phi/\partial R) = v_c^2$, where v_c is the circular speed, to obtain from equation (4-31)

$$\sigma_\phi^2 - \overline{v_R^2} - \frac{R}{\nu}\frac{\partial(\nu\overline{v_R^2})}{\partial R} - R\frac{\partial(\overline{v_R v_z})}{\partial z} = v_c^2 - \bar{v}_\phi^2$$

$$= (v_c - \bar{v}_\phi)(v_c + \bar{v}_\phi) = v_a(2v_c - v_a). \quad (4\text{-}33)$$

If we neglect v_a compared to $2v_c$, we obtain

$$2v_c v_a \simeq \overline{v_R^2}\left[\frac{\sigma_\phi^2}{\overline{v_R^2}} - 1 - \frac{\partial \ln(\nu \overline{v_R^2})}{\partial \ln R} - \frac{R}{\overline{v_R^2}}\frac{\partial(\overline{v_R v_z})}{\partial z}\right]. \tag{4-34}$$

Observations of external galaxies (van der Kruit & Searle 1981; van der Kruit & Freeman 1984) suggest that in these systems $\overline{v_z^2}$ is roughly proportional to ν, so if we assume that the shape of the velocity ellipsoid is constant, we estimate that $[\partial \ln(\nu \overline{v_R^2})/\partial \ln R] \simeq 2(\partial \ln \nu/\partial \ln R)$. The other derivative in equation (4-34) is more problematical. Its value depends on the orientation of the velocity ellipsoid at points just above the plane of our Galaxy. Two extreme possibilities are (i) that the ellipsoid's principal axes remain aligned with the coordinate directions of the (R, ϕ, z) system, and (ii) the principal axes rotate to retain alignment with the coordinate directions of spherical coordinates (r, θ, ϕ) centered on the galactic nucleus. Orbit integrations (Binney & Spergel 1983) suggest that the truth lies nearly midway between these two possibilities. In the first case $\overline{v_R v_z}$ is independent of z, and in the second $\overline{v_R v_z} \simeq (\overline{v_R^2} - \overline{v_z^2})(z/R)$ (see Problem 4-5). We have therefore

$$\frac{2v_c v_a}{\overline{v_R^2}} \simeq \left[\frac{\sigma_\phi^2}{\overline{v_R^2}} - \frac{3}{2} - 2\frac{\partial \ln \nu}{\partial \ln R} + \frac{1}{2}\frac{\overline{v_z^2}}{\overline{v_R^2}} \pm \frac{1}{2}\left(\frac{\overline{v_z^2}}{\overline{v_R^2}} - 1\right)\right], \tag{4-35}$$

where the sign ambiguity covers the range of possible behavior of the velocity ellipsoid near the Sun. If we further assume that $\sigma_\phi^2 \simeq \overline{v_z^2} \simeq 0.45\overline{v_R^2}$ [cf. eq. (3-76)], that the disk of our galaxy is exponential, $\nu = \nu_0 \exp(-R/R_d)$, with $R_0/R_d = 2.4$ (Table 1-2), and that $v_c = 220\,\text{km s}^{-1}$, we may use equation (4-35) to find $v_a \simeq \overline{v_R^2}/(110\pm7\,\text{km s}^{-1})$, in good agreement with the empirical value $D \simeq 120\,\text{km s}^{-1}$.

(b) The mass density in the solar neighborhood   Equation (4-29c) enables us to estimate the density of the galactic disk in the solar neighborhood. We drop the first term in the equation on the assumption that the disk is in a steady state. We also neglect the second and fourth terms, since we have just seen that these terms are unlikely to be larger than $\simeq (\overline{v_R^2} - \overline{v_z^2})z/(RR_d)$, which is a factor $\simeq z^2/(RR_d)$ smaller than the third and fifth terms in equation (4-29c). With these assumptions, equation (4-29c) may be rewritten

$$\frac{1}{\nu}\frac{\partial(\nu \overline{v_z^2})}{\partial z} = -\frac{\partial \Phi}{\partial z}. \tag{4-36}$$

We now recall from §2.2.3 [see eq. (2-57)] that near the plane of a highly flattened system, Poisson's equation can be approximated by

$$\frac{\partial^2 \Phi}{\partial z^2} = 4\pi G \rho. \tag{4-37}$$

Thus, eliminating Φ between equations (4-36) and (4-37), we have

$$\frac{\partial}{\partial z}\left[\frac{1}{\nu}\frac{\partial(\nu\overline{v_z^2})}{\partial z}\right] = -4\pi G \rho. \tag{4-38}$$

Hence, if we can measure as functions of height z the number density ν and the mean-square vertical velocity $\overline{v_z^2}$ of any population of stars in the solar neighborhood, we may use equation (4-38) to calculate the local mass density ρ. Any such estimate of ρ will be very uncertain because it is founded on a triple differentiation of star counts: once to yield $\nu(z)$ (see §4-2 of MB) and twice more to form the left side of equation (4-38). But we may obtain independent estimates of ρ from several stellar populations, such as F stars or K giants, and then average the results. In this way, Oort (1932, 1965) concluded that $\rho_0 \equiv \rho(R_0, z = 0) \simeq 0.15\,\mathrm{M_\odot\,pc^{-3}}$, and ρ_0 is generally called the **Oort limit** in his honor. The column density within one or two tracer scale heights of the plane is rather less uncertain because it requires only a double differentiation of the star counts:

$$\Sigma(z) \equiv \int_{-z}^{z} \rho(z')dz' = -\frac{1}{2\pi G\nu}\frac{\partial(\nu\overline{v_z^2})}{\partial z}. \tag{4-39}$$

Of course, neither $\Sigma(z)$ nor $\rho(z)$ can be reliably determined for z greater than a few tracer scale heights. Oort finds $\Sigma(700\,\mathrm{pc}) \simeq 90\,\mathrm{M_\odot\,pc^{-2}}$.

Bahcall (1984a,b) has endeavored to reduce the uncertainty introduced by multiple differentiation of observational data by proceeding as follows. (i) He assigns all known contributors to the mass density ρ to one of $K \simeq 16$ isothermal components, that is, components for which $\overline{v_z^2}$ is independent of z. By equation (4-36) the density ν_k of the kth such component satisfies

$$\nu_k(z) = \nu_k(0)\exp\left[\frac{\Phi(0) - \Phi(z)}{\overline{v_z^2}(k)}\right]. \tag{4-40a}$$

(ii) Bahcall then uses general astrophysical considerations to estimate the relative mass-to-light ratios (Υ_j/Υ_k) of these components, and uses these ratios and equation (4-40a) to convert equation (4-37) into a differential equation for $\Phi(z)$ alone:

$$\frac{\partial^2 \Phi}{\partial z^2} = 4\pi G \Upsilon_1 \sum_{k=1}^{K} \left(\frac{\Upsilon_k}{\Upsilon_1}\right) j_k(0) \exp\left[\frac{\Phi(0) - \Phi(z)}{\overline{v_z^2}(k)}\right], \qquad (4\text{-}40b)$$

where we have identified ν_k with the luminosity density j_k. (iii) Finally, Bahcall finds the value of Υ_1 for which the solution $\Phi(z)$ of this equation generates a profile $j_1(z)$ [via eq. (4-40a)] in good accord with the observed density profile of one of his components, which he uses as a tracer. For F stars Bahcall finds $\rho_0 \simeq 0.19\,\mathrm{M_\odot\,pc^{-3}}$ and $\Sigma(200\,\mathrm{pc}) \simeq 40\,\mathrm{M_\odot\,pc^{-2}}$; for K giants, which have a larger scale height, $\rho_0 \simeq 0.21\,\mathrm{M_\odot\,pc^{-3}}$ and $\Sigma(700\,\mathrm{pc}) \simeq 75\,\mathrm{M_\odot\,pc^{-2}}$. Combining these estimates, we arrive at a total local mass density of

$$\rho_0 = 0.18 \pm 0.03\,\mathrm{M_\odot\,pc^{-3}}. \qquad (4\text{-}41)$$

The luminosity density near the Sun is $j_V \simeq 0.067\,\mathrm{L_\odot\,pc^{-3}}$ (see Table 1-1). Thus the mass-to-light ratio in the solar neighborhood is

$$\Upsilon_V \equiv \frac{\rho_0}{j_V} \simeq 2.7\,\Upsilon_\odot. \qquad (4\text{-}42)$$

Since different stellar populations have different thicknesses and these thicknesses evolve (see §7.5), a more fundamental quantity is Υ_V averaged over a column through the disk. Unfortunately, the average can only be carried out to $z \approx 700\,\mathrm{pc}$ since the density distribution cannot yet be determined beyond this point. The total surface brightness is $I \simeq 15\,\mathrm{L_\odot\,pc^{-2}}$ (see Table 1-1 and §2.7), and is almost all contained within 700 pc of the plane. Taking Bahcall's surface density $\Sigma(700\,\mathrm{pc}) \simeq 75\,\mathrm{M_\odot\,pc^{-2}}$, we find

$$\Upsilon_V(700\,\mathrm{pc}) \equiv \frac{\Sigma(700\,\mathrm{pc})}{I} \simeq 5\,\Upsilon_\odot. \qquad (4\text{-}43)$$

This is a lower limit to the disk mass-to-light ratio since there may be more mass at $|z| > 700\,\mathrm{pc}$.

It is interesting to compare the surface density derived from equation (4-42) with the surface density Σ_{rot} that would be required to keep the circular speed in our Galaxy constant at $220\,\mathrm{km\,s^{-1}}$, as is found to be the case near the Sun (see §8-2 of MB). If we assume $R_0 = 8.5\,\mathrm{kpc}$, we have from equations (2-161) and (2-164) that

$$\Sigma_{\mathrm{rot}} = \frac{v_c^2}{2\pi G R_0} \simeq 210\,\mathrm{M_\odot\,pc^{-2}}. \qquad (4\text{-}44)$$

Since this value is almost three times as large as the estimates of $\Sigma(700\,\text{pc})$ obtained by Oort and Bahcall, it is natural to conclude that at least half the mass in a column through the Sun is contained in a component that has a total scale height much greater than 700 pc, and may be approximately spherical.

How does the Oort limit ρ_0 compare with estimates of the density of matter near the Sun that are obtained by counting stars and gas clouds? From star counts such as were described in Chapter 4 of MB, one estimates the mass of visible stars to be $\rho_\star \simeq 0.044\,\text{M}_\odot\,\text{pc}^{-3}$. To this must be added the mass of all stellar remnants, principally white dwarfs, that have formed in the lifetime of the Galaxy. Models of the star-forming history of the solar neighborhood (see §9.2) suggest that the density contributed by these remnants is $\rho_{\text{wd}} \simeq 0.028\,\text{M}_\odot\,\text{pc}^{-3}$. Finally, from Table 9-1 of MB we have that the gas near the Sun contributes $\rho_{\text{gas}} \simeq 0.042\,\text{M}_\odot\,\text{pc}^{-3}$. Hence the total mass density near the Sun cannot be less than

$$\rho_{\text{min}} = \rho_\star + \rho_{\text{wd}} + \rho_{\text{gas}} \simeq 0.114\,\text{M}_\odot\,\text{pc}^{-3}. \qquad (4\text{-}45)$$

Since the densities of stellar remnants and molecular gas are very uncertain, the observed constituents of the solar neighborhood could easily weigh as much as $\rho_{\text{min}} \simeq 0.14\,\text{M}_\odot\,\text{pc}^{-3}$. This is still significantly less than the dynamically determined mass estimate (4-41).

(c) The shape of Schwarzschild's velocity ellipsoid   Table 7-1 of MB shows that for most classes of stars in the solar neighborhood, the velocity dispersion in the radial direction is about twice as great as that in the tangential direction, and in §8-2 of MB this piece of information was used to estimate Oort's constant B. The formula (3-76) required for this important step has already been derived from an examination of the orbits of individual stars. In this subsection we rederive equation (3-76) from the collisionless Boltzmann equation.

The collisionless Boltzmann equation for a steady-state axisymmetric galaxy reads [cf. eq. (4-17)]

$$v_R\frac{\partial f}{\partial R} + v_z\frac{\partial f}{\partial z} + \left(\frac{v_\phi^2}{R} - \frac{\partial\Phi}{\partial R}\right)\frac{\partial f}{\partial v_R} - \frac{v_R v_\phi}{R}\frac{\partial f}{\partial v_\phi} - \frac{\partial\Phi}{\partial z}\frac{\partial f}{\partial v_z} = 0. \qquad (4\text{-}46)$$

If we multiply this equation by $v_R v_\phi$ and integrate over all velocities, we obtain in close analogy with the derivation of the Jeans equations

$$\frac{\partial(\nu\overline{v_R^2 v_\phi})}{\partial R} + \frac{\partial(\nu\overline{v_R v_z v_\phi})}{\partial z} - \frac{\nu}{R}\left(\overline{v_\phi^3} - \overline{v}_\phi R\frac{\partial\Phi}{\partial R}\right) + \frac{2\nu}{R}\overline{v_R^2 v_\phi} = 0. \qquad (4\text{-}47)$$

Now suppose that the velocity ellipsoid is always aligned so that one principal axis lies in the azimuthal direction. Then

$$\overline{v_R^2 \left(v_\phi - \overline{v}_\phi\right)} = 0 \quad \text{and} \quad \overline{v_R v_z \left(v_\phi - \overline{v}_\phi\right)} = 0, \qquad (4\text{-}48a)$$

which imply

$$\overline{v_R^2 v_\phi} = \overline{v_R^2}\,\overline{v}_\phi \quad \text{and} \quad \overline{v_R v_z v_\phi} = \overline{v_R v_z}\,\overline{v}_\phi. \qquad (4\text{-}48b)$$

If we use these relations in equation (4-47) and subtract $\overline{v}_\phi$ times the Jeans equation (4-29a) for a steady state galaxy, we obtain

$$\nu \overline{v_R^2}\frac{\partial \overline{v}_\phi}{\partial R} + \nu \overline{v_R v_z}\frac{\partial \overline{v}_\phi}{\partial z} - \frac{\nu}{R}\left(\overline{v_\phi^3} - \overline{v}_\phi \overline{v_\phi^2}\right) + \frac{\nu}{R}\overline{v_R^2}\,\overline{v}_\phi = 0. \qquad (4\text{-}49)$$

In the plane we have by symmetry that $\overline{v_R v_z} = 0$, and if the velocity ellipsoid is symmetric about $v_\phi = \overline{v}_\phi$, we have

$$0 = \overline{\left(v_\phi - \overline{v}_\phi\right)^3} = \overline{v_\phi^3} - 3\overline{v_\phi^2}\,\overline{v}_\phi + 2\overline{v}_\phi^3 = \left(\overline{v_\phi^3} - \overline{v}_\phi \overline{v_\phi^2}\right) - 2\overline{v}_\phi \overline{\left(v_\phi - \overline{v}_\phi\right)^2}, \qquad (4\text{-}50)$$

so eliminating $\left(\overline{v_\phi^3} - \overline{v}_\phi \overline{v_\phi^2}\right)$ from equation (4-49), and dividing the result through by ν, we find

$$\overline{v_R^2}\left(\frac{\partial \overline{v}_\phi}{\partial R} + \frac{\overline{v}_\phi}{R}\right) - \frac{2}{R}\overline{v}_\phi \overline{\left(v_\phi - \overline{v}_\phi\right)^2} = 0. \qquad (4\text{-}51)$$

If the asymmetric drift of the population under study is small, we have $\overline{v}_\phi \simeq v_c$, and with the defining equations (1-9) and (3-62) of Oort's constants A and B and the local epicycle frequency κ_0, equation (4-51) becomes

$$\frac{\sigma_\phi^2}{\sigma_R^2} \equiv \frac{\overline{\left(v_\phi - \overline{v}_\phi\right)^2}}{\overline{v_R^2}} = \frac{-B}{A - B} = \frac{\kappa_0^2}{4\Omega_0^2} = \frac{4B^2}{\kappa_0^2}, \qquad (4\text{-}52)$$

in agreement with equation (3-76). Notice that the approximation that the asymmetric drift is small is equivalent to the use of the epicycle equations, since by equation (4-34) we have that $v_a/v_c \approx \overline{v_R^2}/v_c^2$.

(d) Velocity dispersions in spherical systems   The applications just described used the cylindrical form of the Jeans equations to understand aspects of the structure of the solar neighborhood. Let us now use equation (4-30) to study the dynamics of spherical, or nearly spherical systems.

Consider a galaxy in which both the density and the velocity structures are invariant under rotations about the galactic center. Thus the galaxy does not rotate and

$$\overline{v_\theta^2} = \overline{v_\phi^2}. \tag{4-53a}$$

Hence the velocity ellipsoids are spheroids with their symmetry axes pointing to the galaxy center. We write

$$\beta \equiv 1 - \frac{\overline{v_\theta^2}}{\overline{v_r^2}}, \tag{4-53b}$$

where $\beta(r)$ describes the degree of anisotropy of the velocity distribution at each point. Numerical models of galaxies, and our experience of the high velocity stars in the solar neighborhood (see §7-2 of MB), suggest that $\overline{v_r^2} \geq \overline{v_\theta^2}$ and thus that $\beta \geq 0$.

With these assumptions equation (4-30) becomes

$$\frac{1}{\nu}\frac{d(\nu\overline{v_r^2})}{dr} + 2\frac{\beta\overline{v_r^2}}{r} = -\frac{d\Phi}{dr}. \tag{4-54}$$

Suppose we were able to measure $\overline{v_r^2}$, β, and ν as functions of radius for some stellar population in a spherical galaxy. Then, on setting $(d\Phi/dr) = GM(r)/r^2$, equation (4-54) would enable us to determine the mass $M(r)$ and the circular speed $v_c(r)$ through

$$v_c^2 = \frac{GM(r)}{r} = -\overline{v_r^2}\left(\frac{d\ln\nu}{d\ln r} + \frac{d\ln\overline{v_r^2}}{d\ln r} + 2\beta\right). \tag{4-55}$$

In principle, it is possible to use equation (4-55) to investigate the possibility that the large fraction of the galactic mass that Oort's analysis suggests is not in the disk, is contained in a nearly spherical component. The Galaxy contains a number of approximately spherical tracer populations such as metal-poor globular clusters and RR Lyrae stars. By careful analysis of number counts and line-of-sight velocities of these tracers, one can estimate $\nu(r)$, $\overline{v_r^2}(r)$, and $\beta(r)$. Unfortunately, there are a number of practical difficulties in carrying out this program. (i) Because equation (4-55) involves gradients of $\nu(r)$ and $\overline{v_r^2}$, the derived run of $M(r)$ will be very sensitive to noise in the data; (ii) it is extremely difficult to measure β at radii much larger than the solar radius, since the component of the transverse velocity along the line of sight is then very small; (iii) it is difficult to find a satisfactory tracer population— there are too few globular clusters, and most Population II stars are very faint. For the application of equation (4-55) to the globular cluster system, see Hartwick and Sargent (1978) and Frenk and White (1980). For an alternative approach to this problem, see §10.1.4.

Can we apply equation (4-55) to an external stellar system, taking ν to be the luminosity density? As external observers of a spherical galaxy, we measure only the surface brightness at each projected radius $I(R)$ and the line-of-sight velocity dispersion $\sigma_p(R)$. Since we can measure only two functions where equation (4-55) relates the mass $M(r)$ to the three undetermined functions $\overline{v_r^2}(r)$, $\beta(r)$, and $\nu(r)$, it is clear that we cannot hope to derive a unique mass model of an external galaxy from measurements of I and σ_p alone. However, if we assume a particular form for one of the free functions in the problem, we can derive a unique mass model from the observational data.

The simplest assumption to make is that the velocity ellipsoids are spherical throughout the galaxy, and therefore $\beta = 0$. Equation (4-55) then becomes

$$M(r) = -\frac{r\overline{v_r^2}}{G}\left(\frac{d\ln\nu}{d\ln r} + \frac{d\ln\overline{v_r^2}}{d\ln r}\right). \tag{4-56}$$

Simple geometry (Figure 2-3) shows that the luminosity density and the velocity dispersion $\overline{v_r^2}$ are related to $I(R)$ and $\sigma_v^2(R)$ by

$$I(R) = 2\int_R^\infty \frac{\nu r\,dr}{\sqrt{r^2 - R^2}} \quad ; \quad I(R)\sigma_p^2(R) = 2\int_R^\infty \frac{\nu\overline{v_r^2}r\,dr}{\sqrt{r^2 - R^2}}. \tag{4-57}$$

These are Abel integral equations for the quantities $\nu(r)$ and $\nu(r)\overline{v_r^2}$. Using the techniques described in Appendix 1.B.4, their solutions are found to be

$$\nu(r) = -\frac{1}{\pi}\int_r^\infty \frac{dI}{dR}\frac{dR}{\sqrt{R^2 - r^2}} \tag{4-58a}$$

$$\nu(r)\overline{v_r^2}(r) = -\frac{1}{\pi}\int_r^\infty \frac{d(I\sigma_p^2)}{dR}\frac{dR}{\sqrt{R^2 - r^2}}. \tag{4-58b}$$

With ν and $\overline{v_r^2}$ determined from equations (4-58), it is a simple matter to determine $M(r)$ from equation (4-56).

The left panel of Figure 4-3 shows plots of $-(d\ln\nu/d\ln r)$ and $-(d\ln\overline{v_r^2}/d\ln r)$ as functions of radius in the E1 galaxy M87, for the case $\beta = 0$. From the figure we see that the first term in the parentheses on the right of equation (4-56) is much more important than the second. The right panel of Figure 4-3 shows the mass per unit luminosity within radius r, $\Upsilon(r) = \left[M(r)/4\pi\int_0^r \nu r^2\,dr\right]$, which equation (4-56) predicts for this galaxy.

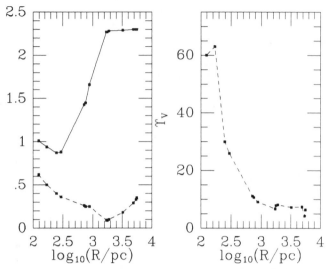

Figure 4-3. Left panel: plots of the logarithmic gradients appearing in equation (4-56) for the giant elliptical galaxy M87: full curve $|d\ln\nu/d\ln r|$; dashed curve $|d\ln\overline{v_r^2}/d\ln r|$. A distance to the galaxy of $16\,\mathrm{Mpc}$ has been assumed, and $\overline{v_r^2}$ has been obtained from the observed dispersion under the assumption $\beta = 0$. Right panel: the ratio $M(r)/L(r)$ derived from these data and equation (4-56). (After Sargent et al. 1978.)

We obtained equation (4-56) under the assumption that $\beta = 0$ at all radii. Unfortunately, there is no good theoretical or observational reason for adopting this assumption. So let us now investigate an alternative assumption, namely, that the mass density $\rho(r)$ is some constant multiple Υ of the luminosity density:

$$\rho(r) = \Upsilon\nu(r). \tag{4-59}$$

We shall see in §4.3 that the mass-to-light ratio Υ may be determined from the observations through the virial theorem [eq. (4-88)]. Furthermore, since the luminosity density $\nu(r)$ is still determined by the observed brightness profile $I(R)$ through equation (4-58a), the assumption (4-59) of a constant mass-to-light ratio determines a unique mass model of the galaxy. However, before we can take this mass model seriously, we have to ensure that it corresponds to physically realizable values of $\overline{v_r^2}$ and β. In particular, we have to ensure that at every radius in our model, $\beta(r) \leq 1$, and therefore by equation (4-53b) $\overline{v_\theta^2} \geq 0$.

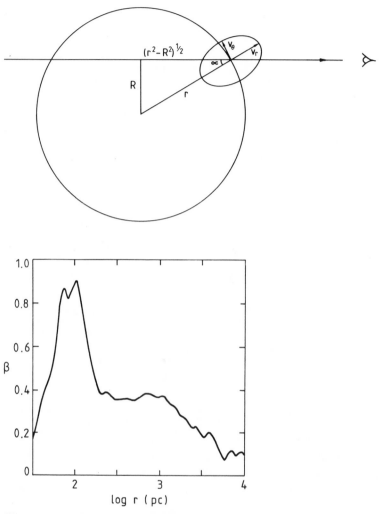

Figure 4-4. Top: the geometry involved in calculating the line-of-sight velocity dispersion at projected radius R in a spherical galaxy. Bottom: the radial variation of the anisotropy parameter β in M87 if the observations on which Figure 4-3 is based are to be interpreted in terms of a constant mass-to-light ratio. (After Binney & Mamon 1982.)

From Figure 4-4 it will be seen that when $\beta \neq 0$, the observed dispersion σ_v is related to $\overline{v_r^2}$ and β through

$$
\begin{aligned}
I(R)\sigma_p^2(R) &= 2\int_R^\infty \overline{(v_r\cos\alpha - v_\theta\sin\alpha)^2}\,\frac{\nu r\,dr}{\sqrt{r^2 - R^2}} \\
&= 2\int_R^\infty \left(\overline{v_r^2}\cos^2\alpha + \overline{v_\theta^2}\sin^2\alpha\right)\frac{\nu r\,dr}{\sqrt{r^2 - R^2}} \qquad (4\text{-}60) \\
&= 2\int_R^\infty \left(1 - \beta\frac{R^2}{r^2}\right)\frac{\nu\overline{v_r^2}r\,dr}{\sqrt{r^2 - R^2}}.
\end{aligned}
$$

If we now use equation (4-55) to eliminate β from this equation, we obtain

$$
I\sigma_p^2 = \int_R^\infty \left[2 + \frac{GMR^2}{r^3\overline{v_r^2}} + \frac{R^2}{r^2}\frac{d\ln\left(\nu\overline{v_r^2}\right)}{d\ln r}\right]\frac{\nu\overline{v_r^2}r\,dr}{\sqrt{r^2 - R^2}}, \qquad (4\text{-}61)
$$

which on rearrangement may be written

$$
I\sigma_p^2 - R^2\int_R^\infty \frac{\nu GM(r)dr}{r^2\sqrt{r^2 - R^2}} = R^2\int_R^\infty \left[2\nu\overline{v_r^2} + \frac{R^2}{r}\frac{d\left(\nu\overline{v_r^2}\right)}{dr}\right]\frac{r\,dr}{\sqrt{r^2 - R^2}}.
$$
$$
(4\text{-}62)
$$

The observations and our mass model completely determine the left side of this equation, while the right side is linear in the unknown quantity $\nu\overline{v_r^2}$. Hence equation (4-62) is a linear integro-differential equation for $\nu\overline{v_r^2}$. It may be solved analytically (Binney & Mamon 1982) and the resulting form of $\overline{v_r^2}$ used in equation (4-55) to evaluate $\beta(r)$. In Figure 4-4 we show the form of $\beta(r)$ that is required if we wish to interpret the same observations of M87 that led to Figure 4-3 in terms of a model with constant mass-to-light ratio. Since $0 < \beta < 1$ everywhere within the galaxy, we conclude that the model is self-consistent within the context of the Jeans equations. This model, together with the isotropic velocity dispersion model, shows that for spherical galaxies, radically different models can be consistent with both the Jeans equations and the observations. However, it has not been established that either model is consistent with the requirement that the DF be non-negative. This can be assured only by constructing a complete solution to the collisionless Boltzmann equation.

We can proceed in the way we have just described from a set of observations of a spherical system to particular mass models only when very high quality data are available. If the data are noisy, the functions $\beta(r)$ and $\overline{v_r^2}(r)$ that we obtain from the data by equations like (4-58) become very uncertain because the right sides of these equations involve derivatives of observed quantities. Therefore, it may be more useful to compare poor quality data with a few simple model galaxies than to build particular models around the data. In §4.4 we shall describe several models that derive from exact solutions of the collisionless Boltzmann equation. An alternative approach is to choose $\beta(r)$, $\nu(r)$, and $\rho(r)$ in some sensible way and then to solve equation (4-55) as a differential equation for $\overline{v_r^2}$. Models of this type are described by Bailey and MacDonald (1981).

(e) Spheroidal components with isotropic velocity dispersion

It is natural to expect a rotating galaxy to be flattened and this leads us to ask whether elliptical galaxies whose images are not circular are flattened systems and, if so, whether rotation is responsible for their flattening. In this subsection we investigate what we can learn from the Jeans equations about the second of these questions.

Earlier in this section we explained that in order to solve the Jeans equations for the dynamical structure of any system, we have to make some assumption about the form of the velocity-dispersion tensor $\boldsymbol{\sigma}^2$. The simplest assumption we can make is to suppose that $\boldsymbol{\sigma}^2$ takes the special **isotropic** form

$$\sigma_{ij}^2 = \overline{(v_i - \overline{v}_i)(v_j - \overline{v}_j)} = \sigma^2 \delta_{ij}. \tag{4-63}$$

In general, there is no reason to assume that the dispersion tensor is isotropic, but isotropic models of stellar systems provide useful points of reference against which observations of real galaxies can be compared. In particular, a system may be said to be flattened by rotation if its rotation velocities $\overline{v}_\phi(R, z)$ are similar to those of the isotropic model that has the same mass distribution.

The Jeans equations for a steady-state axisymmetric system in which $\boldsymbol{\sigma}^2$ is isotropic and the only streaming motion is in the azimuthal direction become [see eqs. (4-29)]

$$\frac{\partial(\nu\sigma^2)}{\partial R} - \nu\left(\frac{\overline{v}_\phi^2}{R} - \frac{\partial\Phi}{\partial R}\right) = 0, \tag{4-64a}$$

$$\frac{\partial(\nu\sigma^2)}{\partial z} + \nu\frac{\partial\Phi}{\partial z} = 0. \tag{4-64b}$$

These equations are formally identical with Euler's equation (1E-8) in a fluid of density ν and pressure $\nu\sigma^2$.

Suppose we know the forms $\nu(R, z)$ and $\Phi(R, z)$ of the luminosity density and the gravitational potential of some galaxy. Then, at any radius R we may integrate equation (4-64b) from $z = \infty$ to arbitrary z. Since $\nu(R, z) \to 0$ as $z \to \infty$, we obtain σ^2 as

$$\sigma^2(R, z) = \frac{1}{\nu} \int_z^\infty \nu \frac{\partial \Phi}{\partial z} dz. \tag{4-65}$$

Inserting this into equation (4-64a), we have that $\bar{v}_\phi$ is given by

$$\bar{v}_\phi^2(R, z) = R \frac{\partial \Phi}{\partial R} + \frac{R}{\nu} \frac{\partial}{\partial R} \int_z^\infty \nu \frac{\partial \Phi}{\partial z} dz. \tag{4-66}$$

As an illustration of the use of this formula, suppose Φ and ν are given by

$$\Phi = \Phi_L(R, z) \equiv \tfrac{1}{2} v_0^2 \ln \left(R_c^2 + R^2 + \frac{z^2}{q_\Phi^2} \right), \tag{4-67a}$$

$$\nu = \nu_L(R, z) \equiv \nu_0 R_c^2 \left(R_c^2 + R^2 + \frac{z^2}{q_\nu^2} \right)^{-1}. \tag{4-67b}$$

From equation (2-55b) we see that if $r \gg R_c$ and $1 \gg (1 - q_\Phi) \simeq 0.3(1 - q_\nu)$, the luminosity density (4-67b) is approximately proportional to the mass density that generates the potential (4-67a). We have

$$\int_z^\infty \nu_L \frac{\partial \Phi_L}{\partial z} dz$$

$$= \tfrac{1}{2} \nu_0 (R_c v_0 q_\nu)^2 \int_{z^2}^\infty \frac{dz^2}{[(R^2 + R_c^2)q_\Phi^2 + z^2][(R^2 + R_c^2)q_\nu^2 + z^2]}$$

$$= \frac{\tfrac{1}{2} \nu_0 R_c^2 v_0^2}{(R^2 + R_c^2)[(q_\Phi/q_\nu)^2 - 1]} \ln \left[\frac{(R^2 + R_c^2)q_\Phi^2 + z^2}{(R^2 + R_c^2)q_\nu^2 + z^2} \right].$$
$$\tag{4-68}$$

If we evaluate this equation at $z = 0$ and then insert the right side into equation (4-66), we obtain

$$\bar{v}_\phi^2(R, 0) = \frac{v_0^2 R^2}{R^2 + R_c^2} \left[1 - \frac{2 \ln(q_\Phi/q_\nu)}{(q_\Phi/q_\nu)^2 - 1} \right]. \tag{4-69}$$

If we write

$$\frac{q_\Phi}{q_\nu} = \frac{1 - \epsilon_\Phi}{1 - \epsilon_\nu} = 1 + (\epsilon_\nu - \epsilon_\Phi) + \dots, \tag{4-70}$$

where ϵ_ν and ϵ_Φ are the ellipticities of the isodensity and the isopotential surfaces, respectively, we may expand equation (4-69) in powers of $(\epsilon_\nu - \epsilon_\Phi)$ to find

$$\bar{v}_\phi^2(R, 0) = \frac{v_0^2 R^2}{R^2 + R_c^2} [(\epsilon_\nu - \epsilon_\Phi) + \dots]. \tag{4-71}$$

Comparison of this equation with equation (2-54c) shows that the rotation speed that is required to flatten the system described by equations (4-67) is a constant multiple $\sqrt{\epsilon_\nu - \epsilon_\Phi}$ of the circular speed $v_c(r)$ in the potential of the system. Furthermore, since for small ϵ_ν, $(\epsilon_\nu - \epsilon_\Phi) \simeq 0.7\epsilon_\nu$, the rotation speed that is required to flatten a typical stellar system to a small ellipticity ϵ_ν decreases only as the square root of ϵ_ν. Therefore even a small ellipticity requires an appreciable rotation speed for its support. For example, if the system is an E1 galaxy $(\epsilon_\nu = 0.1)$, $\bar{v}_\phi/v_c \simeq 0.26$. From equations (4-65) and (4-68) it may be shown that $\sigma \simeq v_c/1.4$ for this system, so $\bar{v}_\phi/\sigma \simeq 0.36$. We shall see in the next section that the rotation speeds of low-luminosity spheroidal components, especially the spheroids of disk galaxies, are comparable to the rotation speeds predicted by this type of analysis, while the rotation speeds of giant elliptical galaxies are a good deal smaller than those predicted by equation (4-66). This indicates that the velocity dispersion tensors $\boldsymbol{\sigma}^2$ of low-luminosity spheroidal components *may* be isotropic, but that in giant elliptical galaxies $\boldsymbol{\sigma}^2$ *must* be markedly anisotropic.

Equation (4-66) enables us to calculate the full velocity field of a galaxy of known potential and luminosity density, and not just the rotation velocity in the equatorial plane. Satoh (1980) has described such velocity fields for some of the potential-density pairs that were described in §2.2. He finds that $\bar{v}_\phi(R, z)$ decreases as $|z|$ increases, especially at large R. This is compatible with the observed velocity fields of many (but not all) spheroidal components (Kormendy & Illingworth 1982).

4.3 The Virial Equations

We obtained the Jeans equation (4-24a) by multiplying the collisionless Boltzmann equation by v_j and integrating over all velocities. In this process an equation in the six phase-space coordinates for a single scalar quantity f was reduced to three partial differential equations in the three spatial coordinates for ν and the velocity moments. We now multiply equation (4-24a) by x_k and integrate over all positions, thus converting these differential equations into a simple tensor equation relating global properties of the galaxy, such as total kinetic energy and mean-square streaming velocity.

We identify ν with the mass density ρ and multiply equation (4-24a) by x_k. Then integrating over the spatial variables, we find

$$\int x_k \frac{\partial(\rho\bar{v}_j)}{\partial t} d^3\mathbf{x} = -\int x_k \frac{\partial(\rho\overline{v_i v_j})}{\partial x_i} d^3\mathbf{x} - \int \rho x_k \frac{\partial\Phi}{\partial x_j} d^3\mathbf{x}. \qquad (4\text{-}72)$$

The second term on the right side of equation (4-72) is the potential energy tensor $\mathbf{W}$ that we introduced in §2.5 [see eq. (2-123)]. The first term on the right side of equation (4-72) can be rewritten with the aid of the divergence theorem:

$$\int x_k \frac{\partial(\rho \overline{v_i v_j})}{\partial x_i} d^3\mathbf{x} = -\int \delta_{ki}\rho\overline{v_i v_j}d^3\mathbf{x} = -2K_{kj}, \qquad (4\text{-}73a)$$

where we have assumed that ρ vanishes at large radii and have defined a new tensor, the **kinetic energy tensor K**, by

$$K_{jk} \equiv \tfrac{1}{2} \int \rho\overline{v_j v_k}d^3\mathbf{x}. \qquad (4\text{-}73b)$$

With the help of equation (4-26) we split $\mathbf{K}$ up into the contributions from ordered and random motion:

$$K_{jk} = T_{jk} + \tfrac{1}{2}\Pi_{jk}, \qquad (4\text{-}74a)$$

where

$$T_{jk} \equiv \tfrac{1}{2} \int \rho\overline{v}_j\overline{v}_k d^3\mathbf{x} \quad ; \quad \Pi_{jk} \equiv \int \rho\sigma_{jk}^2 d^3\mathbf{x}. \qquad (4\text{-}74b)$$

The derivative with respect to time in equation (4-72) may be taken outside the integral sign because x_k does not depend on time. Finally, averaging the (k,j) and the (j,k) components of equation (4-72), we obtain

$$\tfrac{1}{2}\frac{d}{dt} \int \rho\left(x_k\overline{v}_j + x_j\overline{v}_k\right)d^3\mathbf{x} = 2T_{jk} + \Pi_{jk} + W_{jk}. \qquad (4\text{-}75)$$

Here we have exploited the symmetry under exchange of indices of $\mathbf{T}$, $\mathbf{\Pi}$ [see eqs. (4-74)] and $\mathbf{W}$ [see eq. (2-126)].

The left side of equation (4-75) may be brought to a more intuitive form if we define the **moment of inertia tensor I** by

$$I_{jk} \equiv \int \rho x_j x_k d^3\mathbf{x}. \qquad (4\text{-}76)$$

Differentiating $\mathbf{I}$ with respect to time, we have

$$\tfrac{1}{2}\frac{dI_{jk}}{dt} = \tfrac{1}{2} \int \frac{\partial\rho}{\partial t}x_j x_k d^3\mathbf{x}. \qquad (4\text{-}77a)$$

With the continuity equation (4-21), the right side of this equation becomes

$$-\frac{1}{2}\int \frac{\partial(\rho\bar{v}_i)}{\partial x_i}x_j x_k d^3\mathbf{x} = \frac{1}{2}\int \rho\bar{v}_i\left(x_k\delta_{ji} + x_j\delta_{ki}\right)d^3\mathbf{x},$$

where the equality follows by an application of the divergence theorem. Substituting this expression back into equation (4-77a) yields

$$\frac{1}{2}\frac{dI_{jk}}{dt} = \frac{1}{2}\int \rho\left(\bar{v}_j x_k + \bar{v}_k x_j\right)d^3\mathbf{x}. \tag{4-77b}$$

We now combine equations (4-75) and (4-77b) to obtain the **tensor virial theorem**:

$$\frac{1}{2}\frac{d^2 I_{jk}}{dt^2} = 2T_{jk} + \Pi_{jk} + W_{jk}. \tag{4-78}$$

Equation (4-78) enables us to relate the gross kinematic and morphological properties of galaxies.[4] In many applications the left side is simply zero since $\mathbf{I}$ is time-independent.

The trace [see eq. (2-127)] of equation (4-78) is particularly interesting. Equations (4-74b) show that $\text{trace}(\mathbf{T}) + \frac{1}{2}\text{trace}(\mathbf{\Pi}) \equiv K$ is the total kinetic energy of the system, and in §2.5 we saw that $\text{trace}(\mathbf{W})$ is the system's total potential energy W. Thus, if the system is in a steady state, $\ddot{\mathbf{I}} = 0$, and equation (4-78) becomes

$$2K + W = 0. \tag{4-79}$$

Equation (4-79) is a statement of the **scalar virial theorem**.[5] The kinetic energy of a stellar system with mass M is just $K = \frac{1}{2}M\langle v^2\rangle$, where $\langle v^2\rangle$ is the mean-square speed of the system's stars. Hence the virial theorem states that

$$\langle v^2\rangle = \frac{|W|}{M} = \frac{GM}{r_g}, \tag{4-80a}$$

where r_g is the gravitational radius defined by equation (2-132). One often wishes to estimate $\langle v^2\rangle$ without going to the trouble of calculating r_g. A quantity that is very easy to calculate for any system is the **median**

---

[4] Equation (4-78) has here been derived from the collisionless Boltzmann equation, which is only valid for a collisionless system, but we shall find in Chapter 8 that an analogous result is valid for any system of N mutually gravitating particles. However, it should be noted that (4-78) applies only to *self-gravitating* systems. A similar, but less elegant, result may be derived for the case of a massive system embedded in an externally generated gravitational field; see Problem 4-14.

[5] First proved by R. Clausius in 1870. Clausius called the integral $\frac{1}{2}\int \rho\mathbf{x}\cdot\nabla\Phi\, d^3\mathbf{x}$ the "virial". This was one of his successful attempts at new notation. He also coined the term "entropy" and called $-\Phi$ the "ergal".

radius r_h, which is defined to be the radius within which lies half the system's mass. Spitzer (1969) has observed that in many simple stellar systems, the median radius r_h is related to r_g by $r_h \simeq 0.4 r_g$. Hence, a useful approximation is

$$\langle v^2 \rangle = \frac{|W|}{M} \simeq 0.4 \frac{GM}{r_h}. \tag{4-80b}$$

If E is the energy of the system, we have from equation (4-79) that

$$E = K + W = -K = \tfrac{1}{2}W. \tag{4-81}$$

Thus if a system forms by its material collecting together from a state of rest at infinity (in which state, $K = W = E = 0$), and then settles by any process into an equilibrium condition, it invests half of the gravitational energy that is released by the collapse in kinetic form, and in some way disposes of the other half in order to achieve a **binding energy** $E_b = -E$ equal to its kinetic energy. For example, suppose that our Galaxy formed by aggregating from an initial radius r_{in} that was much larger than its present characteristic radius, so that r_{in} may be taken to be infinite. Then, since most of the galactic material is now moving at about $v_c \simeq 220\,\mathrm{km\,s^{-1}}$, whether in the disk or on eccentric and highly inclined halo orbits, we have that $E_b = K \approx \tfrac{1}{2}M_g v_c^2$ of energy must have been released when the Galaxy formed, where M_g is the mass of the Galaxy. This argument suggests that as they form, galaxies may radiate a fraction $\tfrac{1}{2}(v_c/c)^2 \simeq 3 \times 10^{-7}$ of their rest-mass energy.

(a) Mass-to-light ratios of spherical systems   We may use equation (4-79) to evaluate the mass-to-light ratio Υ of a nonrotating spherical galaxy under the assumption that Υ is independent of radius. We choose a coordinate system in which the line of sight to the galaxy center coincides with the x-axis. Then the kinetic energy associated with motion in the x-direction is

$$K_{xx} = \tfrac{1}{2} \int \rho \overline{v_x^2} \, d^3\mathbf{x}. \tag{4-82}$$

With equation (4-59) this may be rewritten in terms of the luminosity density ν as

$$K_{xx} = \tfrac{1}{2}\Upsilon \int dy \int dz \int \nu v_x^2 \, dx. \tag{4-83}$$

The innermost integral in this expression yields the luminosity-weighted dispersion of the line-of-sight velocities. Hence K_{xx} may be expressed in terms of the surface brightness and the line-of-sight velocity dispersion σ_p as

$$K_{xx} = \tfrac{1}{2}\Upsilon \int \int I(y,z)\sigma_p^2(y,z)dydz. \tag{4-84}$$

Since the galaxy is assumed to be spherical and nonrotating, the total kinetic energy is

$$K = 3K_{xx} = \Upsilon J, \tag{4-85a}$$

where we have defined

$$J \equiv 3\pi \int_0^\infty I(R)\sigma_p^2(R)R\,dR. \tag{4-85b}$$

On the other hand, from equations (4-58a) and (4-59) we have

$$\rho(r) = -\frac{\Upsilon}{\pi} \int_r^\infty \frac{dI}{dR} \frac{dR}{\sqrt{R^2 - r^2}}. \tag{4-86}$$

If we use this relation in equation (2-130),[6] we obtain W as

$$W = \Upsilon^2 \widetilde{J}, \tag{4-87}$$

where $\widetilde{J}$ is an integral that depends only on $I(R)$. Using these results in equation (4-79), we obtain finally

$$\Upsilon = -\frac{2J}{\widetilde{J}}. \tag{4-88}$$

Thus Υ may be obtained from measurements of $I(R)$ and $\sigma_p(R)$. Once we have found Υ, we may solve for the radial and tangential components of velocity dispersion as was indicated in §4.2.1(d). For example, from observations of the E1 galaxy M87, one derives a visual-band mass-to-light ratio $\Upsilon_V \simeq 9.2h\,M_\odot/L_\odot$ [Binney & Mamon 1982, where h is defined in equation (1-12) and the distance to M87 is taken to be $13.2h^{-1}\,\mathrm{Mpc}$]. Table 4-2 shows that $\Upsilon = 9.2h$ is similar to the mass-to-light ratios that have been obtained from optical observations of the central regions of other elliptical galaxies (Schechter 1980; Michard 1980; Lauer 1985).

---

[6] Alternatively, W can be obtained from the strip brightness distribution; see Problem 2-10.

(b) Rotation of elliptical galaxies   The tensor virial theorem enables us to deduce valuable information about the internal motions of elliptical galaxies from a knowledge of their shapes and speeds of rotation. For example, consider an axisymmetric system that rotates about its symmetry axis (the z-axis) and is seen edge-on. We may assume that the line of sight to the center of the system coincides with the x-axis, and we have from the symmetry of the problem that

$$W_{xx} = W_{yy} \quad ; \quad W_{ij} = 0 \qquad (i \neq j), \tag{4-89}$$

with similar relations for $\mathbf{\Pi}$ and $\mathbf{T}$. With these assumptions the only independent, nontrivial virial equations [eqs. (4-78)] are

$$2T_{xx} + \Pi_{xx} + W_{xx} = 0 \quad ; \quad 2T_{zz} + \Pi_{zz} + W_{zz} = 0. \tag{4-90}$$

Dividing the first of these equations by the second, we obtain

$$\frac{2T_{xx} + \Pi_{xx}}{2T_{zz} + \Pi_{zz}} = \frac{W_{xx}}{W_{zz}}. \tag{4-91}$$

If the only streaming motion is rotation about the z-axis, $T_{zz} = 0$, and

$$2T_{xx} = \tfrac{1}{2} \int \rho \overline{v}_\phi^2 d^3\mathbf{x} = \tfrac{1}{2} M v_0^2, \tag{4-92}$$

where M is the mass of the system and v_0^2 is the mass-weighted mean-square rotation speed. We may also write

$$\Pi_{xx} = M\sigma_0^2, \tag{4-93}$$

where σ_0^2 is the mass-weighted mean-square random velocity along the line of sight to the galaxy, and

$$\Pi_{zz} \equiv (1 - \delta)\Pi_{xx} = (1 - \delta)M\sigma_0^2, \tag{4-94}$$

where $\delta < 1$ is a parameter that measures the anisotropy of the galaxy's velocity-dispersion tensor. With these definitions equation (4-91) may be written

$$\frac{v_0^2}{\sigma_0^2} = 2(1 - \delta)\frac{W_{xx}}{W_{zz}} - 2. \tag{4-95}$$

In §2.5 we noted that for a system whose isodensity surfaces are similar concentric ellipsoids, any ratio of terms like W_{xx}/W_{zz} depends only on the ellipticity ϵ of these surfaces, and not at all on the radial density structure. Thus equation (4-95), which is an exact equation, states that

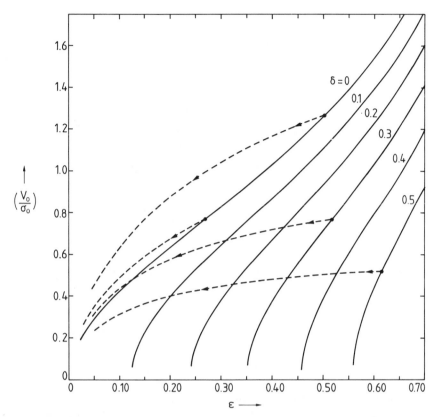

Figure 4-5. The relationship between the rotation parameter v/σ and ellipticity ϵ predicted by (4-95) for elliptical galaxies whose isodensity surfaces are similar coaxial oblate spheroids. The dashed curves show the movement of the point corresponding to the observable quantities $\tilde{v}/\tilde{\sigma}$ and $\tilde{\epsilon}_a$ when the galaxy's inclination angle i is decreased from $i = 90°$.

for such a system v_0/σ_0 depends on ϵ and δ only. Figure 4-5 shows this relationship in the $(v/\sigma, \epsilon)$ plane. Along each curve the anisotropy parameter δ is constant. Notice that the ellipticity of the isodensity surfaces is almost independent of v_0/σ_0 when the latter is small.

In practice we cannot expect to view a given elliptical galaxy from its equatorial plane, so consider how the apparent rotation speed and the apparent ellipticity of a galaxy will vary as the inclination i of the galaxy is slowly changed from edge-on ($i = 90°$). At each point within the galaxy, the line-of-sight component of the rotation velocity becomes a fraction $\sin i$ of its value at edge-on orientation. Therefore, the *apparent*

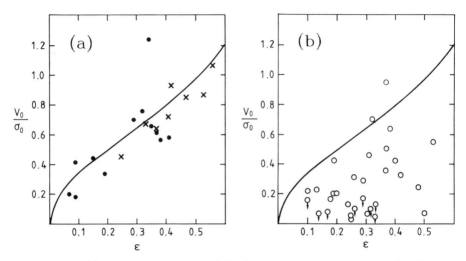

Figure 4-6. (a) The positions in the $(v/\sigma, \epsilon)$ plane of elliptical galaxies (dots), and of spheroids (crosses), that have luminosities smaller than $L = 2.5 \times 10^{10}\,\mathrm{L}_\odot$. (b) The same as (a) but for elliptical galaxies brighter than $L = 2.5 \times 10^{10}\,\mathrm{L}_\odot$. (After Davies et al. 1983.)

mass-weighted RMS rotation velocity at inclination i is

$$\tilde{v}(i) = v_0 \sin i. \tag{4-96a}$$

Similarly, the mean-square random velocity along the line of sight varies as

$$\tilde{\sigma}^2(i) = \sigma_0^2 \sin^2 i + (1 - \delta)\sigma_0^2 \cos^2 i = \sigma_0^2(1 - \delta \cos^2 i). \tag{4-96b}$$

On the other hand, if the true axial ratio is $(1 - \epsilon_t)$ and the apparent axial ratio is $(1 - \epsilon_a)$, we have by equation (5-21) of MB that

$$(1 - \epsilon_a)^2 = (1 - \epsilon_t)^2 \sin^2 i + \cos^2 i, \tag{4-97a}$$

and therefore that

$$\epsilon_a(2 - \epsilon_a) = \epsilon_t(2 - \epsilon_t)\sin^2 i. \tag{4-97b}$$

Figure 4-5 shows how the point with coordinates $[\tilde{v}(i)/\tilde{\sigma}(i), \epsilon_a(i)]$ moves in the $(v/\sigma, \epsilon)$ plane as i is varied for a number of different values of δ and ϵ_t. Notice that when $\delta = 0$, these trajectories run almost parallel to the curve along which $\delta = 0$. Therefore, the representative points in the $(v/\sigma, \epsilon)$ plane of galaxies that are flattened by rotation lie close to the curve $\delta = 0$, independent of the inclination i. This is fortunate, since we often have no way of estimating i for an elliptical galaxy.

From high-quality spectra at a grid of points covering the surface of an elliptical galaxy, it is in principle possible to calculate $\tilde{v}$ and $\tilde{\sigma}$ exactly. In practice the data are less complete than this and a number of assumptions and approximations are involved in estimates of these quantities (Binney 1978, Illingworth 1981). Furthermore, in real galaxies the ellipticity often varies with radius (see §5-2 of MB), so that before one can plot a galaxy in the $(v/\sigma, \epsilon)$ plane one must first choose a characteristic value of ϵ_a. However, these approximations notwithstanding, Figure 4-6a, which shows the position of 12 low-luminosity elliptical galaxies in the $(v/\sigma, \epsilon)$ plane, indicates that for these systems $(\tilde{v}/\tilde{\sigma})$ and ϵ_a are related in much the same way as is predicted by our simple theory in the case $\delta \simeq 0$. Figure 4-6b displays $(\tilde{v}/\tilde{\sigma})$ and ϵ_a for a sample of 31 giant elliptical galaxies. For these galaxies, by contrast, $(\tilde{v}/\tilde{\sigma})$ and ϵ_a are not related in this way; indeed, they are not even well-correlated. Evidently, giant elliptical galaxies are not oblate spheroidal bodies in which $\delta \simeq 0$. The anisotropy of the velocity-dispersion tensors of giant ellipticals that is demonstrated by Figure 4-6b may be an important clue to understanding how these systems formed.

The tensor virial theorem can be similarly applied to triaxial galaxies; see Problem 4-15.

4.4 The Jeans Theorems and Spherical Systems

We remarked above that in the general case, the collisionless Boltzmann equation (4-13) cannot be solved because it involves no less than seven independent variables. In the preceding two sections, we reduced the complexity of the problem posed by this equation by taking moments; in this way we obtained information about general properties of solutions of equation (4-13) without actually recovering any solutions. In this section we obtain certain exact solutions of the collisionless Boltzmann equation. These solutions are important because they develop our understanding of the nature of solutions of the collisionless Boltzmann equation. However, we shall find that the exact solutions of this section usually describe only a restricted subset of all possible stellar-dynamical equilibria. By contrast, "models" based on the Jeans or the virial equations describe a much wider range of systems, but we have no guarantee that a model of this type corresponds to any realizable equilibrium configuration, because we cannot be sure of the validity of the assumption about the form of the velocity-dispersion tensor on which it is based. Thus the exact solutions of this section complement the treatment of the previous two sections.

1 Jeans Theorems

In §3.1.1 we introduced the concept of an integral of motion in a given potential $\Phi(\mathbf{x})$. According to equation (3-38), a function of the phase-space coordinates $I(\mathbf{x}, \mathbf{v})$ is an integral if and only if

$$\frac{d}{dt} I[\mathbf{x}(t), \mathbf{v}(t)] = 0 \qquad (4\text{-}98)$$

along all orbits. With the equations of motion this becomes

$$\frac{dI}{dt} = \boldsymbol{\nabla} I \cdot \frac{d\mathbf{x}}{dt} + \frac{\partial I}{\partial \mathbf{v}} \cdot \frac{d\mathbf{v}}{dt} = 0, \quad \text{or} \quad \mathbf{v} \cdot \boldsymbol{\nabla} I - \boldsymbol{\nabla} \Phi \cdot \frac{\partial I}{\partial \mathbf{v}} = 0. \qquad (4\text{-}99)$$

Comparing this with equation (4-13c), we see that the condition for I to be an integral is identical with the condition for I to be a steady-state solution of the collisionless Boltzmann equation. This leads to the following theorem.

Jeans Theorem *Any steady-state solution of the collisionless Boltzmann equation depends on the phase-space coordinates only through integrals of motion in the galactic potential, and any function of the integrals yields a steady-state solution of the collisionless Boltzmann equation.*

Proof: Suppose f is a steady-state solution of the collisionless Boltzmann equation. Then, as we have just seen, f *is* an integral, and the first part of the theorem is proved. Conversely, if I_1 to I_n are n integrals, and if f is any function of n variables, then

$$\frac{d}{dt} f[I_1(\mathbf{x}, \mathbf{v}), \ldots, I_n(\mathbf{x}, \mathbf{v})] = \sum_{m=1}^{n} \frac{\partial f}{\partial I_n} \frac{dI_n}{dt} = 0 \qquad (4\text{-}100)$$

and f is seen to satisfy the collisionless Boltzmann equation.◁

Much of the work in this section will be based on the second proposition stated by the Jeans theorem, namely, that any function of integrals solves the collisionless Boltzmann equation. However, the first of Jeans's propositions, the assurance that the DF of any steady-state galaxy must be a function of integrals, is not at all helpful since the examples discussed in §3.2, §3.3, and §3.4 indicate that we rarely know the forms of more than three integrals, while, as we saw in §3.1, orbits in any time-independent potential necessarily admit five independent integrals. Thus the first proposition of the Jeans theorem simply tells us that the DF, the form of which we do not know, is a function of at least two integrals whose form is likewise unknown to us.

Fortunately the time averages theorem of §3.5.2(a) can be used to show that if all orbits in a galaxy are regular, we can forget about any non-isolating integrals. This leads to:

Strong Jeans Theorem *The* DF *of a steady-state galaxy in which almost all orbits are regular with incommensurable frequencies may be presumed to be a function only of three independent isolating integrals.*

The proof of this theorem will be found in Appendix 4.A.

In summary, the Jeans theorem tells us that if $I_1, \ldots, I_5$ are five independent integrals in a given potential, then any DFs of the forms $f(I_1)$, $f(I_1, I_2), \ldots f(I_1, \ldots, I_5)$ are solutions of the collisionless Boltzmann equation. The strong Jeans theorem tells us that if the potential is regular, for all practical purposes any time-independent galaxy may be represented by a solution of the form $f(I_1, I_2, I_3)$, where $I_1, \ldots, I_3$ are any three independent isolating integrals.

2 Jeans Theorems Applied to Spherical Systems

In §3.1 we saw that any orbit in a spherical potential necessarily admits four isolating integrals, namely, the energy E and the three components of the angular momentum vector $\mathbf{L}$. Thus, by the Jeans theorem, any non-negative function of these integrals can serve as the DF of a spherical stellar system. On the other hand, we cannot immediately apply the strong Jeans theorem, because, as we saw in §3.5, two of the three frequencies ω_k of any orbit in a spherical potential are always equal—which, of course, is why there are four, not three, isolating integrals. However, there is a simple extension of the strong Jeans theorem to spherical systems (Lynden-Bell 1962b), which permits us to conclude that the DF of any[7] steady-state spherical system can be expressed as a function $f(E, \mathbf{L})$. If the system is spherically symmetric in all its properties,[8] f cannot depend on the direction of $\mathbf{L}$ but only on its magnitude L, and we have that $f = f(E, L)$.

The most interesting and important case is when the stellar system itself provides the potential Φ. If we regard f as the mass DF, then we have

$$\nabla^2 \Phi = 4\pi G\rho = 4\pi G \int f d^3\mathbf{v}, \qquad (4\text{-}101\text{a})$$

---

[7] We exclude potentials such as the Keplerian potential $\Phi \propto 1/r$ in which all the frequencies ω_k are commensurable.

[8] There is no logical reason why the spatial distribution of the stars should not be spherical, while their velocity distribution is not. For example, a spherical system can rotate about a particular axis (Lynden-Bell 1960). However, it is unlikely that this type of model is of practical importance.

or, using spherical symmetry,

$$\frac{1}{r^2}\frac{d}{dr}\left(r^2\frac{d\Phi}{dr}\right) = 4\pi G \int f\left(\tfrac{1}{2}v^2 + \Phi, |\mathbf{r}\times\mathbf{v}|\right)d^3\mathbf{v}. \qquad (4\text{-}101\text{b})$$

Equation (4-101b) is the fundamental equation governing spherical equilibrium stellar systems.

It is convenient for many of the calculations that follow to define a new gravitational potential and a new energy. If Φ_0 is some constant, then let the **relative potential** Ψ and the **relative energy** $\mathcal{E}$ of a star be defined by

$$\Psi \equiv -\Phi + \Phi_0 \quad \text{and} \quad \mathcal{E} \equiv -E + \Phi_0 = \Psi - \tfrac{1}{2}v^2. \qquad (4\text{-}102)$$

In practice, we generally choose Φ_0 to be such that $f > 0$ for $\mathcal{E} > 0$ and $f = 0$ for $\mathcal{E} \leq 0$. The relative potential of an isolated system satisfies Poisson's equation in the form $\nabla^2\Psi = -4\pi G\rho$, subject to the boundary condition $\Psi \to \Phi_0$ as $|\mathbf{x}| \to \infty$.

3 Systems with Isotropic Dispersion Tensors

The simplest spherical models are those with DFs that depend on $\mathcal{E}$ only; $f = f(\Psi - \tfrac{1}{2}v^2)$. In these systems the velocity dispersion $\overline{v_r^2}$ in the radial direction is given by

$$\overline{v_r^2} = \frac{1}{\rho}\int dv_r dv_\theta dv_\phi v_r^2 f\left[\Psi - \tfrac{1}{2}\left(v_r^2 + v_\theta^2 + v_\phi^2\right)\right], \qquad (4\text{-}103\text{a})$$

where the integral extends over all velocities. Similarly, one of the tangential components of velocity dispersion, for example $\overline{v_\theta^2}$, is given by

$$\overline{v_\theta^2} = \frac{1}{\rho}\int dv_r dv_\theta dv_\phi v_\theta^2 f\left[\Psi - \tfrac{1}{2}\left(v_r^2 + v_\theta^2 + v_\phi^2\right)\right]. \qquad (4\text{-}103\text{b})$$

Since equations (4-103) differ only in the labeling of one of the variables of integration, we see that $\overline{v_r^2} = \overline{v_\theta^2} = \overline{v_\phi^2}$, and the velocity-dispersion tensor is everywhere isotropic. As we shall see later in this section, when f depends on L as well as on $\mathcal{E}$, $\overline{v_\theta^2} = \overline{v_\phi^2} \neq \overline{v_r^2}$. Thus the essential distinction between systems whose DFs are functions of $\mathcal{E}$ only, and those for which $f = f(\mathcal{E}, L)$, is that the former have isotropic velocity-dispersion tensors and the latter do not.

When $f = f(\mathcal{E})$ and we have chosen the constant Φ_0 involved in the definition (4-102) of Ψ such that $f(\mathcal{E}) = 0$ for $\mathcal{E} < 0$, equation (4-101b) becomes

$$
\begin{aligned}
\frac{1}{r^2}\frac{d}{dr}\left(r^2\frac{d\Psi}{dr}\right) &= -16\pi^2 G \int_0^{\sqrt{2\Psi}} f\left(\Psi - \tfrac{1}{2}v^2\right) v^2 dv \\
&= -16\pi^2 G \int_0^{\Psi} f(\mathcal{E})\sqrt{2(\Psi - \mathcal{E})}d\mathcal{E}.
\end{aligned}
\tag{4-104}
$$

This equation may be regarded either as a nonlinear equation for $\Psi(r)$ given $f(\mathcal{E})$, or as a linear equation for $f(\mathcal{E})$ given $\Psi(r)$. We discuss each of these approaches in turn. We start by adopting simple forms for f and solving for Ψ (and hence ρ).

(a) Polytropes and Plummer's model   A very simple form for the DF f is

$$
f(\mathcal{E}) = \begin{cases} F\mathcal{E}^{n-\frac{3}{2}}, & (\mathcal{E} > 0); \\ 0, & (\mathcal{E} \le 0), \end{cases}
\tag{4-105}
$$

where once again $\mathcal{E} = \Psi - \tfrac{1}{2}v^2$ is the relative energy. With this form of f we have for the density ρ at radii where $\Psi > 0$

$$
\rho = 4\pi \int_0^\infty f(\Psi - \tfrac{1}{2}v^2)v^2 dv = 4\pi F \int_0^{\sqrt{2\Psi}} \left(\Psi - \tfrac{1}{2}v^2\right)^{n-\frac{3}{2}} v^2 dv.
\tag{4-106}
$$

If we make the substitution $v^2 = 2\Psi\cos^2\theta$, this becomes

$$
\rho = c_n \Psi^n \qquad (\Psi > 0),
\tag{4-107a}
$$

where

$$
c_n \equiv 2^{\frac{7}{2}} F\pi\left[\int_0^{\frac{\pi}{2}} \sin^{2n-2}\theta d\theta - \int_0^{\frac{\pi}{2}} \sin^{2n}\theta d\theta\right] = \frac{(2\pi)^{\frac{3}{2}}(n-\frac{3}{2})!F}{n!}.
\tag{4-107b}
$$

If c_n is to be finite, we must have $n > \tfrac{1}{2}$.

In these models the density rises as the nth power of the relative potential when $\Psi > 0$ and is, of course, zero when $\Psi \le 0$. No polytropic stellar system is homogeneous, for this would correspond to $\rho \propto \Psi^0$ or $n = 0$, which would violate the constraint $n > \tfrac{1}{2}$.

When we use equation (4-107a) to eliminate ρ from Poisson's equation, we find

$$
\frac{1}{r^2}\frac{d}{dr}\left(r^2\frac{d\Psi}{dr}\right) + 4\pi G c_n \Psi^n = 0.
\tag{4-108a}
$$

If we eliminate r and Ψ from equation (4-108a) in favor of the rescaled radial variables,

$$s \equiv \frac{r}{b} \quad \text{and} \quad \psi \equiv \frac{\Psi}{\Psi_0}, \quad \text{where} \quad b \equiv (4\pi G \Psi_0^{n-1} c_n)^{-\frac{1}{2}} \quad \text{(4-108b)}$$

and $\Psi_0 = \Psi(0)$, then equation (4-108a) takes the simple form

$$\frac{1}{s^2} \frac{d}{ds} \left(s^2 \frac{d\psi}{ds} \right) = \begin{cases} -\psi^n, & \psi > 0; \\ 0 & \psi \le 0. \end{cases} \quad \text{(4-108c)}$$

Equation (4-108c) is known as the **Lane-Emden equation** after H. Lane and R. Emden who studied it in connection with self-gravitating polytropic gas spheres. The natural boundary conditions to impose on equation (4-108) at $r = 0$ are: (i) $\psi = 1$ by definition; (ii) $(d\psi/ds) = 0$, since there should be no gravitational force at the center of a self-gravitating star cluster.

Polytropic gases have an equation of state $p = K\rho^\gamma$, where K is a constant [cf. eq. (1E-27)]. Thus the equation of hydrostatic equilibrium for a self-gravitating sphere of polytropic gas [cf. eq. (1E-8)],

$$\frac{dp}{dr} = -\rho \frac{d\Phi}{dr}, \quad \text{(4-109a)}$$

becomes

$$K\gamma\rho^{\gamma-2} \frac{d\rho}{dr} = \frac{d\Psi}{dr}. \quad \text{(4-109b)}$$

If we set the constant involved in the definition of Ψ such that $\Psi = 0$ on the edge of the system, equation (4-109b) yields on integration

$$\rho^{\gamma-1} = \frac{\gamma-1}{K\gamma} \Psi. \quad \text{(4-110)}$$

Equation (4-110) is the same as equation (4-107a) with c_n replaced by $[(\gamma-1)/K\gamma]^{1/(\gamma-1)}$ and $\gamma = 1 + \frac{1}{n}$. Hence *the density distribution of a stellar polytrope of index n is the same as that of a polytropic gas sphere with $\gamma = 1 + \frac{1}{n}$.* A full account of gaseous polytropes will be found in Chandrasekhar (1939).

For general n, (4-108c) cannot be solved in terms of elementary functions. However, there are two special cases for which simple analytical solutions are available. (i) When $n = 1$, equation (4-108c) becomes the linear Helmholtz equation familiar from the theory of spherical waves (see Problem 4-16); and (ii) when $n = 5$, we obtain a model discovered by Schuster (1883) that is worth describing in some detail because it provides the simplest plausible model of a self-consistent stellar system.

Consider the function

$$\psi = \frac{1}{\sqrt{1 + \frac{1}{3}s^2}}. \tag{4-111}$$

Differentiating with respect to s we find that

$$\frac{1}{s^2}\frac{d}{ds}\left(s^2\frac{d\psi}{ds}\right) = -\frac{1}{3s^2}\frac{d}{ds}\left[\frac{s^3}{(1 + \frac{1}{3}s^2)^{\frac{3}{2}}}\right] = -\frac{1}{(1 + \frac{1}{3}s^2)^{\frac{5}{2}}} = -\psi^5. \tag{4-112}$$

Therefore ψ is a solution of equation (4-108c) with $n = 5$. Since ψ also satisfies the central boundary conditions, it represents a physically acceptable potential. According to equation (4-107a), the corresponding density is

$$\rho = c_5\psi^5 = \frac{c_5\Psi_0^5}{(1 + \frac{1}{3}s^2)^{\frac{5}{2}}}. \tag{4-113}$$

Notice that the density is everywhere non-zero. The total mass is finite, however, with value

$$M_\infty = \frac{1}{G}\left(r^2\frac{d\Phi}{dr}\right)_{r\to\infty} = -\frac{b}{G}\left(s^2\frac{d\Psi}{ds}\right)_{s\to\infty} = \frac{\sqrt{3}\Psi_0 b}{G}. \tag{4-114a}$$

Plummer (1911) showed that the density distribution given by equation (4-113) provides a moderately good fit to the observations of some globular clusters, and hence equation (4-113) is known as **Plummer's law**. This distribution fails as a model of elliptical galaxies because its density falls too rapidly in the outer parts; from equation (4-113) we see that at large radii $\rho \propto r^{-5}$, whereas in §5-2 of MB it was shown that the outer envelopes of elliptical galaxies fall off less steeply than $\rho \propto r^{-4}$.

In general, the extent of the outer parts of a polytropic model increases with n; for $n < 5$ the density goes to zero at a finite radius, for $n = 5$ the density is non-zero everywhere but the total mass is finite, and for $n > 5$ the density falls off so slowly at large r that the mass is infinite. However, rather than pursuing the investigation of models with finite n, we now leap from $n = 5$ to the extreme case of infinite polytropic index. For the latter model turns out to have a special significance that sets it aside from the polytropic models of finite index—it is in fact the **isothermal sphere**.

(b) The isothermal sphere   We have just seen that to every polytropic gas sphere with adiabatic index $\gamma < 3$, there corresponds a stellar-
dynamical polytrope with index $n = 1/(\gamma - 1)$. Thus stellar polytropes
with large n correspond to gaseous polytropes for which $\gamma \simeq 1$. Hence
in the limit $n \to \infty$, the corresponding gaseous system has $\gamma = 1$, which
implies that $p = K\rho$. This is the equation of state of an isothermal
body of gas. As $n \to \infty$ the Lane-Emden equation (4-108c) ceases to
be well-defined, but we may derive the equation for the structure of the
$n = \infty$ polytrope by consideration of the equation for the hydrostatic
balance of a self-gravitating isothermal sphere of ideal gas.

The equation of hydrostatic support of an isothermal gas reads

$$\frac{dp}{dr} = \frac{k_B T}{m}\frac{d\rho}{dr} = -\rho\frac{GM(r)}{r^2}, \tag{4-115a}$$

where k_B is Boltzmann's constant, p and T are the pressure and temperature of the gas, and m is the mass per particle. $M(r)$ is the total
mass interior to radius r. Multiplying equation (4-115a) through by
$(r^2 m/\rho k_B T)$ and then differentiating with respect to r, we obtain

$$\frac{d}{dr}\left(r^2\frac{d\ln\rho}{dr}\right) = -\frac{Gm}{k_B T}4\pi r^2\rho, \tag{4-115b}$$

where we have used the relationship $(dM/dr) = 4\pi r^2\rho$.

Now suppose we have a stellar-dynamical system whose DF f is

$$f(\mathcal{E}) = \frac{\rho_1}{(2\pi\sigma^2)^{\frac{3}{2}}}e^{\mathcal{E}/\sigma^2} = \frac{\rho_1}{(2\pi\sigma^2)^{\frac{3}{2}}}\exp\left(\frac{\Psi - \frac{1}{2}v^2}{\sigma^2}\right). \tag{4-116}$$

Then, integrating over all velocities, we find

$$\rho = \rho_1 e^{\Psi/\sigma^2}. \tag{4-117}$$

Poisson's equation for this system reads

$$\frac{1}{r^2}\frac{d}{dr}\left(r^2\frac{d\Psi}{dr}\right) = -4\pi G\rho, \tag{4-118}$$

or, with equation (4-117)

$$\frac{d}{dr}\left(r^2\frac{d\ln\rho}{dr}\right) = -\frac{4\pi G}{\sigma^2}r^2\rho. \tag{4-119a}$$

For future reference, note that if we eliminate ρ rather than Ψ between equations (4-117) and (4-118), we obtain

$$\frac{d}{dr}\left(r^2\frac{d\Psi}{dr}\right) = -4\pi G\rho_1 r^2 e^{\Psi/\sigma^2}. \tag{4-119b}$$

Comparison of equations (4-115b) and (4-119a) shows that if we set

$$\sigma^2 = \frac{k_B T}{m}, \tag{4-120}$$

these equations become identical. Therefore *the structure of an isothermal self-gravitating sphere of gas is identical with the structure of a collisionless system of stars whose density in phase space $f(\mathcal{E})$ is given by equation (4-116).*

A little thought shows why there is this correspondence between the gaseous and stellar-dynamical isothermal spheres. The distribution of velocities at each point in the stellar-dynamical isothermal sphere is the Maxwellian distribution

$$F(v) = Ne^{-\frac{1}{2}v^2/\sigma^2}. \tag{4-121}$$

However, kinetic theory (e.g., Jeans 1940) tells us that this is also the equilibrium Maxwell-Boltzmann distribution which would obtain if the stars were allowed to bounce elastically off each other like the molecules of a gas. Therefore, if the DF of a system is given by equation (4-116), it is a matter of indifference whether the particles of the system collide with one another or not.

Notice that the correspondence between a gaseous polytrope with $\gamma > 1$ and the corresponding stellar-dynamical model is not as close as that between the two isothermal systems; the two kinds of polytrope do *not* have similar DFs, and a stellar polytrope would be drastically altered if elastic collisions were allowed to occur between its stars.

The mean-square speed of the stars at a point in the isothermal sphere is

$$\overline{v^2} = \frac{\displaystyle\int_0^\infty \exp\left(\frac{\Psi - \frac{1}{2}v^2}{\sigma^2}\right)v^4 dv}{\displaystyle\int_0^\infty \exp\left(\frac{\Psi - \frac{1}{2}v^2}{\sigma^2}\right)v^2 dv} = 2\sigma^2\frac{\int_0^\infty e^{-x^2}x^4 dx}{\int_0^\infty e^{-x^2}x^2 dx} = 3\sigma^2. \tag{4-122}$$

Thus $\overline{v^2}$ is independent of position. It is straightforward to verify that the dispersion in any one component of velocity, for example $(\overline{v_r^2})^{1/2}$, is equal to σ.

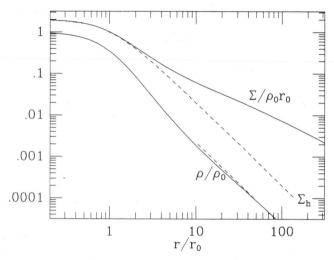

Figure 4-7. Volume (ρ/ρ_0) and projected $(\Sigma/\rho_0 r_0)$ mass densities of the isothermal sphere. The dashed line at bottom right shows the density profile of the singular isothermal sphere. The dashed curve labeled Σ_h shows the surface density of the modified Hubble law (4-128).

It is easy to find one solution of equation (4-119a). If we set $\rho = Cr^{-b}$, the left side of the equation is found to equal $-b$, while the right side equals $-(4\pi G/\sigma^2)Cr^{2-b}$. Therefore, we must set $b = 2$ and $C = (\sigma^2/2\pi G)$, which yields

$$\rho(r) = \frac{\sigma^2}{2\pi Gr^2}. \tag{4-123}$$

This solution describes a model known as the **singular isothermal sphere**.

Unfortunately, the singular isothermal sphere has infinite density at $r = 0$. To obtain a solution that is well behaved at the origin, it is convenient to define new dimensionless variables $\tilde{\rho}$ and $\tilde{r}$ to replace ρ and r in equations (4-119). Let $\tilde{\rho}$ and $\tilde{r}$ be defined in terms of the central density ρ_0 and the **King radius** r_0 by

$$\tilde{\rho} \equiv \frac{\rho}{\rho_0} \quad \text{and} \quad \tilde{r} \equiv \frac{r}{r_0}, \tag{4-124a}$$

where

$$r_0 \equiv \sqrt{\frac{9\sigma^2}{4\pi G\rho_0}}. \tag{4-124b}$$

We shall find that r_0 is the radius at which the projected density of the isothermal sphere falls to roughly half (in fact, 0.5013) of its central

Table 4-1. Density and projected density of the isothermal sphere

$\dfrac{r}{r_0}$	$\log\left(\dfrac{\rho}{\rho_0}\right)$	$\log\left(\dfrac{\Sigma}{r_0\rho_0}\right)$	$\dfrac{r}{r_0}$	$\log\left(\dfrac{\rho}{\rho_0}\right)$	$\log\left(\dfrac{\Sigma}{r_0\rho_0}\right)$
0.0	0.0	0.3049	10.0	−2.7291	−1.2154
0.1	−0.0065	0.3007	20.0	−3.3217	−1.4933
0.2	−0.0256	0.2880	30.0	−3.6489	−1.6511
0.3	−0.0564	0.2676	50.0	−4.0592	−1.8558
0.5	−0.1469	0.2081	70.0	−4.3346	−1.9970
0.7	−0.2640	0.1319	100.0	−4.6337	−2.1532
1.0	−0.4620	0.0053	200.0	−5.2347	−2.4750
2.0	−1.0716	−0.3648	300.0	−5.5939	−2.6732
3.0	−1.5077	−0.6093	500.0	−6.0480	−2.9363
5.0	−2.0560	−0.8931	700.0	−6.3457	−3.1242
7.0	−2.3946	−1.0576	1000.0	−6.6591	−3.3503

value, and because of this some authors call r_0 the core radius in analogy with the usual observational definition (MB p. 307). In terms of our new variables, equations (4-119) become

$$\frac{d}{d\tilde{r}}\left(\tilde{r}^2\frac{d\ln\tilde{\rho}}{d\tilde{r}}\right) = -9\tilde{r}^2\tilde{\rho} \tag{4-125a}$$

or

$$\frac{d}{d\tilde{r}}\left[\tilde{r}^2\frac{d(\Psi/\sigma^2)}{d\tilde{r}}\right] = -9\tilde{r}^2\exp\left[\frac{\Psi(r) - \Psi(0)}{\sigma^2}\right]. \tag{4-125b}$$

In Figure 4-7 we show the function $\tilde{\rho}(\tilde{r})$ obtained by numerically integrating equation (4-125b) from $\tilde{r} = 0$ outward starting from the boundary conditions $\tilde{\rho}(0) = 1$ and $(d\tilde{\rho}/d\tilde{r}) = 0$. Notice that by about $\tilde{r} = 15$, $\tilde{\rho}(\tilde{r})$ is declining as a straight line in the log-log plot of Figure 4-7; in fact, the solution is approaching the singular isothermal sphere of equation (4-123), which in these variables has the form $\tilde{\rho} = \frac{2}{9}\tilde{r}^{-2}$. This is shown as a dashed line in the figure.

In Figure 4-7 we plot the surface density $\Sigma(R)$ of the isothermal sphere in units of $\rho_0 r_0$. These results are tabulated in Table 4-1. From equation (4-123) one may easily show that for $R \gg r_0$,

$$\Sigma(R) = \frac{\sigma^2}{2GR} = \frac{2}{9}\pi\rho_0 r_0\left(\frac{r_0}{R}\right). \tag{4-126}$$

If $M(r)$ is the mass interior to r, we have that the circular speed at r, v_c, is given by

$$v_c^2(r) = \frac{GM(r)}{r}. \tag{4-127a}$$

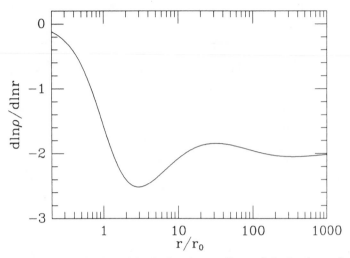

Figure 4-8. The logarithmic density gradient of the isothermal sphere.

But on integrating equation (4-119a) we find that

$$v_c^2 = -\sigma^2 \frac{d\ln\rho}{d\ln r}. \tag{4-127b}$$

In Figure 4-8 we plot $(d\ln\rho/d\ln r)$ for the isothermal sphere, which for $r \gg r_0$ tends to -2. Thus the circular speed at large r is constant at $v_c = \sqrt{2}\sigma$.

At $\tilde{r} \lesssim 2$ $(r \lesssim 2r_0)$ a useful approximation to $\tilde{\rho}(\tilde{r})$ is the modified Hubble law introduced in §2.1.2(d),

$$\tilde{\rho}(\tilde{r}) \approx \tilde{\rho}_h(\tilde{r}) \equiv \frac{1}{(1+\tilde{r}^2)^{\frac{3}{2}}}. \tag{4-128a}$$

Differentiating, we find

$$\frac{d}{d\tilde{r}}\left(\tilde{r}^2\frac{d\ln\tilde{\rho}_h}{d\tilde{r}}\right) + 9\tilde{r}^2\tilde{\rho}_h = -\left[\frac{1+\frac{1}{3}\tilde{r}^2}{\sqrt{1+\tilde{r}^2}}-1\right]9\tilde{r}^2\tilde{\rho}_h. \tag{4-129}$$

The quantity in the square bracket on the right side of this equation is less than 0.05 for $\tilde{r} < 2$, and the error in using equation (4-128a) as a solution to equation (4-119a) is less than 5% in this range. The surface density to which $\tilde{\rho}_h$ gives rise is

$$\Sigma_h(\tilde{R}) = \frac{2}{1+\tilde{R}^2}, \tag{4-128b}$$

where $\tilde{R} \equiv R/r_0$.

Table 4-2. Velocity dispersions and mass-to-light ratios of some elliptical galaxies

NGC	$-M_B$	$\dfrac{\sigma}{\mathrm{km\,s^{-1}}}$	Υ_V	NGC	$-M_B$	$\dfrac{\sigma}{\mathrm{km\,s^{-1}}}$	Υ_V
720	19.9	240	10	4406	20.6	262	12
741	21.2	309	18	4472	21.3	317	13
1395	19.5	245	16	4552	19.9	276	11
1407	20.0	288	22	4636	20.1	227	16
1600	21.1	342	19	4649	20.8	361	18
3379	19.4	227	10	5846	20.4	255	18
4261	20.8	321	13	6086	21.1	331	15
4365	20.0	263	16	7619	20.6	304	12
4374	20.4	292	11	7626	20.5	275	14

NOTES: A Hubble constant $H_0 = 100\,\mathrm{km\,s^{-1}\,Mpc^{-1}}$ has been assumed in estimating distances to the galaxies. Mass-to-light ratios Υ_V are in solar units. From data published in Lauer (1985).

$\widetilde{\rho}_h$ does not fit the isothermal profile well at $\widetilde{r} \gtrsim 3$ because it settles asymptotically to a logarithmic slope equal to -3 and not -2 as is required by the isothermal profile. On the other hand, at large radii, $\widetilde{\rho}_h$ has another use: when $\widetilde{r} \gg 1$ the projected density (4-128b) to which it gives rise is very similar to the Hubble law [eq. (1-14)], which fits the observed surface-brightness profiles of many elliptical galaxies quite well (MB §5-1). Thus $\widetilde{\rho}_h$ provides a simple analytical approximation to the inner parts of an isothermal sphere, or to the outer parts of a Hubble-law galaxy. It does *not* fit the outer parts of an isothermal sphere or the inner parts of the Hubble law. This dual application of $\widetilde{\rho}_h$ has produced a certain amount of confusion in the literature.

The surface-brightness profiles of many elliptical galaxies are well fitted by an isothermal sphere out to a few core radii. This fact makes possible a simple and effective method for determining the core mass-to-light ratios of these galaxies. We determine the King radius r_0 from fitting the photometry, measure the line-of-sight velocity dispersion, which can be identified with the parameter σ, and use equation (4-124b) to determine the central density ρ_0. Then, from Table 4-1, the central luminosity density is $j_0 = 0.495I(0)/r_0$, and the mass-to-light ratio is $\Upsilon = \rho_0/j_0$ (see Richstone & Tremaine 1986 for more detail). Table 4-2 contains a list of central mass-to-light ratios of elliptical galaxies determined in this manner, which is called **core fitting** or **King's method**.

From the astrophysical point of view, the isothermal sphere has a very serious defect: its mass is infinite. Thus from equations (4-127)

and Figure 4-7, we have that $M \simeq 2\sigma^2 r/G$ at large r. Clearly no real astrophysical system can be modeled over more than a limited range of radii with a divergent mass distribution. On the other hand, the rotation curves of spiral galaxies (§10.1.6 and MB §8-3) *are* often remarkably flat out to great radii, and this suggests that we try to construct models that deviate from the isothermal sphere only far from their cores.

(c) Lowered isothermal models   We seek a model that resembles the isothermal sphere at small radii, where the majority of stars have large values of $\mathcal{E}$, and is less dense than the isothermal sphere at large radii, where stars have smaller values of $\mathcal{E}$. We may obtain the DF f_K of such a model by simply diminishing the DF of the isothermal sphere at small values of $\mathcal{E}$. Thus we modify the DF (4-116) of the isothermal sphere in such a way that $f_K = 0$ for $\mathcal{E} \leq \mathcal{E}_0$. We may exploit the arbitrary constant Φ_0 in the definition (4-102) of $\mathcal{E}$ to set the critical relative energy $\mathcal{E}_0 = 0$. Therefore $f_K(\mathcal{E})$ should be of the same form as equation (4-116) for $\mathcal{E} \gg 0$ and be zero for $\mathcal{E} < 0$. A suitable function is

$$f_K(\mathcal{E}) = \begin{cases} \rho_1 (2\pi\sigma^2)^{-\frac{3}{2}} \left(e^{\mathcal{E}/\sigma^2} - 1 \right) & \mathcal{E} > 0; \\ 0 & \mathcal{E} \leq 0. \end{cases} \qquad (4\text{-}130)$$

This DF defines the family of **King models** that were described briefly in Chapter 5 of MB.[9] We now derive the density profiles and other properties of these models.

We proceed much as in the case of the isothermal sphere. Substituting into equation (4-130) for $\mathcal{E}$ from equation (4-102) and integrating over all velocities, we obtain the density at any radius as

$$\begin{aligned} \rho_K(\Psi) &= \frac{4\pi\rho_1}{(2\pi\sigma^2)^{\frac{3}{2}}} \int_0^{\sqrt{2\Psi}} \left[\exp\left(\frac{\Psi - \frac{1}{2}v^2}{\sigma^2} \right) - 1 \right] v^2 \, dv \\ &= \rho_1 \left[e^{\Psi/\sigma^2} \operatorname{erf}\left(\frac{\sqrt{\Psi}}{\sigma} \right) - \sqrt{\frac{4\Psi}{\pi\sigma^2}} \left(1 + \frac{2\Psi}{3\sigma^2} \right) \right], \end{aligned} \qquad (4\text{-}131)$$

where $\operatorname{erf}(x)$ is the error function (Appendix 1.C.3). Poisson's equation for Ψ may therefore be written

$$\frac{d}{dr}\left(r^2 \frac{d\Psi}{dr} \right) = -4\pi G \rho_1 r^2 \left[e^{\Psi/\sigma^2} \operatorname{erf}\left(\frac{\sqrt{\Psi}}{\sigma} \right) - \sqrt{\frac{4\Psi}{\pi\sigma^2}} \left(1 + \frac{2\Psi}{3\sigma^2} \right) \right].$$

$$(4\text{-}132)$$

---

[9] DFs of the form (4-130) were actually introduced by Michie (1963) and studied in detail by Michie and Bodenheimer (1963), but it was King's (1966) paper that made them well known—see King (1981) for a discussion of their history.

Equation (4-132) is an ordinary differential equation for $\Psi(r)$ which may be integrated numerically once we have chosen suitable boundary conditions. At $r = 0$ we have, as usual, $(d\Psi/dr) = 0$. The second boundary condition is the value of Ψ at $r = 0$. The value $\Psi(0)$ from which we start the integration determines the central potential $\Phi(0)$ in the following implicit way. As we integrate equation (4-132) outward, $(d\Psi/dr)$ decreases, because initially $(d\Psi/dr) = 0$ and $(d^2\Psi/dr^2) < 0$. As Ψ decreases toward zero, the range $(0, \sqrt{2\Psi})$ within which the speeds of stars at a given radius must lie narrows, and the density of stars at that radius drops. Eventually at some radius r_t, when Ψ becomes equal to zero, the density vanishes. We call r_t the **tidal radius**. Since we know the mass $M(r_t)$ inside r_t, we can now determine $\Phi(r_t)$ as

$$\Phi(r_t) = -\frac{GM(r_t)}{r_t}. \qquad (4\text{-}133)$$

Clearly, $\Phi(0) = \Phi(r_t) - \Psi(0)$. The bigger the value $\Psi(0)$ from which we start our integration of equation (4-132), the greater will be r_t, $M(r_t)$, and $|\Phi(0)|$.

Figure 4-9a shows the density profiles of King models obtained by integrating equation (4-132) from several values of $\Psi(0)$. The radial coordinate is marked in units of the King radius r_0 that is defined by equation (4-124b). Figure 4-9b shows the projected density profiles $\Sigma_K(R)$ of the King models of Figure 4-9a. Notice that for some of these models r_0 is appreciably larger than the half-brightness, or core radius[10] r_c defined by the condition $\Sigma_K(r_c)/\Sigma_K(0) = \frac{1}{2}$ (see MB p. 307).

The ratio of the tidal radius r_t of a King model to the King radius r_0 defines the **concentration** c through

$$c \equiv \log_{10}(r_t/r_0). \qquad (4\text{-}134)$$

In Chapter 5 of MB it was shown that models with c between 0.75 and 1.75 fit globular clusters very well, and models having $c \gtrsim 2.2$ fit the observed brightness profiles of some elliptical galaxies moderately well. Figure 4-9b shows that the density profile of the model with central potential $\Psi(0) \simeq 8\sigma^2$ is fairly well fitted by the modified Hubble profile (4-128).

King models form a simple sequence that may be parametrized in terms of either c or $\Psi(0)/\sigma^2$. Figure 4-10 gives the relationship between c and $\Psi(0)/\sigma^2$. In the limit $c \to \infty$, $\Psi(0)/\sigma^2 \to \infty$, the sequence of

---

[10] Unfortunately, in the literature r_0 is often called the core radius, but we shall reserve this name for r_c.

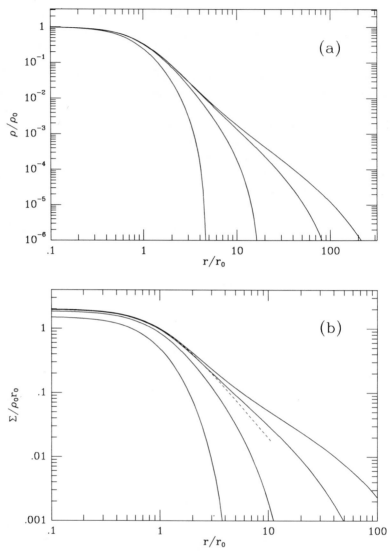

Figure 4-9. (a) Density profiles of four King models: from top to bottom the central potentials of these models satisfy $\Psi(0)/\sigma^2 = 12, 9, 6, 3$. (b) The projected mass densities of these models (full curves), and the projected modified Hubble law (dashed curve).

King models goes over into the isothermal sphere. At each point on the sequence of King models, there is a two-parameter family of systems that are related to each other by changes of scale. Thus for any value of c there are models having any given value of r_0. Among these are

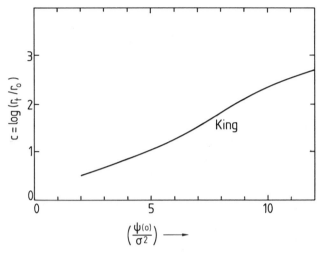

Figure 4-10. The relationship between the concentration c defined by equation (4-134) and the central potential $\Psi(0)$ from which equation (4-132) is integrated.

models with large central densities and thus, by equation (4-124b), large velocity-dispersion parameters σ, as well as models with low densities and low velocity dispersions.

The parameter σ that occurs in the relations we have given for King models must not be confused with the actual velocity dispersion $(\overline{v^2})^{1/2}$ of the stars of the system. From equation (4-130) we have that in a King model the velocity dispersion $\overline{v^2} = 3\overline{v_r^2}$ is just

$$\overline{v^2}(r) = J_2/J_0 \quad \text{where} \quad J_n \equiv \int_0^{\sqrt{2\Psi}} \left[\exp\left(\frac{\Psi - \frac{1}{2}v^2}{\sigma^2}\right) - 1\right] v^{n+2}dv.$$

$$(4\text{-}135)$$

Figure 4-11 is a plot of $[\overline{v^2}(r)/3\sigma^2]^{1/2}$ for several King models. One sees that in all these models the velocity dispersion falls monotonically from the center outward, reaching zero at r_t. The velocity dispersion of the stars at r_t is zero because the potential energy of these stars is already equal to the largest energy allowed to any star.

The King DF (4-130) is only one of several possible modified isothermal DFs. Woolley and Dickens (1961) discussed models for which f is given by (4-116) for $\mathcal{E} > 0$ and is zero otherwise, while Wilson (1975) (see also Hunter 1977a) considered models generated by DFs of the form

$$f_W = \text{constant} \times \begin{cases} [e^{\mathcal{E}/\sigma^2} - (\mathcal{E}/\sigma^2) - 1], & \text{for } \mathcal{E} > 0; \\ 0 & \text{otherwise.} \end{cases} \quad (4\text{-}136)$$

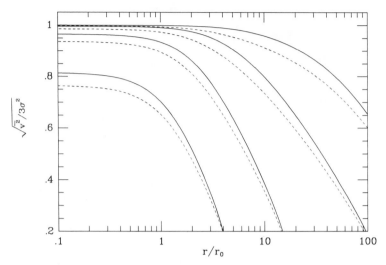

Figure 4-11. The RMS velocity at a given spatial radius r (full curves) and the RMS line-of-sight velocity at projected radius R (dashed curves) for the King models shown in Figures 4-9. Velocities are given in units of $\sqrt{3}\sigma$.

(d) Distribution function from the density profile   We have described a number of cases in which we assume a plausible form of the DF f and solve Poisson's equation for the density profile $\rho(r)$. We now examine the inverse problem, namely how to derive the DF $f(\mathcal{E})$ that generates a spherically symmetric star cluster of any given mass density $\rho(r)$.

If we write the integral of f over all velocities as an integral over $\mathcal{E}$, we find [cf. eq. (4-104)]

$$\rho(r) = 4\pi \int_0^{\Psi} f(\mathcal{E})\sqrt{2(\Psi - \mathcal{E})}d\mathcal{E}. \qquad (4\text{-}137)$$

Since Ψ is a monotonic function of r, we can regard ρ as a function of Ψ. Thus

$$\frac{1}{\sqrt{8\pi}}\rho(\Psi) = 2 \int_0^{\Psi} f(\mathcal{E})\sqrt{\Psi - \mathcal{E}}d\mathcal{E}. \qquad (4\text{-}138)$$

Differentiating both sides with respect to Ψ, we obtain

$$\frac{1}{\sqrt{8\pi}}\frac{d\rho}{d\Psi} = \int_0^{\Psi} \frac{f(\mathcal{E})d\mathcal{E}}{\sqrt{\Psi - \mathcal{E}}}. \qquad (4\text{-}139)$$

Equation (4-139) is an Abel integral equation having solution (see Appendix 1.B.4)

$$f(\mathcal{E}) = \frac{1}{\sqrt{8}\pi^2} \frac{d}{d\mathcal{E}} \int_0^{\mathcal{E}} \frac{d\rho}{d\Psi} \frac{d\Psi}{\sqrt{\mathcal{E} - \Psi}}. \tag{4-140a}$$

An equivalent formula is

$$f(\mathcal{E}) = \frac{1}{\sqrt{8}\pi^2} \left[\int_0^{\mathcal{E}} \frac{d^2\rho}{d\Psi^2} \frac{d\Psi}{\sqrt{\mathcal{E} - \Psi}} + \frac{1}{\sqrt{\mathcal{E}}} \left(\frac{d\rho}{d\Psi} \right)_{\Psi=0} \right]. \tag{4-140b}$$

Thus, given a spherically symmetric density distribution, we may in principle recover a DF that depends on the phase-space coordinates only through the energy, and that generates a model with the given density. However, we have no guarantee that the solution $f(\mathcal{E})$ to equations (4-140) will satisfy the physical requirement that it be nowhere negative. Indeed, we may conclude from equation (4-140a) that *a spherical density distribution $\rho(r)$ can be that of a system whose DF depends only on E if and only if*

$$\int_0^{\mathcal{E}} \frac{d\rho}{d\Psi} \frac{d\Psi}{\sqrt{\mathcal{E} - \Psi}}$$

is an increasing function of $\mathcal{E}$. If $\rho(r)$ does not satisfy this requirement, the "model" that is obtained by setting the anisotropy parameter $\beta = 0$ and integrating the Jeans equation (4-54) is unphysical.

Equation (4-140b) is due to Eddington (1916), and we shall call it **Eddington's formula**. As an example of its use, we determine the DF f_J that self-consistently generates the Jaffe model that was introduced in Problem 2-3. The calculation is simplified by the introduction of dimensionless variables $\widetilde{\rho}$ and $\widetilde{\Psi}$ that are related to the model's mass M, scale radius r_J, density ρ, and potential Φ by

$$\widetilde{\rho} \equiv \frac{4\pi r_J^3}{M}\rho = \frac{r_J^4}{r^2(r + r_J)^2} \quad \text{and} \quad \widetilde{\Psi} \equiv -\frac{r_J}{GM}\Phi = -\ln\left(\frac{r}{r + r_J}\right). \tag{4-141}$$

Eliminating r between these equations yields

$$\widetilde{\rho} = \left(e^{\frac{1}{2}\widetilde{\Psi}} - e^{-\frac{1}{2}\widetilde{\Psi}}\right)^4. \tag{4-142a}$$

Thus $(d\widetilde{\rho}/d\widetilde{\Psi})_0 = 0$, and

$$\frac{d^2\widetilde{\rho}}{d\widetilde{\Psi}^2} = 4\left(e^{2\widetilde{\Psi}} - e^{\widetilde{\Psi}} - e^{-\widetilde{\Psi}} + e^{-2\widetilde{\Psi}}\right). \tag{4-142b}$$

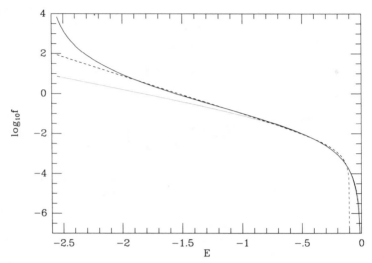

Figure 4-12. Distribution functions of the $R^{1/4}$ model (full curve), the King model with $\Psi(0) = 10.9\sigma^2$ (dashed curve), and Jaffe's model (dotted curve). All three models have unit mass. The $R^{1/4}$ model has $R_e = 1$ and the Jaffe model $r_J = 1.31$. The King model is scaled to have central velocity dispersion equal to the peak velocity dispersion of the $R^{1/4}$ model.

Substituting these expressions into equation (4-140b), and changing to a new variable of integration, $x \equiv \sqrt{\widetilde{\mathcal{E}} - \widetilde{\Psi}}$, where $\widetilde{\mathcal{E}} \equiv (-Er_J/GM)$, yields (Jaffe 1983)

$$f(E) = \frac{M}{2\pi^3 (GMr_J)^{\frac{3}{2}}} \left[F_-\left(\sqrt{2\widetilde{\mathcal{E}}}\right) - \sqrt{2}F_-\left(\sqrt{\widetilde{\mathcal{E}}}\right) \right.$$
$$\left. - \sqrt{2}F_+\left(\sqrt{\widetilde{\mathcal{E}}}\right) + F_+\left(\sqrt{2\widetilde{\mathcal{E}}}\right) \right], \tag{4-143a}$$

where $F_\pm(z)$ is Dawson's integral,[11]

$$F_\pm(x) \equiv e^{\mp x^2} \int_0^x e^{\pm x'^2} dx'. \tag{4-143b}$$

In Figure 4-12 we plot this DF alongside a typical King function and the DF that generates the $R^{1/4}$ model.

---

[11] Note that $F_-(x) = \frac{1}{2}\sqrt{\pi}e^{x^2} \operatorname{erf}(x)$.

Another useful model for which the integral of equation (4-140b) can be evaluated analytically is Hénon's isochrone [eq. (2-32)]. Hénon (1960b) shows that this model is generated by the DF

$$f_I = \frac{M}{\sqrt{2}(2\pi)^3(GMb)^{\frac{3}{2}}} \frac{\sqrt{\tilde{\mathcal{E}}}}{[2(1-\tilde{\mathcal{E}})]^4} \left[27 - 66\tilde{\mathcal{E}} + 320\tilde{\mathcal{E}}^2 - 240\tilde{\mathcal{E}}^3 \right.$$
$$\left. + 64\tilde{\mathcal{E}}^4 + 3(16\tilde{\mathcal{E}}^2 + 28\tilde{\mathcal{E}} - 9)\frac{\arcsin\sqrt{\tilde{\mathcal{E}}}}{\sqrt{\tilde{\mathcal{E}}(1-\tilde{\mathcal{E}})}} \right],$$

$$(4\text{-}144)$$

where $\tilde{\mathcal{E}} \equiv (-Eb/GM)$.

4 Systems with Anisotropic Dispersion Tensors

At the beginning of this section we saw that a general spherical system has a DF $f(E, L)$ that depends on the energy E and the absolute value of the angular momentum L, and we showed that whenever f depends on E only, the velocity dispersions in the radial and tangential directions must be equal. We now examine some systems generated by DFs that depend on both E and L, with the result that their velocity-dispersion tensors are anisotropic. An obvious way to proceed is to generate simple DFs $f(E, L)$ by multiplying some of the functions $f(E)$ studied above by simple functions of L, such as $L^{-2\beta}$ or $e^{-\alpha L^2}$ (see Problems 4-18 and 4-21). However, a more instructive course is to obtain an analog of Eddington's formula (4-140b), from which we can derive a whole family of DFs $f(E, L)$ that generate any given density profile $\rho(r)$, and then to investigate how these models differ dynamically as a result of their varying degrees of velocity anisotropy.

We first express the integral on the right side of (4-101a) in terms of polar coordinates (v, η, ψ) in velocity space. If we orient this coordinate system such that

$$v_r = v\cos\eta \quad ; \quad v_\theta = v\sin\eta\cos\psi \quad ; \quad v_\phi = v\sin\eta\sin\psi, \quad (4\text{-}145a)$$

and use the relative energy and potential $\mathcal{E}$ and Ψ, which are defined by equations (4-102), we have

$$\rho(r) = \int f(\mathcal{E}, L) \, d^3\mathbf{v} = 2\pi \int_0^\pi \sin\eta \, d\eta \int_0^\infty f\left(\Psi - \tfrac{1}{2}v^2, |rv\sin\eta|\right) v^2 dv.$$

$$(4\text{-}145b)$$

(a) Osipkov-Merritt models   We investigate the consequences of f depending on $\mathcal{E}$ and L only through the variable (Osipkov 1979; Merritt 1985a,b)

$$Q \equiv \mathcal{E} - \frac{L^2}{2r_a^2} = \Psi - \tfrac{1}{2}v^2\left(1 + \frac{r^2}{r_a^2}\sin^2\eta\right). \qquad (4\text{-}146a)$$

If we replace the integration variable v in equation (4-145b) with Q, we have at constant r, $dQ = -\left[1 + (r/r_a)^2\sin^2\eta\right]v\,dv$, and thus

$$\rho(r) = 2\pi \int_0^\pi \sin\eta\,d\eta \int_0^\Psi f(Q)\frac{\sqrt{2(\Psi - Q)}dQ}{\left[1 + (r/r_a)^2\sin^2\eta\right]^{\frac{3}{2}}}, \qquad (4\text{-}146b)$$

where we have imposed the condition $f(Q) = 0$ for $Q \le 0$. If we now interchange the order of integrations in equation (4-146b), the inner integral becomes

$$\int_0^\pi \frac{\sin\eta\,d\eta}{\left[1 + (r/r_a)^2\sin^2\eta\right]^{\frac{3}{2}}} = \frac{2}{1 + (r/r_a)^2}. \qquad (4\text{-}147)$$

Hence

$$\left(1 + \frac{r^2}{r_a^2}\right)\rho(r) = 4\pi \int_0^\Psi f(Q)\sqrt{2(\Psi - Q)}dQ. \qquad (4\text{-}148)$$

But the right side of this equation is identical with the right side of equation (4-137) with $\mathcal{E}$ replaced by Q. Hence by equation (4-140b) we have

$$f(Q) = \frac{1}{\sqrt{8}\pi^2}\left[\int_0^Q \frac{d^2\rho_Q}{d\Psi^2}\frac{d\Psi}{\sqrt{Q - \Psi}} + \frac{1}{\sqrt{Q}}\left(\frac{d\rho_Q}{d\Psi}\right)_{Q=0}\right], \qquad (4\text{-}149a)$$

where

$$\rho_Q(r) \equiv \left(1 + \frac{r^2}{r_a^2}\right)\rho. \qquad (4\text{-}149b)$$

Notice that in the limit $r_a \to \infty$, $Q \to \mathcal{E}$ and equation (4-149a) goes over into (4-140b).

Let us use equations (4-149) to recover anisotropic analogs of Jaffe's DF (4-143a). In terms of the variables $\widetilde{\rho}$ and $\widetilde{\Psi}$ defined by equations (4-141), we have

$$\widetilde{\rho}_Q = \left(e^{\frac{1}{2}\widetilde{\Psi}} - e^{-\frac{1}{2}\widetilde{\Psi}}\right)^4 + \frac{r_J^2}{r_a^2}e^{-\Psi}\left(e^{\frac{1}{2}\widetilde{\Psi}} - e^{-\frac{1}{2}\widetilde{\Psi}}\right)^2. \qquad (4\text{-}150a)$$

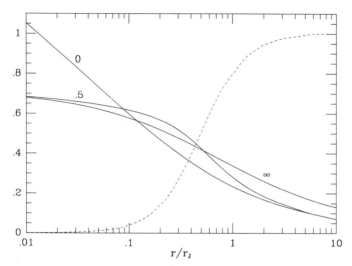

Figure 4-13. Radial bias in the velocity-dispersion tensor steepens the radial gradient in the line-of-sight velocity dispersion σ_p. Each full curve shows $\sigma_p(R)$ for a Jaffe model with the given ratio r_a/r_J of the anisotropy radius to the scale radius. The model with $r_a/r_J = 0$ is composed of stars on radial orbits, while that with $r_a/r_J \to \infty$ has an isotropic velocity-dispersion tensor. The dashed curve shows the anisotropy parameter $\beta(r)$ in the model with $r_a/r_J = 0.5$.

Hence $(d\widetilde{\rho}_Q/d\widetilde{\Psi})_0 = 0$, and

$$\frac{d^2\widetilde{\rho}_Q}{d\widetilde{\Psi}^2} = 4\left(e^{2\widetilde{\Psi}} - e^{\widetilde{\Psi}} - e^{-\widetilde{\Psi}} + e^{-2\widetilde{\Psi}}\right) - 2\frac{r_J^2}{r_a^2}\left(e^{-\widetilde{\Psi}} - 2e^{-2\widetilde{\Psi}}\right). \quad (4\text{-}150b)$$

By analogy with equation (4-143a), the DF we seek is therefore

$$f(Q) = \frac{M}{2\pi^3(GMr_J)^{\frac{3}{2}}}\left[F_-\left(\sqrt{2\widetilde{Q}}\right) - \sqrt{2}F_-\left(\sqrt{\widetilde{Q}}\right)\right.$$
$$\left. - \sqrt{2}\left(1 + \frac{r_J^2}{2r_a^2}\right)F_+\left(\sqrt{\widetilde{Q}}\right) + \left(1 + \frac{r_J^2}{r_a^2}\right)F_+\left(\sqrt{2\widetilde{Q}}\right)\right],$$
$$(4\text{-}151)$$

where F is again defined by equation (4-143b) and $\widetilde{Q} \equiv (Qr_J/GM)$.

In §4.2 we used the Jeans equation (4-54) to show that given measurements of the luminosity density $\nu(r)$ and the line-of-sight velocity dispersion σ_p in a spherical galaxy can be interpreted as arising from different mass distributions $\rho(r)$, depending on what one takes to be the functional form of the anisotropy parameter $\beta(r)$. In the light of that

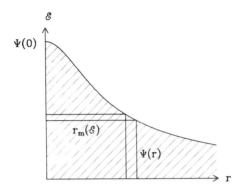

Figure 4-14. Diagram illustrating the change of variables in equation (4-154).

discussion it is interesting to use the family of self-consistent Jaffe models furnished by equation (4-151) for different values of r_a, to see how σ_p depends on $\beta(r)$ for a given density profile $\rho(r)$. In Figure 4-13 we plot σ_p and β for several of these modified Jaffe models. It will be seen that the overall effect of diminishing the anisotropy radius r_a is to increase σ_p at small projected radius R, and to decrease it at large R. By the virial theorem, the mass-weighted mean of σ_p^2 is the same in all models.

(b) Michie models   A natural extension of King models to include velocity anisotropy is the family of **Michie models**, which is defined by the DF

$$f_M(\mathcal{E}, L) = \begin{cases} \rho_1(2\pi\sigma^2)^{-3/2}e^{-L^2/(2r_a^2\sigma^2)}\left[e^{\mathcal{E}/\sigma^2} - 1\right], & \mathcal{E} > 0, \\ 0, & \mathcal{E} \leq 0. \end{cases} \quad (4\text{-}152)$$

In the limit $r_a \to \infty$ this DF reduces to the DF (4-130) of the King models. In contrast to models with DFs of the form $f(Q)$, the DF in a Michie model is not exactly constant on spheroids in velocity space. However, in both families of models the velocity distribution is isotropic at the center, nearly radial in the outer parts, and the transition occurs near the anisotropy radius r_a.

5 The Differential Energy Distribution

When analyzing an N-body model of the type described in §4.7, a natural diagnostic quantity to calculate is the mass dM of stars in each of a number of energy ranges $(\mathcal{E}, \mathcal{E} + d\mathcal{E})$. We shall call the resulting function $(dM/d\mathcal{E})$ the **differential energy distribution**. It is interesting to compute the differential energy distributions of some of the models discussed in this section for later comparison with the energy distributions of typical N-body models.

At radius r, stars of relative energy $\mathcal{E}$ move at speed $v = \sqrt{2(\Psi - \mathcal{E})}$. We change the integration variable in equation (4-145b) from v to $\mathcal{E}$ to obtain

$$\rho(r) = 2\pi \int_0^\pi \sin\eta \, d\eta \int_0^{\Psi(r)} f\left(\mathcal{E}, rv|\sin\eta|\right) v \, d\mathcal{E}. \qquad (4\text{-}153)$$

We next interchange the order of the η and $\mathcal{E}$ integrations, multiply by $4\pi r^2 dr$, and integrate over all radii to obtain the total mass M of the model as

$$M = 8\pi^2 \int_0^\infty r^2 dr \int_0^{\Psi(r)} v \, d\mathcal{E} \int_0^\pi f\left(\mathcal{E}, rv|\sin\eta|\right) \sin\eta \, d\eta. \qquad (4\text{-}154)$$

Interchanging the order of the r and $\mathcal{E}$ integrations, equation (4-154) becomes (see Figure 4-14)

$$M = 8\pi^2 \int_0^{\Psi(0)} d\mathcal{E} \int_0^{r_m(\mathcal{E})} v r^2 dr \int_0^\pi f\left(\mathcal{E}, rv|\sin\eta|\right) \sin\eta \, d\eta, \qquad (4\text{-}155)$$

where $r_m(\mathcal{E})$ is the largest radius that can be reached by a star of relative energy $\mathcal{E}$, and is defined by $\Psi(r_m) = \mathcal{E}$. Equation (4-155) expresses the total mass of the system as a sum over contributions from each range of energies. Thus the differential energy distribution is just the integrand of the outermost integral:

$$\frac{dM}{d\mathcal{E}} = 8\pi^2 \int_0^{r_m(\mathcal{E})} v r^2 dr \int_0^\pi f\left(\mathcal{E}, rv|\sin\eta|\right) \sin\eta \, d\eta, \qquad (4\text{-}156)$$

where $v = \sqrt{2(\Psi - \mathcal{E})}$. If f is a function $f(\mathcal{E})$ of $\mathcal{E}$ only, we may immediately do the integral over η and take f outside the remaining integral, to find

$$\frac{dM}{d\mathcal{E}} = f(\mathcal{E})g(\mathcal{E}), \qquad (4\text{-}157a)$$

where the "density of states" function $g(\mathcal{E})$—the phase-space volume per unit interval in $\mathcal{E}$—is defined by

$$g(\mathcal{E}) \equiv 16\pi^2 \int_0^{r_m(\mathcal{E})} \sqrt{2(\Psi - \mathcal{E})} r^2 dr. \qquad (4\text{-}157b)$$

In Figure 4-15a we plot $(dM/d\mathcal{E})$ for a typical King model and for the $R^{1/4}$ model. Note that the dependence of $(dM/d\mathcal{E})$ on $\mathcal{E}$ is qualitatively different from the way f depends on $\mathcal{E}$: while f generally grows roughly exponentially with $\mathcal{E}$, the density of states function, g, declines so rapidly

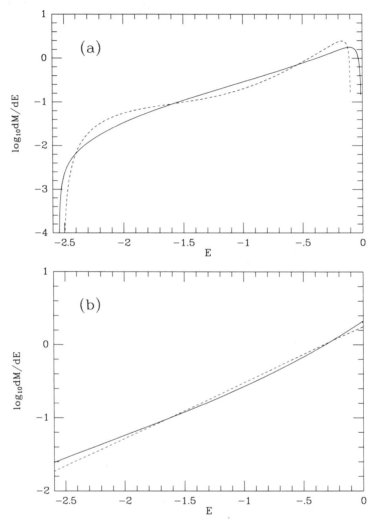

Figure 4-15. (a) (dM/dE) for the $R^{1/4}$ model (full curve) and the King model shown in Figure 4-12 (dashed curve); (b) (dM/dE) for the two extreme Jaffe models: the model with isotropic velocity dispersion (full curve) and the model in which all stars are on perfectly radial orbits (dashed curve).

with $\mathcal{E}$ that $(dM/d\mathcal{E})$ peaks at a *small* value of $\mathcal{E}$, where the DF is very small. As a consequence, most of the mass of a typical stellar system is made up of rather loosely bound stars.

Figure 4-15b shows that the anisotropic Jaffe models that are generated by equation (4-151) all have very similar differential energy distributions. This indicates that $(dM/d\mathcal{E})$ is principally determined by the density profile $\rho(r)$, and is almost independent of the galaxy's internal kinematics. We shall find that the extension of this observation to flattened and triaxial galaxies greatly facilitates the choice of DFs for these systems.

4.5 Axisymmetric Systems

It is likely that few, if any, galaxies are quite spherical. Therefore it is time we moved on from our instructive but somewhat artificial study of spherical systems to the consideration of flattened and triaxial systems. We first consider axisymmetric systems such as normal disk galaxies.

From the work of §3.2 we know that most orbits in flattened potentials admit three isolating integrals, E, L_z, and some third integral I_3. Hence by the strong Jeans theorem, we expect the DFs of flattened galaxies to be functions of three variables. Finding the particular function of three variables that describes any given galaxy is no simple matter. In fact, this task has proved so daunting that only in the last few years, three-quarters of a century after Jeans's (1915) paper posed the problem, has the serious quest for the DF of even our own Galaxy got underway. Hence it is prudent to start our investigation of axisymmetric systems by discussing the simpler subproblem posed by models of razor-thin stellar disks.

1 Planar Systems

Since nearly all the light of a late-type spiral galaxy is emitted by a thin stellar disk, models of planar galaxies are of more than academic interest. However, even the dynamics of apparently isolated disks are probably affected by the gravitational fields of fatter, dark components. Hence the assumption we shall make, that the confining gravitational potentials of our models are entirely generated by the disks themselves, is likely to be something of an idealization. Yet studies of the dynamics of such idealized models will illustrate an important general point: while the total kinetic energy of an equilibrium model is set, via the virial theorem, by the distribution of mass within the system, as galaxy builders we generally have a good deal of freedom as to whether we wish to invest this kinetic energy in random or ordered form. Thus the DF f can rarely be uniquely determined by a knowledge of the mass distribution alone; we shall find that f can be determined only if we know something of the galaxy's internal kinematics.

In §3.3 we saw that planar orbits generally admit two isolating integrals, say E and I_2. Thus application of the strong Jeans theorem to such systems shows that we may assume the DF is of the form $f(E, I_2)$. If the system is axisymmetric, we may take $I_2 = L_z$, and have $f(E, L_z)$. We consider two simple forms of f.

(a) Distribution function for Mestel's disk   In §2.6.3(a) we found that a disk with surface density

$$\Sigma(R) = \frac{\Sigma_0 R_0}{R} \tag{4-158}$$

has a circular speed v_c that is independent of radius and given by

$$v_c^2 = -R\frac{\partial \Psi}{\partial R} = 2\pi G \Sigma_0 R_0. \tag{4-159}$$

We set the arbitrary constant involved in the definition of the relative potential such that $\Psi(R = 1) = 0$, and integrate equation (4-159) with respect to R, to find

$$\Psi(R) = -v_c^2 \ln R. \tag{4-160}$$

Now consider the DF

$$f(\mathcal{E}, L_z) = \begin{cases} FL_z^q e^{\mathcal{E}/\sigma^2}, & L_z > 0. \\ 0, & L_z \leq 0, \end{cases} \tag{4-161}$$

where F, q, and σ are all constants. Inserting equation (4-160) into equation (4-161) and integrating over all velocities in the plane, we find that the surface density produced by this DF in the potential of Mestel's disk is

$$\begin{aligned}
\Sigma'(R) &= FR^q \int_{v_\phi > 0} v_\phi^q dv_\phi \int_{-\infty}^{\infty} \exp\left[-\frac{v_c^2}{\sigma^2}\ln R - \frac{v_R^2 + v_\phi^2}{2\sigma^2} \right] dv_R \\
&= FR^{(q - v_c^2/\sigma^2)} \int_{v_\phi > 0} v_\phi^q e^{-\frac{1}{2}v_\phi^2/\sigma^2} dv_\phi \int_{-\infty}^{\infty} e^{-\frac{1}{2}v_R^2/\sigma^2} dv_R \\
&= 2^{\frac{q}{2}}\sqrt{\pi}\left(\tfrac{q-1}{2}\right)! F\sigma^{(q+2)} R^{(q - v_c^2/\sigma^2)}.
\end{aligned} \tag{4-162}$$

Comparing equations (4-158) and (4-162), we see that the DF of equation (4-161) will self-consistently generate the Mestel disk if we set

$$q = \frac{v_c^2}{\sigma^2} - 1 \quad \text{and} \quad F = \frac{\Sigma_0 R_0}{2^{\frac{q}{2}}\sqrt{\pi}\left(\tfrac{q-1}{2}\right)!\sigma^{(q+2)}}. \tag{4-163}$$

The parameter q that appears in the DF (4-161) of the Mestel disk is a measure of the degree to which the disk is centrifugally supported: from equation (4-161) one may show that σ is the velocity dispersion $\overline{v_R^2}^{1/2}$ in the radial direction, and in the limit $q \to -1$ of maximal pressure support, the system tends to the two-dimensional analog of the envelope of an Eddington model (see Problem 4-21).

We may explicitly calculate the mean azimuthal streaming velocity of the self-consistent Mestel disk as

$$\overline{v}_\phi = \frac{\int v_\phi f(\mathcal{E}, L_z) dv_\phi dv_R}{\int f(\mathcal{E}, L_z) dv_\phi dv_R} = \frac{\int v_\phi^{(q+1)} \exp\left(-\frac{v_\phi^2}{2\sigma^2}\right) dv_\phi}{\int v_\phi^q \exp\left(-\frac{v_\phi^2}{2\sigma^2}\right) dv_\phi} = \frac{\sqrt{2} \left(\frac{1}{2}q\right)!}{\left(\frac{q-1}{2}\right)!} \sigma.$$

(4-164)

When q becomes large, $(\overline{v}_\phi/\sigma)$ increases like $\sqrt{q}$. In the limit $q \to \infty$, all stars are on circular orbits, and $\overline{v}_\phi = v_c$.

(b) Kalnajs disks From Tables 2-1 and 2-2 we have that the potential $\Phi(R)$ at radius R in the equatorial plane of a homogeneous oblate spheroid with eccentricity e, density ρ, and semi-axes of length a and $a_3 = a\sqrt{1 - e^2}$ is

$$\Phi(R) = \frac{\pi G \rho a_3}{ae^2} \left(\frac{\arcsin e}{e} - \sqrt{1 - e^2}\right) R^2 + \text{constant}. \qquad (4\text{-}165)$$

If we now flatten this spheroid down to a disk by letting $e \to 1$, while holding $\Sigma_0 \equiv 2\rho a_3$ constant, we obtain that

$$\Phi(R) = \frac{\pi^2 G \Sigma_0}{4a} R^2 + \text{constant} = \tfrac{1}{2}\Omega_0^2 R^2 + \text{constant}, \qquad (4\text{-}166)$$

where $\Omega_0 \equiv \sqrt{\tfrac{1}{2}\pi^2 G\Sigma_0/a}$ is the angular speed of a circular orbit. By equation (2-135) the surface density of our disk is

$$\Sigma(R) = \Sigma_0 \sqrt{1 - \frac{R^2}{a^2}}. \qquad (4\text{-}167)$$

Now consider the density distribution that arises from combining the DF

$$f(\mathcal{E}, L_z) = \begin{cases} F\left[(\Omega_0^2 - \Omega^2)a^2 + 2(\mathcal{E} + \Omega L_z)\right]^{-\frac{1}{2}} & \text{for } [\ldots] > 0, \\ 0 & \text{for } [\ldots] \leq 0, \end{cases}$$

(4-168)

with the potential of our disk. Since $\mathcal{E} = \Psi - \frac{1}{2}(v_\phi^2 + v_R^2)$ and $L_z = Rv_\phi$, we can also write the argument of the radical in equation (4-168) as

$$(\Omega_0^2 - \Omega^2)a^2 - (v_\phi - \Omega R)^2 - v_R^2 + 2\Psi + \Omega^2 R^2.$$

Hence at any radius R the DF (4-168) depends on the velocities only in the combination $v_R^2 + (v_\phi - \Omega R)^2$. Consequently, the distribution of azimuthal velocities in a model generated by this DF is symmetrical about $v_\phi = \Omega R$, which is therefore the mean azimuthal velocity at R. We choose the arbitrary constant involved in the definition of the relative potential such that

$$\Psi(R) = -\tfrac{1}{2}\Omega_0^2 R^2. \tag{4-169}$$

Substituting this form of Ψ into equation (4-168) and integrating over all velocities, we find the surface density $\Sigma'(R)$ generated by this DF in the potential of our disk to be

$$\Sigma'(R) = F \int_{v_{\phi 1}}^{v_{\phi 2}} dv_\phi \int_{v_{R1}}^{v_{R2}} \frac{dv_R}{\sqrt{(\Omega_0^2 - \Omega^2)(a^2 - R^2) - (v_\phi - \Omega R)^2 - v_R^2}}. \tag{4-170}$$

The limits v_{R1}, v_{R2} of the inner integral in equation (4-170) are just the values of v_R for which the integrand's denominator vanishes. Hence equation (4-170) is of the form

$$\Sigma'(R) = F \int_{v_{\phi 1}}^{v_{\phi 2}} dv_\phi \int_{-b}^{b} \frac{dx}{\sqrt{b^2 - x^2}} = \pi F \int_{v_{\phi 1}}^{v_{\phi 2}} dv_\phi = \pi F\left[v_{\phi 1} - v_{\phi 2}\right]. \tag{4-171}$$

But $v_{\phi 1}$ and $v_{\phi 2}$ are just the roots of the quadratic equation

$$(v_\phi - \Omega R)^2 - (\Omega_0^2 - \Omega^2)(a^2 - R^2) = 0, \tag{4-172}$$

so

$$\Sigma'(R) = 2\pi Fa\sqrt{\Omega_0^2 - \Omega^2}\sqrt{1 - \frac{R^2}{a^2}}. \tag{4-173}$$

Comparing equations (4-167) and (4-173) we see that if we set

$$F = \frac{\Sigma_0}{2\pi a\sqrt{\Omega_0^2 - \Omega^2}}, \tag{4-174}$$

the DF of equation (4-168) in the potential (4-166) of our disk generates the density of our disk, and hence yields a self-consistent stellar-dynamical model of a flat disk galaxy.

It is straightforward to verify the following important properties of these **Kalnajs disks**:

(i) The mean angular velocity of the stars, $(\overline{v}_\phi/R) = \Omega$, is independent of radius R. Thus these disks rotate like rigid bodies.

(ii) At any radius R in the disk, the dispersion $\overline{(v_\phi - \overline{v}_\phi)^2}$ in the azimuthal velocities is equal to the dispersion $\overline{v_R^2}$ of the radial velocities.

(iii) These disks range from "hot" systems with $\Omega \ll \Omega_0$, in which the support against self-gravity comes from random motions, to "cold" systems with $\Omega \approx \Omega_0$, in which all stars move on nearly circular orbits and the random velocities are small.

The Kalnajs disks and the Mestel disk show that many different DFs $f(E, L_z)$ generate a given run of surface density $\Sigma(R)$. Some of these DFs correspond to models in which stars make large excursions in the radial direction, while in other models the stars move on nearly circular orbits. Kalnajs (1976) presents several other examples of this phenomenon.

The most interesting case from the point of view of studies of the solar neighborhood is that in which most of the stars move on nearly circular orbits (see §7.5). Shu (1969) and Zweibel (1978) have discussed parameterizations of $f(E, L_z)$ that are motivated by the idea that at each radius R the distribution of the random velocities of stars forms an approximate Maxwellian about the local standard of rest at R. Villumsen and Binney (1985) have derived DFs for many-particle simulations of this situation.

2 Three-Dimensional Systems with $f(\mathcal{E}, L_z)$

Let us now turn from infinitely thin stellar disks to models of axisymmetric galaxies that have finite thickness perpendicular to their equatorial planes. Although the strong Jeans theorem tells us that the DF of a generic axisymmetric system involves three integrals, functions of two integrals such as $f(\mathcal{E}, L_z)$ generate perfectly valid stellar systems, albeit members of a restricted subclass of all possible models. Of course we cannot expect galaxies to be accurately modeled by such systems, unless there is some good physical reason why Nature should have restricted herself to systems of this type. However, all manner of calculations are greatly simplified when f is a function of only the two familiar integrals, and valuable insight into the structure of more general models can be gained by studying these simple systems.

We start by proving an important property of any model for which $f = f(\mathcal{E}, L_z)$. The ratio of the squared velocity dispersions $\overline{v_R^2}$ and $\overline{v_z^2}$ in the radial and vertical directions is

$$\frac{\overline{v_R^2}}{\overline{v_z^2}} = \frac{\int_{-\infty}^{\infty} dv_\phi \int_{-\infty}^{\infty} dv_z \int_{-\infty}^{\infty} f \left[\Psi - \frac{1}{2}(v_R^2 + v_z^2 + v_\phi^2), Rv_\phi \right] v_R^2 dv_R}{\int_{-\infty}^{\infty} dv_\phi \int_{-\infty}^{\infty} dv_R \int_{-\infty}^{\infty} f \left[\Psi - \frac{1}{2}(v_R^2 + v_z^2 + v_\phi^2), Rv_\phi \right] v_z^2 dv_z}.$$

(4-175)

It is easy to see that the lower integral is equal to the upper one. Hence $\overline{v_R^2} = \overline{v_z^2}$ whenever $f = f(\mathcal{E}, L_z)$. Table 7-5 of MB shows that in the solar neighborhood $\overline{v_R^2}$ is actually greater than $\overline{v_z^2}$ for all classes of stars, so we may immediately conclude that *the DF of our Galaxy is not of the form $f(\mathcal{E}, L_z)$.*

(a) Distribution functions from $\rho(R, z)$ and $\overline{v}_\phi(R, z)$ We have seen that given any spherical density distribution $\rho(r)$, we can find the unique—but not necessarily non-negative—DF $f(\mathcal{E})$ that self-consistently generates ρ. Is it possible, given an axisymmetric density distribution $\rho(R, z)$, to find a DF $f(\mathcal{E}, L_z)$ that self-consistently generates $\rho(R, z)$? The answer to this question is "yes"; in fact, we now show that one can in principle find *infinitely many* DFs $f(\mathcal{E}, L_z)$ that generate a given $\rho(R, z)$.

The density is

$$\rho = \int_{v^2 < 2\Psi} f(\Psi - \tfrac{1}{2}v^2, Rv_\phi) d^3\mathbf{v},$$

(4-176)

where we have imposed the usual condition $f = 0$ for $\mathcal{E} < 0$. Let $\mathbf{v}_m$ be the component of $\mathbf{v}$ in the (R, z) plane. Then, if we define cylindrical coordinates (v_m, v_ϕ, ψ) in velocity space by

$$v_R = v_m \cos \psi \quad ; \quad v_z = v_m \sin \psi,$$

(4-177a)

we have that $d^3\mathbf{v} = v_m dv_m dv_\phi d\psi$, and that the integrand of equation (4-176) is independent of ψ. Integrating out ψ yields

$$\rho = 2\pi \int_0^{\sqrt{2\Psi}} v_m dv_m \int_{v_\phi^2 < (2\Psi - v_m^2)} f \left[\Psi - \tfrac{1}{2}(v_m^2 + v_\phi^2), Rv_\phi \right] dv_\phi.$$

(4-177b)

Changing the variables of integration from (v_m, v_ϕ) to $(\mathcal{E}, L_z)$, equation (4-177b) becomes

$$
\begin{aligned}
\rho &= \frac{2\pi}{R} \int_0^\Psi d\mathcal{E} \int_{L_z^2 < 2(\Psi - \mathcal{E})R^2} f(\mathcal{E}, L_z) dL_z \\
&= \frac{2\pi}{R} \int_0^\Psi d\mathcal{E} \int_0^{R\sqrt{2(\Psi - \mathcal{E})}} [f(\mathcal{E}, L_z) + f(\mathcal{E}, -L_z)] \, dL_z \qquad (4\text{-}178a) \\
&= \frac{4\pi}{R} \int_0^\Psi d\mathcal{E} \int_0^{R\sqrt{2(\Psi - \mathcal{E})}} f_+(\mathcal{E}, L_z) dL_z,
\end{aligned}
$$

where f_+ is the part of f that is even in L_z:

$$
\begin{aligned}
f(\mathcal{E}, L_z) &= f_+(\mathcal{E}, L_z) + f_-(\mathcal{E}, L_z) \\
\text{where} \quad f_\pm(\mathcal{E}, L_z) &\equiv \tfrac{1}{2} [f(\mathcal{E}, L_z) \pm f(\mathcal{E}, -L_z)].
\end{aligned} \qquad (4\text{-}178b)
$$

Thus the density depends only on the even part of f. Physically this result arises because the density contributed by any star does not depend on the star's sense of rotation about the galaxy's symmetry axis.

Since the density depends only on f_+, it follows that given any DF f_0 that generates ρ, we can construct infinitely many other DFs that generate ρ by adding to f_0 any function f_- odd in L_z. While f_- has no influence on the density distribution, it is responsible for the mean streaming velocity,

$$
\overline{v}_\phi = \frac{4\pi}{\rho R^2} \int_0^\Psi d\mathcal{E} \int_0^{R\sqrt{2(\Psi - \mathcal{E})}} f_-(\mathcal{E}, L_z) L_z dL_z, \qquad (4\text{-}178c)
$$

which is directly proportional to f_-.

Lynden-Bell (1962a), and subsequently Hunter (1975) and Dejonghe (1986), showed that just as equation (4-138) may be solved for $f(\mathcal{E})$ in terms of $\rho(\Psi)$, so equation (4-178a) may be solved for $f_+(\mathcal{E}, L_z)$ given $\rho(R, \Psi)$. Similarly, equation (4-178c) may be solved[12] for f_- in terms of $\rho \overline{v}_\phi(R, \Psi)$. These results have been used to derive DFs from analytic density profiles (Lynden-Bell 1962a; Lake 1978). However, a fatal obstacle obstructs the direct application of the formal solutions of equations (4-178) to observational data. The snag is that these solutions presume knowledge of ρ and $\rho \overline{v}_\phi$ as analytic functions of *complex* radius R! Any observations will yield ρ and $\overline{v}_\phi$ only at real values of R, so application of the formal solutions of equations (4-178) to observational data

---

[12] Note that equation (4-178c) establishes exactly the same relation between $R\rho\overline{v}_\phi$ and $L_z f_-$ as equation (4-178a) does between ρ and f_+.

involves fitting the data with analytic functions, which can be continued into the complex plane. Dejonghe (1986) has shown that the problem of thus extending ρ and $\overline{v}_\phi$ to complex R is "ill-conditioned"; that is, any small errors in ρ and $\overline{v}_\phi$ at real R will give rise to large variations in the values of the fitted analytic functions at pure imaginary R. These variations in the fitted functions in turn generate unphysical oscillations in f. Consequently, the formal solutions of equations (4-178) cannot be used to recover DFs directly from observational data.

Although the formal solutions of equations (4-178) cannot help us to derive f from observations, in §4.5.3 we shall describe several practical numerical schemes for recovering DFs that provide adequate fits to noisy data.

Probably the most profitable use of equations (4-178) is in the construction of self-consistent models generated by predetermined DFs $f(\mathcal{E}, L_z)$. The literature contains descriptions of only a handful of such models. Wilson (1975), building on the pioneering work of Prendergast and Tomer (1970), generated some models designed to fit observations of elliptical galaxies. However, Wilson's models differed from observed galaxies in (i) not having reasonably flat rotation curves, and (ii) being always rather round at the center. Toomre (1982) describes a series of elegant scale-free models, in which $\rho \propto r^{-2}$ along any ray from the center. In limiting cases these systems go over into (i) the isothermal sphere, (ii) Spitzer's isothermal sheet (Problem 4-25), and (iii) cold Mestel disks. The main limitations of Toomre's models are that (i) their densities always vanish along the minor axis, and (ii) they have infinite central densities.

3 Axisymmetric Systems Involving Third Integrals

We now take up the challenge of developing stellar-dynamical models in which the DF depends on the kinds of non-classical integrals that we encountered in §3.2. As we saw in the last section, the DF of our Galaxy f_G cannot be of the form $f_G(\mathcal{E}, L_z)$ because $\overline{v_R^2} \neq \overline{v_z^2}$ in the solar neighborhood. In the light of the work of §3.2, which showed that most orbits in axisymmetric potentials admit an effective third integral I_3, we should not be surprised to learn that f_G depends on I_3, but historically it came as a great shock to astronomers to discover this fact. Indeed, Jeans (1915) argued that the inequality of $\overline{v_R^2}$ and $\overline{v_z^2}$ in the solar neighborhood implied that the stars near the Sun could not have settled to a steady state—in this case the Jeans theorems would not apply, and f would not be a function of integrals of motion. In

Jeans's day very little was known about the large-scale structure or the age of our Galaxy, and it was not yet evident that our Galaxy must be in an almost steady state. Furthermore, for several decades after Jeans's paper, many astronomers doubted that any third integral I_3 existed. Not until electronic computers came into widespread use about 1960 did numerical orbit calculations convince astronomers of the existence of I_3 and its importance for galactic dynamics.

We have seen that we can account for any observed $\rho(R, z)$ and $\overline{v}_\phi(R, z)$ with a DF $f(\mathcal{E}, L_z)$ that involves only two integrals. Consequently, we expect *many* DFs of the less restricted form $f(\mathcal{E}, L_z, I_3)$ to be compatible with given functions ρ and $\overline{v}_\phi$. Thus, if we wish to be guided toward a general DF $f(\mathcal{E}, L_z, I_3)$ by observations, we require knowledge of some further moments of f than just ρ and $\overline{v}_\phi$; for example, we might know the distribution over the galaxy of the line-of-sight velocity dispersion σ_p.

We illustrate the role of additional velocity information in determining our choice of the correct DF by outlining two alternative explanations for the low rotation velocities characteristic of giant elliptical galaxies. Recall from Figure 4-6b that giant elliptical galaxies are rotating much more slowly than they would be if they had isotropic velocity distributions. Thus we may safely assume that the velocity distribution at most points inside a typical giant elliptical galaxy must be anisotropic. From equation (4-95) we see that the sense of the anisotropy that is required by the observations is that $\frac{1}{2}(\overline{v_R^2} + \overline{v_\phi^2}) > \overline{v_z^2}$. Now ask whether it is $\overline{v_R^2}$ or $\overline{v_\phi^2}$ that is primarily responsible for this inequality. Consider two extreme cases: (i) Anisotropy generated by large $\overline{v_R^2}$. In our Galaxy, we have for spheroid stars near the Sun $\overline{v_R^2} \gtrsim 2\overline{v_z^2} \approx \overline{(v_\phi - \overline{v}_\phi)^2}$, i.e., the velocity ellipsoid is prolate and points in the radial direction (see Table 7-5 of MB). We know that this elongation arises because the galactic DF depends on a third integral I_3. This suggests that non-rotating elliptical galaxies may also have DFs $f_3(\mathcal{E}, L_z, I_3)$ that cause $\overline{v_R^2}$ to exceed $\overline{v_z^2}$. (ii) Anisotropy generated by large $\overline{v_\phi^2}$. On the other hand, elliptical galaxies may have DFs $f_2(\mathcal{E}, L_z)$ that depend only on $\mathcal{E}$ and L_z. In this case, $\overline{v_R^2} = \overline{v_z^2}$, and the galaxies would be flattened by large values of $\overline{v_\phi^2}$— i.e., the velocity ellipsoids would be prolate and point in the azimuthal direction. This situation could be generated in our Galaxy simply by reversing the rotation directions of half the disk stars. The real situation probably lies somewhere between these two extremes. Indeed, any linear combination $f = sf_2(\mathcal{E}, L_z) + (1-s)f_3(\mathcal{E}, L_z, I_3)$ of these DFs would also be a valid DF. Only an empirical knowledge of the ratio $\overline{v_R^2}/\overline{v_\phi^2}$ of the radial and azimuthal dispersions will enable us to choose a particular

DF from this set.

To make further progress we must develop schemes for the construction of DFs $f(\mathcal{E}, L_z, I_3)$ that generate axisymmetric models with specified observational properties. There are a number of difficulties to be overcome: (i) in general, there is no simple analytical form for the third integral $I_3(\mathbf{x}, \mathbf{v})$; (ii) the required DF is a function of three variables, and as such is much harder to determine than the DFs of one or two variables that were adequate for spherical systems; (iii) in most cases, the density profile $\rho(R, z)$ of the galaxy to be modeled may be taken to be known, but as we have seen, velocity information is also required for the determination of f, and this is usually limited in both accuracy and spatial extent.

For simplicity, we initially restrict ourselves to constructing a galaxy with a potential $\Phi(R, z)$ in which all orbits are regular and may therefore be characterized by the three actions $(J_r, J_a, J_l) = \mathbf{J}$ introduced in §3.5. Let us recall that in §4.4 we constructed models of spherical galaxies in two ways: (i) by choosing a plausible DF and then solving Poisson's equation to determine the spatial structure of the associated model; (ii) by inverting an integral equation for the DF $f(\mathcal{E})$ that generates a model with specified $\rho(r)$. Similarly, there are two routes to models of axisymmetric galaxies involving three integrals: (i) choose $f(\mathbf{J})$ and solve for $\rho(R, z)$; (ii) choose $\rho(R, z)$ and seek $f(\mathbf{J})$. We now discuss each approach in turn.

(a) From f to ρ   The procedure for evaluating the density $\rho(\mathbf{x})$ predicted by a DF of the form $f(\mathbf{J})$ is as follows. (i) Guess the form of the potential Φ_0 of the model. (ii) Establish for this potential the relationships $\mathbf{J}(\mathbf{x}, \mathbf{v})$ between the actions and the Cartesian phase-space cordinates. (iii) At each $\mathbf{x}$ evaluate $\rho(\mathbf{x}) = \int f d^3\mathbf{v}$. (iv) Calculate the potential Φ generated by this density distribution, and, if it differs significantly from that originally assumed, return to step (i) with $\Phi_0 = \Phi$ and repeat all steps until convergence is attained.

This approach is only fruitful if we have made a judicious choice of $f(\mathbf{J})$. One approach to choosing f is based on the differential energy distribution. At the end of §4.4 we saw that members of a particular sequence of spherical models, which have differing DFs $f(\mathcal{E}, L)$ but a common density profile $\rho_J(r)$, all have very similar differential energy distributions $(dM/d\mathcal{E})$ (Figure 4-15). Thus the density profiles of these models appear to be largely determined by $(dM/d\mathcal{E})$. This example suggests that DFs $f(\mathbf{J})$ that give rise to a common differential energy distribution $(dM/d\mathcal{E})$ may generate models having similar overall density profiles even when the shapes and internal kinematics of the models are very dissimilar. Thus, to construct a variety of axisymmetric models with similar radial density profiles but different shapes and internal dynamics, we first compute $(dM/d\mathcal{E})$ for the spherical model with $f = f_0(\mathcal{E})$ and the desired density profile (§4.4.5). We then modify $f_0(\mathcal{E})$ by shifting stars over the energy hypersurfaces in action space. For example, if we push the stars toward the J_a-axis, we flatten the system, while if we shift them toward the J_l-axis, the system becomes prolate. Shifting stars toward the J_r-axis enhances the radial velocity dispersion but has no direct effect on the shape. For further detail, see Binney (1987).

(b) From ρ to f   The route from the density to the DF of an axisymmetric system is most easily understood by considering how to obtain f numerically. In any numerical scheme we will have to be satisfied with determining $f(\mathbf{J})$ on a grid of N points in action space $\mathbf{J}_\alpha$, $\alpha = 1, \ldots, N$. Associated with this grid is a partition of action space into N cells, with cell α having volume $\Delta^3 \mathbf{J}_\alpha$. In the crudest numerical scheme, we assume that all of the mass in cell α, $M_\alpha = (2\pi)^3 f(\mathbf{J}_\alpha)\Delta^3 \mathbf{J}_\alpha$, is concentrated at the point $\mathbf{J}_\alpha$. Similarly, we partition real space into K cells, centered on positions $\mathbf{x}_\beta$, and with associated volumes $\Delta^3 \mathbf{x}_\beta$. Since we can determine the potential from the density, we can compute the probability $P_{\beta\alpha}$ that a star on orbit α is found in spatial cell β. Thus the stars in cell α of action space contribute mass $P_{\beta\alpha}M_\alpha$ to cell β of real space. If the required density in cell β of real space is $\rho_\beta \equiv \rho(\mathbf{x}_\beta)$, then a satisfactory model will solve the K linear equations,

$$
\begin{aligned}
\rho_\beta \Delta^3 \mathbf{x}_\beta &= \sum_{\alpha=1}^{N} P_{\beta\alpha} M_\alpha \\
&= (2\pi)^3 \sum_{\alpha=1}^{N} P_{\beta\alpha} \Delta^3 \mathbf{J}_\alpha f(\mathbf{J}_\alpha).
\end{aligned}
\tag{4-179}
$$

The obvious next step is to set $K = N$ and solve the system of equations (4-179) by standard numerical techniques. However, there is no guarantee that the resulting values $f(\mathbf{J}_\alpha)$ will be non-negative, as is physically required. In addition, we have seen that in general there are many DFs consistent with a given $\rho(R, z)$, and thus we should treat with reserve *any* algorithm that yields a unique DF.

In practice, it is best to choose $K > N$, that is, to have more grid points in action space than in real space. In this case, equations (4-179) are likely to have many solutions—and many of these may be physically acceptable in that they involve no negative $f(\mathbf{J}_\alpha)$. To proceed further, we need a numerical scheme that selects only physically acceptable solutions, and, furthermore, isolates one particular solution out of the many available. A number of such schemes have been applied to this problem. They include:

(i) Linear programming (Schwarzschild 1979, 1982). This technique finds the unique solution of equations (4-179) that satisfies the constraints $f(\mathbf{J}_\alpha) \geq 0$ and maximizes a "profit function" $P = \sum_\alpha p_\alpha f(\mathbf{J}_\alpha)$, where the constants p_α are chosen by the programmer. Suitable profit functions include L_z, the velocity dispersion in any direction at any point, or indeed any moment of the DF. For further details, see Richstone (1980, 1982), Richstone and Tremaine (1984), Pfenniger (1984b), and Statler (1986).

(ii) Lucy's (1974) algorithm. This is an iterative procedure for generating approximate solutions to sets of linear equations of the form (4-179), subject to the restriction $f(\mathbf{J}_\alpha) \geq 0$. The method yields a DF that approximates an exact solution more and more closely as the number of iterations is increased. The particular solution toward which the iteration converges is determined by the trial solution from which the iteration begins. The method is simple to program and generates smooth DFs; also, since it runs quickly it is possible to use finer grids than one can with linear programming. A disadvantage is that there is no simple characterization of the criterion by which the particular solution is chosen. For details, see Newton and Binney (1984) and Newton (1986).

(iii) Maximum entropy. In this approach we find the unique solution that satisfies equations (4-179) and maximizes a nonlinear profit function $P = \sum_\alpha C_\alpha[f(\mathbf{J}_\alpha)]$, where the C_α are functions chosen by the programmer. One natural profit function is the entropy $P_B \equiv -\text{const} \times \int f \ln f \, d^3\mathbf{J}$; in this case $C_\alpha(y) = -y \ln y \, \Delta^3 \mathbf{J}_\alpha$. As in the case of Lucy's method, the maximum-entropy solution must be determined iteratively, and there is no guarantee that a solution exists. Even if a solution is found, it may not be physically acceptable because f may somewhere be negative. However, by choosing the functions $C_\alpha(y)$ to be strongly negative for $y < 0$, we can generally ensure that the maximum-profit solution is physically acceptable. For details see Richstone (1987).

4.6 Triaxial Systems

There are at least two lines of evidence that triaxial stellar systems are common. First, at least half of spiral galaxies show clear signs of a bar.

Second, it appears that many—possibly most—elliptical galaxies are triaxial (see MB, §5-2). Constructing models of triaxial systems is yet more difficult than building axisymmetric models, and much remains to be done in this area. However, the Jeans theorem enables us to construct one special kind of triaxial system that is comparatively straightforward to understand.

In a nonrotating triaxial potential, the only isolating integral that is always available is the energy E. It can be shown that all self-consistent systems with $f = f(E)$ must be spherical.[13] In a triaxial potential that is stationary in a frame rotating with uniform angular velocity, the only generally available integral is the Jacobi constant E_J [eq. (3-83b)]. It is natural to ask whether useful models can be obtained by assuming that the DF is a function $f(E_J)$ of Jacobi's integral. We shall see that the answer to this question is a qualified "yes". However, it will help us to understand why the answer is only a qualified yes, if we first step out of the realm of stellar dynamics into a very old branch of applied mathematics—the investigation of the possible configurations of a rotating, self-gravitating mass of incompressible fluid, such as water.

1 Equilibrium Bodies of Incompressible Fluid

The study of rotating fluid bodies was initiated by Isaac Newton himself. Newton treated the Earth as a homogeneous, incompressible fluid body in order to estimate the expected flattening of the earth at the poles.[14] His treatment was only valid for small rotation speeds, but in 1742 the Scottish mathematician C. Maclaurin found an exact solution for the equilibrium of a rotating body. In Maclaurin's solutions, now called **Maclaurin spheroids**, the fluid surface is a flattened axisymmetric figure. The surprise came in 1834 when a German, C. G. Jacobi, showed that, rather than forming a Maclaurin spheroid, a rotating fluid mass might form a triaxial body. It is a measure of the unexpectedness of this result that Jacobi's discovery came over ninety years after Maclaurin's. The relationship of the **Jacobi ellipsoids**, as these configurations are now called, to the simpler Maclaurin spheroids remained obscure until the

---

[13] This is a straightforward extension of Lichtenstein's theorem for barotropic stars; see Tassoul (1978), §4.3.

[14] Unfortunately, the best astronomical evidence available at the time, and for years after Newton's death, suggested that the Earth was actually prolate, a point of view that was vigorously supported by I. Cassini. The controversy was settled only in 1736 when the Academy of Sciences in Paris sent surveying parties to the Arctic and the Equator. When the leader of the Arctic party, Maupertuis, returned to Paris after suffering hunger and shipwreck with proof that the Earth is oblate, Voltaire congratulated him on having "flattened the poles and Cassini".

end of the nineteenth century, when H. Poincaré and others showed that the Jacobi ellipsoids are actually the preferred configurations of rapidly rotating fluid bodies because they have lower energy for fixed angular momentum and mass. Meanwhile, however, Riemann went one step further than Jacobi by showing that even the Jacobi ellipsoids are only special members of a much larger family of triaxial equilibrium configurations, the **Riemann ellipsoids**.

The importance of the Jacobi ellipsoids for galactic dynamics is that their very existence suggests that a rapidly rotating galaxy may not remain axisymmetric. The Riemann ellipsoids are important because they draw attention to a distinction between the rate at which the *matter* of a rotating triaxial body streams and the rate at which the *figure* of the body rotates. This distinction is very important for the bars of disk galaxies because the pattern speed governs the interaction of the bar with the rest of the galaxy via its rotating potential, while the streaming velocity of the bar's stars determines the observed velocities.

(a) Maclaurin spheroids   In equilibrium, the velocity $\mathbf{v}$, the pressure p, the density ρ, and the gravitational potential Φ of a fluid body are related by Euler's equation [see eq. (1E-8)],

$$(\mathbf{v} \cdot \nabla)\mathbf{v} = -\nabla\Phi - \frac{1}{\rho}\nabla p. \tag{4-180}$$

When we are dealing with an incompressible fluid it is useful to define a **pseudo-potential** η by

$$\eta = \Phi + \frac{p}{\rho}, \tag{4-181}$$

in terms of which equation (4-180) becomes

$$(\mathbf{v} \cdot \nabla)\mathbf{v} = -\nabla\eta. \tag{4-182}$$

Now assume that the body rotates with constant angular speed Ω about the z-axis of the cylindrical coordinates (R, ϕ, z) and the Cartesian coordinates (x, y, z). Then

$$\mathbf{v} = \Omega(-y\hat{\mathbf{e}}_x + x\hat{\mathbf{e}}_y), \tag{4-183}$$

and hence

$$(\mathbf{v} \cdot \nabla)\mathbf{v} = -\Omega^2(x\hat{\mathbf{e}}_x + y\hat{\mathbf{e}}_y) = -\tfrac{1}{2}\Omega^2\nabla R^2. \tag{4-184}$$

Equation (4-182) now becomes

$$\nabla\left(\eta - \tfrac{1}{2}\Omega^2 R^2\right) = 0, \tag{4-185}$$

or on integrating,

$$\eta - \tfrac{1}{2}\Omega^2 R^2 = \text{constant} = \left(\Phi - \tfrac{1}{2}\Omega^2 R^2\right)_{\text{surface}}, \tag{4-186}$$

where the second equality follows because the pressure p must vanish on the surface. From this equality it follows that the surface of the body is a level surface of the effective potential $\Phi_{\text{eff}} \equiv \left(\Phi - \tfrac{1}{2}\Omega^2 R^2\right)$ in the rotating frame in which the fluid is stationary [see eq. (3-89a)].[15]

Maclaurin's inspiration was to assume that the body's surface forms an oblate spheroid. Table 2-2 shows that in this case the potential inside the body is

$$\Phi(R, z) = \pi G\rho(A_1 R^2 + A_3 z^2) + \text{constant}, \tag{4-187}$$

where $A_1(e)$ and $A_3(e)$ are functions of the ellipticity e of the surface. Thus the effective potential of the body is given by

$$\frac{\Phi_{\text{eff}}(R, z)}{\pi G\rho} = R^2 \left(A_1 - \frac{\Omega^2}{2\pi G\rho}\right) + A_3 z^2 + \text{constant}. \tag{4-188}$$

If the semi-major axis of the surface is of length a, the equation of the surface is

$$R^2 + \frac{z^2}{1 - e^2} = a^2, \tag{4-189}$$

so that the condition (4-186) that a level surface of the effective potential coincides with the fluid surface is

$$\left(A_1 - \frac{\Omega^2}{2\pi G\rho}\right) \Big/ A_3 = (1 - e^2). \tag{4-190a}$$

Hence

$$\frac{\Omega^2}{2\pi G\rho} = A_1(e) - (1 - e^2)A_3(e). \tag{4-190b}$$

If we take $A_1(e)$ and $A_3(e)$ from Table 2-1, this becomes

$$\frac{\Omega^2}{2\pi G\rho} = \frac{\sqrt{1 - e^2}}{e^3}\left[(3 - 2e^2)\arcsin e - 3e\sqrt{1 - e^2}\right], \tag{4-191}$$

which is the defining equation of the Maclaurin spheroids. In the limit $e \to 0$ we have

$$\Omega^2 \simeq \frac{8\pi}{15}G\rho e^2, \tag{4-192}$$

---

[15] Equation (4-186) could equally well have been derived by requiring hydrostatic equilibrium in the frame in which the fluid is at rest.

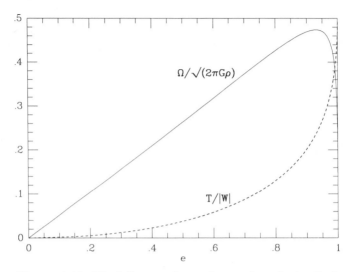

Figure 4-16. The full curve shows the angular velocity Ω of a Maclaurin spheroid as a function of the spheroid's eccentricity $e = \sqrt{1 - b^2/a^2}$. The dashed curve shows the ordered kinetic energy parameter $t \equiv T/|W|$ for Maclaurin spheroids.

which is Newton's original result relating the Earth's rotation to its oblateness.

In Figure 4-16 we plot the relationship between Ω and e given by equation (4-191). Notice that Ω attains a maximum value $\Omega_{\max} = 0.474\sqrt{2\pi G\rho}$ at $e = 0.930$ (which corresponds to $\epsilon = 0.632$), and then falls off to zero as $e \to 1$. Ω peaks in this way because the surface density of the spheroid falls off as a grows, and this diminishes the gravitational self-attraction of the system.

A convenient measure of the importance of rotation for the structure of any self-gravitating body is the ratio t of the rotational kinetic energy T to the body's self-gravitational potential energy W;

$$t \equiv \frac{T}{|W|}. \tag{4-193}$$

By the virial theorem we have $0 \le t \le \frac{1}{2}$. For a Maclaurin spheroid,

$$T = \tfrac{1}{2}\rho\Omega^2 \int R^2 d^3\mathbf{x} = \frac{4\pi}{15}\rho\Omega^2\sqrt{1-e^2}\,a^5, \tag{4-194}$$

while W is given by Table 2-2. Combining T and W we obtain

$$t = \frac{3}{2e^2} - 1 - \frac{3\sqrt{1-e^2}}{2e\arcsin e}. \tag{4-195}$$

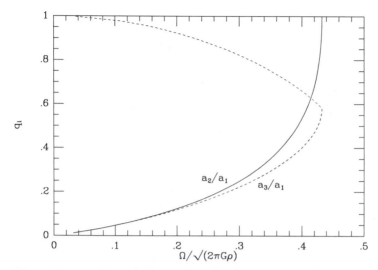

Figure 4-17. The axis ratios of Jacobi ellipsoids as a function of the angular velocity Ω with which their figures rotate. The upper branch of the dashed curve shows the axis ratio of the Maclaurin spheroids as a function of the angular velocity with which their material streams.

For a sphere $t = 0$, and for a flat disk $t = \frac{1}{2}$, as is required by the virial theorem. The value of t for arbitrary e is shown in Figure 4-16.

(b) Jacobi ellipsoids   Jacobi realized that the assumption of axisymmetry in the derivation we have just presented is unnecessary. The effective potential of an ellipsoidal fluid mass in the corotating frame is given by

$$\frac{\Phi_{\text{eff}}(x,y,z)}{\pi G \rho} = x^2\left(A_1 - \frac{\Omega^2}{2\pi G \rho}\right) + y^2\left(A_2 - \frac{\Omega^2}{2\pi G \rho}\right) + z^2 A_3 + \text{constant},$$

(4-196)

where Table 2-1 gives the A_i as functions of the lengths a_i of the body's semi-axes. Thus, if one of the level surfaces of Φ_{eff} is to coincide with the bounding surface of the fluid, we require

$$\frac{A_1 - \dfrac{\Omega^2}{2\pi G \rho}}{A_3} = \frac{a_3^2}{a_1^2} \quad ; \quad \frac{A_2 - \dfrac{\Omega^2}{2\pi G \rho}}{A_3} = \frac{a_3^2}{a_2^2}.$$

(4-197)

Eliminating Ω from these equations, we obtain

$$A_3 = \frac{a_1^2 a_2^2}{a_3^2(a_2^2 - a_1^2)}(A_1 - A_2).$$

(4-198)

Given any value $q_3 \equiv (a_3/a_1)$ of the ratio of the lengths of two of the semi-axes, we may solve equation (4-198) for the other axial ratio $q_2 \equiv (a_2/a_1)$. Without loss of generality we may assume that $a_1 \geq a_2 \geq a_3$, and hence confine q_2 to the range $1 \geq q_2 \geq q_3$. Of course, a solution $q_2 = 1$, corresponding to a Maclaurin spheroid, always exists, but when $q_3 < 0.583$ a second solution to equation (4-198) for q_2 exists. One says that the Jacobi sequence **bifurcates** from the Maclaurin sequence at axial ratio $q_3 = 0.583$ (which corresponds to $e = 0.813$). The axial ratios q_i are plotted as a function of Ω in Figure 4-17.

(c) Riemann ellipsoids   The Jacobi ellipsoids can be generalized by permitting motion of the fluid relative to the ellipsoidal boundary. In the simplest case, the flow vectors are perpendicular to the angular velocity vector Ω. Riemann showed that for any pair of axis ratios q_2 and q_3 for which a Jacobi bar exists, it is possible to construct a one-parameter family of such models. This family extends from the Jacobi ellipsoids, which have rotating boundaries with respect to which the fluid is stationary, to the **Dedekind ellipsoids**, in which the fluid flows inside a stationary bounding surface. For a full account of the theory of rotating fluid masses, see Chandrasekhar (1969).

2 Systems with Distribution Functions $f(E_J)$

Let us now ask whether we can construct a triaxial stellar system with a DF of the form $f(E_J)$. What would be the properties of such a system? Since $f(E_J) = f(\frac{1}{2}v^2 + \Phi_{\text{eff}})$, where v and $\Phi_{\text{eff}} = \Phi - \frac{1}{2}\Omega^2 R^2$ are the velocity and effective potential (3-89a) in the rotating frame, we will find $\overline{\mathbf{v}} = 0$ and $\sigma_{ij}^2 = \sigma\delta_{ij}$. Hence in the rotating frame, the Jeans equations for the stellar system will be identical with the equation of hydrostatic support that is satisfied by a Jacobi ellipsoid, and the existence of the latter suggests that analogous stellar systems are possible.

A simple form of f that will give us insight into what we can achieve by taking f to be of the form $f(E_J)$ is the power-law

$$f(E_J) = \begin{cases} F(\Psi_{\text{eff}} - \frac{1}{2}v^2)^{(n-\frac{3}{2})}, & (\Psi_{\text{eff}} > \frac{1}{2}v^2) \\ 0 & (\Psi_{\text{eff}} \leq \frac{1}{2}v^2), \end{cases} \tag{4-199}$$

where $\Psi_{\text{eff}} \equiv -\Phi_{\text{eff}} + \Phi_0$ in the usual way. Of course this form of f is just the rotating analog of the DF (4-105) of the polytropes. Thus the density inside the bar is by analogy with equations (4-107)

$$\rho = c_n \Psi_{\text{eff}}^n, \tag{4-200}$$

and inside the bar Poisson's equation reads

$$\nabla^2 \Psi_{\text{eff}} = -4\pi G c_n \Psi_{\text{eff}}^n + 2\Omega^2. \qquad (4\text{-}201)$$

We choose the constant Φ_0 such that $\Psi_{\text{eff}} = 0$ at the edge of the bar [see eq. (4-186)].

Equation (4-201) has been studied, among others, by Jeans (1919), James (1964), and Ipser and Managan (1984), because it is the equation satisfied by the relative potential of a uniformly rotating body of polytropic gas, and is therefore of importance in the theory of rotating stars (e.g., Tassoul 1978). It cannot be solved by analytical means when $n \neq 0$, but its solutions have been calculated numerically. It is found that non-axisymmetric solutions exist for $n \leq 0.808$, but not for $n > 0.808$. Since we require $n > 0.5$ for c_n to be finite [see eq. (4-107b)], we conclude that triaxial stellar-dynamical polytropic models are possible for n in the range $0.5 < n \leq 0.808$ (Vandervoort 1980).

None of these polytropic models exhibits strong central concentration; even in the limiting case $n = 0.808$, the central density is only 3.12 times the mean density. Furthermore, no triaxial stellar model whose DF depends only on E_J will be much more inhomogeneous than this, as the following argument shows.

In Chapter 2 we saw that the equipotential surfaces generated by a distribution of matter are always more spherical than the matter distribution itself. The Jacobi ellipsoids nonetheless manage to have level surfaces of the *effective* potential that are the same shape as their matter distributions, by maintaining a delicate balance between the gravitational and the centrifugal components of the effective potential. Thus, for a homogeneous ellipsoid,

$$\begin{aligned}
\Phi_{\text{eff}}(\mathbf{x}) &= \Phi(\mathbf{x}) - \tfrac{1}{2}\Omega^2(x^2 + y^2) \\
&= \tfrac{1}{2}\left[x^2\left(\Omega_x^2 - \Omega^2\right) + y^2\left(\Omega_y^2 - \Omega^2\right) + z^2\Omega_z^2 \right] + \text{constant},
\end{aligned}$$
$$(4\text{-}202)$$

where $\Omega_x < \Omega_y < \Omega_z$ are frequencies determined by the axis ratios a_y/a_x and a_z/a_x and the density of the ellipsoid. An appropriate choice of Ω makes the ratio $\left[(\Omega_x^2 - \Omega^2)/(\Omega_y^2 - \Omega^2)\right]^{1/2}$ equal to the ratio a_y/a_x of the lengths of the x and y semi-axes and thus ensures that Φ_{eff} is constant on the surface of the ellipsoid. Now consider whether we can repeat this trick for an inhomogeneous ellipsoid. Suppose such a body were to exist. Then $\Phi(\mathbf{x})$, which would no longer be quadratic in the coordinates, would rise at small radii more rapidly than the centrifugal component $\tfrac{1}{2}\Omega^2 R^2$ of the effective potential, while at large radii $\Phi(\mathbf{x})$ would rise less rapidly than $\tfrac{1}{2}\Omega^2 R^2$. Therefore, when we set Ω so as to ensure that

Φ_{eff} is constant on the ellipsoidal surface of the body, these two terms would make comparable contributions to the effective potential near the surface, and at small radii the effective potential would be dominated by Φ. Hence, near the center $\Phi_{\text{eff}} \approx \Phi$, and since $\rho = \rho(\Phi_{\text{eff}})$, the isodensity surfaces would nearly coincide with the surfaces of constant Φ. On the other hand, if the body is really centrally concentrated, the shape of the central isopotential surfaces would be dominated by the distribution of matter at the center, and the central surfaces of constant Φ would be markedly more spherical than the isodensity surfaces. From this contradiction we conclude that there can be no triaxial configurations of a very inhomogeneous stellar system with a DF $f(E_J)$ that depends on Jacobi's integral alone.

Hence, models with $f = f(E_J)$ cannot satisfactorily represent elliptical galaxies. However, the bars of disk galaxies do appear to be less centrally concentrated than the other components of galaxies (Kormendy 1982). Thus models of this type may capture some of the essential physics of real bars. But the complexity of rapidly rotating stellar bars is such that it may be many years before we have a complete picture of how they work.

3 Freeman's Analytic Bars

Freeman (1966a,b) developed a series of models that are generalizations of the Kalnajs disks to the non-axisymmetric domain. Since Freeman's models are complicated and somewhat artificial, we shall not give a full account of them here. However, they are the only analytic bar models in the literature, so a brief summary of their properties is in order. For more information, see Hunter (1970).

Freeman's bars are based on the simple orbital structure of a two-dimensional harmonic oscillator potential,

$$\Phi(\mathbf{x}) = \tfrac{1}{2}\Omega_x^2 x^2 + \tfrac{1}{2}\Omega_y^2 y^2 + \text{constant}, \tag{4-203}$$

which rotates at angular frequency Ω. In §2.6.1 we saw that this potential is generated by a flat elliptical disk with surface density

$$\Sigma(\mathbf{x}) = \Sigma_0 \sqrt{1 - \frac{x^2}{a^2} - \frac{y^2}{b^2}}. \tag{4-204}$$

In §3.3.2 we showed that prograde orbits in the potential (4-203) have two isolating integrals and may be decomposed into retrograde motion around an epicycle, superposed on prograde motion of the guiding center. The guiding center moves around an ellipse that is centered on the same point as the potential and shares the potential's major axis. In the notation of equations (3-98), the guiding center's orbit has semi-axes of lengths X_1 and Y_1, and period $2\pi/\alpha$. The retrograde epicyclic motion has period $2\pi/\beta$. It takes place on an ellipse that is elongated parallel to the short axis of the potential and has semi-axes of lengths X_2 and Y_2. The ratios X_1/Y_1 and X_2/Y_2 are fixed by Ω and the potential.

Retrograde orbits are similarly superpositions of elliptic motions at frequencies α and β, the only difference being that now the guiding center moves in a retrograde sense at frequency β, while the epicyclic motion is prograde at frequency α.

For either prograde or retrograde orbits, the two isolating integrals may be taken to be X_1 and X_2, and Freeman constructed self-consistent bars by seeking DFs $f(X_1, X_2)$ that generate the surface density (4-204). The following are his main results.

(i) Models can be constructed with all possible axis ratios a/b, so long as $\Omega/\Omega_y < 1$ and $\Omega/\Omega_x < 1$, i.e., so long as the effective potential

$$\Phi_{\text{eff}} \equiv \Phi - \tfrac{1}{2}\Omega^2 R^2 = \tfrac{1}{2}(\Omega_x^2 - \Omega^2)x^2 + \tfrac{1}{2}(\Omega_y^2 - \Omega^2)y^2 + \text{constant} \tag{4-205}$$

has a minimum at the center.

(ii) Models can be constructed with pattern speeds Ω that range from zero to the smaller of Ω_x and Ω_y.

(iii) In the circular limit $b = a$, the models are Kalnajs disks.

(iv) In the frame that rotates with the bar, stars may stream in the retrograde sense. Indeed, this retrograde motion can be so strong that even in an inertial frame, the mean stellar rotation velocity can be in the direction opposite to that of the figure's rotation. This effect is most pronounced along the minor axis of the bar. Furthermore, the retrograde streaming causes the total angular momentum vector of some of Freeman's bars to be anti-aligned with the angular velocity vector Ω.

4 General Triaxial Systems

The triaxial systems we have examined so far are very special. The structure of a general triaxial system depends strongly on the speed at which the figure of the model rotates (the pattern speed). Consider first the case of zero pattern speed. The action space of such a system is partitioned between the major orbit families described in §3.4. The orbits of each of the three major families occupy the volume of action space adjacent to the axis on which lie the closed orbits that sire that family: the box orbits adjoin the J_r axis, the short-axis tubes the J_a axis, and the long-axis tubes the J_l axis. The proportions of action space that are assigned to each family depend on the shape of the potential. In a highly flattened, nearly axisymmetric potential, most of action space is assigned to the short-axis tubes. If the potential is nearly axisymmetric about its longest axis, the long-axis tubes dominate. Boxes are favored by potentials that have no two axes of similar length.

Just as in the case of the axisymmetric systems investigated in §4.5.3, we can construct models starting either from specified $\rho(\mathbf{x})$ or from $f(\mathbf{J})$. To date all investigators have taken the density distribution to be given, and have used linear programming (Schwarzschild 1979; Merritt 1980) or Lucy's method (Statler 1986) to find orbit distributions consistent with $\rho(\mathbf{x})$. An alternative approach—which in fact led to the first self-consistent triaxial model galaxy—is to use an N-body program to follow the collapse and relaxation toward equilibrium of a collection of stars from triaxial initial conditions (Aarseth & Binney 1978; Wilkinson & James 1981).

When the pattern speed is non-zero, the situation becomes more complicated. If the pattern speed is large enough that resonances like the Lindblad and corotation resonances (§3.3.3) lie within the body of the system, additional orbit families appear and stochastic orbits begin to occupy an appreciable portion of phase space (Contopoulos 1986). In these circumstances, the Jeans theorem is of limited value, and the only rapidly rotating triaxial systems in the literature are the endpoints of N-body calculations (Hohl & Zang 1979).

4.7 The Choice of Equilibrium

We have seen that the range of equilibrium configurations accessible to a collisionless stellar system is large. The question we now have to address is, what determines the particular configuration to which a given stellar system settles. Two classes of explanation are in principle possible: (i) The configuration actually adopted is favored by some fundamental physical principle in the same way that the velocity distribution of an ideal gas always relaxes toward the Maxwell-Boltzmann distribution. (ii) The present configuration of the galaxy is simply a reflection of the particular initial conditions that gave rise to the galaxy's formation, in the same way that the shape of a particular stone in a field is due to particular circumstances rather than to any general physical principle. These two classes of explanation are not mutually exclusive. For example, the meandering course of the Mississippi River has been determined more by chance than by any fundamental principle, yet many characteristics of the river channel—typical sizes of elbows and the formation of oxbow lakes—can be understood in terms of simple physical arguments. Yet it is profitable to analyze the question of how we may understand the large-scale structures of galaxies in terms of first one and then the other of these points of view. We start by asking whether the present states of galaxies are simply more probable than any other configurations.

1 The Principle of Maximum Entropy

In the 1890s J. W. Gibbs discovered that the standard relations between the thermodynamic variables of simple systems could be derived by hypothesizing that the probability dp that the system would be found to be in any small volume $d\tau$ of its phase space is proportional to $e^{-\beta H}d\tau$, where β is a parameter he identified as the inverse of the system's temperature, and H is the system's Hamiltonian. Since Gibbs's day there have been almost as many attempts to explain why this hypothesis correctly predicts experimental results as there have been books written on statistical mechanics. Unfortunately, after all these decades of debate, scientists are still far from agreement on this question. However, it is generally agreed that Gibbs's hypothesis can be derived from an alternative principle, that of **maximum entropy**: the correct thermodynamic relations for any physical system may be derived by seeking the probability density p that maximizes the **entropy**

$$S \equiv -\int_{\text{phase space}} p\ln p\,d\tau, \tag{4-206}$$

subject to all relevant constraints. Can we derive the structures of galaxies from this principle?

In §4.1.1 we said that the DF f is best thought of as a probability density. It is therefore tempting to substitute $p = f$ and $d\tau = d^3\mathbf{x}d^3\mathbf{v}$ in equation (4-206) and seek the form of f that maximizes S subject to given values of the galaxy's mass M and energy E.[16] Clearly M and E

---

[16] Strictly, the phase space of a galaxy of N stars is $6N$-dimensional, and the infinitesimal $d\tau$ in equation (4-206) refers to an element of this phase space rather than an element of the phase space of a single star. However, we may neglect correlations between the particles of a collisionless system (see §8.1.3), so the probability $pd\tau$ associated with a range of configurations in the $6N$-dimensional phase space of the whole galaxy is just the product of factors $fd^3\mathbf{x}d^3\mathbf{v}$ associated with individual stars. It is straightforward to show that S is then simply N times the entropy $-\int f\ln f d^3\mathbf{x}d^3\mathbf{v}$ associated with an individual star.

are determined by f, so this is a well-defined procedure. Ogorodnikov (1965) and Lynden-Bell (1967a) show that this calculation leads to the conclusion that S is extremized if and only if f is the DF (4-116) of the isothermal sphere. However, the isothermal sphere is a system with infinite mass and energy. Hence this calculation shows that the maximization of S subject to fixed M and E leads to a DF that is incompatible with finite M and E. From this contradiction it follows that *no* DF that is compatible with finite M and E maximizes S: if we constrain only M and E, configurations of arbitrarily large entropy can be constructed by suitable rearrangements of the galaxy's stars.[17] The reason why this is so is easily explained.

Suppose we have a spherical galaxy of total mass M and binding energy $|E|$. We mentally divide the system into a main body of mass M_1 and gravitational radius r_1, and an outer envelope of mass $M_2 \ll M$ and radius r_2. By the virial theorem (4-81) the binding energy $|E_1|$ of the main body is of order GM_1^2/r_1, and the binding energy of the envelope is $|E_2| \approx GM_1M_2/r_2$. Then we shrink the main body by a small amount ϵ, so that its radius changes from r_1 to $(1-\epsilon)r_1$. The shrinkage releases an energy $\Delta E \approx \epsilon GM_1^2/r_1$. We deposit this energy in the outer envelope, which swells in response to a radius r_2' given by $|E_2'| = |E_2 + \Delta E| \approx GM_1M_2/r_2'$.

The velocity dispersion in the swollen envelope is $\sigma_2' \approx \sqrt{GM_1/r_2'}$. Thus the volume $V \approx \sigma_2'^3 r_2'^3$ of the region of phase space over which the representative points of the envelope's stars are distributed is $V \approx (GM_1r_2')^{3/2}$. Finally, by equation (4-206) the entropy S of the envelope is

$$S = -\int f \ln f \, d^3\mathbf{x} \, d^3\mathbf{v} + \text{constant} \approx -\frac{N_2}{V} \ln\left(\frac{N_2}{V}\right) V + \text{constant}$$

$$\approx \tfrac{3}{2} N_2 \ln(r_2') + \text{constant} \approx -\tfrac{3}{2} N_2 \ln(|E_2 + \Delta E|) + \text{constant},$$

(4-207)

where N_2 is an unimportant normalization constant. Thus the entropy of the envelope tends to infinity as ΔE tends to $|E_2|$. It is easy to see that the entropy of the main body changes by only a finite amount as a result of the energy transfer. Hence the entropy of the combined system increases without limit. It follows that we can always increase the entropy of a self-gravitating system of point masses at fixed total mass and energy by increasing the system's degree of central concentration.

---

[17] It has sometimes been argued that the principle of increase of entropy dooms the Universe to a "heat death", in which all matter is in a uniform, isothermal, maximum-entropy state. This argument is clearly invalid once gravitational forces are included, since then there is no maximum-entropy state.

From this discussion we conclude that galaxies, unlike cold white dwarfs, for example, are not in long-term thermodynamic equilibrium, but rather resemble main-sequence stars in constantly evolving to states of higher central concentration and entropy. Consequently, if we wish to understand their present configurations, we must investigate the initial conditions from which they started in life, and the rates at which various dynamical processes occur within them. In the following, we discuss two important relaxation processes and then we draw on our knowledge of cosmology to ask by what path galaxies may have evolved to their present configurations.

2 Phase Mixing and Violent Relaxation

We do not yet know how galaxies formed. Observations of the large-scale structure of the Universe (Peebles 1980b) and of the cosmic microwave background (Weinberg 1972) suggest that for many thousand years after the Big Bang the Universe was remarkably homogeneous. Since then, part or all of the matter in the Universe has broken up into the gravitationally bound clumps that we call galaxies. It is probable that the radiative properties of matter played an important role in this process. However, in view of the complexity of this problem, a worthwhile first step is to investigate how galaxies might have formed in the simplest case, where each element of matter moved as a point mass under the mean gravitational potential that is generated by all neighboring matter.

The most powerful method of investigating how systems of point particles evolve under the influence of gravity is direct numerical simulation, and we shall describe some simulations a little later. First we isolate two processes that, between them, dominate the early evolution of a system of point particles.

(a) Phase mixing Consider a collection of N independent pendulums that are all of the same length L, and therefore all have the same dynamical properties. Initially all the pendulums are swung back so that they make angles θ with the vertical that are uniformly distributed in the interval $\theta_0 \pm \frac{1}{2}\Delta\theta$, where $\Delta\theta \ll \theta_0$. Each pendulum is given a small random velocity so that the momenta $p = L^2\dot{\theta}$ are uniformly distributed in the range $\pm\Delta p$. When the pendulums are released, they start to oscillate. This situation is shown schematically in Figure 4-18—of course, in reality the phase trajectories will be closed curves of constant energy,

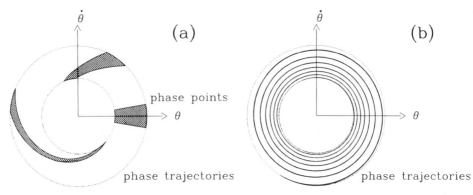

Figure 4-18. Schematic representation of how an initially compact group of phase points winds up into a larger region of lower coarse-grained phase-space density. (a) Initially the phase points fill the wedge on the $\dot\theta = 0$ axis. As time passes, and the phase-points move on circles like those shown, this wedge is drawn out into a band of ever-decreasing width. (b) After several dynamical times the coarse-grained phase-space density is approaching uniformity in the annulus shown.

but not circles. The period of each pendulum depends on its amplitude, and therefore on its energy. The more energetic pendulums, which in Figure 4-18 move on circles of larger radius, oscillate slowest. Thus the patch formed in the figure by the phase-space points of the pendulums is gradually sheared out into a spiral of ever-diminishing pitch angle, which remains confined to the area between the minimum and maximum energy curves.

The evolution of the whole system of pendulums may be described by the collisionless Boltzmann equation. According to this equation, the density of phase points in an infinitesimal volume around the phase point of any particular pendulum is constant. Consequently, the density of phase points in the spiral into which the occupied patch in Figure 4-18a is sheared is the same as the density in the original patch.

A macroscopic observer estimates the coarse-grained DF $\overline{f}$ introduced in §4.1.1 by counting how many pendulums have phase-space coordinates in each of a number of cells of finite size. Initially $\overline{f} = f$, but at late times the spiral in which the phase points of the pendulums lie has been wound so tightly that any of the macroscopic observer's phase-space cells that intersect the spiral will contain both strips of the spiral and strips that are empty of phase points. When the observer works out the mean density within each of his cells, he finds that $\overline{f}$ is constant throughout the annulus in phase space between the limiting energy curves. Since the area of the annulus is larger than the area of the small patch within which the phase points originally lay, at late times $\overline{f}$ is smaller than f. If the initial spread in momenta Δp is small, then $\overline{f}$ will be much smaller than f.

The process that causes $\overline{f}$ to decrease in this way as the pendulums get out of phase with one another is called **phase mixing**. An example is given in Problem 5-3.

The significance of the collisionless Boltzmann equation for the relaxation of galaxies to equlibrium configurations is therefore rather subtle: it does not ensure that the empirically measurable DF $\overline{f}$ is constant along stellar orbits, but rather it prevents this phase-space density actually *increasing* along an orbit. One can always diminish the coarse-grained phase-space density by mixing the incompressible fluid of phase points with "air" in the form of unpopulated regions of phase-space, but one cannot in this or any other way compress the phase fluid.

The entropy defined by equation (4-206) is time-independent since f is constant along every orbit. However, if in equation (4-206) we replace f by $\overline{f}$, we obtain an entropy $\overline{S}$ that in general increases in time; it can be shown that any decrease in the value of $\overline{f}$ along orbits, such as occurs during phase mixing, causes $\overline{S}$ to increase, just as the entropies of familiar thermodynamic systems increase when their different parts come into thermal equilibrium. In particular, we may state that no isolated collisionless stellar system A could evolve into a system B for which the entropy $\overline{S}(B) < \overline{S}(A)$. For discussion of this idea, see Tremaine et al. (1986).

(b) Violent relaxation   Phase mixing plays an important role during the relaxation of a galaxy of stars toward a steady state. But another relaxation process, known as **violent relaxation**, is also at work (Lynden-Bell 1967a). While phase mixing changes the coarse-grained phase-space density near the phase point of each star, violent relaxation changes the energies of the stars themselves. When a star moves in a fixed potential Φ, its energy $E = \frac{1}{2}v^2 + \Phi$ is constant. But if Φ is a function $\Phi(\mathbf{x}, t)$ of both space and time, E is not constant. In fact, using d/dt to denote the derivative of a quantity associated with a star,

$$\frac{dE}{dt} = \frac{1}{2}\frac{dv^2}{dt} + \frac{d\Phi}{dt} = \mathbf{v} \cdot \frac{d\mathbf{v}}{dt} + \frac{\partial \Phi}{\partial t} + \mathbf{v} \cdot \nabla\Phi = \frac{\partial \Phi}{\partial t}\bigg|_{\mathbf{x}(t)}, \qquad (4\text{-}208)$$

where the last equality follows because the star's equation of motion is $\dot{\mathbf{v}} = -\nabla\Phi$.

As a simple example of the way in which a time-varying gravitational field gives rise to changing stellar energy, consider a star that is at rest at the center of a collapsing spherical protogalaxy. As the protogalaxy collapses, the potential well at its center becomes deeper. On the other hand, the velocity of the central star remains zero. Therefore the energy of this star decreases.

Other stars in the collapsing protogalaxy will gain energy. For example, consider a star that is initially located near the half-mass radius of the system and is moving slowly radially outward. This star will be slow to respond to the overall collapse of the system, and by the time it starts to fall toward the center, the system as a whole will be approaching its most compact configuration. Hence the star will acquire a lot of kinetic energy as it falls into the deep potential well at the center of the system. Later, by the time the star has passed close to the center and is on its way out again, the system will have reexpanded significantly and the potential well out of which it has to climb will be less deep than that into which it fell. Consequently, it will reach the potential at which it originally started with more kinetic energy than it had originally.

Clearly, the change in the energy of a particular star depends in a complex way on the initial position and velocity of that star, but the overall effect is to widen the range of energies of the stars. In this respect, a time-varying potential provides a relaxation mechanism analogous to collisions in a gas. However, there is an important distinction between relaxation in a gas and violent relaxation. Since the mass of the star whose energy is being followed does not appear in equation (4-208) or in the equation of motion that determines $\mathbf{x}(t)$, violent relaxation changes the energy per unit mass of a star that has a given initial position and velocity in a way that is independent of the star's mass. In contrast, it is well known that collisional relaxation tends to pump energy from the most particles to the least massive particles, thus establishing equipartition of energy. It is sometimes desirable to be able to check that collisional relaxation through gravitational encounters is not playing an important role in a numerical simulation of a stellar system, and the best way of doing this is to check that there is no tendency for the distribution of energies of stars of different masses to evolve in different ways.

It is important to realize that the collisionless relaxation processes that we have described—phase mixing and violent relaxation—are quite distinct effects. For example, the phase-space density of an ensemble of non-interacting pendulums is radically reduced by phase mixing although the energies of the individual pendulums are exactly constant. By contrast, while the phase-space density in the neighborhood of the phase point of a star that is at rest at the center of a collapsing protogalaxy cannot change during the collapse (the motion of any star that does not move more than infinitesimally from the center of the proto-galaxy is unaffected by the collapse), we have seen that violent relaxation does change the energy of this star.

3 Numerical Simulation of the Relaxation Process

Modern computers make it possible to study the relaxation of stellar systems experimentally. We assign initial positions and velocities to a large number N of particles and then integrate their equations of motion numerically. Once the forces acting on each particle have been determined, a variety of simple numerical methods enable us to integrate the equations of motion with adequate accuracy. However, as we saw in §2.8, direct summation of the forces acting on all the particles becomes prohibitively costly when there are more than a few thousand particles in the system. Consequently, when one sets up a numerical simulation of a stellar system, one has to strike some sort of compromise between the number of particles and the accuracy with which the force on each particle is calculated. In discussing the results of these simulations, we should always bear in mind the possibility that the results have been distorted by the particular compromise struck. For the reader who wishes to experiment with numerical simulations, an N-body program is reproduced in Appendix 4.B (Aarseth 1985).

There are clearly an immense number of ways in which we may assign positions and velocities to N particles, and each such assignment will lead to a different simulation. However, many published simulations start from initial conditions that may be grouped into four broad categories.

(i) At $t = 0$ the particles are distributed through an ellipsoidal volume. The body formed by the distribution may be rotating rigidly and each particle will have a random velocity chosen from a Maxwellian distribution. We shall refer to these simulations as **collapse simulations**.

(ii) At $t = 0$ the particles form an axisymmetric disk. Each particle moves in the tangential direction with a velocity that is close to that required for centrifugal support of this disk, and has a small random radial motion. The particles of these **disk simulations** are often constrained to remain in the plane $z = 0$.

(iii) At $t = 0$ the particles are grouped into two galaxies. The velocities of the particles are chosen so that each galaxy is in a roughly steady state, and is approaching the other galaxy. Thus the galaxies are set to collide with one another.

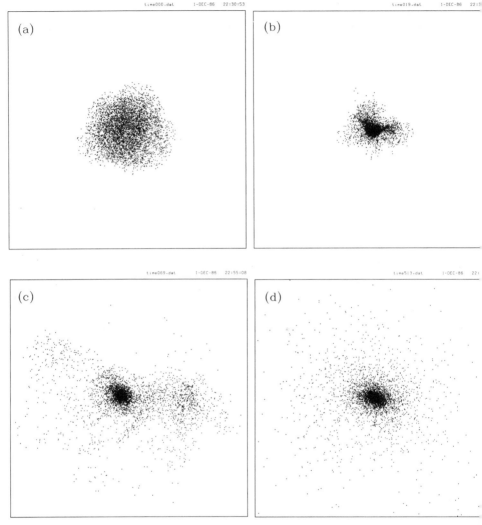

Figure 4-19. Four stages of a collapse simulation. (a) Initially the 5,000 particles have small random velocities and are approximately homogeneously distributed within a rough spherical boundary. (b) Gravity causes the system to fragment into sublumps that fall together to form a tight minimum configuration. (c) The kinetic energy built up during t infall phase enables the system to reexpand to a centrally concentrated configuration larg than the initial configuration. (d) After a series of pulsations of ever-decreasing intensity the system settles to a quasi-steady state. (Plots kindly supplied by T. S. van Albada: f details see van Albada 1982.)

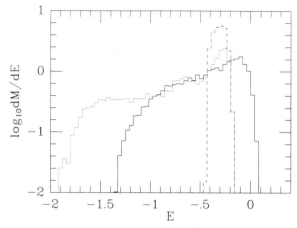

Figure 4-20. The evolution of the differential energy distribution of the model shown in Figure 4-19. The corresponding times are: $t = 0$ (dashed and then dash-dot curve); after one initial free-fall time (dotted and then dash-dot curve); after violent relaxation has ceased (full curve). Energy is measured in units of GM/R_e, where M is the system's mass and R_e is the effective radius of the $R^{1/4}$ model that best fits the final configuration (see Figure 4-22). (After van Albada 1982.)

(iv) At $t = 0$ the particles are nearly homogeneously distributed throughout a spherical volume, and are receding from the sphere's center with velocities that are approximately proportional to radius. These **cosmological simulations** model galaxy formation by gravitational clustering in an expanding Universe.

In this section we shall discuss only collapse simulations. One motivation for these simulations is that they may mimic the process by which galaxies formed. Disk simulations will be discussed in §6.3 and collision simulations in §7.4. Only very recently have cosmological simulations become feasible in which the individual particles have masses smaller than those of galaxies, and it is still unclear to what degree even the best current simulations provide adequate models of the internal dynamics of galaxies—for more details, see Efstathiou et al. (1985).

Figures 4-19, 4-20, and 4-21 show three general features of collapse simulations. Figure 4-19 shows the spatial structure of an initially cold and nonrotating system at four evolutionary stages. The system first contracts to a very compact configuration and then partially reexpands. After the elapse of two or three times the time required for the system

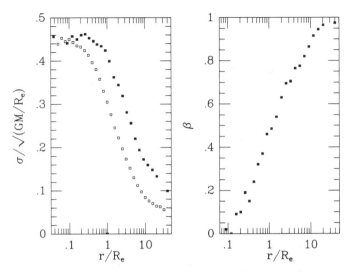

Figure 4-21. Left panel: radial velocity dispersion (filled squares), and projected velocity dispersion (open squares) in the final configuration of the simulation shown in Figure 4-19. Right panel: anisotropy parameter $\beta \equiv 1 - \sigma_\theta^2/\sigma_r^2$ in this system. The velocity dispersions have been scaled to those of the best-fitting $R^{1/4}$ model (see Figure 4-22). (After van Albada 1982.)

to reach its minimum-radius configuration, the system has settled to a nearly steady state. Initially cold and slowly rotating systems form more compact minimum-radius configurations than hot or rapidly rotating systems, and this is reflected in the central density of the final equilibrium system, which tends to be comparable to the density of the minimum-radius configuration. Note that the equilibrium system shown in Figure 4-19d extends well beyond the outer boundary of the initial configuration.

Figure 4-20 shows the evolution of the differential energy distribution (dM/dE) that was introduced in §4.4.5. Initially the energies of the particles lie in a narrow range, but this range is rapidly extended as the violent relaxation of the system causes different particles to gain and lose energy. In the final configuration the most densely populated energies lie near the escape energy $E = 0$. This is just what we found to be the case in equilibrium models that are based on the Jeans theorem.

As Figure 4-21 shows, in the outer parts of the final equilibrium systems the mean-square radial velocity, $\overline{v_r^2}$, tends to be larger than the mean-square tangential velocities, $\overline{v_\theta^2}$ and $\overline{v_\phi^2}$. This situation arises

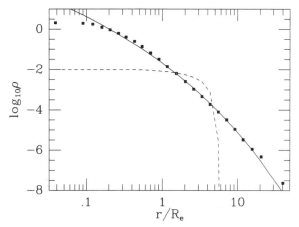

Figure 4-22. The radial density profile of the final configuration of the simulation shown in Figure 4-19 (squares), together with the simulation's initial density profile (dashed line) and the density profile of the best-fitting $R^{1/4}$ model (full curve). The initial and final simulation profiles have been scaled to the effective radius R_e of the latter model. (After van Albada 1982.)

because the particles that populate the outer regions of the final system were accelerated onto their present orbits by fluctuations in the strong gravitational field that prevails near the galactic center. Consequently, most of these particles are on highly elongated orbits that pass close to the center of the system.

Figure 4-22 shows the intial (dashed curve) and final (squares) density profiles of the simulation of Figure 4-19 together with the density profile of the $R^{1/4}$ model. This particular simulation, which started from cold and somewhat clumpy initial conditions, generates a system that fits the $R^{1/4}$ model extremely well. Simulations that start from warmer or more homogeneous initial conditions generate systems that fit the $R^{1/4}$ law less well (van Albada 1982).

If the initial configuration is ellipsoidal rather than spherical, the final system is also ellipsoidal, although it will generally be less flattened or elongated than was the initial configuration (Binney 1976, Aarseth & Binney 1978). Figure 4-23 illustrates this result. The velocity-dispersion tensors of these ellipsoidal models are everywhere anisotropic, in contrast to the case of spherical systems, in which the dispersion tensor is isotropic at small radii. When the initial configuration is slowly rotat-

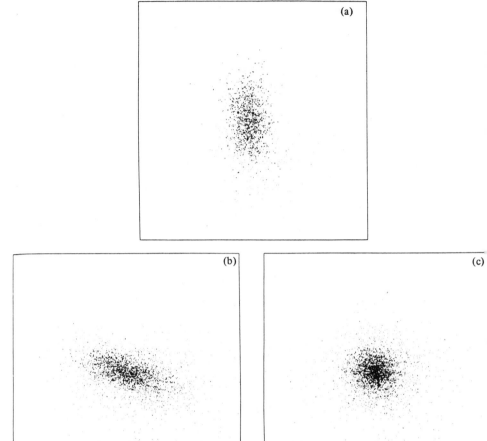

Figure 4-23. Three orthogonal projections of the final configuration to which 25,000 particles settled after they had been released from a cold, elliptical initial configuration (From Wilkinson & James 1982.)

ing, the figure of the final configuration also rotates (Wilkinson & James 1982).

If the initial configuration is spherical but slowly rotating, the final system will be an oblate figure of rotation. When the initial rotation is rapid, the final state will be a prolate triaxial bar (Hohl & Zang 1979; Miller & Smith 1979). The axisymmetric bodies that are formed in this way have ellipticity $\epsilon \lesssim 0.2$. In §5-2 of MB it was shown that if elliptical galaxies are oblate spheroids, the commonest true ellipticity must lie near $\epsilon = 0.35$. This suggests that if elliptical galaxies formed by violent relaxation, then they are not rotationally flattened oblate spheroids.

Problems

4-1. [1] From equations (1D-60) and (4-13) show that the collisionless Boltzmann equation can be written in the form $(\partial f/\partial t) = [H, f]$, where $[\ ,\ ]$ is the Poisson bracket.

4-2. [2] Show that in a frame that rotates with constant angular velocity $\mathbf{\Omega}$, the collisionless Boltzmann equation is

$$\frac{\partial f}{\partial t} + \mathbf{v} \cdot \nabla f - [\nabla \Phi_{\text{eff}} + 2(\mathbf{\Omega} \times \mathbf{v})] \cdot \frac{\partial f}{\partial \mathbf{v}} = 0, \tag{4P-1}$$

where $\Phi_{\text{eff}} \equiv \Phi - \frac{1}{2}|\mathbf{\Omega} \times \mathbf{r}|^2$.

4-3. [2] By analogy with the derivation of equation (4-17), show that in spherical coordinates the collisionless Boltzmann equation is

$$0 = \frac{\partial f}{\partial t} + v_r \frac{\partial f}{\partial r} + \frac{v_\theta}{r} \frac{\partial f}{\partial \theta} + \frac{v_\phi}{r \sin \theta} \frac{\partial f}{\partial \phi} + \left(\frac{v_\theta^2 + v_\phi^2}{r} - \frac{\partial \Phi}{\partial r} \right) \frac{\partial f}{\partial v_r}$$
$$+ \frac{1}{r} \left(v_\phi^2 \cot \theta - v_r v_\theta - \frac{\partial \Phi}{\partial \theta} \right) \frac{\partial f}{\partial v_\theta} - \frac{1}{r} \left[v_\phi (v_r + v_\theta \cot \theta) + \frac{1}{\sin \theta} \frac{\partial \Phi}{\partial \phi} \right] \frac{\partial f}{\partial v_\phi}. \tag{4P-2}$$

4-4. [2] Consider an infinite homogeneous system of collisionless test particles in n-dimensional space. The particles have an isotropic velocity distribution $f(v)$. Initially the particles are subject to no forces. At $t = 0$ a gravitational potential well suddenly appears in a finite region of the space. Show that as $t \to \infty$, the density of unbound particles traveling through the well is smaller than the background density if $n < 2$, and larger if $n > 2$.

4-5. [1] Suppose the principal axes of the velocity ellipsoid near the Sun are always parallel to the unit vectors of spherical coordinates. Then show that for $|z|/R$ small, $\overline{v_R v_z} \simeq (\overline{v_r^2} - \overline{v_\theta^2})(z/R)$. (Hint: Write v_R and v_z in terms of v_r and v_θ, and then average $v_R v_z$ using $\overline{v_r v_\theta} = 0$.)

4-6. [1] Suppose that the Oort limit has been calculated as described in §4.2.1(b) from observations of tracers whose distances have been systematically overestimated by a factor λ. By what factor is the derived local mass density in error?

4-7. [2] Consider a hypothetical disk galaxy in which all the mass is contained in a central point mass. The disk density is negligible; more precisely, the disk consists of a population of stars of zero mass with RMS z-velocity σ_z that is independent of z. At radius R, the number density of these stars as a function of z is $\nu(z) = \nu(0) \exp(-\frac{1}{2}z^2/z_0^2)$, where $z_0 \ll R$ is a constant. (a) What is the relation between σ_z and z_0? (b) What does equation (4-38) predict for the local mass density if these stars are used as tracers? Why is the wrong answer obtained?

4-8. [1] The stars of a spherically symmetric cluster have mean-square proper motions relative to the cluster mean $\overline{\mu^2}$, and mean-square radial velocity relative to the mean $\overline{v_r^2}$. What is the cluster distance? Does this result depend on an assumption that the velocity ellipsoid is everywhere isotropic?

4-9. [1] In a singular isothermal sphere with an isotropic dispersion tensor, the RMS speed is $\sqrt{\frac{3}{2}}v_c$, where v_c is the circular speed. In a system with the same mass distribution, but with all stars on randomly oriented circular orbits, the RMS speed is v_c. Thus, two systems have identical density distributions but different amounts of kinetic energy per star. How is this consistent with the virial theorem?

4-10. [1] In a spherical stellar system with mass profile $M(r)$, a stellar population with number density $\nu(r)$ has anisotropy parameter $\beta \equiv 1 - (\overline{v_\theta^2}/\overline{v_r^2})$ of the form $\beta(r) = r^2/(r_a^2 + r^2)$, where r_a is a constant. Show that

$$\overline{v_r^2}(r) = \frac{G \int_r^\infty [(r_a/r')^2 + 1]\nu(r')M(r')\,dr'}{(r_a^2 + r^2)\nu(r)}. \qquad (4P\text{-}3)$$

4-11. [2] The velocity dispersion in some axisymmetric stellar system is isotropic and a function $\sigma(\rho)$ of the density alone. Show that the azimuthal streaming velocity must be a function $\overline{v}_\phi(R)$ of the cylindrical radius R only. Is this configuration physically plausible?

4-12. [2] A static, isotropic and spherically symmetric stellar system is confined by a spherical vessel of radius r_b. Show that $2K + W = 4\pi r_b^3 p$, where K and W are the system's kinetic and potential energies, and p is the pressure exerted by the system on the vessel's walls.

4-13. [1] Show that the part of equation (4-72) that is antisymmetric in j and k is equivalent to the law of conservation of angular momentum.

4-14. [1] Show that in the presence of an externally generated gravitational potential Φ_{ext}, the right side of equation (4-78) acquires an extra term:

$$V_{jk} \equiv -\frac{1}{2} \int \left(x_k \frac{\partial \Phi_{\text{ext}}}{\partial x_j} + x_j \frac{\partial \Phi_{\text{ext}}}{\partial x_k} \right) \rho\, d^3\mathbf{x}. \qquad (4P\text{-}4)$$

4-15. [2] Consider a triaxial system whose density distribution is stationary in a frame that rotates with angular frequency Ω about the x_3-axis, which coincides with a principal axis of the moment of inertia tensor $\mathbf{I}$. Show that at an instant when $I_{12} = 0$, the left side of equation (4-78) is Ω^2 times the diagonal tensor with components $(I_{22} - I_{11})$, $(I_{11} - I_{22})$, and 0 along the diagonal. Hence show that

$$\Omega^2 = -\frac{(W_{11} - W_{22}) + 2(T_{11} - T_{22}) + (\Pi_{11} - \Pi_{22})}{2(I_{11} - I_{22})}, \qquad (4P\text{-}5a)$$

and if $T_{33} = 0$,

$$\frac{v_0^2}{\sigma_0^2} = (1 - \delta)\frac{W_{11} + W_{22}}{W_{33}} - 2, \qquad (4P\text{-}5b)$$

where $v_0^2 \equiv \frac{2}{M}(T_{11} + T_{22})$, $\sigma_0^2 \equiv \frac{1}{2M}(\Pi_{11} + \Pi_{22})$ and $(1 - \delta)(\Pi_{11} + \Pi_{22}) \equiv 2\Pi_{33}$.

4-16. [1] Solve the Lane-Emden equation (4-108c) for the case $n = 1$, to show that

$$\psi = \begin{cases} \dfrac{\sin s}{s} & \text{for } s < \pi; \\ \left(\dfrac{\pi}{s} - 1\right) & \text{for } s \geq \pi. \end{cases} \tag{4P-6}$$

Show that the model's total mass is $M = \frac{1}{2}\Psi_0 G^{-\frac{3}{2}}\sqrt{\pi/c_1}$, where c_1 is defined by equation (4-107b).

4-17. [1] For a Maxwellian distribution of velocities with one-dimensional dispersion σ, show that: (a) the mean speed is $\overline{v} = \sqrt{8/\pi}\sigma$; (b) the mean-square speed is $\overline{v^2} = 3\sigma^2$; (c) the mean-square of one component of velocity is $\overline{v_x^2} = \sigma^2$; (d) the mean-square relative speed of any two particles is $\overline{v_{rel}^2} = 6\sigma^2$; (e) the fraction of particles with $v^2 > 4\overline{v^2}$ is 0.00738.

4-18. [1] Show that if the DF of a spherical system is of the form $f(\mathcal{E}, L) = L^{-2\beta}f_0(\mathcal{E})$, where $\beta < 1$, then at any point in the system $(\overline{v_\theta^2}/\overline{v_r^2}) = 1 - \beta$.

4-19. [1] Show that the DF

$$f(\mathcal{E}, L) = \begin{cases} F\delta(L^2)(\mathcal{E} - \mathcal{E}_0)^{-\frac{1}{2}} & \text{for } \mathcal{E} > \mathcal{E}_0 \\ 0 & \text{for } \mathcal{E} \leq \mathcal{E}_0, \end{cases} \tag{4P-7a}$$

where F and $\mathcal{E}_0$ are constants, self-consistently generates a model with density

$$\rho(r) = \begin{cases} Cr^{-2} & \text{for } r < r_0 \\ 0 & \text{for } r \geq r_0, \end{cases} \tag{4P-7b}$$

where C is a constant and the relative potential at r_0 satisfies $\Psi(r_0) = \mathcal{E}_0$. This is the only analytic stellar system known to us in which all stars are on perfectly radial orbits (Fridman & Polyachenko 1984).

4-20. [2] The aim of this problem is to determine the asymptotic behavior of the density $\rho(r)$ in a Michie model [eq. (4-152)].

(a) At large radii, the density in a Michie model is dominated by stars with $\mathcal{E}/\sigma^2 \ll 1$. In this case, show that the density can be written in the form

$$\rho \propto \frac{1}{r^2} \int_0^{r^2\Psi/(r_a\sigma)^2} e^{-x}\left(\Psi - \frac{r_a^2\sigma^2 x}{r^2}\right)^{\frac{3}{2}} dx. \tag{4P-8}$$

(b) The outer boundary of the model is at the tidal radius r_t where the relative potential $\Psi = 0$. If r_t is sufficiently large, show that at large radii

$$\rho \propto \frac{\Psi^{\frac{3}{2}}}{r^2}. \tag{4P-9}$$

Hence argue that in an isolated cluster $(r_t \to \infty)$ $\rho \propto r^{-7/2}$ at large radii, while in a cluster with finite tidal radius $\rho \propto (r_t - r)^{3/2}r^{-7/2}$.

4-21. [3] Eddington (see Shiveshwarkar 1936) took f to be of the form

$$f(\mathcal{E}, L) = \frac{\rho_1}{(2\pi\sigma^2)^{3/2}} \exp\left(\frac{\mathcal{E}}{\sigma^2}\right) \exp\left[-\frac{1}{2}\left(\frac{L}{r_a\sigma}\right)^2\right]. \tag{4P-10a}$$

Show that the density of Eddington's model may be written,

$$\rho = \frac{\rho(0)}{1 + r^2/r_a^2} \exp\left[\frac{\Psi(r) - \Psi(0)}{\sigma^2}\right]. \tag{4P-10b}$$

Show further that $\overline{v_r^2} = \sigma^2$ everywhere, and $\overline{v_\theta^2} = \overline{v_\phi^2} = \sigma^2 r_a^2/(r_a^2 + r^2)$.

Deduce that for $r \lesssim r_a$ the model resembles the isothermal sphere. Show that for $r \gg r_a$, $\rho \propto r^{-2}(\log r)^{-1}$, $(v_c/\sigma) \propto (\log r)^{-1/2}$. Interpret physically the fact that ρ varies approximately as r^{-2} at large r, in terms of a stellar traffic jam.

4-22. [2] Consider a spherical system with DF $f(\mathcal{E}, L)$. Let $N(\mathcal{E}, L)$ be the number of stars with $\mathcal{E}$ and L in the ranges $(\mathcal{E}, \mathcal{E} + d\mathcal{E})$ and $(L, L + dL)$.

(a) Show that

$$N(\mathcal{E}, L) = 8\pi^2 L f(\mathcal{E}, L) T_r(\mathcal{E}, L), \tag{4P-11}$$

where T_r is the radial period defined by equation (3-16).

(b) A collection of test particles orbiting a point mass has DF $f(\mathcal{E})$. Show that the number of particles with eccentricities in the range $(e, e + de)$ is proportional to $e\,de$.

4-23. [3] Consider a spherical system in which at every point in real space the star density in velocity space is constant on ellipsoidal figures of rotation. Show that the DF has the form $f = f(Q)$, where $Q(\mathcal{E}, L)$ is defined by equation (4-146a).

4-24. [3] Every star in a spherical system loses mass slowly and isotropically. If the initial DF is of the form $f_0(\mathcal{E})$, what is the DF f_p after every star has been reduced to a fraction p of its original mass? Is f_p a function of $\mathcal{E}$ alone? How has the gravitational radius of the system changed? (Hint: See Richstone & Potter 1982).

4-25. [2] We may study the vertical structure of a thin axisymmetric disk by neglecting all radial derivatives and adopting the form $f = f(E_z)$ for the DF, where $E_z \equiv \frac{1}{2}v_z^2 + \Phi(z)$. Show that if $f = \rho_0(2\pi\sigma_z^2)^{-1/2}\exp(-E_z/\sigma_z^2)$, the approximate form (2-57) of Poisson's equation may be written

$$2\frac{d^2\phi}{d\varsigma^2} = e^{-\phi}, \quad \text{where} \quad \phi \equiv \frac{\Phi}{\sigma_z^2}, \quad \varsigma \equiv \frac{z}{z_0}, \quad \text{and} \quad z_0 \equiv \frac{\sigma_z}{\sqrt{8\pi G\rho_0}}. \tag{4P-12a}$$

By solving this equation subject to the boundary conditions $\phi(0) = d\phi/d\varsigma|_0 = 0$, show that the density ρ in the disk is given by (Spitzer 1942)

$$\rho(z) = \rho_0 \operatorname{sech}^2(\tfrac{1}{2}z/z_0). \tag{4P-12b}$$

5 Stability of Collisionless Systems

Having discussed the equilibria of stellar systems, we now turn to the question of whether these equilibria are sufficiently stable to correspond to real, physical systems. Obviously, it is rare in nature to find an equilibrium that is so delicately balanced, like a pencil on its point, that the slightest perturbation will cause it to evolve rapidly away from its initial state. Hence the study of the stability of stellar equilibria is almost as important for galactic astronomy as the study of the equilibria themselves, and we should test any proposed stellar-dynamical configuration for stability before we employ it as a model of a real dynamical system. In fact, it turns out that many equilibrium stellar systems *are* unstable. For example, the simplest model of a disk galaxy consists of a self-gravitating disk in which the stars move on precisely circular orbits—but any such "cold" disk, with no random motions, is violently unstable. It appears that a minimum level of random motion [eq. (6-53)] is necessary for a stable disk.

Stability analyses can do more than simply discard otherwise satisfactory models. An additional hope is that stability requirements will constrain the nature and distribution of the matter in galaxies. A classic example of such an application of stability studies is furnished by Laplace's (1802) work on Saturn's rings. Laplace showed that Saturn's rings could not be rigid bodies, as most astronomers then believed, because the motion of such rings would be unstable. A more recent and still controversial example concerns the structure of disk galaxies. Models of our own and other spiral galaxies that are designed to match known rotation curves and velocity dispersion measurements appear to have enough random motion to suppress axisymmetric instabilities; however, many of the models exhibit a fierce non-axisymmetric instability that results in the growth of a large barlike structure in the central regions. Since these galaxies do not appear to be unstable in nature, it has been argued that the bulk of their mass must lie in invisible spheroidal components that dominate the overall gravitational potential (see §6.3.1).

A related aim of stability studies is to explain some of the striking irregularities of galaxies, such as spiral structure, as arising from instabilities in an initially featureless system. All instability phenomena in galactic disks, including both the bar instability and spiral structure, are strongly influenced by differential rotation (the variation of angular speed with distance from the galactic center). Differential rotation shears out any unstable structure since the components of the structure at different radii are carried around the galaxy at different rates. We shall defer this complication to Chapter 6, which is largely devoted to instabilities in differentially rotating disks. In this chapter we examine the stability of both static and uniformly rotating stellar systems.

Unfortunately it is much easier to generate equilibria than to test them for stability, and we shall be able to give only an incomplete and limited account of stability theory. Even so, the complexity of the mathematics will sometimes be considerable. We have arranged the material so that subsequent chapters are largely independent of the material in this chapter and the next, and in this chapter the reader may wish to skip §5.2 and §5.3 on a first reading.

In Chapter 3 we discussed the stability of orbits in a variety of fixed gravitational potentials. The instabilities that we shall encounter here are of a different and rather more serious type. They are caused by cooperative effects, in which a density perturbation gives rise to extra gravitational forces, which deflect the stellar orbits in such a way that the original density perturbation is enhanced. The task of analyzing these instabilities is made much easier by the many similarities between stellar systems and two other kinds of systems whose instabilities have been thoroughly studied:

(i) *Self-gravitating fluid systems.* As we saw in Chapter 4, certain stellar systems have close fluid analogs. The analogy arises because a fluid system is supported against gravity by gradients in the pressure p, while a stellar system is supported by gradients in the stress tensor $-\nu\sigma_{ij}^2$ [see discussion after eq. (4-27)]. Similarly, in this chapter we shall find it useful to draw analogies between the instabilities of self-gravitating fluid and stellar systems. A fluid system resists the gravitational collapse of a local density enhancement through pressure gradients, while a stellar system is resistant to collapse because the spread $\overline{(v_i - \overline{v}_i)(v_j - \overline{v}_j)} = \sigma_{ij}^2$ in stellar velocities at every point tends to disperse any density enhancement before it has had time to grow. Since the fluid systems are simpler (and important in their own right for other aspects of galactic astronomy such as the interstellar medium and star formation), we shall analyze fluid systems alongside the analogous stellar-dynamical ones.

(ii) *Electrostatic plasmas.* Rarefied plasmas share with collisionless stellar systems the property that the mean field of the system is more important than the fields of individual nearby particles. Furthermore, the stability theory of plasmas is very well developed. Hence many of the techniques of plasma physics can be used in stellar dynamics. However, there is a fundamental difference between plasmas and stellar systems. Plasmas have both positive and negative charges, so they are neutral on large scales and can form static homogeneous equilibria. But since gravity is always an attractive force, gravitating systems never form static *homogeneous* equilibria. This essential inhomogeneity of self-gravitating systems greatly complicates the study of their stability, and much less is known about the stability of stellar systems than of plasmas.

1 Mathematical Preliminaries

We have seen that the dynamics of self-gravitating collisionless stellar systems are described by the coupled collisionless Boltzmann and Poisson equations, (4-13) and (2-10),[1]

$$\frac{\partial f}{\partial t} + \mathbf{v} \cdot \frac{\partial f}{\partial \mathbf{x}} - \boldsymbol{\nabla}\Phi \cdot \frac{\partial f}{\partial \mathbf{v}} = 0, \tag{5-1}$$

$$\nabla^2 \Phi(\mathbf{x}, t) = 4\pi G \int f(\mathbf{x}, \mathbf{v}, t)\, d^3 \mathbf{v}. \tag{5-2}$$

An equilibrium stellar system is described by a time-independent DF $f_0(\mathbf{x}, \mathbf{v})$ and potential $\Phi_0(\mathbf{x})$ that are solutions of (5-1) and (5-2), that is,

$$\mathbf{v} \cdot \frac{\partial f_0}{\partial \mathbf{x}} - \boldsymbol{\nabla}\Phi_0 \cdot \frac{\partial f_0}{\partial \mathbf{v}} = 0 \quad ; \quad \nabla^2 \Phi_0 = 4\pi G \int f_0\, d^3 \mathbf{v}. \tag{5-3}$$

Now consider a small perturbation to this equilibrium, which we write as

$$\begin{aligned}
f(\mathbf{x}, \mathbf{v}, t) &= f_0(\mathbf{x}, \mathbf{v}) + \epsilon f_1(\mathbf{x}, \mathbf{v}, t), \\
\Phi(\mathbf{x}, t) &= \Phi_0(\mathbf{x}) + \epsilon \Phi_1(\mathbf{x}, t),
\end{aligned} \tag{5-4}$$

where $\epsilon \ll 1$. Substituting into equations (5-1) and (5-2), we find that the terms that are independent of ϵ sum to zero by virtue of (5-3). Dropping the terms proportional to ϵ^2 since $\epsilon \ll 1$, we have

$$\frac{\partial f_1}{\partial t} + \mathbf{v} \cdot \frac{\partial f_1}{\partial \mathbf{x}} - \boldsymbol{\nabla}\Phi_0 \cdot \frac{\partial f_1}{\partial \mathbf{v}} - \boldsymbol{\nabla}\Phi_1 \cdot \frac{\partial f_0}{\partial \mathbf{v}} = 0, \tag{5-5}$$

---

[1] Furthermore, these equations are also applicable to systems like globular clusters in which two-body relaxation has been important, so long as the growth time of the instabilities that are predicted is much shorter than the two-body relaxation time.

$$\nabla^2 \Phi_1 = 4\pi G \int f_1 \, d^3\mathbf{v}. \tag{5-6}$$

The first of these equations is called the **linearized collisionless Boltzmann equation**; the second is simply Poisson's equation (5-2), except that it relates Φ_1 and f_1, instead of Φ and f. In this chapter we shall be mainly concerned with analyzing solutions of these equations.

As we mentioned earlier, we shall examine the stability of fluid systems in parallel with the stability of stellar systems. Therefore we now consider the fluid-dynamical analogs of equations (5-5) and (5-6). A fluid with density $\rho(\mathbf{x}, t)$, pressure $p(\mathbf{x}, t)$, and velocity $\mathbf{v}(\mathbf{x}, t)$ in a potential field $\Phi(\mathbf{x}, t)$ obeys the continuity equation (1E-3),

$$\frac{\partial \rho}{\partial t} + \nabla \cdot (\rho \mathbf{v}) = 0, \tag{5-7}$$

Euler's equation (1E-8),

$$\frac{\partial \mathbf{v}}{\partial t} + (\mathbf{v} \cdot \nabla)\mathbf{v} = -\frac{1}{\rho}\nabla p - \nabla \Phi, \tag{5-8}$$

and Poisson's equation,

$$\nabla^2 \Phi = 4\pi G\rho. \tag{5-9}$$

We must also specify the equation of state relating p and ρ. Since our main objective is to use the fluid systems as analogs to stellar systems, it is sufficient to use a very simple equation of state. Thus, we will consider barotropic fluids [eq. (1E-9)], in which the pressure is a function of the density only:

$$p(\mathbf{x}, t) = p[\rho(\mathbf{x}, t)]. \tag{5-10}$$

The equilibrium fluid system is described by the time-independent density, pressure, velocity, and potential distributions $\rho_0(\mathbf{x})$, $p_0(\mathbf{x})$, $\mathbf{v}_0(\mathbf{x})$, and $\Phi_0(\mathbf{x})$, which are solutions of the fluid equations (5-7) to (5-10). Then we consider a small perturbation in these quantities:

$$\begin{aligned} \rho(\mathbf{x}, t) = \rho_0(\mathbf{x}) + \epsilon\rho_1(\mathbf{x}, t) \quad &; \quad p(\mathbf{x}, t) = p_0(\mathbf{x}) + \epsilon p_1(\mathbf{x}, t), \\ \mathbf{v}(\mathbf{x}, t) = \mathbf{v}_0(\mathbf{x}) + \epsilon\mathbf{v}_1(\mathbf{x}, t) \quad &; \quad \Phi(\mathbf{x}, t) = \Phi_0(\mathbf{x}) + \epsilon\Phi_1(\mathbf{x}, t), \end{aligned} \tag{5-11}$$

where $\epsilon \ll 1$. Substituting equations (5-11) into the fluid equations (5-7) to (5-10), we find that the terms that are independent of ϵ sum to zero, and discarding terms proportional to ϵ^2, we obtain

$$\frac{\partial \rho_1}{\partial t} + \nabla \cdot (\rho_0 \mathbf{v}_1) + \nabla \cdot (\rho_1 \mathbf{v}_0) = 0, \tag{5-12a}$$

$$\frac{\partial \mathbf{v}_1}{\partial t} + (\mathbf{v}_0 \cdot \nabla)\mathbf{v}_1 + (\mathbf{v}_1 \cdot \nabla)\mathbf{v}_0 = \frac{\rho_1}{\rho_0^2}\nabla p_0 - \frac{1}{\rho_0}\nabla p_1 - \nabla\Phi_1 \tag{5-12b}$$
$$= -\nabla h_1 - \nabla\Phi_1,$$

$$\nabla^2\Phi_1 = 4\pi G\rho_1, \tag{5-12c}$$

$$p_1 = \left(\frac{dp}{d\rho}\right)_0 \rho_1 \equiv v_s^2\rho_1. \tag{5-12d}$$

Here we have introduced two new quantities, the sound speed $v_s(\mathbf{x})$ [see eq. (1E-17b)], defined by

$$v_s^2(\mathbf{x}) \equiv \left[\frac{dp(\rho)}{d\rho}\right]_{\rho_0} \equiv \left(\frac{dp}{d\rho}\right)_0, \tag{5-13}$$

and the specific enthalpy [see eq. (1E-10)],

$$h(\mathbf{x},t) \equiv \int_0^{\rho(\mathbf{x},t)} \frac{dp(\rho)}{\rho}. \tag{5-14}$$

The perturbation of the enthalpy is

$$h_1 = \left(\frac{dp}{d\rho}\right)_0 \frac{\rho_1}{\rho_0} = v_s^2\frac{\rho_1}{\rho_0}. \tag{5-15}$$

Equations (5-12a) to (5-12c), along with (5-12d) or (5-15), form a complete set of linear equations that govern the response of a fluid system to small perturbations.

5.1 The Jeans Instability

As we have mentioned, an infinite homogeneous gravitating system cannot be in static equilibrium. Nevertheless, the mathematical convenience of supposing that a gravitating system, like an electrostatic plasma, can form an infinite homogeneous medium is so great that it is useful to consider its stability properties as if it were an equilibrium state. By this means we introduce some of the basic techniques of stability analyses, and, more importantly, we obtain some results that turn out to be applicable in other, more realistic contexts.[2]

We construct our fictitious infinite homogeneous equilibrium by perpetrating what we shall call the **Jeans swindle** after Sir James Jeans,

---

[2] A full account of the stability of homogeneous self-gravitating stellar systems is given in Lynden-Bell (1967b).

who studied this problem in 1902 (Jeans 1929). Mathematically, the difficulty we must overcome is that if the density and pressure of the medium ρ_0, p_0 are constant, and the mean velocity $\mathbf{v}_0$ is zero, it follows from Euler's equation (5-8) that $\nabla\Phi_0 = 0$. On the other hand, Poisson's equation (5-9) requires that $\nabla^2\Phi_0 = 4\pi G\rho_0$. These two requirements are inconsistent unless $\rho_0 = 0$. Physically, there are no pressure gradients in a homogeneous medium to balance gravitational attraction. A similar inconsistency arises in an infinite homogeneous stellar system whose DF is independent of position. We remove the inconsistency by the *ad hoc* assumption that Poisson's equation describes only the relation between the perturbed density and the perturbed potential, while the unperturbed potential is zero. This assumption constitutes the Jeans swindle; it is a swindle, of course, because in general there is no formal justification for discarding the unperturbed gravitational field. However, there are circumstances in which the swindle is justified. For example,

(i) Consider the evolution of ordinary sound waves. The atmosphere has vertical density and pressure gradients that balance the earth's gravitational field; yet we can reliably compute the dispersion relation for sound waves using the assumption of an infinite homogeneous medium with no equilibrium density or pressure gradients. This approximation is valid because the wavelength of the sound waves is much smaller than the scale over which the equilibrium density and pressure vary. For similar reasons, the Jeans swindle should be valid for the analysis of small-scale instabilities.

(ii) In a uniformly rotating, homogeneous system, the equilibrium gravitational field may be balanced by centrifugal force rather than pressure gradients; in this case a homogeneous system can be in static equilibrium in the rotating frame and no Jeans swindle is necessary (although the stability properties are somewhat modified from those of the nonrotating medium because of Coriolis forces. See problem 5-2).

Clearly, the consistency of the Jeans swindle has to be checked with every application, but with the encouragement that there are at least *some* systems for which the swindle is justified, let us go on to the analysis.

1 Physical Basis of the Jeans Instability

Consider an infinite homogeneous fluid of density ρ_0 and pressure p_0, with no internal motions, so that $\mathbf{v}_0 = 0$. Now draw a sphere of radius r around any point and suppose that we compress this spherical region by reducing its volume V to $(1-\alpha)V$, where $\alpha \ll 1$. We will deal only with order-of-magnitude arguments in this subsection, and so the details of how the fluid is compressed and of the exact shape of the perturbed density distribution are not important. To order of magnitude, the density perturbation is $\rho_1 \approx \alpha\rho_0$, and the pressure perturbation is $p_1 \approx (dp/d\rho)_0\alpha\rho_0 = \alpha v_s^2\rho_0$. The pressure force per unit mass is $\mathbf{F}_p = -\nabla p/\rho$, and our compressive perturbation therefore leads to an extra outward pressure force $\mathbf{F}_{p1}$ of magnitude roughly $|\mathbf{F}_{p1}| = |\nabla p_1/\rho_0| \approx p_1/(\rho_0 r) \approx \alpha v_s^2/r$ (we have replaced the gradient ∇ by $1/r$). Similarly, the enhanced density of the perturbation gives rise to an extra inward gravitational force $\mathbf{F}_{G1} = -\nabla\Phi_1$. Thus $|\mathbf{F}_{G1}| \approx GM\alpha/r^2$, where $M = \frac{4}{3}\pi\rho_0 r^3$ is the mass originally within r; in other words, $|\mathbf{F}_{G1}| \approx G\rho_0 r\alpha$. If the net force $\mathbf{F}_{p1} + \mathbf{F}_{G1}$ is outward, the compressed fluid reexpands and the perturbation is stable; if the net force is inward, the fluid continues to contract and the perturbation is unstable. Thus, there is instability if $|\mathbf{F}_{G1}| > |\mathbf{F}_{p1}|$, or if

$$G\rho_0 r\alpha \gtrsim \alpha v_s^2/r,$$

that is, if

$$r^2 \gtrsim \frac{v_s^2}{G\rho_0}. \tag{5-16}$$

Thus, *perturbations with a scale longer than* $\approx v_s/\sqrt{G\rho_0}$ *are unstable.* An equivalent statement is that the system is unstable on a scale r if the dynamical time t_{dyn} [eq. (2-30)] is less than the time needed for a sound wave to traverse this scale, $t_s \approx r/v_s$. We now derive more exact forms of this stability criterion for fluids and for stellar systems.

2 The Jeans Instability for a Fluid

We use the linearized fluid equations (5-12a) to (5-12c) and (5-15). The equilibrium state is $\rho_0 = constant$, $\mathbf{v}_0 = 0$, and the Jeans swindle lets us set $\Phi_0 = 0$. We then have

$$\frac{\partial\rho_1}{\partial t} + \rho_0\nabla\cdot\mathbf{v}_1 = 0, \tag{5-17a}$$

$$\frac{\partial\mathbf{v}_1}{\partial t} = -\nabla h_1 - \nabla\Phi_1, \tag{5-17b}$$

$$\nabla^2\Phi_1 = 4\pi G\rho_1, \tag{5-17c}$$

$$h_1 = v_s^2\rho_1/\rho_0. \tag{5-17d}$$

By taking the time derivative of equation (5-17a) and the divergence of equation (5-17b) and eliminating $\mathbf{v}_1$, Φ_1, and h_1 with the aid of the other equations, we can combine equations (5-17) into the single equation

$$\frac{\partial^2 \rho_1}{\partial t^2} - v_s^2 \nabla^2 \rho_1 - 4\pi G\rho_0 \rho_1 = 0. \tag{5-18}$$

Because the medium is homogeneous, the coefficients of the partial derivatives in equation (5-18) are independent of position $\mathbf{x}$ and time t. Equations of this type are easy to solve; in fact, if we substitute into equation (5-18) a trial solution of the form[3]

$$\rho_1(\mathbf{x}, t) = Ce^{i(\mathbf{k}\cdot\mathbf{x}-\omega t)}, \tag{5-19}$$

where C is a constant, then the equation is satisfied, provided ω and $k = |\mathbf{k}|$ satisfy the dispersion relation

$$\omega^2 = v_s^2 k^2 - 4\pi G\rho_0. \tag{5-20}$$

Furthermore, it can be shown that the solutions (5-19) are complete, that is, the general solution of equation (5-18) is a superposition of solutions of the form (5-19):

$$\rho_1(\mathbf{x}, t) = \int C(\mathbf{k})e^{i[\mathbf{k}\cdot\mathbf{x}-\omega(k)t]}\, d^3\mathbf{k}, \tag{5-21}$$

where the function $\omega(k)$ is defined by equation (5-20). Hence the behavior of the system is entirely determined by the trial functions (5-19) and the dispersion relation (5-20), and it is these that we shall study.

When the density ρ_0 or the wavelength $\lambda = 2\pi/k$ is small, the dispersion relation (5-20) reduces to that of a sound wave, $\omega^2 = v_s^2 k^2$ [eq. (1E-21)]. As the wavelength is increased, the frequency ω decreases and the wave becomes more and more sluggish, until eventually ω^2 becomes negative. When $\omega^2 < 0$, say $\omega^2 = -\gamma^2$, the temporal dependence of the solution (5-19) is proportional to $\exp(\pm\gamma t)$, corresponding to exponentially growing and decaying solutions. The presence of the growing solution implies that the system is unstable. Hence from (5-20) the system is unstable when

$$k^2 < k_J^2 \equiv \frac{4\pi G\rho_0}{v_s^2}, \tag{5-22}$$

---

[3] The physical density perturbation must, of course, be real. However, if we find a complex function ρ_1 that satisfies equation (5-18), then clearly $\mathrm{Re}(\rho_1)$ also satisfies the same equation. Hence we shall allow ρ_1 to be complex with the understanding that the physical density is given by its real part.

where k_J is the **Jeans wavenumber** for the fluid. In terms of the wavelength, this result states that the perturbation is unstable if λ exceeds the **Jeans length** $\lambda_J = 2\pi/k_J$, that is, if[4]

$$\lambda^2 > \lambda_J^2 = \frac{\pi v_s^2}{G\rho_0}. \tag{5-23}$$

We define the **Jeans mass** M_J as the mass originally contained within a sphere of diameter λ_J:

$$M_J = \frac{4\pi}{3}\rho_0(\tfrac{1}{2}\lambda_J)^3 = \tfrac{1}{6}\pi\rho_0\left(\frac{\pi v_s^2}{G\rho_0}\right)^{\frac{3}{2}}. \tag{5-24}$$

The Jeans instability in a fluid has a simple interpretation in terms of energy. The energy density of an ordinary sound wave is positive. However, the gravitational energy density of a sound wave is negative, because the enhanced attraction in the compressed regions outweighs the reduced attraction in the dilated regions. The Jeans instability sets in at the wavelength λ_J where the net energy density becomes negative, so that the system can evolve to a lower energy state by allowing the wave to grow.

3 The Jeans Instability for a Stellar System

We use the linearized collisionless Boltzmann and Poisson equations (5-5) and (5-6). Since the equilibrium state is assumed to be homogeneous and time-independent, we set $f_0(\mathbf{x}, \mathbf{v}, t) = f_0(\mathbf{v})$, and use the Jeans swindle to set $\Phi_0 = 0$. We have

$$\frac{\partial f_1}{\partial t} + \mathbf{v}\cdot\frac{\partial f_1}{\partial \mathbf{x}} - \boldsymbol{\nabla}\Phi_1\cdot\frac{\partial f_0}{\partial \mathbf{v}} = 0 \quad ; \quad \nabla^2\Phi_1 = 4\pi G\int f_1\,d^3\mathbf{v}. \tag{5-25}$$

These equations have a solution of the form $f_1(\mathbf{x}, \mathbf{v}, t) = f_a(\mathbf{v})\exp[i(\mathbf{k}\cdot\mathbf{x} - \omega t)]$, $\Phi_1(\mathbf{x}, t) = \Phi_a\exp[i(\mathbf{k}\cdot\mathbf{x} - \omega t)]$, provided $f_a(\mathbf{v})$ and Φ_a satisfy[5]

$$(\mathbf{k}\cdot\mathbf{v} - \omega)f_a - \Phi_a\mathbf{k}\cdot\frac{\partial f_0}{\partial \mathbf{v}} = 0 \quad ; \quad -k^2\Phi_a = 4\pi G\int f_a\,d^3\mathbf{v}. \tag{5-26}$$

---

[4] Note that we can write the dispersion relation (5-20) as $\omega^2 = v_s^2(k^2 - k_J^2)$, which is similar to the dispersion relation for electrostatic plasma waves (e.g., Krall & Trivelpiece 1973, §4.4), $\omega^2 = v_s^2(k^2 + k_P^2)$, where $k_P^2 = 4\pi ne^2/mv_s^2$, n is the electron number density and e and m are the electron charge and mass. In a plasma all wavelengths are stable because the dispersion relation involves $+k_P^2$ instead of $-k_J^2$, because of the repulsion of like charges.

[5] As in the previous subsection, we shall allow f_1 and Φ_1 to be complex with the understanding that only their real parts are physically meaningful.

Combining these equations, we obtain

$$1 + \frac{4\pi G}{k^2} \int \frac{\mathbf{k} \cdot \partial f_0/\partial \mathbf{v}}{\mathbf{k} \cdot \mathbf{v} - \omega} d^3\mathbf{v} = 0. \tag{5-27}$$

This equation is the required dispersion relation since it relates $\mathbf{k}$ and ω.

For illustrative purposes assume that f_0 is Maxwellian,

$$f_0(\mathbf{v}) = \frac{\rho_0}{(2\pi\sigma^2)^{3/2}} e^{-\frac{1}{2}v^2/\sigma^2}, \tag{5-28}$$

where ρ_0 is the density. When f_0 is of this form, the integral over all velocities in equation (5-27) can be done in rectangular coordinates (v_x, v_y, v_z), where the v_x-axis is chosen to lie in the direction of $\mathbf{k}$. The integrals over v_y and v_z are simple, using $\int_{-\infty}^{\infty} \exp(-\frac{1}{2}v^2/\sigma^2) \, dv = \sqrt{2\pi\sigma^2}$, and equation (5-27) becomes

$$1 - \frac{2\sqrt{2\pi}G\rho_0}{k\sigma^3} \int_{-\infty}^{\infty} \frac{v_x e^{-\frac{1}{2}v_x^2/\sigma^2}}{kv_x - \omega} \, dv_x = 0. \tag{5-29}$$

By analogy with the fluid case, we expect the boundary between stable and unstable solutions to occur at $\omega = 0$. At $\omega = 0$ the integral in (5-29) is evaluated easily, and we have

$$k^2(\omega = 0) \equiv k_J^2 = \frac{4\pi G\rho_0}{\sigma^2}, \tag{5-30}$$

where k_J is the Jeans wavenumber for the stellar system. Thus the formula for the Jeans length of a collisionless system is the same that obtained for fluids, equation (5-22), except that the velocity dispersion σ is substituted for the sound speed v_s.

Pursuing the analogy with fluids, we suspect that all perturbations with wavelengths $\lambda > \lambda_J = 2\pi/k_J$ will be unstable in the stellar system. To check this, we set $\omega = i\gamma$, where γ is real and positive,[6] and substitute into the dispersion relation (5-29). Using the relation $\int_0^{\infty} x^2 \exp(-x^2) \, dx/(x^2 + \beta^2) = \frac{1}{2}\sqrt{\pi} - \frac{1}{2}\pi\beta \exp(\beta^2)[1 - \text{erf}(\beta)]$, where $\text{erf}(\beta)$ denotes the error function [eq. (1C-11)], we find

$$k^2 = k_J^2 \left\{ 1 - \frac{\sqrt{\pi}\gamma}{\sqrt{2}k\sigma} \exp\left(\frac{\gamma^2}{2k^2\sigma^2}\right) \left[1 - \text{erf}\left(\frac{\gamma}{\sqrt{2}k\sigma}\right) \right] \right\}. \tag{5-31}$$

---

[6] In principle, it is also possible to have unstable modes with $\omega = \omega_R + i\gamma$, where ω_R is real and non-zero. These are sometimes called **overstable** modes. In Appendix 5.A we show that there are no overstable modes in the system we are examining here.

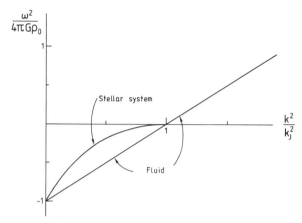

Figure 5-1. The dispersion relation for infinite homogeneous fluid and stellar systems, equations (5-20) and (5-31). Only the unstable branch of the dispersion relation for the stellar system is plotted.

This relation is plotted in Figure 5-1, along with the dispersion relation (5-20) for a fluid. As expected, both the fluid and the stellar system are unstable if $\lambda > \lambda_J$, although the growth rates are generally quite different in the two kinds of system.

The resemblance of fluid and stellar systems breaks down at smaller wavelengths $\lambda < \lambda_J$. Here the fluid supports waves that can simply be regarded as gravity-modified sound waves. The analysis of the stellar system is more complicated. Because of the singularity in the integrand of (5-29) that occurs when ω is real, it is difficult to determine whether modes with real ω exist. A careful analysis is given in Appendix 5.A, along the lines of Landau's (1946; see also Lifshitz & Pitaevskii 1981) treatment of the analogous problem in plasma physics. We find that all perturbations with wavelengths $\lambda < \lambda_J$ are damped [Im $(\omega) < 0$]. There are no nonsingular modes with real ω. The damping of perturbations with $\lambda < \lambda_J$ is an example of the interesting phenomenon of damping in a time-reversible system; in other words, there is dissipation even though there are no dissipative terms in the collisionless Boltzmann equation itself. Phase mixing, discussed in §4.7.2, is another example of this kind of dissipation; however, the damping process described here is more closely related to **Landau damping** in collisionless plasmas. The relation of Landau damping to phase mixing is explored in more detail in Appendix 5.A.

To summarize, the stability of an infinite homogeneous stellar system is closely related to the stability of the analogous fluid system: in both cases there is instability if and only if the wavenumber of the disturbance is less than the Jeans wavenumber k_J [eqs. (5-22) and (5-30)]. However, the oscillations of the two systems are quite different: the fluid supports short-wavelength sound waves and the stellar system does not.

(a) **Example**   An infinite homogeneous stellar system with a Maxwellian DF [eq. (5-28)] is subject to an impulsive perturbation of the form

$$\Phi(\mathbf{x}, t) = -\sigma^2 T \delta(t) \cos(\mathbf{k} \cdot \mathbf{x}), \qquad (5\text{-}32)$$

where T is a short time and $\delta(t)$ is the Dirac delta function (Appendix 1.C). Using linear perturbation theory and the Jeans swindle, find the resulting density perturbation in the system for all $t > 0$.[7]

We shall leave the step-by-step analysis to Problem 5-3. The answer is that for $t > 0$ the density perturbation is given by

$$\rho_1(\mathbf{x}, t) = \rho_0 k \sigma T Y(k \sigma t) \cos(\mathbf{k} \cdot \mathbf{x}), \qquad (5\text{-}33)$$

where $k = |\mathbf{k}|$ and the function $Y(\tau)$ satisfies the integral equation

$$Y(\tau) = \tau e^{-\frac{1}{2}\tau^2} + \frac{k_J^2}{k^2} \int_0^\tau Y(\tau - \tau'') \tau'' e^{-\frac{1}{2}\tau''^2} d\tau''. \qquad (5\text{-}34)$$

The results of solving equation (5-34) numerically are shown in Figure 5-2. As we might have expected, the perturbation grows when $k^2 < k_J^2$ and damps for $k^2 > k_J^2$. In the limit $k_J^2/k^2 = 0$ the damping is entirely due to phase mixing (cf. §4.7) since the effects of self-gravity vanish.

It is instructive to contrast the response of the stellar system to the response of the equivalent plasma. The plasma obeys equation (5-34), but with k_J^2/k^2 replaced by $-k_P^2/k^2$ [see footnote following eq. (5-22)]. As k_P^2/k^2 increases from zero, damping becomes less important, and eventually the plasma exhibits slowly decaying waves.

4 Discussion

All of the results that we have obtained so far rest on two foundations: (i) the assumption that the system is infinite and homogeneous; (ii) the Jeans swindle of neglecting the gravitational potential of the

---

[7] Suggested by A. Toomre.

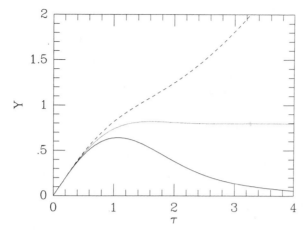

Figure 5-2. Evolution of an infinite homogeneous
stellar system in response to an impulsive perturbation
with wavenumber k. Curves are for $k_J/k = 0.5$ (full
curve), 1.0 (dotted curve), and 1.25 (dashed curve).
The dimensionless density Y satisfies equation (5-34).

unperturbed system. It is time for us to pause and investigate the consequences of these assumptions.

The virial theorem [eq. (4-79)] tells us that a stellar system of mass M and mean square velocity $\overline{v^2}$ has a characteristic size $\lambda_0 \approx GM/\overline{v^2}$. In terms of the mean density $\rho \approx M/\lambda_0^3$, we have $\lambda_0^2 \approx \overline{v^2}/G\rho$. However, from equation (5-30) the Jeans length is given by $\lambda_J^2 \approx \overline{v^2}/G\rho \approx \lambda_0^2$. Thus the Jeans length is comparable to the size of the system, and the assumption of homogeneity is not valid. Accordingly, the Jeans analysis does not establish that there is a real physical instability in any isolated stellar system. Nevertheless, the analysis is a cornerstone of the stability theory of self-gravitating systems, for the following reasons.

First, as we have already argued at the beginning of this section, the homogeneity assumption is valid and the Jeans swindle is legitimate on scales $\lambda \ll \lambda_0$ (because the effects of the self-gravity of the equilibrium system are small on scales much smaller than λ_0). Therefore, we can conclude from our analysis that stationary stellar systems are generally stable on small scales. This is a very important and general conclusion.

Second, the fundamental correctness of Jeans's analysis and the significance of the Jeans length can be verified by analyzing the stability of homogeneous systems that are expanding or collapsing. For these systems one does not have to invoke the Jeans swindle; the unperturbed gravitational field simply causes deceleration of the expansion.

Just as before, one finds that λ_J, now a function of time, roughly divides the stable from the unstable perturbations. The only difference is that perturbations of wavelength $\lambda \gg \lambda_J$ grow as a power of the time ($\rho_1/\rho_0 \propto t^{2/3}$) rather than exponentially in time (Peebles 1971; Weinberg 1972; Peebles 1980b). This is one of the most important applications of the Jeans instability, since precisely this kind of instability may have led to the formation of galaxies from the homogeneous "cosmic soup" that filled the early expanding Universe.

Third, the Jeans swindle can be avoided in certain homogeneous, uniformly rotating systems, and in these the usual Jeans criterion is found to remain valid (see Problem 5-2 and Chandrasekhar 1961).

Fourth, instabilities that are closely related to the Jeans instability play an important role in the dynamics of rotating disks (see §5.3.1 and §6.2.3.)

Finally, the Jeans analysis is useful because with simple modifications it can be used to investigate whether more complicated DFs are unstable on small scales (see Lynden-Bell 1967b).

5.2 The Stability of Spherical Systems

The stability analysis of realistic, finite, inhomogeneous systems is much more difficult than the analysis of homogeneous systems like those of §5.1. Only for spherical systems is the theory in a satisfactory state; many of the known results for these systems are contained in two important papers by V. A. Antonov (1960, 1962b). Antonov's results mostly concern spherical systems in which the unperturbed DF f_0 depends only on energy, $f_0 = f_0(E) = f_0[\frac{1}{2}v^2 + \Phi_0(r)]$. The more general spherical systems in which f_0 depends on energy and angular momentum, $f_0 = f_0(E, L)$, are discussed briefly in §5.2.2(c).

The simplest and most physical stability criteria are based on energy. For example, a ball resting at the bottom of a bowl is stable because all nearby positions of the ball have higher energy. A similar stability criterion can be established for stellar systems: they are certain to be stable if all neighboring configurations with the same total mass have higher energy. However, energy criteria are generally sufficient but not necessary for stability. Satisfaction of an energy criterion will be necessary for stability only if we can be sure that all neighboring configurations are dynamically accessible. As an example of a system that cannot reach certain low-energy configurations, consider a ball that rolls around a bowl at constant height, maintaining a balance between centrifugal and gravitational forces. The ball's trajectory is stable (in the absence of friction) despite the existence of states of lower energy, because the ball cannot attain these states while conserving its angular momentum.

To investigate the stability of spherical stellar systems, we shall once again exploit the analogy between stellar systems and fluids. First we discuss the stability of spherical bodies of barotropic gas, using energy arguments. Next, following Antonov, we show how to relate the stability of gaseous bodies to the stability of the analogous stellar systems. Using this apparatus, we can appeal to results from stellar stability theory to demonstrate the stability of many spherical stellar systems. Finally, we discuss stability criteria that are not based on energy arguments and have no fluid analog. Some readers may prefer to pass directly to the discussion of Antonov's work which follows in the next subsection, and to return to this section for the relevant results about fluid systems as they are required.

1 Stability of Barotropic Stars

We assume that the unperturbed star is static ($\mathbf{v} = 0$), that its density decreases outward ($d\rho_0/dr < 0$), and that the density is zero at the surface of the star r_m [$\rho_0(r_m) = 0$]. Euler's equation (5-8) for a spherical system reads

$$\frac{dp_0}{dr} = -\rho_0 \frac{d\Phi_0}{dr}, \tag{5-35}$$

or

$$\left(\frac{dp}{d\rho}\right)_0 = -\rho_0 \left(\frac{d\Phi}{d\rho}\right)_0 = \rho_0 \left|\frac{d\Phi}{d\rho}\right|_0, \tag{5-36}$$

where the last line follows because $(d\Phi/d\rho)_0 < 0$ (since Φ_0 must increase outward and we have assumed that ρ_0 decreases outward). The linearized fluid equations (5-12) become

$$\frac{\partial \rho_1}{\partial t} + \nabla \cdot (\rho_0 \mathbf{v}_1) = 0, \tag{5-37a}$$

$$\frac{\partial \mathbf{v}_1}{\partial t} = -\nabla h_1 - \nabla \Phi_1, \tag{5-37b}$$

$$\nabla^2 \Phi_1 = 4\pi G \rho_1, \tag{5-37c}$$

$$h_1 = \left(\frac{dp}{d\rho}\right)_0 \frac{\rho_1}{\rho_0}. \tag{5-37d}$$

The boundary conditions on equations (5-37) are (see Cox 1980, §17.6 for more detail): (i) ρ_1 and $\mathbf{v}_1$ are finite at $r = 0$; (ii) $p_1 = 0$ at $r = r_m$;[8] (iii) Φ_1 is finite at $r = 0$; (iv) at $r = r_m$, Φ_1 matches onto a solution of Laplace's equation that decays to zero as $r \to \infty$. In some cases we shall replace equation (5-37c) by the integral form of Poisson's equation,

$$\Phi_1(\mathbf{r}, t) = -G \int \frac{\rho_1(\mathbf{r}', t)\, d^3\mathbf{r}'}{|\mathbf{r} - \mathbf{r}'|}, \tag{5-38}$$

in which case only boundary conditions (i) and (ii) are needed.

The coefficients of equations (5-37) and the boundary conditions are independent of the time t. Hence we may solve these equations in terms of **normal modes** that have the form

$$\rho_1(\mathbf{r}, t) = \mathrm{Re}\left[\rho_a(\mathbf{r})e^{-i\omega_a t}\right] \quad ; \quad \mathbf{v}_1(\mathbf{r}, t) = \mathrm{Re}\left[\mathbf{v}_a(\mathbf{r})e^{-i\omega_a t}\right],$$
$$\Phi_1(\mathbf{r}, t) = \mathrm{Re}\left[\Phi_a(\mathbf{r})e^{-i\omega_a t}\right] \quad ; \quad h_1(\mathbf{r}, t) = \mathrm{Re}\left[h_a(\mathbf{r})e^{-i\omega_a t}\right]. \tag{5-39}$$

With these definitions, equations (5-37) become

$$-i\omega_a\rho_a + \boldsymbol{\nabla} \cdot (\rho_0\mathbf{v}_a) = 0, \tag{5-40a}$$

$$-i\omega_a\mathbf{v}_a = -\boldsymbol{\nabla}h_a - \boldsymbol{\nabla}\Phi_a, \tag{5-40b}$$

$$\nabla^2\Phi_a = 4\pi G\rho_a, \tag{5-40c}$$

$$h_a = \left(\frac{dp}{d\rho}\right)_0 \frac{\rho_a}{\rho_0}, \tag{5-40d}$$

or, using equation (5-36),

$$h_a = -\left(\frac{d\Phi}{d\rho}\right)_0 \rho_a = \left|\frac{d\Phi}{d\rho}\right|_0 \rho_a. \tag{5-41}$$

These equations can be combined into a single fourth-order equation (Ledoux & Walraven 1958, §78); however, for our purposes it is simpler to proceed by a different method. We do not need to know the exact functional forms of ρ_a, $\mathbf{v}_a$, Φ_a, but only the possible values of ω_a, since the system is unstable if and only if $\mathrm{Im}(\omega_a) > 0$. In these circumstances

---

[8] Strictly, we should take into account that the boundary of the star moves, and it is on the perturbed boundary that the pressure should vanish. Thus the correct boundary condition is $p_1 + \boldsymbol{\xi}\cdot\boldsymbol{\nabla}p_0 = 0$, where $\boldsymbol{\xi}$ is the displacement of a fluid element (see Chandrasekhar 1969 or Cox 1980 for more detail). However, $\boldsymbol{\nabla}p_0 = -\rho_0(d\Phi_0/dr)\hat{\mathbf{e}}_r$ by equation (5-35) and $\rho_0(r_m) = 0$ by assumption, so that $\boldsymbol{\xi}\cdot\boldsymbol{\nabla}p_0 = 0$ at the surface.

it is often simplest to attempt to find a **variational principle** for the set of equations, and this is what we now do.

Let us denote a second mode by ω_b, ρ_b, $\mathbf{v}_b$, etc. We multiply equation (5-40b) by $\rho_0 \mathbf{v}_b^*$, where $\mathbf{v}_b^*$ is the complex conjugate of $\mathbf{v}_b$. Integrating the resulting equation over the volume of the star, we obtain

$$i\omega_a \int \rho_0 \mathbf{v}_b^* \cdot \mathbf{v}_a \, d^3\mathbf{r} = \int \rho_0 \mathbf{v}_b^* \cdot \boldsymbol{\nabla}(h_a + \Phi_a) \, d^3\mathbf{r}. \qquad (5\text{-}42)$$

If we integrate the right side by parts [as in eq. (1B-43)], the boundary term is proportional to $\rho_0 \mathbf{v}_b^*(h_a + \Phi_a)$, which equals $\mathbf{v}_b^*(p_a + \rho_0 \Phi_a)$ by equations (5-15) and (5-12d). This term vanishes because both p_a and ρ_0 vanish on the surface. Thus

$$i\omega_a \int \rho_0 \mathbf{v}_b^* \cdot \mathbf{v}_a \, d^3\mathbf{r} = - \int \boldsymbol{\nabla} \cdot (\rho_0 \mathbf{v}_b^*)(h_a + \Phi_a) \, d^3\mathbf{r}, \qquad (5\text{-}43)$$

and using equation (5-40a) to eliminate $\boldsymbol{\nabla} \cdot (\rho_0 \mathbf{v}_b^*)$, we obtain

$$i\omega_a \int \rho_0 \mathbf{v}_b^* \cdot \mathbf{v}_a \, d^3\mathbf{r} = i\omega_b^* \int \rho_b^*(h_a + \Phi_a) \, d^3\mathbf{r}. \qquad (5\text{-}44)$$

The integral form of Poisson's equation (5-38) enables us to eliminate Φ_a, and we eliminate h_a with the aid of equation (5-41):

$$\frac{\omega_a}{\omega_b^*} \int \rho_0 \mathbf{v}_b^* \cdot \mathbf{v}_a \, d^3\mathbf{r} = \int \left| \frac{d\Phi}{d\rho} \right|_0 \rho_b^* \cdot \rho_a \, d^3\mathbf{r} - G \iint \frac{d^3\mathbf{r} \, d^3\mathbf{r}'}{|\mathbf{r} - \mathbf{r}'|} \rho_b^*(\mathbf{r}) \rho_a(\mathbf{r}').$$
$$(5\text{-}45)$$

Finally, we take the complex conjugate of equation (5-45), interchange the labels a and b, and subtract the resulting equation from (5-45). The result is

$$\left(\frac{\omega_a}{\omega_b^*} - \frac{\omega_b^*}{\omega_a} \right) \int \rho_0 \mathbf{v}_b^* \cdot \mathbf{v}_a \, d^3\mathbf{r} = 0. \qquad (5\text{-}46)$$

If $a = b$, the integral is positive definite and we conclude that $\omega_a^2 = (\omega_a^*)^2 = (\omega_a^2)^*$. Hence *all modes have real ω^2.* If $\omega^2 \geq 0$, then ω is real, the perturbations in equations (5-39) oscillate with frequency ω, and the mode is stable. If $\omega^2 < 0$, say $\omega^2 = -\gamma^2$, where γ is real, then the temporal dependence of the perturbations is proportional to $\exp(\pm\gamma t)$, one component of the solution grows exponentially, and the mode is unstable. Furthermore, if $a \neq b$ then

$$\text{either} \quad \omega_a = \omega_b \quad \text{or} \quad \int \rho_0 \mathbf{v}_b^* \cdot \mathbf{v}_a \, d^3\mathbf{r} = 0. \qquad (5\text{-}47)$$

In the language of mathematical physics we say that either the modes are degenerate or they are orthogonal.

When we set $a = b$ in equation (5-45) we obtain

$$\frac{\omega_a^2}{|\omega_a|^2} \int \rho_0 |\mathbf{v}_a|^2 \, d^3\mathbf{r} = \int \left|\frac{d\Phi}{d\rho}\right|_0 |\rho_a|^2 d^3\mathbf{r} - G \iint \frac{d^3\mathbf{r}\, d^3\mathbf{r}'}{|\mathbf{r} - \mathbf{r}'|} \rho_a^*(\mathbf{r})\rho_a(\mathbf{r}').$$

(5-48)

Since $\int \rho_0 |\mathbf{v}_a|^2 \, d^3\mathbf{r}$ and $|\omega_a|^2$ are positive, we may conclude that a mode is unstable if and only if the quantity in square brackets on the right side of equation (5-48) is negative. This result permits us to state a sufficient condition for stability:

Chandrasekhar's Variational Principle *A barotropic star with $d\rho_0/dr < 0$ and $\rho_0(r_m) = 0$ is stable if the quantity*

$$\mathcal{E}[\rho_1] \equiv \int \left|\frac{d\Phi}{d\rho}\right|_0 \rho_1^2 d^3\mathbf{r} - G \iint \frac{d^3\mathbf{r}\, d^3\mathbf{r}'}{|\mathbf{r} - \mathbf{r}'|} \rho_1(\mathbf{r})\rho_1(\mathbf{r}') \qquad (5\text{-}49)$$

is non-negative for all real functions $\rho_1(\mathbf{r})$ that conserve the total mass of the star ($\int \rho_1 \, d^3\mathbf{r} = 0$).

Proof: Assume that the star is unstable and that $\mathcal{E}[\rho_1] \geq 0$ for all functions ρ_1 that conserve mass, and look for a contradiction. Since the star is unstable, there is some mode with $\omega_a^2 < 0$, and for this mode the quantity on the right of equation (5-48) is negative. This quantity is just $\mathcal{E}[\text{Re}(\rho_a)] + \mathcal{E}[\text{Im}(\rho_a)]$; hence at least one of these two terms is negative. Suppose the first term is negative; then set $\rho_1 = \text{Re}(\rho_a)$. The continuity equation (5-40a) guarantees that $\int \rho_a \, d^3\mathbf{r} = 0$, and taking the real part of this equation yields $\int \rho_1 \, d^3\mathbf{r} = 0$. Hence $\mathcal{E}[\rho_1] < 0$ and ρ_1 conserves mass, which contradicts our assumptions. A similar contradiction appears if $\mathcal{E}[\text{Im}(\rho_a)] < 0.\lhd$

This variational principle was first stated by Chandrasekhar (1963, 1964; see also Lebovitz 1965), who gave a more general stability criterion for adiabatic perturbations to a spherical body of gas with an arbitrary equation of state (i.e., not just a barotropic gas). The requirement that $\mathcal{E}[\rho_1]$ be non-negative can also be shown to be necessary for stability (Laval et al. 1965; Cox 1980, §15.2), but we shall not need these more general results.

Chandrasekhar's variational principle—which is really a stability criterion based on energy, as is shown in Appendix 5.B—can be used to prove a remarkably general result that was derived independently by Antonov (1962b) and Lebovitz (1965):

The Antonov-Lebovitz Theorem *Any barotropic star with $d\rho_0/dr < 0$ and $\rho_0(r_m) = 0$ is stable to all perturbations that are not spherically symmetric.*

Proof: See Appendix 5.C.◁

This theorem shows that only radial perturbations are dangerous for a spherical star. To investigate the stability of radial perturbations, we return to the basic perturbation equations (5-40). Since the perturbation is radial, we set $\mathbf{v}_a = v_a \hat{\mathbf{e}}_r$, $\nabla h_a = (dh_a/dr)\hat{\mathbf{e}}_r$, etc., and use equation (2-23a) to write (5-40b) as

$$-i\omega_a v_a = -\frac{dh_a}{dr} - \frac{4\pi G}{r^2} \int_0^r \rho_a r^2 \, dr. \tag{5-50}$$

Using (5-40a) and the definition of divergence in spherical coordinates (1B-46), we obtain

$$\int_0^r \rho_a r^2 \, dr = \frac{1}{i\omega_a} \int_0^r \frac{d}{dr}(r^2 \rho_0 v_a) \, dr = \frac{r^2 \rho_0 v_a}{i\omega_a}, \tag{5-51}$$

and after substituting for h_a from equations (5-41) and (5-40a), equation (5-50) becomes

$$\frac{d}{dr}\left[\frac{1}{r^2}\left|\frac{d\Phi}{d\rho}\right|_0 \frac{d}{dr}(r^2 \rho_0 v_a)\right] + (\omega_a^2 + 4\pi G\rho_0)v_a = 0. \tag{5-52}$$

To simplify this equation we make two new definitions. We define the local adiabatic index $\gamma(r)$ by

$$\gamma = \left(\frac{d\ln p}{d\ln \rho}\right)_0; \tag{5-53}$$

using equation (5-35) it is easy to show that an alternative definition is

$$\left|\frac{d\Phi}{d\rho}\right|_0 = \gamma\frac{p_0}{\rho_0^2}. \tag{5-54}$$

We also replace v_a by $\varsigma_a \equiv iv_a/(\omega_a r)$. Physically, $|\varsigma_a|$ is the amplitude of the fractional displacement in radius of a fluid element.

After a number of algebraic manipulations using equations (5-53) and (5-54), equation (5-52) can be cast in the form

$$\frac{d}{dr}\left(p_0\gamma r^4 \frac{d\varsigma_a}{dr}\right) + \left\{\omega_a^2 \rho_0 r^4 + r^3\frac{d}{dr}[(3\gamma - 4)p_0]\right\}\varsigma_a = 0. \tag{5-55}$$

This equation was first derived by Eddington in 1918 (see Eddington 1926, and Cox 1980, Chapter 8). We can derive a variational principle by multiplying by ς_a^* and integrating from $r = 0$ to $r = r_m$. After performing an integration by parts, we find

$$\omega_a^2 \int_0^{r_m} \rho_0 r^4 |\varsigma_a|^2 \, dr = \int_0^{r_m} \left\{ p_0 \gamma r^4 \left| \frac{d\varsigma_a}{dr} \right|^2 - r^3 |\varsigma_a|^2 \frac{d}{dr} [(3\gamma - 4)p_0] \right\} dr.$$
(5-56)

This equation leads us to a theorem on the stability of radial modes that is analogous to Chandrasekhar's variational principle:

Eddington's Variational Principle *A barotropic star with* $d\rho_0/dr < 0$ *and* $\rho_0(r_m) = 0$ *is stable to radial oscillations if the quantity*

$$\mathcal{F}[\varsigma] = \int_0^{r_m} \left\{ p_0 \gamma r^4 \left(\frac{d\varsigma}{dr} \right)^2 - r^3 \varsigma^2 \frac{d}{dr} [(3\gamma - 4)p_0] \right\} dr$$
(5-57)

is non-negative for all functions $\varsigma(r)$.

Proof: Follows from equation (5-56) in almost the same way that Chandrasekhar's variational principle follows from equation (5-48).◁

It can also be shown that the condition that $\mathcal{F}[\varsigma] \geq 0$ for all functions $\varsigma(r)$ is necessary for stability (see Cox 1980, Chapter 8).

The simplest model stars are polytropes, composed of barotropic gases with $\gamma(r) = constant$, so that the equation of state is $p = K\rho^\gamma$ [cf. eq. (1E-27)]. For polytropes we have

$$\mathcal{F}[\varsigma] = \gamma \int_0^{r_m} p_0 r^4 \left(\frac{d\varsigma}{dr} \right)^2 dr - (3\gamma - 4) \int_0^{r_m} r^3 \varsigma^2 \frac{dp_0}{dr} \, dr.$$
(5-58)

The first integral is positive, and the second integral is negative since dp_0/dr is negative. Thus if $\gamma \geq \frac{4}{3}$ then $\mathcal{F}[\varsigma]$ is non-negative and the star is stable. Hence *a polytropic star is stable if* $\gamma \geq \frac{4}{3}$. It can also be shown that the star is unstable if $\gamma < \frac{4}{3}$.

This result has been known since the nineteenth century and is due to A. Ritter (see Chandrasekhar 1939, p. 178). It can also be derived by simple and elegant arguments involving only equilibrium configurations (e.g., Weinberg 1972). A general barotrope may have $\gamma(r) < \frac{4}{3}$ over a limited range of radii and still be stable; in such cases the stability of the star depends on whether or not a suitably weighted average of $(3\gamma - 4)$ is positive.

2 Stability of Stellar Systems

(a) Antonov's variational principle   We consider the stability of a stellar system in which the unperturbed DF $f_0(E)$ depends only on energy $E = \frac{1}{2}v^2 + \Phi(r)$ and satisfies $df_0/dE < 0$. The King, Wilson, and $R^{1/4}$ models that we discussed in §4.4.3 are all of this type.

We start from the linearized collisionless Boltzmann equation (5-5), which can be written

$$\frac{\partial f_1}{\partial t} + D_0 f_1 - \nabla\Phi_1 \cdot \frac{\partial f_0}{\partial \mathbf{v}} = 0, \qquad (5\text{-}59)$$

where the operator

$$D_0 \equiv \mathbf{v} \cdot \nabla - \nabla\Phi_0 \cdot \frac{\partial}{\partial \mathbf{v}}. \qquad (5\text{-}60)$$

If we now write

$$f_1(\mathbf{r}, \mathbf{v}, t) = \mathrm{Re}[f_\omega(\mathbf{r}, \mathbf{v})e^{-i\omega t}], \qquad (5\text{-}61)$$

and eliminate Φ_1 using the integral form of Poisson's equation (5-38), we obtain

$$-i\omega f_\omega + D_0 f_\omega + G\frac{df_0}{dE}\mathbf{v} \cdot \nabla \int \frac{d^6\mathbf{w}'}{|\mathbf{r} - \mathbf{r}'|} f_\omega(\mathbf{r}', \mathbf{v}') = 0, \qquad (5\text{-}62)$$

where $d^6\mathbf{w}' \equiv d^3\mathbf{r}' \, d^3\mathbf{v}'$. The key to proceeding further is a trick introduced by Antonov (1960). We can always write

$$f_\omega(\mathbf{r}, \mathbf{v}) = f_+(\mathbf{r}, \mathbf{v}) + f_-(\mathbf{r}, \mathbf{v}), \qquad (5\text{-}63)$$

where

$$f_\pm(\mathbf{r}, \mathbf{v}) = \frac{1}{2}\left[f_\omega(\mathbf{r}, \mathbf{v}) \pm f_\omega(\mathbf{r}, -\mathbf{v})\right]. \qquad (5\text{-}64)$$

Thus f_+ is symmetric in $\mathbf{v}$ and f_- is antisymmetric in $\mathbf{v}$. Note that the operator D_0 turns an antisymmetric function into a symmetric one, and vice versa. Substituting (5-63) into (5-62) and separating the symmetric and antisymmetric components, we find

$$-i\omega f_+ + D_0 f_- = 0,$$
$$-i\omega f_- + D_0 f_+ + G\frac{df_0}{dE}\mathbf{v}\cdot\nabla \int \frac{d^6\mathbf{w}'}{|\mathbf{r} - \mathbf{r}'|} f_+(\mathbf{r}', \mathbf{v}') = 0. \qquad (5\text{-}65)$$

Eliminating f_+ from this pair of equations, we have

$$\omega^2 f_- + D_0^2 f_- + G\frac{df_0}{dE}\mathbf{v} \cdot \nabla \int \frac{d^6\mathbf{w}'}{|\mathbf{r} - \mathbf{r}'|} D_0 f_-(\mathbf{r}', \mathbf{v}') = 0. \qquad (5\text{-}66)$$

We now multiply through by $f_-^*/(df_0/dE)$ and integrate over phase space:

$$\omega^2 \int \frac{d^6\mathbf{w}}{df_0/dE}|f_-|^2 + \int \frac{d^6\mathbf{w}}{df_0/dE} f_-^* D_0^2 f_-$$
$$+ G\int d^6\mathbf{w}\, f_-^*\mathbf{v} \cdot \nabla \int \frac{d^6\mathbf{w}'}{|\mathbf{r} - \mathbf{r}'|} D_0 f_-(\mathbf{r}', \mathbf{v}') = 0. \qquad (5\text{-}67)$$

We can rewrite the second term by noting that $D_0(E) = 0$; hence $D_0[1/(df_0/dE)] = 0$, and the denominator df_0/dE can be taken inside the operator D_0. The second term becomes

$$
\int \frac{d^6\mathbf{w}}{df_0/dE} f_-^* D_0^2 f_- = \int d^6\mathbf{w}\, f_-^* D_0 \left[\frac{D_0 f_-}{df_0/dE} \right]
$$
$$
= \int d^6\mathbf{w}\, f_-^* \left[\mathbf{v} \cdot \frac{\partial}{\partial \mathbf{r}} \left(\frac{D_0 f_-}{df_0/dE} \right) - \nabla\Phi_0 \cdot \frac{\partial}{\partial \mathbf{v}} \left(\frac{D_0 f_-}{df_0/dE} \right) \right].
$$

(5-68)

We now integrate each term by parts. The boundary terms vanish because $f_- \to 0$ as $|\mathbf{r}|, |\mathbf{v}| \to \infty$ and we find

$$
\int \frac{d^6\mathbf{w}}{df_0/dE} f_-^* D_0^2 f_- = -\int d^6\mathbf{w} \left(\mathbf{v} \cdot \frac{\partial f_-^*}{\partial \mathbf{r}} - \nabla\Phi_0 \cdot \frac{\partial f_-^*}{\partial \mathbf{v}} \right) \left(\frac{D_0 f_-}{df_0/dE} \right)
$$
$$
= -\int \frac{d^6\mathbf{w}}{df_0/dE} |D_0 f_-|^2.
$$

(5-69)

Doing the integral over $\mathbf{r}$ in the third term of (5-67) by parts, this term becomes

$$
-G \int d^6\mathbf{w}\, (\mathbf{v} \cdot \nabla) f_-^* \int \frac{d^6\mathbf{w}'}{|\mathbf{r} - \mathbf{r}'|} D_0 f_-(\mathbf{r}', \mathbf{v}'). \tag{5-70}
$$

Since integration by parts shows that $\int d^3\mathbf{v}\, \nabla\Phi_0 \cdot (\partial f_-^*/\partial \mathbf{v}) = 0$, we have $\int d^3\mathbf{v}\, D_0 f_-^* = \int d^3\mathbf{v}\, \mathbf{v} \cdot \nabla f_-^*$. Hence we can write (5-70) as

$$
-G \iint d^6\mathbf{w}\, d^6\mathbf{w}' \frac{D_0 f_-^*(\mathbf{r}, \mathbf{v}) D_0 f_-(\mathbf{r}', \mathbf{v}')}{|\mathbf{r} - \mathbf{r}'|}. \tag{5-71}
$$

Combining these results, equation (5-67) yields

$$
\omega^2 \int \frac{d^6\mathbf{w}}{df_0/dE} |f_-|^2 = \int \frac{d^6\mathbf{w}}{df_0/dE} |D_0 f_-|^2
$$
$$
+ G \iint \frac{d^6\mathbf{w}\, d^6\mathbf{w}'}{|\mathbf{r} - \mathbf{r}'|} D_0 f_-^*(\mathbf{r}, \mathbf{v}) D_0 f_-(\mathbf{r}', \mathbf{v}').
$$

(5-72)

The right side of equation (5-72) is real; hence we conclude that *all modes have real ω^2; stable modes have $\omega^2 \geq 0$ and unstable modes have $\omega^2 < 0$.* Using equation (5-72) we can derive a necessary and sufficient condition for stability:

Antonov's Variational Principle *A spherical stellar system with $f_0 = f_0(E)$ and $df_0/dE < 0$ is stable if and only if the quantity*

$$
\mathcal{G}[f_1] \equiv \int \frac{d^6\mathbf{w}}{|df_0/dE|} (D_0 f_1)^2 - G \iint \frac{d^6\mathbf{w}\, d^6\mathbf{w}'}{|\mathbf{r} - \mathbf{r}'|} D_0 f_1(\mathbf{r}, \mathbf{v}) D_0 f_1(\mathbf{r}', \mathbf{v}')
$$

(5-73)

is non-negative for all real functions $f_1(\mathbf{r}, \mathbf{v})$.

Proof: To show that $\mathcal{G}[f_1] \geq 0$ is sufficient for stability we proceed in the same manner as in the demonstration of Chandrasekhar's variational principle for fluids, except that we start with equation (5-72) instead of (5-48). The proof that $\mathcal{G}[f_1] \geq 0$ is necessary for stability is more delicate. Antonov's original proof assumes that the stellar system has a complete set of normal modes, i.e., that every perturbation can be written as a sum of normal modes, but this assumption is difficult to justify for a stellar system. (In Appendix 5.A, for example, we show that the normal modes with real ω of an infinite homogeneous stellar system are all singular.) An alternative and more satisfactory proof is given by Kulsrud and Mark (1970).◁

(b) Antonov's laws   The analogy between Antonov's variational principle for stellar systems and Chandrasekhar's variational principle for stars can be made exact by the following theorem:

Antonov's First Law *A spherical stellar system with* $f_0 = f_0(E)$ *and* $df_0/dE < 0$ *is stable if the barotropic star with the same equilibrium density distribution is stable.*

Proof: Consider the function

$$\rho_1(\mathbf{r}) = \int D_0 f_1 \, d^3\mathbf{v}. \tag{5-74}$$

Note that $\int \rho_1 \, d^3\mathbf{r} = \int d^6\mathbf{w} [\mathbf{v} \cdot \nabla f_1 - \nabla \Phi_0 \cdot (\partial f_1/\partial \mathbf{v})]$, and integration by parts shows that each of these terms vanishes since $f_1 \to 0$ as $|\mathbf{r}|, |\mathbf{v}| \to \infty$. Hence $\int \rho_1 d^3\mathbf{r} = 0$, and if ρ_1 is interpreted as a density, the total mass is conserved in the perturbation.

Now according to Schwarz's inequality (e.g., Arfken 1970), for any two real functions A and B

$$\int A^2 \, d^3\mathbf{v} \int B^2 \, d^3\mathbf{v} \geq \left(\int AB \, d^3\mathbf{v} \right)^2. \tag{5-75}$$

Thus if we set $A = |df_0/dE|^{1/2}$ and $B = D_0 f_1/|df_0/dE|^{1/2}$, we have

$$\int \frac{(D_0 f_1)^2 d^3\mathbf{v}}{|df_0/dE|} \geq \frac{\left(\int D_0 f_1 d^3\mathbf{v} \right)^2}{\int |df_0/dE| d^3\mathbf{v}} = \frac{\rho_1^2}{\int |df_0/dE| d^3\mathbf{v}}. \tag{5-76}$$

The denominator on the right side of (5-76) can be related to the unperturbed density $\rho_0 = \int f_0(\frac{1}{2}v^2 + \Phi_0) \, d^3\mathbf{v}$, by differentiating both sides of the latter expression with respect to Φ_0:

$$\left(\frac{d\rho}{d\Phi} \right)_0 = \int \frac{df_0}{dE} d^3\mathbf{v} = -\int \left| \frac{df_0}{dE} \right| d^3\mathbf{v}. \tag{5-77}$$

Combining equations (5-73), (5-76), and (5-77), we obtain

$$\mathcal{G}[f_1] \geq \int \left| \frac{d\Phi}{d\rho} \right|_0 \rho_1^2 d^3\mathbf{r} - G \iint \frac{d^3\mathbf{r}\, d^3\mathbf{r}'}{|\mathbf{r} - \mathbf{r}'|} \rho_1(\mathbf{r})\rho_1(\mathbf{r}'). \qquad (5\text{-}78)$$

If the barotropic star with density $\rho_0(r)$ is stable, then the right side is non-negative by Chandrasekhar's variational principle. Hence $\mathcal{G}[f_1] \geq 0$ and the stellar system is stable by Antonov's variational principle. This completes the proof of Antonov's first law.◁

An immediate consequence is:

Antonov's Second Law *A spherical stellar system with $f_0 = f_0(E)$ and $df_0/dE < 0$ is stable to all nonradial perturbations.*

Proof: Follows from the Antonov-Lebovitz theorem and Antonov's first law.◁

Unfortunately, Antonov's first law is not very helpful when we consider the stability of stellar systems to *radial* perturbations, because it provides a sufficient but not necessary condition for stability. Therefore it can only be used to prove stability, and then only if the barotropic analog of the stellar system of interest is stable. We saw in §4.4.3 that the stellar polytropes of greatest interest are those with index $n \geq 5$ ($\gamma \leq \frac{6}{5}$), and we showed in the previous subsection that the barotropic analogs of these systems are unstable. There is a simple physical reason for the instability of the barotropic systems. The molecules of a gaseous polytrope of index n have $(2n - 3)$ degrees of internal freedom that are excited when the system is compressed adiabatically. Thus, when n is large, most of the gravitational energy that is released during a contraction is fed into internal degrees of freedom rather than the three translational degrees of freedom that contribute to the pressure. Consequently, the pressure does not rise rapidly enough to balance the increased gravitational self-attraction, and collapse ensues. In contrast, when a stellar polytrope contracts, all of the released gravitational energy goes into translational motions that resist the increased self-attraction. Hence we expect stellar systems to be more stable than gaseous ones. This expectation is confirmed by the following two powerful theorems:

Antonov's Third Law *A spherical stellar system with $f_0 = f_0(E)$ and $df_0/dE < 0$ is stable to radial perturbations if its density $\rho_0(r)$ and potential $\Phi_0(r)$ satisfy everywhere the inequality*

$$\frac{d^3\rho_0}{d\Phi_0^3} \leq 0. \qquad (5\text{-}79)$$

Antonov's third law is a special case of an even more general result:

Doremus-Feix-Baumann Theorem *Any spherical stellar system with* $f_0 = f_0(E)$ *and* $df_0/dE < 0$ *is stable to radial perturbations.*

Proof: Both these theorems are based on Antonov's variational principle, but the proofs are complicated and tedious. We refer the reader to Antonov's (1962b) paper for the first theorem, and to Sygnet et al. (1984) and Kandrup and Sygnet (1985), who have considerably simplified the original proof of the second (Doremus et al. 1971).◁

The Doremus-Feix-Baumann theorem, combined with Antonov's second law, shows that almost all realistic spherical systems with $f = f_0(E)$ are stable to both radial and nonradial perturbations. Thus, for example, all of the polytropic stellar systems (§4.4) with $n > \frac{3}{2}$ are stable.[9] In particular, Plummer's model ($n = 5$) is stable; this result is important because the density distribution in the Plummer model, $\rho(r) \propto (r^2 + a^2)^{-5/2}$, is similar to the density distribution in many real stellar systems.

The DF for the isochrone model [eq. (2-32)] is given by equation (4-144); it is straightforward to show that $df_0/dE < 0$ and hence the isochrone is stable. The DFs of the King models [eq. (4-130)] also satisfy $df_0/dE < 0$ and hence all King models are also stable.

(c) Systems with $f = f(E, L)$ In many spherical systems the DF is of the form $f_0 = f_0(E, L)$, i.e., it depends both on energy and angular momentum, and the velocity-dispersion tensor is anisotropic. In such systems Antonov's laws cannot be used to establish stability to nonradial perturbations. However, a version of Antonov's variational principle still applies to radial perturbations. To show this, write $\nabla \Phi_1 = \hat{\mathbf{e}}_r d\Phi_1/dr$, and substitute this into the left side of the linearized collisionless Boltzmann equation (5-59):

$$-\nabla \Phi_1 \cdot \frac{\partial f_0}{\partial \mathbf{v}} = -\frac{\partial \Phi_1}{\partial r} \frac{\partial f_0}{\partial v_r} = -\frac{\partial \Phi_1}{\partial r} v_r \frac{\partial f_0}{\partial E}. \qquad (5\text{-}80)$$

Notice that partial derivatives of the form $\partial f_0/\partial L$ do not appear, and hence all of the steps leading to Antonov's variational principle can be carried out by simply replacing df_0/dE by $\partial f_0/\partial E$ wherever it appears. Similarly, Antonov's third law becomes:

---

[9] Polytropes with $\frac{1}{2} < n < \frac{3}{2}$ are somewhat unrealistic because they have $df_0/dE > 0$, with an integrable singularity in f_0 at the boundary $E = E_0$. However, numerical experiments (Hénon 1973b; Barnes et al. 1986) indicate that these systems are stable.

Antonov's Fourth Law *A spherical stellar system with $\partial f_0(E,L)/\partial E <$ 0 is stable to radial perturbations if the function*

$$\rho_0(\Phi_0, L) = \int d^3\mathbf{v} f_0(\tfrac{1}{2}v^2 + \Phi_0, L), \qquad (5\text{-}81)$$

where L is regarded as a constant, satisfies $\partial^3 \rho_0(\Phi_0, L)/\partial \Phi_0^3 \leq 0$ for all L.

Proof: See Antonov (1962b).◁

From these results, one can establish the stability of many anisotropic spherical systems to *radial* perturbations. In particular, *all* Michie models can be proved stable in this limited sense.

The stability of anisotropic spherical systems to nonradial perturbations is presently under active investigation (see Merritt 1987b for a review). On the one hand, Gillon et al. (1976) have argued that any spherical system with $\partial f_0(E,L)/\partial E < 0$ and $\partial f_0(E,L)/\partial L < 0$ is stable to both radial and nonradial modes. On the other hand, Antonov (1973) argued that any spherical system composed entirely of stars on radial orbits was unstable, and there is an extensive Soviet literature on instabilities in anisotropic systems with mainly radial orbits (see Fridman & Polyachenko 1984). The apparent contradiction between the two schools has not yet been fully resolved.

In this situation, numerical experiments are an indispensable guide (Hénon 1973b; Merritt & Aguilar 1985; Barnes 1986; Barnes et al. 1986). Figure 5-3 shows the evolution of an initially spherical Jaffe model with anisotropy radius $r_a = 0.1 r_J$ [see eqs. (4-141) and (4-151)]. There is a strong nonradial instability, which leads to a triaxial or barlike final state, and persists in models with r_a as large as $0.3 r_J$. Some of the unstable numerical models would be stable according to the Gillon et al. criterion, which suggests that there is an error in their proof.

The instability appears in models with predominantly radial orbits, and hence is called the **radial-orbit instability**. It appears that the instability arises because near-radial orbits are almost closed (the angle between successive apocenters, $\Delta\psi$, [eq. (3-17)], is nearly π). Thus nonradial forces can in some cases realign the orientation of the orbits, distorting the galaxy into a triaxial configuration (see Palmer & Papaloizou 1987 for more detail).

(d) Summary   We have described a set of powerful theorems that establish that all spherical stellar systems with DF $f_0 = f_0(E)$

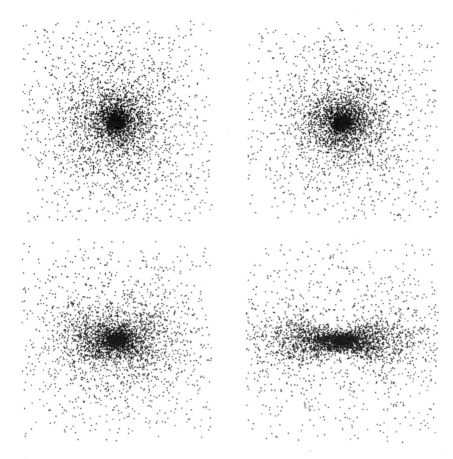

Figure 5-3. Instability in an anisotropic Jaffe model with $r_a = 0.1r_J$ [cf. eqs. (4-141) and (4-151)]. From Merritt (1987b), courtesy of the International Astronomical Union.

and $df_0/dE < 0$ are stable. The situation is less satisfactory for systems whose DFs depend on both energy and angular momentum, $f_0 = f_0(E, L)$. Antonov's theorems cannot be used to decide the stability of these systems to nonradial perturbations. Numerical experiments show that systems with predominantly radial orbits are unstable to a barlike mode. The nature and consequences of this radial-orbit instability are only beginning to be explored in the literature.

5.3 The Stability of Uniformly Rotating Systems

Stability analyses are still more difficult for rotating systems than for spherical systems. First, since rotating systems are usually flattened rather than spherically symmetric, the Poisson equation and dynamical equations are geometrically much more awkward than in the nonrotating case. Second, and more fundamental, the system now has a reservoir of rotational kinetic energy to feed any possible unstable modes.

Because of these complications, we have relatively few general results of the kind that are available for nonrotating systems. Instead, we must rely on specific simple models of rotating fluid and stellar systems. In this chapter we restrict ourselves to uniformly rotating systems, and defer the discussion of differential rotation to the next chapter. Just as in the case of spherical systems, the results for rotating fluids will provide a valuable guide when we come to consider the more complex problem of rotating stellar systems. For more details on fluid systems, and for the application of the theory to rotating stars, see Tassoul's (1978) book, *Theory of Rotating Stars*. The stability of homogeneous fluid systems is discussed thoroughly and elegantly—though by a very different approach than we use here—in Chandrasekhar's (1969) classic work, *Ellipsoidal Figures of Equilibrium*.

1 The Uniformly Rotating Sheet

We begin by investigating an extremely simple model system that exhibits the effects of rotation and a flattened geometry. Our model consists of an infinite fluid disk or sheet of zero thickness and constant surface density Σ_0. The sheet occupies the plane $z = 0$, and is uniform in the x- and y-directions. The sheet rotates with constant angular velocity $\mathbf{\Omega} = \Omega \hat{\mathbf{e}}_z$.

We consider the stability of the sheet to disturbances in its own plane. We do not examine bending or corrugation modes since these turn out to be always stable (see Problem 5-4). The analysis is easiest if we work in a frame that rotates with the unperturbed sheet at Ω; thus the continuity equation (5-7), the Euler equation (5-8), and Poisson's equation (5-9) read

$$\frac{\partial \Sigma}{\partial t} + \mathbf{\nabla} \cdot (\Sigma \mathbf{v}) = 0, \tag{5-82a}$$

$$\frac{\partial \mathbf{v}}{\partial t} + (\mathbf{v} \cdot \mathbf{\nabla})\mathbf{v} = -\frac{\mathbf{\nabla} p}{\Sigma} - \mathbf{\nabla}\Phi - 2\mathbf{\Omega} \times \mathbf{v} + \Omega^2(x\hat{\mathbf{e}}_x + y\hat{\mathbf{e}}_y), \tag{5-82b}$$

$$\nabla^2 \Phi = 4\pi G \Sigma \delta(z). \tag{5-82c}$$

where $\delta(z)$ is the Dirac delta function (see Appendix 1.C), $\mathbf{v}(x,y,t) = v_x(x,y,t)\hat{\mathbf{e}}_x + v_y(x,y,t)\hat{\mathbf{e}}_y$ is the velocity in the rotating frame, $\Sigma(x,y,t)$ is the surface density, and the last two terms on the right side of equation (5-82b) are the Coriolis and centrifugal forces [eq. (1D-42)]. Note that equations (5-82a) and (5-82b) are defined only in the (x,y) plane but equation (5-82c) must hold throughout three-dimensional space. In a two-dimensional system of this kind, the pressure p is assumed to act only in the plane of the sheet and has dimensions of force per unit length. We assume that the equation of state is barotropic and write [cf. eq. (5-10)]

$$p(x,y,t) = p[\Sigma(x,y,t)]. \tag{5-83}$$

In the unperturbed state $\Sigma = \Sigma_0$, $\mathbf{v} = 0$, and $p = p_0 = p(\Sigma_0)$. Equation (5-82a) is satisfied trivially, and equations (5-82b) and (5-82c) read

$$\nabla\Phi_0 = \Omega^2(x\hat{\mathbf{e}}_x + y\hat{\mathbf{e}}_y), \tag{5-84}$$

$$\nabla^2\Phi_0 = 4\pi G\Sigma_0\delta(z). \tag{5-85}$$

Since the sheet is uniform there is no preferred direction in the (x,y) plane. Hence the gravitational field $\nabla\Phi_0$ must point in the z-direction.[10] Thus equation (5-84) cannot be satisfied as it stands: there are no pressure gradients or gravitational forces to balance the centrifugal force. To proceed further we must perpetrate a version of the Jeans swindle: we assume that the centrifugal force is balanced by a gravitational force that is produced by some unspecified mass distribution. The nature of this mass distribution does not concern us, since its only function is to ensure centrifugal balance in the equilibrium state.

We now consider a small perturbation of the form $\Sigma(x,y,t) = \Sigma_0 + \epsilon\Sigma_1(x,y,t)$, $\mathbf{v}(x,y,t) = \epsilon\mathbf{v}_1(x,y,t)$, etc., where $\epsilon \ll 1$. Keeping only terms linear in ϵ in equations (5-82) and (5-83), we have

$$\frac{\partial\Sigma_1}{\partial t} + \Sigma_0\nabla\cdot\mathbf{v}_1 = 0, \tag{5-86a}$$

$$\frac{\partial\mathbf{v}_1}{\partial t} = -\frac{v_s^2}{\Sigma_0}\nabla\Sigma_1 - \nabla\Phi_1 - 2\Omega\times\mathbf{v}_1, \tag{5-86b}$$

$$\nabla^2\Phi_1 = 4\pi G\Sigma_1\delta(z), \tag{5-86c}$$

---

[10] Although we do not need the explicit analytic form of Φ_0, it is easy to derive. When $z \neq 0$, $\nabla^2\Phi_0 = 0$ and hence Φ_0 must be a linear function of z. By symmetry, Φ_0 is even in z and hence $\Phi_0 = a_0 + a_1|z|$. The zero of the potential can be chosen so that $a_0 = 0$. Integrating (5-85) from $z = -\varsigma$ to $z = \varsigma$ yields $d\Phi_0/dz(\varsigma) - d\Phi_0/dz(-\varsigma) = 4\pi G\Sigma_0$, which requires $a_1 = 2\pi G\Sigma_0$. Thus $\Phi_0 = 2\pi G\Sigma_0|z|$.

where we have introduced the sound speed v_s defined by [cf. eq. (5-13)]

$$v_s^2 = \left[\frac{dp(\Sigma)}{d\Sigma}\right]_{\Sigma_0}. \tag{5-87}$$

These equations are very similar to equations (5-17) except that (i) there is a Coriolis force term in Euler's equation arising from the rotation; (ii) Poisson's equation is modified to apply to a two-dimensional system.

To solve equations (5-86) we write $\Sigma_1(x, y, t) = \Sigma_a \exp[i(\mathbf{k} \cdot \mathbf{x} - \omega t)]$, $\mathbf{v}_1(x, y, t) = (v_{ax}\hat{\mathbf{e}}_x + v_{ay}\hat{\mathbf{e}}_y) \exp[i(\mathbf{k} \cdot \mathbf{x} - \omega t)]$, and $\Phi_1(x, y, z = 0, t) = \Phi_a \exp[i(\mathbf{k} \cdot \mathbf{x} - \omega t)]$. With no loss of generality, we can choose the x-axis to be parallel to $\mathbf{k}$, so that $\mathbf{k} = k\hat{\mathbf{e}}_x$. First consider Poisson's equation (5-86c). For $z \neq 0$ we have $\nabla^2\Phi_1 = 0$ but when $z = 0$, $\Phi_1 = \Phi_a \exp[i(kx - \omega t)]$. The only continuous function that satisfies both constraints and that approaches zero far from the sheet has the form

$$\Phi_1(x, y, z, t) = \Phi_a e^{i(kx - \omega t) - |kz|}. \tag{5-88}$$

To relate Φ_a to Σ_a we integrate equation (5-86c) from $z = -\varsigma$ to $z = \varsigma$, where ς is a positive constant, and then let $\varsigma \to 0$. Since $\partial^2\Phi_1/\partial x^2$ and $\partial^2\Phi_1/\partial y^2$ are continuous at $z = 0$, but $\partial^2\Phi_1/\partial z^2$ is not, we have

$$\lim_{\varsigma \to 0} \int_{-\varsigma}^{\varsigma} \frac{\partial^2\Phi_1}{\partial z^2}\, dz = \lim_{\varsigma \to 0} \left.\frac{\partial\Phi_1}{\partial z}\right|_{-\varsigma}^{\varsigma} = 4\pi G\Sigma_1 \int_{-\varsigma}^{\varsigma} \delta(z)\, dz = 4\pi G\Sigma_1. \tag{5-89}$$

Hence $-2|k|\Phi_a = 4\pi G\Sigma_a$ or

$$\Phi_1(x, y, z, t) = -\frac{2\pi G\Sigma_a}{|k|} e^{i(kx - \omega t) - |kz|}. \tag{5-90}$$

Substituting for Σ_1, $\mathbf{v}_1$, and Φ_1 in equations (5-86) we obtain

$$-i\omega\Sigma_a = -ik\Sigma_0 v_{ax},$$

$$-i\omega v_{ax} = -\frac{v_s^2 ik\Sigma_a}{\Sigma_0} + \frac{2\pi Gi\Sigma_a k}{|k|} + 2\Omega v_{ay}, \tag{5-91}$$

$$-i\omega v_{ay} = -2\Omega v_{ax}.$$

This set of three homogeneous equations in the three variables Σ_a, v_{ax}, v_{ay} has nontrivial solutions only when

$$\omega^2 = 4\Omega^2 - 2\pi G\Sigma_0|k| + k^2 v_s^2. \tag{5-92}$$

This is the dispersion relation for the uniformly rotating sheet. The sheet is stable if $\omega^2 \geq 0$ and unstable if $\omega^2 < 0$.

Equation (5-92) has several interesting features. First consider the case in which the sheet is not rotating. If $\Omega = 0$, then the sheet is unstable if $v_s^2 k^2 - 2\pi G\Sigma_0 |k| < 0$, in other words if $|k| < k_J = 2\pi G\Sigma_0 / v_s^2$, where k_J is the Jeans wavenumber for the sheet. Despite the change in geometry, there is evidently a direct analog to the classical Jeans instability [eq. (5-22)] for a three-dimensional infinite homogeneous medium: in both cases long wavelengths are subject to a gravitational instability. However, there is one important difference. In an infinite homogeneous medium with sound speed $v_s = 0$, the instability grows as $\exp(\gamma t)$ where $\gamma^2 = -\omega^2 = 4\pi G\rho_0$, independent of the scale of the perturbation. In a sheet with zero sound speed and zero angular speed, the growth rate is given by $\gamma^2 = 2\pi G\Sigma_0 |k|$. Thus, in the sheet, the growth rate increases and the instability becomes more violent as the wavelength of the perturbation decreases. As the perturbation's wavelength $\lambda = 2\pi / |k| \to 0$, the growth rate $\gamma \to \infty$.

This violent instability at short wavelengths is not suppressed by rotation. For a rotating sheet with zero sound speed, the dispersion relation (5-92) states that perturbations with wavenumber $|k| > 2\Omega^2 / \pi G\Sigma_0$ are unstable, and their growth rate is given by $\gamma^2 = -\omega^2 = 2\pi G\Sigma_0 |k| - 4\Omega^2$, which again diverges as $\lambda \to 0$.

Clearly neither rotation nor pressure is able by itself to stabilize the sheet. A rotating sheet with zero sound speed is unstable at small wavelengths, and a nonrotating sheet with a given non-zero sound speed is unstable at large wavelengths. However, rotation and pressure working together *can* stabilize the sheet: if both effects are present, the right side of the dispersion relation (5-92) is quadratic in k, with a minimum at the "most unstable wavenumber" $|k| = \pi G\Sigma_0 / v_s^2 = \frac{1}{2} k_J$. The sheet is stable at all wavelengths if the minimum is positive, which requires that

$$\frac{v_s \Omega}{G\Sigma_0} \geq \frac{\pi}{2} = 1.5708. \tag{5-93}$$

Toomre (1964) has given a simple physical interpretation of this stability criterion. Consider a small circular patch located anywhere in the sheet. The radius of the patch is ΔR and its mass $M = \pi\Sigma_0(\Delta R)^2$. Now suppose that the patch's area is reduced to a fraction $(1 - \alpha)$ of its original value, where $\alpha \ll 1$. The pressure perturbation will be $p_1 \approx \alpha p_0 \approx \alpha v_s^2 \Sigma_0$. The pressure force per unit mass is $\mathbf{F}_p = -\nabla p / \Sigma$, so the extra outward pressure force has magnitude $|\mathbf{F}_{p1}| \approx \alpha v_s^2 / \Delta R$. Similarly, the compression leads to an extra inward gravitational force $\mathbf{F}_{G1}$, where $|\mathbf{F}_{G1}| \approx GM\alpha / (\Delta R)^2 \approx G\Sigma_0\alpha$. In the absence of other effects, the sheet is expected to be stable if $|\mathbf{F}_{p1}|$ exceeds $|\mathbf{F}_{G1}|$, that is, if

$$\Delta R \lesssim \frac{v_s^2}{G\Sigma_0} \equiv \Delta R_l. \tag{5-94}$$

(Compare the analysis of the Jeans instability in §5.1.1.) There are also internal motions in the perturbed region, which arise from the rotation of the sheet. If we neglect external influences, the compressed region will tend to conserve spin angular momentum around its own center. The typical spin angular momentum per unit mass is $S \approx \Omega \Delta R^2$, where Ω is the angular speed of the sheet. The outward centrifugal force per unit mass is given by $|\mathbf{F}_c| \approx \Omega^2 \Delta R \approx S^2/\Delta R^3$. If S is conserved, the centrifugal force felt by each element is increased by the compression; the amount of the increase is $|\mathbf{F}_{c1}| \approx \alpha S^2/\Delta R^3 \approx \alpha \Omega^2 \Delta R$. Thus, there is stability if $|\mathbf{F}_{c1}|$ exceeds $|\mathbf{F}_{G1}|$, which requires

$$\Delta R \gtrsim \frac{G\Sigma_0}{\Omega^2} \equiv \Delta R_u. \tag{5-95}$$

Equations (5-94) and (5-95) show that both small and large regions are stable, the one through pressure forces and the other through centrifugal force. The instability at intermediate radii is suppressed if $\Delta R_u \lesssim \Delta R_l$, which requires

$$\frac{v_s \Omega}{G\Sigma_0} \gtrsim 1, \tag{5-96}$$

an order of magnitude statement of the stability criterion (5-93).

   Although equation (5-93) was derived for a fluid sheet, a very similar stability criterion applies to the analogous stellar system. A razor-thin sheet of stars with a Maxwellian velocity distribution is stable if (Toomre 1964)

$$\frac{\sigma \Omega}{G\Sigma_0} \geq 1.68, \tag{5-97}$$

where σ is the one-dimensional velocity dispersion of the stars. We defer the derivation of this important result to §6.2.3, where it appears as a special case of Toomre's local stability criterion for differentially rotating disks. Notice that the number on the right of equation (5-97) differs by less than 7% from the number in the fluid stability criterion (5-93): once again we find a close analogy between stellar systems and fluids.

   The approximation that the sheet is razor-thin has greatly simplified the stability analysis. Nevertheless, it is possible to investigate the stability of more realistic sheets with three-dimensional structure. These sheets or disks are still uniform in the x and y directions, but have an equilibrium vertical structure $\rho_0(z)$ that is determined by the equation of state and the equation of hydrostatic equilibrium. Experience shows that the stability criteria derived by such analyses are generally very similar to equation (5-93). For example, Goldreich and Lynden-Bell (1965) have analytically determined the stability of a uniformly rotating isothermal disk (equation of state $p = v_s^2 \rho$, where v_s is a constant). They find that the disk is stable if

$$\frac{v_s \Omega}{G \Sigma_0} \geq 1.06, \tag{5-98}$$

a result only about 30% smaller than equation (5-93). The reason that the idealized two-dimensional sheet works so well is that the most unstable wavelength is several times the characteristic disk thickness. For example, if we define the characteristic thickness to be $T = (\int \rho_0 \, dz)^2 / \int \rho_0^2 \, dz$, then the most unstable wavelength is about $4.5T$. The behavior of perturbations with such relatively long wavelengths is insensitive to the details of the vertical structure.

In conclusion, the uniformly rotating sheet exhibits three important features:

(i) A cold sheet is violently unstable.

(ii) The sheet can be completely stabilized by a sound speed v_s or velocity dispersion σ that satisfies the stability criteria (5-93) or (5-97), respectively.

(iii) The stability properties of fluid and stellar sheets are very similar.

We shall encounter all of these features again in more realistic models of both uniformly rotating and differentially rotating disks.

2 Maclaurin Disks

In §4.5 we introduced a family of models of uniformly rotating stellar systems called Kalnajs disks. This family provides one of the few examples of a set of rotating, self-gravitating systems for which a complete stability analysis is possible. The results from this analysis are discussed in §5.3.3. However, as we have seen, fluid systems share many of the stability properties of stellar systems and are simpler to analyze. Hence we shall first examine the stability of the two-dimensional fluid analogs of the Kalnajs disks—the **Maclaurin disks**.

The unperturbed surface density in a Maclaurin disk satisfies [cf. eq. (4-167)]

$$\Sigma_0(R) = \begin{cases} \Sigma_c \sqrt{1 - \dfrac{R^2}{a^2}}, & R \leq a, \\ 0 & R > a, \end{cases} \tag{5-99}$$

where a is the radius of the disk edge. The mass of the disk is

$$M = \tfrac{2}{3}\pi\Sigma_c a^2. \tag{5-100}$$

The potential in the plane of the disk is [see eq. (4-166)]

$$\Phi_0(R) = \tfrac{1}{2}\Omega_0^2 R^2 + \text{constant}, \tag{5-101}$$

where

$$\Omega_0^2 = \frac{\pi^2 G\Sigma_c}{2a}. \tag{5-102}$$

The equation of state is $p = K\Sigma^3$, where p is assumed to act only in the plane of the disk, and the equilibrium rotation speed Ω is given by

$$\Omega^2 = \Omega_0^2 - \frac{3K\Sigma_c^2}{a^2}. \tag{5-103}$$

Disks of a given mass and radius form a one-parameter family specified by the rotation speed, $0 \leq \Omega \leq \Omega_0$.

We now turn to the stability analysis. The linearized continuity equation is

$$\frac{\partial\Sigma_1}{\partial t} + \boldsymbol{\nabla}\cdot(\Sigma_0\mathbf{v}_1) + \boldsymbol{\nabla}\cdot(\Sigma_1\mathbf{v}_0) = 0, \tag{5-104}$$

where $\Sigma_1(R,\phi,t)$, $\mathbf{v}_1(R,\phi,t)$ are the perturbed surface density and velocity, and the unperturbed velocity $\mathbf{v}_0 = \Omega R\hat{\mathbf{e}}_\phi$. The linearized Euler equation is

$$\frac{\partial\mathbf{v}_1}{\partial t} + (\mathbf{v}_0\cdot\boldsymbol{\nabla})\mathbf{v}_1 + (\mathbf{v}_1\cdot\boldsymbol{\nabla})\mathbf{v}_0 = -\boldsymbol{\nabla}h_1 - \boldsymbol{\nabla}\Phi_1, \tag{5-105}$$

where the perturbed enthalpy is [see eq. (5-15)]

$$h_1 = 3K\Sigma_0\Sigma_1. \tag{5-106}$$

We search for modes of the form

$$\begin{aligned}
\mathbf{v}_1 &= \text{Re}\{[v_R(R)\hat{\mathbf{e}}_R + v_\phi(R)\hat{\mathbf{e}}_\phi]e^{i(m\phi-\omega t)}\}, \\
\Sigma_1 &= \text{Re}[\Sigma_a(R)e^{i(m\phi-\omega t)}], \\
h_1 &= \text{Re}[h_a(R)e^{i(m\phi-\omega t)}], \\
\Phi_1 &= \text{Re}[\Phi_a(R)e^{i(m\phi-\omega t)}],
\end{aligned} \tag{5-107}$$

where $m \geq 0$.[11] Using equation (1B-21) and the equation for divergence in cylindrical coordinates (1B-45), the continuity equation becomes

$$-i\omega_r \Sigma_a + \frac{1}{R}\frac{d}{dR}(R\Sigma_0 v_R) + \frac{im}{R}\Sigma_0 v_\phi = 0, \qquad (5\text{-}108)$$

where[12]

$$\omega_r \equiv \omega - m\Omega. \qquad (5\text{-}109)$$

Similarly, the two components of Euler's equation may be written

$$-i\omega_r v_R - 2\Omega v_\phi = -\frac{d}{dR}(h_a + \Phi_a),$$
$$-i\omega_r v_\phi + 2\Omega v_R = -\frac{im}{R}(h_a + \Phi_a). \qquad (5\text{-}110)$$

Defining

$$\eta(R) \equiv h_a(R) + \Phi_a(R), \qquad (5\text{-}111)$$

we can solve for v_R, v_ϕ as

$$v_R = i\frac{\omega_r d\eta/dR - 2m\Omega\eta/R}{4\Omega^2 - \omega_r^2},$$
$$v_\phi = -\frac{m\omega_r \eta/R - 2\Omega d\eta/dR}{4\Omega^2 - \omega_r^2}. \qquad (5\text{-}112)$$

Substituting these expressions into the continuity equation (5-108), we obtain

$$-\omega_r \Sigma_a + \frac{1}{4\Omega^2 - \omega_r^2}\left\{\frac{1}{R}\frac{d}{dR}\left[\Sigma_0 \omega_r R\frac{d\eta}{dR} - 2m\Omega\Sigma_0\eta\right]\right.$$
$$\left. - m^2\omega_r\frac{\Sigma_0\eta}{R^2} + 2m\Omega\frac{\Sigma_0}{R}\frac{d\eta}{dR}\right\} = 0. \qquad (5\text{-}113)$$

We now change to the new radial coordinate

$$\xi = \sqrt{1 - \frac{R^2}{a^2}}. \qquad (5\text{-}114)$$

---

[11] We do not have to consider modes with $m < 0$ since $\mathrm{Re}\{\Sigma_a(R)\exp[i(m\phi - \omega t)]\} = \mathrm{Re}\{\Sigma_a^*(R)\exp[i(-m\phi + \omega^* t)]\}$, i.e., every mode with $m < 0$ described by an eigenfunction $\Sigma_a(R)$ and eigenfrequency ω is also a mode with $m > 0$ and eigenfunction $\Sigma_a^*(R)$ and eigenfrequency $-\omega^*$.

[12] In the rotating frame the azimuthal angle is $\phi_r = \phi - \Omega t$. Thus the perturbation is proportional to $\exp[i(m\phi - \omega t)] = \exp[i(m\phi_r + m\Omega t - \omega t)] = \exp[i(m\phi_r - \omega_r t)]$, which shows that ω_r is the apparent frequency seen by an observer from the rotating frame.

We have $\Sigma_0(R) = \Sigma_c \xi$ and $d/dR = -(R/a^2\xi)d/d\xi$; thus (5-113) yields

$$\frac{d}{d\xi}\left[(1-\xi^2)\frac{d\eta}{d\xi}\right] - \eta\left(\frac{m^2\xi^2}{1-\xi^2} - \frac{2m\Omega}{\omega_r}\right) - a^2\xi[4\Omega^2 - \omega_r^2]\frac{\Sigma_a}{\Sigma_c} = 0.$$

$$(5\text{-}115)$$

This ordinary differential equation links the perturbed density Σ_a to the potential variations Φ_a involved in the definition of η. The frequencies of the normal modes are determined by requiring that this relation be consistent with Poisson's equation. In §2.6.4(a) we saw that Poisson's equation may be conveniently solved using the oblate spheroidal coordinates introduced in equation (2-58). The potential and surface density can be expanded in terms of the functions Φ_{lm} and Σ_{lm} given by equations (2-177a) and (2-179a). Adjusting those equations to the notation of the present section, we have that if

$$\Phi_a(\xi) = \sum_{l=m}^{\infty} b_{lm} P_l^m(\xi),$$

$$(5\text{-}116)$$

then

$$\Sigma_a(\xi) = -\frac{2}{\pi^2 Ga} \sum_{l=m}^{\infty} \frac{b_{lm} P_l^m(\xi)}{g_{lm}\xi},$$

$$(5\text{-}117)$$

where g_{lm} is defined by equation (2-179b) and P_l^m is an associated Legendre function (Appendix 1.C.5). The summation begins at $l = m$ because P_l^m is defined only for $l \geq |m|$. Since Φ_a is defined only for $\xi \geq 0$, we may assume that Φ_a is even in ξ. Thus its expansion requires only the P_l^m's that are even in ξ, and these all have $l - m$ even.

Using this relation and equations (5-106) and (5-111), we obtain

$$\eta = 3K\Sigma_0\Sigma_a + \Phi_a = \sum_{l=m}^{\infty}\left(-\frac{6K\Sigma_c}{\pi^2 Gag_{lm}} + 1\right)b_{lm}P_l^m(\xi).$$

$$(5\text{-}118)$$

If we substitute (5-117) and (5-118) into (5-115), we find

$$\sum_{l=m}^{\infty} b_{lm}\left\{\frac{d}{d\xi}\left[(1-\xi^2)\frac{dP_l^m(\xi)}{d\xi}\right]\right.$$

$$- P_l^m(\xi)\left[\frac{m^2\xi^2}{1-\xi^2}\frac{2m\Omega}{\omega_r} + \frac{a^2(4\Omega^2 - \omega_r^2)}{\Sigma_c(3K\Sigma_c - \frac{1}{2}\pi^2 Gag_{lm})}\right]\right\} = 0.$$

$$(5\text{-}119)$$

The associated Legendre functions satisfy the differential equation (1C-17):

$$\frac{d}{d\xi}\left[(1-\xi^2)\frac{dP_l^m(\xi)}{d\xi}\right] + \left[l(l+1) - \frac{m^2}{1-\xi^2}\right]P_l^m(\xi) = 0.$$

$$(5\text{-}120)$$

Hence, (5-119) simplifies to

$$\sum_{l=m}^{\infty} b_{lm} P_l^m(\xi) \left[l(l+1) - m^2 - \frac{2m\Omega}{\omega_r} + \frac{a^2(4\Omega^2 - \omega_r^2)}{\Sigma_c(3K\Sigma_c - \frac{1}{2}\pi^2 G a g_{lm})} \right] = 0.$$

$$(5\text{-}121)$$

Since the associated Legendre functions are linearly independent, the coefficient in braces must vanish for each l and m. Thus, using (5-102) and (5-103) we can write

$$\omega_r^3 - \omega_r\{4\Omega^2 + (l^2 + l - m^2)[\Omega_0^2(1 - g_{lm}) - \Omega^2]\}$$
$$+ 2m\Omega[\Omega_0^2(1 - g_{lm}) - \Omega^2] = 0. \qquad (5\text{-}122)$$

This is the characteristic equation for the modes of a Maclaurin disk (Takahara 1976; Smith 1979).[13] The Maclaurin disks are virtually the only rotating, self-gravitating fluid systems for which the characteristic equation can be written in closed form. Each mode is specified by two "quantum numbers" l and m that must satisfy the constraints $0 \le m \le l$ and $l - m$ even; each mode has $\frac{1}{2}(l - m)$ radial nodes in the range $0 < R \le a$, and m azimuthal nodes in the range $0 \le \phi \le 2\pi$. Schematic diagrams of the first few modes are shown in Figure 5-4.

Let us now examine the solutions of equation (5-122) for increasing values of l. The mode $l = m = 0$ is unphysical, since it corresponds to a surface density $\Sigma_1(R, \phi) \propto \xi^{-1} = (1 - R^2/a^2)^{-1/2}$, which has the same sign everywhere and hence cannot conserve mass. For $l = 1$ the only mode has $m = 1$; in this case the characteristic equation is

$$\omega^2(\omega - 3\Omega) = 0. \qquad (5\text{-}123)$$

The root $\omega = 3\Omega$ is spurious, arising because we multiplied (5-113) by $4\Omega^2 - \omega_r^2$ to get (5-115), and for $m = 1$, $\omega = 3\Omega$ is a root of this factor. The roots arising when $\omega^2 = 0$ represent a uniform translation of the whole disk.[14]

The first modes corresponding to a real oscillation of the disk are the $l = 2$ modes. In this case no $m = 1$ mode exists since $l - m$ must

---

[13] The careful reader might wonder whether our perturbation analysis is valid near the edge of the disk, since the perturbed surface density Σ_a diverges at the disk edge $\xi = 0$ according to (5-117). This divergence is an artifact that arises because of the sharp edge of the disk, and it can be removed in a more careful treatment (see Hunter 1963) without affecting the characteristic equation (5-122).

[14] The proof is simple. The associated Legendre polynomial $P_1^1(\xi) = (1-\xi^2)^{1/2} = R/a$. Hence $\eta \propto R$, and (5-112) yields $v_R = -ik$, $v_\phi = k$, where k is a constant. The space velocity is thus $\mathbf{v}_1 = \mathrm{Re}[k(-i\hat{\mathbf{e}}_R + \hat{\mathbf{e}}_\phi)\exp(i\phi)]$. In Cartesian coordinates, using $\hat{\mathbf{e}}_R = \hat{\mathbf{e}}_x\cos\phi + \hat{\mathbf{e}}_y\sin\phi$, $\hat{\mathbf{e}}_\phi = -\hat{\mathbf{e}}_x\sin\phi + \hat{\mathbf{e}}_y\cos\phi$, we have $\mathbf{v}_1 = \hat{\mathbf{e}}_x\mathrm{Im}(k) + \hat{\mathbf{e}}_y\mathrm{Re}(k)$, corresponding to uniform translation of the whole disk.

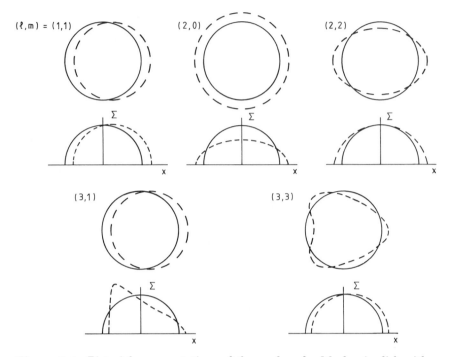

Figure 5-4. Pictorial representations of the modes of a Maclaurin disk with $l \leq 3$. For each mode, the upper diagram shows the distortion of the boundary of the disk, and the lower diagram shows the surface density profile along the horizontal axis: full curves show unperturbed quantities, and dashed curves the configuration of the unperturbed system. The $l = m = 1$ mode corresponds to a translation of the whole disk; the $l = 2$, $m = 0$ mode is a self-similar pulsation of the disk; and the $l = 2$, $m = 2$ mode is a barlike or ellipsoidal deformation.

be even, but we have interesting candidates both for $m = 0$ and $m = 2$. The $m = 0$ mode has either

$$\omega = 0 \qquad \text{or} \qquad \omega^2 = 3\Omega_0^2 - 2\Omega^2. \tag{5-124}$$

The root at $\omega = 0$ simply represents a distortion into a slightly larger or smaller equilibrium disk. The remaining roots represent a stable pulsation of the disk, sometimes called the "breathing mode".

For $m = 2$, on the other hand, we obtain

$$\omega_r^3 - \omega_r(2\Omega^2 + \tfrac{1}{2}\Omega_0^2) + \Omega(\Omega_0^2 - 4\Omega^2) = 0. \tag{5-125}$$

This cubic equation has a spurious root at $\omega_r = 2\Omega$, which again arises because we have multiplied by $4\Omega^2 - \omega_r^2$; when this root is factored out, there remains the quadratic

$$\omega_r^2 + 2\Omega\omega_r - (\tfrac{1}{2}\Omega_0^2 - 2\Omega^2) = 0. \tag{5-126}$$

Table 5-1. Onset of instability in Maclaurin disks

(l, m)	Ω^2/Ω_0^2	(l, m)	Ω^2/Ω_0^2
$(2, 2)$	0.5000	$(4, 4)$	0.6042
$(3, 1)$	0.8380	$(5, 1)$	0.8638
$(3, 3)$	0.5625	$(5, 3)$	0.8244
$(4, 0)$	0.8984	$(5, 5)$	0.6348
$(4, 2)$	0.8201		

Solving, we find

$$\omega_r = -\Omega \pm \sqrt{\tfrac{1}{2}\Omega_0^2 - \Omega^2}, \qquad (5\text{-}127)$$

thus the disk is dynamically unstable if

$$\frac{\Omega^2}{\Omega_0^2} > \tfrac{1}{2}. \qquad (5\text{-}128)$$

According to equation (5-127), the frequency of the unstable mode has non-zero real and imaginary parts (this is true both for the frequency in the rotating frame ω_r and for the frequency in the inertial frame ω). Growing modes of this kind are often called **overstabilities** to distinguish them from **instabilities** in which the frequency is purely imaginary. The presence of overstable modes is related to the rotation of the system: recall that all modes in spherical systems have purely real or purely imaginary frequencies.

To repeat, if the angular speed exceeds $1/\sqrt{2}$ of the circular angular speed, the disk is unstable. The instability has the form

$$\Sigma_1 \propto \text{Re}\left[\frac{P_2^2(\xi)}{\xi}e^{i(2\phi - \omega t)}\right] = 3\frac{R^2}{a\sqrt{a^2 - R^2}}\cos(2\phi - \omega t), \qquad (5\text{-}129)$$

which represents a rotating elliptical deformation of the disk (cf. Fig. 5-4). This is sometimes called a "bar" mode, since it deforms the disk into a shape reminiscent of the bars in disk galaxies. We shall see below, and again in Chapter 6, that the bar instability is an important and common feature of a wide range of both fluid and stellar disks.

To determine the stability for arbitrary (l, m) requires solving the cubic equation (5-122). By doing just that, it can be shown that the nonrotating "hot" ($\Omega = 0$) Maclaurin disk is stable to all modes, while the most rapidly rotating "cold" ($\Omega = \Omega_0$, zero pressure) disk is unstable to all modes except $l = 2$, $m = 0$ (Hunter 1963). The first instability to occur as Ω^2 increases from zero is the $l = m = 2$ mode. The critical values of Ω^2/Ω_0^2 at which other instabilities set in are listed in Table 5-1 for $(l, m) \leq (5, 5)$.

The Maclaurin disks have many analogies to the rotating sheet discussed in §5.3.1. In particular, both require a minimum pressure or sound speed to be stable. For the Maclaurin disks, the sound speed v_s is given by

$$v_s^2 = \left(\frac{dp}{d\Sigma}\right)_0 = 3K\Sigma_0^2 = 3K\Sigma_c^2\xi^2. \tag{5-130}$$

According to equation (5-103) we have

$$v_s^2 = (\Omega_0^2 - \Omega^2)a^2\xi^2, \tag{5-131}$$

and using (5-99) and (5-102) we may write

$$\frac{v_s\Omega_0}{G\Sigma_0} = \frac{\pi^2}{2}\sqrt{1 - \frac{\Omega^2}{\Omega_0^2}}. \tag{5-132}$$

All modes are stable if $\Omega^2/\Omega_0^2 < \frac{1}{2}$; thus stability requires

$$\frac{v_s\Omega_0}{G\Sigma_0} \geq \frac{\pi^2}{2\sqrt{2}} = 3.4894. \tag{5-133}$$

This result is very similar to the stability criterion for the uniformly rotating sheet, equation (5-93), except that the sound speed required for stability is larger by a factor of 2.2. The difference arises because the most unstable mode in the Maclaurin disks is the $l = 2$, $m = 2$ mode, which is a large-scale or **global** mode. This mode does not arise in the sheet because it extends uniformly to infinity. However, the stability analysis of the sheet does correctly predict the behavior of the Maclaurin disks to small-scale or **local** modes. Local modes have $l \gg m$: in this limit $P_l^m(\xi)$ oscillates rapidly, the large-scale gradients are unimportant, and the characteristic equation for the Maclaurin disk reduces to the dispersion relation for the rotating sheet (see Problem 5-6). These results suggest that global modes are much more dangerous for rotating disks than local modes.

3 Kalnajs Disks

This family comprises the simplest rotating self-gravitating stellar disks [§4.5.1(b)]. Kalnajs disks have exactly the same surface density and potential [eqs. (5-99) and (5-101)] as the Maclaurin disks. The mean angular speed Ω of the stars in a Kalnajs disk is independent of position, and relative to this mean speed the stars have isotropic velocity dispersion in the disk plane, of magnitude

$$\overline{v_x^2} = \overline{v_y^2} = \tfrac{1}{3}a^2(\Omega_0^2 - \Omega^2)(1 - R^2/a^2). \tag{5-134}$$

This is exactly analogous to the relation between pressure and rotation rate for Maclaurin disks [eqs. (5-99) and (5-103), on setting $\overline{v_x^2} = \overline{v_y^2} = p/\Sigma_0$].

Kalnajs (1972b) used the linearized collisionless Boltzmann equation to analyze the stability of the disks named in his honor. Just as for the Maclaurin disks, the normal modes of Kalnajs disks have potential distributions that are described by associated Legendre functions $P_l^m(\xi)$. However, the characteristic equations for the frequencies of the modes are very different in the two cases. In particular, for a given pair of indices (l, m), a Kalnajs disk has up to $l + 1$ different modes with different frequencies, while the Maclaurin disk has only three. In view of this added complexity, we will consider only the $l = m = 2$ barlike mode, since this was the dominant instability in Maclaurin disks. This mode can be analyzed relatively simply, by a technique that does not even require the use of the collisionless Boltzmann equation (Kalnajs & Athanassoula-Georgala 1974).

Suppose the disk is subject to a perturbing potential

$$\epsilon\Phi_1(R, \phi, t) = \epsilon\mathrm{Re}[R^2 e^{i(2\phi - \omega t)}] = \epsilon\mathrm{Re}[(x + iy)^2 e^{-i\omega t}], \qquad (5\text{-}135)$$

where $\epsilon \ll 1$. Since R^2 is proportional to $P_2^2(\xi)$, where as usual $\xi^2 \equiv 1 - R^2/a^2$, we know from equations (5-116) and (2-179b) that the corresponding surface density perturbation is

$$
\begin{aligned}
\epsilon\Sigma_1(R, \phi, t) &= -\epsilon\frac{8}{3\pi^2 Ga}\mathrm{Re}\left[\frac{R^2 e^{i(2\phi - \omega t)}}{\sqrt{1 - R^2/a^2}}\right] \\
&= -\epsilon\frac{8}{3\pi^2 Ga}\mathrm{Re}\left[\frac{(x + iy)^2 e^{-i\omega t}}{\sqrt{1 - R^2/a^2}}\right].
\end{aligned}
\qquad (5\text{-}136)
$$

The unperturbed potential is given by equation (5-101). The unperturbed equations of motion,

$$\ddot{x}_0 = -\left(\frac{\partial\Phi_0}{\partial x}\right)_{\mathbf{R}_0(t)} = -\Omega_0^2 x_0 \quad ; \quad \ddot{y}_0 = -\left(\frac{\partial\Phi_0}{\partial y}\right)_{\mathbf{R}_0(t)} = -\Omega_0^2 y_0, \qquad (5\text{-}137)$$

have solution vector

$$\mathbf{R}_0(t) \equiv [x_0(t), y_0(t)] = \mathrm{Re}\left(X e^{i\Omega_0 t}, Y e^{i\Omega_0 t}\right), \qquad (5\text{-}138)$$

where X and Y are complex constants. This is the equation of an ellipse centered on the origin. If we now add the potential perturbation (5-135), the perturbed orbit $\mathbf{R}(t) = [x(t), y(t)]$ is governed by the equations

$$\ddot{x} = -\left[\frac{\partial}{\partial x}(\Phi_0 + \epsilon\Phi_1)\right]_{\mathbf{R}(t)} \quad ; \quad \ddot{y} = -\left[\frac{\partial}{\partial y}(\Phi_0 + \epsilon\Phi_1)\right]_{\mathbf{R}(t)}. \qquad (5\text{-}139)$$

Since $\epsilon \ll 1$, we may work to first order in ϵ and set $\mathbf{R}(t) = \mathbf{R}_0(t) + \epsilon\mathbf{R}_1(t)$, where $\mathbf{R}_1(t) = [x_1(t), y_1(t)]$. The right sides of equations (5-139) must be evaluated at $\mathbf{R}(t)$; to first order in ϵ we have

$$
\begin{aligned}
\left[\frac{\partial}{\partial x}(\Phi_0 + \epsilon\Phi_1)\right]_{\mathbf{R}(t)} &= \left[\frac{\partial}{\partial x}(\Phi_0 + \epsilon\Phi_1)\right]_{\mathbf{R}_0(t)+\epsilon\mathbf{R}_1(t)} \\
&= \left[\frac{\partial\Phi_0}{\partial x}\right]_{\mathbf{R}_0(t)+\epsilon\mathbf{R}_1(t)} + \epsilon\left[\frac{\partial\Phi_1}{\partial x}\right]_{\mathbf{R}_0(t)} + O(\epsilon^2) \\
&= \left[\frac{\partial\Phi_0}{\partial x}\right]_{\mathbf{R}_0(t)} + \epsilon\left[\frac{\partial^2\Phi_0}{\partial x^2}\right]_{\mathbf{R}_0(t)} x_1 \\
&\quad + \epsilon\left[\frac{\partial^2\Phi_0}{\partial x\partial y}\right]_{\mathbf{R}_0(t)} y_1 + \epsilon\left[\frac{\partial\Phi_1}{\partial x}\right]_{\mathbf{R}_0(t)} + O(\epsilon^2),
\end{aligned}
$$
(5-140)

with a similar equation in y. By equation (5-101), $(\partial^2\Phi_0/\partial x\partial y) = 0$, so when we subtract the unperturbed equations (5-137) and substitute for Φ_1 from (5-135), the perturbed equations of motion become

$$
\begin{aligned}
\ddot{x}_1 + \Omega_0^2 x_1 &= -2\mathrm{Re}\left[(x_0 + iy_0)e^{-i\omega t}\right], \\
\ddot{y}_1 + \Omega_0^2 y_1 &= -2\mathrm{Re}\left[i(x_0 + iy_0)e^{-i\omega t}\right].
\end{aligned}
$$
(5-141)

These are the equations of motion of driven harmonic oscillators. Since we are seeking the response of the disk to a perturbing potential of frequency ω, we discard the homogeneous solution. Substituting for x_0, y_0 from (5-138), the right sides of equations (5-141) can be written

$$
\mathrm{Re}\left[-s(X + iY)e^{i(\Omega_0 - \omega)t} - s(X^* + iY^*)e^{-i(\Omega_0 + \omega)t}\right],
$$

where $s = (1, i)$ in the upper and lower equations. The particular solution is

$$
x_1(t) = \mathrm{Re}\left[s\frac{(X + iY)e^{i(\Omega_0 - \omega)t}}{\omega^2 - 2\Omega_0\omega} + s\frac{(X^* + iY^*)e^{-i(\Omega_0 + \omega)t}}{\omega^2 + 2\Omega_0\omega}\right], \quad (5\text{-}142)
$$

with a similar result for y_1. Equation (5-138) yields the identities

$$
\begin{aligned}
2x_0(t) &= Xe^{i\Omega_0 t} + X^* e^{-i\Omega_0 t} \\
2\dot{x}_0(t) &= i\Omega_0\left(Xe^{i\Omega_0 t} - X^* e^{-i\Omega_0 t}\right),
\end{aligned}
$$
(5-143)

which can be solved for X:

$$
X = \left[x_0(t) + \frac{\dot{x}_0(t)}{i\Omega_0}\right]e^{-i\Omega_0 t}, \quad (5\text{-}144)
$$

with a similar equation for Y. Substituting into equation (5-142), we find

$$
x_1(t) = \text{Re}\left\{ se^{-i\omega t}\left[\frac{x_0 + iy_0 - i(\dot{x}_0 + i\dot{y}_0)/\Omega_0}{\omega^2 - 2\Omega_0\omega} \right.\right.
$$
$$
\left.\left. + \frac{x_0 + iy_0 + i(\dot{x}_0 + i\dot{y}_0)/\Omega_0}{\omega^2 + 2\Omega_0\omega} \right]_t \right\}, \tag{5-145}
$$

with a similar expression for y_1. Equation (5-145) gives the perturbed position (x_1, y_1) of an individual star as a function of the unperturbed position in phase space $(\mathbf{x}_0, \mathbf{v}_0)$.

To first order in the perturbation, the equation of continuity may be written in the form [cf. eq. (1E-4)]

$$
\Sigma_1 + \boldsymbol{\nabla} \cdot (\Sigma_0 \boldsymbol{\xi}) = 0, \tag{5-146}
$$

where $\boldsymbol{\xi}$ is the mean displacement of the stars at a given position. We may write $\boldsymbol{\xi} = \overline{x_1}\hat{\mathbf{e}}_x + \overline{y_1}\hat{\mathbf{e}}_y$, where the bar, as usual, denotes the average over velocities of the stars at a given position. The unperturbed mean velocity at (x, y) is

$$
\overline{v}_x = \overline{\dot{x}_0} = -\Omega y \quad ; \quad \overline{v}_y = \overline{\dot{y}_0} = \Omega x. \tag{5-147}
$$

The mean displacement is obtained by averaging equation (5-145) over all velocities at a given point, using equation (5-147), and replacing (x_0, y_0) by (x, y):

$$
\overline{x}_1 = \left(\frac{1 + \Omega/\Omega_0}{\omega^2 - 2\Omega_0\omega} + \frac{1 - \Omega/\Omega_0}{\omega^2 + 2\Omega_0\omega} \right) \text{Re}\left[s(x + iy)e^{-i\omega t} \right], \tag{5-148}
$$

with a similar expression for $\overline{y}_1$. Substituting into the continuity equation (5-146), we obtain

$$
\Sigma_1 = \frac{2\Sigma_c}{a^2\sqrt{1 - R^2/a^2}} \frac{\omega + 2\Omega}{\omega(\omega^2 - 4\Omega_0^2)} \text{Re}\left[(x + iy)^2 e^{-i\omega t} \right]. \tag{5-149}
$$

The characteristic frequencies can be determined by the requirement that equations (5-136) and (5-149) be consistent. Eliminating Σ_1 from these equations and using the relation $\Omega_0^2 = \pi^2 G\Sigma_c/2a$ [eq. (5-102)], we find the characteristic equation

$$
\omega^3 - \tfrac{5}{2}\Omega_0^2\omega + 3\Omega\Omega_0^2 = 0. \tag{5-150}
$$

To analyze stability we rewrite (5-150) as $\omega^3 - \tfrac{5}{2}\Omega_0^2\omega = -3\Omega\Omega_0^2$. The cubic polynomial on the left side has a local minimum of $-2(\tfrac{5}{6})^{3/2}\Omega_0^3$

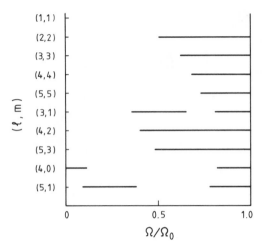

Figure 5-5. Stability of Kalnajs disks as a function of the degree of centrifugal support Ω/Ω_0. The solid lines represent zones of instability.

at $\omega = (\frac{5}{6})^{1/2}\Omega_0$. If the right side is less than this local minimum, then there can be only one real root, and one of the pair of complex roots must correspond to an unstable mode. Hence the disk is stable if and only if $-2(\frac{5}{6})^{3/2}\Omega_0^3 < -3\Omega\Omega_0^2$, or if

$$\frac{\Omega^2}{\Omega_0^2} < \frac{125}{486} = 0.2572. \tag{5-151}$$

Thus we see that, like the fluid Maclaurin disks, the stellar-dynamical Kalnajs disks are unstable to a bar-making mode if they rotate too rapidly.

Figure 5-5 shows the regions of instability for modes with $l, m \leq 5$, as determined by Kalnajs (1972b) using the linearized collisionless Boltzmann equation. As l and m increase, the behavior becomes increasingly complicated: for given l and $m \neq 0$ there are $l + 1$ independent modes (l modes if $m = 0$) with the same spatial dependence but different functional forms in velocity space. The modes have been ordered by the number of radial nodes $\frac{1}{2}(l - m)$. Note that all Kalnajs disks are unstable; however, Kalnajs was also able to construct stable composite systems consisting of superimposed Kalnajs disks.

4 Maclaurin Spheroids

The presence of a bar instability in both the rapidly rotating Maclaurin disks and in the Kalnajs disks suggests that such instabilities are a fairly general feature of rotating self-gravitating systems. Additional support for this suggestion comes from the presence of a classic instability of the Maclaurin spheroids introduced in §4.6.1.

The Maclaurin spheroids are a family of uniformly rotating three-dimensional masses of incompressible fluid. Their surface is a spheroid with eccentricity e; the non-rotating spheroid is spherical ($e = 0$), and the most rapidly rotating spheroid is a zero-thickness disk ($e = 1$)— in fact, it is precisely the most rapidly rotating Maclaurin disk as described already in §5.3.2. For studies of the stability of rotating stars and other fluid bodies, the Maclaurin spheroids are to be preferred over the Maclaurin disks because they more accurately represent the full three-dimensional structure of these systems. However, as analogs of rotating stellar systems, where the velocity dispersion in the equatorial plane is not directly related to the vertical dispersion, the Maclaurin disks appear to be just as realistic as the spheroids and are much easier to analyze. Hence we shall only quote the results of the stability analysis for the spheroids, which in fact predate the discussions of disk systems by many decades.

The complete spectrum of normal modes of the Maclaurin spheroids was first analyzed by Bryan (1888). More recent discussions appear in Lyttleton (1953), Chandrasekhar (1969), and Tassoul (1978). Since the spheroid is incompressible, the perturbations consist of displacements of its surface.

The stability characteristics are very similar to those of Maclaurin disks. The non-rotating spheroid ($\Omega^2 = 0$) is stable to all modes. As Ω^2 increases, the first unstable mode is a bar mode with radial and vertical displacements ξ_R and ξ_z given by

$$\begin{aligned} \xi_R &= k\cos(2\phi - \omega t), \\ \xi_z &= 0, \end{aligned} \tag{5-152}$$

where k is a constant and the frequency ω is given by

$$\omega = \Omega \pm \sqrt{2\pi G\rho(2A_1 - B_1) - \Omega^2}. \tag{5-153a}$$

The constant A_1 is given in Table 2-1, and

$$B_1 = -\frac{(3 + 2e^2)(1 - e^2)}{2e^4} + \frac{3\sqrt{1 - e^2}}{2e^5}\arcsin e. \tag{5-153b}$$

The spheroid is unstable if the quantity inside the radical is negative, which occurs when the eccentricity $e > 0.9529$. This result was first derived by Riemann in 1860.

The modes described by equation (5-152) are closely related to the Riemann ellipsoids [§4.6.1(c)]. If the modes are stable (i.e., if ω is real), then they appear stationary in a reference frame that rotates at the **pattern speed** $\frac{1}{2}\omega$. In this frame, the spheroid looks like a stationary, slightly triaxial ellipsoid, and in fact it turns out to be exactly a Riemann ellipsoid.[15] As equation (5-152) indicates, there are two pattern speeds for each Maclaurin spheroid, corresponding to two distinct Riemann ellipsoids. In particular, when

$$\pi G \rho (2A_1 - B_1) = \Omega^2, \tag{5-154}$$

corresponding to an eccentricity $e = 0.8127$, the pattern speeds are $0, \Omega$, and the Riemann ellipsoids are simply the Dedekind and Jacobi ellipsoids introduced in §4.6.

5 Comparison of Stellar and Fluid Systems

Comparison of the stability of Maclaurin spheroids, Maclaurin disks, and Kalnajs disks provides insight into the similarities and differences between fluid and stellar systems.

It is convenient to express our results in terms of the parameter $t = T/|W|$ [used already in eq. (4-193)], the ratio of the rotational kinetic energy[16] T to the absolute value of the self-gravitational energy W. For Maclaurin disks and Kalnajs disks, t is related to the rotation speed Ω by

$$t = \frac{\Omega^2}{2\Omega_0^2}, \tag{5-155}$$

while for Maclaurin spheroids t is related to the eccentricity e by equation (4-195). The most rapidly rotating members of all three families have $t = \frac{1}{2}$. In fact, in this limit all three families are identical: in each case, the most rapidly rotating member is a cold disk with each fluid element or star in a circular orbit, and with zero pressure or zero random motion.

As we have seen, the Maclaurin disks are unstable when $\Omega^2 > \frac{1}{2}\Omega_0^2$; this corresponds to $t > t_{\mathrm{crit}} = 0.25$. The Kalnajs disks are unstable when $\Omega^2 > \frac{125}{486}\Omega_0^2$, corresponding to $t > t_{\mathrm{crit}} = 0.1286$. Finally, the Maclaurin spheroids are unstable to the barlike mode when $e > 0.9529$, corresponding to $t > t_{\mathrm{crit}} = 0.2738$.

---

[15] Note that the pattern speed of the Riemann ellipsoids was denoted by Ω in §4.6 but is denoted $\frac{1}{2}\omega$ here; the pattern speed does not equal the mean angular speed of the fluid particles in the adjacent Maclaurin spheroid, which is the sense in which Ω is used here.

[16] Defined by $T \equiv \frac{1}{2} \int \rho \bar{v}_\phi^2 \, d^3\mathbf{x}$.

Evidently the thickness has little effect on the onset of the instability, since $t_{\rm crit}$ changes only from 0.25 to 0.2738 between Maclaurin disks and spheroids. However, the fluid systems are substantially more stable than the stellar system, with $t_{\rm crit}$ for the Maclaurin disks and spheroids exceeding $t_{\rm crit}$ for the Kalnajs disks by about a factor of two. This is an interesting contrast to the case of spherical systems (§5.2) where the stellar systems are *more* stable than their fluid analogs.

Despite these differences, it is clear that the bar instability is a very common disease in uniformly rotating self-gravitating systems.

6 Secular Instabilities

The instabilities that we have discussed so far all share the property that their characteristic equations are real, so that if there is a mode with frequency ω there is also a mode with frequency ω^*. Behavior of this kind is characteristic of dissipationless systems, since it implies that growing and decaying modes occur in pairs. Instabilities associated with such complex-conjugate frequency pairs are often called **dynamical instabilities**. A system may also exhibit **secular instability**, an instability that only appears in the presence of a dissipative force and has a growth rate proportional to the strength of the dissipative force (see Lyttleton 1953 and Hunter 1977b for more detailed discussion). A simple example of a secular instability is given in Problem 5-5.

We can use the properties of the Riemann ellipsoids to determine the secular stability of the Maclaurin spheroids. Suppose that the fluid in a Maclaurin spheroid has a small viscosity, so that shearing motions dissipate energy. A perturbation of the form (5-152) carries the spheroid into a neighboring Riemann ellipsoid with the same total angular momentum. The ellipsoid has internal shear so that energy is slowly lost. The energy loss leads to damping and a return to the unperturbed state—that is, the spheroid is secularly stable—if and only if all these ellipsoids have a higher energy than the original spheroid. Examination of the Riemann ellipsoids shows that this condition is satisfied only for Maclaurin spheroids that rotate more slowly than the spheroid that satisfies (5-154), i.e., the spheroid at the beginning of the Jacobi sequence. Spheroids rotating less rapidly than the first member of the Jacobi sequence are secularly stable; more rapidly rotating spheroids are secularly unstable. Note that secular instability sets in at $e = 0.8127$, well before the onset of dynamical instability at $e = 0.9529$.

Energy arguments only isolate the point where secular instability sets in. Analytic and numerical solutions of the Navier-Stokes equations for a viscous incompressible fluid (Chandrasekhar 1969; Press &

Teukolsky 1973) allow one to follow the evolution of an unstable Maclaurin spheroid. It progresses through a sequence of Riemann ellipsoids of constant angular momentum and decreasing energy until it reaches the Jacobi sequence. Since the Jacobi ellipsoid has no internal shear there is no viscous dissipation, and there the evolution stops.

There is a simple physical basis for the instability. The kinetic energy of a uniformly rotating body with no internal motions is $\frac{1}{2}L^2/I$, where L is the angular momentum and I is the moment of inertia. The deformation of the spheroid into an ellipsoid increases this moment of inertia and thus lowers the rotational kinetic energy. The deformation also raises the potential energy; however, for sufficiently large angular momentum the loss of rotational kinetic energy is greater than the gain of potential energy and the spheroid is secularly unstable.

A similar secular instability is present in Maclaurin disks. Like the Maclaurin spheroids, the Maclaurin disks become unstable at the point of bifurcation of a Jacobi sequence. The first member of the Jacobi sequence can be regarded as arising from a Maclaurin spheroid or disk by an infinitesimal distortion that is stationary in the frame rotating with the spheroid (since the Jacobi ellipsoid has no internal shear). Thus, the point of onset of secular instability is marked by a mode whose frequency is zero in the frame rotating with the spheroid or disk. Setting $\omega_r = 0$ in equation (5-122) for the barlike ($l = m = 2$) mode, we find

$$\Omega^2 = \tfrac{1}{4}\Omega_0^2. \tag{5-156}$$

Hence for $\Omega^2 > \frac{1}{4}\Omega_0^2$ we expect that the Maclaurin disks are secularly unstable. This expectation can be verified by computation of the energy of neighboring Riemann disks with the same angular momentum as a given Maclaurin disk. Once again, secular instability sets in well before the onset of dynamical instability at $\Omega^2 = \frac{1}{2}\Omega_0^2$.

Are stellar-dynamical systems like the Kalnajs disks subject to secular instabilities as well? The answer to this question is not yet well-understood. There is no simple analog to viscosity in stellar systems. On a microscopic level, collisionless stellar systems are clearly reversible and therefore inviscid. However, macroscopically, the phenomenon of phase mixing (§4.7) is irreversible and has many features in common with viscosity; in particular, phase mixing converts ordered motion into random motion. In Chapter 7 we shall encounter other examples of macroscopically irreversible behavior in stellar systems, such as dynamical friction, tidal heating, and mergers. Thus we should beware of assuming that all instabilities in stellar systems are dynamical.

5.4 Summary

(i) Jeans's stability analysis of homogeneous stellar systems shows that all perturbations with wavelengths less than the Jeans length λ_J are stable, and perturbations with wavelengths exceeding λ_J are unstable. However, if we calculate the Jeans length associated with the density distribution and random velocities of a self-gravitating system in equilibrium, we find that it is comparable to the size of the system as given by the virial theorem. Therefore the Jeans analysis does not describe any real instability in an isolated, finite system. However, it does suggest that any stellar system with approximately isotropic velocity dispersion is stable on all scales much shorter than the smallest dimension of the system. The Jeans length is also an important concept in theories of galaxy and star formation.

(ii) All spherical stellar systems in which the DF is a decreasing function of energy alone are stable. However, anisotropic spherical systems with a preponderance of highly eccentric orbits are unstable to deformation into a triaxial shape.

(iii) Modes that are stable in spherical, nonrotating systems may become either secularly or dynamically unstable when the system is set in rotation. The most important unstable mode in the Maclaurin spheroids, the Maclaurin disks, and the Kalnajs disks is the fundamental $m = 2$ mode (the "bar" mode). It becomes secularly unstable in the fluid systems and dynamically unstable in the stellar systems when the ratio of kinetic energy in rotation to potential energy $T/|W|$ exceeds 0.14 (approximately).

These results are encouraging but incomplete. Our conclusion that spherical systems are generally stable is consistent with the smooth and regular structure of globular clusters and elliptical galaxies. Correspondingly, the widespread instabilities that we find in rapidly rotating systems may be connected with some of the large-scale departures from axial symmetry that we find in disk galaxies; in particular, it is tempting to identify the unstable bar mode with the bars observed in both S0 and spiral galaxies.

However, as we stressed in Chapter 4, real stellar systems do not rotate uniformly. To obtain a complete understanding of the instabilities of disk galaxies we must analyze differentially rotating systems. Since the behavior of these systems also appears to be intimately connected with the problem of the origin of spiral structure, we shall begin Chapter 6 with a discussion of spiral structure; the stability analysis itself will be taken up in §6.2.

Problems

5-1. [1] At typical sea-level conditions ($p = 1.01 \times 10^6$ dyne cm^{-2} and $T = 15°C$), the density of air is 1.23×10^{-3} g cm^{-3} and the speed of sound is 3.40×10^4 cm sec^{-1}. Find (a) the fractional change in frequency due to the self-gravity of the air, for a sound wave with wavelength 1 meter; (b) the Jeans length.

5-2. [2] The Jeans instability can be analyzed exactly, without invoking the Jeans swindle, in certain cylindrical rotating systems. Consider a homogeneous, self-gravitating fluid of density ρ_0, contained in an infinite cylinder of radius R_0. The cylinder walls and the fluid rotate at angular speed $\mathbf{\Omega} = \Omega\hat{\mathbf{e}}_z$, where $\hat{\mathbf{e}}_z$ lies along the axis of the cylinder.

(a) Show that the gravitational force per unit mass inside the cylinder is

$$-\nabla\Phi_0 = -2\pi G\rho_0(x\hat{\mathbf{e}}_x + y\hat{\mathbf{e}}_y). \tag{5P-1}$$

(b) Using Euler's equation in a rotating frame [cf. eq. (5-82b)], find the condition on Ω so that the fluid is in equilibrium with no pressure gradients.

(c) Now let $R_0 \to \infty$, or, what is equivalent, consider wavelengths $\lambda \ll R_0$, so that the boundary condition due to the wall can be neglected. Working in the rotating frame, find the dispersion relation analogous to equation (5-20) for (i) waves propagating perpendicular to $\mathbf{\Omega}$; (ii) waves propagating parallel to $\mathbf{\Omega}$. Show that waves propagating perpendicular to $\mathbf{\Omega}$ are always stable, while waves propagating parallel to $\mathbf{\Omega}$ are stable if and only if the usual Jeans criterion (5-22) is satisfied. [This problem is discussed by Chandrasekhar (1961), §120, who also shows that the usual Jeans criterion applies to *all* waves except those propagating perpendicular to $\mathbf{\Omega}$. See Lynden-Bell (1962c) for the analysis of the analogous homogeneous rotating stellar system.]

5-3. [2] (Suggested by A. Toomre) This problem investigates the transient response of an infinite homogeneous stellar system with a Maxwellian velocity distribution $f_0(\mathbf{v})$ [eq. (5-28)]. Note that parts (d) and (e) require numerical computations.

(a) Consider an external potential perturbation of the form

$$\Phi(\mathbf{x},t) = \epsilon p(t)\cos(\mathbf{k}\cdot\mathbf{x}), \tag{5P-2}$$

where $\epsilon \ll 1$ and $p(t) \to 0$ as $t \to -\infty$. Using linear perturbation theory and the Jeans swindle, show that the perturbed DF and density resulting from this external potential may be written in the form

$$f_1(\mathbf{x},\mathbf{v},t) = \text{Re}[f_b(\mathbf{v},t)e^{i\mathbf{k}\cdot\mathbf{x}}] \quad ; \quad \rho_1(\mathbf{x},t) = \text{Re}[\rho_b(t)e^{i\mathbf{k}\cdot\mathbf{x}}], \tag{5P-3}$$

where

$$f_b(\mathbf{v},t) = i\mathbf{k}\cdot\frac{\partial f_0}{\partial\mathbf{v}}\int_{-\infty}^t \Phi_b(t')e^{i\mathbf{k}\cdot\mathbf{v}(t'-t)}\,dt'$$

$$\rho_b(t) = \int f_b(\mathbf{v},t)\,d^3\mathbf{v} \quad ; \quad \Phi_b(t) = \epsilon p(t) - \frac{4\pi G}{k^2}\rho_b(t). \tag{5P-4}$$

(b) Using equation (5-28), show that the fractional density perturbation $y(t) = \rho_b(t)/\rho_0$ satisfies the integral equation

$$y(t) = \int_{-\infty}^{t} [\epsilon k^2 p(t') - 4\pi G \rho_0 y(t')](t' - t) e^{-\frac{1}{2}k^2 \sigma^2 (t'-t)^2} \, dt'. \qquad (5\text{P-}5)$$

Now consider the response to an impulsive perturbation, $\epsilon p(t) = -\sigma^2 T \delta(t)$, where T is a short time and δ denotes the Dirac delta function.

(c) In the limit where the effects of self-gravity are negligible $(G\rho_0 \to 0)$, the damping is entirely due to phase mixing. Evaluate $y(t)$ analytically in this limit.

(d) By introducing the dimensionless time $\tau = k\sigma t$ and replacing $y(t)$ by $Y(\tau) = y(t)/(k\sigma T)$, show that $Y(\tau)$ satisfies the integral equation

$$Y(\tau) = \begin{cases} \tau e^{-\frac{1}{2}\tau^2} + \dfrac{k_J^2}{k^2} \displaystyle\int_0^{\tau} Y(\tau - \tau'')\tau'' e^{-\frac{1}{2}\tau''^2} \, d\tau'', & \tau > 0, \\ 0, & \tau < 0, \end{cases} \qquad (5\text{P-}6)$$

where $k_J = \sqrt{4\pi G \rho_0}/\sigma$ is the Jeans wavenumber. Solve this equation numerically for $k_J/k = 0.5$, 1.0, 1.25 to show that the disturbance is damped when $k_J/k < 1$ and grows when $k_J/k > 1$ (see Figure 5-2).

(e) If we formally set k_J^2/k^2 negative, then our system behaves like a collisionless plasma. Solve for $Y(\tau)$ when $k_J^2/k^2 = -4$. You should find slowly decaying (Landau-damped) waves.

5-4. [2] In this problem we investigate the stability of an infinite self-gravitating thin sheet to bending or corrugation modes, in which the vertical position or height of the sheet $z_s(x, t)$ oscillates according to

$$z_s(x, t) = \epsilon z_0 e^{i(kx - \omega t)}, \qquad (5\text{P-}7)$$

where $\epsilon \ll 1$. We neglect any horizontal motions or external vertical potential so that the equation of motion is simply

$$\frac{\partial^2 z_s(x, t)}{\partial t^2} = F_v(x, t), \qquad (5\text{P-}8)$$

where F_v is the vertical force due to the self-gravity of the perturbed sheet.

(a) Using Gauss's theorem, show that the potential due to an infinite straight wire with mass per unit length ς, oriented parallel to the y-axis at location (x', z'), is

$$\Phi(x, z) = 2G\varsigma \ln \left[\sqrt{(x - x')^2 + (z - z')^2} \right]. \qquad (5\text{P-}9)$$

(b) By considering the sheet as a collection of wires, show that the potential of the sheet is

$$\Phi(x, z, t) = 2G\Sigma_0 \int_{-\infty}^{\infty} \ln \sqrt{(x - x')^2 + [(z - z_s(x', t)]^2} \, dx', \qquad (5\text{P-}10)$$

where Σ_0 is the surface density of the unperturbed sheet.

(c) Show that the vertical force per unit mass on an element of the sheet may be written

$$F_v(x,t) = -2\pi G\Sigma_0|k|z_s(x,t), \tag{5P-11}$$

where we have used the identity $\int_0^\infty (1-\cos u)du/u^2 = \frac{1}{2}\pi$, and kept only the dominant term in ϵ.

(d) Show that the dispersion relation for bending modes is

$$\omega^2 = 2\pi G\Sigma_0|k|, \tag{5P-12}$$

and hence that the bending modes are stable.

5-5. [2] This problem investigates a simple example of secular stability. We consider a spherical bowl of radius R, which rotates about a vertical axis with angular speed Ω. A particle slides inside the bowl. The particle is subject to a frictional force $\mathbf{F}_{\mathrm{fr}} = -k(\mathbf{v} - \mathbf{v}_b)$, where $\mathbf{v}$ is the velocity of the particle and $\mathbf{v}_b$ is the velocity of the bowl. The coefficient of friction k may be assumed to be very small. The particle is initially at rest, at the bottom of the bowl, and then is given a small displacement.

(a) Prove that the particle returns to rest at the bottom of the bowl (secular stability) if and only if $\Omega < \sqrt{g/R}$, where g is the acceleration due to gravity.

(b) If the motion is secularly unstable, what is the final fate of the particle?

5-6. [2] The characteristic equation (5-122) for the Maclaurin disks is particularly simple for axisymmetric modes with $l \gg 1$ (small-scale or local modes).

(a) For axisymmetric modes, show that the characteristic equation can be written in the form

$$\omega = 0 \quad \text{or} \quad \omega^2 = 4\Omega^2 + (l^2 + l)(\Omega_0^2 - \Omega^2) - (l^2 + l)g_{l0}\Omega_0^2. \tag{5P-13}$$

Does the root $\omega = 0$ correspond to a real mode, or is it spurious?

(b) Using Stirling's approximation (1C-9) for $n!$, show that $g_{l0} \simeq 4/(\pi l)$ for $l \gg 1$.

(c) Using the asymptotic form of $P_l(\xi)$ for large l [eq. (1C-26)], show that the surface density of an axisymmetric mode with $l \gg 1$ can be locally approximated as a sinusoid with the large wavenumber $k \simeq l/(\xi a)$.

(d) Using equation (5-131) to eliminate $\Omega_0^2 - \Omega^2$ in favor of the sound speed v_s^2, show that the characteristic equation derived in part (a) tends to the dispersion relation (5-92) for the rotating sheet, as $l \to \infty$.

5-7. [2] A simple way to stabilize the Maclaurin disks is to imagine that they are embedded in a fixed axisymmetric potential field (say, due to the halo or spheroid of the galaxy). Suppose that the halo potential is $\Phi_h(R) = \frac{1}{2}f\Omega_0^2 R^2$, i.e., the halo supplies a fraction f of the potential gradient from the disk (5-101).

(a) Show that the equilibrium relation (5-103) becomes

$$\Omega^2 = \Omega_0^2(1+f) - \frac{3K\Sigma_c}{a^2}. \tag{5P-14}$$

Thus, a cold disk would rotate at $\Omega = \Omega_0\sqrt{1+f}$.

(b) Show that the dispersion relation (5-122) becomes

$$\omega_r^3 - \omega_r\{4\Omega^2 + (l^2 + l - m^2)[\Omega_0^2(1 + f - g_{lm}) - \Omega^2]\} \\ + 2m\Omega[\Omega_0^2(1 + f - g_{lm}) - \Omega^2] = 0. \tag{5P-15}$$

(c) Show that the disk is stable to the $l = m = 2$ mode if

$$\frac{\Omega^2}{\Omega_0^2} < \tfrac{1}{2} + 2f. \tag{5P-16}$$

Thus, a cold disk is stable to the $l = m = 2$ mode if $f > \frac{1}{2}$.

(d) Is it possible to stabilize *all* of the modes of a cold disk by a sufficiently large value of f? [Hint: Use Stirling's approximation (1C-9) to evaluate g_{lm} for large l.]

5-8. [2] Just as in the case of the Maclaurin disks, we can stabilize the Kalnajs disks by embedding them in a fixed axisymmetric halo potential $\Phi_h(R) = \frac{1}{2}f\Omega_0^2 R^2$. Show that a Kalnajs disk embedded in such a halo is stable to the $l = m = 2$ modes if and only if

$$\frac{\Omega^2}{\Omega_0^2} < \frac{(5 + 8f)^3}{486}. \tag{5P-17}$$

5-9. [2] The onset of secular instability in a Maclaurin disk occurs at nearly the same point as the onset of dynamical instability in a Kalnajs disk. In this problem we investigate whether this agreement persists when the disks are embedded in a fixed halo potential of the type seen in the previous two problems.

(a) The onset of secular instability is marked by a mode whose frequency is zero in the frame rotating with the disk. By setting $\omega_r = 0$ in the dispersion relation derived in Problem 5-7, argue that the $l = m = 2$ mode of a Maclaurin disk embedded in a halo is secularly unstable if

$$\frac{\Omega^2}{\Omega_0^2} > \tfrac{1}{4} + f. \tag{5P-18}$$

(b) What halo fraction f is needed to provide secular stability to a cold Maclaurin disk? From the results of Problem 5-8, what halo fraction is needed to provide dynamical stability to a cold Kalnajs disk?

(c) For a cold Maclaurin or Kalnajs disk embedded in a halo, the parameter $t = T/|W|$ has the value $\frac{1}{2}(1 + f)$ (in defining t, only the self-gravitational energy of the disk is counted in W, not the potential arising from the interaction of the disk with the halo). Plot $t/(1 + f)$ against f for the Maclaurin disks at the point of onset of secular instability, and for the Kalnajs disks at the point of onset of dynamical instability. Comment on the argument by Ostriker and Peebles (1973) that the onset of dynamical instability in rotating stellar systems will generally correspond closely to the onset of secular instability in the analogous fluid systems.

6 Disk Dynamics and Spiral Structure

The majestic sweep of spiral arms across the face of a galaxy like M51 (Figure 6-1) is one of the most spectacular sights in the sky. Although this beauty alone is perhaps reason enough to study spiral structure, features such as these are more than just a scenic frosting on the galactic cake. As the primary sites of star formation, spiral arms play a central role in determining the chemical composition, density, and thermal balance of the interstellar gas. Their strength and shape provide important clues to the dynamics of the gas and stars in the galactic disk. Moreover, spiral arms may be the main force that drives the secular dynamical evolution of galactic disks, through such processes as gravitational scattering of stars off the arms (see §7.5.2) and angular momentum transport by spiral gravitational fields.

Spiral structure has proven to be one of the more obstinate problems in astrophysics. The first major attack was made by the Swedish astronomer Bertil Lindblad, who struggled with this problem from 1927 until his death in 1965. Lindblad correctly recognized that spiral structure arises through the interaction between the orbits and the gravitational forces of the stars of the disk, and thus should be investigated using stellar dynamics. In this view he stood almost alone; at his death most astronomers believed that spiral structure is caused by the interstellar magnetic field, which we now know is too weak to give rise to spiral structure [see §6.4.3(a)]. However, Lindblad's methods were not well suited to quantitative analysis, and his work has now been superseded by investigations based on more powerful analytical and numerical tools.

Shortly before Lindblad's death, C. C. Lin and Frank Shu made the crucial step that elucidated the theory toward which Lindblad had been groping. They recognized that spiral structure in a stellar disk could be regarded as a **density wave**, a wavelike oscillation that propagates through the disk in much the same way that waves propagate through violin strings or over the ocean surface. Lin and Shu also realized that many of the powerful mathematical techniques of wave mechanics could be applied to deduce the properties of density waves in differentially rotating stellar disks. They combined these insights with a bold hypothesis already made by Lindblad: that the spiral patterns in galaxies are long-lasting, in other words, that the appearance of the pattern remains unchanged (except for overall rotation) over many orbital periods. This **quasi-stationary spiral structure** hypothesis, combined with the idea that spiral structure is a density wave, leads to what we shall call the **Lin-Shu hypothesis**, that *spiral structure is a quasi-stationary density wave*. The Lin-Shu hypothesis has been central to the development of spiral structure theory over the past two decades, mainly because it enables theorists to make a broad variety of quantitative predictions for comparison with observations of spiral galaxies. Although many of these predictions have been successful, the validity of the Lin-Shu hypothesis is still the subject of intense debate among dynamicists.

The study of wave mechanics in differentially rotating disks has been developed into an extensive formalism called **density-wave theory**. Density-wave theory is intimately related to the study of the stability of differentially rotating stellar disks. In fact, perhaps its most important application has been to the interpretation of instabilities in galaxy models rather than to the explanation of spiral structure. Hence, after a brief introduction to the phenomenology of spiral structure in §6.1, we will discuss density-wave theory and the stability of differentially rotating disks in §6.2 and §6.3. We describe the current status of spiral structure theory in §6.4, and §6.5 and §6.6 are devoted to bars and warps, respectively.

6.1 Introduction

1 Observations

There is an intimate relation between spiral structure and the other large-scale properties of a galaxy. This relation is reflected in Hubble's classification of normal spirals, which was briefly described in §1.1.3 (see Hubble 1936, Sandage 1961, or MB §5-3 for more details). Hubble's classification tells us that the properties of the spiral arms (how tightly they are wound, how well they are resolved into individual stars and HII regions, etc.) are strongly correlated with many apparently unrelated properties of the galaxy (such as the luminosity of the spheroid relative to the disk, and the fractional mass in interstellar gas). One of the ultimate goals of spiral structure theory is to explain the origin of these correlations.

Figure 6-1. The Sc galaxy M51 (NGC 5194) together with its companion galaxy NGC 5195. Courtesy of A. R. Sandage.

(a) Photographs   Despite the presence of these general correlations, there is a great deal of variety in the spiral structure exhibited by galaxies, even galaxies of the same Hubble type. Hence the best introduction to spiral structure is to inspect a variety of photographs of nearby spiral galaxies (Sandage 1961; Sandage & Brucato 1979; Sandage

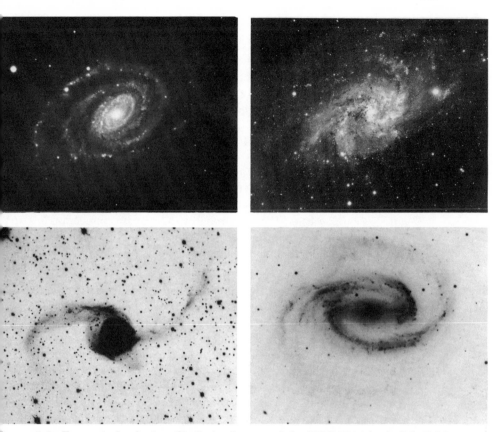

Figure 6-2. Four spiral galaxies. Clockwise from top left: NGC 5364 (type Sc); NGC 598 = M33 (type Sc); NGC 1300 (type SBb); NGC 3256 [type Sb(pec)]. Photograph of NGC 5364 courtesy of A. R. Sandage; others courtesy of F. Schweizer.

& Tammann 1981). To suggest what can be learned from the study of these pictures, consider the four galaxies shown in Figure 6-2:

(i) NGC 5364: This galaxy contains two very regular spiral arms, which can be traced for $1\frac{1}{4}$ to $1\frac{1}{2}$ revolutions around the galaxy's center. Since the arms are fairly open and there is only a small central spheroid, the galaxy is classified as Sc. This is one of the most symmetrical galaxies in the sky; spirals exhibiting this high degree of symmetry and regularity in the arms are rare. They are often called **grand-design** spirals. Presumably, they have been formed by some large-scale *global* process that involves the whole galaxy. There are dark strips on the inside of each arm, although they are not obvious on this photograph. These strips are called **dust lanes** because they are believed to be caused by absorption of the galaxy's starlight in dense clouds of gas and dust (see MB §3-11). Similar lanes are seen in many other Sb and Sc spirals, usually located on the inside of the spiral arms.

(ii) NGC 598: Also known as M33, this is a member of the Local Group of galaxies. It has little or no central spheroid, and its arms are broken up into stars and HII regions; hence it is classified Sc. Although there is a definite spirality to the galaxy, the arms are much broader and less distinct than in NGC 5364. The arms appear to branch or divide, and it is difficult to trace the arms unambiguously across the galaxy. A more extreme example of a galaxy with many short, fragmented arms is NGC 2841 (see Fig. 6-24). In galaxies like M33 or NGC 2841, it is not necessary that the arms be formed by a global process; rather, there is probably little or no connection between the arms on opposite sides of the galaxy, and a *local*, rather than global, origin seems more likely.

(iii) NGC 1300: Here is one of the most dramatic barred spirals. It is classified SBb (the letter B stands for "bar"). The two spiral arms are very symmetrical, though not perfectly so. They can easily be followed through 180°, and on the best plates through almost a full circle. Careful examination of Figure 6-2 shows that sharp straight dust lanes extend from the sides of the central nucleus out to the end of the bar. At the start of each spiral arm there is a cluster of HII regions (which look like bright stars on this photograph). Straight dust lanes in the bar and a cluster of HII regions at the start of the spiral arms are fairly common features of barred spiral galaxies. Note that the spiral arms start at the tips of the bar, which suggests that the bar is closely related to the spiral structure.

(iv) NGC 3256: This galaxy is classified as Sb(pec), where "pec" denotes "peculiar". The galaxy is peculiar indeed, with very little symmetry and a lopsided appearance. It may be the result of a recent merger of two galaxies (see §7.4). Its picture is included here as a reminder that not all galaxies exhibit the relatively well-formed, symmetrical patterns of Figures 6-1 and the other panels of 6-2.

(b) Spiral arms at other wavelengths   Photographs like those in Figure 6-2 are dominated by the blue light from luminous, young O and B stars and HII regions, which delineate narrow, sharply defined spiral arms. In Baade's (1963) apt phrase, the HII regions are "strung out like pearls along the arms". Since O and B stars live less than 10^7 yr (compared to a galactic age of 10^{10} yr), they are seen only in regions of

Figure 6-3. Radio continuum map of M51 (Mathewson et al. 1972).

recent star formation, and we are led to conclude that *spiral arms are regions of rapid, active star formation.*

Since photographs are so strongly affected by the young stellar population, it is natural to ask how spiral arms would look at other wavelengths. A radio continuum map of M51 is shown in Figure 6-3. The most striking features of the map are the strong ridges of radio emission,

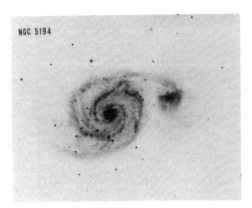

NGC 5194

Figure 6-4. Matched pictures of M51 in blue light (left) and red light (right). This is one example from an atlas of photographs by D. M. Elmegreen (1981), reprinted by permission of *The Astrophysical Journal*.

which follow the spiral arms. Although it is not obvious in Figure 6-3, the ridges actually follow the dust lanes on the inner edges of the optical spiral arms rather than the optical arms themselves. This displacement shows that the radio emission does not come from the stars that mark the optical arms; a completely separate mechanism is needed.

A different view is obtained in the 21-cm line of neutral hydrogen (HI). The intensity of the 21-cm line directly measures the HI surface density in the galaxy. Figure 6-23 (see also Figures 8-24 and 9-10 of MB) shows the 21-cm map of the spiral galaxy M81. The HI is strongly concentrated into spiral arms, and these arms lie on the inner (concave) edge of the optical arms (just as for the continuum observations). This displacement of the 21-cm radio line and the continuum spiral arms inside the optical arms appears to be a common feature of many galaxies.

Figure 6-23 also displays contours of constant radial velocity. Their overall shape resembles the usual contour field for an axisymmetric spiral galaxy (MB, Figure 8-17) but there are marked deviations. In particular, there are sharp kinks and bulges in the contours as they pass over the spiral arms. These distortions indicate that velocity perturbations of up to $20 \, \text{km s}^{-1}$ accompany the arms.

Finally, it is worthwhile to look at galaxies in red light, where the emission is dominated by the older, more typical disk stars rather than by young blue stars. Figure 6-4 shows that in M51 the red stars exhibit a spiral pattern similar to the blue stars, but with smoother, broader arms with relative amplitudes of 10%–20% (see also Figures 5-32 and 5-33 of MB). These features are common to many spiral galaxies, and suggest that the enhanced star-formation rate that produces the young

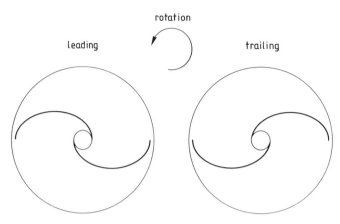

Figure 6-5. Leading and trailing arms.

stars is associated with changes in the overall density of the old disk stars. In other words, *the entire disk participates in the spiral pattern.*

The principal features that distinguish lenticular or S0 galaxies from spirals are the absence of interstellar gas and the absence of spiral arms, and these two features are strongly correlated, i.e., there are practically no examples of gas-free galaxies that exhibit spiral arms. The few known exceptions (e.g., Strom et al. 1976) may have recently lost their gas, perhaps through a collision with another galaxy. Thus, even though spiral structure is present in the old disk stars, *the interstellar gas seems to be an essential ingredient of spiral structure.*

(c) The number of arms   In many cases the shapes of spiral galaxies are approximately invariant under a rotation about their centers. A galaxy that looks identical after a rotation through an angle of $2\pi/m$ radians is said to have m-fold symmetry. A galaxy with m-fold symmetry usually has m dominant spiral arms. Most spiral galaxies have two arms and approximate twofold symmetry.

(d) Leading and trailing arms   Spiral arms can be classified by their orientation relative to the direction of rotation of the galaxy. A **trailing** arm is one whose outer tip points in the direction opposite to galactic rotation, while the outer tip of a **leading** arm points in the direction of rotation (see Fig. 6-5).

It is usually difficult to determine observationally whether the arms of a given galaxy are leading or trailing. Consider the two galaxies A

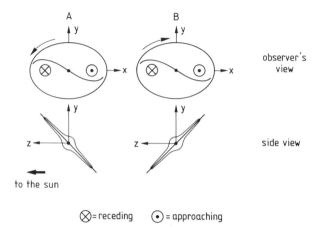

Figure 6-6. The appearance of leading and trailing arms. Galaxy A has leading arms, while galaxy B has trailing arms, but both exhibit the same pattern on the sky and the same radial velocity field.

and B in Figure 6-6. In both cases the (x, y) plane is the celestial sphere and the z-axis points toward the Sun. Galaxy A is inclined so that the side nearest the Sun is in the half plane $y > 0$, while galaxy B has its near side in the half plane $y < 0$. We have marked a spiral pattern and a rotation direction on both galaxies; the spiral in Galaxy A is leading and in Galaxy B is trailing. Despite this difference the appearance of both galaxies as seen from the Sun is the same; as we show in Figure 6-6, in both systems the spiral pattern appears to curve in an anti-clockwise direction as one moves out from the center, and the side with $x > 0$ has radial velocity toward the Sun. Thus radial velocity measurements and photographs of the spiral pattern cannot by themselves distinguish leading and trailing spirals.

To determine whether a particular galaxy leads or trails, we must therefore determine which side of the galaxy is closer to us. One way to do this is to count the numbers of objects such as novae and globular clusters that are seen on either side of the apparent major axis. If the objects are heavily obscured by dust in the central plane, then it is clear from Figure 6-7 that fewer objects will be seen on the near side of the galaxy. The near side can also be determined by studying the pattern of absorption in the dust lanes (see de Vaucouleurs 1959, MB §5-2, or the discussion of NGC 7331 in Sandage 1961). In almost all cases where the answer is unambiguous the spiral arms trail. The arms in our own Galaxy trail as well (§4.3 of MB). We conclude that in general *spiral arms are trailing.*

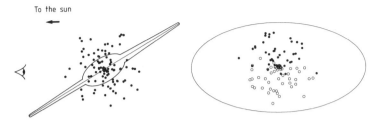

To the sun

Figure 6-7. Distinguishing near and far sides of a disk galaxy. The dots represent objects such as novae or globular clusters. There is an obscuring dust layer in the central plane of the disk which is shown as a line in the side view at left. In the observer's view, at right, objects behind the dust layer are fainter and are shown as open circles.

2 The Winding Problem

Suppose that at some initial time $t = 0$ we paint a stripe radially outward across the disk of the galaxy. The initial equation of the stripe is $\phi = \phi_0$, where ϕ is the azimuthal angle (Figure 6-8). The disk rotates with an angular speed $\Omega(R)$, where R is the distance from the center of the disk. The disk is said to be in **differential rotation** if $\Omega(R)$ is not independent of R; when the disk is in differential rotation the stripe will not remain radial as the disk rotates. The equation of the stripe $\phi(R, t)$ obeys

$$\phi(R, t) = \phi_0 + \Omega(R)t. \tag{6-1}$$

The **pitch angle** i of the arm at any radius r is the angle between the tangent to the arm and the circle $r = constant$ (see Figure 6-8); by definition $0 < i < 90°$. Thus

$$\cot i = \left| R \frac{\partial \phi}{\partial R} \right|. \tag{6-2}$$

For the arm of equation (6-1)

$$\cot i = Rt \left| \frac{d\Omega}{dR} \right|. \tag{6-3}$$

When the pitch angle is small, it can be related to the separation ΔR between adjacent arms at a given azimuth. If the locations of the arms at azimuth ϕ are R and $R + \Delta R$, we have from (6-1)

$$2\pi = |\Omega(R + \Delta R) - \Omega(R)|t, \tag{6-4}$$

so that when $\Delta R \ll R$

$$\Delta R = \frac{2\pi R}{\cot i}. \tag{6-5}$$

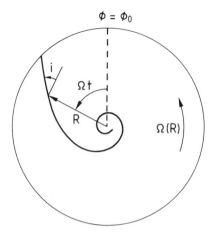

Figure 6-8. How a material arm winds up in a differentially rotating disk. The rotation law is $\Omega(R) \propto R^{-1}$.

For a typical galaxy with a flat rotation curve, $\Omega R = v_c = 220\,\mathrm{km\,s^{-1}}$, $R = 10\,\mathrm{kpc}$, and $t = 10^{10}\,\mathrm{yr}$, the pitch angle would be $i = 0.25°$, and the interarm separation would be $\Delta R = 0.28\,\mathrm{kpc}$, implying that the spiral is much too tightly wound to be observable. This difficulty is called the **winding problem**: if the material originally making up a spiral arm remains in the arm, the differential rotation of the galaxy winds up the arm in a time short compared with the age of the galaxy. A remarkably clear statement of the winding problem was given already by Wilczynski (1896).

There are a number of possible resolutions of the winding problem.

(i) It may be that the spiral pattern is statistically in a steady state, but that any individual spiral arm is quite young. If, for example, we continuously dribble cream into a freshly stirred cup of coffee, each new droplet briefly takes on a spiral form before it is stretched out and disappears. Similarly, if clumpy features are continuously produced in a galactic disk (say, by local gravitational instability leading to a burst of star formation) each feature will be sheared out into a spiral, which lasts until the bright young stars die off. In some galaxies the spiral structure may simply be an aggregate of these local features. This hypothesis is particularly plausible in the cases of galaxies like NGC 2841 that have many short arms but no global pattern. We shall discuss this possibility further in §6.4.2(a).

(ii) The spiral pattern may be a temporary phenomenon resulting from a recent violent disturbance such as an encounter with another galaxy. Not all spirals are made in this way since events of this kind are relatively rare, but they may be common enough to account for many of the most striking grand-design spirals, such as M51.

A more subtle concept is that the spiral arms are some sort of wave pattern that rotates through the galactic disk. The nature of the wave differs considerably in different theories:

(iii) It is possible that the spiral arm is a **detonation wave** of star formation that propagates around the disk; in this picture, new stars that explode as supernovae induce further star formation in adjacent regions [see §6.4.3(b)].

(iv) Another possibility is that the wave pattern is present in the density and hence the gravitational potential of the disk itself. As the interstellar gas travels over this "gravitational washboard", it is periodically compressed. Rapid star formation in these strongly compressed regions creates the young, bright stars that mark the visible spiral arms. If the Lin-Shu hypothesis is correct, the density wave is neutrally stable, so that the spiral pattern remains unchanged over many revolutions of the galaxy. The great attraction of this hypothesis is that it yields a broad range of explicit, testable predictions for the shape and kinematics of the spiral pattern, based solely on the properties of the underlying disk. As a result, theories based on the Lin-Shu hypothesis have become the standard against which all other explanations of spiral structure are tested.

The historical development of the theory and observations of spiral structure can be traced from the proceedings of four major conferences held in the 1960s and 1970s (Woltjer 1962; Becker & Contopoulos 1970; Weliachew 1975; Berkhuisen & Wielebinski 1978), and from reviews by Toomre (1977a) and Athanassoula (1984).

6.2 Wave Mechanics of Differentially Rotating Disks

1 Preliminaries

(a) **Kinematic density waves**   The radius of an orbit in the equatorial plane of an axisymmetric galaxy is a periodic function of time with period T_r [see eq. (3-16)]. During the interval T_r the azimuthal angle increases by an amount $\Delta\phi$ [eq. (3-17b)]. These quantities are related to the radial and azimuthal oscillation frequencies ω_r and ω_a by $\omega_r = 2\pi/T_r$ and $\omega_a = \Delta\phi/T_r$. In general, $\Delta\phi/(2\pi)$ is irrational, so that the orbit forms a rosette figure of the kind shown in Figure 3-1.

Now suppose that we view the orbit from a frame rotating at angular speed Ω_p. In this frame the azimuthal angle increases in one radial

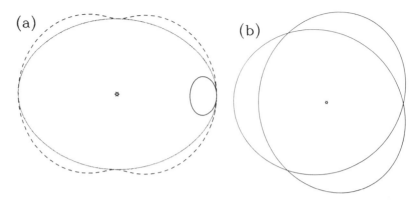

Figure 6-9. The appearance of elliptical orbits in a frame rotating at $\Omega_p = \Omega - n\kappa/m$. (a) $(n,m) = (0,1)$, solid line; $(1,2)$, dotted line; $(1,-2)$, dashed line; (b) $(n,m) = (2,3)$.

period by $\Delta\phi_p = \Delta\phi - \Omega_p T_r$. Therefore we can choose Ω_p so that the orbit is closed; in particular, if $\Delta\phi_p = 2\pi n/m$, where m and n are integers, the orbit closes after m radial oscillations. In this case

$$\Omega_p = \omega_a - \frac{n\omega_r}{m} \simeq \Omega - \frac{n\kappa}{m}, \tag{6-6}$$

where in the last equality we have approximated ω_a and ω_r by their values for nearly circular orbits, the circular frequency Ω and the epicycle frequency κ [see eqs. (3-58)]. The appearance of the closed orbits in the rotating frame is shown in Figure 6-9.

In Figure 6-10 we show the behavior of $\Omega - n\kappa/m$ in the epicycle approximation for several values of m and n. The curves are plotted for two representative galactic rotation curves, the Bahcall-Soneira model for our Galaxy (§2.7) and the isochrone potential [eq. (2-32)].

This diagram exhibits an interesting fact noted by Lindblad many years ago: while most of the $\Omega - n\kappa/m$ curves vary rapidly with radius, the curve for $n = 1, m = 2$ (or $n = 2, m = 4$, etc.) is relatively constant across much of the galaxy.[1] To understand the significance of constant $\Omega - n\kappa/m$, let us suppose for the moment that $\Omega - \frac{1}{2}\kappa$ were exactly constant and equal to some number Ω_p. Then in a frame rotating at Ω_p the orbits of the type shown as a dotted line in Figure 6-9a would be

---

[1] This result is related to the shape of galaxy rotation curves in their inner parts. In most galaxies the rotation curve rises linearly from the center with a steep slope. Thus, both Ω and κ are large near the galactic center, so that in general $|\Omega - n\kappa/m|$ is large near the center. However, in the central region where the rotation speed is roughly proportional to radius, $\Omega \simeq \frac{1}{2}\kappa$ [see eq. (3-59)]. Thus $\Omega - \frac{1}{2}\kappa$ is much smaller than $\Omega - n\kappa/m$ for values of n and m other than 1 and 2.

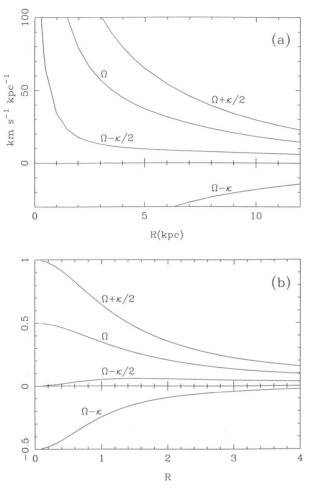

Figure 6-10. Behavior of $\Omega - n\kappa/m$ in: (a) the Bahcall-Soneira model for our Galaxy; (b) the isochrone potential [eq. (2-32)].

exactly closed. Hence we can set up a nested, aligned set of these orbits covering a range of radii, as shown in Figure 6-11a. If we fill up these orbits with stars we create a barlike wave pattern or density wave. Of course, we are not restricted to bars; by rotating the axes of the ellipses we can create leading or trailing spiral density waves as in Figures 6-11b and c. As viewed from an inertial frame the pattern rotates at angular speed Ω_p, which we call the **pattern speed.**

In a real galaxy $\Omega - \frac{1}{2}\kappa$ is not exactly constant. Hence, no matter

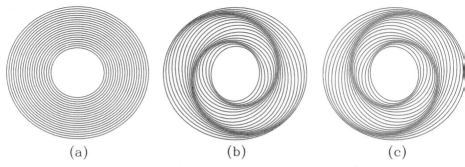

Figure 6-11. Arrangement of closed orbits in a galaxy with $\Omega - \frac{1}{2}\kappa$ independent of radius, to create bars and spiral patterns (after Kalnajs 1973).

what the value of Ω_p, most orbits are not exactly closed. The orientations of different orbits drift at slightly different speeds, so the pattern tends to twist or wind up. This is a modified version of the winding problem which we have already discussed—but now applying to waves rather than material arms—and the rate of winding can be calculated in a similar way. Let $\phi_p(R, t)$ be the angle of the major axis of the pattern, as viewed in the frame rotating at the pattern speed. Let us suppose the major axes are aligned along $\phi_p = \phi_0$ at time $t = 0$; thus $\phi_p(R, 0) = \phi_0$. The drift rate is $\partial \phi_p / \partial t = \Omega - \frac{1}{2}\kappa - \Omega_p$; thus

$$\phi_p(R, t) = \phi_0 + [\Omega(R) - \tfrac{1}{2}\kappa(R) - \Omega_p]t. \tag{6-7}$$

Equation (6-2) now gives the pitch angle as

$$\cot i = Rt \left| \frac{d(\Omega - \frac{1}{2}\kappa)}{dR} \right|. \tag{6-8}$$

In the Bahcall-Soneira model the average value of $\left| Rd(\Omega - \frac{1}{2}\kappa)/dR \right|$ is about $4.0\,\mathrm{km\,s^{-1}\,kpc^{-1}}$ between 5 and 10 kpc, and after $t = 10^{10}\,\mathrm{yr}$ the pitch angle in this region is about $i = 1.4°$. For comparison we computed after equation (6-5) that a material arm would have $i = 0.25°$ in a galaxy with a similar rotation curve. Thus, the wave pattern winds up more slowly than the material arm by about a factor of five. Although the winding time is still shorter than the age of the galaxy, we have come some way toward resolving the winding problem. Thus, we conclude that in galaxies with rotation curves similar to our own, $n = 1, m = 2$ *density waves can resist the winding process better than material arms.* This result suggests a natural explanation for the prevalence of two-armed spirals, if only we could find a way to adjust the slow drift rates of all the orbits to a common standard.

Density waves of the type described above are called **kinematic density waves** because they involve only the kinematics of orbits in a central potential. In fact, the orbits will deviate from the paths we have assumed because the spiral pattern itself produces a non-axisymmetric component of the gravitational field. The non-axisymmetric gravitational forces have a strong influence on the differential precession and the winding rate of the density wave. Since the mid-1960s, a major goal of spiral structure theorists has been to determine whether these gravitational forces can coordinate the drift rates of the orbits in such a way as to produce long-lived spiral patterns.

(b) Resonances   Stellar orbits, like springs, drums, and bridges, have natural resonant frequencies. If the gravitational field generated by spiral structure perturbs a stellar orbit at or near one of its resonant frequencies, then the response of the orbit will be large, even when the perturbing field is weak. To investigate the effect of non-axisymmetric forces on stars in the disk, a natural first step is to find the resonant stellar orbits.

An arbitrary gravitational potential in the disk plane can be written $\Phi_1(R, \phi, t)$. The simplest and most important potentials are those that are stationary in a rotating frame; these can be written in the form $\Phi_1(R, \phi, t) = \Phi(R, \phi - \Omega_p t)$, where Ω_p is the pattern speed of the potential. Examples of systems that generate potentials of this form include the rotating bars seen at the centers of many disk galaxies, a satellite galaxy in a circular orbit in the disk plane, and any spiral structure with a well-defined pattern speed. More complicated potentials can be regarded as superpositions of potentials with different pattern speeds.

We now specialize to the effect of a weak potential on a disk composed of stars on circular or near-circular orbits. Since the potential is periodic in $(\phi - \Omega_p t)$, it can be decomposed into a series of terms proportional to $\cos[m(\phi - \Omega_p t) + constant]$. We studied orbits in potentials of this form in §3.3.3, and found that resonances occurred when the circular frequency Ω and the epicycle frequency κ in the unperturbed orbit satisfied one of two conditions: $\Omega = \Omega_p$ (corotation resonance) or $m(\Omega - \Omega_p) = \pm\kappa$ (Lindblad resonances). The location and even the existence of these resonances depend on the circular-speed curve and the pattern speed. Inspection of Figure 6-10 shows that a typical galaxy can have zero, one, or two inner Lindblad resonances.

Note that when $n = \pm 1$, the condition (6-6) for a stationary kinematic density wave in a frame rotating at Ω_p is identical to the condition for a Lindblad resonance. This is a reasonable result: near resonance a weak perturbation with m-fold symmetry can produce a strong response with the same symmetry, hence we expect that when the Lindblad resonance condition is satisfied exactly, a stationary m-fold wave pattern can be present even in the absence of any perturbing force.

2 The Dispersion Relation for Tightly Wound Spiral Arms

(a) The tight-winding approximation   We analyze the behavior of density waves in disks by a three-step process. First, we use Poisson's equation to calculate the gravitational potential of an assumed surface density pattern. Second, we determine how this potential affects the stellar orbits and thus alters the surface density in the galaxy. Finally, we match this response surface density to the input surface density to obtain a self-consistent density wave.

This is a complicated procedure. Since the gravitational force is long-range, in general all parts of the galaxy are strongly coupled together, and the wave patterns must be determined numerically.

Fortunately, there is one important limit in which the analysis is much simpler. In the early 1960s a number of workers, notably A. J. Kalnajs, C. C. Lin, and A. Toomre, realized that for tightly wound density waves (i.e., waves whose radial wavelength is much less than the radius), the long-range coupling is negligible, the response is determined locally, and the relevant solutions are analytic. As we shall see, this **tight-winding** or **WKB approximation**[2] is an indispensable tool for understanding the origin and evolution of density waves in galaxies.

How tightly wound *are* spiral arms? To address this question it is useful to think of the center of each arm as a mathematical curve in the plane of the galaxy, which we write in the form $\phi + g(R,t) = constant$. We assume that the galaxy has $m > 0$ identical spiral arms. Then the arm pattern is invariant under a rotation of $2\pi/m$, and the equation for the kth arm is $\phi + 2\pi(k-1)/m + g(R,t) = constant$, where $k = 1,\ldots,m$. A more convenient form of this equation, which defines the locations of all m arms, is

$$m\phi + f(R,t) = \text{constant} \quad (\text{mod } 2\pi), \tag{6-9}$$

where $f(R,t) = mg(R,t)$ is the **shape function**. The radial separation between adjacent arms at a given azimuth is ΔR, where $|f(R+\Delta R,t) -$

---

[2] Named after the closely related Wentzel-Kramers-Brillouin approximation of quantum mechanics.

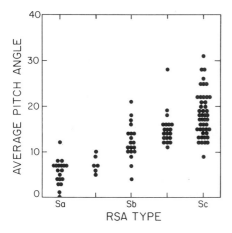

Figure 6-12. Measured pitch angle as a function of Hubble type for 113 galaxies (Kennicutt 1981). Reprinted by permission of *The Astronomical Journal.*

$f(R,t)| = 2\pi$. If the arms are tightly wound we may replace $f(R+\Delta R, t)$ by $f(R,t) + (\partial f/\partial R)\Delta R$ so that $(\partial f/\partial R)\Delta R = 2\pi$. In this case ΔR is identical to the **radial wavelength**

$$\lambda(R,t) \equiv \frac{2\pi}{|\partial f(R,t)/\partial R|}. \qquad (6\text{-}10)$$

It is also useful to introduce the **radial wavenumber**

$$k(R,t) \equiv \frac{\partial f(R,t)}{\partial R}. \qquad (6\text{-}11)$$

Note that $\lambda = 2\pi/|k|$, but k can be positive or negative. The sign of k determines whether the arms lead or trail. If we assume that the galaxy rotates in the direction of increasing ϕ, then

$$\text{leading arms} \Leftrightarrow k < 0 \quad ; \quad \text{trailing arms} \Leftrightarrow k > 0. \qquad (6\text{-}12)$$

The pitch angle i is given by equation (6-2) as

$$\cot i = \left| \frac{kR}{m} \right|. \qquad (6\text{-}13)$$

The condition for tight winding is that the pitch angle is small; in other words,

$$\cot i = \left| \frac{kR}{m} \right| \gg 1. \qquad (6\text{-}14)$$

Figure 6-12 shows measured pitch angles as a function of Hubble type. The correlation of pitch angle with type is expected, since openness of spiral arms is one of the criteria used to assign the Hubble type. For our purposes, the important result is that the median pitch angle is about $13°-14°$, so that $\cot i \simeq 4$. Thus the WKB approximation is satisfied at the median pitch angle in our sample, but not by very much.

Evidently the results of analyses based on the WKB approximation must be viewed with caution. We shall buttress these analyses with numerical experiments wherever possible, using the WKB results more as a guide to interpreting the experiments than as a definitive theory.

(b) Potential of a tightly wound spiral pattern   The surface density in a zero-thickness disk can be represented mathematically as the sum of an axisymmetric or unperturbed surface density $\Sigma_0(R)$, and a perturbed surface density $\Sigma_1(R, \phi, t)$, which represents the spiral pattern. For a tightly wound spiral it is convenient to write Σ_1 in a form that separates the rapid variations in density as one passes between arms from the slower variation in the strength of the spiral pattern as one moves along an arm. We may accomplish this by writing

$$\Sigma_1(R, \phi, t) = H(R, t)e^{i[m\phi + f(R,t)]}, \qquad (6\text{-}15)$$

where $f(R, t)$ is the shape function of equation (6-9), and $H(R, t)$ is a smooth function of radius that gives the amplitude of the spiral pattern. As usual, the physical surface density is given by the real part of equation (6-15). This expression assumes that the surface density variation is approximately sinusoidal in radius, which proves to be correct for the linear perturbation theory that we carry out below; in any event, more complicated surface density variations can be Fourier decomposed into a sum of sinusoidal variations.

The next step is to determine the gravitational potential due to the pattern (6-15). Since the surface density oscillates rapidly around zero mean, there will be nearly complete cancelation of the contribution from the distant parts of the pattern to the local potential; in other words, the perturbed potential at a given location will almost entirely be determined by the properties of the pattern within a few wavelengths of that location. Thus, to determine the potential in the neighborhood of a point (R_0, ϕ_0), we may replace the shape function $f(R, t)$ by the first two terms in its Taylor series, $f(R_0, t) + k(R_0, t)(R - R_0)$. Hence

$$\Sigma_1(R, \phi, t) \simeq \Sigma_a e^{ik(R_0, t)(R-R_0)}, \quad \text{where } \Sigma_a = H(R_0, t)e^{i[m\phi_0 + f(R_0, t)]}. \qquad (6\text{-}16)$$

(We have neglected variations with angle ϕ since these are much slower than radial variations whenever the wave is tightly wound.) Equation

(6-16) shows that in the vicinity of (R_0, ϕ_0), the spiral wave closely resembles a plane wave with wavevector $\mathbf{k} = k\hat{\mathbf{e}}_R$. The potential of a plane wave in a razor-thin disk was determined in §5.3.1. According to equation (5-90),

$$\Phi_1(R, \phi, t) \simeq \Phi_a e^{ik(R_0, t)(R-R_0)}, \quad \text{where} \quad \Phi_a = -\frac{2\pi G \Sigma_a}{|k|}. \quad (6\text{-}17)$$

We are now free to set $R = R_0$, thereby obtaining our final result for the potential due to the surface density (6-15)

$$\Phi_1(R, \phi, t) = -\frac{2\pi G}{|k|} H(R, t) e^{i[m\phi + f(R,t)]}. \quad (6\text{-}18)$$

The fractional error in this result is $O(|kR|^{-1})$. An alternative relation can be obtained by differentiating equation (6-18) with respect to radius and neglecting the derivative of $H(R, t)$ compared to the derivative of $f(R, t)$ (which also involves an error of order $|kR|^{-1}$). We find[3]

$$\Sigma_1(R, \phi, t) = \frac{i \, \text{sign}(k)}{2\pi G} \frac{d}{dR} \Phi_1(R, \phi, t), \quad (6\text{-}19)$$

again with fractional error $O(|kR|^{-1})$. Since this result involves the wavenumber k only through its sign, it is valid for any tightly wound spiral whose Fourier decomposition involves predominantly either leading [$\text{sign}(k) < 0$] or trailing [$\text{sign}(k) > 0$] waves, rather than a mixture of both.

(c) **The dispersion relation for gaseous disks**   We now determine the response of the galactic disk to a tightly wound potential perturbation. Since the response of a gaseous disk exhibits most of the important features that also occur in the stellar case, we shall find it helpful to consider spiral structure in gaseous disks before we tackle the more complicated stellar disks.

We again neglect the thickness of the disk, and assume that the pressure p acts only in the disk plane. In this approximation the motion

---

[3]   A more rigorous derivation of equation (6-18) is given by Shu (1970), who also presents a more accurate version whose fractional error is only $O(|kR|^{-2})$:

$$\Sigma_1(R, \phi, t) = \frac{i \, \text{sign}(k)}{2\pi G \sqrt{R}} \frac{d}{dR} \left[\sqrt{R} \Phi_1(R, \phi, t) \right].$$

is confined to the plane $z = 0$. Using equation (1B-54) for $(\mathbf{v} \cdot \boldsymbol{\nabla})\mathbf{v}$ in cylindrical coordinates, Euler's equations (1E-8) can be written

$$\frac{\partial v_R}{\partial t} + v_R \frac{\partial v_R}{\partial R} + \frac{v_\phi}{R} \frac{\partial v_R}{\partial \phi} - \frac{v_\phi^2}{R} = -\frac{\partial \Phi}{\partial R} - \frac{1}{\Sigma} \frac{\partial p}{\partial R}, \qquad (6\text{-}20a)$$

$$\frac{\partial v_\phi}{\partial t} + v_R \frac{\partial v_\phi}{\partial R} + \frac{v_\phi}{R} \frac{\partial v_\phi}{\partial \phi} + \frac{v_\phi v_R}{R} = -\frac{1}{R} \frac{\partial \Phi}{\partial \phi} - \frac{1}{\Sigma R} \frac{\partial p}{\partial \phi}, \qquad (6\text{-}20b)$$

where we have replaced the volume density ρ by the surface density Σ since we are dealing with a two-dimensional disk. Since the gaseous disk is only intended to serve as a heuristic model of a real stellar disk, we are free to choose a simple equation of state. Thus, suppose that

$$p = K\Sigma^\gamma. \qquad (6\text{-}21)$$

With this equation of state, sound waves in a disk with surface density Σ_0 propagate at a speed v_s given by [see eq. (1E-17)]

$$v_s^2 = \frac{dp}{d\Sigma}_0 = \gamma K \Sigma_0^{\gamma-1}. \qquad (6\text{-}22)$$

The equations of motion (6-20) are simplified if we replace p by the specific enthalpy [see eq. (1E-10)]

$$h = \frac{\gamma}{\gamma - 1} K \Sigma^{\gamma-1}. \qquad (6\text{-}23)$$

For example, the right side of equation (6-20a) then becomes

$$-\frac{\partial \Phi}{\partial R} - \frac{1}{\Sigma} \frac{\partial p}{\partial R} = -\frac{\partial \Phi}{\partial R} - \gamma K \Sigma^{\gamma-2} \frac{\partial \Sigma}{\partial R} = -\frac{\partial}{\partial R}(\Phi + h),$$

with a similar simplification in equation (6-20b).

We now assume that the spiral wave is only a small perturbation on the axisymmetric disk, so that we can linearize the equations of motion. Denoting quantities in the unperturbed axisymmetric disk by the subscript "0" we have $v_{R0} = 0$ and $\partial\Phi_0/\partial\phi = \partial p_0/\partial\phi = 0$. Euler's equations for the unperturbed disk become simply

$$\frac{v_{\phi0}^2}{R} = \frac{d}{dR}(\Phi_0 + h_0), \qquad (6\text{-}24)$$

representing a balance of the centrifugal force on the left side with gravitational and pressure forces on the right. In the cases of interest to us, the sound speed v_s is much smaller than the rotation speed $v_{\phi0}$—for example, in the interstellar gas $v_s \simeq 10\,\mathrm{km\,s^{-1}}$ while $v_{\phi 0} \simeq 220\,\mathrm{km\,s^{-1}}$—and it is easy to show that the term dh_0/dR in equation (6-24) is smaller than $d\Phi_0/dR$ by a factor of order $(v_s/v_{\phi 0})^2$. Hence

$$v_{\phi 0} \simeq \sqrt{R\frac{d\Phi_0}{dR}} = R\Omega(R), \qquad (6\text{-}25)$$

where $\Omega(R)$ is the circular frequency.

We now write $v_R = v_{R1}$, $v_\phi = v_{\phi 0} + v_{\phi 1}$, $h = h_0 + h_1$, $\Phi = \Phi_0 + \Phi_1$, where quantities labeled with the subscript "1" are assumed to be small. Keeping only terms that are first order in small quantities, equations (6-20) become

$$\frac{\partial v_{R1}}{\partial t} + \Omega\frac{\partial v_{R1}}{\partial \phi} - 2\Omega v_{\phi 1} = -\frac{\partial}{\partial R}(\Phi_1 + h_1), \qquad (6\text{-}26a)$$

$$\frac{\partial v_{\phi 1}}{\partial t} + \left[\frac{d(\Omega R)}{dR} + \Omega\right]v_{R1} + \Omega\frac{\partial v_{\phi 1}}{\partial \phi} = -\frac{1}{R}\frac{\partial}{\partial \phi}(\Phi_1 + h_1). \qquad (6\text{-}26b)$$

We shall find it useful to introduce a new variable

$$B(R) = -\tfrac{1}{2}\left[\frac{d(\Omega R)}{dR} + \Omega\right] = -\Omega - \tfrac{1}{2}R\frac{d\Omega}{dR}. \qquad (6\text{-}27)$$

Note that in the solar neighborhood, $B(R)$ is simply Oort's B constant [eq. (3-61c)]. We also use the epicycle frequency $\kappa(R)$ defined by [eq. (3-59)]

$$\kappa^2(R) = R\frac{d\Omega^2}{dR} + 4\Omega^2 = -4B\Omega. \qquad (6\text{-}28)$$

Any solution of equations (6-26) can be written as a sum of terms of the form

$$v_{R1} = \mathrm{Re}[v_{Ra}(R)e^{i(m\phi - \omega t)}] \quad ; \quad v_{\phi 1} = \mathrm{Re}[v_{\phi a}(R)e^{i(m\phi - \omega t)}],$$

$$\Phi_1 = \mathrm{Re}[\Phi_a(R)e^{i(m\phi - \omega t)}] \quad ; \quad h_1 = \mathrm{Re}[h_a(R)e^{i(m\phi - \omega t)}], \qquad (6\text{-}29)$$

$$\Sigma_1 = \mathrm{Re}[\Sigma_a(R)e^{i(m\phi - \omega t)}].$$

As usual we assume $m > 0$. Substituting these definitions into equations (6-26) and solving for v_{Ra} and $v_{\phi a}$ we find

$$v_{Ra} = -\frac{i}{\Delta}\left[(m\Omega - \omega)\frac{d}{dR}(\Phi_a + h_a) + \frac{2m\Omega}{R}(\Phi_a + h_a)\right], \qquad (6\text{-}30a)$$

$$v_{\phi a} = \frac{1}{\Delta}\left[-2B\frac{d}{dR}(\Phi_a + h_a) + \frac{m(m\Omega - \omega)}{R}(\Phi_a + h_a)\right], \qquad (6\text{-}30b)$$

where

$$\Delta \equiv \kappa^2 - (m\Omega - \omega)^2. \tag{6-31}$$

By equations (6-22) and (6-23), the linearized version of the equation of state is

$$h_a = \gamma K \Sigma_0^{\gamma-2} \Sigma_a = v_s^2 \Sigma_a / \Sigma_0, \tag{6-32}$$

where the sound speed $v_s(\Sigma_0)$ is a function of the unperturbed density.

So far our analysis has closely paralleled the analysis of the normal modes of Maclaurin disks in §5.3.2. In the Maclaurin disk $\Omega = constant$, $\kappa = 2\Omega$, and with these substitutions equations (6-30) are identical to equation (5-112).[4] In a differentially rotating disk, however, Ω and κ are not constant, and if ω is real, there may be radii at which $\Delta = 0$ so that equations (6-30) diverge. The origin of the singularity can be clarified by writing the exponent in equations (6-29) as $i(m\phi - \omega t) = im(\phi - \Omega_p t)$, where $\Omega_p = \omega/m$ is the pattern speed. Thus the singularity in the perturbed velocities arises when

$$\Omega_p = \Omega \pm \frac{\kappa}{m}, \tag{6-33}$$

which we recognize as the condition for a Lindblad resonance. This is also the condition for the existence of a kinematic density wave with $n = \pm 1$; see equation (6-6).[5] Our analysis breaks down near these resonances and a separate, more careful treatment is required, which we will not give here (see Goldreich & Tremaine 1979).

We now proceed with the analysis of the response of the gaseous disk. The perturbed surface density is related to the perturbed velocities by the equation of continuity (1E-3), which we write in cylindrical coordinates using equation (1B-45). Keeping only terms linear in small quantities, we have

$$\frac{\partial \Sigma_1}{\partial t} + \frac{1}{R}\frac{\partial}{\partial R}(R\Sigma_0 v_{R1}) + \Omega\frac{\partial \Sigma_1}{\partial \phi} + \frac{\Sigma_0}{R}\frac{\partial v_{\phi 1}}{\partial \phi} = 0. \tag{6-34}$$

With equations (6-29) this becomes

$$i(m\Omega - \omega)\Sigma_a + \frac{1}{R}\frac{d}{dR}(R\Sigma_0 v_{Ra}) + \frac{im\Sigma_0}{R}v_{\phi a} = 0. \tag{6-35}$$

---

[4] Note, however, that in this chapter the mean angular frequency and the circular frequency are nearly equal and are both denoted by Ω, while in Chapter 5 they are denoted separately as Ω and Ω_0.

[5] Resonances with $|n| > 1$ are unimportant in gaseous disks because they correspond to orbits that are self-intersecting, as in Figure 6-9b, and hence cannot be present in a fluid.

Equations (6-30), (6-32), and (6-35) provide four constraints on the five variables Σ_a, v_{Ra}, $v_{\phi a}$, h_a, and Φ_a. Thus, they determine the dynamical *response* Σ_a of the disk to an *imposed* potential Φ_a—and then to obtain a self-consistent density wave we require that Σ_a and Φ_a are related by Poisson's equation. These equations can be solved numerically to yield the *global* forms of the self-consistent density waves in a given disk (Bardeen 1975; Aoki et al. 1979). In this section, however, we concentrate on a simpler task: we use the WKB approximation to obtain analytic *local* solutions for density waves. As usual, the potential of a tightly wound wave can be written in the form

$$\Phi_a(R) = F(R)e^{if(R)} = F(R)e^{i\int^R k\,dR}, \qquad (6\text{-}36)$$

where $k = df(R)/dR$ [cf. eq. (6-15)] and $|kR| \gg 1$. The potential and surface density are related by Poisson's equation (6-17), which holds with fractional error $O(|kR|^{-1})$. Thus, Σ_a and—through equation (6-32)—h_a share with Φ_a the factor $\exp[if(R)]$, which varies rapidly with radius. Hence, in equation (6-30) we may drop the terms proportional to $(\Phi_a + h_a)/R$ relative to those involving $d(\Phi_a + h_a)/dR$, without increasing the fractional error beyond $O(|kR|^{-1})$. We may also write $d(\Phi_a+h_a)/dR = ik(\Phi_a + h_a)$ to the same level of accuracy. Thus equations (6-30) simplify to

$$v_{Ra} = \frac{(m\Omega - \omega)k(\Phi_a + h_a)}{\Delta} \quad ; \quad v_{\phi a} = -\frac{2Bik(\Phi_a + h_a)}{\Delta}. \qquad (6\text{-}37)$$

Similarly, in equation (6-35) we replace $d(R\Sigma_0 v_{Ra})/dR$ by $ikR\Sigma_0 v_{Ra}$. Since equations (6-37) show that v_{Ra} and $v_{\phi a}$ are of the same order, this term dominates over the third term in equation (6-35) by $O(|kR|)$, and hence we drop the latter term. The continuity equation thus has the form

$$(m\Omega - \omega)\Sigma_a + k\Sigma_0 v_{Ra} = 0. \qquad (6\text{-}38)$$

We eliminate v_{Ra} using equation (6-37a), eliminate h_a using (6-32), and eliminate Φ_a using (6-17). We find

$$\Sigma_a \left(1 + \frac{k^2 v_s^2}{\Delta} - \frac{2\pi G\Sigma_0 |k|}{\Delta}\right) = 0. \qquad (6\text{-}39)$$

The quantity in brackets must vanish, yielding *the dispersion relation for a gaseous disk in the tight-winding limit*:

$$(m\Omega - \omega)^2 = \kappa^2 - 2\pi G\Sigma|k| + k^2 v_s^2, \qquad (6\text{-}40)$$

where for brevity we have dropped the subscript on Σ_0. In the case of uniform rotation (constant angular speed), $\kappa = 2\Omega$, and equation (6-40) reduces to the dispersion relation for a rotating sheet that was derived in §5.3.1 [eq. (5-92)]. Before discussing the consequences of this relation, we shall derive the analogous dispersion relation for collisionless stellar-dynamical disks.

(d) The dispersion relation for stellar disks   The dispersion relation for a stellar disk may be calculated by the same general technique that we used to obtain the dispersion relation for a gaseous disk: we use the equations of motion to calculate the surface density perturbation Σ_1 arising from a potential perturbation Φ_1 of the form (6-29), and then require that Σ_1 and Φ_1 be related by Poisson's equation. The hardest step in this calculation is determining the perturbation $\overline{v}_{R1}$ in the mean radial velocity of the stars at a given point (R, ϕ) that is induced by Φ_1. If the disk were quite "cold", that is, if the unperturbed orbits were all circular, we could obtain $\overline{v}_{R1}$ from equation (6-37) with $h_a = 0$, because the disk would be dynamically identical to a gas disk with zero pressure. Thus for a cold stellar disk,

$$\overline{v}_{Ra} = \frac{m\Omega - \omega}{\Delta} k\Phi_a, \tag{6-41}$$

where Δ is defined by equation (6-31).

Unfortunately, this expression is accurate only if the typical stellar epicycle amplitude is much smaller than the wavelength $2\pi/k$ of the imposed spiral pattern: if this condition is not fulfilled, stars passing through a given location (R, ϕ) at a given time, which come from a range of radii of width equal to twice the epicycle amplitude, have sampled entirely different parts of the spiral potential. This leads to a partial cancelation of the effects of the spiral potential on the mean velocity perturbation, which can be formally incorporated by rewriting equation (6-41) in the form

$$\overline{v}_{Ra} = \frac{m\Omega - \omega}{\Delta} k\Phi_a \, \mathcal{F}, \tag{6-42}$$

where $\mathcal{F} \leq 1$ is called the **reduction factor**. For a given potential perturbation, the reduction factor describes how much the response to a spiral perturbation is reduced below the value for a cold disk.

Once we have $\overline{v}_{Ra}$, it is straightforward to calculate the response density Σ_a, since the Jeans equation (4-21) is identical to the continuity equation of the gaseous disk. Thus the analog of equation (6-38) is

$$(m\Omega - \omega)\Sigma_a + k\Sigma_0\overline{v}_{Ra} = 0. \tag{6-43}$$

The final step is to eliminate $\overline{v}_{Ra}$ between equations (6-42) and (6-43), and combine the resulting equation with the WKB form of Poisson's equation (6-17) to obtain the dispersion relation.

In Appendix 6.A we evaluate the reduction factor $\mathcal{F}$ for a two-dimensional disk with the **Schwarzschild distribution function**

$$f_0(R, v_R, v_\phi) = \frac{\Sigma(R)}{2\pi\sigma_R\sigma_\phi} \exp\left\{ -\frac{v_R^2}{2\sigma_R^2(R)} - \frac{[v_\phi - v_c(R)]^2}{2\sigma_\phi^2(R)} \right\}, \tag{6-44}$$

where $v_c(R) = R\Omega(R)$ is the circular speed, and $\sigma_R(R)$ and $\sigma_\phi(R)$ are the radial and tangential velocity dispersions. As required by the epicycle approximation, we assume that $\sigma_R \ll \overline{v}_\phi$, $\sigma_\phi \ll \overline{v}_\phi$, and that the ratio σ_R/σ_ϕ satisfies equation (4-52).[6] The DF (6-44), which was introduced by K. Schwarzschild (1907) to fit the kinematics of the solar neighborhood, is a solution of the collisionless Boltzmann equation in the epicycle approximation (§7.5).

For the DF (6-44), the reduction factor can be written in the form [eqs. (6A-29) and (6A-25)]

$$\mathcal{F}\left(\frac{\omega - m\Omega}{\kappa}, \frac{k^2\sigma_R^2}{\kappa^2}\right) \equiv \mathcal{F}(s, \chi) = \frac{2}{\chi}(1 - s^2)e^{-\chi}\sum_{n=1}^{\infty}\frac{I_n(\chi)}{1 - s^2/n^2} \quad (6\text{-}45\text{a})$$

$$= \frac{1 - s^2}{\sin \pi s}\int_0^\pi e^{-\chi(1+\cos\tau)}\sin s\tau \sin\tau \, d\tau. \quad (6\text{-}45\text{b})$$

Here $I_n(\chi)$ is a modified Bessel function [see eq. (1C-48)]. These formulae were derived independently by Lin and Shu (1966) and Kalnajs (1965). A tabulation of $\mathcal{F}(s, \chi)$ is given by Lin et al. (1969). Note that the definition of $\mathcal{F}$ implies that $\mathcal{F}(s, 0) = 1$, a relation that is easily checked using equation (6-45b).

Equation (6-45), together with equations (6-17), (6-42), and (6-43), can be combined to yield the dispersion relation for tightly wound density waves in a stellar disk, the analog of equation (6-40) for a gaseous disk:

$$(m\Omega - \omega)^2 = \kappa^2 - 2\pi G\Sigma|k|\mathcal{F}\left(\frac{\omega - m\Omega}{\kappa}, \frac{k^2\sigma_R^2}{\kappa^2}\right). \quad (6\text{-}46)$$

The dispersion relations (6-40) and (6-46) are the key equations in the analytic study of density waves in disks. Strictly speaking, the WKB approximation is only marginally valid in most galactic disks. Nevertheless, when properly reinforced by numerical work, these dispersion relations provide an invaluable guide to the stability, evolution, and shape of density waves in galaxies. Like any local dispersion relations, they do not establish that a permanent standing wave pattern can be set up in the disk—this requires more input physics, including the boundary conditions at the center and outer edge of the disk, and some

---

[6] Note that this DF is apparently inconsistent with the asymmetric drift equation (4-34), because it implies that the mean speed $\overline{v}_\phi(R)$ is equal to the circular speed. However, the asymmetric drift $v_c - \overline{v}_\phi$ is only of order σ_R^2/v_c, which is second order in the small parameter σ_R/v_c and hence vanishes in the first-order approximation that we use here.

understanding of how the wave behaves at the Lindblad and corotation resonances, where the local dispersion relations break down. Rather, they establish the relation between wavenumber and frequency that is satisfied by a traveling wave as it propagates across the disk.

3 Local Stability of Differentially Rotating Disks

A simple and important application of the dispersion relations (6-40) and (6-46) for gaseous and stellar disks is to determine whether a given disk is locally stable to axisymmetric perturbations.[7] (All of the analysis we have done so far is for tightly wound non-axisymmetric disturbances, that is, for $|kR/m| \gg 1$. However, it is a simple exercise to show that the dispersion relations (6-40) and (6-46) also hold for axisymmetric disturbances $(m = 0)$ so long as $|kR| \gg 1$.)

Consider first the case of a gaseous disk. Equation (6-40) becomes

$$\omega^2 = \kappa^2 - 2\pi G\Sigma|k| + k^2 v_s^2. \tag{6-47}$$

Since the quantities on the right side of equation (6-47) are all real, ω^2 must also be real. If $\omega^2 > 0$, then ω is real and the disk is stable. If, on the other hand, $\omega^2 < 0$, say $\omega^2 = -p^2$, then $\omega = \pm ip$, and $\exp(-i\omega t) = \exp(\pm pt)$. Hence for $\omega^2 < 0$, there is a perturbation whose amplitude grows exponentially, and the disk is unstable. The line of neutral stability is therefore

$$\kappa^2 - 2\pi G\Sigma|k| + k^2 v_s^2 = 0. \tag{6-48}$$

The gaseous disk is stable to all axisymmetric perturbations if there is no solution of equation (6-48) for any positive value of $|k|$. Since the equation is quadratic, it is easily solved, and we find that axisymmetric stability requires

$$Q \equiv \frac{v_s \kappa}{\pi G\Sigma} > 1 \qquad \text{for gas.} \tag{6-49}$$

The line of neutral stability defined by equation (6-48) is drawn in Figure 6-13 in terms of the dimensionless ratios Q and $\lambda/\lambda_{\text{crit}}$, where $\lambda = 2\pi/|k|$ and[8]

$$\lambda_{\text{crit}} \equiv \frac{4\pi^2 G\Sigma}{\kappa^2}. \tag{6-50}$$

---

[7] It can be shown that gaseous and stellar disks are stable to all local non-axisymmetric disturbances (Goldreich & Lynden-Bell 1965; Julian & Toomre 1966).

[8] λ_{crit} is the longest unstable wavelength in a zero-pressure disk.

The stability criterion for a uniformly rotating sheet [eq. (5-93)] is a special case of (6-49) when $\kappa = 2\Omega$. Of course, in a general disk v_s, κ, and Σ are all functions of radius, so that Q is also a function of radius. In this case $Q(R) < 1$ only implies *local* axisymmetric instability near radius R, in the sense that a traveling wave that crosses a region with $Q(R) < 1$ will grow while it is in that region.

The analysis of the stability of a stellar disk is similar. By analogy we expect that the boundary between stable and unstable axisymmetric waves is given by $\omega = 0$, just as in the case of a gaseous disk.[9] Thus from equation (6-46) the stability boundary is

$$\kappa^2 - 2\pi G\Sigma |k| \mathcal{F}(0, k^2\sigma_R^2/\kappa^2) = 0, \qquad (6\text{-}51)$$

or, using equation (6-45a) and the identity (1C-55),

$$\frac{|k|\sigma_R^2}{2\pi G\Sigma} = \left[1 - e^{-k^2\sigma_R^2/\kappa^2} I_0\left(\frac{k^2\sigma_R^2}{\kappa^2}\right)\right], \qquad (6\text{-}52)$$

a relation first derived by Toomre (1964). There is no solution to equation (6-52), and thus the disk is stable, if

$$Q \equiv \frac{\sigma_R \kappa}{3.36 G\Sigma} > 1 \qquad \text{for stars.} \qquad (6\text{-}53)$$

The stability boundary (6-52) is plotted in Figure 6-13 as a function of the dimensionless ratios Q and $\lambda/\lambda_{\mathrm{crit}}$. Note the close analogy between gaseous and stellar disks: the stability criterion for stellar disks (6-53) is obtained from the criterion for gaseous disks (6-49) simply by replacing the sound speed v_s by the radial velocity dispersion σ_R, and the coefficient $\pi \simeq 3.14$ by 3.36. The inequality (6-49) or (6-53) is known as **Toomre's stability criterion**;[10] its physical interpretation is discussed in the context of the uniformly rotating sheet in §5.3.1. Toomre's Q serves as a thermometer for galactic disks. "Hot" disks have large velocity dispersion and high Q, while "cool" disks have low dispersion and Q, and "cold" disks have zero dispersion and $Q = 0$.

The most reliable evidence on the value of Q in a real galactic disk comes from the solar neighborhood. From Table 1-1 the local surface density is $\Sigma \simeq 75 \, M_\odot \, \mathrm{pc}^{-2}$. From Table 1-2, the epicycle frequency $\kappa \simeq 36 \, \mathrm{km \, s^{-1} \, kpc^{-1}}$. For σ_R we take the radial velocity dispersion of late-type (G, K, and M) dwarfs, since these are the largest known

---

[9] The rigorous proof of this statement is complicated. See Julian and Toomre (1966).

[10] An approximate version of equation (6-49) dates back to Safronov (1960).

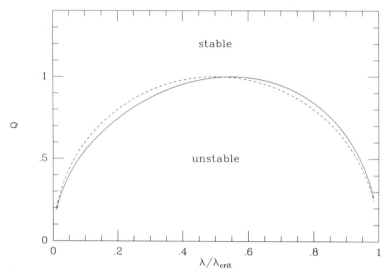

Figure 6-13. Neutral stability curves for short-wavelength axisymmetric perturbations in a gaseous disk [dashed line, from eq. (6-48)] and a stellar disk [solid line, from eq. (6-52)].

contributors to the surface density in the solar neighborhood.[11] Table 7-1 of MB lists a local value $\sigma_R \simeq 30\,\mathrm{km\,s^{-1}}$ for these objects, but a more appropriate value is the average of σ_R over a vertical column through the Sun, about $45\,\mathrm{km\,s^{-1}}$ with an uncertainty of at least 20% (Wielen 1977). We compute Q using equation (6-53), with the numerical coefficient reduced from 3.36 to 2.9 to account for the non-zero disk thickness and the influence of interstellar gas (Toomre 1974). With these numbers we estimate $Q = 1.7$, with a possible range from about 1 to 3. Despite the many uncertainties in the calculations, it appears that the solar neighborhood is safely stable in the local sense.

Is the WKB approximation valid in the solar neighborhood? As Q drops below unity, instability first appears at a single wavelength, which we write as

$$\lambda(\text{most unstable}) \equiv p\lambda_{\mathrm{crit}}, \qquad (6\text{-}54)$$

where the constant p is 0.5 or 0.55 for zero-thickness gas or stellar disks, respectively (see Figure 6-13). With our parameters for the solar neighborhood, $\lambda_{\mathrm{crit}} \simeq 10\,\mathrm{kpc}$ and $\lambda(\text{most unstable}) \simeq 6\,\mathrm{kpc}$, compared to the solar radius $R_0 = 8.5\,\mathrm{kpc}$. These wavelengths are so large as to cast serious doubt on the validity of the WKB approximation, for which we

---

[11] White dwarfs make a substantial (but very uncertain) contribution and probably increase σ_R (see Wielen 1977).

would like to have $\lambda/R \ll 1$. On the other hand, a more natural criterion for the validity of the WKB approximation for axisymmetric waves is $|kR| \gg 1$, which implies $\lambda/R \ll 2\pi$, a significantly less stringent condition. A useful rule of thumb based on numerical experiments is that the WKB results for axisymmetric waves are reasonably accurate when $\lambda/R \lesssim 2$, which suggests that the WKB approximation is reliable in this case.

4 Long Waves, Short Waves, and Group Velocity

The dispersion relations (6-40) and (6-46) for gaseous and stellar disks relate the wavenumber k to the frequency ω or pattern speed $\Omega_p = \omega/m$. We have plotted these relations in Figure 6-14a, where the disks have been characterized by Toomre's Q [eqs. (6-49) or (6-53)], and by $k_{\text{crit}} \equiv 2\pi/\lambda_{\text{crit}}$ [eq. (6-50)]. We write the dimensionless frequency

$$s = \frac{m(\Omega_p - \Omega)}{\kappa}. \tag{6-55}$$

The corotation resonance is at $s = 0$, and the Lindblad resonances are at $s = \pm 1$. Since the axes are labeled with k/k_{crit} and s, the curves apply to any disk, but the relation between s and radius depends on the disk model. For example, in a Mestel disk $s = 2^{-1/2}m(r/r_{\text{CR}} - 1)$ and the Lindblad resonances are at $r/r_{\text{CR}} = 1 \pm \sqrt{2}/m$.

The following features of the dispersion relations should be noted:

(i) The dispersion relations for leading $(k < 0)$ and trailing $(k > 0)$ waves [see eq. (6-12)] are identical. (This is a consequence of the anti-spiral theorem described in §6.4.)

(ii) In all disks with $Q > 1$ there is a forbidden region around corotation in which tightly wound density waves are evanescent.[12] The width of the forbidden region increases as Q increases.

(iii) Between corotation and the Lindblad resonances $s = \pm 1$ there are two branches of the dispersion relation. The **long-wave** branch (the one with larger λ or smaller $|k|$) begins at $k = 0$, $|s| = 1$, and $|s|$ decreases as $|k|$ increases. Sufficiently near the Lindblad resonances the long wave dispersion relation is independent of Q and is the same for gaseous and stellar disks:

$$|k| = \frac{2\pi}{\lambda_{\text{crit}}}(1 - s^2). \tag{6-56}$$

---

[12] An **evanescent** wave has a complex wavevector k and decays exponentially with radius.

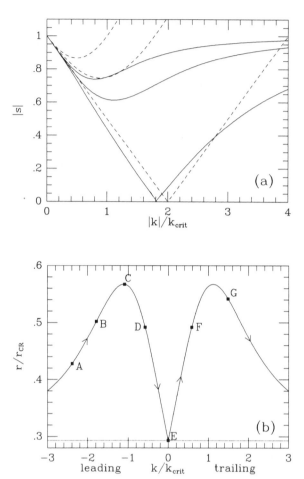

Figure 6-14. (a) The dispersion relation for tightly wound disturbances in gaseous [eq. (6-40), dashed lines] and stellar [eq. (6-46), solid lines] disks. The curves shown are (bottom to top) $Q = 1$, 1.5, and 2. Since only $|s|$ and $|k|$ are shown, there is no distinction between leading and trailing waves, or waves inside and outside corotation. (b) Dispersion relation in the form of wavenumber versus radius for an $m = 2$ tightly wound wave in a stellar Mestel disk with $Q = 1.5$. The radial scale is in units of the corotation radius r_{CR}, and only waves inside corotation are shown. The inner Lindblad resonance is at $r = 0.293r_{CR}$. The direction of the group velocity is shown by arrows.

This result is easily proved from either of the dispersion relations (6-40) or (6-46). The behavior of the stellar and gaseous disks in this regime is similar because the dispersion relation is determined solely by Coriolis and centrifugal forces and by the self-gravity of the disk.

(iv) Along the **short-wave** branch (the one with smaller λ or larger $|k|$), $|s|$ increases as $|k|$ increases. In stellar disks, $|k| \to \infty$ as $|s| \to 1$, so that the short-wave branch terminates at the Lindblad resonance $|s| = 1$, while in gaseous disks the short-wave branch passes smoothly through the Lindblad resonance at wavenumber $|k| = 2\pi G\Sigma/v_s^2$.

There is a simple physical basis for many of these results. The dimensionless frequency s represents the ratio of the forcing frequency seen by a particle, $m(\Omega_p - \Omega)$, to its natural radial frequency, κ. If no perturbing forces are present, a steady wave can be set up only if the two frequencies are equal, which occurs at the Lindblad resonances, $s = \pm 1$. Thus in a stellar disk with $Q \gg 1$, in which gravitational and pressure forces are negligible, a stationary disturbance can be present only when $s = \pm 1$. As Q decreases, the attractive self-gravity of the disk becomes more important. The self-gravity reduces the natural radial frequency below κ, so that waves can be present when $|s| < 1$. As the wavenumber $|k|$ increases, the gravitational forces become more important, so that the natural frequency $|s|$ decreases with increasing $|k|$ on the long-wave branch. Eventually the repulsive pressure forces in a gas disk, or the reduction factor $\mathcal{F}$ in a stellar disk, begin to dominate over the self-gravity, so that $|s|$ increases again on the short-wave branch. In a gas disk, the pressure forces can actually increase the natural radial frequency above κ, so that waves can also be present when $|s| > 1$; for $|s| \gg 1$ the density wave is essentially a sound wave. A stellar disk has no pressure forces and therefore cannot support waves with $|s| > 1$.

The waves described by Figure 6-14a are traveling waves, and they propagate with a **group velocity**. In Appendix 1.E we show that the group velocity of a wave packet in a dispersive medium is $v_g = d\omega(k)/dk$; similarly, it can be shown (Toomre 1969; Whitham 1974) that in an inhomogeneous medium where the frequency depends on position, $\omega = \omega(k, R)$, and the group velocity is

$$v_g = \frac{\partial \omega(k, R)}{\partial k}, \tag{6-57}$$

so long as the distance over which ω varies is much larger than the wavelength. Toomre (1969) first pointed out that this relation could be applied to Figure 6-14a to determine the evolution of a tightly wound density wave packet.[13]

For example, the dispersion relation for a gaseous disk (6-40) yields

$$v_g(R) = \text{sign}(k)\frac{|k|v_s^2 - \pi G\Sigma}{\omega - m\Omega}. \tag{6-58}$$

In this equation v_s, Σ, and Ω are functions of radius determined by the properties of the unperturbed disk, and k is determined by the dispersion relation. A wave packet localized around a radius R propagates radially outward if $v_g(R) > 0$, and inward if $v_g(R) < 0$. As the packet propagates, the wavenumber k changes so that the packet continues to satisfy the dispersion relation with fixed ω. The direction of propagation is toward corotation on the long-wave branch and away from corotation on the short-wave branch. The evaluation of the group velocity in stellar disks is more tedious (Toomre 1969); however, we can obtain an approximate estimate of v_g by graphical methods, directly from Figure 6-14a. The axes in this figure are $x \equiv k/k_{\text{crit}}$ and $s = (\omega - m\Omega)/\kappa$ (to within a sign). Thus $d\omega|_R = \kappa ds$ and $dk|_R = k_{\text{crit}}dx$; hence $v_g = (\partial\omega/\partial k)_R = (\kappa/k_{\text{crit}})(ds/dx)$. In other words, to within a sign the group velocity is simply the slope of the curves in Figure 6-14a times the characteristic velocity $\kappa/k_{\text{crit}} = 2\pi G\Sigma/\kappa$ [see eq. (6-50)].

In the solar neighborhood, $\kappa/k_{\text{crit}} \simeq 57\,\text{km s}^{-1}$ (based on $\Sigma = 75\,M_\odot\,\text{pc}^{-2}$, $\kappa = 36\,\text{km s}^{-1}\,\text{kpc}^{-1}$; see Tables 1-1 and 1-2). It is instructive to estimate the time required for a wave packet to propagate across the Galaxy. Since the slopes of the curves in Figure 6-14a are of order unity, this time will not be very different from $10\,\text{kpc}/57\,\text{km s}^{-1} \approx 2 \times 10^8\,\text{yr}$. For comparison, the rotation period at the solar radius is $2\pi/\Omega \simeq 2.4 \times 10^8$ yr. It is evident that any WKB wave packet will propagate across a substantial fraction of the galactic disk within at most a few rotation times; more precisely, it will propagate into either a Lindblad resonance or the forbidden zone around the corotation resonance. What happens to a density wave at the resonances?

With the help of Figure 6-14b, which shows the leading and trailing branches of the dispersion relation separately, let us follow a packet of leading waves inside corotation in a $Q = constant$ stellar Mestel disk, starting at point A on the short branch of the dispersion relation. The group velocity is positive and the packet propagates outward. As the

---

[13] In a one-dimensional medium, the group velocity is also the velocity at which momentum and energy are carried by the wave. In a disk, waves carry energy and angular momentum. The group velocity determines the direction and rate of radial transport of angular momentum in the disk. See Toomre (1969) and Lynden-Bell and Kalnajs (1972).

radius increases (point B), $|k|$ decreases. Eventually, at point C, the edge of the forbidden region, the group velocity changes sign, and the packet begins to propagate inward as a leading wave on the long branch (point D). The change in direction of group velocity can be regarded as a reflection of the packet off the forbidden zone, in the same way that wave packets in quantum mechanics reflect off potential barriers. The group velocity eventually carries the packet into the inner Lindblad resonance (point E). Here the WKB theory is suspect because $k = 0$, but more detailed calculations (Goldreich & Tremaine 1978, 1979) show that so long as the parameter

$$X \equiv \frac{k_{\text{crit}} R}{m} \qquad (6\text{-}59)$$

greatly exceeds unity, the wave simply reflects off the resonance and propagates out again as a trailing wave (point F). At the forbidden zone it is reflected again, this time into a short-branch trailing wave, and then propagates inward (point G). From now on its wavelength becomes shorter and shorter without limit; more detailed analysis (Mark 1974) shows that in a stellar disk the wave is eventually absorbed at the Lindblad resonance.[14]

The evolution that we have described here is reminiscent of the winding-up of material spiral arms, since a tightly wound leading wave becomes first a loosely wound leading wave, then a loosely wound trailing wave, and finally a trailing wave that becomes ever more tightly wound. The rate of winding can be worked out quantitatively using the dispersion relation and the group velocity. For the simple example of a gaseous, $Q = constant$ Mestel disk, we show in Problem 6-4 that a trailing wave winds up at a rate

$$\frac{d}{dt}(\cot i) = \Omega_p. \qquad (6\text{-}60)$$

For comparison, according to equation (6-3), a material arm in a Mestel disk would wind up at a rate

$$\frac{d}{dt}(\cot i) = \left| R\frac{d\Omega}{dR} \right| = \Omega, \qquad (6\text{-}61)$$

where the last equation follows because $\Omega \propto R^{-1}$ in a Mestel disk. In this example, wave packets wind up at a rate comparable to material

---

[14] In a gas disk the wave would propagate straight through the resonance and eventually reach the the center or edge of the disk.

arms: waves outside corotation ($\Omega < \Omega_p$) wind up faster while waves inside corotation ($\Omega > \Omega_p$) wind up slower. In most disks the winding rate is slowest for wave packets inside corotation; these packets typically wind up in a few rotation times, at rates comparable to the winding rates for the kinematic density waves of §6.2.1 (see Toomre 1969 and 1977a for examples).

In this example, any tightly wound disturbance, whether leading or trailing, will wind up and disppear into a Lindblad resonance within a few rotation times. This suggests that such a disk would be unable to sustain *any* tightly wound spiral wave pattern.

However, the behavior of galactic disks can be far more complex than this simple example would suggest. The principal reason is that the WKB approximation usually fails badly for waves on the long branch of dispersion relation. The reason for this failure is clear from the pitch angle of the waves. As a convenient reference, we use the parameter X defined in equation (6-59). Figure 6-14 shows that in general $|k| \lesssim k_{\mathrm{crit}}$ for waves on the long branch; hence for these waves $|kR/m| \lesssim X$, and equation (6-14) then shows that the WKB approximation fails for the *entire* long branch unless $X \gg 1$. In the solar neighborhood, for example, we have seen in §6.2.3 that $\lambda_{\mathrm{crit}} \simeq 10\,\mathrm{kpc}$. This is probably not large enough to invalidate the WKB analysis of axisymmetric disturbances, since the most unstable axisymmetric wavelength is only $0.55\lambda_{\mathrm{crit}} \simeq 6\,\mathrm{kpc}$. However, almost all of the long waves of the non-axisymmetric dispersion relation have wavelengths that exceed λ_{crit}, and so the accuracy and even the physical reality of the long wave branch of the WKB dispersion relation are very doubtful. This difficulty is reflected in the value of X, which is $\simeq 5/m$ in the solar neighborhood: thus even when $m = 2$, the condition $X \gg 1$ is marginal, and for $m > 2$ it is clearly not satisfied.[15]

Thus the fate of tightly wound leading waves is far less certain than the fate of trailing waves. A trailing disturbance becomes ever more tightly wrapped, thus fulfilling the requirements of the WKB approximation better and better. A tightly wound leading disturbance unwinds and inevitably becomes a loosely wound disturbance whose evolution cannot be followed within the context of the WKB approximation. To investigate the fate of leading disturbances and their influence on disk stability, we must turn to numerical experiments.

6.3 Global Stability of Differentially Rotating Disks

In the last section we studied the way in which tightly wound spiral

---

[15] There are, of course, disk systems other than galaxies in which the WKB approximation works extremely well. For example, in Saturn's rings, $X \approx 10^7/m$.

disturbances propagate through galactic disks. We derived the relation between frequency and wavenumber of such disturbances, from which we deduced Toomre's criterion for local stability, and the group velocity with which these disturbances propagate.

Unfortunately, the WKB analysis does not give a complete picture of disk dynamics, because it does not apply to loosely wound structures. There are no analytic methods that can determine the stability of a general galactic disk to arbitrary perturbations. Hence we begin this section with a description of numerical experiments on disk dynamics. We shall find that many of the results of these experiments can be understood by treating the disk as a resonant cavity, within which WKB disturbances rattle to and fro.

1 Numerical Work on Disk Stability

A wide variety of N-body simulations of galactic disks have been carried out since 1970. Both direct N-body codes and fast Fourier transform (FFT) codes (see §2.8) have been used, although FFT codes are usually the method of choice for simulating collisionless disks.

One of the earliest and most influential studies was carried out by Hohl (1971) using a 100,000 body FFT code on a 256×256 grid. The initial surface density and potential of the 100,000 stars were those of a Kalnajs disk [eqs. (4-167) and (4-166)]:

$$\Sigma(R) = \Sigma_0 \sqrt{1 - \frac{R^2}{a^2}} \quad ; \quad \Phi_0(R) = \tfrac{1}{2}\Omega_0^2 R^2, \qquad (6\text{-}62)$$

where $\Omega_0 = \sqrt{\tfrac{1}{2}\pi^2 G\Sigma_0/a}$. The epicycle frequency is therefore $\kappa = 2\Omega_0$. Hohl chose the initial DF of the stars to be the Schwarzschild DF [eq. (6-44)]. The radial velocity dispersion $\sigma_R(R)$ was chosen so that $Q(R) = 1$ at all radii, the tangential dispersion $\sigma_\phi(R)$ was set equal to $\sigma_R(R)$ in order to satisfy equation (4-52), and the mean rotation speed $\overline{v}_\phi(R)$ was chosen to satisfy the asymmetric drift equation (4-33). These conditions yield a mean rotation speed that increases linearly with radius,

$$\overline{v}_\phi(R) = 0.808\Omega_0 R. \qquad (6\text{-}63)$$

Thus the disk is initially in rigid rotation at about 80% of the circular angular speed.

This DF is not an exact equilibrium solution of the collisionless Boltzmann equation [the exact solution is the Kalnajs DF (4-168)], but

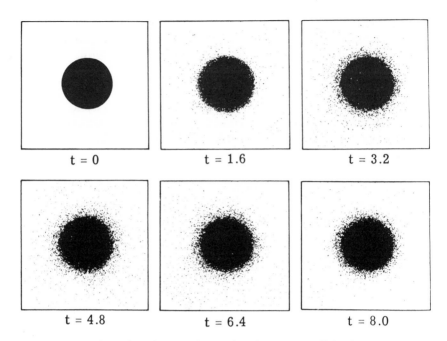

Figure 6-15. Initial evolution of a uniformly rotating disk of 100,000 stars with $Q = 1$ that is constrained to remain axisymmetric. The time unit is the constant rotation period of the stars in the initial disk. Reprinted from Hohl (1971), by permission of *The Astrophysical Journal*.

it avoids the unphysical—but integrable—singularity in the Kalnajs DF and more closely resembles the velocity distribution in the solar neighborhood. To obtain an exact equilibrium, Hohl simply ran his program for several orbital times while constraining the gravitational field to remain exactly radial. The resulting evolution of the spatial distribution of the stars is shown in Figure 6-15. The time unit is the initial circular orbital period, $2\pi/\Omega_0$. Apart from a blurring of the sharp outer edge of the initial distribution, there is rather little change. In particular, there is no visible change between $t = 3.2$ and $t = 8$, suggesting that the disk has settled into equilibrium. The velocity dispersion of the stars does not rise noticeably, so that Q remains near unity, and there is no sign of any instability. These results suggest that in this case Toomre's *local* stability criterion $Q > 1$ is also sufficient for *global* stability to axisymmetric modes.

At $t = 8$, with the disk now in equilibrium, Hohl removed the constraint that the gravitational field should remain radial. The resulting evolution, shown in Figure 6-16, is dramatically different. In less than two rotations, the disk evolves into a barlike structure. At later times

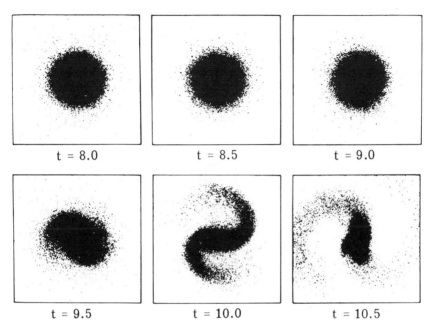

t = 8.0   t = 8.5   t = 9.0

t = 9.5   t = 10.0   t = 10.5

Figure 6-16. Evolution of the disk in Figure 6-15 without the constraint that the disk remain axisymmetric. The initial condition is the disk in Figure 6-15 at $t = 8.0$. Reprinted from Hohl (1971), by permission of *The Astrophysical Journal*.

the model settles into a nearly axisymmetric disk with large random velocities surrounding a slowly rotating oval structure. There is evidently a strong instability to bar formation, which was not predicted by the local analysis of the previous section.

Hohl's results are consistent with the analytic study of the normal modes of the Kalnajs disks reported in §5.3.3. We found there that the Kalnajs disks are unstable to a bar mode if $\Omega > 0.507\Omega_0$ [eq. (5-151)].[16] The investigations by Kalnajs and Hohl are mutually reinforcing: Kalnajs's analytic work shows that the strong bar instability found by Hohl is not likely to be a numerical artifact, while Hohl's work shows that the instability is not connected with the singularity in the Kalnajs DF.

Hohl also examined whether the bar instability was present in a differentially rotating disk. He constructed an axisymmetric disk from

---

[16] Axisymmetric ($m = 0$) instabilities are also present in some of the Kalnajs disks in Figure 5-5. In rapidly rotating disks these arise because $Q < 1$, while in slowly rotating Kalnajs disks they are probably connected with the integrable singularity in his DF; thus neither instability would be present in a model like Hohl's.

the final state of his first simulation, by redistributing the stars at random in longitude. The new disk was axisymmetric and exhibited strong differential rotation and large random motions (typically, $Q \approx 4$). Hohl evolved this disk for six rotations and found it to be completely stable.

The crucial difference between this stable disk and the earlier unstable disk is the velocity dispersion. Hohl's results suggest that disks with large velocity dispersion ("hot" disks) are not subject to a bar instability, while cold disks are unstable. To test this conjecture, Hohl cooled the stable, hot disk by choosing 10% of the stars at random each rotation and placing them on circular orbits. He found that after about three rotations the bar instability reappeared and began to heat the disk. Further cooling only caused the instability to heat the disk as fast as it was being cooled; quantitatively, Toomre's Q could never be reduced below about 2.5.

Ostriker and Peebles (1973) confirmed and extended Hohl's results on the importance of the bar instability in differentially rotating disks. They studied a direct 300-body simulation of a truncated Mestel disk with $Q = 1$, and found, like Hohl, that their model was "rapidly and grossly unstable" to a bar mode. Two particularly important new conclusions emerged from the Ostriker-Peebles experiment:

(i) The models were parametrized by the dimensionless number $t = T/|W|$, where T is the kinetic energy of rotation and W is the potential energy. Introducing also the kinetic energy in random motions $\frac{1}{2}\Pi$ [see eq. (4-74)], we have from the virial theorem $T + \frac{1}{2}\Pi = -\frac{1}{2}W$ or $\Pi/T = t^{-1} - 2$. Since $\Pi/T > 0$ we have $0 \leq t \leq \frac{1}{2}$. Ostriker and Peebles noticed that their N-body models were stable if and only if $t < t_{\mathrm{crit}} = 0.14 \pm 0.02$ and that this limit was similar to the stability criterion for barlike modes in Kalnajs disks, $t < \frac{1}{2}\Omega^2/\Omega_0^2 = 0.1286$ (see §5.3.3.). Hence they proposed that $t \lesssim 0.14$ or $\Pi/T \gtrsim 5$ was a necessary condition for stability of any stellar system to barlike modes. We shall see below that in fact the physics of the bar instability is only indirectly related to $T/|W|$. Nevertheless, this **Ostriker-Peebles criterion** provides a surprisingly useful empirical guide for identifying systems that are likely to be unstable.

(ii) What is more important, Ostriker and Peebles were the first to stress the grave consequences of the bar instability for our own Galaxy and other disk galaxies. In a column through the solar neighborhood, the RMS random velocity of the late-type dwarfs that dominate the stellar mass contribution is about $60\,\mathrm{km\,s^{-1}}$ (Wielen 1977), compared to a rotation speed of about $220\,\mathrm{km\,s^{-1}}$. If these values are roughly constant across the disk, then $\Pi/T \simeq 0.15$, much less than the value of 5.1 needed for stability according to the Ostriker-Peebles criterion. They argued that the most promising way to stabilize the Galaxy was to add an unseen component with $\Pi/T \gg 1$: either an invisible spherical component (a "massive halo")[17] or a "hot disk" that was either invisible or else concentrated well inside the solar radius. Thus they made the bold hypothesis that the Galaxy might require a major, previously undetected component, possibly containing even more mass than the visible disk, in order to be stable against bar formation. (See §10.1.7 for an assessment of the current status of this hypothesis.)

By now N-body simulations of a wide range of disk models have been carried out (e.g., Sellwood 1981; Zang & Hohl 1978; Efstathiou et al. 1982). These simulations have been supplemented by linear normal mode calculations, notably by Zang (1976) and Kalnajs (1978). The simulations confirm the two main conclusions of Hohl's classic paper:

(i) Toomre's *local* stability criterion $Q > 1$ is generally a fairly accurate predictor of stability to axisymmetric modes of all wavelengths.

(ii) If most of the kinetic energy of a disk is in rotational rather than random motion, then the disk is usually strongly unstable to large-scale barlike modes.

The second conclusion leads directly to the question: why are disk galaxies apparently stable? This issue has dominated the study of disk dynamics since the early 1970s. Its importance for our understanding of disk galaxies is difficult to overemphasize. From a pessimist's point of view, it represents a major obstacle, because it is impossible to investigate almost any aspect of disk dynamics with confidence without first understanding why the disk is stable. The optimist's point of view is that the bar instability may provide a clue to the existence of a previously unrecognized component of disk galaxies, which may even contain most of the mass of the galaxy.

2 Swing Amplifier and Feedback Loops

A qualitative understanding of the dynamics of the bar instability is gradually emerging. Remarkably, it appears that most features of the instability can be understood by augmenting the WKB dispersion relations we have derived already with two new physical concepts: feedback loops and the swing amplifier.

---

[17] A similar halo was invoked by Kalnajs (1972b) to stabilize the bar mode in his disk models, but only as a mathematical device rather than a plausible component of real galaxies.

(a) The swing amplifier   In §6.2.4 we argued that any leading disturbance in a disk inevitably unwinds. If the parameter X defined by equation (6-59) is of order unity or smaller, the WKB approximation is unable to follow the evolution of the disturbance once it reaches the long branch of the dispersion relation. In this section we use numerical experiments to follow the evolution of loosely wound leading disturbances.

Following Toomre and Zang (Toomre 1981) we consider a stellar Mestel disk with constant Q. We assume that a fraction $f/(1+f)$ of the total radial force on the unperturbed disk arises from a rigid, fixed "halo" component; thus the disk surface density $\Sigma(R) = \Sigma_0 R_0/R$ and the circular angular speed Ω are related by [see eq. (2-164)]

$$\Omega^2 = \frac{2\pi G \Sigma_0 R_0}{R^2}(1+f). \qquad (6\text{-}64)$$

Toomre and Zang use linear perturbation theory to follow numerically the evolution of a leading wave packet in a disk with $Q = 1.5$, $f = 1$. The evolution is shown in Figure 6-17. Not surprisingly, within four rotation periods (of a particle at the initial corotation radius) the wave unwinds into a relatively open pattern (frame 3) and then into a trailing pattern that becomes more and more tightly wound (frame 9). The remarkable and striking feature of these calculations, however, is that the amplitude of the trailing wave in frame 9 is about twenty times larger than the amplitude of the initial leading wave in frame 1, and that at intermediate stages (frames 4, 5, and 6) an even stronger *transient* spiral pattern is formed.

These results are a manifestation of **swing amplification**, a phenomenon that is not captured by the WKB approximation.[18] The reason for the failure of the WKB approximation in this disk is clear: for a Mestel disk obeying equation (6-64), using the relation $\kappa = \sqrt{2}\Omega$ [see eq. (3-59)], we find

$$X = \frac{2}{m}(1+f). \qquad (6\text{-}65)$$

For the disk used by Toomre and Zang, $X = 2$. With such a small value of X, the WKB approximation is invalid throughout the long branch of the dispersion relation.

To follow the evolution of a wave packet without the WKB approximation generally requires numerical programs like the one used

---

[18] Mark (1976) has shown that a weak version of swing amplification (giving amplifications of up to a factor of two) is present in the WKB approximation if Q is very close to unity. Mark's results have an elegant physical interpretation in terms of tunneling across the forbidden region at corotation. See also Goldreich and Tremaine (1978).

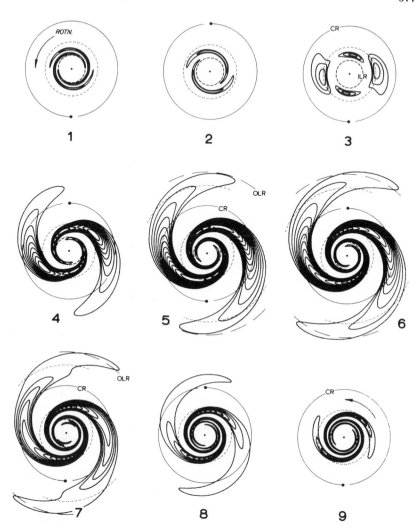

Figure 6-17. Evolution of a packet of leading waves in a stellar Mestel disk with $Q = 1.5$ and $f = 1$. Contours represent fixed fractional excess surface densities; since the calculations are based on linear perturbation theory, the amplitude normalization is arbitrary. Contours in regions of depleted surface density are not shown to minimize confusion. The time interval between diagrams is one-half of a rotation period at corotation. From Toomre (1981), by permission of Cambridge University Press.

by Toomre and Zang.[19] However, there is a simple heuristic explanation that suggests the basic reason for the amplification (Toomre 1981). Consider a material arm described by equation (6-1). For the purposes of this argument we redefine the quadrant of the pitch angle so that $0 < i < 90°$ for trailing arms and $90° < i < 180°$ for leading arms. Thus equation (6-3) is written as

$$\cot i = -Rt\frac{d\Omega}{dR} = 2At, \tag{6-66}$$

where $A = -\frac{1}{2}R(d\Omega/dR)$ is a generalization of Oort's A constant [eq. (3-61a)]. The rate of change of pitch angle is

$$\frac{di}{dt} = \frac{2A}{1 + 4A^2t^2}. \tag{6-67}$$

When the arm is tightly wound, its rotation rate di/dt is slow, but as it swings from leading to trailing it reaches a maximum rotation rate of $2A$. This maximum rate of rotation is comparable to the average angular speed of stars around their epicycles, κ (for a Mestel disk, $2A = \Omega$ and $\kappa = \sqrt{2}\Omega$). Moreover, both the unwinding of the arm and the rotation of the stars around their epicycles are in the same sense, opposite to the direction of rotation (see Figure 6-18). Thus, for corotating stars, there is a temporary near-match between the epicyclic motion and the rotating spiral feature that enhances the effect of the gravitational force from the spiral on the stellar orbit—and the contribution of the star's own gravity to the spiral perturbation. This enhancement can lead to rapid growth in the strength of the arm over the interval of about one radian when the arm is most open. An alternative view that leads to the same conclusion is based on transforming to the frame that rotates with the arm: in this frame there is a temporary cancelation of the stabilizing influence of the Coriolis force, which leads to a transitory gravitational instability in the arm. A quantitative version of this argument is presented by Toomre (1981).

Strong amplification requires that Q is not too large—in order to assure that the disk is susceptible to gravitational instability. Numerical results show that the amplification factor during the swing from leading to trailing is extremely sensitive to the value of Q when Q is near unity.

---

[19] An alternative method—by which swing amplification was originally discovered by Goldreich and Lynden-Bell (1965) and Julian and Toomre (1966)—is to carry out a short-wavelength analysis based on the assumption that both the azimuthal wavenumber m and the radial wavenumber k are large, but that the pitch angle $i = \operatorname{arccot}|kR/m|$ is not (see also Goldreich & Tremaine 1978).

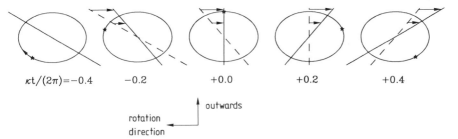

Figure 6-18. Schematic diagram showing the reason for swing amplification. The panels show the motion of a star in its epicycle and the motion of an unwinding material arm in a Mestel disk. The dashed and solid straight lines show the movement of the material arm between panels. The interval between panels is 0.2 times the epicycle period. Note the temporary similarity of angular speeds of the arm and the star: between $\kappa t/(2\pi) = -0.2$ and $+0.2$ the arm and the star swing around at roughly the same rate, so that the gravitational field of the arm can steadily attract the star.

For example, in Mestel disks there is usually an increase of a factor of five in the amplification as Q is decreased from 1.5 to 1.2. Strong amplification also requires that X is not too large. If $X \gg 1$, the wave will propagate away from corotation on the long branch of the dispersion relation before swinging through $i = 90°$, so that corotating stars are not coupled to the swinging spiral field. Figure 6-19 shows the gain of the swing amplifier as a function of the parameters X and Q, in a disk with a flat rotation curve. This figure suggests that $X \lesssim 3$, $Q \lesssim 3$ are necessary and sufficient conditions for strong swing amplification in such disks.

(b) Feedback loops   Swing amplification of a single leading disturbance is not sufficient by itself to destabilize a galactic disk. However—as anyone who has set up a public address system knows—a strong amplifier together with a small degree of feedback from output to input can give rise to a rapidly growing instability. Thus, any mechanism that can turn trailing waves into leading waves is liable to initiate instability in a disk where the swing amplifier works well.

There are at least two possibilities for trailing→leading feedback:

(i) Suppose that the disk has a sharp outer edge that lies outside the forbidden zone around corotation but inside the outer Lindblad resonance. Trailing waves that approach this edge can reflect off the edge, in the same way that waves reflect off the end of a hanging chain or an organ pipe. The reflection reverses the sign of the wavevector k and hence reflects the trailing waves into leading waves

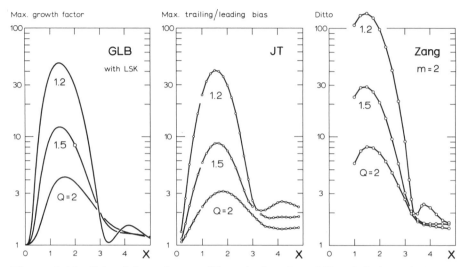

Figure 6-19. Gain of the swing amplifier as a function of X and Q, for a disk with a flat rotation curve. These calculations are based on the Goldreich and Lynden-Bell (1965) model of a local patch of a shearing gas sheet (left); the Julian and Toomre (1966) model of a shearing sheet of stars (center); and the Zang (1976) model of a Mestel disk (right). Note that all three models give similar results. Reprinted from Toomre (1981), by permission of Cambridge University Press.

with the same wavelength. This is perhaps the simplest example of feedback, but it is probably not present in galactic disks because their outer edges are not sufficiently sharp.

(ii) In some cases the disk may have no inner Lindblad resonance, because the maximum value of $\Omega - \frac{1}{2}\kappa$ is less than the pattern speed. For example, in the isochrone potential, Figure 6-10b shows that there is no inner Lindblad resonance when $\Omega_p > 0.0593$. If there is no inner Lindblad resonance, waves can propagate right in to the center of the disk. It can be shown that a trailing wave propagating into the center emerges as a leading wave propagating outward (see Figure 6-20 for a heuristic argument in support of this statement).

(c) Physical interpretation of the bar instability   These results suggest a compelling physical interpretation of the strong bar instability that is so common in numerical simulations of differentially rotating disks. Any minor leading disturbance unwinds and is then swing amplified into a short trailing disturbance, which propagates through the disk center and emerges as a short leading disturbance, which then unwinds and is amplified further.

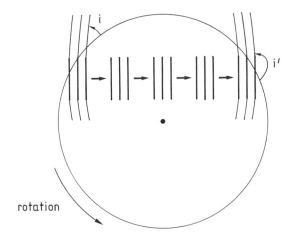

Figure 6-20. A graphical argument that suggests why trailing waves that propagate through the center of a disk emerge as leading waves. A small patch of three incoming trailing waves with inclination $i < 90°$ is shown on the left. The patch propagates through the center as a plane wave and emerges with a pitch angle $i' = 180° - i$. Since $i' > 90°$ the merging wave is leading.

One striking clue that supports this point of view is shown in Figure 6-21. The figure shows an $m = 2$ unstable mode for a Gaussian disk, with surface density $\Sigma(R) = \Sigma_0 \exp(-\frac{1}{2}R^2/R_0^2)$. A fraction $f/(1 + f)$ of the radial force on the unperturbed disk arises from a rigid "halo" component; $f = 0$ in the model on the left and $f = \frac{1}{2}$ on the right. The details of the disk model are described in Toomre (1981); the main feature to notice here is the "lumpy" structure of the mode, which is much more prominent in the right panel. This lumpy structure is naturally explained as the result of interference between leading and trailing waves of nearly equal amplitude, propagating through the disk center. In the model on the left, which has a large growth rate, the swing-amplified trailing waves have larger amplitude than the leading waves, so that the mode looks like a trailing spiral even though a weak leading wave is present. In the model on the right, which has a much lower growth rate because of the rigid halo, the amplitudes of the leading and trailing waves are nearly equal, so that the interference pattern is much more pronounced.

In practice, a galaxy disk can be stabilized by one or more of the following procedures:

(i) Increase Q. Large random motions reduce the susceptibility of a

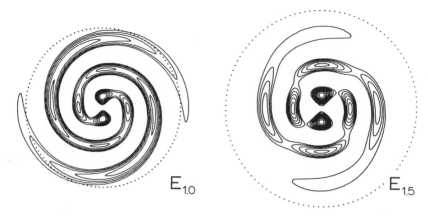

Figure 6-21. Shapes of unstable modes in a Gaussian disk (from Toomre 1981). Left: no rigid halo; right: one-third of the central force arises from a rigid halo. The growth rate is 17% of the pattern speed in the first case but only 3% of the pattern speed in the second. This calculation used a cold disk and softened gravity to model a disk with non-zero Q (see Problem 6-5). Dots mark the corotation circle. Reprinted by permission of Cambridge University Press.

disk to gravitational instability and thus inhibit the swing amplifier. Figure 6-19 shows that for a disk with a flat rotation curve, $Q \lesssim 3$ is necessary for strong swing amplification. The "hot" axisymmetric disk constructed by Hohl from the debris of his original bar-unstable disk was stable because $Q \simeq 4$.

(ii) Increase X. The presence of a halo or some other hot component that does not respond to the perturbations will lead to more tightly wound waves. For example, in a disk with a flat rotation curve, Figure 6-19 shows that $X \lesssim 3$ is needed for strong swing amplification. Equation (6-65) shows that in such a disk, $X < 3$ for an $m = 2$ wave if more than $\frac{1}{3}$ of the equilibrium radial force arises in the disk.

(iii) Cut off the feedback. The commonest feedback loop involves the propagation of waves through the disk center, which requires that there is no inner Lindblad resonance. Thus a mode can be stabilized by a readjustment of the mass distribution in the inner disk so as to increase the central angular speed and thus create an inner Lindblad resonance.

Which of these mechanisms is the explanation for the stability of disk galaxies? We do not know. It is possible that the stars in the inner parts of galactic disks have sufficiently high random motions to stabilize the disk by increasing Q. Alternatively, there may be a massive halo that contains a mass comparable to the disk mass, thereby increasing X. Both these possibilities, proposed by Ostriker and Peebles (1973), have profound implications because they imply the existence of a major new component of disk galaxies. A third possibility, which requires a less drastic revision of our views of galactic structure, is that galactic rotation curves have an inner Lindblad resonance for all pattern speeds at which there is substantial swing amplification. Toomre (1981) suggests an appealing scenario in which the strong spiral perturbations generated in an unstable galaxy rearrange the galaxy mass, shifting material inward and thereby raising $\Omega - \frac{1}{2}\kappa$ until a Lindblad resonance appears and the instability is cut off. In this scenario the bar instability would be self-regulating.

3 Summary

(i) Analyses based on the WKB approximation show that disks are stable to short-wavelength perturbations if $Q > 1$ [eqs. (6-49) and (6-53)]. Numerical experiments suggest that $Q > 1$ implies stability to *all* axisymmetric perturbations, but not to non-axisymmetric ones.

(ii) In most disks, the strongest non-axisymmetric instability is an $m = 2$ barlike instability. The instability appears to be due to a feedback loop in which trailing waves propagate through the galactic center, emerging as leading waves that are swing amplified into stronger trailing waves.

(iii) A disk that is stable to the $m = 2$ mode generally has either: (a) large random motions, possibly supplied by a "hot disk" component in the inner regions; (b) a halo containing a mass comparable to the disk's; or (c) a rotation curve that has an inner Lindblad resonance for all pattern speeds at which strong swing amplification can occur (i.e., there is a sharp peak in $\Omega - \frac{1}{2}\kappa$ at small radii). We do not yet know which of these features is most responsible for the stability of real galaxy disks.

6.4 Theories of Spiral Structure

Newton's equations of motion and the law of gravitation are time-reversible. Thus, if we make a movie of the trajectories of N point masses interacting through their mutual gravity, the trajectories seen when we run the movie backward are also dynamically possible. Similarly, if the dynamics of a spiral galaxy are governed by Newton's equations, and the galaxy is in a steady state, then a time-reversed movie of the galaxy should also represent a possible steady-state solution of the equations. However, time reversal changes trailing spirals into leading spirals, by changing the sign of all velocities without changing the instantaneous luminosity density. This argument is the basis of the **anti-spiral theorem**, which states that *if a steady-state solution of a time-reversible set of equations has the form of a trailing spiral, then there must be an identical solution in the form of a leading spiral* (Lynden-Bell & Ostriker 1967).

The anti-spiral theorem implies that, in contrast to the stellar systems discussed in Chapter 4, spiral galaxies cannot be understood simply as steady-state solutions of the collisionless Boltzmann equation and Newton's law of gravity—the prevalence of trailing spirals implies that there is an additional ingredient in the explanation. The most likely causes are (i) the spirals are not in a steady state (for example, they may be growing in amplitude because they arise from an unstable normal mode, or they may be a result of a recent disturbance); or (ii) the spiral form is influenced by processes that are not time-reversible (such as dissipation in the interstellar gas).

1 The Lin-Shu Hypothesis

The Lin-Shu (1964, 1966) hypothesis is that spiral structure consists of a quasi-steady density wave, that is, a density wave that is maintained in a steady state over many rotation periods. Lin and Shu did not at first address the question of the origin of the spiral structure, but the most natural theory is that the spiral pattern is the most unstable normal mode of the galactic disk. As the wave amplitude builds, energy dissipation in the interstellar medium leads to damping [see Kalnajs (1972a)]. The damping rate increases as the wave amplitude increases, so that eventually the wave reaches a stable, finite amplitude, which is the state in which spiral galaxies are found today.

This scenario is appealing because it provides a number of concrete theoretical predictions for comparison with observations. For example, the prevalence of trailing spirals in real galaxies [§6.1.1(d)] is a consequence of the swing amplifier, which amplifies density waves as they swing from leading to trailing and thus ensures that the trailing arms in the feedback cycle always have the larger amplitude. The prevalence of two-armed spirals is in part due to the fact that the region between the Lindblad resonances where waves can propagate is larger for $m = 2$

than for higher m. In addition, $m = 2$ waves are most likely to be unstable, as a result of a competition between two effects: waves with $m = 1$ have twice as large a value of X [eq. (6-65)] and hence swing amplification is much weaker; while waves with $m > 2$ almost always have an inner Lindblad resonance so that feedback through the disk center cannot take place. A third point of comparison with observation is the surface brightness in red light, which is dominated by old disk stars. In this waveband, galaxies usually exhibit smooth broad arms, with relative amplitudes of up to 20% (Figure 6-4). Since these stars are also the dominant stellar contributor to the mass of the disk, the surface brightness is directly related to the surface density. Thus, the presence of a spiral pattern in red light is direct evidence that a real *density* wave is present, i.e., that the whole mass of the disk participates in the spiral wave. In addition, the relatively low amplitude reassures us that the linearized theory that we have developed is an adequate tool for describing real galaxies.

The most striking comparisons of the Lin-Shu hypothesis with observation have come from analysis of the response of the interstellar gas to a density wave. The overall surface density in the disk varies with azimuth roughly as $1 + \epsilon \cos(m\phi)$, where ϵ is the fractional amplitude of the wave. However, the young, bright stars and interstellar gas define much narrower and stronger arms, with gas density contrasts of a factor of two or three between arm and interarm regions. One of the triumphs of the Lin-Shu hypothesis is that it led to the prediction of precisely such narrow, high-density gaseous arms by Fujimoto (1968) and W. W. Roberts (1969). Once these arms can be explained, there is also a natural explanation for the narrow arms defined by the bright young stars: the high gas density triggers rapid star formation, and the lifetimes of the brightest stars are so short that they cannot drift far from their formation sites before they die. Thus the angular width of the arms is determined by the lifetime t_* of the stars created in the arm; the width $\Delta\theta = |\Omega - \Omega_p|t_*$, where Ω is the circular frequency and Ω_p is the pattern speed. Typical values are $\Omega = 2\Omega_p$, $\Omega_p = 10 \text{ km s}^{-1} \text{ kpc}^{-1}$, and $t_* = 2 \times 10^7 \text{ yr}$ (for a $9 M_\odot$ star, corresponding to spectral class B3; see tables 3-6 and 3-9 of MB). We find $\Delta\theta = 12°$, in qualitative agreement with the observed width.[20]

In general, calculations of spiral structure in the interstellar gas like those of Fujimoto and Roberts must be done numerically, since the linear

---

[20] This argument can be inverted to estimate the pattern speed Ω_p, for example, by measuring the widths of OB associations, or by measuring color gradients across the arms (larger associations or redder colors imply older stars). Most studies of this kind suggest that the arms lie inside corotation ($\Omega > \Omega_p$), but the data are not generally good enough to determine Ω_p reliably (see Schweizer 1975).

approximation we have used so far is clearly invalid when high density contrasts are present. However, by invoking some simple approximations we can reduce the problem to a simple analog that is easy to understand.

(a) Response of the interstellar gas to a density wave

The interstellar medium contains only a small fraction of the total mass in the galactic disk (in the solar neighborhood, this fraction is about 10% according to Table 1-1). Hence to a first approximation we can assume that the gas moves in the potential field of the stars alone. We write the potential as the sum of the unperturbed axisymmetric potential $\Phi_0(R)$ and the perturbed potential due to the stellar density wave,

$$\Phi_1(R, \varphi) = \text{Re}[\Phi_a(R)e^{im\varphi}], \qquad (6\text{-}68)$$

where φ is now the azimuthal angle in a frame rotating at the pattern speed Ω_p. We assume that the wave is tightly wound, so that in the neighborhood of radius R_0, we have [see eq. (6-36)]

$$\Phi_a(R) \simeq Fe^{ikR}, \qquad (6\text{-}69)$$

where F is a constant. We can choose the zero point of azimuth so that F is real; thus

$$\Phi_1(R, \varphi) = F \cos(kR + m\varphi). \qquad (6\text{-}70)$$

We shall concentrate our attention on interstellar clouds (see MB §3-11) since these contain the bulk of both the atomic and molecular gas. In a first approximation, the clouds can be regarded as test particles moving in the potential of the stellar disk. Their motion can be calculated using the equations of motion for a star in a weak non-axisymmetric field (§3.3.3); equations (3-114) become

$$\ddot{R}_1 + \left(\frac{d^2\Phi_0}{dR^2} - \Omega_0^2\right) R_1 - 2R_0\Omega_0\dot{\varphi}_1 = kF \sin(kR + m\varphi), \quad (6\text{-}71\text{a})$$

$$\ddot{\varphi}_1 + 2\Omega_0\frac{\dot{R}_1}{R_0} = \frac{mF}{R_0^2} \sin(kR + m\varphi), \quad (6\text{-}71\text{b})$$

where Ω_0, R_1, and φ_1 are defined by equations (3-109) and (3-111). Notice that we have not replaced R by its unperturbed value R_0 in the argument of the sines—because the wavenumber k is large, and we want to include the possibility that kR_1 is of order unity.

Since the waves are assumed to be tightly wound ($|kR_0/m| \gg 1$), the tangential force [the right side of equation (6-71b) times R_0] is

a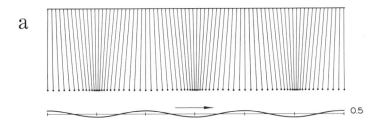

b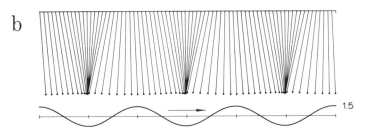

Figure 6-22. A simple model for the response of the interstellar gas. The diagram depicts a row of pendulums exposed to a force field of the form $C\sin(kx + \omega t)$. If we set $\omega = m(\Omega_0 - \Omega_p)$, regard x as radial distance in the galactic disk, and let the natural frequency of the pendulums be the epicycle frequency κ_0, then the pendulums obey equation (6-73). The motion is shown for forcing amplitudes of $0.5g$ and $1.5g$ [eq. (6-75)] and $\omega = 0.75\kappa_0$. All collisions are assumed to be inelastic. From Toomre (1977a), based on unpublished work by A. J. Kalnajs. Reproduced by permission from *Annual Review of Astronomy and Astrophysics*.

smaller than the radial force [the right side of (6-71a)], by the large factor kR_0/m. Hence the right side of (6-71b) can be dropped, and equation (6-71b) integrates immediately to an approximate statement of the conservation of angular momentum

$$\dot{\varphi}_1 + \frac{2\Omega_0 R_1}{R_0} = \text{constant}. \tag{6-72}$$

Since a readjustment of R_0 simply changes R_1 by a constant, we can always choose R_0 so that the constant in (6-72) is zero; then, eliminating φ_1 from (6-71a) and replacing φ in the argument of the sine by its unperturbed value $(\Omega_0 - \Omega_p)t$, we obtain

$$\ddot{R}_1 + \kappa_0^2 R_1 = kF\sin[k(R_0 + R_1) + m(\Omega_0 - \Omega_p)t], \tag{6-73}$$

where the epicycle frequency κ_0 is defined in (3-118b).

A. J. Kalnajs has discussed a simple system that obeys equation (6-73). Consider the endless row of identical pendulums shown in Figure 6-22. Each pendulum acts like a simple harmonic oscillator of frequency κ_0. The equilibrium position of a pendulum is described by the coordinate R_0 and its horizontal displacement from equilibrium by R_1. Each pendulum is subject to a horizontal force per unit mass $kF \sin(kR + \omega t)$. The equation of motion is obviously just (6-73) with $\omega = m(\Omega_0 - \Omega_p)$.

In general, equation (6-73) must be solved numerically for the displacement $R_1(R_0, t)$. If the forcing F is sufficiently strong, adjacent pendulums may collide. A collision takes place if two pendulums whose supports are separated by a small value ΔR_0 occupy the same position, i.e., if $R_0 + R_1(R_0, t) = R_0 + \Delta R_0 + R_1(R_0 + \Delta R_0, t)$. Letting ΔR_0 shrink to zero, we find that the condition for collision is $\partial R_1 / \partial R_0 < -1$. To obtain a crude estimate of the level of forcing that leads to collisions, we revert for the moment to the linear approximation, where kR_1 can be dropped from the argument of the sine. In the linear approximation we find

$$R_1 = \frac{kF}{\kappa_0^2 - \omega^2} \sin(kR_0 + \omega t). \qquad (6\text{-}74)$$

Hence in the linear approximation collisions occur if

$$g \equiv \frac{k^2 F}{|\kappa_0^2 - \omega^2|} > 1. \qquad (6\text{-}75)$$

Of course, this condition has not been derived self-consistently, since it requires that $|kR_1| > 1$ at some point in the cycle, so that the approximation of dropping kR_1 from the argument of the sine is not valid. Nevertheless, numerical experiments with the full nonlinear equation (6-73) show that in many cases the crude criterion $g \gtrsim 1$ provides a fairly accurate criterion for when collisions occur, and g proves to be a useful dimensionless parameter for describing the effects of spiral waves on the interstellar gas.

Figure 6-22 (taken from Toomre 1977a) shows the results of numerical integrations of equation (6-73) with $\omega = 0.75\kappa_0$. In case (a), $g = 0.5$, no collisions occur, and the linear approximation is fairly accurate. The whole pattern travels to the right at speed ω/k. As the forcing is increased, collisions first occur at $g = 0.98$. Case (b) shows the behavior of the row of pendulums at even stronger forcing, $g = 1.5$. The collisions are assumed to be completely inelastic. The most prominent features are the "traffic jams" where several pendulums are in contact. Each pendulum lingers for about 20% of the cycle in the traffic jam and then swings away to the left until it enters the next jam.

The motion of the row of pendulums provides a simple model for the motion of clouds in a spiral density wave. Each pendulum can be regarded as a cloud. The traffic jams correspond to the narrow regions of high density shocked gas found by Fujimoto and by Roberts. Equation (6-73) and Figure 6-22 show convincingly that dense, narrow gaseous arms are a natural consequence of the response of the interstellar gas to a tightly wound density wave.

Other features of the spiral pattern can also be explained by this model. The concentration of dust is also increased in the traffic jams, leading to the dust lanes commonly associated with spiral arms. The gas compression is accompanied by a strong enhancement in the magnetic field strength, since the field lines are frozen into the gas, and this leads to enhanced synchrotron emission from relativistic electrons, just as Figure 6-3 shows in M51. All of these features (the HI arms, the dust lanes, and the radio continuum arms) will be displaced from the optical arms due to the time delay required for star formation. If the wave is inside corotation and trailing, the time delay implies that the optical arms should lie outside the other arms, which is consistent with observations of most spiral galaxies.

We can also test the Lin-Shu hypothesis by comparing the expected HI velocity field with observations. Figure 6-23 shows contours of constant velocity superimposed on a map of the density distribution of HI in M81. The solid lines are observed contours and the symbols mark theoretical contours obtained from a model constructed by Visser (1980). Notice the kinks in the contours as they cross the inner edge of the most prominent spiral arm (the effect is most easily seen near $\alpha = 9^{\rm h}52^{\rm m}, \delta = 69°20'$). The kinks are caused by the discontinuous change in velocity as the gas clouds enter the traffic jam. The similarity of the kinks in the theoretical and observed contours strongly suggests that the gas is passing through traffic jams consistent with those predicted by the Lin-Shu hypothesis.

The picture we have just described is based on a grossly oversimplified model of the interstellar medium (Shu 1978). In reality, most of interstellar space is occupied by a hot $(T \approx 10^6{\rm K})$ ionized medium, which hardly responds at all to the spiral potential because its internal pressure is so high. The interstellar clouds occupy only about 2%–4% of the disk volume (§3-11 of MB). An additional complication is that the clouds have random velocities $v_{\rm rms} \approx 6\,{\rm km\,s}^{-1}$; hence the cloud motion is not perfectly ordered and clouds can collide even outside the traffic jams (the typical time between collisions is $t_{\rm coll} \approx 10^7\,{\rm yr}$). Therefore our model based on the pendulum chain is oversimplified because it implies that collisions occur exclusively in traffic jams; in reality the traffic jams

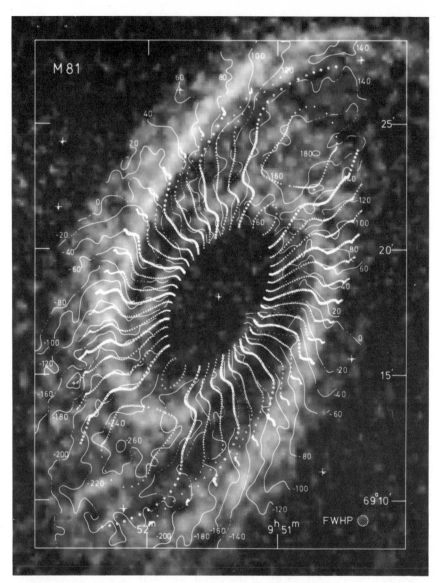

Figure 6-23. Constant-velocity contours of HI in the spiral galaxy M81. Solid lines represent observations made at the Westerbork Synthesis Radio Telescope, while chains of symbols represent predictions of a model based on the Lin-Shu hypothesis (Visser 1980). The contours are superimposed on an artificial photograph in which brightness is proportional to HI column density. The shaded circle at lower right represents the spatial resolution of the observations.

are only regions of greatly enhanced collision rates. A better model would involve treating the clouds as atoms of an imperfect fluid with mean free path $v_{rms}t_{coll} \approx 100\,pc$ and sound speed v_{rms}. In these more realistic models the traffic jam is spread out into a high-density region of small but finite width, corresponding to a shock wave in the imperfect fluid. An additional complication arises from the self-gravity of the gas. In linearized theory we can neglect the self-gravity of the gas because on average the fractional mass in gas is small. However, in the traffic jams the gas and star densities may be comparable and self-gravity may play an important role in determining the strength of spiral structure in the gas.

To sum up, we can demonstrate convincingly that interstellar clouds respond in a strongly nonlinear fashion to the imposed spiral potential from a density wave. The resulting density and velocity fields can be adjusted to resemble closely the 21-cm maps of real spiral galaxies. It seems very plausible that star formation will proceed much more rapidly in regions of high density, thus producing the young stars that delineate the optical arms. The presence of dust lanes and radio continuum arms that lie slightly inside the optical arms can also be explained.

(b) The present status of the Lin-Shu hypothesis   The observational tests that we have described represent a qualified success for the Lin-Shu hypothesis. They clearly establish that in the galaxies studied, the spiral arm is a density wave, and that the behavior of the interstellar gas and the shape of the spiral arms are roughly consistent with the assumption that the pattern is a quasi-stationary density wave. However, other models in which the spiral structure is a transitory density wave, such as the tidal models described below, can often produce equally good agreement with the observations. Moreover, even though the most regular and well-formed "grand-design" spirals are generally chosen for testing the Lin-Shu hypothesis, it has proved to be frustratingly difficult to determine unambiguously the basic parameters of the models, such as the pattern speed or the location and number of the Lindblad resonances.

2 Other Density Wave Theories

(a) Chaotic spiral arms   Other theories concentrate on explaining the more ragged spiral galaxies, which have many short, fragmented

Figure 6-24. The Sb galaxy NGC 2841 (Sandage 1961). Note that no single spiral arm can be traced through more than about 30°. This is the prototype of the class of galaxies for which chaotic spiral arm theories may be appropriate.

arms and no clear two-armed symmetry. A typical example is NGC 2841 (Fig. 6-24). For these galaxies "a swirling hotch-potch of spiral arms is a reasonably apt description" (Goldreich & Lynden-Bell 1965). Although ragged spirals are less dramatic than grand-design spirals, they are more common.

In a chaotic theory the pieces of arms are constantly forming and dying. In the simplest picture, local gravitational instability in the interstellar gas creates a patch of new stars, which are sheared by the differential rotation (with some help from swing amplification) into an appearance of spirality. As time passes, the arm is sheared more and more while the brightest stars die. Both effects lead to the gradual disappearance of the arm fragment. Meanwhile new arms form elsewhere.

We can estimate the lengths of the arm fragments from our earlier analysis of gravitational instability in a gas disk. The most unstable wavelength for axisymmetric perturbations in a gas disk with $Q = 1$ is $\lambda(\text{most unstable}) = \frac{1}{2}\lambda_{\text{crit}} = 2\pi^2 G\Sigma_g/\kappa^2$ [eqs. (6-50) and (6-54)], where Σ_g is the gas surface density and κ is the epicycle frequency. The most unstable wavelength for non-axisymmetric perturbations is similar. A reasonable estimate for the length of a feature is therefore $L \simeq \frac{1}{2}\lambda(\text{most unstable}) = \pi^2 G\Sigma_g/\kappa^2$. Taking values from the solar neighborhood $\Sigma_g = 5\,\mathrm{M_\odot\,pc^{-2}}$, $\kappa = 36\,\mathrm{km\,s^{-1}\,kpc^{-1}}$ (see Tables 1-1 and 1-2), we find $L \simeq 0.2\,\mathrm{kpc}$, in reasonable agreement with the lengths of arm fragments in external galaxies. (This analysis is somewhat oversimplified because it neglects the dynamics of the stars, which may also contribute to the instability.)

One appealing aspect of the theory is that star formation is self-regulating. Cooling and infall of new material reduce Q in the interstellar gas, leading to gravitational instability and star formation. Star formation produces supernovae that heat the interstellar gas, thus raising its velocity dispersion and Q, and inhibiting the subsequent growth of the instabilities that create new stars.

A separate self-regulating mechanism may be present for the stars: infall of gas continually leads to the formation of new stars with low velocity dispersion, thereby reducing the Q of the disk stars. The lowering of Q boosts the gain of the swing amplifier and thus increases the strength of the spiral arms. Stochastic heating of the disk stars by gravitational scattering off the spiral arms (§7.5.2) then increases Q and reduces the susceptibility of the disk to further spiral-making. Since the surface density of the stars is higher than the surface density of the gas, instabilities in the stellar disk tend to produce larger-scale spiral features (since λ_{crit} is bigger). A self-regulating instability in the stellar disk may be able to produce large-scale, open spiral structure similar to that seen in most Sc galaxies (Sellwood & Carlberg 1984).

(b) Tidal arms   Many of the most beautiful spirals have nearby companion galaxies. The best example is M51 (Fig. 6-1), whose companion galaxy NGC 5195 is located near the tip of one of its two main spiral arms. Another example is M81 (Fig. 6-23); its brightest and nearest companion is M82, separated by 40′ in angle or 40 kpc in distance.

Can the spiral patterns in M51 and M81 be caused by a recent encounter with their companions? Toomre and Toomre (1972) investigated the M51 system with this possibility in mind. They modeled each galaxy as a disk of test particles surrounding a central point mass. Their best case (Fig. 6-25) adequately reproduces the outer arms but cannot explain the fact that these arms can be followed more or less continuously right into the galactic center. This shortcoming is probably due to the absence of self-gravity in the test particle disk (Toomre 1981). A tidal model of the M81 spiral has also been constructed by Toomre (1981). In these models the spiral is a density wave; however, the wave is a transitory one rather than the long-lived wave envisaged by Lin and Shu. Tidal models can successfully reproduce most of the features explained by the Lin-Shu hypothesis, such as the location and strength of dust lanes, radio continuum arms and HI arms, and the kinks in HI velocity contours across the arms. However, they cannot account for all spirals because encounters with massive companion galaxies in favorable orbits are not common enough.

3 Other Theories

(a) Magnetohydrodynamic theories   In the 1950s most astronomers suspected that spiral structure was somehow the result of a complicated interaction between the interstellar gas and magnetic field. The following simple energy argument allows us to assess the viability of such theories.

The energy density due to a magnetic field B is $B^2/(8\pi)$ in Gaussian units, or $B^2/(2\mu_0)$ in MKS units. The kinetic energy density in a perturbation that imparts a velocity $\Delta\mathbf{v}$ to gas of density ρ is $\frac{1}{2}\rho\Delta\mathbf{v}^2$. The velocity perturbations associated with spiral arms are of order $|\Delta\mathbf{v}| \simeq 20\,\mathrm{km\,s^{-1}}$. Setting $\rho = 0.042\,\mathrm{M_\odot\,pc^{-3}}$ (the local gas density according to Table 1.1), we find $\frac{1}{2}\rho\Delta\mathbf{v}^2 \simeq 6 \times 10^{-12}\mathrm{erg\ cm^{-3}}$. If spiral structure is caused by magnetic fields, the magnetic energy density must be at least this large; hence $B \gtrsim 1 \times 10^{-5}\mathrm{gauss} = 1 \times 10^{-9}\mathrm{tesla}$. However, Faraday rotation and dispersion measures of pulsars indicate a mean field of only $B \approx 2 \times 10^{-6}\mathrm{gauss}$ (Spitzer 1978), too small by a factor of 5 in the field or 25 in energy density. Thus it appears that the interstellar magnetic field plays no major role in spiral structure.

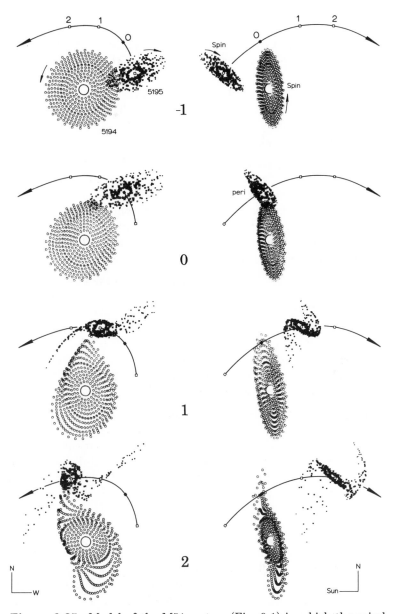

Figure 6-25. Model of the M51 system (Fig. 6-1) in which the spiral arms are caused by a recent passage of NGC 5195. The scenes on the left are viewed from the Sun, and those on the right are viewed from the side. See Toomre and Toomre (1972) and Toomre (1978).

(b) Spiral detonation waves   An alternative wave theory of spiral structure has been proposed by Mueller and Arnett (1976). Their theory is based on the hypothesis that star formation is self-propagating, i.e., that the star-formation rate is enhanced by the presence of young stars nearby. The hypothesis is at least plausible from a theorist's point of view: star formation could be triggered by shock waves driven by supernova explosions or expanding HII regions. There are known to be several examples of prominent star-forming regions close to such objects (B. G. Elmegreen & Lada 1977).

Mueller and Arnett argued that a wave of star formation could propagate through the galaxy, very much as a detonation wave propagates through an exploding stick of dynamite or a fire through a forest. The differential rotation would shear the wavefront into a trailing spiral form. An isolated detonation wave ultimately evolves into a spiral with a fixed shape and pattern speed (Cowie & Rybicki 1982; Balbus 1984). Thus the winding problem is eliminated.

Both Mueller and Arnett and subsequent workers (see Seiden & Gerola 1982 for a review) have concentrated on numerical simulations of self-propagating star formation. They use Monte Carlo codes that divide the galaxy into differentially rotating cells and assign both spontaneous and induced star formation probabilities to each cell. The induced rate depends on whether stars were recently formed in an adjacent cell. The codes include a "regeneration" time, a minimum interval between successive bursts of star formation in a given cell.

The principal difficulty with self-propagating star formation is that good spiral formation requires ad hoc fine-tuning of the induced star-formation rate to within ±10%. Also, it is not clear that this theory can reproduce features like the broad arms seen in the old disk component (Fig. 6-4), the regular patterns seen in galaxies like NGC 5364 (Fig. 6-2), and the sharp velocity variations across spiral arms (Fig. 6-23).

(c) Driving by bars and oval distortions   In barred spiral galaxies the spiral arms usually start at the end of the bar, suggesting that the bar and the spiral are related. Can the bar drive spiral structure without help from density waves in the disk?

We first consider a simple model investigated by Sanders and Huntley (1976). They constructed a uniform disk of gas with no self-gravity. The gas was initially in circular orbits under the influence of a fixed central force field. Then an additional potential due to a rigidly rotating bar was slowly introduced. Once the gas had settled again into a steady

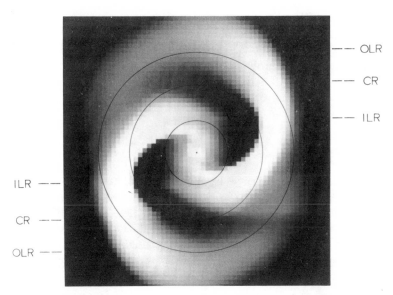

Figure 6-26. Spiral structure driven by a central bar in a differentially rotating gas disk with no self-gravity. The circles marked ILR, CR, and OLR mark the locations of the inner Lindblad resonance, the corotation resonance, and the outer Lindblad resonance. From Sanders and Huntley (1976), by permission of *The Astrophysical Journal.*

state it exhibited the strong trailing spiral pattern shown in Figure 6-26. Note that (i) the gas response is spiral even though the bar forcing is not; (ii) the gas response is very strong (the density is enhanced by a factor of about 2.5 in the spiral arms) even though the bar is rather weak (the maximum tangential force was only 13% of the local axisymmetric force). The spirality of the response is a consequence of viscosity in the gas disk; a simulation with a disk of test particles does not yield a spiral pattern (Sanders 1977). This result is consistent with the anti-spiral theorem: a reversible, steady-state system of test particles can show no preference for leading or trailing spirals, while in the gas disk, viscosity breaks the time-reversal symmetry and leads to a trailing response.

This simulation demonstrates that the formation of spiral arms requires only the presence of a bar and a dissipative interstellar medium. The spiral density waves in the stellar component, which are the heart of Lin and Shu's theory, are not needed. This prompts the question: could all unbarred spiral galaxies have a weak oval distortion in their centers, which is strong enough to drive spiral structure but too weak to appear barlike on photographic plates? The answer is probably no, for two reasons: (i) bar forcing seems to be unable to produce tightly wound spirals; (ii) we have seen, for example in Figure 6-4, that a spiral pattern is often present in the old disk stars as well as in the young stars and gas. Like the tidal models, the Sanders-Huntley model is attractive for some, but not all spiral galaxies.

4 Summary

In the previous subsections we have described plausible models for spiral structure in barred galaxies, galaxies with close companions, and galaxies with chaotic spiral arms. None of these satisfy the Lin-Shu hypothesis, that spiral structure is a quasi-stationary density wave in the stars of the galactic disk. The skeptic might ask at this point, is there any need for quasi-stationary density waves, that is, are there *any* regular, grand-design spirals with neither bars nor companions?

This question has been studied by Kormendy and Norman (1979), who examined 54 spiral galaxies with published rotation curves. They found that 25 had bars, 8 more had close companions (including M51 and M81; Figs. 6-1 and 6-23), and 9 of the remainder had no clear global spiral pattern (including NGC 2841, Fig. 6-24). The remaining 12 galaxies (including M33; Fig. 6-2) are candidates for quasi-stationary density waves. However, the spirals in this class are disappointingly ragged compared to some of the galaxies with bars and companions. In addition, 10 of the 12 galaxies exhibit almost rigid rotation throughout the spiral pattern; often the coherent pattern ends where differential rotation begins. Thus the winding problem is not severe, so that these spirals might represent material arms rather than density-wave patterns.

To summarize, modern spiral structure theory offers a number of mechanisms that can explain much of the wide variety of spiral structure that is seen in different galaxies. The common thread of several of these mechanisms is that *because of the swing amplifier, galactic disks respond with remarkable vigor to a wide variety of perturbations*, whether these be tidal forces, gravitational instability of some local patch of gas or stars, or fresh leading density waves. In some cases there is clear evidence that Lindblad's original conception of the spiral arm as a density wave is correct. However, there is little or no direct evidence for the hypothesis that the spiral pattern is stationary (i.e., that it will look the same in 10^9 yr or so). Thus, ironically, the Lin-Shu hypothesis, which has proved to be so fruitful in advancing our understanding of disk dynamics, may still prove to be largely irrelevant to spiral structure.

6.5 Bars

Roughly half of all disk galaxies contain bars. These structures are among the most common examples of triaxial stellar systems. They are the only triaxial systems whose principal axis directions can be determined directly (since the plane of two of the axes lies in the surrounding disk). Consequently, the study of bars is closely linked to the study of triaxial elliptical galaxies. This section reviews our current—and still seriously incomplete—understanding of the dynamics of bars. Many of the properties of triaxial systems have already been presented in Chapters 3 and 4, and only the most relevant ones will be collected and summarized here. We begin with a brief review of the observations (de Vaucouleurs & Freeman 1972; Kormendy 1981, 1982).

It may be that our own Galaxy is the nearest barred spiral, but, paradoxically, it is much more difficult to identify a bar in our Galaxy than it is in many other galaxies. Part of the difficulty is that the Sun is located near the galactic plane, and the characteristic non-axisymmetric structure of a bar cannot be seen in any edge-on galaxy. In addition, dust in the galactic plane obscures our direct view of the central regions of the Galaxy, so that we cannot identify a bar through its effects on stellar kinematics. The best evidence comes from kinematic features in 21-cm maps of the central regions of the Galaxy (W. W. Roberts 1979; Liszt & Burton 1980). In particular, at longitude $\ell = 0$ the HI radial velocity would be zero if the gas were on circular orbits; in fact, the gas exhibits streaming motions with line-of-sight velocities up to $\pm 200\,\mathrm{km\,s^{-1}}$ (Oort 1977). These velocities could be the result of a recent explosion in the galactic center, or, more plausibly, they may be evidence for radial streaming motion in a bar.

The next two nearest galaxies, the Large and Small Magellanic Clouds, are barred irregular galaxies. The nearest giant spiral galaxy, M31, contains an oval distortion which B. Lindblad interpreted as a bar (see Stark 1977), although most observers still classify M31 as a normal rather than a barred spiral. Figure 6-27 contains photographs of two other barred galaxies.

Bars are generally quite elongated. The axis ratios in their equatorial planes range from about 2.5:1 to 5:1. Their thickness is less well determined since it is difficult to recognize an edge-on barred galaxy. However, since half of all edge-on disk galaxies are presumably barred, we can set some limits from photometry of edge-on disks. It appears that bars are quite flat, possibly as flat as disks.

Bars are not very centrally condensed. The surface brightness along the major axis is nearly constant, while the surface brightness falls off

Figure 6-27. Two examples of barred galaxies. Top: NGC 2523, classified SBb. Bottom: NGC 5383, classified SBb. Note the two nearly straight dust lanes parallel to the bar in NGC 5383. Similar dust lanes are visible in NGC 1300 (Fig. 6-2). Photographs from Sandage (1961).

sharply along the minor axis. The bar emits up to about $\frac{1}{3}$ of the galaxy's total luminosity. Since the bar is concentrated toward the center of the galaxy, it makes an even larger relative contribution to the luminosity (and hence presumably the mass) of the inner region.

Because the bar is straight, the bar pattern is presumed to rotate rigidly, at some pattern speed Ω_b. In the frame in which the pattern is stationary, there may be streaming motions of the stars and gas inside the bar, and stars can flow into or out of the bar.

Late-type barred spiral galaxies exhibit complicated gas motions in the vicinity of the bar. Many bars contain long straight dust lanes that are displaced toward the leading edge of the bar (on the assumption that the spiral arms trail and that the pattern speed of the bar and the spiral are the same). Examples of this phenomenon are seen in NGC 1300 (Fig. 6-2) and NGC 5383 (Fig. 6-27). Many barred galaxies contain a rather narrow, well defined ring near the outer edge of the bar (e.g., NGC 2523 in Fig. 6-27).

Apart from the presence of the bar there are few systematic differences between barred spiral and S0 galaxies and their unbarred counterparts. In particular, the rotation curves of barred galaxies are similar to those of other disk galaxies. The rotation speed rises linearly out to a "turnover" radius R_t and is constant beyond R_t. The length of the bar is generally close to R_t, so that bars appear to be associated with regions of nearly uniform angular speed.

One major difference between bars and elliptical galaxies is that bars are flat. Thus, most of the three-dimensional galaxy models that we discussed in Chapter 4 cannot be used as bar models. Two-dimensional models, in which the bar lies entirely in a plane, are more appropriate as simple models of bars.

1 Bars as Stellar Systems

(a) Weak bars   In Chapter 3 we investigated orbits in a rotating, nearly axisymmetric gravitational field. That analysis provides a useful first approach to the study of bar dynamics, even though the deviation from axisymmetry in most barred galaxies is too large to be treated by perturbation theory.

In a nearly axisymmetric potential, almost all of the stars are on loop orbits sired by the nearly circular, closed, loop orbit described by equation (3-120a) with $C_1 = 0$, or, in a continuum description, by equations (6-30) with $h_a = 0$. Equations (6-30) can be combined with Poisson's equation and the linearized continuity equation (6-35) to construct self-consistent weak bars. However, we can obtain some insight into the properties of these bars by a simpler route. We showed in §3.3.3 that the closed loop orbit is aligned parallel to the bar whenever the quantity C_2 defined by equation (3-120b) is positive. Stars on orbits that are aligned parallel to the bar generally contribute to the barlike nature of the overall gravitational field, whereas stars on orbits that are aligned perpendicular to the bar tend to cancel the gravitational field of the bar. Hence *any self-consistent weak bar must be composed mainly of orbits with $C_2 > 0$*. Let us examine the behavior of C_2 as we march out from the galaxy center to the corotation radius R_{CR} defined by $\Omega(R_{\mathrm{CR}}) = \Omega_b$. Since the angular speed decreases outward in almost all galaxies, for $R < R_{\mathrm{CR}}$ we have $\Omega > \Omega_b$, and the coefficient of the term Φ_b/R in C_2 is $2\Omega/(\Omega - \Omega_b) > 2$. This is sufficiently large that for most reasonable bar potentials, the term proportional to Φ_b swamps the term proportional to $d\Phi_b/dR$. Since $\Phi_b < 0$ by definition, we find that $C_2 > 0$ if and only if $\Delta > 0$. In other words, *a self-consistent weak bar can only be present inside corotation in regions where $\Omega_b > \Omega - \frac{1}{2}\kappa$*. If the bar is to extend continuously from the center of the galaxy to some maximum radius R_b—as the bars in real galaxies seem to do—then the pattern speed must exceed $\Omega - \frac{1}{2}\kappa$ throughout the region $0 < R < R_b$. Since the maximum of $\Omega - \frac{1}{2}\kappa$ occurs at small radii in most galaxies, the presence of a continuous weak bar of reasonable extent implies that Ω_b exceeds the maximum of $\Omega - \frac{1}{2}\kappa$. Also, since C_2 is always negative just outside corotation, a weak bar must always stop at or before corotation, that is, $R_b < R_{\mathrm{CR}}$.[21]

There is a natural connection between these results and the theory of disk stability discussed in §6.3. We argued that the bar instability arises when short trailing waves can propagate through the galactic center, emerging as short leading waves that can be swing amplified back to short trailing waves. This process can occur only if there is no inner Lindblad resonance. Thus the pattern speed of the bar Ω_b must exceed the maximum of $\Omega - \frac{1}{2}\kappa$, just as we found above by a different argument.

(b) Freeman's analytic bars   These models are flat elliptical disks with surface density and potential given by equations (4-204) and (4-203) of §4.6.3. The family of Freeman bars is very rich, including bars with all

---

[21] Sufficiently far outside corotation, C_2 may once again become positive. However, bars in this region are much less interesting, since they do not extend continuously from the center as the observations require.

possible axis ratios and a wide variety of pattern speeds. In the circular limit $b/a \to 1$, the Freeman bars are examples of weak bars formed from perturbed Kalnajs disks.

The Freeman bars are somewhat artificial because the potential is exactly quadratic, and because the DF contains an integrable singularity. However, they do resemble real bars in that they are not strongly centrally condensed. Thus, there is some hope that many of the features of Freeman bars may be present in real bars. However, the reliability of this analogy and the ultimate usefulness of the Freeman bars will not be known until a more complete survey of realistic bar models is available.

(c) Numerical bar models N-body experiments allow us to investigate much more realistic bar models than the ones we have described so far. One attractive feature of N-body models is that the bars can be allowed to form naturally from instabilities in an initially axisymmetric disk.

Sellwood (1981) has carried out an instructive series of experiments using an N-body code (typically $N = 20,000$) in which Poisson's equation was solved using an FFT program based on a two-dimensional polar grid (§2.8). Sellwood allowed a bar to develop in his models through the bar instability in a differentially rotating disk. He found that the initial pattern speed of the bar always exceeded the maximum of $\Omega - \frac{1}{2}\kappa$ and that the bar ended inside corotation, consistent with the predictions we developed for weak bars. He also found evidence of slow evolution of the bars after they formed: the bar pattern speeds slowed, and the bars became stronger and longer, apparently because there was a steady transfer of angular momentum from the bars to the disk stars. A similar angular momentum transfer to halo stars can also occur (Sellwood 1980). Sellwood's experiments raise the interesting possibility that there has been substantial secular evolution of the angular momentum distribution and other global properties of barred galaxies over their lifetimes.

2 Gas Dynamics in Barred Spiral Galaxies

Detailed simulations of the gas flow near a rotating bar can be directly compared with observations to help constrain the pattern speed and other parameters of bars in real galaxies (see W. W. Roberts 1979 for a review).

Figure 6-28 shows a photographic representation of a simulation of steady-state gas flow in a bar+disk potential (W. W. Roberts et al.

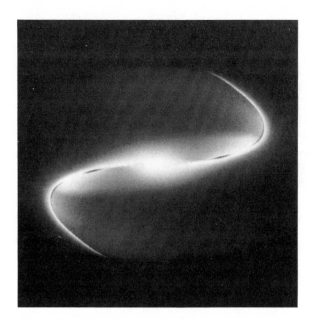

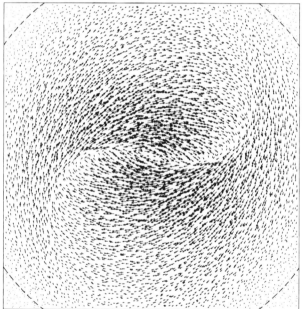

Figure 6-28. A simulation of steady-state gas flow in a bar potential. Top: The brightness is proportional to the gas surface density and the lines represent shocks. The orientation has been chosen to match that of NGC 1300 (Fig. 6-2). Bottom: The velocity field. Reprinted from W. W. Roberts et al. (1979), by permission of *The Astrophysical Journal.*

1979). The picture has been oriented and tipped so as to resemble NGC 1300 (Fig. 6-2). The locations of shock waves in the gas flow have been shaded black. The locations of the shocks are also visible on the map of the velocity field shown in Figure 6-28. One striking feature is that the shock waves strongly resemble the dust lanes in NGC 1300. This result suggests that the dust lanes, like those in spirals, arise in "traffic jams" containing a greatly enhanced density of interstellar material. Just as in the spirals, the traffic jams arise when the trajectories of interstellar clouds intersect.

Even rather weak bars can drive large noncircular motions. For example, the strongly perturbed velocity field in Figure 6-28 results from a non-axisymmetric force that is only about 20% of the axisymmetric force; the strength of the perturbation in the flow field reflects the fact that most of the gas is close to an inner Lindblad resonance and hence responds strongly to even a weak imposed force.

Over many rotation periods, the bar has important secular effects on the disk gas. Gas orbiting through the inner parts of the bar sinks to the galactic center as it loses energy and angular momentum on each passage through the shock. Gas orbiting through the outer parts of the bar appears to be concentrated into a ring near the end of the bar (W. W. Roberts et al. 1979). Gas orbiting inside the outer Lindblad resonance tends to collect into a ring at the resonance (Schwarz 1981). It may be that these rings can be identified with the rings seen in many barred galaxies (Kormendy 1982).

3 Summary

Possibly the most important feature of the bar models we have discussed is that they exist. A wide variety of numerical experiments have shown that bars are straightforward to construct, stable over many orbital times, and, most important, that fairly realistic bars are naturally and inevitably formed in unstable disks.

Barred stellar systems are complicated, and our understanding is far from complete. For example, we are only beginning to obtain reliable observational estimates of bar pattern speeds from gas kinematics. However, both extensive observations and realistic theoretical models are now routinely available. Therefore there are good prospects for rapid progress in the future.

Finally, the discussion in this section has neglected one of the potentially most fundamental and least understood aspects of barred galaxies: the various secular evolution processes driven by the bar. These include rearrangement of the disk gas, which may sweep the central regions clear of gas and form more distant gas into rings; transfer of angular momentum among the disk, the bar, and the halo components, which may cause changes in the strength, axis ratio, and length of the bar itself; and the possible formation of lenses, elliptical features of nearly constant surface brightness found between the spheroid and disk in roughly half of all barred galaxies (Kormendy 1982).

The timescales for these and several other processes (Kormendy 1982) are very uncertain but often appear to be less than a Hubble time. We must therefore consider the possibility that barred galaxies have undergone substantial evolution in their lifetimes. Even some unbarred galaxies with strong non-axisymmetric features (such as spiral structure or oval distortions) may evolve. Fortunately it is now beginning to be possible to run accurate N-body models of galaxies for a Hubble time, and these processes will soon be subject to direct experimental scrutiny.

6.6 Warps

In visible light, disk galaxies are usually remarkably thin and flat. However, the HI disk, which generally extends to much larger radii than the visible disk, sometimes shows a noticeable warp. Warps are most clearly visible in external galaxies that are seen edge-on but have also been detected in our own Galaxy (Fig. 6-29), and are inferred from abnormalities in the velocity fields of external galaxies that are inclined to the line of sight (see MB §8-4 for a fuller discussion). Warped disks are extremely common: for example, all of the spiral galaxies in the Local Group (our own Galaxy, M31, and M33) are warped. Photometry of edge-on galaxies shows that the optical disk ends rather abruptly at a radius that we shall call R_{max} (van der Kruit & Searle 1982). The warp generally begins at or near R_{max}; thus for most galaxies it is not known whether or not the optical disk (and thus the stars) participates with the interstellar gas in the warp. However, there are several known examples of galaxies with an optical warp, such as M31 (Innanen et al. 1982).

In general, a warped galaxy is twisted upward at one side and downward at the other; these are sometimes called "integral sign" warps since from an appropriate viewing angle such a disk resembles an integral sign.

Warps are one of the long-standing unresolved puzzles in galactic dynamics. In this section we will sketch the nature of the problem presented by warps and some attempts at its solution. For recent reviews of the warp problem, see van Woerden (1979) and Toomre (1983).

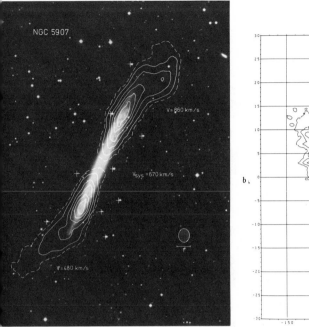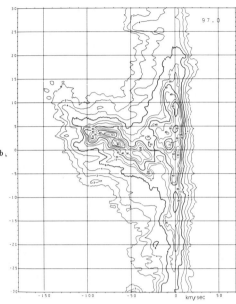

Figure 6-29. Left: A photograph of NGC 5907, with HI column density contours superimposed (Sancisi 1976). HI with velocity near the mean velocity of the system has been excluded, to show more clearly the distribution of HI in the outer parts. Right: HI brightness temperature in our Galaxy as a function of velocity and galactic latitude at longitude $\ell = 97°$. Note that the line of maximum brightness temperature is warped, reaching $b = 3°$ at $v = -90 \, \mathrm{km \, s^{-1}}$ and even higher at more negative velocities (Weaver & Williams 1974).

(a) Kinematics of warps   In the usual cylindrical coordinates (R, ϕ, z) a warped but very thin disk can be described by its height $z(R, \phi, t)$ above the galactic plane at position (R, ϕ) and time t. Suppose that initially the warp has the simple form

$$z(R, \phi, t = 0) = z_0(R) \cos m\phi, \qquad m \geq 0, \qquad (6\text{-}76)$$

and that the vertical velocity at every point is zero. An integral sign warp like those seen in most warped galaxies is represented by setting $m = 1$. We now ask how the warp evolves with time. Using the epicycle approximation [see eqs. (3-67), (3-69), and (3-66)] we have

$$R = R_g + X \cos(\kappa t + \psi), \qquad (6\text{-}77\mathrm{a})$$

$$\phi = \Omega_g t + \phi_0 - \frac{2\Omega_g X}{\kappa R_g} \sin(\kappa t + \psi), \qquad (6\text{-}77\mathrm{b})$$

$$z = Z \cos(\nu t + \varsigma), \qquad (6\text{-}77\mathrm{c})$$

where ψ, ς, X, and Z are independent of time, and the epicycle and vertical frequencies κ and ν are defined by equations (3-58). We neglect the epicyclic motion in R and ϕ, an approximation that is valid if the scale of the warp is larger than the size of the horizontal epicycle. Thus $R = constant$ and $\phi = \Omega(R)t + \phi_0$. In order to match the assumed vertical displacement and speed at $t = 0$, the parameters in equation (6-77c) must be $Z = z_0(R)\cos m\phi_0$ and $\xi = 0$. Therefore the disk height at time t is described by the relation

$$z(R,\phi,t) = \tfrac{1}{2}z_0(R)\big(\cos\{[m\Omega(R) - \nu(R)]t - m\phi\} \\ + \cos\{[m\Omega(R) + \nu(R)]t - m\phi\}\big). \qquad (6\text{-}78)$$

We see that the warp runs around the circle $R = constant$ in two waves. The waves are analogous to the kinematic density waves discussed in §6.2.1, and are called kinematic **bending waves**. The angular phase velocity or pattern speed Ω_p of the two waves is $\Omega(R) + \nu(R)/m$ (the "fast" wave) or $\Omega(R) - \nu(R)/m$ (the "slow" wave). In a flattened galaxy $\nu > \Omega$, and the pattern speed of an $m = 1$ slow wave is retrograde (i.e., the pattern rotates in the opposite direction to the galaxy). From now on, we consider only the slow wave, since the fast wave will disappear rapidly as a result of the winding process discussed below.

If only the slow wave is present, the warped disk attains its maximum height at $\phi(R,t) = [\Omega(R) - \nu(R)]t$. Because $\Omega - \nu$ is not exactly constant with radius, the warp will tend to wind up. Physically, the winding occurs because every ring in the warped disk precesses at a different rate. Our analysis in §6.1.2 of the winding problem in spiral arms can be applied without change, and we find that the pitch angle of the line of maximum height is [eq. (6-2)]

$$\cot i = Rt\left|\frac{d}{dR}(\Omega - \nu)\right|. \qquad (6\text{-}79)$$

Thus warps present precisely the same winding problem that is presented by kinematic density waves (Kahn & Woltjer 1959).

The severity of the winding problem depends on the azimuthal and vertical frequencies Ω and ν and their dependence on radius. From equations (3-58) we have

$$\Omega^2 = \frac{1}{R}\frac{\partial \Phi}{\partial R}\bigg|_{z=0} \quad ; \quad \nu^2 = \frac{\partial^2 \Phi}{\partial z^2}\bigg|_{z=0}. \qquad (6\text{-}80)$$

For a spherical galaxy, in which $\Phi = \Phi(\sqrt{R^2 + z^2})$, it is simple to prove that $\Omega = \nu$ and hence that no winding occurs. Physically, the warp does not wind up because there is no tendency to precess in a spherical field. In a real galaxy, the warp will wind up due to the nonspherical potential field from two sources, the halo and the disk. We now make separate estimates of the size of each of these effects.

First consider the effect of a nonspherical halo. In this estimate we neglect the effect of the disk entirely, and consider the halo to be described by the logarithmic potential of equation (2-54a),

$$\Phi_L = \tfrac{1}{2} v_0^2 \ln \left(R_c^2 + R^2 + \frac{z^2}{q_\Phi^2} \right), \tag{6-81}$$

where we shall set the core radius R_c to zero for simplicity. In this potential

$$\Omega = \frac{v_0}{R}, \quad \nu = \frac{v_0}{q_\Phi R}, \tag{6-82}$$

and hence equation (6-79) yields

$$\cot i = \frac{v_0 t}{R} \left| \frac{1}{q_\Phi} - 1 \right|. \tag{6-83}$$

Taking values from our own Galaxy, we use $v_0 = 220 \,\text{km s}^{-1}$, $R = 12 \,\text{kpc}$, and $t = 10^{10} \,\text{yr}$. The pitch angle i is less certain, but relief maps of galactic HI show no observable twist in the warp, suggesting that conservatively we may assume $i \gtrsim 45°$. Then equation (6-83) yields $|q_\Phi^{-1} - 1| < 0.005$. It is highly improbable that any galactic halo is so accurately spherical.

Next consider the effect of the disk potential. According to Problem 2-6, at large radii the disk potential can be written approximately as the sum of a monopole and a quadrupole term,

$$\Phi_d = -\frac{GM_d}{\sqrt{R^2 + z^2}} - \frac{G}{4} \frac{(R^2 - 2z^2)}{(R^2 + z^2)^{5/2}} \int_0^\infty 2\pi R'^3 \Sigma(R') \, dR'. \tag{6-84}$$

We assume that the disk is exponential, $\Sigma(R) = \Sigma_0 \exp(-R/R_d)$, and add the disk potential (6-84) to the potential of a spherical halo of the form (6-81) with $q_\Phi = 1$. Substituting into (6-80) we find

$$\Omega^2 = \frac{v_0^2}{R^2} + \frac{GM_d}{R^3}\left(1 + \frac{9R_d^2}{2R^2}\right) \quad ; \quad \nu^2 = \frac{v_0^2}{R^2} + \frac{GM_d}{R^3}\left(1 + \frac{27R_d^2}{2R^2}\right). \tag{6-85}$$

Taking $M_d = 6 \times 10^{10} \, M_\odot$, $R_d = 3.5 \,\text{kpc}$ (Table 1-2), and the remaining parameters from the example above, we find a present pitch angle of only 0.7°, which is far too small.

Both these arguments imply that the warp in our Galaxy cannot be primordial. It is possible to avoid the winding problem by assuming that the oblateness of the halo changes with radius in just such a way as to counterbalance the differential precession due to the disk, but such models are contrived and in any case do not explain how the warp formed in the first place. We must find a mechanism to excite and maintain the warp.

(b) Bending waves with self-gravity   In view of the analogy between bending and density waves, it is natural to wonder whether the winding problem could be solved by the self-gravity of the disk. Let us imagine a tightly wound bending wave in which the height above the galactic plane is given by [cf. eq. (6-29)]

$$z(R, \phi, t) = \text{Re}[z_a(R)e^{im(\phi - \Omega_p t)}], \quad \text{where} \quad z_a(R) = Z(R)e^{i \int k \, dR},$$
$$(6\text{-}86)$$

$m \geq 0$, and $|kR| \gg 1$. Just as in the case of tightly wound density waves, the effect of the self-gravity is dominated by the rapid radial corrugations, and the resulting vertical restoring force is shown in Problem 5-4 to be $F_z = -2\pi G\Sigma|k|z$, where Σ is the surface density. Thus the total restoring force is increased from $-\nu^2 z$ to $-(\nu^2 + 2\pi G\Sigma|k|)z$, the vertical frequency is increased from ν to $\sqrt{\nu^2 + 2\pi G\Sigma|k|}$, and the pattern speed Ω_p is changed from $\Omega \pm \nu/m$ to $\Omega \pm \frac{1}{m}\sqrt{\nu^2 + 2\pi G\Sigma|k|}$. This can be rewritten as a dispersion relation in the form (Hunter & Toomre 1969)

$$(m\Omega - \omega)^2 = \nu^2 + 2\pi G\Sigma|k|, \qquad (6\text{-}87)$$

where $\omega = m\Omega_p$. Note the similarity to the dispersion relation for tightly wound density waves in a cold disk [eq. (6-46) with $\mathcal{F} = 1$]. The only differences are that the epicycle frequency κ is replaced by the vertical frequency ν, and that the minus sign in front of the term $2\pi G\Sigma|k|$ in the density-wave dispersion relation is replaced by a plus sign (because self-gravity increases the stiffness of the disk to bending waves, while it decreases the stiffness to density waves). Thus there are **vertical resonances** at $m(\Omega - \Omega_p) = \pm\nu$, which are analogous to the Lindblad resonances at $m(\Omega - \Omega_p) = \pm\kappa$; the sign change implies that bending waves with $\Omega_p > 0$ exist in the region outside an outer vertical resonance and inside an inner resonance, in contrast to density waves, which exist inside an outer Lindblad resonance and outside an inner resonance (Fig. 6-30). However, a retrograde wave ($\Omega_p < 0$) such as the $m = 1$ slow wave exists in the region outside an inner vertical resonance.

Since the right side of equation (6-87) is always positive, ω is always real, and we conclude that *tightly wound bending waves are always stable.*

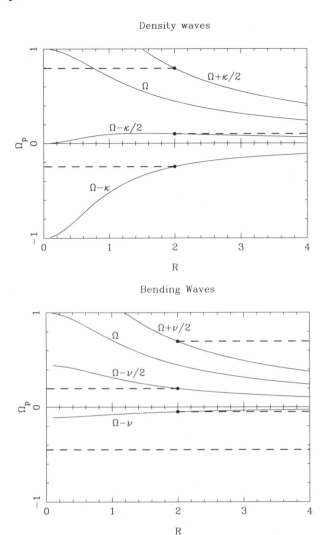

Figure 6-30. The regions where tightly wound density waves (top) and bending waves (bottom) exist in an isochrone potential are marked by heavy dashed lines. The potential is a flattened logarithmic potential of the form (6-81) with $v_0 = R_c = 1$ and $q_\Phi = 0.9$. The Lindblad and vertical resonances are marked by dots. For density waves, we assume that $Q = 1$ so that there is no forbidden region around corotation (cf. Fig. 6-14). The wave regions shown are for prograde ($\Omega_p > 0$) $m = 2$ waves and retrograde ($\Omega_p < 0$) $m = 1$ waves.

(c) The care and feeding of warps   Could galaxies be susceptible to a bending instability analogous to the bar instability which we have found to be so common? The answer appears to be "no". Hunter and Toomre (1969) have shown that a thin, self-gravitating disk in centrifugal equilibrium is stable to all vertical perturbations with $m = 0$ or $m = 1$. For $m > 1$ there is no analytic proof of stability, but numerical normal mode analyses have never revealed any instabilities. In any event, the observed warps have $m = 1$ and so are definitely stable.

It has sometimes been argued that warps represent a stable $m = 1$ bending mode, which was excited at the time of galaxy formation. However, numerical calculations of bending modes of galactic disks (Hunter & Toomre 1969; Sparke 1984) show that in most cases, free modes damp out in less than 5–10 rotation times, essentially because bending waves, like density waves, have a group velocity that tends to carry them to the edge of the disk where they are absorbed.[22] Thus bending wave packets would disappear in a few times 10^9 yr from the radii where the warps are most prominent.

Again arguing by analogy with spiral arms, the warp might represent a response by the disk to a recent close encounter with a companion galaxy, rather like the intense spiral structure induced in M51 by its encounter with NGC 5195. For our Galaxy the best candidate is the Large Magellanic Cloud. The effects of the Large Cloud on the Galaxy can be computed fairly accurately, since the orbit of the Cloud can be determined from the kinematics of the Magellanic Stream [§7.1.1(b)]. It appears that the Large Cloud never approaches closer to the Galaxy than about 50 kpc; at this distance the Cloud's tidal field is not nearly strong enough to cause the observed galactic warp (Hunter & Toomre 1969).

Further evidence that tidal forcing is not the major cause of warps comes from a survey of warped galaxies carried out with the Westerbork Synthesis Radio Telescope (van Woerden 1979). Three out of four edge-on galaxies with warps have no companions with luminosity greater than 15% of the luminosity of the warped galaxy within a projected distance of about 200 kpc. Thus, even if the companions are on nearly radial orbits so that there was a close encounter in the past, any such encounter must have taken place at least 10^9 yr ago (assuming a relative velocity of $200 \, \mathrm{km \, s^{-1}}$)—comparable to the time in which bending waves excited by the encounter would die away. Thus it is very unlikely that these galaxies have tidally excited warps.

---

[22] More precisely, the retrograde $m = 1$ waves propagate inward to the inner vertical resonance, where they reflect into leading waves, which then propagate out to the disk edge. See Problem 6-2.

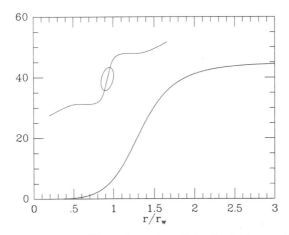

Figure 6-31. The orientation of the Laplacian surface in an exponential disk with surface density $\Sigma(R) = M_d \exp(-R/R_d)/(2\pi R_d^2)$ and a halo with potential $\Phi_h = \frac{1}{2} v_0^2 (R'^2 + z'^2/q_\Phi^2)/R_c^2$ whose symmetry axis is inclined with respect to that of the disk by $I = 45°$. The curve shows the solution of equation (6P-11) for the case $q_\Phi = 0.9$. The horizontal axis is plotted in units of the characteristic radius $r_w \equiv (9GM_d R_c^2 R_d^2/v_0^2)^{1/5}$. Interior to r_w the Laplacian surface almost coincides with the disk, while beyond $2r_w$ it closely follows the equatorial plane of the halo. The insert at top left shows how a galaxy with gas in such a Laplacian surface might appear on the sky. The spheroid has major axis $0.5r_w$. See Problem 6-6.

Many explanations of warps have been proposed (Toomre 1983). Most of them are flawed, or at least have not been supported by realistic numerical calculations. Perhaps the most plausible suggestion is that the symmetry axes of the galactic halo are misaligned with the disk. Gas in the outer parts of the disk would then tend to settle into the symmetry plane of the halo rather than the symmetry plane of the inner disk. This model is explored quantitatively in Problem 6-6 and Figure 6-31.

Problems

6-1. [1] By analogy with Jacobi ellipsoids, [§4.6.1(b)], a **Jacobi-type bar** is defined to be a bar with zero streaming motions, that is, a bar in which the mean velocity at every point is $\overline{\mathbf{v}} = \mathbf{\Omega}_p \times \mathbf{x}$, where $\mathbf{\Omega}_p$ is the pattern speed. A simple measurement can be used to determine whether the bar in a given galaxy is Jacobi-type, even without knowing the pattern speed. Consider a barred galaxy that is axisymmetric at large distances. Measure the mean line-of-sight velocity of the bar (relative to the center of the galaxy), along a line that coincides with the apparent minor axis of the disk at large radii. Prove that if the bar is Jacobi-type and Ω_p lies along the symmetry axis of the disk, then this velocity will be zero at every point on the line.

6-2. [1] Show from equation (6-87) that the group velocity of bending waves is

$$v_g = -\text{sign}(k)\frac{2\pi G\Sigma}{m(\Omega - \Omega_p)}. \qquad (6P\text{-}1)$$

Hence show that $m = 1$ retrograde bending waves propagate inward when they are trailing, and outward when they are leading.

6-3. [2] Show that the group velocity of density waves in a gaseous disk with $Q = 1$ is equal (within a sign) to the sound speed.

6-4. [2] The rate of change of pitch angle i in a trailing wavepacket is determined by the equation

$$\frac{d}{dt}(\cot i) = \left.\frac{\partial \cot i}{\partial R}\right|_\omega v_g, \qquad (6P\text{-}2)$$

where v_g is the group velocity. For a gaseous Mestel disk with $Q = constant$, show that

$$\frac{d}{dt}(\cot i) = \frac{\omega}{m} = \Omega_p. \qquad (6P\text{-}3)$$

[Hint: Show first that for a disk of this form $\kappa, \Omega \propto R^{-1}$, $\Sigma \propto R^{-1}$, $v_s \propto R^0$. Then use the dispersion relation (6-40), the definition of group velocity (6-57), and equation (6-13).]

6-5. [2] A useful model for exploring the properties of differentially rotating disks is based on the softened point-mass potential [eq. (2-194)]. In this model the usual potential between two particles of mass m at separation d, $\Phi = -Gm^2/d$, is replaced by $\Phi_\epsilon = -Gm^2/\sqrt{d^2 + \epsilon^2}$, where ϵ is the softening parameter (Miller 1971).

(a) Consider a tightly wound surface density distribution in the $z = 0$ plane,

$$\Sigma_1(R, \phi) = \Sigma_a e^{i[m\phi + f(R)]}. \qquad (6P\text{-}4)$$

Argue that the softened potential at $z = 0$ due to this surface density distribution is equal to the usual Newtonian potential created by the same density distribution at a height $z = \epsilon$. Hence, by an extension of the arguments given in §6.2.2(b), show that the softened potential due to the surface density (6P-4) is

$$\Phi_\epsilon(R, \phi, z = 0) = \Phi(R, \phi, z = \epsilon) = -\frac{2\pi G e^{-k\epsilon}}{|k|}\Sigma_a e^{i[m\phi + f(R)]}, \qquad (6P\text{-}5)$$

where $k = \partial f/\partial R$.

(b) Show that the WKB dispersion relation for a cold disk with softened gravity is

$$(m\Omega - \omega)^2 = \kappa^2 - 2\pi G\Sigma|k|\mathcal{F}, \qquad (6P\text{-}6)$$

where the reduction factor $\mathcal{F} = \exp(-k\epsilon)$.

(c) Show that a cold disk is stable to short-wavelength axisymmetric disturbances if

$$\epsilon > \frac{2\pi G\Sigma}{\kappa^2 e}. \qquad (6P\text{-}7)$$

The reduction factor due to softened gravity mimics the reduction factor due to velocity dispersion [eq. (6-46)]. In particular, disks satisfying equation (6P-7) are stable to axisymmetric WKB disturbances, just like stellar disks with $Q > 1$. Thus, cold disks with softened gravity provide close analogs to stellar disks and have the advantage that they are much easier to investigate numerically (see Figure 6-21 for one example of their use).

6-6. [3] One possible explanation of the warps seen in the outer parts of many disks is that the halo and disk are both axisymmetric but their symmetry axes are misaligned. In systems of this kind, dissipative material such as the gas in the galactic disk—and hence the stars that form from the gas—settles into a configuration in which the net torque from halo plus disk on each ring of material is zero. At each radius, there is a unique plane, called the **Laplacian surface**, in which the net torque vanishes. The object of this problem is to compute the orientation of the Laplacian surface as a function of radius in a simple system.

(a) Consider a galaxy containing an axisymmetric disk and halo, in which the halo symmetry axis is tipped relative to the disk symmetry axis by an inclination I. Let the disk axis be the z-axis, and the intersection of the equatorial planes of the disk and halo mark the y-axis (see Figure 6-31). We model the halo potential Φ_h by the logarithmic potential of equation (6-81) in the limit $R^2, z^2 \ll R_c^2$; thus

$$\Phi_h = \frac{v_0^2}{2R_c^2}\left(x'^2 + y^2 + \frac{z'^2}{q_\Phi^2}\right) + \text{constant}, \qquad (6P\text{-}8)$$

where $x' = x\cos I + z\sin I$ and $z' = z\cos I - x\sin I$. Consider the torque per unit mass $\mathbf{N}_h$ due to the halo on a ring of radius r_0, which has inclination i and whose symmetry axis, like that of the halo, lies in the x–z plane. Prove that $\mathbf{N}_h$ is directed along the y-axis and has magnitude

$$N_{hy} = \frac{v_0^2 r_0^2}{4R_c^2}\left(\frac{1}{q_\Phi^2} - 1\right)\sin 2(i - I). \qquad (6P\text{-}9)$$

(b) We model the disk potential Φ_d using equation (6-84). Prove that the torque on the ring due to the disk is also directed along the y-axis and has magnitude

$$N_{dy} = \frac{3G}{8r_0^3}\left[\int_0^\infty 2\pi R'^3\Sigma(R')dR'\right]\sin 2i. \qquad (6P\text{-}10)$$

(c) For an exponential disk, $\Sigma(R) = \Sigma_0 \exp(-R/R_d)$, with total mass $M_d = 2\pi\Sigma_0 R_d^2$, show that the inclination i of the Laplacian surface at radius r is given implicitly by $N_{hy} + N_{dy} = 0$ or

$$\frac{v_0^2 r^5}{9GM_d R_c^2 R_d^2}\left(\frac{1}{q_\Phi^2} - 1\right) = \frac{\sin 2i}{\sin 2(I - i)}. \qquad (6P\text{-}11)$$

A sketch of the solution to equation (6P-11) is shown in Figure 6-31.

6-7. [3] Consider a flat, non-axisymmetric disk galaxy. Assume that the distance to the galaxy is known, and that the galaxy is stationary in a frame rotating with the pattern speed Ω_p. Also assume that surface brightness is conserved, i.e., stars are neither created nor destroyed (in practice, this assumption is only correct for S0 galaxies, since bright, short-lived stars are created in the spiral arms of spiral galaxies). Show that measurements of the line-of-sight velocity field and surface brightness distribution of the stars can be used to yield a direct determination of Ω_p (Tremaine & Weinberg 1984b).

7 Collisions and Encounters of Stellar Systems

The velocities of galaxies are not perfectly ordered. Rather, the velocity of each galaxy is the combination of the Hubble velocity [see eq. (1-11) or Appendix 9.A] appropriate to that galaxy's position, and a residual, or **peculiar velocity**. The peculiar velocities of neighboring galaxies typically are of order $100 \, \mathrm{km \, s^{-1}}$ (Bean et al. 1983). Thus galaxies closer than $100 \, \mathrm{km \, s^{-1}}/H_0 \approx 1h^{-1}$ Mpc are frequently found to be approaching one another. For example, our Galaxy's neighbor, M31 (the Andromeda galaxy) (Figure 1-5), which is at a distance of about 700 kpc, is approaching us at about $100 \, \mathrm{km \, s^{-1}}$. Thus in about $10^{10} \, \mathrm{yr}$ our Galaxy and M31 will probably collide. In this chapter we investigate what happens during such collisions.

A collision between our Galaxy and M31 would have devastating consequences for the gas in both systems. If a gas cloud from M31 encountered a galactic cloud, shock waves would be driven into the clouds, heating and compressing the gas. In the denser parts of the clouds, the compressed post-shock gas would cool rapidly and fragment into new stars. The more massive of these stars would heat and photo-ionize much of the remaining gas and ultimately explode as supernovae, thereby shock-heating the gas still further. Since astronomers do not yet have a good understanding of this complex chain of events, we shall not pursue these ideas further in this book (but see Larson & Tinsley 1978). Instead we concentrate on the stellar-dynamical aspects of galaxy encounters.

In contrast to the gas, stars emerge unscathed from a galaxy collision. To see this, consider what would happen to the solar neighborhood in a collision with the disk of M31. According to Table 4-16 of MB, by far the commonest stars in the solar neighborhood are dwarf M stars, of which there are about 20 $\mathrm{pc^{-2}}$. Table 3-7 of MB indicates that all these stars have radii smaller than $R_{\mathrm{M}} = 10^{-0.2} \, \mathrm{R_\odot} = 4.4 \times 10^{10} \, \mathrm{cm}$. Hence the fraction of the area of the galactic disk that is filled by the disks of these stars is of order $20\pi(R_{\mathrm{M}}/1 \, \mathrm{pc})^2 = 1.3 \times 10^{-14}$. Thus even if M31 were to score a direct hit on our Galaxy, the probability that even one of the 10^{11}–10^{12} stars in M31 would actually hit any star in our Galaxy would be small.[1]

However, the distribution and dynamics of the stars in the two galaxies would be radically changed by such a collision, because the gravitational field of M31 would deflect the stars of our Galaxy from their original orbits and vice versa for the stars of M31. The first part of this chapter is concerned with this type of gravitational interaction between stellar systems.

The most straightforward way to investigate what happens when galaxies collide or pass close by one another is to simulate the encounter numerically using an N-body program. Figure 7-1 shows various stages in a parabolic encounter of two identical N-body "galaxies". Initially each galaxy is a King model with $\Psi(0)/\sigma^2 = 7$ [see §4.4.3(c)]. The two dotted curves in Figure 7-1 show the trajectories that the centers of these models would follow if the galaxies were point masses. At the point of closest approach, the centers would be separated by 1.2 times the median radius r_h of the models, and would be moving relative to one another at a speed equal to about three times the initial RMS speed within each model. The N-body simulation shows that the galaxies actually move on trajectories that are quite unlike the marked trajectories. In particular, the galaxies do not separate at late times as they would if they remained rigid King models, but return, crash into each other, and finally merge into a single galaxy. In fact, in this encounter, these models behave more like soft deformable lumps of putty than like billiard balls.[2]

Since every star in these model galaxies moves according to Newton's laws of motion, the total energy of the system is strictly conserved. Nevertheless, the kinetic energy associated with the relative motion of the two galaxies diminishes during the encounter because the internal energies of the galaxies increase. For some purposes it is useful to think of the galaxies as made of a viscous fluid that absorbs energy when it is deformed.

Not every close encounter between galaxies leads to a merger like the one shown in Figure 7-1. To see this, let v_∞ be the speed at which galaxy A initially approaches galaxy B and consider how the energy that is gained by a star at position vector $\mathbf{r}$ in galaxy A depends on v_∞. The star's orbit with respect to the center of A gains energy at a rate $\mathbf{v} \cdot \mathbf{g}[\mathbf{r}(t)]$, where $\mathbf{g}$ is the difference between the gravitational attraction of B at $\mathbf{r}$ and at the center of A, and $\mathbf{v}$ is the star's velocity with respect to the center of A. As we increase v_∞, the time t_0 required for the two

---

[1] We have neglected gravitational focusing, which enhances the collision probability by a factor of about five [eq. (8-123)].

[2] Indeed, we shall see that galaxies behave like the toy putty that is elastic at high impact speeds, but soft and inelastic at low speeds.

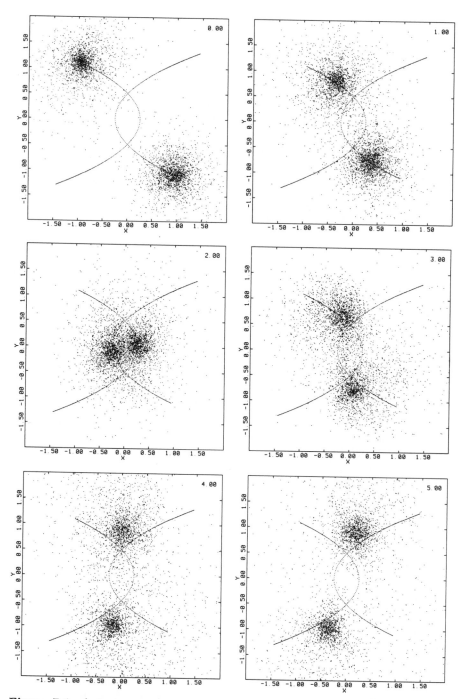

Figure 7-1. An encounter between two King models. The dotted curves show the trajectories their centers would follow if the galaxies were point masses. In fact energy is transferred from the mutual orbit to the internal energies of the two galaxies, with the result that the systems do not separate fully after the encounter: the last panel shows them near the point of maximum post-encounter separation. (After Barnes & Hut 1986.)

galaxies to reach the point of closest approach diminishes. Hence as v_∞ increases, the energy increment $\Delta E(t_0) = \int_0^{t_0} \mathbf{v} \cdot \mathbf{g}[\mathbf{r}(t)]dt$ diminishes and the star's orbit through A withdraws less energy from the relative orbit of the galaxies. Thus there is a critical speed v_f such that for $v_\infty > v_f$ the galaxies reach the point of closest approach with sufficient orbital energy to make good their escape to infinity. If $v_\infty < v_f$, the galaxies merge, while if $v_\infty \gg v_f$ the encounter alters both the orbits and the internal structures of the galaxies only slightly. In §7.2 we shall see that when $v_\infty \gg v_f$ the effect of the encounter can be handled analytically.

Unfortunately, many problems of astrophysical interest lie in the more complex regime, $v_\infty \lesssim v_f$. Such encounters can be discussed analytically only when one of the galaxies is much larger and more massive than the other. We start our study of interactions between galaxies by considering this special case of interaction between systems of very unequal size. The results we obtain are valid for all values of v_∞/v_f.

7.1 Dynamical Friction

We consider the motion of a body of mass M through a population of stars of individual mass m. The body M may, in principle, be extended, but for simplicity we shall temporarily assume that it is a point mass. We first study the effect on M of an encounter with a single star, and then add the effects of successive encounters with different stars. Let $(\mathbf{x}_M, \mathbf{v}_M)$ and $(\mathbf{x}_m, \mathbf{v}_m)$ be the positions and velocities of M and m, respectively; let $\mathbf{r} = \mathbf{x}_m - \mathbf{x}_M$ and $\mathbf{V} = \dot{\mathbf{r}}$. Then, as we saw in Appendix 1.D.2, the separation vector $\mathbf{r}$ obeys the equation,

$$\left(\frac{mM}{m+M}\right)\ddot{\mathbf{r}} = -\frac{GMm}{r^2}\hat{\mathbf{e}}_r. \tag{7-1}$$

This is the equation of motion of a fictitious particle, called the reduced particle, in the Keplerian potential of a fixed body of mass $m + M$. If $\Delta\mathbf{v}_m$ and $\Delta\mathbf{v}_M$ are the changes in the velocities of m and M during the encounter, we have

$$\Delta\mathbf{v}_m - \Delta\mathbf{v}_M = \Delta\mathbf{V}. \tag{7-2a}$$

Furthermore, since the velocity of the center of mass of the two bodies is unaffected by the encounter (see Appendix 1.D.2), we also have

$$m\Delta\mathbf{v}_m + M\Delta\mathbf{v}_M = 0. \tag{7-2b}$$

Eliminating $\Delta\mathbf{v}_m$ between equations (7-2) we obtain $\Delta\mathbf{v}_M$ as

$$\Delta\mathbf{v}_M = -\left(\frac{m}{m+M}\right)\Delta\mathbf{V}. \tag{7-3}$$

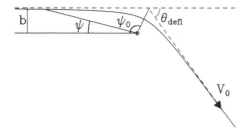

Figure 7-2. The motion of the reduced particle during a hyperbolic encounter.

We now evaluate $\Delta \mathbf{V}$.

Let the component of the initial separation vector that is perpendicular to the initial velocity vector $\mathbf{V}_0 = \mathbf{V}(t = -\infty)$ have length b (see Figure 7-2). We call b the **impact parameter** of the encounter. Then the conserved angular momentum per unit mass associated with the motion of the reduced particle is

$$L = bV_0. \tag{7-4}$$

Equation (3-21), which relates the radius and azimuthal angle of a particle in Keplerian orbit, reads in this case,

$$\frac{1}{r} = C\cos(\psi - \psi_0) + \frac{G(M+m)}{b^2 V_0^2}, \tag{7-5}$$

where the angle ψ is shown in Figure 7-2. The constants C and ψ_0 are determined by the initial conditions. Differentiating equation (7-5) with respect to time we obtain,

$$\begin{aligned} \frac{dr}{dt} &= Cr^2 \dot{\psi} \sin(\psi - \psi_0) \\ &= CbV_0 \sin(\psi - \psi_0), \end{aligned} \tag{7-6}$$

where the second line follows because $r^2\dot{\psi} = L$. If we define the direction $\psi = 0$ to point toward the particle as $t \to -\infty$, we find on evaluating equation (7-6) at $t = -\infty$,

$$-V_0 = CbV_0 \sin(-\psi_0). \tag{7-7a}$$

On the other hand, evaluating equation (7-5) at this time we have

$$0 = C\cos\psi_0 + \frac{G(M+m)}{b^2 V_0^2}. \tag{7-7b}$$

Eliminating C between these equations, we obtain

$$\tan\psi_0 = -\frac{bV_0^2}{G(M+m)}. \tag{7-8}$$

But from equation (7-5) or (7-6) we see that the point of closest approach is reached when $\psi = \psi_0$. Since the orbit is symmetrical about this point, the angle through which the reduced particle's velocity is deflected, is $\theta_{\text{defl}} = 2\psi_0 - \pi$ (see Figure 7-2). By conservation of energy, the relative speed after the encounter equals the initial speed V_0. Hence the components $\Delta V_{\parallel}$ and $\Delta V_{\perp}$ of ΔV parallel and perpendicular to the original relative velocity vector V_0 are given by

$$|\Delta V_{\perp}| = V_0 \sin \theta_{\text{defl}} = V_0 |\sin 2\psi_0| = \frac{2V_0 |\tan \psi_0|}{1 + \tan^2 \psi_0}$$

$$= \frac{2bV_0^3}{G(M+m)} \left[1 + \frac{b^2 V_0^4}{G^2(M+m)^2} \right]^{-1}, \qquad (7\text{-}9a)$$

$$|\Delta V_{\parallel}| = V_0[1 - \cos \theta_{\text{defl}}] = V_0(1 + \cos 2\psi_0) = \frac{2V_0}{1 + \tan^2 \psi_0}$$

$$= 2V_0 \left[1 + \frac{b^2 V_0^4}{G^2(M+m)^2} \right]^{-1}. \qquad (7\text{-}9b)$$

$\Delta V_{\parallel}$ always points in the direction opposite to V_0. By equation (7-3) we obtain the components of Δv_M as

$$|\Delta v_{M\perp}| = \frac{2mbV_0^3}{G(M+m)^2} \left[1 + \frac{b^2 V_0^4}{G^2(M+m)^2} \right]^{-1}, \qquad (7\text{-}10a)$$

$$|\Delta v_{M\parallel}| = \frac{2mV_0}{M+m} \left[1 + \frac{b^2 V_0^4}{G^2(M+m)^2} \right]^{-1}. \qquad (7\text{-}10b)$$

$\Delta v_{M\parallel}$ always points in the same direction as V_0. Notice that in the limit of large impact parameter b, equation (7-10a) agrees with our rough estimate (4-3) of the same quantity.

We now imagine that the mass M is traveling through an infinite homogeneous sea of stars. We invoke the Jeans swindle (§5.1) to neglect the gravitational potential generated by this sea, so that the motion of each star is determined solely by the gravitational force from M. In this approximation, the changes $\Delta v_{M\perp}$ sum to zero. The changes $\Delta v_{M\parallel}$ are, by contrast, all parallel to V_0 and form a non-zero resultant. In consequence, the mass M suffers a steady deceleration, which is said to be due to **dynamical friction**.

If the phase-space number density of stars is $f(\mathbf{v})$, the rate at which M encounters stars that have velocities in the velocity-space element $d^3 \mathbf{v}_m$ at impact parameters between b and $b + db$, is

$$2\pi b \, db \times V_0 \times f(\mathbf{v}_m) \, d^3 \mathbf{v}_m. \qquad (7\text{-}11)$$

Table 7-1. Typical values of the Coulomb logarithm

	Open Cluster	Globular Cluster	Elliptical Galaxy	Solar Nbd.	Cluster of Galaxies
Center	5.8	10.1	20.9	18.1	6.1
r_h	6.5	12.0	23.9	–	8.6

NOTES: Column labeled "Solar Nbd." refers to the solar neighborhood. Results are based on the log of equation (7-13b) for $M = m$, where $m = 1\,M_\odot$ for stellar systems and $1.2 \times 10^{11}\,M_\odot$ for galaxy systems [characteristic luminosity $L_\star = 1 \times 10^{10}\,L_\odot$ from equation (1-15) times typical mass-to-light ratio of $12\Upsilon_\odot$ from Table 4-2]. System parameters taken from Tables 1-2, 1-3, and 1-4, using b_{max} equal to the core radius and median radius for the first and second rows, respectively, and setting v_{typ} to $\sqrt{3}$ times the one-dimensional velocity dispersion.

Hence the net rate of change of $\mathbf{v}_M$ due to these encounters is

$$\frac{d\mathbf{v}_M}{dt}\bigg|_{\mathbf{v}_m} = \mathbf{V}_0 f(\mathbf{v}_m)\, d^3\mathbf{v}_m \int_0^{b_{max}} \frac{2mV_0}{M+m} \left[1 + \frac{b^2 V_0^4}{G^2(M+m)^2}\right]^{-1} 2\pi b\, db,$$
(7-12)

where b_{max} is the largest impact parameter that need be considered. This is roughly the distance at which the stellar density $\int f(\mathbf{v}_m) d^3\mathbf{v}_m$ becomes much smaller than it is in the neighborhood of M. When we perform the integral over b in equation (7-12), we find

$$\frac{d\mathbf{v}_M}{dt}\bigg|_{\mathbf{v}_m} = 2\pi \ln(1+\Lambda^2) G^2 m(M+m) f(\mathbf{v}_m)\, d^3\mathbf{v}_m \frac{(\mathbf{v}_m - \mathbf{v}_M)}{|\mathbf{v}_m - \mathbf{v}_M|^3},$$
(7-13a)

where

$$\Lambda \equiv \frac{b_{max} V_0^2}{G(M+m)}.$$
(7-13b)

In typical applications of equation (7-13a), Λ is very large. For example, we shall shortly apply relations based on equation (7-13a) to the motion of a globular cluster of mass $M \approx 10^6\,M_\odot$ at speed $V_0 \approx 100\,\mathrm{km\,s^{-1}}$ through a galaxy of effective radius $b_{max} \approx 2\,\mathrm{kpc}$ in which the stars have mass $m \approx M_\odot$; in this case $\Lambda \approx 4.6 \times 10^3$ and $\ln(1+\Lambda^2) \approx 17$. Typical values for Λ in other applications of equation (7-13a) are given in Table 7-1. Hence we usually have $\frac{1}{2}\ln(1+\Lambda^2) \simeq \ln\Lambda$, which is called the **Coulomb logarithm** by analogy with the equivalent logarithm that occurs in the theory of plasmas. In the following discussion we shall everywhere replace $\frac{1}{2}\ln(1+\Lambda^2)$ by $\ln\Lambda$, and replace the factor V_0 in Λ by v_{typ}, a typical speed. Neither b_{max} nor v_{typ} is precisely defined, but a fractional uncertainty δ in either quantity produces a fractional error in the friction of only $\delta/\ln\Lambda$, which is generally much smaller than δ.

In the approximation $\ln \Lambda = constant$, equation (7-13a) states that the stars that have velocity $\mathbf{v}_m$ exert a force on M that acts parallel to $\mathbf{v}_m - \mathbf{v}_M$ and is inversely proportional to the square of this vector. Thus the problem of integrating the acceleration $d\mathbf{v}_M/dt|_{\mathbf{v}_m}$ over all field-star velocities $\mathbf{v}_m$ is equivalent to the problem of finding the gravitational field at the point with position vector $\mathbf{v}_M$ that is generated by the "mass density" $\rho(\mathbf{v}_m) \equiv 4\pi \ln \Lambda Gm(M+m)f(\mathbf{v}_m)$. If the stars move isotropically,[3] the density distribution is spherical and by Newton's first theorem (§2.1.1), the total acceleration of M is simply equal to (G/v_M^2) times the total "mass" that lies at $v_m < v_M$. Hence for an isotropic distribution of stellar velocities,

$$\frac{d\mathbf{v}_M}{dt} = -16\pi^2 \ln \Lambda G^2 m(M+m) \frac{\int_0^{v_M} f(v_m)v_m^2 dv_m}{v_M^3} \mathbf{v}_M; \qquad (7\text{-}14)$$

i.e., only stars moving slower than M contribute to the force. Like an ordinary frictional drag, the force described by equation (7-14) always opposes the motion. Equation (7-14) is usually called the **Chandrasekhar dynamical friction formula** (Chandrasekhar 1943).

For sufficiently small v_M we may replace $f(v_m)$ in the integral of equation (7-14) by $f(0)$ to find

$$\frac{d\mathbf{v}_M}{dt} \simeq -\frac{16\pi^2}{3} \ln \Lambda G^2 f(0)m(M+m)\mathbf{v}_M \qquad (v_M \text{ small}). \qquad (7\text{-}15)$$

Thus in this case, the drag is proportional to v_M, as in Stokes's law for the drag on a marble falling through honey. For sufficiently large v_M, the integral in equation (7-14) converges to a definite limit, and the frictional force therefore falls like v_M^{-2}.

If $f(\mathbf{v}_m)$ is Maxwellian with dispersion σ, then

$$f = \frac{n_0}{(2\pi\sigma^2)^{3/2}} \exp(-\tfrac{1}{2}v^2/\sigma^2), \qquad (7\text{-}16)$$

and equation (7-14) becomes

$$\frac{d\mathbf{v}_M}{dt} = -\frac{4\pi \ln \Lambda G^2(M+m)n_0 m}{v_M^3}\left[\mathrm{erf}(X) - \frac{2X}{\sqrt{\pi}}e^{-X^2}\right]\mathbf{v}_M, \qquad (7\text{-}17)$$

where $X \equiv v_M/(\sqrt{2}\sigma)$, n_0 is the total number density of stars, and erf is the error function [eq. (1C-11)]. In the limit $M \gg m$, this formula

---

[3] If the velocity distribution is ellipsoidal, one can show that the component of $\mathbf{v}_M$ that decays fastest is that in the direction of the smallest principal velocity dispersion of the background stars (see Problem 7-10).

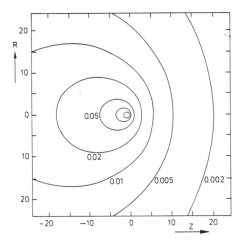

Figure 7-3. A mass travels from left to right at speed v through a homogeneous Maxwellian distribution of stars with one-dimensional dispersion $\sigma = v$. Deflection of the stars by the mass enhances the stellar density downstream more than upstream. Contours of equal stellar density are labeled with the corresponding fractional density enhancement. (From Mulder 1983.)

illustrates two important characteristics of dynamical friction. These are:

(i) The frictional drag is proportional to the mass density $(n_0 m)$ of the stars being scattered, but independent of the mass of each individual star. In particular, if we replace $n_0 m$ in equation (7-17) by the overall background density ρ, we obtain a formula that is equally valid for a background made up of a spectrum of different masses:

$$\frac{d\mathbf{v}_M}{dt} = -\frac{4\pi \ln \Lambda G^2 \rho M}{v_M^3} \left[\operatorname{erf}(X) - \frac{2X}{\sqrt{\pi}} e^{-X^2} \right] \mathbf{v}_M \qquad (M \gg m).$$
(7-18)

(ii) When $M \gg m$, the frictional acceleration is proportional to M and thus the frictional force must be proportional to M^2. It is instructive to consider why this is so. Stars are deflected by M in such a way that the density of background stars behind M is greater than in front of it (see Figure 7-3). The amplitude of this density enhancement is proportional to M and the gravitational force that it exerts on M is proportional to M times its amplitude. Hence the force is proportional to M^2.

Does an extended body such as a globular cluster or a small galaxy experience a similar drag to that experienced by a point particle of equal mass? If M is an extended body with median radius r_h, say, then equation (7-10b) for $\Delta \mathbf{v}_{M\parallel}$ will seriously overestimate $\Delta \mathbf{v}_{M\parallel}$ for encounters with pericenter distances r_p—computed as if M were a point mass—that satisfy $r_p \lesssim r_h$. The pericenter distance r_p is related to V_0 and b by the energy equation of the reduced particle at pericenter:

$$E = \tfrac{1}{2} V_0^2 = -\frac{G(M + m)}{r_p} + \tfrac{1}{2} \left(\frac{bV_0}{r_p} \right)^2.$$
(7-19)

Setting $r_p = r_h$ and $V_0 = v_{\text{typ}}$, we obtain the smallest impact parameter b_h for which equation (7-10b) gives a reasonably reliable estimate of $\Delta \mathbf{v}_{M\parallel}$ as

$$b_h = r_h \sqrt{1 + \frac{2G(M+m)}{r_h v_{\text{typ}}^2}}. \tag{7-20}$$

The fraction of the integral in equation (7-12) that is contributed by impact parameters less than b_h is

$$\frac{\ln(1 + \Lambda_h^2)}{\ln(1 + \Lambda^2)}, \tag{7-21a}$$

where, by analogy with equation (7-13b),

$$\Lambda_h \equiv \frac{b_h v_{\text{typ}}^2}{G(M+m)} = \Lambda \sqrt{\frac{r_h^2}{b_{\text{max}}^2} + \frac{2r_h}{\Lambda b_{\text{max}}}}. \tag{7-21b}$$

Since $\ln(1 + \Lambda^2)$ typically lies in the range 10 to 20, the fraction (7-21a) is small so long as $\Lambda_h \lesssim 1$. Thus, when $r_h \lesssim b_{\text{max}}/\sqrt{\Lambda} = \sqrt{G(M+m)}b_{\text{max}}/V_0$, the formulae we have derived for the case where M is a point mass will only slightly overestimate the drag experienced on an extended body.

Although the Chandrasekhar dynamical friction formula (7-14) was derived for a mass moving through an infinite homogeneous background, it can be employed to estimate the drag on a small body traveling through a much larger stellar system. In such applications we replace $f(v)$ by the value of the DF in the vicinity of the small body, v_{typ} by the local velocity dispersion, and b_{max} by the distance from the body over which the density of the larger system changes by a factor of two or so. When employed in this way, Chandrasekhar's formula suffers from several shortcomings:

(i) The choice of b_{max} is rather arbitrary.

(ii) It neglects the self-gravity of the wake. Thus equation (7-14) takes into account the mutual attraction of M and the background stars, but neglects the attraction of the background stars for each other.

(iii) We obtained equation (7-14) in the approximation that stars move past M on Keplerian hyperbolae. Orbits in the combined force fields of M and the larger system would really be more complex.

These deficiencies become especially worrisome when M is so large as to be comparable to the mass of the larger system that lies interior to M's orbit. For example, the center of the Coma galaxy cluster is dominated by two supergiant galaxies that are thought to form a binary system. The orbital motion of this binary is probably resisted by the gravitational field of the wake the supergiants induce in the rest of the cluster. But we cannot expect equation (7-14) to yield an accurate estimate of the magnitude of this drag because the supergiants contain much of the mass of the cluster's core.

Notwithstanding these difficulties, simulations have shown that Chandrasekhar's formula often provides a remarkably accurate description of the drag experienced by a body orbiting in a stellar system (White 1976; D. N. C. Lin & Tremaine 1983; Bontekoe & van Albada 1987). As a rough guide, Chandrasekhar's formula is fairly accurate provided that (i) the mass M does not exceed 20% of the mass of the larger system; (ii) M's orbit is not confined either within the central core or beyond the outer boundary of the larger system. A more sophisticated discussion of dynamical friction is given by Tremaine and Weinberg (1984a).

1 Applications of Dynamical Friction

(a) Decay of globular cluster orbits   As a globular cluster orbits through a galaxy, it is subject to dynamical friction. This drag causes the cluster to lose energy and spiral in toward the galaxy center. We now estimate the time $t_{\text{fric}}(r_i)$ required for a cluster that is initially on a circular orbit of radius r_i to reach the center.

The flatness of many observed rotation curves (see §10.1.6) suggests that we approximate the density interior to r_i with the density distribution

$$\rho(r) = \frac{v_c^2}{4\pi G r^2} \tag{7-22}$$

of the singular isothermal sphere with circular speed v_c and velocity dispersion $\sigma = v_c/\sqrt{2}$ [see eq. (4-123)].[4] Equation (7-18) then gives the frictional force on a cluster of mass M moving at speed v_c at radius r as

$$\begin{aligned} F &= -\frac{4\pi \ln \Lambda G^2 M^2 \rho(r)}{v_c^2} \left[\text{erf}(1) - \frac{2}{\sqrt{\pi}} e^{-1} \right] \\ &= -0.428 \ln \Lambda \frac{GM^2}{r^2}. \end{aligned} \tag{7-23}$$

---

[4] Equation (7-22) overestimates the density inside the galaxy's core, but this leads to a negligible underestimate in the calculated time for the cluster's orbit to decay because the orbital decay is slowest near r_i, where the density is smallest and equation (7-22) is most accurate.

If we assume $b_{max} \approx 2\,\text{kpc}$, $M = 10^6\,M_\odot$, and $v_{typ} \approx v_c = 250\,\text{km s}^{-1}$, we have by equation (7-13b) that $\ln \Lambda \simeq 10$.

The force (7-23) is tangential and thus causes the cluster to lose angular momentum per unit mass L at a rate

$$\frac{dL}{dt} = \frac{Fr}{M} \simeq -0.428 \frac{GM}{r} \ln \Lambda. \qquad (7\text{-}24)$$

Since the cluster continues to orbit at speed v_c as it spirals to the center, its angular momentum per unit mass at radius r is at all times $L = rv_c$. Substituting the time derivative of this expression into equation (7-24), we obtain

$$r\frac{dr}{dt} = -0.428 \frac{GM}{v_c} \ln \Lambda. \qquad (7\text{-}25)$$

Solving this differential equation subject to the initial condition $r(0) = r_i$, we find that the cluster reaches the center after a time[5]

$$t_{\text{fric}} = \frac{1.17}{\ln \Lambda} \frac{r_i^2 v_c}{GM} = \frac{2.64 \times 10^{11}}{\ln \Lambda} \left(\frac{r_i}{2\,\text{kpc}} \right)^2 \left(\frac{v_c}{250\,\text{km s}^{-1}} \right) \left(\frac{10^6\,M_\odot}{M} \right) \text{ yr.}$$
$$(7\text{-}26)$$

It is interesting to apply this formula to the Andromeda galaxy, M31. The most luminous globular cluster in M31 has luminosity $L \simeq 2.6 \times 10^6\,L_\odot$, and thus if the mass-to-light ratio of the cluster is the same ($\Upsilon_V \approx 2\Upsilon_\odot$) as of those in our Galaxy (Table 1-3), its mass is $M \approx 5 \times 10^6\,M_\odot$. Since the circular speed close to the center of M31 is of order $250\,\text{km s}^{-1}$, equation (7-26) with $\ln \Lambda = 10$ predicts that any clusters as massive as this cluster whose original orbits lay within $r_i \approx 3\,\text{kpc}$ would already have spiraled into the center of the galaxy. Tremaine et al. (1975) have extended this analysis to include a spectrum of initial cluster masses, and concluded that the nucleus of M31 may represent the debris produced when twenty to thirty massive globular clusters spiraled to the center and then merged to form a single massive system.

(b) Fate of the Magellanic Clouds   The Small and Large Magellanic Clouds lie $21°$ from one another in the southern sky. They are $63\,\text{kpc}$ and $50\,\text{kpc}$, respectively, from the Sun and have masses $M_{SMC} \approx 2 \times 10^9\,M_\odot$ and $M_{LMC} \approx 2 \times 10^{10}\,M_\odot$. Since their galactocentric radial velocities are similar, and they are enveloped in a common

---

[5] In reality, some mass will be stripped from the cluster by the galaxy's tidal field. However, for most globular clusters this process will not greatly lengthen t_{fric}.

cloud of hydrogen, it is generally believed that they are—or at least recently have been—in orbit around one another. The center of mass of the combined Clouds must in turn be orbiting in the gravitational field of our Galaxy. If our Galaxy has a massive halo (see §10.1) that reaches out to the Clouds, we may use equation (7-26) to estimate the time for the decay of an object on a circular orbit as

$$
t_{\rm fric} \approx \frac{1.0 \times 10^{10}}{\ln \Lambda} \left(\frac{r}{60\,{\rm kpc}} \right)^2 \left(\frac{v_c}{220\,{\rm km\,s^{-1}}} \right) \left(\frac{2 \times 10^{10}\,{\rm M_\odot}}{M} \right) \ {\rm yr}.
$$

$$(7\text{-}27)$$

For the Clouds the Coulomb logarithm is $\ln \Lambda \approx 3$. Thus, if the rotation curve of our Galaxy is flat out to $r \gtrsim 60\,{\rm kpc}$ and the Clouds are currently on a nearly circular orbit, the Clouds will spiral to the center of our Galaxy within a fraction of a Hubble time (Tremaine 1976).

How sensitive is this interesting conclusion to the simplifying assumptions on which it is based?

If the mass distribution of our Galaxy does not reach as far as the Clouds, $t_{\rm fric}$ will be greater than the value given by equation (7-27). Unfortunately, it is not easy to estimate $t_{\rm fric}$ analytically for a truncated galactic mass distribution, and no numerical calculations of this case are available; nor would they provide a definite answer without better data on the extent of the galactic halo.

The second important simplifying assumption underlying equation (7-27) is that the Clouds are on a circular orbit. If the orbit is eccentric, $t_{\rm fric}$ will be either smaller or greater than the estimate (7-27) according as the Clouds are currently near apocenter or pericenter. Within the next decade it may become possible to measure the proper motions of the Clouds, and thus determine (for any assumed galactic potential) their current orbits, but at present we are able to measure only their radial velocities. Nevertheless, a valuable constraint on the transverse velocities is provided by the **Magellanic Stream** that was described in §9-2 of MB. The Stream consists of a long thin chain of neutral hydrogen clouds that stretches for at least 110° across the sky, starting near the Clouds. The velocity of the material in the Stream varies smoothly with position and near the Clouds coincides with the mean velocity of the Clouds. There seems little doubt that the Stream consists of material torn out of the Clouds by the Galaxy's tidal field, and that the Stream lies in the orbital plane of the Clouds. Since the Stream very nearly passes through the south galactic pole, we infer that the Clouds are orbiting in a plane that is nearly perpendicular to the galactic plane. The velocity vector of the center of mass of the Clouds must lie in this plane.

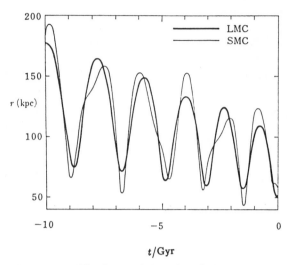

Figure 7-4. The decay of the orbit of the Magellanic Clouds around our Galaxy according to the model of Murai and Fujimoto (1980).

Murai and Fujimoto (1980) have followed many trial orbits for the Clouds. In these calculations, the galactic potential was assumed to be spherical with constant circular speed $v_c = 250\,\mathrm{km\,s^{-1}}$, and the Clouds experienced dynamical friction according to equation (7-17). They found that the unknown transverse velocities of the Clouds are tightly constrained by the requirement that the Clouds remain bound to one another when their equations of motion are integrated backward in time. In particular, they conclude that the Clouds must now be near their point of closest approach to our Galaxy because all orbits that allow the Clouds to pass within 30 kpc of the galactic center cause the Clouds to separate quickly when the equations of motion are integrated backward. Murai and Fujimoto chose a particular orbit for the Clouds from among those that are compatible with the Clouds having been a binary system in the past, by assuming that the Magellanic Stream is material that has been stripped out of one or other of the Clouds, and fitting models of this process to the observations. In the model that fits the data best, the Stream is generated about $2 \times 10^8\,\mathrm{yr}$ before the present, when the two Clouds came within 3 kpc of each other, and the Large Cloud tidally stripped gas from the disk of the Small Cloud. D. N. C. Lin and Lynden-Bell (1982) have independently reached a very similar conclusion.

Figure 7-4 shows the galactocentric distances of the two Clouds over the last $10^{10}\,\mathrm{yr}$ in the preferred model of Murai and Fujimoto. The greatest distance reached by the Clouds on their orbit around our Galaxy has decreased by about 50% over this epoch, and the Clouds will spiral to the galactic center in about another ten billion years.

(c) Galactic cannibalism   Our discussion of the probable fate of the Magellanic Clouds prompts us to ask whether large galaxies make a practice of gobbling up their smaller neighbors, and if so, how bloated they become in consequence. Since galactic cannibalism will be an important process only if giant galaxies have extended mass distributions, we shall again assume that the density associated with a giant galaxy can be approximated by (7-22), where v_c is the galaxy's constant circular speed.

Studies of the distribution of galaxies on the sky [see Peebles (1980b) for references] suggest that the number density $n(r, L)\delta L$ of galaxies with luminosity in the range $(L, L + \delta L)$ near any galaxy falls with distance r from the center of the galaxy as

$$n(r, L)\delta L = \phi(L)\left[\left(\frac{r_0}{r}\right)^{1.8} + 1\right]\delta L. \tag{7-28}$$

Here $\phi(L)$ is the **luminosity function** of galaxies that was discussed in §1.1.3, and $r_0 \approx 3h^{-1}\,\text{Mpc}$ is the **correlation length** of the galaxy distribution. Since we are interested only in $n(r, L)$ for $r \ll r_0$, we shall retain only the first term in the square bracket of equation (7-28).

Let us imagine that all the galaxies near a giant galaxy have mass-to-light ratio Υ and are moving on circular orbits in the extended mass distribution of the giant galaxy. Then the orbital radius r of a galaxy of mass $M = \Upsilon L$ diminishes according to equation (7-25). Hence the inward current of galaxies with luminosities in the range $(L, L + \delta L)$ through radius r would be

$$\begin{aligned}
\delta F(r, L) &= 4\pi r^2 \left|\frac{dr}{dt}\right| n(r, L)\delta L \\
&\simeq 5.4 \frac{G\Upsilon L r_0 \phi(L) \ln \Lambda}{v_c} \left(\frac{r_0}{r}\right)^{0.8} \delta L.
\end{aligned} \tag{7-29}$$

Thus the flux decreases with increasing r and at any radius more galaxies are lost to smaller radii than are gained from larger radii. Consequently, in this naive picture the density of galaxies around any giant galaxy is constantly decreasing, and one expects a deficit of galaxies in the vicinity of luminous galaxies. Attempts to test this prediction observationally have unfortunately not produced conclusive results (Ostriker & Turner 1979; White & Valdes 1980).

We obtain a rough estimate of the number of galaxies that a giant galaxy has eaten in a Hubble time $t_H \equiv H_0^{-1}$ by integrating $n(r, L)$ from $r = 0$ out to the radius $r_i(t_H, L)$ at which $t_{\rm fric} = t_H$. With equations (7-26) and (7-28) we obtain

$$
\begin{aligned}
\delta N(L) &= 4\pi\delta L \int_0^{r_i} n(r, L) r^2 dr \\
&= 9.5 \left(\frac{G\Upsilon L \ln \Lambda}{H_0 r_0^2 v_c} \right)^{0.6} r_0^3 \phi(L)\delta L.
\end{aligned}
\tag{7-30}
$$

Integrating L times equation (7-30) over all L, we obtain the amount ΔL by which the giant's luminosity has swollen at the expense of former neighbors. Using equation (1-15) for the luminosity function $\phi(L)$, we find

$$
\begin{aligned}
\Delta L &= \int_0^{\infty} L\delta N(L) \approx 9.5 \left(\frac{G\Upsilon L_\star \ln \Lambda}{H_0 r_0^2 v_c} \right)^{0.6} r_0^3 n_\star L_\star \int_0^{\infty} x^{1.6+\alpha} e^{-x} dx \\
&= 9.5 \left(\frac{G\Upsilon L_\star \ln \Lambda}{H_0 r_0^2 v_c} \right)^{0.6} r_0^3 n_\star L_\star (1.6 + \alpha)!,
\end{aligned}
\tag{7-31}
$$

where we neglect the slow variation of the Coulomb logarithm with mass (and hence with luminosity). If the victim galaxies do not have extended mass distributions, Table 4-2 suggests that their mass-to-light ratios $\Upsilon \approx 12\,h\,{\rm M_\odot/L_\odot}$. Finally setting $v_c \approx 300\,{\rm km\,s^{-1}}$ appropriate to a giant galaxy, $r_0 = 3h^{-1}\,{\rm Mpc}$, $\ln \Lambda = 3$, and taking values for $L_\star$, $n_\star$ and α from §1.1.3, we obtain

$$
\Delta L \approx 0.13 L_\star
\tag{7-32}
$$

independent of h. Thus this formula suggests that a typical giant galaxy has eaten one or two satellites of luminosity $\approx 0.1 L_\star$.

7.2 High-Speed Encounters

In the last section we obtained a number of analytic results for encounters of galaxies of very unequal size. Another important kind of encounter that can be treated analytically is a collision of two stellar systems at very high speed; as we saw at the beginning of this chapter, the effects of an encounter on the internal structures of the galaxies *decrease* as the encounter speed increases. Hence high-speed encounters can be considered as generating small perturbations of otherwise steady-state systems.

Let the colliding systems have masses M_1 and M_2 and median radii r_1 and r_2, and suppose that at the instant of closest approach, their centers are separated by distance b' and have relative speed V. Then the effective duration of the encounter may be crudely estimated as

$$t_{enc} \approx \frac{\max(r_1, r_2, b')}{V}. \tag{7-33}$$

If the systems have internal velocity dispersions of order σ_i ($i = 1, 2$), the characteristic crossing times of the majority of their stars will be comparable to $t_i \equiv r_i/\sigma_i$. If $t_i \gg t_{enc}$, that is, if

$$V \gg \sigma_i \frac{\max(r_1, r_2, b')}{r_i} \qquad i = 1, 2, \tag{7-34}$$

the majority of stars in each system will, in the course of the encounter, barely move from their initial locations with respect to their galactic centers. Therefore, if condition (7-34) is satisfied, we can assume that the stars do not significantly change their positions with respect to the center of their own system during the encounter. A wide variety of numerical experiments (e.g., Aguilar & White 1985) have demonstrated that this **impulse approximation** yields remarkably accurate results, often even when the condition (7-34) is not strictly satisfied.

In the impulse approximation, the density distribution in each system is unchanged during the encounter, so we can obtain approximate expressions for the motion of the centers by treating the two systems as extended rigid bodies. We now show that whenever the impulse approximation is valid, the centers travel at nearly uniform velocity throughout the encounter. For simplicity we consider only encounters in which neither center penetrates inside the median radius of the other system; $r_i \lesssim b'$. The relative motion of the centers may then be approximated as that of a reduced particle in the Keplerian potential of a fixed mass $M_1 + M_2$.

The reduced particle travels at constant velocity so long as its potential energy $-G(M_1+M_2)/r$ is at all times much smaller in magnitude than its initial kinetic energy $\frac{1}{2}V_\infty^2$, or, equivalently, $G(M_1 + M_2)/r \ll \frac{1}{2}V^2$. We have

$$\frac{G(M_1 + M_2)}{r} \lesssim \frac{G(M_1 + M_2)}{b'} \simeq \frac{r_1\sigma_1^2 + r_2\sigma_2^2}{b'}, \tag{7-35a}$$

where we have approximated the masses $M_i \approx r_i\sigma_i^2/G$ by the virial theorem. Since we are assuming that $b' > \max(r_1, r_2)$, equation (7-34) yields $\sigma_i^2 \ll V^2 r_i^2/b'^2$; thus

$$\frac{G(M_1 + M_2)}{r} \ll V^2 \frac{r_1^3 + r_2^3}{b'^3} < 2V^2, \tag{7-35b}$$

and we conclude that the relative motion of the two centers is at nearly uniform velocity. From this it follows that b' almost coincides with the impact parameter b defined by Figure 7-2. Hence in a suitably oriented coordinate system, the position vector $\mathbf{X}(t)$ of the center of M_2 with respect to the center of M_1 may be approximated as

$$\mathbf{X}(t) \simeq (0, b, Vt). \tag{7-36}$$

We now focus attention on one system, say M_1, which we call the perturbed system, and study how its structure is changed by the passage of the other system, M_2, which we call the perturbing system. Let $\Delta\mathbf{v}'_\alpha$ be the change in the velocity of the αth star of the perturbed system. We break $\Delta\mathbf{v}'_\alpha$ into two components. The component

$$\Delta\overline{\mathbf{v}} \equiv \frac{\sum_\beta m_\beta \Delta\mathbf{v}'_\beta}{\sum_\beta m_\beta}, \tag{7-37a}$$

where m_β is the mass of the βth star, reflects the change in the center of mass velocity of the entire system, while the component

$$\Delta\mathbf{v}_\alpha \equiv \Delta\mathbf{v}'_\alpha - \Delta\overline{\mathbf{v}} \tag{7-37b}$$

is the change in the velocity of the αth star with respect to the systemic velocity. We next estimate $\Delta\mathbf{v}_\alpha$.

Now that we have an approximate expression (7-36) for the position of the perturber's center with respect to the center $\mathbf{r} = 0$ of the perturbed system, we may consider known the gravitational potential $\Phi(\mathbf{r}, t)$ generated by the perturber at each point $\mathbf{r}$ in the perturbed system. The resulting rate of change of the velocity of the αth star is

$$\dot{\mathbf{v}}'_\alpha = -\boldsymbol{\nabla}\Phi(\mathbf{r}_\alpha, t), \tag{7-38}$$

and since the impulse approximation implies that $\mathbf{r}_\alpha$ is constant during the encounter, we obtain

$$\Delta\mathbf{v}'_\alpha = -\int_{-\infty}^{\infty} \boldsymbol{\nabla}\Phi(\mathbf{r}_\alpha, t)dt. \tag{7-39}$$

Equation (7-37b) may now be written

$$\Delta\mathbf{v}_\alpha = -\int_{-\infty}^{\infty}\left[\boldsymbol{\nabla}\Phi(\mathbf{r}_\alpha, t) - \frac{1}{M_1}\sum_\beta m_\beta\boldsymbol{\nabla}\Phi(\mathbf{r}_\beta, t)\right]dt. \tag{7-40}$$

In the impulse approximation, the potential energy of the system does not change during the encounter. Hence the change in the internal energy is simply the change in the internal kinetic energy. This is

$$\Delta E = \frac{1}{2} \sum_\alpha m_\alpha \left[(\mathbf{v}_\alpha + \Delta \mathbf{v}_\alpha)^2 - \mathbf{v}_\alpha^2 \right]$$
$$= \frac{1}{2} \sum_\alpha m_\alpha \left[|\Delta \mathbf{v}_\alpha|^2 + 2 \mathbf{v}_\alpha \cdot \Delta \mathbf{v}_\alpha \right]. \tag{7-41}$$

In any axisymmetric system, $\sum_\alpha m_\alpha \mathbf{v}_\alpha \cdot \Delta \mathbf{v}_\alpha = 0$ by symmetry (see Problem 7-1). Hence the internal energy of the entire system changes by

$$\Delta E = \frac{1}{2} \sum_\alpha m_\alpha |\Delta \mathbf{v}_\alpha|^2. \tag{7-42}$$

Note that ΔE is necessarily positive.

How does the system react to the injection of energy by an impulsive encounter? There are two important aspects to the system's response: (i) relaxation to a new dynamical equilibrium, and (ii) mass loss. We consider each in turn, along with a number of other aspects of the response of stellar systems to high-speed encounters.

(a) Return to equilibrium   After the increments (7-40) have been added to the velocities of all the stars of M_1, the system is no longer in virial equilibrium. Hence the passage of the perturber initiates a period of readjustment to a new equilibrium configuration. An interesting feature of this process is that it changes the internal kinetic energy of the system more than did the encounter itself.

To show this, let the internal kinetic and total internal energies be T_0 and E_0, respectively. Then we have by the virial theorem

$$T_0 = -E_0. \tag{7-43}$$

If the encounter increases the kinetic energy by δT, the final energy E_1 is

$$E_1 = E_0 + \delta T. \tag{7-44}$$

Applying the virial theorem to the final equilibrium, we have that when the relaxation is complete, the kinetic energy is

$$T_1 = -E_1 = -(E_0 + \delta T) = T_0 - \delta T. \tag{7-45}$$

Thus during the relaxation process to which the encounter gives rise, the kinetic energy decreases by $2\delta T$ from $T_0 + \delta T$ to $T_0 - \delta T$.

(b) Mass loss   Equation (7-41) shows that the encounter not only increases the system's energy by the overall amount ΔE, but also redistributes a portion of the system's orginal energy stock between stars; the encounter behaves like the kind of croupier gamblers dream about, one who feeds extra chips in from the bank at the same time as he redistributes chips among those at the gaming table. But notwithstanding the bank's largesse, it can happen that as the night wears on in this stellar casino, the players at the table get steadily poorer. How come? In *this* casino, gamblers who get lucky are more likely to quit the table than those who are down on their luck; stars that gain a lot of energy are likely to become unbound and drift off, and the steady departure of the system's richest and most fortunate denizens gradually impoverishes all that remain. Numerical experiments show that (i) the rate at which the term (7-42) pumps energy into the system is typically roughly balanced by the rate at which the system loses energy with escaping stars; (ii) the precise balance between energy loss and gain depends on the details of the perturber's force field (Richstone 1975; Dekel et al. 1980; Aguilar & White 1985).

It is possible to estimate the lifetimes of repeatedly perturbed stellar systems from a knowledge of the overall rate of energy input, without investigating how quickly the system loses mass: the timescale for the system's mass and luminosity to diminish significantly will always be on the order of the time required to pump into the system an amount of energy equal to the system's original binding energy. All that has to be determined by a detailed calculation of the rate of mass loss is whether the system initially uses the energy input to swell while retaining its stars, and then dissolves rather suddenly, or whether mass is lost steadily.

(c) Adiabatic invariance   Condition (7-34) ensures that the impulse approximation is valid for the majority of stars in the perturbed system. However, the orbital time of stars at the center of the system will be very short, so the impulse approximation is unlikely to apply there. Indeed, sufficiently close to the center, the orbital times of most stars may be so short that their orbits deform adiabatically as the perturber approaches (cf. §3.6). In this case, changes that occur in the structure of orbits as the perturber approaches will be reversed as the perturber departs, and the encounter will effect no net change in the structure of the central region.

(d) Tidal approximation   If the ratios r_1/b' and r_2/b' of the median radii r_i to the distance of closest approach b' are small, we can obtain a convenient expression for the velocity impulse (7-40). We first rewrite this expression in terms of a continuous density distribution $\rho(\mathbf{x})$;

$$\Delta\mathbf{v}(\mathbf{x}) = -\int_{-\infty}^{\infty}\left[\nabla\Phi(\mathbf{x},t) - \frac{1}{M_1}\int\rho(\mathbf{x}')\nabla\Phi(\mathbf{x}',t)\,d^3\mathbf{x}'\right]dt. \quad (7\text{-}46)$$

Next we place the origin of the coordinates at the center of mass of the perturbed system and expand $\nabla_k\Phi(\mathbf{x},t)$ in a Taylor series about the origin:

$$\nabla_k\Phi(\mathbf{x},t) = \Phi_k^{(1)} + \sum_j\Phi_{kj}^{(2)}x_j + O(|\mathbf{x}|^2), \quad (7\text{-}47a)$$

where

$$\Phi_k^{(1)} \equiv \left.\frac{\partial\Phi}{\partial x_k}\right|_{\mathbf{x}=0} \quad;\quad \Phi_{kj}^{(2)} \equiv \left.\frac{\partial^2\Phi}{\partial x_k\partial x_j}\right|_{\mathbf{x}=0}. \quad (7\text{-}47b)$$

Hence

$$\int\rho(\mathbf{x}')\nabla_k\Phi(\mathbf{x}',t)\,d^3\mathbf{x}' = M_1\Phi_k^{(1)} + \sum_j\Phi_{kj}^{(2)}\int\rho(\mathbf{x}')x_j'\,d^3\mathbf{x}' + O(\overline{r^2})$$

$$= M_1\Phi_k^{(1)} + O(\overline{r^2}),$$
$$(7\text{-}48)$$

where $\overline{r^2}$ is the mean-square radius of the system, and we have exploited our choice $\mathbf{x} = 0$ for the position of the center of mass of the system. Inserting equations (7-47a) and (7-48) into equation (7-46), we find

$$\Delta v_k(\mathbf{x}) = -\int_{-\infty}^{\infty}\sum_j\Phi_{kj}^{(2)}x_j\,dt + O(\overline{r^2})$$
$$= -\int_{-\infty}^{\infty}\mathbf{x}\cdot[\nabla(\nabla_k\Phi)]_{\mathbf{x}=0}\,dt + O(\overline{r^2}). \quad (7\text{-}49)$$

Since we are considering only encounters for which $r_2 \ll b'$ we may assume that

$$\Phi(\mathbf{x},t) = -\frac{GM_2}{|\mathbf{x}-\mathbf{X}(t)|}, \quad (7\text{-}50)$$

where $\mathbf{X}$ is the position vector of the center of the perturbing system. Differentiating equation (7-50) we obtain

$$(\mathbf{x}\cdot\nabla)\nabla_k\Phi = GM_2\left[\frac{x_k}{|\mathbf{x}-\mathbf{X}|^3} - 3(x_k-X_k)\frac{\mathbf{x}\cdot(\mathbf{x}-\mathbf{X})}{|\mathbf{x}-\mathbf{X}|^5}\right]. \quad (7\text{-}51)$$

When we substitute this expression to order $|\mathbf{x}|$ into equation (7-49), we find

$$\Delta\mathbf{v}(\mathbf{x}) = -GM_2\int_{-\infty}^{\infty}\left[\frac{\mathbf{x}}{|\mathbf{X}|^3} - 3\frac{\mathbf{X}(\mathbf{x}\cdot\mathbf{X})}{|\mathbf{X}|^5}\right]dt + O(\overline{r^2}). \quad (7\text{-}52)$$

Substituting from equation (7-36) for $\mathbf{X}$ yields

$$
\begin{aligned}
\Delta\mathbf{v}(\mathbf{x}) &= -GM_2 \int_{-\infty}^{\infty} \frac{\mathbf{x}[b^2 + (Vt)^2] - 3(0, b, Vt)(yb + zVt)}{[b^2 + (Vt)^2]^{5/2}} dt + \mathrm{O}(\overline{r^2}) \\
&= -\frac{GM_2}{b^2 V} \left[x \int_{-\infty}^{\infty} \frac{du}{(1 + u^2)^{3/2}}, \ y \int_{-\infty}^{\infty} \frac{u^2 - 2}{(1 + u^2)^{5/2}} du, \right. \\
&\qquad \left. z \int_{-\infty}^{\infty} \frac{1 - 2u^2}{(1 + u^2)^{5/2}} du \right] + \mathrm{O}(\overline{r^2}),
\end{aligned}
$$

(7-53)

where we have made the substitution $u = Vt/b$. Evaluating the integrals in equation (7-53), we obtain finally

$$
\Delta\mathbf{v}(\mathbf{x}) = \frac{2GM_2}{b^2 V}(-x, y, 0) + \mathrm{O}(\overline{r^2}). \tag{7-54}
$$

The velocity increments given by the leading term in equation (7-54) tend to deform a sphere of stars into an ellipsoid whose long axis lies in the direction of the perturber's point of closest approach. This distortion is reminiscent of the way in which the Moon raises tides on the surface of the oceans.

By equations (7-42) and (7-54) the total energy change ΔE in the tidal approximation is (Spitzer 1958)

$$
\begin{aligned}
\Delta E &= \frac{2G^2 M_2^2}{b^4 V^2} \int \rho(\mathbf{x})(x^2 + y^2) d^3\mathbf{x} \\
&= \frac{4G^2 M_2^2 M_1}{3b^4 V^2} \overline{r^2},
\end{aligned}
$$

(7-55)

where $\overline{r^2}$ is the mean-square radius of the perturbed system. Aguilar and White (1985) show that equation (7-55) gives tolerably accurate results for impact parameters $b \gtrsim 5r_h$, where r_h is the perturbed system's median radius. If the perturber is also extended, with median radius r_h', then the correct criterion is $b \gtrsim 5\max(r_h, r_h')$.

(e) Penetrating encounters   According to equation (7-55), the internal, random energy gained in a fast, distant encounter increases sharply as the impact parameter b of the encounter decreases. This suggests that a few close encounters may pump more energy into a typical system than a large number of distant encounters. Hence it is important to obtain an estimate of the energy gained from encounters for which b/r_i is not large, and thus equation (7-55) is not valid.

One way in which we might seek to improve on equation (7-55) is to retain more terms from the Taylor series (7-47). A much simpler and more instructive procedure is to calculate the energy input at zero impact parameter, $b = 0$. The analysis of such head-on encounters is greatly simplified by two observations:

(i) In the impulse approximation we may neglect changes in the velocities of the perturbed system's stars that lie along the perturber's trajectory. This conclusion follows because, during the approach and recession phases of the perturber's motion, the perturbed system's stars receive equal and opposite impulses parallel to the perturber's trajectory.[6]

(ii) By symmetry, the velocity increments $\Delta \mathbf{v}$ generated by a head-on encounter point toward the perturber's line of motion.

Let (R, z) be cylindrical coordinates such that the z-axis coincides with the perturber's trajectory $(R = 0, z = Z_p(t) = Vt)$, and $(R = 0, z = 0)$ defines the center of the perturbed system. Defining $r(R, z, t)$ to be the distance from (R, z) to the center of the perturber, and supposing for simplicity that the perturber is spherical with potential $\Phi(r)$, the R-component of the perturber's gravitational field at (R, z) is

$$-\frac{d\Phi}{dr}\frac{R}{r}.$$

Hence the only non-zero component of the velocity increment $\Delta \mathbf{v}$ of the star at (R, z) is

$$\Delta v_R = -\int_{-\infty}^{\infty}\frac{d\Phi}{dr}\frac{R}{r}dt = -\frac{2R}{V}\int_0^{\infty}\frac{d\Phi}{dr}\frac{dZ_p}{r}. \tag{7-56}$$

In this expression $r = \sqrt{(Z_p - z)^2 + R^2}$ is a function of Z_p.

According to equation (7-56), Δv_R depends only on the distance R of the star from the perturber's axis. We can therefore write down a simple expression for the energy gained in a head-on encounter by a system that is axisymmetric about the perturber's line of motion:

$$\Delta E = \pi \int_0^{\infty}[\Delta v_R(R)]^2 \Sigma(R)R dR, \tag{7-57}$$

where $\Sigma(R)$ is the perturbed system's surface density when projected parallel to the perturber's line of motion. For example, the velocity increment generated by a Plummer model $\Phi = -GM/\sqrt{r^2 + a^2}$ [cf. eq. (4-111)] is

$$\Delta v_R = -\frac{2GMR}{V(R^2 + a^2)}. \tag{7-58}$$

---

[6] This is a general property of impulsive encounters rather than a peculiarity of head-on collisions; see also equation (7-54).

If the perturbed system is an identical Plummer model, equations (7-57) and (7-58) give its energy gain in a head-on encounter as

$$\Delta E = \frac{G^2 M^3}{3V^2 a^2}. \tag{7-59}$$

Comparing equations (7-55) and (7-59) we see that the ratio $b^4/\overline{r^2}$ of the tidal approximation is replaced by $4a^2$ in the formula for a head-on encounter.

The head-on formula (7-59) complements the tidal formula (7-55), and a smooth interpolation between them yields an adequate formula for the energy gain in any high-speed encounter.

1 Applications of the Impulse Approximation

(a) Disruption of open clusters   The distribution of matter in the galactic plane is by no means smooth. The most conspicuous irregularities are (i) open clusters (see §3-5 of MB); (ii) giant molecular clouds (see §9-3 of MB); and (iii) spiral arms. In this subsection and in §7.5 we shall find that these irregularities have far-reaching consequences for the structure and evolution of the disk as a whole. Here we concentrate on the action of giant molecular clouds on open clusters.

The masses of open clusters lie in the range $10^2 \, \mathrm{M}_\odot \lesssim M_{\text{cluster}} \lesssim 10^3 \, \mathrm{M}_\odot$, and their characteristic radii $r_{\text{cluster}} \approx 1 \, \text{pc}$ (Table 1-3). Much of the hydrogen in our Galaxy is concentrated into a few thousand **giant molecular cloud complexes** of mass $M_{\text{cloud}} \gtrsim 10^5 \, \mathrm{M}_\odot$ and radius $r_{\text{cloud}} \approx 10 \, \text{pc}$ (see §9-3 of MB). Both open clusters and molecular cloud complexes move through the galactic disk with random velocities of order $7 \, \text{km s}^{-1}$. If we assume that the distribution of these velocities is approximately Maxwellian, with dispersion σ, then it follows (see Problem 7-3) that the distribution of the relative velocities is also Maxwellian, but with dispersion $\sqrt{2}\sigma$. Thus if $\sigma \approx 7 \, \text{km s}^{-1}$, typical cluster-cluster or cluster-cloud encounters occur at relative speeds that substantially exceed the velocity dispersion $\sigma_\star \lesssim 1 \, \text{km s}^{-1}$ of the cluster stars. Hence condition (7-34) indicates that we may use the impulse approximation to study the effect on clusters of close encounters with either other clusters or molecular cloud complexes.

According to equation (7-55), the energy input from an impulsive encounter is proportional to the square of the perturber's mass. Cloud complexes are nearly as numerous and about a thousand times more massive than open clusters. Hence cluster-cloud encounters must pump nearly a million times more energy into clusters than cluster-cluster encounters, and we need consider only cluster-cloud encounters. For simplicity we shall neglect the possibility that the orbits of stars in the cluster's core deform adiabatically (but see Spitzer 1958).

Consider the rate at which a given cluster encounters clouds at relative speeds in the range $(V, V+dV)$ and impact parameters in the range $(b, b+db)$. With our assumption of Maxwellian velocity distributions, the probability that any given cluster-cloud pair has relative speed in the given range is (Problem 7-3)

$$dP = \frac{4\pi V^2 dV}{[2\pi(\sqrt{2}\sigma)^2]^{3/2}} \exp\left(-\frac{V^2}{2(\sqrt{2}\sigma)^2}\right). \tag{7-60}$$

Hence the rate at which a given cluster encounters clouds with speed V and impact parameter b is

$$\dot{C} = n_{\text{cloud}} \times V \times 2\pi b\, db \times dP = \frac{n_{\text{cloud}} 8\pi^2 b\, db}{[2\pi(\sqrt{2}\sigma)^2]^{3/2}} \exp\left(-\frac{V^2}{4\sigma^2}\right) V^3 dV, \tag{7-61}$$

where n_{cloud} is the number density of clouds. If we assume that each such encounter increases the internal energy of the cluster according to equation (7-55), we have that encounters with impact parameters in the range b_{min} to b_{max} increase the cluster's energy at a rate

$$\dot{E} = \frac{32\pi^2 G^2 (M^2 n)_{\text{cloud}} (M\overline{r^2})_{\text{cluster}}}{3[2\pi(\sqrt{2}\sigma)^2]^{3/2}} \int_0^\infty \exp\left(-\frac{V^2}{4\sigma^2}\right) V\, dV \int_{b_{\text{min}}}^{b_{\text{max}}} \frac{db}{b^3}$$

$$= \frac{4\sqrt{\pi} G^2 (M^2 n)_{\text{cloud}} (M\overline{r^2})_{\text{cluster}}}{3\sigma b_{\text{min}}^2}, \tag{7-62}$$

where we have assumed that $b_{\text{min}}^2/b_{\text{max}}^2$ is negligible. There are two conditions involved in the determination of b_{min}:

(i) Since $r_{\text{cluster}} < r_{\text{cloud}}$, the tidal approximation, upon which equation (7-62) is based, is valid for encounters with $b \gtrsim 5 r_{\text{cloud}}$. Furthermore, we have seen that the energy input in a head-on encounter is comparable to that predicted by the tidal approximation when $b \approx r_{\text{cloud}}$ [eqs. (7-55) and (7-59)]. Therefore we will set $b_{\text{min}} = r_{\text{cloud}}$ and multiply the heating rate (7-62) by a factor g of order unity to account for encounters in which the tidal approximation fails. Here we use $g = 3$.[7]

_____

[7] We obtain $g = 3$ by assuming that $\dot{E}$ varies as b^{-3} for $b > r_{\text{cloud}}$ and is constant for $b < r_{\text{cloud}}$.

(ii) We also need to be sure that there is not a significant contribution to $\dot{E}$ from encounters so close that the cluster flies apart with a significant post-encounter energy; each cluster can only be disrupted once! Equation (7-55) can be rewritten in the form

$$\frac{\Delta E}{E_{\text{bind}}} = 6.7 \left(\frac{r_{\text{cloud}}}{b} \right)^4 \left(\frac{GM_{\text{cloud}}}{V^2 r_{\text{cloud}}} \right) \left(\frac{M_{\text{cloud}}}{M_{\text{cluster}}} \right) \left(\frac{\overline{r^2} r_h}{r_{\text{cloud}}^3} \right), \quad (7\text{-}63)$$

where the binding energy is related to the median radius r_h by [eq. (4-80b)]

$$E_{\text{bind}} \simeq 0.2 \frac{GM_{\text{cluster}}^2}{r_h}. \quad (7\text{-}64)$$

Thus

$$\frac{\Delta E}{E_{\text{bind}}} \lesssim 1 \qquad \text{provided} \qquad b > f r_{\text{cloud}}, \quad (7\text{-}65a)$$

where

$$f \equiv 1.6 \left(\frac{GM_{\text{cloud}}}{V^2 r_{\text{cloud}}} \right)^{\frac{1}{4}} \left(\frac{M_{\text{cloud}}}{M_{\text{cluster}}} \right)^{\frac{1}{4}} \left(\frac{\overline{r^2} r_h}{r_{\text{cloud}}^3} \right)^{\frac{1}{4}}. \quad (7\text{-}65b)$$

For typical cloud and cluster parameters $f \approx 0.5$. Thus, when we adopt $b_{\min} = r_{\text{cloud}}$, condition (7-65a) is not significantly violated. Equation (7-62) now yields

$$\dot{E} = \frac{4\sqrt{\pi}g}{3\sigma} G^2 \left(\frac{M^2 n}{r^2} \right)_{\text{cloud}} (M\overline{r^2})_{\text{cluster}}. \quad (7\text{-}66)$$

When we divide the cluster binding energy (7-64) by equation (7-66), we obtain the time t_d required for distant encounters to dissolve the cluster:

$$t_d \approx \frac{E_{\text{bind}}}{\dot{E}} = \frac{0.085\sigma}{gG} \left(\frac{M}{\overline{r^2} r_h} \right)_{\text{cluster}} \left(\frac{r^2}{M^2 n} \right)_{\text{cloud}}. \quad (7\text{-}67)$$

The cloud parameters M_{cloud}, n_{cloud}, and r_{cloud} are all poorly determined. Fortunately, they enter equation (7-67) in terms of the observationally accessible combinations $\Sigma_{\text{cloud}} \equiv (M/\pi r^2)_{\text{cloud}}$, the mean surface density of a cloud, and $\rho_{\text{mol}} \equiv (Mn)_{\text{cloud}}$, the mean density of gas in molecular clouds. We adopt $\Sigma_{\text{cloud}} = 250 \, M_\odot \, \text{pc}^{-2}$ and $\rho_{\text{mol}} = 0.025 \, M_\odot \, \text{pc}^{-3}$ (see Hut & Tremaine 1985), roughly half of the total gas density (see Table 1-1). Taking $\sigma = 7 \, \text{km s}^{-1}$ and $g = 3$, we have

$$t_d \approx 5.7 \times 10^8 \left(\frac{M_{\text{cluster}}}{250 \, M_\odot} \right) \left(\frac{1 \, \text{pc}}{r_h} \right)^3 \left(\frac{r_h^2}{\overline{r^2}} \right) \text{yr}. \quad (7\text{-}68)$$

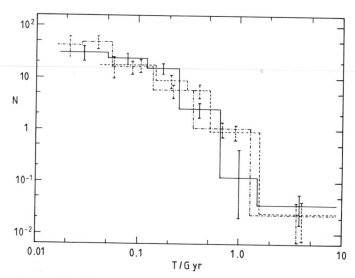

Figure 7-5. The age distribution of open clusters within 1 kpc of the Sun. The y-axis gives the number of clusters per square kpc per 10^8 yr. The full line shows data from Becker and Fenkart (1971), while the broken lines show two age calibrations of Lindoff (1968). (From Wielen 1971.)

Clearly, this result is very uncertain, in part because of uncertainties in the molecular cloud parameters, but also because t_d depends on the cube of the cluster radius and the ill-determined ratio of the mean-square radius to the median radius. However, the estimate (7-68) does suggest that we should not expect to see many open clusters older than about 10^9 yr. In fact, it was the observation of Oort (1958) and von Hörner (1958) that there are very few open clusters with ages $\gtrsim 10^9$ yr that prompted Spitzer (1958) to derive equation (7-67), thus showing that clusters might be dissolved by the very clouds that bring them into the world. The available data on the age distribution of open clusters have been reviewed by Wielen (1971), from whose paper we take Figure 7-5. Uncertainties in the ages of clusters and the wide range of cluster structures at birth make it difficult to determine the half-life $t_{1/2}$ of a typical cluster at all exactly, but the data suggest $t_{1/2} \simeq 2 \times 10^8$ yr. In view of the many simplifying assumptions we have employed, this figure is in satisfying agreement with our estimate (7-68). Two-body relaxation (§8.4) also makes an important contribution to the dissolution of open clusters. For a comprehensive review of this subject, see Wielen (1985).

(b) Disruption of wide binaries   Binary stars in the disk of our Galaxy can be thought of as open clusters with just two members. Like clusters, binaries can be disrupted by encounters with passing perturbers. Obviously the vulnerability of a binary to disruption is an increasing function of the separation s or semi-major axis a of its components. Since the orbital periods of wide binaries may be 10^6 yr or even longer, their existence must be inferred from the presence of pairs of stars in close proximity in space, line-of-sight velocity, and proper motion. The orbital motion cannot be directly detected.

First we consider disruption of wide binaries by molecular clouds. We specialize to a population of binaries with a one-dimensional velocity dispersion of $30 \, \mathrm{km \, s^{-1}}$, and assume that each binary is composed of two solar-type stars on a circular orbit of radius a. To sufficient accuracy, we may treat the binary as a cluster with mass $M_{\mathrm{cluster}} = 2 \, M_\odot$ and radius $r_h = \sqrt{\overline{r^2}} = a$. For typical cloud parameters, the quantity f defined by equation (7-65b) is approximately 0.3, so that equation (7-67) for the disruption time remains valid. In our earlier calculation, we assumed that the open clusters and the molecular clouds had the same velocity dispersion $\sigma = 7 \, \mathrm{km \, s^{-1}}$. Generalizing to the present case of unequal dispersions, we take $\sigma = \sqrt{(30^2 + 7^2)/2} \, \mathrm{km \, s^{-1}} \simeq 22 \, \mathrm{km \, s^{-1}}$. With the usual parameters for the molecular clouds, we find $t_d \approx 1.4 \times 10^{10} (0.1 \, \mathrm{pc}/a)^3$ yr. This result is subject to the same substantial uncertainties that surround equation (7-68), and some additional complications as well. In particular, (i) typical binaries spend much of their time above or below the thin molecular cloud layer, during which time they are not subject to tidal forces from the clouds; (ii) the gas density in the Galaxy may have been higher in the past (see §9.2), which would enhance the cumulative disruptive effects of the clouds; (iii) in contrast to open clusters, binary stars are also disrupted by encounters with individual field stars, at a rate that is comparable to and possibly exceeds the rate of disruption by clouds. For a more complete discussion of the disruption of wide binaries, see Bahcall et al. (1985), Hut and Tremaine (1985), and Problem 8-3.

It is interesting that our crude estimate of the disruption time indicates that binaries with semi-major axes in excess of about 0.1 pc would not survive for a Hubble time. The widest known binaries do indeed have separations of about 0.1 pc (Latham et al. 1984); however, it is still controversial whether the apparent cutoff in binary separations near 0.1 pc arises from observational selection, the conditions of binary formation, or the destructive effects of encounters with clouds or other stars (Wasserman & Weinberg 1987).

(c) **Disk shocking of globular clusters**   In the solar neighborhood the galactic disk has density $\rho \simeq 0.18 \, M_\odot \, \mathrm{pc}^{-3}$, and the scale-height on which this density decreases with increasing distance from the plane is $z_0 \approx 350 \, \mathrm{pc}$. Globular clusters regularly pass through the galactic disk, and since the tidal radii r_t of clusters fall in the range $30 \, \mathrm{pc} \lesssim r_t \lesssim 100 \, \mathrm{pc}$ (see Table 1-3 or Table 7-6 of MB), there are times when any cluster is wholly immersed in the disk. At these times the attractive gravitational field of the cluster is enhanced by the gravitational field of the layers of the disk that pass through the cluster. The fractional increase in the gravitational attraction is most pronounced for stars near the cluster's tidal radius: the mean density $\bar{\rho}_c \equiv 3m(r)/(4\pi r^3)$ of cluster stars interior to cluster radius r decreases strongly with increasing r, while the density of the disk material is independent of r. Hence for sufficiently large r the disk stars may dominate the force on cluster stars so long as the cluster is within the disk. In fact, we shall see in §7.3 [eq. (7-84)] that $\bar{\rho}_c(r_t)$ is approximately $3\bar{\rho}_g(R)$, where $\bar{\rho}_g(R) \equiv 3M(R)/4\pi R^3$ is the mean density of galactic matter interior to the cluster's orbit. At the solar radius $\bar{\rho}_g(R_0) \approx 0.04 \, M_\odot \, \mathrm{pc}^{-3}$, which is significantly less than the density $\simeq 0.18 \, M_\odot \, \mathrm{pc}^{-3}$ of galactic material in the disk. Hence the additional attraction due to the disk stars is large at r_t, and even though it only acts during the limited time during which the cluster is in the disk, it may have important consequences for the cluster's dynamics.

To estimate the effects of the disk on clusters, we must first determine the mean speed at which clusters cross the disk. If the galactic potential were spherical, the magnitude of the velocity $\mathbf{v}$ of a globular cluster on an inclined circular orbit would be the circular speed v_c, and as the cluster crossed the plane the component of $\mathbf{v}$ perpendicular to the plane would be $v_\perp = v_c \sin\theta$, where θ is the angle between the cluster's angular momentum vector and the Galaxy's symmetry axis. Furthermore, since the globular cluster population of our Galaxy forms a nearly spherical subsystem (see §4-4 of MB), if all globular clusters were on such circular orbits, the $\cos\theta$ values of globular clusters would be nearly uniformly distributed, and hence the mean value of $|v_\perp|$ would be $V_\perp \equiv \overline{|v_\perp|} = v_c \int_0^{\pi/2} \sin^2\theta \, d\theta = \frac{\pi}{4} v_c$. For $v_c = 220 \, \mathrm{km\,s}^{-1}$, $V_\perp \simeq 170 \, \mathrm{km\,s}^{-1}$. Although globular cluster orbits are probably not all circular, this estimate of $V_\perp$ is sufficiently accurate for our purposes.

The internal velocity dispersions of globular clusters are of order $\sigma_\star = 5 \, \mathrm{km\,s}^{-1}$ (see Table 1-3 or Table 7-6 of MB), so during the passage of the cluster through the disk, a typical cluster member moves along its orbit a distance of order $2z_0(\sigma_\star/V_\perp) \approx 20 \, \mathrm{pc}$. Thus orbits that are confined to, say, the innermost 2 pc of the cluster evolve adiabatically during the passage through the disk, and emerge unscathed from the encounter. But the effect of the passage on orbits that lie between about 20 pc of the cluster center and the cluster tidal radius $r_t \approx 50\,\mathrm{pc}$ can be crudely treated by the impulse approximation.[8]

Let (R, Z) be cylindrical coordinates with origin at the galactic center, and let (x, y, z) be a Cartesian coordinate system with origin at the cluster center and oriented so that the z-axis points vertically out of the plane and the x-axis points radially outward. Since the radius of the globular cluster is small compared with the scale-height z_0 of the galactic disk, we may evaluate equation (7-46) for the changes $\Delta \mathbf{v}$ in the velocities of stars with respect to the cluster center by expanding the galactic potential in the Taylor series (7-47) about $\mathbf{x} = 0$. We calculate only Δv_z, which is the largest component of $\Delta \mathbf{v}$. By equation (7-49) we have to first order in $|\mathbf{x}|$,

$$
\begin{aligned}
\Delta v_z(\mathbf{x}) &\simeq - \int \mathbf{x} \cdot [\boldsymbol{\nabla}(\nabla_z \Phi)]_{\mathbf{x}=0}\, dt \\
&= \int \mathbf{x} \cdot (\boldsymbol{\nabla} g_z)_{\mathbf{x}=0} dt,
\end{aligned}
\tag{7-69}
$$

where $\mathbf{g}$ is the Galaxy's gravitational field. Since the field is axisymmetric,

$$
\mathbf{x} \cdot \boldsymbol{\nabla} g_z = x \frac{\partial g_z}{\partial R} + z \frac{\partial g_z}{\partial Z}.
\tag{7-70}
$$

Retaining only the second, larger term in this expression, we find that the change in v_z as the cluster center moves from $Z = -Z_1$ to $Z = Z_1$ is given by

$$
\begin{aligned}
|\Delta v_z(\mathbf{x})| &\simeq \left| z \int_{-Z_1}^{Z_1} \left(\frac{\partial g_z}{\partial Z} \right)_{Z=v_\perp t} dt \right| = \left| \frac{z}{v_\perp} \int_{-Z_1}^{Z_1} \frac{\partial g_z}{\partial Z}\, dZ \right| \\
&\simeq \frac{2z}{v_\perp} |g_z(R, Z_1)|,
\end{aligned}
\tag{7-71}
$$

where we have exploited the fact that g_z is an odd function of Z and again neglected the radial gradient of g_z. If the z-coordinates of a star at successive passages of the cluster through the disk are uncorrelated, the mean-square velocity $\langle v^2 \rangle$ of the star will increase on a timescale

$$
t_{\mathrm{shock}} \equiv \left[\frac{1}{\sigma_\star^2} \frac{d\langle v^2 \rangle}{dt} \right]^{-1} \simeq \frac{\tfrac{1}{2} T_\psi \sigma_\star^2}{\langle |\Delta \mathbf{v}|^2 \rangle} \simeq \frac{T_\psi \sigma_\star^2 V_\perp^2}{8 z^2\, g_z^2},
\tag{7-72}
$$

----

[8] The characteristic speed of stars near the tidal radius is actually less than $\sigma_\star$, so this approximation is actually better than it may seem. However, one may show (see Problem 7-5) that if the cluster stars had periods much longer than the cluster's orbital period around the Galaxy, they would *not* acquire random motions as a result of the cluster passing through the disk.

where $v_\perp$ has been replaced by the average value $V_\perp$, T_ψ is the azimuthal period of the cluster's orbit, and $\overline{z^2}$ and $\overline{g_z^2}$ are the mean-square values of z and $g_z(R, Z_1)$ at each passage. For a spherical shell of stars of radius r, $\overline{z^2} = \frac{1}{3}r^2$, so for stars at the median radius, $\overline{z^2} = \frac{1}{3}r_h^2$. In Table 1-1 we saw that in the solar neighborhood the disk has surface density $\Sigma_0 \simeq 75\,M_\odot\,\mathrm{pc}^{-2}$. An infinite disk with this surface density generates a gravitational field $g_z = 2\pi G\Sigma_0 = 6.6 \times 10^{-9}\,\mathrm{cm\,s}^{-2}$. Above an exponential disk, $g_z \propto \exp(-R/R_d)$, where $R_d \simeq 0.41R_0$ in our Galaxy (see Table 1-2). So we estimate $\overline{g_z^2}$ at $R = 0.5R_0$, where most of the best-observed clusters cross the disk, as $\overline{g_z^2}(0.5R_0) \simeq [\exp(0.5R_0/R_d) \times 6.6 \times 10^{-9}]^2 \simeq 5 \times 10^{-16}\,\mathrm{cm}^2\,\mathrm{s}^{-4}$. Taking $T_\psi = 1 \times 10^8\,\mathrm{yr}$, $\sigma_\star = 5\,\mathrm{km\,s}^{-1}$, $V_\perp = 170\,\mathrm{km\,s}^{-1}$, and $r_h = 10\,\mathrm{pc}$, we obtain

$$t_\mathrm{shock} \approx 6 \times 10^9\,\mathrm{yr}. \tag{7-73}$$

Thus this process, which is known as **disk shocking** (Ostriker et al. 1972), is unimportant for stars that lie close to the cluster center, but it may unbind loosely bound stars. Spitzer and Chevalier (1973) have studied the effect of disk shocks on models of globular clusters of the type described in §8.4. They find that disk shocks substantially enhance the rate at which stars are lost from clusters, and, paradoxically, thereby increase the rate at which two-body relaxation causes the central parts to shrink.

(d) Ring galaxies   About a dozen galaxies are known that consist of prominent rings enclosing either one or two small bright patches. Figure 7-6 shows one such system, II Hz 4. These remarkable systems are thought to be generated by the head-on impact of a disk galaxy with another system; the disk gives rise to the ring, while the two bright patches represent the nuclei of the colliding systems. We can use equation (7-56) to develop an instructive model of ring-galaxy formation, even though the impulse approximation may not apply to all ring galaxies.

Suppose that a singular isothermal sphere with circular speed v_c scores a direct hit at speed $V \gg v_c$ on an identical singular isothermal sphere. The change Δv_R in the velocity of a star at distance R from the intruder's line of motion is by equation (7-56)

$$\Delta v_R = -2R\frac{v_c^2}{V}\int_0^\infty \frac{dZ_p}{R^2 + Z_p^2} = -\frac{\pi v_c^2}{V}. \tag{7-74}$$

Now consider the response of a stellar disk that resides in the target system with its normal parallel to the intruder's line of motion. At time

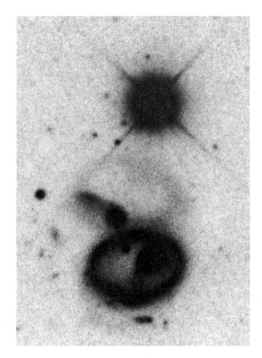

Figure 7-6. The ring galaxy II Hz 4. (Courtesy of National Optical Astronomy Observatories.)

$t = 0$, immediately after the intruder has passed, each star in this disk is moving inward with speed (7-74). Hence the disk first contracts and then expands as the stars bounce off the "centrifugal barriers" set by their non-zero angular momenta. If V is sufficiently large, and therefore $\Delta v_R/v_c$ small, we may neglect the changes in the target's potential that are generated by these oscillations, and assume that stars originally at R_0 execute harmonic radial oscillations at the epicycle frequency $\kappa(R_0) = \sqrt{2}(v_c/R_0)$ [see eq. (3-58)]. Let $R(R_0, t)$ be the radius at time t of a star that was originally at R_0. Then equation (3-67) implies

$$R(R_0, t) = R_0 - \Delta R \sin(\kappa t). \tag{7-75}$$

Differentiating this expression with respect to time, setting $t = 0$, and comparing with equation (7-74), we eliminate ΔR to find,

$$R(R_0, t) = R_0 \left[1 - \frac{\pi v_c}{\sqrt{2}V} \sin\left(\frac{\sqrt{2}v_c t}{R_0} \right) \right]. \tag{7-76}$$

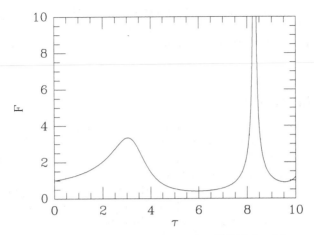

Figure 7-7. A plot of the function $F(\tau)$ defined by equation (7-77c) for the case $v_c/V = 0.1$.

If the undisturbed disk had surface density $\Sigma_0(R_0)$, the surface density at time t is

$$\Sigma(R) = \Sigma_0(R_0)\left[\frac{R}{R_0}\left|\frac{\partial R}{\partial R_0}\right|\right]^{-1} \tag{7-77a}$$

$$= F(\tau)\Sigma_0(R_0),$$

where

$$\tau \equiv \frac{\sqrt{2}v_c t}{R_0} \tag{7-77b}$$

and

$$F(\tau) \equiv \left(1 - \frac{\pi v_c}{\sqrt{2}V}\sin\tau\right)^{-1}\left|1 - \frac{\pi v_c}{\sqrt{2}V}(\sin\tau - \tau\cos\tau)\right|^{-1}. \tag{7-77c}$$

Thus the new surface density $\Sigma(R)$ is equal to the original surface density at a neighboring point times an amplification factor F that depends only on the ratio t/R_0. In Figure 7-7, $F(\tau)$ is plotted for $v_c/V = 0.1$. F is seen to peak first at $\tau_m = 3.1$. Thus for this value of v_c/V, at any time t, Σ is most sharply amplified at $R_m(t) = \sqrt{2}v_c t/\tau_m$. Physically this corresponds to a wave of enhanced density moving outward at speed $\sqrt{2}v_c/\tau_m$. Lynds and Toomre (1976) identify the crests of this wave with the rings of galaxies such as II Hz 4.

Numerical experiments (Lynds & Toomre 1976) show that whenever an intruder passes close to the center of the target disk on a trajectory that is inclined by less than about 30° to the axis of the disk, the physical process described by our simple model generates a striking ring. If the intruder misses the center of the target just slightly, the target nucleus is displaced from the center of the ring, as is observed in II Hz 4. If both intruder and target are disk systems, two rings can be generated.

7.3 Tidal Radii

We have just seen that when globular clusters pass through the galactic plane, the velocities of the cluster stars are randomly perturbed. Similarly, galaxies in clusters are subject to tidal perturbations from neighboring galaxies. Repeated perturbations of the velocities of stars in a stellar system cause the orbits of the stars to diffuse in phase space, just as the Brownian motion of gas molecules causes one gas to diffuse through another.[9] Hence we would expect orbits in systems such as globular clusters and cluster galaxies to be smoothly populated. In particular, orbital diffusion would be expected to erase all sharp features in the spatial structures of these systems. Hence it is at first sight surprising to observe that globular clusters and many cluster galaxies have rather sharp edges. The origin of this apparent conflict of theory and observation lies in our neglect of the large-scale gravitational field of the galaxy or cluster of galaxies in which the system that experiences gravitational encounters orbits. In reality this field, which we shall call the field of the **host system**, prunes the **satellite system**, after each encounter, and thus prevents the satellite from swelling. Systems that are continuously pruned in this way have density profiles that plunge to zero at a finite radius r_t. In practice the **tidal radius** r_t is measured by fitting the projected density profile with one of the Michie-King models discussed in §4.4. Our goal in this section is to find a dynamical basis for this ad hoc fitting procedure.

In the simplest case, the satellite system, mass m, is on a circular orbit beyond the outer edge of the host system. We suppose that both systems are spherical and let their centers be distance D apart. Then if the mass of the host system is M, the angular speed with which the systems orbit around their common center of mass is

$$\Omega = \sqrt{\frac{G(M+m)}{D^3}}. \tag{7-78}$$

The gravitational potential generated by the two systems is stationary when referred to a coordinate system that is centered on the common center of mass, and rotates at speed Ω. We orient this coordinate system

---

[9] We shall develop this idea in §8.3.

so that the centers of the satellite and host systems are at $\mathbf{x}_m = [D(1 + m/M)^{-1}, 0, 0]$ and $\mathbf{x}_M = [-D(1 + M/m)^{-1}, 0, 0]$, respectively. In §3.3.2 we studied orbits in steadily rotating potentials, and found that along any orbit in such a potential, Jacobi's integral [eq. (3-83b)]

$$
\begin{aligned}
E_J &= \tfrac{1}{2}v^2 + \Phi(\mathbf{x}) - \tfrac{1}{2}|\boldsymbol{\Omega} \times \mathbf{x}|^2 \\
&= \tfrac{1}{2}v^2 + \Phi_{\mathrm{eff}}(\mathbf{x})
\end{aligned}
\tag{7-79}
$$

is constant, where $\boldsymbol{\Omega}$ is the vector $(0, 0, \Omega)$. Since $v^2 \geq 0$, a star whose Jacobi integral takes the value E_J will never trespass into a region where $\Phi_{\mathrm{eff}}(\mathbf{x}) > E_J$. Consequently, the surface $\Phi_{\mathrm{eff}}(\mathbf{x}) = E_J$, which we called the zero-velocity surface for stars of Jacobi integral E_J, forms an impenetrable wall for such stars. Figure 7-8 shows the curves in which the equatorial plane of two orbiting bodies cuts the zero-velocity surfaces corresponding to several values of E_J. For simplicity we have drawn Figure 7-8 for two point masses, but the general form of the figure would be similar for any pair of centrally concentrated bodies with a similar mass ratio. From the figure we see that the zero-velocity surfaces near the satellite system m are centered on m, but farther out the zero-velocity surfaces surround both bodies. Hence, at the critical value of E_J corresponding to the last zero-velocity surface to enclose only m, there is a discontinuous change in the region within which Jacobi's integral confines stars. This suggests that we identify the tidal radius of m as the distance r_J between m and the saddle point of Φ_{eff} that lies between the two masses—this saddle point is one of the Lagrange points (cf. Figure 3-13).

We may evaluate r_J by noticing that at $(x_m - r_J, 0, 0)$ the effective potential has a saddle point, so

$$
\left(\frac{\partial \Phi_{\mathrm{eff}}}{\partial x}\right)_{x = x_m - r_J} = 0.
\tag{7-80}
$$

Now for two point masses a distance D apart

$$
\Phi_{\mathrm{eff}}(\mathbf{x}) = -G\left[\frac{M}{|\mathbf{x} - \mathbf{x}_M|} + \frac{m}{|\mathbf{x} - \mathbf{x}_m|} + \tfrac{1}{2}\frac{M+m}{D^3}|\hat{\mathbf{e}}_z \times \mathbf{x}|^2\right].
\tag{7-81}
$$

Hence

$$
0 = \frac{1}{G}\left(\frac{\partial \Phi_{\mathrm{eff}}}{\partial x}\right)_{x_m - r_J} = \frac{M}{(D - r_J)^2} \mp \frac{m}{r_J^2} - \frac{M+m}{D^3}\left(\frac{D}{1 + m/M} - r_J\right),
\tag{7-82}
$$

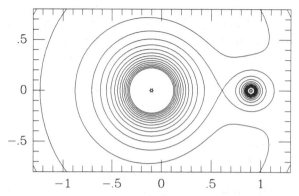

Figure 7-8. Contours of equal effective potential Φ_{eff} defined by equation (7-81) for two point masses, m and $M = 9m$, in circular orbit about one another. The particles are unit distance apart. The center of mass is at the origin, and the central Lagrange point is located near $(0.6,0)$.

where the top (bottom) sign is chosen if r_J is positive (negative). This leads to a cubic equation for r_J. However, if $m \ll M$, then $r_J \ll D$, and we can expand $(D - r_J)^{-2}$ in powers of (r_J/D) to find

$$0 = \frac{M}{D^2}\left(1 + \frac{2r_J}{D} + \cdots\right) \mp \frac{m}{r_J^2} - \frac{M}{D^2} + \frac{M+m}{D^3}r_J. \tag{7-83}$$

Hence to first order in (r_J/D),

$$r_J = \pm D\left[\frac{m}{M\left(3 + m/M\right)}\right]^{\frac{1}{3}} \simeq \pm\left(\frac{m}{3M}\right)^{\frac{1}{3}} D. \tag{7-84}$$

We call radius r_J the **Jacobi limit** of the mass m. The Jacobi limit of an orbiting stellar system provides a useful estimate of the system's tidal radius r_t, but it is important to realize that r_J provides only a crude estimate of r_t. There are several reasons why r_J may differ from r_t:

(i) The zero-velocity surface is not spherical (see Figure 7-8 and Problem 7-4), so it cannot be adequately characterized by a single radius.

(ii) A star will not necessarily escape from the neighborhood of the satellite system just because its zero-velocity surface does not close around m. As we saw in §3.3, most orbits in steady potentials respect non-classical integrals in addition to the classical integral E_J. Numerical orbit calculations show that non-classical integrals can confine stars to the immediate neighborhood of m that would otherwise be free to go. Studies by Hénon (1969a, 1970), Jefferys (1974, 1976), and Keenan (1980) have shown that there can be trapped orbits about m with apocenters as large as $2r_J$ or more. Thus we do not expect the stellar density $\rho(r)$ to drop abruptly to zero at r_J. However, the proportion of phase space occupied by bound orbits diminishes very rapidly beyond r_J, so we do expect ρ to decline steeply near r_J, and we shall not err greatly if we identify r_J with the empirical tidal radius r_t.

(iii) In general, the satellite system m will not be on a circular orbit. When m is on an elongated orbit, there is no frame in which the potential experienced by a test star is stationary, and no analog of Jacobi's integral exists.[10] Thus no direct generalization of our derivation of the Jacobi limit to the case of elongated orbits is possible. King (1962) has argued that in the case of elongated orbits, the tidal limit is determined by conditions at pericenter, while Innanen et al. (1983) argue that the radius of the satellite system pulsates in phase with the system's orbit through the host's field. Theoretical estimates of the tidal radii of globular clusters must be considered very uncertain until these conjectures have been tested by convincing numerical experiments.

(iv) In many cases, the satellite orbits within the body of the host system, so that the point-mass approximation used in deriving equation (7-84) is not accurate.

Thus many factors combine to decide whether a particular star is tidally stripped from a satellite in an elongated orbit in the potential of a massive host system. A characteristic radius such as r_J can never be more than a rough guide to which stars are most vulnerable to tidal stripping, and what value of r_t will be obtained on fitting a Michie-King model to the system's surface density profile.

7.4 Mergers

When two galaxies of similar mass collide at a speed that is comparable with the internal velocities of the galaxies, neither the dynamical friction approximation of §7.1, nor the impulse approximation of §7.2 is valid. In these cases we must have recourse to numerical simulations such as the one shown in Figure 7-1. These simulations suggest that a slow encounter between galaxies often leads to the two systems merging into one. We discussed the origin of this phenomenon from a qualitative point

---

[10] However, analogs of the Lagrange points do exist (Szebehely 1967).

of view at the beginning of the chapter. In this section we shall try to determine: (i) the ranges of initial speeds and impact parameters for which the encounter leads to a merger; (ii) how these ranges depend on the internal structures and orientations of the original galaxies; (iii) the observational properties, such as the density profile, velocity dispersion, and rotation curve of the system formed in the merger.

1 Encounters between Spherical Galaxies

We consider first the simplest case, which is the encounter of two non-rotating spherical galaxies. We assume for simplicity that the galaxies have the same mass M and internal structure, being characterized by median radius r_h and internal mean-square velocities $\langle v^2 \rangle \simeq 0.4GM/r_h$, where we have used equation (4-80b). Let the reduced particle that describes the relative orbit of the two galaxies have mass $\mu = \frac{1}{2}M$, and energy and angular momentum per unit mass E_{orb} and L. Then the encounter may be characterized by the dimensionless parameters

$$\hat{E} \equiv \frac{E_{\mathrm{orb}}}{\frac{1}{2}\langle v^2 \rangle}, \quad \text{and} \quad \hat{L} \equiv \frac{L}{r_h \langle v^2 \rangle^{1/2}}. \qquad (7\text{-}85)$$

(a) Criteria for merging   It is convenient to associate each initial orbit with a point in the $(\hat{E}, \hat{L})$ plane that is shown in Figure 7-9. Points to the right of the line marked "parabolic orbits" in Figure 7-9 correspond to unbound orbits, while the left side of the figure is occupied by bound orbits; these orbits are often called hyperbolic and elliptic orbits by analogy with the Keplerian case, those with $\hat{E} = 0$ being called parabolic orbits. The full curve in Figure 7-9 marks the locus $\hat{L} = \hat{L}_{\mathrm{circ}}(\hat{E})$ of the angular momentum of the circular orbit with energy $\hat{E}$. Since a circular orbit has the maximum possible angular momentum for a given energy, points in Figure 7-9 that lie above the full curve do not correspond to possible orbits.

Any bound orbit $\hat{E} < 0$ will eventually lead to a merger because even the smallest tidal interaction will eventually drain away the orbital energy. But it is clear that low angular momentum or low energy orbits will lead to mergers more quickly than orbits that keep the galaxies well separated. The dashed line in Figure 7-9 crudely divides the region of $(\hat{E}, \hat{L})$ space corresponding to bound orbits into a lower portion, in which galaxies merge within a Hubble time, and an upper region, in which mergers occur more slowly or not at all.

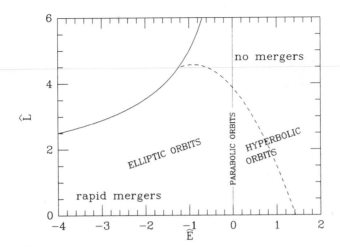

Figure 7-9. The time required for two galaxies to merge
is a function of the initial position of the binary orbit
in the $(\hat{E}, \hat{L})$ plane defined by equations (7-85). Orbits
are only possible below and to the right of the full curve
formed by the circular orbits. In principle all elliptic
orbits $(\hat{E} < 0)$ will eventually lead to a merger, but the
time to merging increases rapidly toward the upper right
portion of the diagram. For typical galactic parameters,
orbits below and to the left of the dashed line evolve to
mergers in about a Hubble time.

It is sometimes useful to parametrize unbound orbits in terms of the
initial speed V and impact parameter b of the reduced particle. Since
$E_{\text{orb}} = \frac{1}{2}V^2$ and $L = bV$, we have in this case

$$\hat{E} = \frac{V^2}{\langle v^2 \rangle} \quad \text{and} \quad \hat{L} = \frac{b}{r_h}\sqrt{\hat{E}}. \qquad (7\text{-}86)$$

Clearly, at any given positive value of $\hat{E}$, and therefore of V, mergers
will not occur when b and therefore $\hat{L}$ are sufficiently great. Hence mergers will not occur in the upper right portion of Figure 7-9. Notice that
when $\hat{L} = 0$, that is, when the encounter is head-on, there is a maximum
energy $\hat{E}_{\text{max}}$, and therefore a maximum encounter speed V_{max}, such that
all encounters that start from $V < V_{\text{max}}$ lead to a merger. Several sim-
ulations have led to the estimates $V_{\text{max}} \simeq 1.2\sqrt{\langle v^2 \rangle} \Rightarrow \hat{E}_{\text{max}} \simeq 1.4$.
When $\hat{L} > 0$, the highest initial speed that leads to a merger is smaller
than V_{max}.

When $0 < \hat{E} < \hat{E}_{\text{max}}$, there is a maximum angular momentum
$0 < \hat{L}_{\text{max}}(\hat{E})$ that leads to a merger. As $\hat{E} \to 0$, $\hat{L}_{\text{max}}$ increases without

limit because when $\hat{E} \simeq 0$ even a wide passage can cause the orbit to become bound, and a merger will then inevitably ensue. However, numerical experiments show that the only parabolic encounters that lead to reasonably rapid merging are those with $\hat{L} \lesssim 3.5$. With this value of $\hat{L}$, the centers of rigid galaxies on a parabolic orbit would pass within about $3r_h$ of each other.

We define the merger cross-section $\mathcal{A}_{\mathrm{merge}}$ of an unbound orbit as

$$\mathcal{A}_{\mathrm{merge}}(\hat{E}) = \pi[b(\hat{L}_{\mathrm{max}}, \hat{E})]^2 = \frac{\hat{L}_{\mathrm{max}}(\hat{E})}{\hat{E}}\pi r_h^2, \qquad (7\text{-}87)$$

where we have made use of the second of equations (7-86). Hence $\mathcal{A}_{\mathrm{merge}}$, which decreases from infinity at $\hat{E} = 0$ to zero at $\hat{E}_{\mathrm{max}} \simeq 1.4$, can be very much greater than the galaxies' geometrical cross-section πr_h^2.

(b) Structure of merger remnants   When two galaxies merge, the first minimum in the separation of the centers is followed by a period during which the density distribution of the combined system fluctuates coherently. The density fluctuations are heavily damped by phase mixing (see §4.7.2), and the system soon settles into a new equilibrium configuration, which we shall call the **remnant**. Since the directions of the initial encounter velocity **V** and impact parameter **b** define special directions, we should not necessarily expect the remnant to be spherical, or even axisymmetric. Furthermore, the remnant may have a radial structure that is quite unlike those of the progenitor galaxies.

Numerical simulations yield the following conclusions (White 1978, 1979; Villumsen 1982, 1983):

(i) Remnants formed from head-on encounters ($\hat{L} = 0$) of spherical nonrotating galaxies are prolate bodies whose longest axes coincide with the initial lines of motion of the galaxy centers. The pattern speeds of these remnants are small or zero.

(ii) A remnant that forms from an orbit that is neither hyperbolic nor highly eccentric is generally oblate and rotating. However, the remnant's flattening is only partly caused by rotation; in addition, the velocity dispersion is greater parallel to the equatorial plane than in the perpendicular direction. The rotation curves are fairly flat out to the median radius, and then drop toward zero.[11]

---

[11] The importance of velocity anisotropy for the dynamics of the remnant depends on the magnitude of the remnant's dimensionless spin $\hat{S}$ defined by equation (7-88). Conservation of energy and angular momentum during the encounter imply $\hat{S} = \frac{1}{8}\hat{L}\sqrt{1 - \frac{1}{4}\hat{E}}$, where we have assumed that the dimensionless ratio $r_h\langle v^2\rangle/(GM)$ is the same for the progenitors and the remnant. Remnants with $\hat{S} \lesssim 0.4$ tend to be prolate (White 1979).

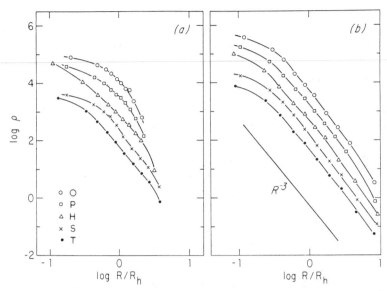

Figure 7-10. The density profiles of remnants tend to the form $\rho \propto r^{-3}$ independent of the initial density profile: (a) five pre-merger density profiles (offset vertically for clarity); (b) the density profiles of the remnants formed by merging pairs of identical systems with the profiles shown in (a). The symbols identify corresponding profiles and mark the radii containing 10%, 20%,... of each system's mass. (From White 1979.)

(iii) Figure 7-10 shows that over a range of about three decades in radius, radial density profiles of merger remnants closely resemble the power law $\rho \propto r^{-3}$. This result is interesting because, as we saw in §2.1.2, if $\rho \propto r^{-3}$, the surface density $\Sigma \propto r^{-2}$, which is reminiscent of the Hubble-Reynolds law (1-14) for the surface-brightness distributions of elliptical galaxies. Unfortunately, the central density profiles of merger remnants, which are observationally of greatest interest, cannot be reliably predicted by simulations with the available spatial resolution.

(iv) If the progenitors had radial composition gradients, for example metallicity gradients, the remnant has a similar, though less pronounced gradient.

2 Mergers of Spinning Galaxies

Since most galaxies rotate, it is important to understand how the merging process is affected by galaxy spin. Unfortunately, mergers between spinning galaxies are much more complicated than mergers of nonrotating galaxies, because six extra numbers are required to specify an encounter between spinning galaxies of equal mass: two to specify the magnitudes of the two spin vectors, and four to specify the orientations of the spin vectors with respect to the initial position and velocity vectors of the reduced particle. We cannot survey the whole of this six-dimensional parameter space in the way that we have surveyed $(\hat{E}, \hat{L})$ space for encounters between nonrotating galaxies. However, two specimen encounters will give an idea of the importance of galaxy spin for the merging process.

Figure 7-11 shows six snapshots of the encounter of two identical spinning galaxies on the orbit $(\hat{E} = -1.16, \hat{L} = 4.49)$. The spin angular momenta $\mathbf{S}$ of these galaxies satisfy[12]

$$\hat{S} \equiv \frac{S}{r_h \langle v^2 \rangle^{1/2}} = 0.6, \qquad (7\text{-}88)$$

where $\langle v^2 \rangle$ now includes both rotational and random internal velocities. Hence these galaxies are spinning more rapidly than the majority of giant elliptical galaxies (see §4.3), but less rapidly than a disk galaxy ($\hat{S} \simeq 1.8$; see Problem 7-6). The spin vectors are aligned parallel to one another and *anti-parallel* to the orbital angular momentum vector. (We shall call this a "retrograde" encounter.) Between each of the top three snapshots, the orbital motion has carried the galaxies about 180° in a clockwise sense around the center of mass, while between the third and fourth snapshots, and between the fourth and fifth snapshots, the orbital angle has increased by almost 360°. Hence by the fifth snapshot, the galaxies have completed three full rotations about one another without losing their identity. Only in the sixth snapshot, taken more than an orbital time later, can the galaxies be said to have merged.

Figure 7-12 shows how the same galaxies interact when we reverse the direction of their orbit about the center of mass so as to make the orbital and spin angular momenta *parallel* (a "prograde" encounter). Notice that the sixth snapshot in Figure 7-12 corresponds to almost the same time as the third snapshot of Figure 7-11. Thus these galaxies merge twice as rapidly in a prograde encounter as in a retrograde encounter.

A crude model of an encounter between spinning galaxies will help us to understand why the mutual alignment of the spin and orbital angular momentum vectors is so important. We replace the two galaxies

---

[12] In the literature, spin is often parametrized by $\lambda \equiv S\sqrt{|E_{\text{bind}}|}/(GM) \simeq 0.4\hat{S}/\sqrt{2}$, where E_{bind} is the system's binding energy per unit mass.

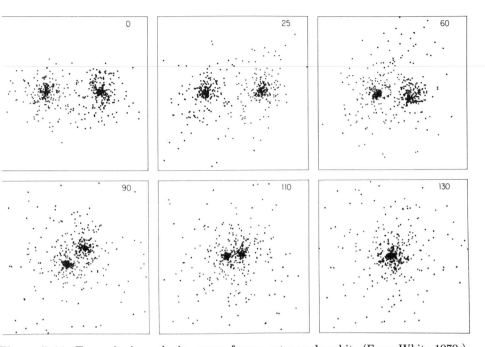

Figure 7-11. Two spinning galaxies merge from a retrograde orbit. (From White 1979.)

by two particles of equal mass M that move on an orbit similar to the
initial orbit of the galaxies in Figure 7-11. We now arrange five rings of
massless particles in circular orbits around one of the massive particles.
We choose the sense of these orbits to be opposite to that of the orbit of
the massive particles (retrograde encounter), and numerically integrate
the equations of motion of these test masses in the time-dependent gravitational field of the two orbiting massive particles. Figure 7-13 shows
that each ring gradually distorts. The degree of distortion increases
with radius, but it remains moderate even for the outermost ring at a
radius equal to 60% of the minimum separation D_{min} of the two massive
particles.

Now consider the behavior of the same system when the massive
particles orbit in the opposite sense (prograde encounter). Figure 7-14
shows that by the time the massive particles have reached the point of
closest approach, the ring system is devastated. Only the innermost
ring, whose initial radius was 20% of D_{min}, remains intact. The other
rings are first pulled into elongated shapes by the gravitational field of
the binary system in which they orbit, and are then snapped off where
they come close to either of the massive particles. The slow-motion
replay in Figure 7-15 of the distortion of the outermost ring (whose

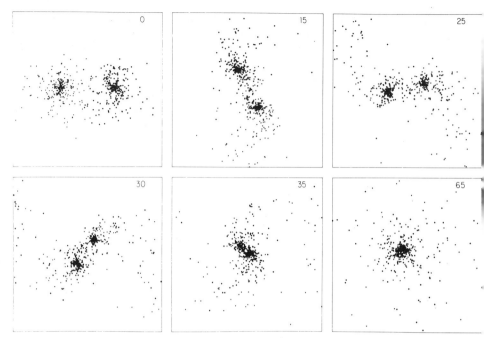

Figure 7-12. Two spinning galaxies merge from a prograde orbit. Notice how much faster the galaxies merge than in Figure 7-11 (times are shown at top right of each panel). (From White 1979.)

initial radius $r = 0.6D_{\min}$) illustrates this phenomenon. At the outset the ring distorts in a way that is reminiscent of the way the Moon raises tides on the surface of the oceans. Later, when the distortion has become large, the half of the ring that has been pushed away from the disturbing mass forms a fairly thin and distinctly curved **tidal tail**, while the half of the ring that has fallen toward the disturbing mass is captured by, and moves away with, the disturber.

Why is the prograde encounter so much more violent than the retrograde one? The orbital frequency of a ring of radius r is $\omega_{\text{ring}} = \sqrt{(GM/r^3)}$, while from equation (3-22b) the angular velocity of the line joining the two massive particles at pericenter is

$$\omega_{\text{orb}} = \sqrt{\left(\frac{2GM}{D_{\min}^3}\right)(1+e)}, \qquad (7\text{-}89)$$

where e is the eccentricity of the orbit. Setting $e = 1$ we have that a test particle on the ring in Figure 7-14 of radius $r = 4^{-2/3}D_{\min} \simeq 0.4D_{\min}$ is in resonance; it is continuously pulled either inward or outward, de-

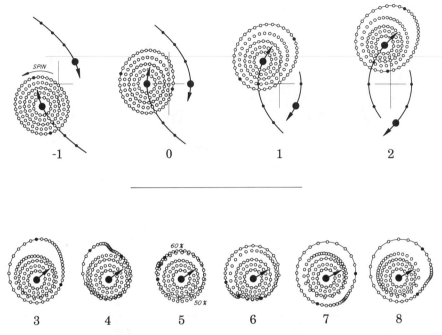

Figure 7-13. Two equal point masses move on a parabolic trajectory. Five rings of test particles form a counter-rotating disk about one of the masses. Reproduced from Toomre and Toomre (1972) by permission of *The Astrophysical Journal.*

pending on its initial position in relation to the disturbing mass. Consequently, it responds violently to the disturber. When, as in Figure 7-13, the rings and the disturber move in opposite senses, each test particle is pulled alternately inward and outward, with little net result.

These heuristic arguments explain why the prograde N-body encounter of Figure 7-12 leads to more rapid merging than does the retrograde encounter shown in Figure 7-11. Notice the two tidal tails, one from each galaxy, which are clearly seen in the third snapshot of Figure 7-12. These tails play a vital role in the merger process. Their gravitational fields tend to brake the galaxies and hence to drain energy and angular momentum from the orbital motion.

Our knowledge of the details of how spinning galaxies merge is still rudimentary, in part because the parameter space involved is so large, and in part because it is difficult to construct a stable numerical model of an isolated, rapidly spinning galaxy. Some results that have emerged are (Gerhard 1981; Farouki & Shapiro 1982; Negroponte & White 1983):

(i) The remnants are always featureless elliptical objects even when the progenitors had prominent disk components. The remnant profile

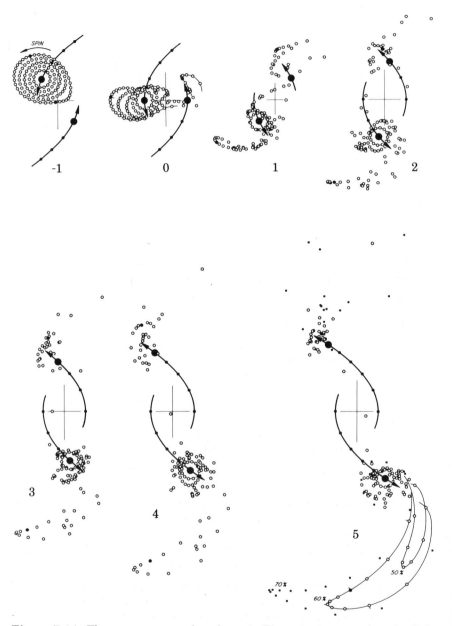

Figure 7-14. The same setup as that shown in Figure 7-13 except that the disk of test particles now corotates with the binary orbit. Reproduced from Toomre and Toomre (1972) by permission of *The Astrophysical Journal*.

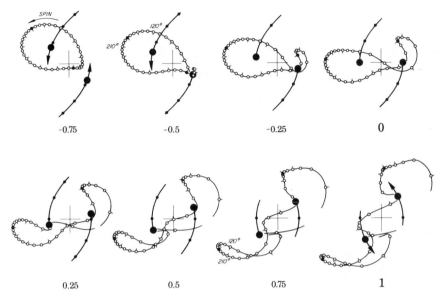

Figure 7-15. A slow-motion replay of the destruction of the outermost ring in Figure 7-14. Reproduced from Toomre and Toomre (1972) by permission of *The Astrophysical Journal.*

is approximately of the form $\rho \propto r^{-3}$ just as when spinless systems merge.

(ii) As in the case of nonrotating progenitors, when spinning galaxies merge, either a prolate or an oblate remnant can form, depending on the value of $\hat{L}$. The roundest systems are formed when the spin vectors are perpendicular to the orbital angular momentum **L**. Most remnants have significantly anisotropic velocity-dispersion tensors.

(iii) The spin of the final merger remnant varies greatly depending on the mutual orientation of the two spin vectors and the orbital angular momentum. However, only a small proportion of all possible initial configurations lead to remnants that rotate as slowly as giant elliptical galaxies. Typical remnant rotation curves resemble the observed rotation curves of elliptical galaxies in shape, but remnants generally lie higher in the $(v/\sigma, \epsilon)$ diagrams of Figure 4-6 than do giant elliptical galaxies.

3 Observations of Galaxy Mergers

Some galaxies appear to be in a highly disturbed state. Figure 7-16, which is a reproduction of a page of H. Arp's *Atlas of Peculiar Galaxies,*

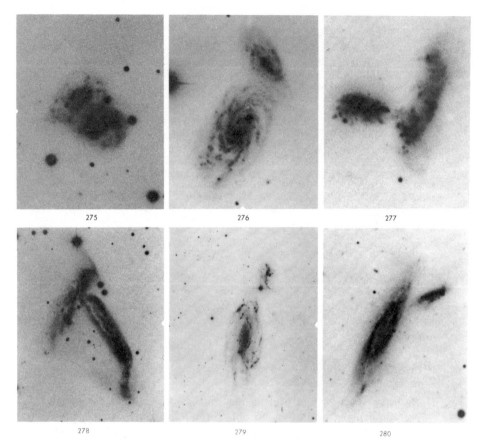

Figure 7-16. Some interacting galaxies. Reproduced from Arp (1966) by permission of *The Astrophysical Journal.*

shows six such systems. Hundreds of other examples can be found in Arp's atlas and in the catalogs of Vorontsov-Velyaminov (1968) and Arp and Madore (1987). At one time it was widely believed that irregular systems of this kind are exploding galaxies, but in the early 1970s it became clear that most are actually colliding systems, and that many of these collisions will result in the formation of a single remnant galaxy.

Most of the interacting galaxies that have been cataloged are distant and therefore difficult to observe. The dozen or so peculiar galaxies that are sufficiently nearby to be studied in detail are therefore especially important. Figure 7-17 is a picture of the NGC 4038/4039 pair of galaxies. This system consists of overlapping blobs of light from which two long curved tails of much lower surface brightness emerge. Were

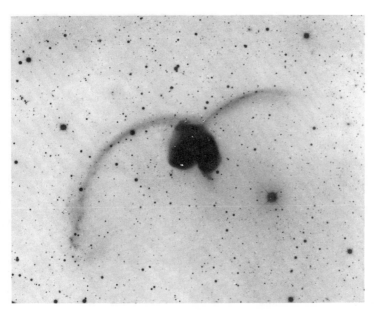

Figure 7-17. The interacting galaxies NGC 4038 and NGC 4039.
Courtesy of D. F. Malin and Kitt Peak National Observatory.

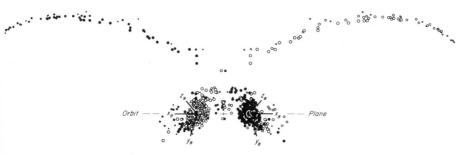

igure 7-18. A model of the NGC 4038/4039 pair by Toomre and Toomre (1972).
eproduced by permission of *The Astrophysical Journal.*

these long, curved tails formed in a collision, like the tidal tails in the
simulation of Figure 7-14? In a classic paper A. and J. Toomre (1972)
argued that this is indeed the case. The Toomres studied encounters between disks of massless particles orbiting around point masses like those
shown in Figure 7-14. They showed that for a suitable choice of orbits
and orientations of these systems, and of the direction from which we
view the encounter, it is possible to obtain a projected mass distribution
that is very similar to Figure 7-17. We show their model in Figure 7-18.

From a model like the Toomres', we can predict the line-of-sight

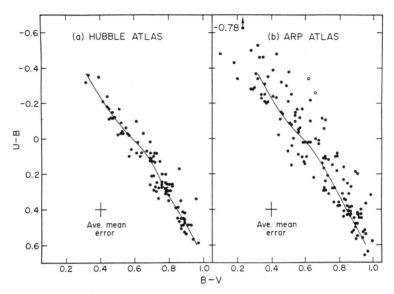

Figure 7-19. Tidally disturbed galaxies often have anomalous colors: the left panel shows the colors of predominantly undisturbed galaxies, while the galaxies of the right panel are all highly disturbed. The full curve is the same in each plot. Reproduced from Larson and Tinsley (1978) by permission of *The Astrophysical Journal.*

velocities[13] at each point in the system, and the observed velocities are in complete agreement with the model (Toomre 1977b). Unfortunately only a few such observational tests of the encounter hypothesis have so far been completed (Huchtmeier & Bohnenstengel 1975; Schweizer 1982).

(a) Photometry of merger remnants   Galaxies showing signs of severe tidal disturbance usually have anomalous $U - B$ and $B - V$ colors. This phenomenon is illustrated by Figure 7-19, which at left shows the two-color distribution of a sample of normal galaxies, and at right shows the same plot for the galaxies in Arp's *Atlas of Peculiar Galaxies.* Larson and Tinsley (1978) have shown that the greater spread of points in the right panel of Figure 7-19 is not due to observational errors, but is expected if galaxies undergo intense bursts of star formation when they are disturbed tidally. In the models of Larson and Tinsley, a typical burst of star formation temporarily doubles the galaxy's luminosity.

[13] Up to the change of sign that occurs when the model is reflected through the plane of the sky.

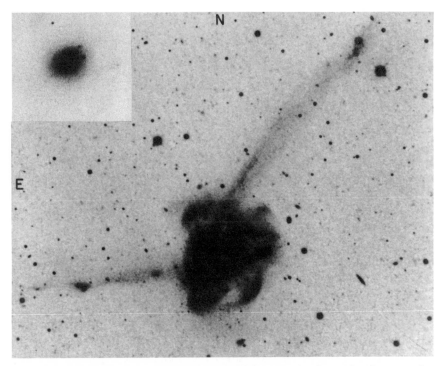

Figure 7-20. Two plates of the galaxy NGC 7252: in the main photograph, a deep exposure shows that at large radii the brightness distribution is highly irregular; in the inset, a shallow exposure at the same scale shows that near its center the galaxy is smooth and elliptical-like. Reproduced from Schweizer (1982) by permission of *The Astrophysical Journal*.

Figure 7-20 shows a deep photograph of the peculiar galaxy NGC 7252. From the two narrow tails that appear in this photograph we suspect that the rather confused main body of NGC 7252 was formed by the collision of two disk galaxies. The short-exposure plate shown in Figure 7-20 shows that the brightness distribution at the center of NGC 7252 is smooth and symmetrical. This suggests that the central portion of NGC 7252 has already relaxed to an equilibrium configuration,[14] and thus that the collision and merger are at a more advanced stage than in the case of NGC 4038/9. Furthermore, Figure 7-21 shows that the brightness distribution throughout the main body of the galaxy is well fitted by an $R^{1/4}$ law, suggesting that at least some galaxies whose spheroids fit the $R^{1/4}$ law may be merger remnants.

[14] The central parts relax faster than the outer parts, because the dynamical time is shortest at the center.

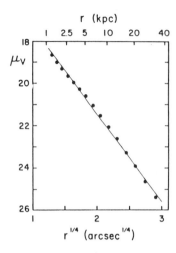

Figure 7-21. The radial brightness profile of the galaxy NGC 7252, shown in Figure 7-20, is quite well fitted by the $R^{1/4}$ law. Reproduced from Schweizer (1982) by permission of *The Astrophysical Journal*.

(b) Frequency of merging   We may estimate the number of galaxies that have experienced a merger as follows (Toomre 1977b). About 11 NGC[15] galaxies display tidal tails and other indications that these galaxies are undergoing a merger. The characteristic lifetime of a tidal tail is of order the orbital time at the Holmberg radius (see §1.1.3), say $5 \times 10^8 h^{-1}$ yr. Hence if mergers have occurred at a constant rate over a Hubble time, $1.0 \times 10^{10} h^{-1}$ yr, any given population of galaxies would contain about 20 merger remnants for every object currently displaying a tidal tail. Thus the population from which the 11 merging NGC galaxies are drawn may contain about 220 remnants. There are about 4000 NGC galaxies, but the population from which the 11 merging objects are drawn is probably rather larger than this; since galaxies become temporarily more luminous when they merge, a magnitude-limited sample of merging galaxies surveys a larger volume than a magnitude-limited sample of quiescent galaxies. In particular, temporarily doubling the luminosity of each merging galaxy makes the merging pair twice as luminous as the remnant will be when it has settled down, so the volume within which the pair satisfies a given magnitude limit is $2^{3/2} \simeq 2.83$ times larger than the corresponding volume for the remnant. Hence if the merger rate has been constant, there should be of order 78 remnants among NGC galaxies. Toomre (1977b) points out that there are about 440 elliptical galaxies in the NGC and argues that all these ellipticals are the product of mergers, which he suggests occurred more frequently in the past. This is an attractive idea, although it may be unable to account for the systematic variations of color and velocity dispersion

---

[15] NGC stands for "New General Catalog", a catalog of galaxies, nebulae, and star clusters compiled by Dreyer in 1890 and containing some 8000 objects.

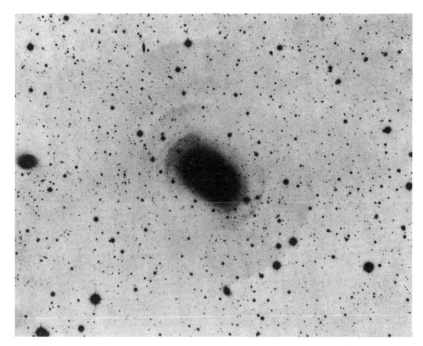

Figure 7-22. The giant elliptical galaxy NGC 3923 is surrounded by faint ripples of brightness. Courtesy of D. F. Malin and the Anglo-Australian Telescope Board.

that are observed within the family of ellipticals (see §5-4 of MB). For more discussion of these questions, see Ostriker (1980) or Aarseth and Fall (1980).

(c) Ripples Figure 7-22 is a photograph of NGC 3923. This photograph has been specially processed to accentuate the arclike **ripples** or "shells" in the density that can be seen on both sides of the major axis. About 10% of bright elliptical galaxies show such ripples, which surface photometry (Fort et al. 1986; Pence 1986) shows to be sudden drops in the surface brightness superimposed on the steady outward decrease in the galaxy's surface brightness. Quinn (1984) has summarized observations of these features.

How do these ripples form? One possibility is that ripples form when a giant elliptical galaxy eats a low-mass companion (Quinn 1984). Figure 7-23 shows the outcome of a numerical experiment in which a small spherical galaxy was allowed to fall into the fixed potential well

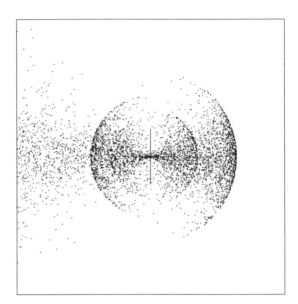

Figure 7-23. Ripples like those shown in Figure 7-22 are formed when a numerical disk galaxy is tidally disrupted by a fixed galaxy-like potential. (See Hernquist & Quinn 1987.)

like that of an elliptical galaxy. Ripples are clearly visible. Hence ripples may be the "smile of the face of the tiger".

7.5 Encounters in Stellar Disks

In a cool stellar disk, like our galactic disk at the solar radius, stars are on nearly circular orbits. Hence the galactocentric velocities of the stars in any small region of the disk are all nearly equal to the velocity $\mathbf{v}_{\rm LSR}$ of an exactly circular orbit at that radius. The inertial frame in which at a given instant $\mathbf{v}_{\rm LSR} = 0$ is called the local standard of rest (see §1.1.2 and §6-5 of MB), and we define the peculiar velocity of a disk star with total velocity $\mathbf{v}'$ to be

$$\mathbf{v} \equiv \mathbf{v}' - \mathbf{v}_{\rm LSR}. \tag{7-90}$$

In the first years of this century, K. Schwarzschild (1907) found that the number density of stars near the Sun with peculiar velocities $\mathbf{v}$ in any velocity range $d^3\mathbf{v}$ is well fitted by the **Schwarzschild distribution function** [see also eq. (6-44)]

$$f(\mathbf{v})d^3\mathbf{v} = \frac{n_0\, d^3\mathbf{v}}{(2\pi)^{\frac{3}{2}}\sigma_R\sigma_\phi\sigma_z} \exp\left[-\left(\frac{v_R^2}{2\sigma_R^2} + \frac{v_\phi^2}{2\sigma_\phi^2} + \frac{v_z^2}{2\sigma_z^2} \right) \right], \tag{7-91}$$

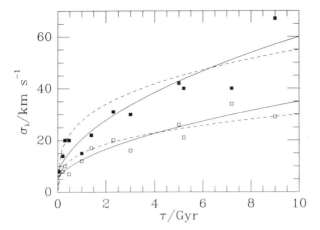

Figure 7-24. In the solar neighborhood, older stellar populations have larger velocity dispersions. Here the mean ages τ of stars of different spectral types are plotted against estimates of the RMS values of v_R (filled squares) and v_ϕ (open squares) for each group. The full curves are fits to these data of the form $v \propto (t + t_0)^{0.5}$, where t_0 is a suitable constant. The dashed curves show fits of the form $v \propto t^{0.3}$. (From data published in Wielen 1977.)

where n_0, σ_R, σ_ϕ, and σ_z are constants. Equation (7-91) states that the density of stars in velocity space is constant on ellipsoids with axis lengths in the ratio $\sigma_R : \sigma_\phi : \sigma_z$. The ellipsoid with semi-axes of length σ_R, σ_ϕ, and σ_z is called the **Schwarzschild velocity ellipsoid** (see §7-1 of MB for details).

In the 1930s and 1940s it was found that the dispersions σ_i ($i = R, \phi, z$) of hot, blue stars are smaller than those of cool, red stars. Similarly, metal-rich stars are associated with smaller values of the dispersions σ_i than are metal-poor stars. As was explained in §7-1 of MB, these facts suggest that as a stellar population ages, it "heats up" in the sense that the dispersions σ_i associated with it increase. As Figure 7-24 shows, the available observations suggest that $\sigma_i \propto t^\alpha$, $\alpha \simeq 0.5$. In this section we investigate the dynamical origin of this effect.

(a) The random energies of disk stars   A star's peculiar velocity **v** constantly changes as the star moves on its orbit through the Galaxy. If the Galaxy's gravitational field were perfectly smooth and axisymmetric, these changes in **v** would be solely due to epicyclic oscillations in R

and z, and there would be no long-term growth in the magnitude of **v**. We wish to study the slow changes in the amplitudes of these oscillations that are caused by irregularities in the galactic gravitational potential. So we start by identifying functions of R, z, and $\dot{\mathbf{v}}$ that are integrals of motion in a time-independent axisymmetric potential, and that can be used as measures of the amplitudes of the current oscillations in R and z.

In §3.2 we studied orbits in axisymmetric potentials, and showed that the **epicycle energy**

$$E_R \equiv \tfrac{1}{2}\left[v_R^2 + \kappa^2(R - R_g)^2\right] \tag{7-92a}$$

is approximately constant along a nearly circular orbit in an axisymmetric potential [eq. (3-64)]. Here the guiding-center radius R_g satisfies

$$R_g v_c(R_g) = L_z \tag{7-92b}$$

and the epicycle frequency κ is given by equation (3-62). E_R is zero for circular orbits and grows with the epicycle amplitude, so E_R provides a natural measure of the amplitude of the random component of a star's motion parallel to the galactic plane. We shall find it useful to eliminate R_g from equation (7-92) in favor of the azimuthal component of the velocity **v** defined by equation (7-90). We have from equation (3-74)[16]

$$v_\phi = 2B(R - R_g) = -\frac{\kappa^2}{2\Omega_g}(R - R_g). \tag{7-93}$$

Substituting equation (7-93) into (7-92a), we have that to leading order in $(R - R_g)/R_g$

$$E_R \simeq \tfrac{1}{2}\left[v_R^2 + \left(\frac{2\Omega_g}{\kappa}\right)^2 v_\phi^2\right] = \tfrac{1}{2}[v_R^2 + \gamma^2 v_\phi^2], \tag{7-94a}$$

where

$$\gamma \equiv \frac{2\Omega_g}{\kappa}. \tag{7-94b}$$

The natural measure of random motion perpendicular to the galactic plane is the z-energy

$$E_z = \tfrac{1}{2}(v_z^2 + \nu^2 z^2) \tag{7-95}$$

defined by equation (3-64). The total random energy is therefore

$$E_{\text{rand}} = E_R + E_z = \tfrac{1}{2}(v_R^2 + \gamma^2 v_\phi^2 + v_z^2) + \tfrac{1}{2}\nu^2 z^2. \tag{7-96}$$

---

[16] Note that the quantity called $v_\phi - v_c(R_0)$ in (3-74) is here called v_ϕ.

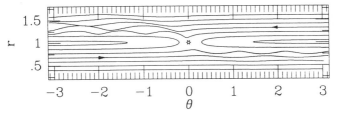

Figure 7-25. A molecular cloud represented by a Plummer model of mass $M = 0.004$ and scale length $a = 0.05$ is on a circular orbit of unit radius and speed about the center of a singular isothermal sphere. The initially circular orbits of stars orbiting in the same plane as the cloud are disturbed as they pass the cloud. Each curve shows the trajectory of such a star in terms of the star's radius and the difference between its azimuth and that of the cloud (measured in radians).

In the epicycle approximation, $\sigma_\phi^2 = \sigma_R^2/\gamma^2$ [eqs. (3-76) and (4-52)] and hence $v_R^2/\sigma_R^2 + v_\phi^2/\sigma_\phi^2 = 2E_R/\sigma_R^2$. Thus if we define

$$f_S(z, \mathbf{v}) \equiv \frac{n_0}{(2\pi)^{\frac{3}{2}}\sigma_R\sigma_\phi\sigma_z} \exp\left[-\left(\frac{E_R}{\sigma_R^2} + \frac{E_z}{\sigma_z^2} \right) \right], \qquad (7\text{-}97)$$

we have that Schwarzschild's empirical distribution (7-91) is $f(\mathbf{v}) = f_S(0, \mathbf{v})$. Furthermore, since f_S depends on the phase-space coordinates only through the integrals E_R and E_z, by the Jeans theorem it is a time-independent solution of the collisionless Boltzmann equation; indeed, it is the *only* time-independent solution consistent with the Schwarzschild distribution and the epicycle approximation.[17] Notice that f_S implies a Gaussian dependence of density on z: $n(z) = \int f_S(z, \mathbf{v}) \, d^3\mathbf{v} = n_0 \exp(-\frac{1}{2}\nu^2 z^2/\sigma_z^2)$. Notice also that f_S is isothermal with respect to both the radial and the vertical energies, although the "temperatures" σ_R^2 and σ_z^2 associated with the two types of independent oscillations, are different. We shall refer to the steady increase of σ_R^2 and σ_z^2 with time as disk "heating".

1 Scattering of Disk Stars by Molecular Clouds

Long before giant molecular cloud complexes were first detected, Spitzer and Schwarzschild (1953) suggested that encounters between disk stars

---

[17] The least satisfactory part of the approximation is that the potential is quadratic in z (see §3.2.3). However, it is straightforward to generalize our results to more general z-dependences.

and massive gas clouds might be responsible for the random velocities of old disk stars. Giant molecular clouds have masses in excess of $10^5\,M_\odot$ and tend to cluster into complexes of mass $M \gtrsim 10^6\,M_\odot$ and diameter $D \lesssim 200\,\mathrm{pc}$ that move on nearly circular orbits. In Figure 7-25 we represent a cloud by a Plummer model with scale length $a = 0.05$ and mass $M = 0.004$, and place it in an orbit of radius $R_c = 1$ in the potential of the singular isothermal sphere with unit circular speed. We complete the system by arranging a number of test particles ("stars") on circular orbits around the center of the isothermal sphere. As Figure 7-25 shows, stars on orbits with radii $R < R_c$ overtake the cloud and are set vibrating in the radial direction as they pass the cloud, while stars on orbits with initial radii $R > R_c$ start to vibrate when they are overtaken by the cloud. Hence our initially cold disk is soon nicely warmed.

We can obtain a rough estimate of the effectiveness of the process illustrated by Figure 7-25 as follows. We view the encounter from the rotating frame in which the cloud is at rest and use the impulse approximation to estimate the radial velocity acquired by a star that is initially on an orbit of radius $R = R_c - b$, where $|b| \ll R_c$. Since b/R_c is small, we can approximate the star's unperturbed orbit as a straight line. Integrating the gravitational attraction of a Plummer model, $\Phi(r) = -GM/\sqrt{r^2 + a^2}$, along this line yields

$$v_R = \frac{2GMb}{V_0(b^2 + a^2)},\tag{7-98}$$

where V_0 is the speed at which the star moves past the cloud on the unperturbed orbit. But

$$V_0 = R(\Omega - \Omega_c),\tag{7-99}$$

where $\Omega_c = \Omega(R_c)$ is the circular frequency of the frame of reference. Expanding Ω in a power series about R_c and using equation (3-61a) we have

$$V_0 \simeq -2A_c(R - R_c) = 2A_c b,\tag{7-100}$$

where A_c is Oort's "constant" at the radius R_c of the cloud. Hence

$$v_R(b) = \frac{GM}{A_c(b^2 + a^2)}.\tag{7-101}$$

In the impulse approximation, the star's angular momentum is unchanged by the encounter, so the star's post-encounter guiding-center radius $R_g = R$, and by equation (7-92a) the radial velocity (7-101) corresponds to epicycle energy

$$E_R = \frac{f^2}{2}\left[\frac{GM}{A_c(b^2 + a^2)}\right]^2,\tag{7-102a}$$

where $f = 1$ according to this derivation. We have introduced the factor f because in fact our derivation contains a subtle error. Since the radial velocity is not conserved in the unperturbed potential—it oscillates with the epicycle frequency κ_0—our derivation is only valid if the duration of the perturbation is much less than the epicycle period. In fact, the duration is $\approx b/V_0 = (2A_c)^{-1}$, which is comparable to the epicycle period $2\pi/\kappa_0$. Hence equation (7-102a) with $f = 1$ always makes an error of order unity in E_R. The correct derivation (Julian & Toomre 1966 and Problem 7-12) yields for $b \gg a$

$$f = \frac{\Omega}{A} K_0 \left(\frac{\kappa}{2A} \right) + \frac{\kappa}{2A} K_1 \left(\frac{\kappa}{2A} \right), \qquad (7\text{-}102b)$$

where K_ν is a modified Bessel function (Appendix 1.C.7) and we have dropped the subscripts on Ω, κ, and A. For a flat rotation curve, $A = \frac{1}{2}\Omega$, $\kappa = \sqrt{2}\Omega$, and $f = 0.9226$; for a Keplerian curve $f = 1.6797$.

Equations (7-102) are valid only for impact parameters b such that $v_R \ll V_0$. Adopting $A_c = 14.5\,\mathrm{km\,s^{-1}\,kpc^{-1}}$ (see Table 1-2), $M = 10^6\,\mathrm{M_\odot}$, and $a = 80\,\mathrm{pc}$, we find that we require $b \gtrsim 200\,\mathrm{pc}$. The predicted velocity gain for $b = 200\,\mathrm{pc}$ is only $v_R = 6\,\mathrm{km\,s^{-1}}$. The velocity perturbation is larger for smaller impact parameters; however, Figure 7-25 shows that encounters at small impact parameters simply switch the star from one circular orbit to another, with no sensible increase in the star's random velocity. Thus we conclude that first encounters such as those shown in Figures 7-25 leave stars with peculiar velocities of up to about $7\,\mathrm{km\,s^{-1}}$. This maximum velocity is to be compared with the radial velocity dispersions σ_R of populations of old disk stars, which exceed $45\,\mathrm{km\,s^{-1}}$ (see Figure 7-24 or Table 7-1 of MB). Clearly, such dispersions have not been established by subjecting stars to single encounters with molecular clouds, so we must next consider the cumulative effects of many encounters.

The speeds with which the stars in Figure 7-25 approach the scattering cloud are due entirely to the differential rotation of the galactic disk. Once a star has acquired non-zero peculiar velocity, we have to consider two types of encounter. For large impact parameters, the approach speed V_0 will still be dominated by the contribution from differential rotation (**shear-dominated** encounters), but at impact parameters smaller than $b_{\min} \approx \sigma_R/2A_c$, the encounter will be determined by the star's peculiar velocity **v** (**dispersion-dominated** encounters). Clearly, the proportion of dispersion-dominated encounters increases as E_{rand} increases, so we now take a close look at such an encounter.

Consider a close encounter between a cloud with a scale length a and a star with epicycle amplitude $\Delta R \gg a$. We assume that the impact parameter b of the encounter also satisfies $b \ll \Delta R$, so that we may treat the star's pre- and post-encounter trajectories as straight. Since the cloud is much more massive than the star, the speed with which the star recedes from the cloud is very nearly equal to the original speed of approach. Furthermore, during a close encounter we can neglect the difference between the cloud's rest frame and the star's instantaneous local standard of rest, and thus conclude that the magnitude of the star's peculiar velocity $\mathbf{v}$ is unchanged by the encounter; that is, the whole effect of such an encounter is to *deflect* the star's peculiar velocity vector $\mathbf{v}$.

Consideration of equation (7-96) shows that such a deflection usually changes the star's random energy: we recall from §3.2.3 that the ratio γ defined by equation (7-94b) varies from unity for a disk that rotates as a solid body, to two for a Kepler potential. In the solar neighborhood, $\gamma \simeq 1.5$. Thus by equation (7-96), a given speed v with respect to the local standard of rest corresponds to about twice as much random energy if the star's motion is directed in the azimuthal direction than it does if the star is moving in the meridional plane. Thus if the star's velocity is deflected toward the azimuthal direction, the star's random energy will go up. Furthermore, as the star moves around its epicycle, v_R and v_ϕ oscillate at angular frequency κ and mutual phase $\frac{\pi}{2}$. Hence at intervals separated by $\frac{1}{2}T_r = \pi/\kappa$, v_ϕ passes through zero while $|v_R|$ is at a maximum. At these instants *any* scattering is bound to increase $|v_\phi|$ and therefore E_{rand}. Thus a star that can arrange to be deflected twice each radial period just as v_ϕ passes through zero can repeatedly increase its random energy, and at no cost to the obliging deflectors! In practice, stars cannot choose the points on their epicycles at which they are deflected, and it will often happen that a scattering will diminish v_ϕ, and therefore E_{rand}. However, it is straightforward to show that these "cooling" deflections are less common than deflections that increase E_{rand} (Problem 7-9). Hence repeated scattering of stars by clouds will heat the disk overall.

Numerous authors have estimated the rate at which star-cloud encounters heat disks. Woolley and Candy (1968) and Icke (1982) estimated encounter cross-sections by numerically integrating the orbits of stars past clouds in differentially rotating disks. This approach enables them to include the effects of shear-dominated encounters such as those portrayed in Figure 7-25. However, neither Woolley and Candy, nor Icke had the resources to integrate as many orbits as are required to obtain a satisfactory estimate of the heating rate. Lacey (1984) followed the seminal paper of Spitzer and Schwarzschild (1953) in concentrating on dispersion-dominated encounters. These encounters almost certainly make the largest contribution to the overall heating rate, and have the advantage that they can be handled analytically. An interesting test of Lacey's predictions is provided by Villumsen (1985), who simulated disk heating by allowing 4000 test particles and 400 clouds of mass $5 \times 10^5 \, M_\odot$ to orbit from cold initial conditions in the potential of a fixed exponential disk. The most important conclusions of this work are:

(i) Lacey predicts $\sigma_R \propto \sigma_z \propto t^{1/4}$ while Villumsen measures $\sigma_R \propto t^{0.25 \pm 0.02}$ and $\sigma_z \propto t^{0.31 \pm 0.02}$.

(ii) At the end of Villumsen's simulation, $\sigma_z/\sigma_R \simeq 0.6$ compared with Lacey's prediction $\sigma_z/\sigma_R \simeq 0.8$.

(iii) Lacey finds that if the masses of individual molecular clouds lie near the upper end of the probable range, $M \approx 10^6 \, M_\odot$, scattering by clouds can account for the observed age-velocity dispersion relation for dispersions less than about $30 \, \mathrm{km \, s^{-1}}$. However, it does not seem likely that the disk populations with the largest dispersions, for example, white dwarfs ($\sigma_R \simeq 50 \, \mathrm{km \, s^{-1}}$) or carbon stars ($\sigma_R \simeq 48 \, \mathrm{km \, s^{-1}}$), owe their present dispersions to star-cloud scattering.

The agreement between Lacey and Villumsen that $\sigma_R \propto t^{0.25}$ is particularly significant because the observations seem to require σ_R to grow at least as fast as $t^{0.3}$ and more like $t^{0.5}$ (see Figure 7-24). In view of this discrepancy, it is important to note that the following simple scaling argument shows that if star-cloud scattering were the only process heating the disk, the exponent α in the relation $\sigma_R \propto t^\alpha$ could not be much greater than 0.25.

We expect the main contribution to disk heating to come from dispersion-dominated encounters. We have seen that in such encounters, the star's peculiar speed v is unchanged. Differencing equation (7-96) at constant v^2 yields

$$
\begin{aligned}
\Delta E_{\mathrm{rand}} &= \tfrac{1}{2}(\gamma^2 - 1)\Delta v_\phi^2 \\
&= \tfrac{1}{2}(\gamma^2 - 1)[2v_\phi \, \Delta v_\phi + (\Delta v_\phi)^2].
\end{aligned} \tag{7-103}
$$

In the impulse approximation, the magnitude of $\Delta \mathbf{v}$ in a single encounter satisfies $|\Delta \mathbf{v}| \propto v^{-1}$, so the average value of $(\Delta v_\phi)^2$ per encounter is $\propto v^{-2}$. Thus the contribution of this term to the rate of change of E_{rand} is $\propto nv^{-1}$, where n is the number density of clouds. It can be shown that the contribution of the term in equation (7-103) proportional to $v_\phi \Delta v_\phi$ is also $\propto nv^{-1}$.[18] Hence

$$\frac{dE_{\text{rand}}}{dt} \propto \frac{n}{v} \propto \frac{n}{\sqrt{E_{\text{rand}}}}. \tag{7-104}$$

If the star's orbit is confined to the cloud layer, $n = constant$ and equation (7-104) yields $E_{\text{rand}} \propto t^{\frac{2}{3}}$. However, this is an unrealistic limit. Most stars travel well above and below the cloud layer, so that the star passes through the cloud layer only twice in each vertical period, irrespective of its energy. Consequently, in this case we have for approximately harmonic oscillations $\langle n \rangle \propto v^{-1} \propto (E_{\text{rand}})^{-1/2}$, and equation (7-104) yields $E_{\text{rand}} \propto t^{\frac{1}{2}}$. Accounting for the decrease of the vertical frequency ν with energy will further decrease the exponent. Since $\sigma_R \propto \sqrt{E_R}$, we see that if star-cloud scattering were the only process heating the disk, $\sigma_R \propto t^\alpha$, with $\alpha \lesssim 0.25$. Thus the results of Lacey and Villumsen are consistent with basic physical arguments and are not likely to be subject to significant error.

We conclude that heating by molecular clouds is unable to explain an exponent α as large as that observed. Actually, it would be rather surprising if star-cloud encounters were alone sufficient to account for the heating of the galactic disk, since there is likely to be some contribution to the overall heating rate from spiral structure.

2 Scattering of Disk Stars by Spiral Arms

The disks of spiral galaxies are far from smooth. Gas, dust, and young stars are always concentrated into spiral arms. These arms are sometimes part of a global spiral structure, but they are more commonly short and mangy, being less than a kiloparsec wide and several kiloparsecs long. Studies of disks at infrared wavelengths suggest that these spiral features are associated with peaks in the density of the old stars that make up most of the mass of galactic disks [see §6.1.1(b)], so it is natural to ask whether the gravitational fields of spiral features, like the fields from molecular clouds, are able to heat galactic disks (Barbanis & Woltjer 1967).

N-body simulations of self-gravitating stellar disks suggest that spiral features may be able to heat disks quite effectively. For example, Figure 7-26 shows the evolution of the radial velocity dispersion σ_R at

---

[18] This follows from Chandrasekhar's dynamical friction formula (7-14) for the case in which M is a star and the field objects m are clouds. Since the cloud velocity dispersion is small, the integral in equation (7-14) is simply $n/(4\pi)$. Thus $\dot{v}_\phi \propto nv_\phi/v^3$, from which it follows that upon averaging over an epicycle, $v_\phi \dot{v}_\phi \propto nv^{-1}$.

several radii in a 20,000 particle simulation. The disk heats more slowly at large radii because the dynamical time is larger there; however, at every radius σ_R shows the same characteristic behavior: a rapid increase near the start of the simulation, which then slows so that σ_R approaches a constant value. Figure 7-27 shows that this behavior is mirrored by the development of spiral structure in the disk: the disk quickly develops several scraggy spiral arms, which gradually become smoother and eventually fade away. This sequence of events can be explained by the following model: initially, the disk is cold, so that Toomre's Q [eq. (6-53)] is small, the swing amplifier is efficient, and strong spiral structure develops. The spiral structure heats the disk, increasing Q and decreasing the efficiency of the swing amplifier, until finally the spiral structure cannot be maintained. Thus the spiral structure is killed by the heat that it injects into the disk, just as yeast in a vat of fermenting beer is killed by the alcohol it creates. The issue of how to maintain the spiral structure despite the disk heating does not concern us here;[19] instead, our aim in this section is to investigate the dynamical process by which spiral structure heats the galactic disk.

Consider a spiral potential with pattern speed Ω_p,

$$\Phi_1(R, \phi, t) = F(R) \cos[f(R) + m\varphi], \tag{7-105}$$

where $\varphi = \phi - \Omega_p t$ is the azimuthal angle in a frame rotating at Ω_p. The effect of this potential on a stellar orbit can be determined using linear perturbation theory. If the orbit is nearly circular, with mean radius R_0, equations (3-114) yield expressions for the orbit in the form $R(t) = R_0 + R_1(t)$, $\varphi(t) = \varphi_1(t) + (\Omega_0 - \Omega_p)t$. We have

$$\ddot{R}_1 + \left(\frac{d^2\Phi_0}{dR^2} - \Omega^2\right)_{R_0} R_1 - 2R_0\Omega_0\dot{\varphi}_1 =$$

$$-\frac{dF(R_0)}{dR}\cos[\eta(R_0, t)] + F(R_0)k(R_0)\sin[\eta(R_0, t)], \tag{7-106a}$$

$$\ddot{\varphi}_1 + 2\Omega_0\frac{\dot{R}_1}{R_0} = m\frac{F(R_0)}{R_0^2}\sin[\eta(R_0, t)], \tag{7-106b}$$

where

$$\eta(R_0, t) \equiv f(R_0) + m(\Omega_0 - \Omega_p)t. \tag{7-106c}$$

In these equations, Φ_0 is the axisymmetric potential, Ω_0 is the circular frequency at R_0 [eq. (3-111)], and $k = df/dR$. We now integrate equation (7-106b) to obtain an equation for $\dot{\varphi}_1$ that can be used to eliminate

---

[19] In one theory, the pattern persists because the disk is continually cooled by infall of new gas [see §6.4.2(a)].

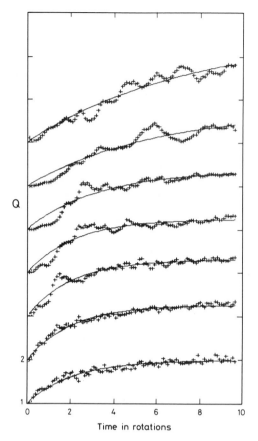

Figure 7-26. The temporal evolution of the radial velocity dispersion in an N-body disk. Each curve is for a different annulus: the radii of the annuli increase from bottom to top and the curves have been offset by one unit vertically for clarity. The radial dispersion σ_R is expressed in terms of Toomre's parameter $Q \equiv \sigma_R \kappa / (3.36 G \Sigma)$ [eq. (6-53)]. Reproduced from Sellwood and Carlberg (1984) by permission of *The Astrophysical Journal.*

$\dot{\varphi}_1$ from equation (7-106b). In so doing, the algebra later in this section will be greatly simplified if we consider only tightly wound spirals, $|kR| \gg 1$. Then all of the forcing terms on the right side of equations (7-106) can be dropped, except for the dominant term proportional to k in equation (7-106a). Thus equation (7-106b) yields

$$\dot{\varphi}_1 = -2\Omega_0 \frac{R_1}{R_0} + \text{constant}, \tag{7-107}$$

which can be substituted into equation (7-106a) to yield

$$\ddot{R}_1 + \kappa_0^2 R_1 = F(R_0)k(R_0)\sin[f(R_0) + m(\Omega_0 - \Omega_p)t] + \text{constant}, \tag{7-108}$$

where κ_0 is the usual epicycle frequency [eq. (3-118b)]. The constant is unimportant since it can be absorbed by a shift $R_1 \to R_1 + constant$.

Equation (7-108) can be solved to yield [cf. eq. (3-119)]

$$R_1(t) = C\cos(\kappa_0 t + \psi) + \frac{F(R_0)k(R_0)}{\Delta}\sin[f(R_0) + m(\Omega_0 - \Omega_p)t], \tag{7-109}$$

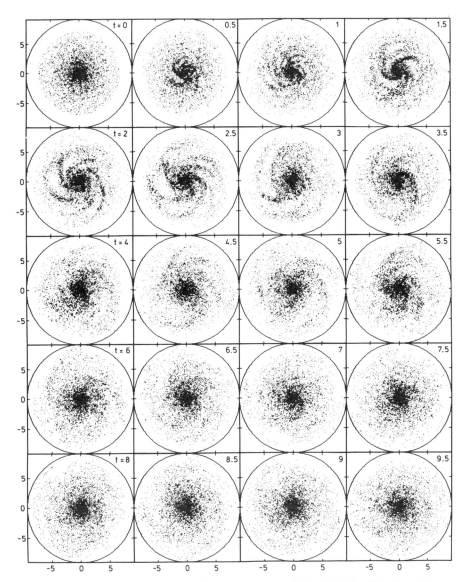

Figure 7-27. Several snapshots of the evolution of the N-body disk for which σ_R is plotted in Figure 7-28. Notice that the disk, which is initially axisymmetric, quickly develops pronounced spiral structure which then slowly fades. Reproduced from Sellwood and Carlberg (1984) by permission of *The Astrophysical Journal*.

where $\Delta = \kappa_0^2 - m^2(\Omega_0 - \Omega_p)^2$, and C and ψ are arbitrary constants. The most important feature of this result for our purposes is that the spiral potential imposes a forced radial oscillation on the star but does not lead to any steady growth in the RMS radial velocity. In other words, *a spiral potential with a fixed pattern speed cannot heat the disk.*[20]

Given this result, how can we explain the steady growth in σ_R seen in Figure 7-26? The answer is that disk heating requires *transitory* rather than steady spiral patterns. To see this, let us multiply the potential of equation (7-105) by a Gaussian function of time with unit area, $p(t) = (2\pi s^2)^{-1/2} \exp(-\frac{1}{2}t^2/s^2)$. Equation (7-108) is thereby modified to read

$$\ddot{R}_1 + \kappa_0^2 R_1 = p(t)F(R_0)k(R_0)\sin[f(R_0) + m(\Omega_0 - \Omega_p)t], \qquad (7\text{-}110)$$

which can be shown to have the solution

$$R_1(t) = C\cos(\kappa_0 t + \psi) + \frac{F(R_0)k(R_0)}{\kappa_0}$$
$$\times \int_{-\infty}^{t} p(t')\sin[f(R_0) + m(\Omega_0 - \Omega_p)t']\sin[\kappa_0(t - t')]\,dt'. \qquad (7\text{-}111)$$

Inserting the chosen form for $p(t)$ and setting the amplitude C of the free oscillation to zero, we obtain

$$R_1(t \to \infty) = \frac{F(R_0)k(R_0)}{2\kappa_0}\left\{\cos[\kappa_0 t - f(R_0)]e^{-\frac{1}{2}s^2[m(\Omega_0-\Omega_p)-\kappa_0]^2}\right.$$
$$\left. - \cos[\kappa_0 t + f(R_0)]e^{-\frac{1}{2}s^2[m(\Omega_0-\Omega_p)+\kappa_0]^2}\right\}. \qquad (7\text{-}112)$$

Thus the transitory spiral pattern has induced a permanent epicyclic oscillation. When the characteristic duration of the transient, s, is much less than the orbital period (an impulsive transient), the arguments of the exponential $s[m(\Omega_0 - \Omega_p) \pm \kappa_0]$ are small and the induced epicycle amplitude is simply $F(R_0)k(R_0)\sin[f(R_0)]/\kappa_0$, which varies slowly across the disk. On the other hand, when s is much greater than the orbital period, the induced epicycle amplitude is strongly peaked near the Lindblad resonances $m(\Omega_0 - \Omega_p) \pm \kappa_0 = 0$. In this latter case most of the disk is not heated—just as in the case of a steady pattern.

Thus the ability of spiral structure to heat the disk is strongly dependent on the nature of spiral structure. According to the Lin-Shu

---

[20] Strictly, we have only established that no heating occurs except at the corotation and Lindblad resonances, since the derivation leading to equation (7-109) fails at these resonances. In general, heating is likely to occur at the Lindblad resonances because the energy carried by the short trailing wave is deposited there.

hypothesis (§6.4.1), in which the spiral structure is a quasi-stationary wave with a single, well-defined pattern speed, disk heating is negligible, except possibly at the Lindblad and corotation resonances. In chaotic models [§6.4.2(a)] the arms are transient, forming and disappearing on an orbital timescale, and in these theories spiral structure can heat the disk efficiently over a wide range of radii.

Which of these models is a more accurate representation of the spiral structure in our Galaxy? We do not yet know. However, assuming that the chaotic models are correct, we may explore their predictions for the evolution of the velocity ellipsoid with time.

Let us suppose that a given star is subjected to N independent transient perturbations. Each transient induces an epicyclic motion which can be written in the form $R_1(t) = C_i \cos(\kappa_0 t + \psi_i)$, $i = 1, \ldots, N$, where C_i and ψ_i are given by equations similar to (7-112) for each transient. After N transients, $R_1(t) = \sum_{i=1}^{N} C_i \cos(\kappa_0 t + \psi_i)$. The mean-square value of R_1, averaged over one epicycle, is

$$
\begin{aligned}
\overline{R_1^2} &\equiv \frac{\kappa_0}{2\pi} \int_0^{2\pi/\kappa_0} R_1^2(t)dt \\
&= \tfrac{1}{2} \sum_{i,j=1}^{N} C_i C_j \cos(\psi_i - \psi_j).
\end{aligned}
\tag{7-113}
$$

Since the transients are uncorrelated, the phases of the epicyclic oscillations that they induce are uncorrelated. Hence on average $\cos(\psi_i - \psi_j)$ will be zero when $i \neq j$, and the only terms in the sum that contibute to $\overline{R_1^2}$ will be those with $i = j$. Thus we have $\overline{R_1^2} = \tfrac{1}{2} N \overline{C^2}$, where $\overline{C^2}$ is the mean-square amplitude induced by a single transient. If the rate of occurrence and the strength of new transients is independent of time, we conclude that $\overline{R_1^2}$, and hence the mean-square radial velocity σ_R^2 should grow linearly with time. In other words, $\sigma_R \propto t^\alpha$, where $\alpha = 0.5$; this is actually an upper limit to α, since the heating rate decreases once the epicycle size becomes comparable to the radial wavelength of the spiral arms (Carlberg & Sellwood 1985), and a more accurate estimate is $0.2 \lesssim \alpha \lesssim 0.5$. This range for α is consistent with the observations presented in Figure 7-24, and provides a substantially better fit than the values predicted in the previous subsection for heating by molecular clouds.

One potential difficulty with this model is that spiral structure only affects the radial and azimuthal velocities of the disk stars. It has no direct effect on their velocities in the z-direction. Thus, scattering by spiral arms cannot explain why all three dispersions σ_R, σ_ϕ, and σ_z grow at roughly the same rate (Figure 7-24). Some mechanism is needed to convert the radial and azimuthal velocities into vertical velocities. Carlberg (1984) has suggested that the vertical velocities may arise from scattering off molecular clouds, even though most of the heating is due to spiral structure.

3 Summary

There is little doubt that imperfections in the Galaxy's gravitational field cause the random velocities of disk stars to increase. Irregularities on many scales may contribute to this phenomenon. At the small-scale end of the spectrum lie molecular clouds, while spiral arms lie at the large-scale end.

Although both molecular clouds and spiral arms appear to be capable of heating the disk at about the observed rate, neither mechanism can explain all of the features of the observed velocity dispersion versus age relation. Molecular clouds appear to yield too low a value for $d \ln \sigma_i / d \ln t$, while spiral arms are unable to heat the stars efficiently in the z-direction.

Clearly, we still do not completely understand the mechanism by which the velocity dispersions of disk stars grow. Perhaps the combined effect of clouds and spiral arms can reproduce the observations. Alternatively, there may be irregularities on intermediate scales that dominate the heating rate. A final possibility is that the heating may arise from some completely different process—one exotic possibility is that the galactic halo is composed of $10^6 \, M_\odot$ black holes and that these are responsible for disk heating (Lacey & Ostriker 1985).

Problems

7-1. [2] At time zero the stellar streaming velocity $\mathbf{v}(\mathbf{x})$ within an axisymmetric galaxy of density $\rho(R, z)$ constitutes circular rotation at angular frequency $\omega(R, z)$. The galaxy is then perturbed by the high-speed passage of a massive system. Show that within the impulse approximation the instantaneous change $\Delta \mathbf{v}$ in $\mathbf{v}$ that is produced by the encounter satisfies

$$\int (\mathbf{v} \cdot \Delta \mathbf{v}) \rho \, d^3\mathbf{x} = 0. \qquad (7P\text{-}1)$$

(Hint: Write $\mathbf{v} = \omega R \, \hat{\mathbf{e}}_\phi$ and exploit the fact that $\Delta \mathbf{v}$ can be derived from a potential.)

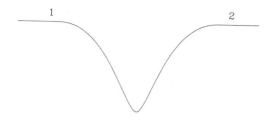

Figure 7-28. An analog for disk shocking of globular clusters: two railroad cars on a track passing through a V-shaped valley.

7-2. [1] Rewrite equation (7-56) in the form

$$\Delta v_R = -\frac{2}{RV} \int_R^\infty \left(r^2 \frac{d\Phi}{dr} \right) \frac{d}{dr} \sqrt{1 - \frac{R^2}{r^2}}\, dr \qquad (7\text{P-}2)$$

and integrate by parts, to show that

$$\Delta v_R = -\frac{2G}{RV} \left[M(\infty) - 4\pi \int_R^\infty \rho \sqrt{1 - \frac{R^2}{r^2}} r^2\, dr \right]$$

$$= -\frac{2G}{RV} \left[M(R) + 2\pi R^3 \int_0^1 \frac{\rho(R/x)}{x^2} \left(1 + \tfrac{1}{4}x^2 + \tfrac{1}{8}x^4 + \cdots \right) dx \right].$$
$$(7\text{P-}3)$$

The term proportional to $M(R)$ represents the contribution to Δv_R from matter interior to R, while the integral represents the contribution from matter exterior to R. Show that the latter is finite provided $\rho(R) < O(R^{-1})$ at large R, and that for the singular isothermal sphere, interior and exterior mass contribute to Δv_R in proportions that are independent of R.

7-3. [1] Show that the probability $P(V)dV$ that two stars drawn from a Maxwellian distribution with one-dimensional dispersion σ have relative speed in the interval $(V, V + dV)$ is

$$P(V)dV = (2\sqrt{\pi}\sigma^3)^{-1} \exp\left(-\frac{V^2}{4\sigma^2} \right) V^2 dV. \qquad (7\text{P-}4)$$

In words, the relative speed distribution is Maxwellian with dispersion $\sqrt{2}\sigma$. (Hint: Write down the probability that one of the stars has velocity in $d^3\mathbf{v}_1$ and the other has velocity in $d^3\mathbf{v}_2$, transform to new variables $\mathbf{v}_{cm} \equiv \frac{1}{2}(\mathbf{v}_1 + \mathbf{v}_2)$ and $\mathbf{V} \equiv \mathbf{v}_1 - \mathbf{v}_2$, and integrate over all $\mathbf{v}_{cm}$ and all directions of $\mathbf{V}$.)

7-4. [2] A satellite system of mass m is in a circular orbit around a host of mass $M \gg m$. Let (x, y, z) be Cartesian coordinates with $\hat{\mathbf{e}}_x$ pointing along the line joining the two masses and $\hat{\mathbf{e}}_z$ normal to the orbital plane. In §7.3, we showed that the distance of the last closed zero-velocity surface from m along the x-axis is r_J [eq. (7-84)]. Show that the distance of this surface from m along the y- and z-axes is $\frac{2}{3}r_J$ and $(3^{2/3} - 3^{1/3})r_J$, respectively. Thus, the zero-velocity surface is not spherical and the tidal radius cannot be characterized by a single number.

7-5. [2] Figure 7-28 shows the track of a frictionless railroad crossing a valley. At $t = 0$, two cars A and B of mass m are released with speed v and separation d from near point 1 on the horizontal stretch of track on the left side of the valley. Show that when the cars emerge onto the horizontal stretch of track on the right at point 2, they have zero relative velocity.

A spring with rest length d and stiffness ω^2/m is now fixed between the cars and they are again released from point 1 with speed v and separation d. Describe qualitatively how their final relative motion depends on the value of ω. Draw an analogy between this system and the disk shocking of globular clusters.

7-6. [1] Show that a centrifugally supported Kuzmin disk (§2.2.1) of mass M and scale length a has median radius $r_h = \sqrt{3}a$, RMS speed $\langle v^2 \rangle^{1/2} = \frac{1}{2}\sqrt{GM/a}$, and spin angular momentum per unit mass $S = \frac{8}{5}\sqrt{GMa}$, and thus dimensionless spin $\hat{S} = \frac{16}{5\sqrt{3}}$ [cf. eq. (7-88)].

7-7. [1] If the Murai and Fujimoto (1980) model of the orbit of the Magellanic Clouds is correct, is there more angular momentum in the orbit of the Magellanic Clouds around our Galaxy or in the spin of the disk of our Galaxy?

7-8. [2] Show that the total spin angular momentum of two disk galaxies is comparable with the maximum orbital angular momentum that is compatible with rapid merging in a parabolic encounter.

7-9. [2] A star moves on a nearly circular orbit through the equatorial plane of a razor-thin galaxy. At time zero, when the star is moving at speed v with respect to the local standard of rest, the star is instantaneously deflected by a molecular cloud that is itself on a perfectly circular orbit, through an angle η (in a clockwise sense) onto a new, nearly circular orbit within the galactic plane. Show that the deflection changes the star's random energy by an amount

$$\Delta E_R = E_R(\gamma^2 - 1)[\sin^2 \eta(\sin^2 \psi - \gamma^{-2}\cos^2 \psi) - \frac{1}{2\gamma}\sin 2\eta \sin 2\psi], \quad (7P\text{-}5)$$

where γ is defined by equation (7-94b) and ψ is the epicycle phase introduced in §3.2.3 [see eqs. (3-67) and (3-69)]. At the radius of the star's orbit, there are n_c Plummer-model clouds per unit area, and each cloud has mass M_c and a scale length a that is much smaller than the star's epicycle radius. Using the impulse approximation, assuming that the relative velocity is dominated by the velocity dispersion of the stars, and assuming that the deflection angle η is small, show that the expectation value of the rate of change of E_R is

$$\dot{E}_R \simeq \frac{\sqrt{2}G^2 M_c^2 n_c}{a\sqrt{E_R}}(\gamma^2 - 1)\int_0^{\pi/2} \frac{\sin^2 \psi - \gamma^{-2}\cos^2 \psi}{(\sin^2 \psi + \gamma^{-2}\cos^2 \psi)^{3/2}}\, d\psi. \quad (7P\text{-}6)$$

Verify that $\dot{E}_R \geq 0$ for $1 \leq \gamma \leq 2$.

7-10. [3] In the core of a certain flattened elliptical galaxy, the mean stellar velocity vanishes and the velocity distribution is Gaussian, with dispersion σ_z parallel to the galaxy's symmetry axis $\hat{\mathbf{e}}_z$, and dispersion $\sigma_\perp = \sigma_z/\sqrt{1-e^2} > \sigma_z$ in directions orthogonal to $\hat{\mathbf{e}}_z$. A globular cluster moves through the core at velocity $\mathbf{v} = v_z \hat{\mathbf{e}}_z + v_\perp \hat{\mathbf{e}}_\perp$, where $\hat{\mathbf{e}}_\perp \cdot \hat{\mathbf{e}}_z = 0$. Show that the frictional drag on the cluster may be written $\mathbf{F} = -\gamma_z v_z \hat{\mathbf{e}}_z - \gamma_\perp v_\perp \hat{\mathbf{e}}_\perp$, where

$$1 < \frac{\gamma_z}{\gamma_\perp} = \frac{I(1,\tfrac{3}{2})}{I(2,\tfrac{1}{2})} \; ; \; I(\mu,\nu) \equiv \int_1^\infty \exp\left[-\tfrac{1}{2}\sigma_\perp^{-2}\left(\frac{v_\perp^2}{\lambda} + \frac{v_z^2}{\lambda - e^2}\right)\right] \frac{d\lambda}{\lambda^\mu (\lambda - e^2)^\nu}. \tag{7P-7}$$

[Hint: See Binney (1977)].

7-11. [3] Marochnik (1968) and Kalnajs (1972c) show how the dynamical friction formula (7-17) may be derived from the collisionless Boltzmann equation, rather than from individual orbits as was done in §7.1:

(a) Show that the potential of a point mass M moving on the trajectory $\mathbf{x}(t) = \mathbf{x}_0 + \mathbf{v}_0 t$ can be written

$$\Phi_M(\mathbf{x},t) = -\frac{GM}{2\pi^2} \int \frac{d^3\mathbf{k}}{k^2} e^{i[\mathbf{k}\cdot(\mathbf{x}-\mathbf{x}_0)-\omega(\mathbf{k})t]}, \tag{7P-8}$$

where $\omega(\mathbf{k}) \equiv \mathbf{k} \cdot \mathbf{v}_0$.

(b) M is surrounded by a sea of background objects of mass m and Maxwellian DF (7-16). Use the linearized collisionless Boltzmann equation (5-5) to show that to first order M changes the background's DF by

$$f_1(\mathbf{x},\mathbf{v},t) = \frac{GMn_0}{(2\pi\sigma^2)^{5/2}\pi} \int \frac{d^3\mathbf{k}}{k^2} \frac{\mathbf{k}\cdot\mathbf{v}}{\mathbf{k}\cdot\mathbf{v} - \omega} e^{-\tfrac{1}{2}v^2/\sigma^2} e^{i[\mathbf{k}\cdot(\mathbf{x}-\mathbf{x}_0)-\omega(\mathbf{k})t]}. \tag{7P-9}$$

(c) Show that the force on M to which the wake described by equation (7P-9) gives rise may be written

$$\mathbf{F} = \frac{4G^2 M^2 n_0 m}{(2\pi\sigma^2)^{3/2}} i \int \frac{d^3\mathbf{k}}{k^3} \mathbf{k} \int_{-\infty}^\infty \frac{e^{-\tfrac{1}{2}u^2/\sigma^2} u \, du}{ku - \omega(\mathbf{k})}, \tag{7P-10}$$

where u is the component of $\mathbf{v}$ parallel to $\mathbf{k}$. By replacing ω in equation (7P-10) by $\omega = \mathbf{k}\cdot\mathbf{v}_0 + i\epsilon$ (which for small positive ϵ corresponds to assuming that M grows slowly as $e^{\epsilon t}$), show that

$$\mathbf{F} = -\frac{4\pi G^2 M^2 n_0 m}{(2\pi\sigma^2)^{3/2}} \omega \int \frac{d^3\mathbf{k}}{k^5} \mathbf{k} \exp\left(-\frac{\omega^2}{2k^2\sigma^2}\right). \tag{7P-11}$$

[Hint: Write $(ku - \omega)^{-1} = [(ku - \mathbf{k}\cdot\mathbf{v}_0) + i\epsilon]/[(ku - \mathbf{k}\cdot\mathbf{v}_0)^2 + \epsilon^2]$ and show that the real part of the double integrand of (7P-10) is antisymmetric under $\mathbf{k} \to -\mathbf{k}$, $u \to -u$. Then take the limit $\epsilon \to 0$ and use equation (1C-2) to evaluate the integral over u.]

(d) Show that on integrating over a finite range of wavevectors, $k_{\min} \le |\mathbf{k}| \le k_{\max}$, equation (7P-11) yields an expression for the acceleration of M that in the limit $m/M \to 0$ is identical with equation (7-17), except that $\ln\Lambda$ is replaced by $\ln(k_{\max}/k_{\min})$. Explain why appropriate choices for $k_{\min}$ and $k_{\max}$ are 2π over the characteristic impact parameters $b_{\max}$ and $b_{\min} = GM/v_0^2$, respectively, and thus argue that $\ln\Lambda$ and $\ln(k_{\max}/k_{\min})$ will generally be very similar.

7-12. [3] The goal of this problem is to determine the epicycle energy induced in a star by a passing molecular cloud.

(a) Let (R_c, ϕ_c) and (R, ϕ) be polar coordinates of the cloud and star, where $|R - R_c| \ll R_c$. Let $x = R - R_c$ and $y = R_c(\phi - \phi_c)$; hence show that the equations of motion (3-114) have the form

$$\ddot{x}_1 + \left(\frac{d^2 \Phi_0}{dR^2} - \Omega^2 \right)_{R_0} x - 2\Omega_0 \dot{y} = -\left. \frac{\partial \Phi_1}{\partial x} \right|_{R_0}, \tag{7P-12a}$$

$$\ddot{y} + 2\Omega_0 \dot{x} = -\left. \frac{\partial \Phi_1}{\partial y} \right|_{R_0}, \tag{7P-12b}$$

where Φ_1 is the potential of the cloud.

(b) If $\Phi_1 = 0$ show that a solution of equations (7P-12) is

$$x(t) = x_0 \quad ; \quad y(t) = -2Ax_0 t, \tag{7P-13}$$

where A is Oort's constant [eq. (3-61a)]. This solution represents a star on a circular orbit.

(c) Assume that the cloud is on a circular orbit, that the star is initially on a circular orbit, and that the cloud potential $\Phi_1 = -GM/r$ is sufficiently weak that the right sides of equations (7P-12) can be evaluated along the unperturbed stellar orbit (7P-13). Show that equations (7P-12) can be combined into a single differential equation,

$$\ddot{x}_1 + \kappa_0^2 x_1 = -\frac{GM\Omega_0}{Ax_0^2} \frac{[1 + A/\Omega) + (2At)^2]}{[1 + (2At)^2]^{3/2}}, \tag{7P-14}$$

where $x(t) = x_0 + x_1(t)$ and $y(t) = -2Ax_0 t + y_1(t)$. [Hint: Since the unperturbed y-coordinate is linear in time, the y-derivative in equation (7P-12b) can be converted to a time derivative.]

(d) Solve equation (7P-14) to show that the epicycle amplitude induced by the cloud is (Julian & Toomre 1966)

$$X = \frac{GM\Omega}{\kappa A^2 x_0^2} \left[K_0 \left(\frac{\kappa}{2A} \right) + \frac{\kappa}{2\Omega} K_1 \left(\frac{\kappa}{2A} \right) \right], \tag{7P-15}$$

where K_ν is a modified Bessel function (Appendix 1.C.7) and we have dropped the subscripts on Ω and κ. Thus, derive the correction factor f in equation (7-102b).

8 Kinetic Theory

In all of our discussions so far, we have concentrated on collisionless systems, that is, systems in which the constituent particles move under the influence of the *mean* potential generated by all the other particles. The fundamental equation describing motion in such a system is the collisionless Boltzmann equation (4-13). However, the collisionless Boltzmann equation is not valid for arbitrarily long times. In its derivation, we assumed that it is always possible to define a volume in phase space that is large enough to contain many stars but small enough so that the gravitational acceleration $-\nabla\Phi$ is the same for all stars in the volume. Individual stellar encounters[1] invalidate this assumption because the accelerations of the two stars that are undergoing an encounter can be very different even though they are immediate neighbors. Encounters gradually perturb stars away from the trajectories they would have taken if the distribution of matter in the system were perfectly smooth. After many such encounters the star eventually loses its memory of its initial orbit. The characteristic time over which the loss of memory occurs is called the relaxation time t_{relax}; over timescales exceeding t_{relax} the collisionless Boltzmann equation is not valid.

We have shown that the relaxation time is of order

$$t_{relax} \approx \frac{0.1N}{\ln N} t_{cross}, \tag{8-1}$$

where t_{cross} is the crossing time and N is the number of stars in the system [eq. (4-9)]. Thus, the relaxation time exceeds the crossing time if $N \gtrsim 35$. Galaxies typically have $N \approx 10^{11}$ and $t_{cross} \approx 10^8\,yr$, so stellar encounters have been entirely unimportant in the galaxy's lifetime of $10^{10}\,yr$. However, encounters may have played a major role in determining the present structure of many other stellar systems, such as globular clusters ($N \approx 10^5$, $t_{cross} \approx 10^5\,yr$, lifetime $10^{10}\,yr$), galactic or open clusters ($N \approx 10^2$, $t_{cross} \approx 10^6\,yr$, lifetime $10^8\,yr$), galactic nuclei ($N \approx 10^8$, $t_{cross} \approx 10^4\,yr$, lifetime $10^{10}\,yr$), and clusters of galaxies ($N \approx 10^3$, $t_{cross} \approx 10^9\,yr$, lifetime $10^{10}\,yr$).

---

[1] In general we use the term "encounter" to denote the gravitational perturbation of the orbit of one star by another, and "collision" to denote actual physical contact between stars. However, to conform with common use, we use the term "collisionless system" to describe a stellar system in which encounters play no role.

Stellar encounters influence the structure of a stellar system in a number of distinct ways:

(i) Relaxation. Each star slowly wanders away from its initial orbit, and the structure of the system accordingly becomes less dependent on the initial conditions. The evolution is toward a state of higher entropy. Although the maximum entropy state of an ideal gas in a box is one of uniform density, we have seen in §4.7.1 that for a stellar system of fixed mass and energy the high entropy states are very inhomogeneous, with a small, dense central core and an extended low-density halo.

(ii) Equipartition. A typical stellar system contains stars whose masses span a wide range. From elementary kinetic theory we know that encounters tend to produce equipartition of kinetic energy: on average, the stars with the largest kinetic energy lose energy to stars with less kinetic energy. If the system forms by violent relaxation, then the position and velocity distribution of the stars is initially independent of the stellar mass. Hence the most massive stars have the largest kinetic energy $\frac{1}{2}mv^2$, and over a time of order t_{relax} they lose energy to less massive stars. Shorn of their original kinetic energy, the massive stars then sink toward the system's center.

(iii) Escape. From time to time an encounter gives enough energy to a star that it can escape from the system. Thus there is a slow but irreversible leakage of stars from the system, and in a sense the only permanent equilibrium state of a stellar system consists of two stars in a Kepler orbit, with all the others having escaped to infinity. The timescale over which the stars "evaporate" in this way can be directly related to the relaxation timescale by the following simple argument (Ambarzumian 1938; Spitzer 1940). From equation (2-25) the escape speed v_e at $\mathbf{x}$ is given by $v_e^2 = -2\Phi(\mathbf{x})$. The mean-square escape speed in a system whose density is $\rho(\mathbf{x})$ is therefore

$$\langle v_e^2 \rangle = \frac{\int \rho(\mathbf{x}) v_e^2 d^3\mathbf{x}}{\int \rho(\mathbf{x}) d^3\mathbf{x}} = -2\frac{\int \rho(\mathbf{x})\Phi(\mathbf{x}) d^3\mathbf{x}}{M} = -\frac{4W}{M}, \qquad (8\text{-}2)$$

where M and W are the total mass and potential energy of the system. According to the virial theorem (4-79), $-W = 2K$, where $K = \frac{1}{2}M\langle v^2 \rangle$ is the total kinetic energy. Hence

$$\langle v_e^2 \rangle = 4\langle v^2 \rangle. \qquad (8\text{-}3)$$

Thus the root mean square (RMS) escape speed is just twice the RMS speed. The fraction of particles in a Maxwellian distribution that have speeds exceeding twice the RMS speed is $\gamma = 7.38 \times$

10^{-3} (Problem 4-17). We can crudely represent the evaporation process as simply removing a fraction γ of the stars every relaxation time. Thus the rate of loss is $dN/dt = -\gamma N/t_{\text{relax}} \equiv -N/t_{\text{evap}}$, where the **evaporation time**, the characteristic time in which the system's stars evaporate, is $t_{\text{evap}} = 136 t_{\text{relax}}$. Thus we expect that evaporation sets an upper limit to the lifetime of any bound stellar system of about $10^2 t_{\text{relax}}$.

(iv) Inelastic encounters. Until now we have treated stars as point masses, but in dense systems we must consider the possibility that two stars will actually pass so close that they raise powerful tides on one another or even suffer a physical collision. In most cases a head-on or nearly head-on collision leads to coalescence of the two stars; thus, collisions lead to evolution of the mass spectrum of the stars. In addition, the dissipation of kinetic energy in the collisions causes the total kinetic energy of the system to decrease. Even if there is no physical contact, the violent tides raised in a close passage (where the minimum separation of the centers is less than a few times the sum of the stellar radii) can dissipate so much kinetic energy that the stars capture one another to form a binary system.

The characteristic timescale on which a star suffers a collision is given approximately by $t_{\text{coll}} \approx (\nu \Sigma v)^{-1}$, where ν is the number density of stars, Σ is the collision cross-section, and v is the RMS velocity of stars in the system. We may write $\nu \approx N/r^3$, where r is the characteristic radius of the system, and $\Sigma \approx \pi(2r_\star)^2$, where $r_\star$ is the stellar radius (neglecting gravitational focusing; a more exact calculation is given in §8.4.5). In terms of the crossing time $t_{\text{cross}} \approx r/v$

$$\frac{t_{\text{coll}}}{t_{\text{cross}}} \approx \frac{r^2}{4\pi N r_\star^2}. \tag{8-4}$$

From the virial theorem we have $v^2 \approx GNm/r$ where m is the stellar mass; it proves convenient to use this relation to eliminate r in favor of v. We also eliminate $r_\star$ in favor of the escape speed from the stellar surface, $v_\star = \sqrt{2Gm/r_\star}$. Thus

$$\frac{t_{\text{coll}}}{t_{\text{cross}}} \approx 0.02 \frac{N v_\star^4}{v^4}. \tag{8-5}$$

In terms of the relaxation time [eq. (8-1)],

$$\frac{t_{\text{coll}}}{t_{\text{relax}}} \approx 0.2 \left(\frac{v_\star}{v}\right)^4 \ln N. \tag{8-6}$$

In systems composed of objects whose escape speed is much smaller than the RMS velocity dispersion, $t_{coll} \ll t_{relax}$, and inelastic collisions are more important than gravitational encounters. An example of such a system is a rich cluster of galaxies in which individual galaxies take the place of stars in the preceding arguments. Here $v_* \approx 300\,\mathrm{km\,s^{-1}}$, while $v \approx 1500\,\mathrm{km\,s^{-1}}$.

(v) Binary formation by three-body encounters. A binary star cannot form in an isolated encounter of two point masses (since the motion is always along a hyperbola). However, an encounter involving three stars can leave two of the participants in a bound Kepler orbit. It is simple to estimate the rate of binary formation by this process. We showed in §4.1 that the velocity perturbation in an encounter of two stars of mass m and velocity v is $\delta v \approx Gm/pv$, where p is the distance of closest approach. If three stars approach each other within a distance p, we expect the velocity perturbations to be of similar magnitude. Thus, to form a binary by a three-body encounter, we must have $\delta v \approx v$, which requires $p \approx Gm/v^2$. For a given star, the time interval between encounters with other stars at separation p or less is of order $(\nu p^2 v)^{-1}$. In each such encounter, there is a probability $p^3 \nu$ that a third star will also lie within a distance p. Hence the time t_3 required for a given star to suffer a triple encounter at separation less than p is $t_3 \approx (\nu^2 p^5 v)^{-1}$. Setting $p \approx Gm/v^2$, we find the time required for a given star to become part of a binary by a triple encounter to be

$$t_3 \approx \frac{v^9}{\nu^2 G^5 m^5}. \tag{8-7}$$

Using the virial theorem, $v^2 \approx GNm/r$, we may express t_3 in terms of the relaxation time [eq. (8-1)]:

$$\frac{t_3}{t_{relax}} \approx 10 N^2 \ln N. \tag{8-8}$$

Hence the total number of binaries formed per relaxation time is only of order

$$\frac{N t_{relax}}{t_3} \approx \frac{0.1}{N \ln N}. \tag{8-9}$$

Since the system dissolves after the evaporation time of about $100 t_{relax}$, the rate of binary formation by triple encounters is completely negligible if N is larger than about 100.

On the basis of these conclusions, we will neglect binary formation by three-body encounters in our initial discussion of the evolution of stellar systems due to encounters. We return to a detailed discussion of binary formation and evolution in §8.4.4.

(vi) Interactions with primordial binaries. The many binary stars found in the solar neighborhood were almost certainly produced at the time of star formation. It is likely that binary stars are also produced at the time of star formation in globular and open clusters. These are called **primordial binary stars** to distinguish them from binaries formed by triple or inelastic encounters long after the cluster was born. Encounters of a binary with other cluster stars lead to exchange of energy between the binary orbit and the random motions in the cluster. Such encounters can increase or decrease the binding energy of the binary, and on occasion can even disrupt the binary (see §8.4.4). By energy conservation, such processes also affect the total kinetic energy of the cluster, and the evolutionary track followed by the cluster can therefore depend sensitively on the fraction of its stars that belong to primordial binary systems. The importance of binary stars can be illustrated by comparing the total binding energy of a cluster to the binding energy of a close binary. Consider, for example, a globular cluster with mass $M = 10^5 M_\odot$ and RMS velocity $\langle v^2 \rangle^{1/2} = 10\,\mathrm{km\,s^{-1}}$. From the virial theorem, its energy is $E = -\frac{1}{2} M \langle v^2 \rangle = -1 \times 10^{50}$ ergs. A binary star consisting of two $1 M_\odot$ stars with a separation of $2 R_\odot$ has a binding energy of 1×10^{48} ergs. Thus, 100 such binaries contain as much binding energy as the whole cluster, and if the binary binding energy is exchanged with passing stars, the structure and evolution of the cluster will be drastically altered.

Since the collisionless Boltzmann equation, which has been our main tool so far, is not valid when encounters are important, our first step toward exploring the consequences of encounters is to find other equations and results that are valid in the presence of encounters (§8.1 and §8.2). Detailed calculations of the evolution of stellar systems require approximate methods (§8.3 and §8.4).

For readable reviews of the subject of this chapter, see Hénon (1973a), Lightman and Shapiro (1978), and Spitzer (1987).

8.1 Exact Results

1 Virial Theorem

In Chapter 4 we proved the tensor virial theorem [eq. (4-78)],

$$\frac{1}{2}\frac{d^2 I_{jk}}{dt^2} = 2K_{jk} + W_{jk}, \tag{8-10}$$

which relates the moment of inertia tensor I_{jk} of an isolated stellar system to the kinetic and potential energy tensors K_{jk} and W_{jk}. Although this theorem was proved using the collisionless Boltzmann equation, we now show that with slightly different definitions the same result can be shown to hold for any system of N mutually gravitating particles.

Consider a system of N particles with masses m_α and positions $\mathbf{x}^\alpha, \alpha = 1, \ldots, N$. The moment of inertia tensor is

$$I_{jk} = \sum_{\alpha=1}^{N} m_\alpha x_j^\alpha x_k^\alpha. \tag{8-11}$$

Its second time derivative is

$$\frac{d^2 I_{jk}}{dt^2} = \sum_{\alpha=1}^{N} m_\alpha \left(\ddot{x}_j^\alpha x_k^\alpha + 2\dot{x}_j^\alpha \dot{x}_k^\alpha + x_j^\alpha \ddot{x}_k^\alpha \right). \tag{8-12}$$

The acceleration of particle α is

$$\ddot{x}_j^\alpha = \sum_{\substack{\beta=1 \\ \beta \neq \alpha}}^{N} \frac{G m_\beta (x_j^\beta - x_j^\alpha)}{|\mathbf{x}^\beta - \mathbf{x}^\alpha|^3}; \tag{8-13}$$

substituting this result and a similar formula for $\ddot{x}_k^\alpha$ into equation (8-12) we find

$$\frac{d^2 I_{jk}}{dt^2} = 2 \sum_{\alpha=1}^{N} m_\alpha \dot{x}_j^\alpha \dot{x}_k^\alpha + \sum_{\substack{\alpha,\beta=1 \\ \beta \neq \alpha}}^{N} \frac{G m_\alpha m_\beta}{|\mathbf{x}^\alpha - \mathbf{x}^\beta|^3} \left\{ (x_j^\beta - x_j^\alpha) x_k^\alpha + (x_k^\beta - x_k^\alpha) x_j^\alpha \right\}.$$

$$\tag{8-14}$$

We identify the first sum on the right side with $4K_{jk}$, where $\mathbf{K}$ is the kinetic energy tensor [eq. (4-73b)]. The second sum is closely related to the potential energy tensor $\mathbf{W}$ that was defined in Chapter 2 for continuous systems. By analogy with equations (2-125a) and (2-126) we define the potential energy tensor for a system of point particles as[2]

$$
\begin{aligned}
W_{jk} &= G \sum_{\substack{\alpha,\beta=1 \\ \beta \neq \alpha}}^{N} m_\alpha m_\beta \frac{x_j^\alpha (x_k^\beta - x_k^\alpha)}{|\mathbf{x}^\alpha - \mathbf{x}^\beta|^3} \\
&= -\tfrac{1}{2} G \sum_{\substack{\alpha,\beta=1 \\ \beta \neq \alpha}}^{N} m_\alpha m_\beta \frac{(x_j^\alpha - x_j^\beta)(x_k^\alpha - x_k^\beta)}{|\mathbf{x}^\alpha - \mathbf{x}^\beta|^3},
\end{aligned}
\tag{8-15}
$$

where the second line is obtained by interchanging the dummy indices α and β in the first line and adding the result to the first line. From the second line we conclude that $\mathbf{W}$ is symmetric, that is, $W_{jk} = W_{kj}$. The second sum on the right side of equation (8-14) is just $W_{kj} + W_{jk} = 2W_{jk}$, and we have therefore arrived at equation (8-10).◁

The most useful form of the virial theorem is obtained by taking the trace of the tensor $\mathbf{I}$, $I \equiv \text{trace}(\mathbf{I}) \equiv \sum_{j=1}^{3} I_{jj}$. Furthermore, we assume that the system is in a steady state, so that $d^2 I/dt^2 = 0$. Equation (8-15) then becomes the **scalar virial theorem**

$$
2K + W = 0,
\tag{8-16}
$$

where

$$
K = \text{trace}(\mathbf{K}) = \tfrac{1}{2} \sum_{\alpha=1}^{N} m_\alpha v_\alpha^2 \quad ; \quad W = \text{trace}(\mathbf{W}) = -\tfrac{1}{2} \sum_{\substack{\alpha,\beta=1 \\ \alpha \neq \beta}}^{N} \frac{G m_\alpha m_\beta}{|\mathbf{x}^\alpha - \mathbf{x}^\beta|}
$$

$$
\tag{8-17}
$$

are the total kinetic and potential energy. If E is the total energy of the system, $E = K + W$, then an alternative statement of the scalar virial theorem is

$$
E = -K = \tfrac{1}{2} W.
\tag{8-18}
$$

The only approximation involved in deriving the scalar virial theorem is that the moment of inertia I is time-independent. In a system with a small number of particles, there are statistical fluctuations in I, and equations (8-16) and (8-18) only hold for the time-averaged values of K and W.

---

[2] Formally, we can derive equation (8-15) by replacing the continuous density $\rho(\mathbf{x})$ in (2-125a) by a sum of Dirac delta functions: $\rho(\mathbf{x}) = \sum_{\alpha=1}^{N} m_\alpha \delta(\mathbf{x} - \mathbf{x}^\alpha)$.

2 Liouville's Theorem

A complete specification of the state of a system of N stars requires a knowledge of the coordinates and velocities of each of the N stars. We represent this state by a point in the $6N$-dimensional space, called Γ-**space**, whose coordinates are the spatial coordinates and velocities of all the stars. This state is called a **microstate** and its representative point a Γ-**point**. In practice, we do not have (and are not interested in) the detailed information that is required to specify a particular microstate. We are really only concerned with the "average" behavior of a system with a given set of macroscopic properties (density distribution, velocity distribution, number of binary stars, etc.). To make the definition of "average" more precise, we imagine that the system in question is replicated a large number of times to form a collection of systems with the same macroscopic properties but distributed over a range of all possible microstates that can produce the given macroscopic properties. Such a collection is called an **ensemble**. The state of the ensemble can be described by the density of Γ-points in Γ-space, and the evolution of the ensemble is described by the evolution of the density distribution in Γ-space.

Consider a system of N particles and denote the position and velocity of the αth particle by $\mathbf{x}_\alpha, \mathbf{v}_\alpha$. Let the six-dimensional vector $\mathbf{w}_\alpha \equiv (\mathbf{x}_\alpha, \mathbf{v}_\alpha)$ denote the location of an individual particle in the six-dimensional phase space. The Γ-point of a given system in the $6N$-dimensional Γ-space is determined by the collection of N six-vectors $\mathbf{w}_1, \ldots, \mathbf{w}_N$. The probability that a Γ-point is found in a unit volume of Γ-space at time t is denoted by $f^{(N)}(\mathbf{w}_1, \ldots, \mathbf{w}_N, t)$; thus the normalization condition is

$$\int f^{(N)}(\mathbf{w}_1, \ldots, \mathbf{w}_N, t)\, d^6\mathbf{w}_1 \ldots d^6\mathbf{w}_N = 1, \text{ where } d^6\mathbf{w}_\alpha \equiv d^3\mathbf{x}_\alpha d^3\mathbf{v}_\alpha.$$
(8-19)

The function $f^{(N)}$ is known as the **N-particle distribution function** or N-particle DF. The derivation of an equation governing the evolution of $f^{(N)}$ is exactly analogous to the derivation of the collisionless Boltzmann equation governing the evolution of the phase-space density f (cf. §4.1). Since the Γ-points drift smoothly through Γ-space, $f^{(N)}$ must satisfy a continuity equation analogous to that satisfied by the density $\rho(\mathbf{x}, t)$ of an ordinary fluid flow (see Appendix 1.E):

$$\frac{\partial f^{(N)}}{\partial t} + \sum_{\alpha=1}^{N} \left\{ \frac{\partial}{\partial \mathbf{x}_\alpha} \left[f^{(N)} \frac{d\mathbf{x}_\alpha}{dt} \right] + \frac{\partial}{\partial \mathbf{v}_\alpha} \left[f^{(N)} \frac{d\mathbf{v}_\alpha}{dt} \right] \right\} = 0. \quad (8\text{-}20)$$

To simplify this equation, note that $d\mathbf{x}_\alpha/dt = \mathbf{v}_\alpha$ and $\partial \mathbf{v}_\alpha/\partial \mathbf{x}_\alpha = 0$ because $\mathbf{x}_\alpha$ and $\mathbf{v}_\alpha$ are independent coordinates in Γ-space. Also, if the forces are conservative, $d\mathbf{v}_\alpha/dt = -\partial\Phi_\alpha/\partial\mathbf{x}_\alpha$, where Φ_α is the potential at particle α due to the other particles. Thus $d\mathbf{v}_\alpha/dt$ depends only on the coordinates $\mathbf{x}_1, \ldots, \mathbf{x}_N$ and not on the velocities, so that its partial derivative $\partial(d\mathbf{v}_\alpha/dt)/\partial\mathbf{v}_\alpha = 0$. Hence equation (8-20) becomes

$$\frac{\partial f^{(N)}}{\partial t} + \sum_{\alpha=1}^{N}\left[\mathbf{v}_\alpha \cdot \frac{\partial f^{(N)}}{\partial\mathbf{x}_\alpha} - \frac{\partial\Phi_\alpha}{\partial\mathbf{x}_\alpha} \cdot \frac{\partial f^{(N)}}{\partial\mathbf{v}_\alpha}\right] = 0. \qquad (8\text{-}21)$$

The left side of this equation is just the convective or Lagrangian derivative in Γ-space (see Appendix 1.E and §4.1). Hence we can write

$$\frac{df^{(N)}}{dt} = 0; \qquad (8\text{-}22)$$

in other words the flow of Γ-points through Γ-space is incompressible; the density of Γ-points $f^{(N)}$ around the Γ-point of a given system always remains constant. This is **Liouville's theorem** and equation (8-22) is **Liouville's equation** (see, e.g., Landau & Lifshitz 1980 or Reichl 1980).[3]

3 The BBGKY Hierarchy*

We can clarify the relation between the Liouville and collisionless Boltzmann equations by integrating over all of the $\mathbf{w}_\alpha$ except one, which, without loss of generality, we may choose to be $\mathbf{w}_1$. We define the **one-particle distribution function** to be

$$f^{(1)}(\mathbf{w}_1, t) = \int f^{(N)}(\mathbf{w}_1, \ldots, \mathbf{w}_N, t)\, d^6\mathbf{w}_2 \ldots d^6\mathbf{w}_N, \qquad (8\text{-}23)$$

and then integrate equation (8-21) over $d^6\mathbf{w}_2 \ldots d^6\mathbf{w}_N$. The term involving $\partial f^{(N)}/\partial t$ simply yields $\partial f^{(1)}/\partial t$. Integration of the term involving $\partial f^{(N)}/\partial\mathbf{x}_\alpha$ yields zero if $\alpha = 2, \ldots, N$ because $\int(\partial f^{(N)}/\partial\mathbf{x}_\alpha)d^3\mathbf{x}_\alpha = 0$ so long as $f^{(N)} \to 0$ sufficiently fast as $|\mathbf{x}_\alpha| \to \infty$. The integration of the term involving $\partial f^{(N)}/\partial\mathbf{v}_\alpha$ yields zero if $\alpha = 2, \ldots, N$ for a similar reason. Thus we obtain

$$\frac{\partial f^{(1)}(\mathbf{w}_1, t)}{\partial t} + \mathbf{v}_1 \cdot \frac{\partial f^{(1)}(\mathbf{w}_1, t)}{\partial\mathbf{x}_1} = \int \frac{\partial\Phi_1}{\partial\mathbf{x}_1} \cdot \frac{\partial f^{(N)}}{\partial\mathbf{v}_1}\, d^6\mathbf{w}_2 \ldots d^6\mathbf{w}_N.$$
$$(8\text{-}24)$$

---

[3] Liouville's equation is in fact not due to Liouville. It was first explicitly derived by Gibbs (1884), two years after Liouville's death. Gibbs was also the first to recognize its potential usefulness in astronomy. It might therefore be better called the Gibbs equation.

* This section contains more advanced material and can be skipped on a first reading.

At this point the derivation is greatly simplified if we assume that $f^{(N)}$ is a symmetric function of $\mathbf{w}_1, \ldots, \mathbf{w}_N$. This assumption is certainly valid if all the stars have the same characteristics (mass, composition, etc.), and it is straightforward to modify the calculations below for a system containing several distinct stellar types. The gravitational potential at the location of star α is

$$\Phi_\alpha = \sum_{\beta \neq \alpha} \Phi_{\alpha\beta}, \quad \text{where} \quad \Phi_{\alpha\beta} = -\frac{Gm}{|\mathbf{x}_\alpha - \mathbf{x}_\beta|}, \tag{8-25}$$

and m is the stellar mass. Since $f^{(N)}$ is symmetric in $\mathbf{w}_1, \ldots, \mathbf{w}_N$, we can write the right side of (8-24) as

$$(N-1) \int \frac{\partial \Phi_{12}}{\partial \mathbf{x}_1} \cdot \frac{\partial f^{(N)}}{\partial \mathbf{v}_1} \, d^6\mathbf{w}_2 \ldots d^6\mathbf{w}_N. \tag{8-26}$$

If we define the **two-particle distribution function** by

$$f^{(2)}(\mathbf{w}_1, \mathbf{w}_2, t) = \int f^{(N)}(\mathbf{w}_1, \ldots, \mathbf{w}_N, t) \, d^6\mathbf{w}_3 \ldots d^6\mathbf{w}_N, \tag{8-27}$$

then we obtain

$$\frac{\partial f^{(1)}}{\partial t} + \mathbf{v}_1 \cdot \frac{\partial f^{(1)}}{\partial \mathbf{x}_1} = (N-1) \int \frac{\partial \Phi_{12}}{\partial \mathbf{x}_1} \cdot \frac{\partial f^{(2)}}{\partial \mathbf{v}_1} \, d^6\mathbf{w}_2. \tag{8-28}$$

Equation (8-28) expresses the rate of change of $f^{(1)}(\mathbf{w}_1, t)$ in terms of $f^{(2)}(\mathbf{w}_1, \mathbf{w}_2, t)$. A similar equation can be obtained for the rate of change of $f^{(2)}(\mathbf{w}_1, \mathbf{w}_2, t)$ in terms of the three-particle DF $f^{(3)}(\mathbf{w}_1, \mathbf{w}_2, \mathbf{w}_3, t)$ by integrating equation (8-21) over $d^6\mathbf{w}_3 \ldots d^6\mathbf{w}_N$. Continuing in this way, we obtain a sequence of equations, each expressing the rate of change of $f^{(n)}$ in terms of $f^{(n+1)}$. This sequence is known as the **BBGKY hierarchy**, after N. N. Bogoliubov, M. Born and H. S. Green, J. G. Kirkwood, and J. Yvon, who all discovered the equations independently between 1935 and 1946. Obtaining an exact solution for the hierarchy is no easier than solving the Liouville equation (8-21) itself, since the solution for $f^{(n+1)}$ must be known before $f^{(n)}$ can be determined. Nevertheless, the BBGKY hierarchy is useful because it is sometimes possible to guess an approximate form for $f^{(n+1)}$ that permits the hierarchy to be closed.

For example, let us write the two-particle DF as

$$f^{(2)}(\mathbf{w}_1, \mathbf{w}_2, t) = f^{(1)}(\mathbf{w}_1, t) f^{(1)}(\mathbf{w}_2, t) + g(\mathbf{w}_1, \mathbf{w}_2, t). \tag{8-29}$$

The function g is symmetric in $\mathbf{w}_1$ and $\mathbf{w}_2$ because $f^{(N)}$ is symmetric in all of the $\mathbf{w}_\alpha$, and is called the **two-particle correlation function** since, loosely speaking, it measures the *excess* probability of finding a particle at $\mathbf{w}_1$ due to the presence of a particle at $\mathbf{w}_2$ (and vice versa). Notice that the one-particle DF $f^{(1)}(\mathbf{w}, t)$ is closely related to the phase-space density $f(\mathbf{w}, t)$ defined in §4.1, since both describe the density of stars in six-dimensional phase space. In fact, $f = N f^{(1)}$, since $\int f(\mathbf{w}, t) \, d^6\mathbf{w} = N$, the total number of stars, while $\int f^{(1)}(\mathbf{w}, t) d^6\mathbf{w} = 1$ by equations (8-19) and (8-23). Hence we can rewrite equation (8-28) as

$$\frac{\partial f(\mathbf{w}_1, t)}{\partial t} + \mathbf{v}_1 \cdot \frac{\partial f(\mathbf{w}_1, t)}{\partial \mathbf{x}_1} = \frac{(N-1)}{N} \frac{\partial f(\mathbf{w}_1, t)}{\partial \mathbf{v}_1}$$

$$\cdot \frac{\partial}{\partial \mathbf{x}_1} \int \Phi_{12} f(\mathbf{w}_2, t) \, d^6\mathbf{w}_2 + N(N-1) \int \frac{\partial \Phi_{12}}{\partial \mathbf{x}_1} \cdot \frac{\partial g(\mathbf{w}_1, \mathbf{w}_2, t)}{\partial \mathbf{v}_1} \, d^6\mathbf{w}_2.$$
(8-30)

Now suppose that the number of stars $N \gg 1$ so that $N - 1 \simeq N$, and note that

$$\int \Phi_{12} f(\mathbf{w}_2, t) \, d^6\mathbf{w}_2 = \int \Phi_{12} f(\mathbf{w}_2, t) \, d^3\mathbf{x}_2 d^3\mathbf{v}_2$$

$$= -G \int \rho(\mathbf{x}_2, t)|\mathbf{x}_1 - \mathbf{x}_2|^{-1} \, d^3\mathbf{x}_2 \qquad (8\text{-}31)$$

$$= \Phi(\mathbf{x}_1),$$

where $\rho(\mathbf{x})$ and $\Phi(\mathbf{x})$ are the total density and potential from all the stars. Then, on dropping the subscript '1' we have

$$\frac{\partial f(\mathbf{x}, \mathbf{v}, t)}{\partial t} + \mathbf{v} \cdot \frac{\partial f(\mathbf{x}, \mathbf{v}, t)}{\partial \mathbf{x}} - \frac{\partial \Phi(\mathbf{x})}{\partial \mathbf{x}} \cdot \frac{\partial f(\mathbf{x}, \mathbf{v}, t)}{\partial \mathbf{v}}$$

$$= -N^2 G m \int \frac{\partial g(\mathbf{x}, \mathbf{v}, \mathbf{x}_2, \mathbf{v}_2, t)}{\partial \mathbf{v}} \cdot \frac{\partial}{\partial \mathbf{x}_2} \left(\frac{1}{|\mathbf{x} - \mathbf{x}_2|} \right) d^3\mathbf{x}_2 d^3\mathbf{v}_2.$$
(8-32)

This equation follows exactly from Liouville's theorem and the assumptions that $N \gg 1$ and that $f^{(N)}$ is a symmetric function of the $\mathbf{w}_\alpha$. However, it can be solved only if we choose some approximate form for the two-particle correlation function $g(\mathbf{w}, \mathbf{w}_2, t)$. The simplest assumption is that $g = 0$, and this yields the collisionless Boltzmann equation [eq. (4-13b)]. Hence we have shown that *the collisionless Boltzmann equation is based on the approximation that the total number of particles $N \gg 1$ and that the two-particle correlation function is zero.* A more accurate approximation, based on a local calculation of the two-particle correlation function using the next equation in the hierarchy, can be used to derive the Fokker-Planck equation, which we will discuss in §8.3 (see Lifshitz & Pitaevskii 1981).

The BBGKY hierarchy helps clarify and formalize the kinetic theory of stellar systems, plasmas, or gases. However, it is generally too complicated to be used for practical calculations of the evolution of a model stellar system. For this purpose, the Fokker-Planck equation described below is more useful.

8.2 The Gravothermal Catastrophe

By analogy with the ideal gas, we can define the temperature T of a self-gravitating system using the relation [see eq. (1E-24)]

$$\tfrac{1}{2}m\overline{v^2} = \tfrac{3}{2}k_B T, \tag{8-33}$$

where m is the stellar mass and k_B is Boltzmann's constant. In general the mean-square velocity and hence the temperature depend on position. The mean temperature is therefore $\overline{T} \equiv \int \rho(\mathbf{x})T\, d^3\mathbf{x} / \int \rho(\mathbf{x})\, d^3\mathbf{x}$, and the total kinetic energy of a system of N stars is then

$$K = \tfrac{3}{2}Nk_B\overline{T}. \tag{8-34}$$

According to the virial theorem in the form (8-18), the total energy is $E = -K$, and hence

$$E = -\tfrac{3}{2}Nk_B\overline{T}. \tag{8-35}$$

The heat capacity of the system is then

$$C \equiv \frac{dE}{d\overline{T}} = -\tfrac{3}{2}Nk_B. \tag{8-36}$$

The heat capacity is therefore *negative*: by losing energy the system automatically grows hotter.

This apparently paradoxical result is not restricted to stellar-dynamical systems. *Any* bound, finite system in which the dominant forces are gravitational exhibits a negative heat capacity. In fact, the stability of nuclear burning in the cores of stars like the Sun is a consequence of their negative heat capacity: the reaction rate is a strongly increasing function of density and temperature, so that if the reactions proceed too fast, the core gains energy and therefore expands and cools, bringing the reaction rate back into equilibrium.

A system with negative heat capacity exhibits behavior quite different from normal laboratory systems. For example, suppose that we place a self-gravitating system in contact with a heat bath. Assume that initially the heat bath and the stellar system have the same temperature T. If a small amount of heat $dQ > 0$ is transferred to the bath, the temperature of the stellar system will change to $T - dQ/C = T + dQ/|C|$.

The self-gravitating system is now hotter than the bath, and heat continues to flow from hot to cold, that is, from the system to the bath. Therefore, the temperature of the system continues to rise without limit. Similarly, if heat begins to flow from the bath to the system, then the temperature of the system will decrease to zero.

To investigate this kind of behavior more quantitatively, we consider a self-gravitating ideal gas[4] contained in a spherical container of radius r_b. The gas is assumed to be thermally conducting, so that heat can flow from one part of the gas to another if there are temperature gradients within the gas, over the thermal diffusion timescale t_{diff}. The equilibrium state of this system is therefore isothermal, and is closely related to the isothermal sphere described in §4.4.3(b). We showed there that the structure of an isothermal self-gravitating sphere of gas is identical with the structure of a collisionless stellar system with DF $f(E) \propto \exp(-E/\sigma^2)$, where $\sigma^2 = k_B T/m$. The presence of the wall at radius r_b does not affect the structure of the system at radii $r < r_b$. Hence the radial density distribution of the isothermal gas, $\rho(r)$, is given in Figure 4-7 [in the normalized units defined in equation (4-124)], while the pressure is related to the density by $p(r) = \rho(r)k_B T/m$, where the temperature T is independent of radius.

The total energy of the gas is $E = K + W$, where the kinetic energy K is given by equation (8-34) and the potential energy W is given by equation (2-19). Thus

$$E = \tfrac{3}{2}Nk_B T + \tfrac{1}{2}\int_0^{r_b} 4\pi r^2 \rho(r)\Phi(r)\,dr. \qquad (8\text{-}37)$$

We let M be the total mass of the gas, and for brevity we introduce the inverse temperature

$$\beta \equiv \frac{m}{k_B T}. \qquad (8\text{-}38)$$

The ideal gas law now reads $p = \rho/\beta$, and we can write the kinetic energy as $K = \tfrac{3}{2}M/\beta$. We could determine the potential energy W by performing the integral in (8-37) using the density distribution of the isothermal sphere. However, the calculation can be simplified by using the virial theorem. In Problem 4-12 we derived a form of the

---

[4] In all of our calculations involving a self-gravitating gas, we implicitly assume that the repulsive forces between the gas molecules dominate over the gravitational forces at short distances. Thus binary systems and other short-range correlations between the molecules do not develop, and the molecules are primarily influenced by the mean gravitational field due to the other parts of the gas rather than the fields of nearby molecules.

virial theorem valid for a collisionless system confined to a spherical container. The same theorem holds for a gaseous system and reads

$$2K + W = 4\pi r_b^3 p(r_b). \qquad (8\text{-}39)$$

Hence

$$E = 4\pi r_b^3 p(r_b) - K = \frac{4\pi r_b^3 \rho(r_b)}{\beta} - \frac{3M}{2\beta}. \qquad (8\text{-}40)$$

We now eliminate the inverse temperature β in favor of the King radius r_0 and central density ρ_0 of the isothermal sphere, using equations (4-120) and (4-124b), which yield

$$\beta = \frac{9}{4\pi G \rho_0 r_0^2}. \qquad (8\text{-}41)$$

We also eliminate the central density in favor of the mass M by the relation

$$M = 4\pi \rho_0 r_0^3 \int_0^{\tilde{r}_b} \tilde{r}^2 \tilde{\rho}(\tilde{r}) \, d\tilde{r} \equiv 4\pi \rho_0 r_0^3 \widetilde{M}(\tilde{r}_b), \qquad (8\text{-}42)$$

where the second line defines a dimensionless mass $\widetilde{M}(\tilde{r})$. We can therefore rewrite equation (8-40) in the form

$$E = \frac{GM^2}{r_b} \left[\frac{\tilde{r}_b^4 \tilde{\rho}(\tilde{r}_b)}{9\widetilde{M}(\tilde{r}_b)^2} - \frac{\tilde{r}_b}{6\widetilde{M}(\tilde{r}_b)} \right], \qquad (8\text{-}43)$$

where $\tilde{r}_b = r_b/r_0$ and $\tilde{\rho}(\tilde{r})$ is defined by the differential equation (4-123a). Similarly, the inverse temperature can be written

$$\beta = 9 \frac{r_b}{GM} \frac{\widetilde{M}(\tilde{r}_b)}{\tilde{r}_b}. \qquad (8\text{-}44)$$

Hence, for a container of a given radius r_b containing a given mass of gas M, we can determine E and β as functions of the parameter $\tilde{r}_b$ and hence as functions of one another. The result is shown in Figure 8-1. We have plotted the dimensionless ratios $r_b/(GM\beta)$ and Er_b/GM^2 so that the graph can be used for containers of any mass M and radius r_b. The subscripts on the labels A–E give the values at the corresponding points on the curve of a parameter that is not $\tilde{r}_b$ but the more physical parameter

$$\mathcal{R} = \frac{\rho_0}{\rho(r_b)} = \frac{1}{\tilde{\rho}(\tilde{r}_b)}, \qquad (8\text{-}45)$$

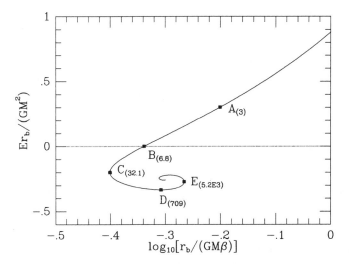

Figure 8-1. The relation between dimensionless temperature $r_b/(GM\beta)$ and dimensionless energy Er_b/GM^2 for a mass M of isothermal gas in a spherical container of radius r_b, at temperature $T = m/k_B\beta$. The curve spirals inward to the point $(\frac{1}{2}, -\frac{1}{4})$ corresponding to the singular isothermal sphere (see Problem 8-10).

which measures the degree of central concentration of the gas in the container.

We now perform the following thought experiment. Suppose that the walls of the container conduct heat, and that the container is surrounded by a heat bath at very high temperature, so that the gravitational potential energy of the gas is small compared with the kinetic energy in random motions (point A on Figure 8-1). Under these conditions, the gas behaves almost like an ideal gas with no self-gravity: the energy $E \simeq \frac{3}{2}M/\beta = \frac{3}{2}Mk_BT/m$, the heat capacity $C = dE/dT \simeq \frac{3}{2}Mk_B/m$, and the gas is nearly homogeneous ($\mathcal{R}$ is near unity). If we now reduce the temperature of the heat bath, energy flows from the gas to the bath, and the system moves down along the curve in Figure 8-1. At point B ($\mathcal{R} = 6.8$) its total energy passes through zero and becomes negative. As we continue to reduce the temperature of the heat bath, the system continues to lose energy and cool, its heat capacity $C = dE/dT$ becomes larger and larger, and finally at point C ($\mathcal{R} = 32.125$) the heat capacity becomes infinite. There is no equilibrium state with $T < T(C) = 0.40GMm/(k_Br_b)$. Systems between points C and D ($32.125 < \mathcal{R} < 708.61$) have negative heat capacity and are unstable for the reasons given at the beginning of this section: if the temperature of the gas momentarily rises above the temperature of the heat bath, then energy flows from the gas to the bath; because of its negative heat capacity the gas becomes hotter as it loses energy; the increased temperature difference leads to even faster energy loss to the heat bath; and the gas is heated without limit. Similarly, if the gas is momentarily cooler than the bath, then energy flows into the gas, which cools continuously as a result.

In systems between points D and E ($708.61 < \mathcal{R} < 5221.5$) the heat capacity is once again positive. However, a more detailed analysis (Horwitz & Katz 1978) shows that these systems are also unstable when in contact with a heat bath. In fact, *an isothermal sphere in contact with a heat bath is unstable whenever the density contrast between center and edge exceeds* $\mathcal{R} = 32.125$. Katz (1978) gives an elegant derivation of this result based on the geometry of the E-β curve.

Now consider a second thought experiment. Suppose that our container is surrounded by a thermally insulating wall (i.e., the stars bounce off the wall without gain or loss of energy). Initially the gas has energy E, mass M, and radius r_{b0}. Suddenly the container is expanded to a radius r_b. Since the expansion is sudden, the gas does no work on the container, and thus E is constant. After the expansion, when the gas has again settled into equilibrium, its temperature can be determined from Figure 8-1. If the energy of the gas is positive, then the value of the vertical coordinate Er_b/GM^2 is increased by a factor r_b/r_{b0}. The value of $\mathcal{R}$ is decreased and the gas is therefore more homogeneous. However, if the energy is negative, then Er_b/GM^2 becomes more negative by a factor of r_b/r_{b0}. If the final value of Er_b/GM^2 is more negative than -0.335 (the value at point D in Figure 8-1), then no equilibrium is possible after the expansion.

Point D has a deeper significance than this. Stability analysis shows that a thermally isolated ($E = constant$) sphere is unstable at all points in the equilibrium sequence of Figure 8-1 beyond point D, on the thermal diffusion timescale t_{diff} (Antonov 1962a; Lynden-Bell & Wood 1968; Katz 1978; Horwitz & Katz 1978). In other words, *an isothermal gas in an insulating spherical container is unstable if the density contrast between center and edge exceeds* $\mathcal{R} = 708.61$. It can be shown that this instability (named the **gravothermal catastrophe** by Lynden-Bell & Wood 1978) arises because the isothermal sphere becomes a local entropy *minimum* at fixed E, M, and r_b, rather than a local entropy *maximum*. Thus the system can reach states of higher entropy by evolving away from isothermality. If at some instant the central core is slightly hotter than its surroundings, then it continues to become hotter and hotter, and denser and denser; heat flows out to the outer parts of the system, but the peripheral temperature cannot rise fast enough to catch up with the runaway temperature of the central core.[5]

The onset of instability can be heuristically explained in terms of the virial theorem. The halo has positive heat capacity C_h since it is not strongly influenced by self-gravity, while the core, which is confined primarily by self-gravity, has negative heat capacity C_c. If the core momentarily becomes hotter than the halo, heat flows from the core to the halo, and the temperatures of *both* the core and halo rise. If $C_h < |C_c|$, the halo temperature rises more than the core temperature so that the heat flow is shut off. If $C_h > |C_c|$, then the halo has so much thermal inertia that it cannot heat up as fast as the core, and the temperature difference between core and halo grows. Of course, the division into a separate core and halo is artificial, but this argument appears to capture the essence of the instability that sets in at $\mathcal{R} = 708.61$.

Is there a gravothermal catastrophe in isothermal stellar systems as well as in gaseous ones? The answer is almost certainly yes. The velocity distribution in an isothermal stellar system is the same as in an isothermal gas (§4.4), and hence the entropy of an isothermal stellar system is the same as the entropy of an isothermal gas with the same temperature and density distribution. If we can approximate the stellar encounters as local, then the entropy of the stellar system cannot decrease with time. Hence instability is likely to arise when the equilibrium state becomes a local entropy minimum instead of a maximum, just as it does for a gaseous system. This argument has been confirmed by numerical calculations of the normal modes of an isothermal stellar system (Inagaki 1980; Ipser & Kandrup 1980). These authors find that instability sets in when $\mathcal{R} > 708.61$, which is Antonov's original criterion for instability in an isothermal gas.

The gravothermal catastrophe in a gas develops through heat conduction and hence the growth time is comparable to the thermal diffusion time t_{diff}. The analog of the diffusion time in a stellar system is the relaxation time t_{relax}, and hence the gravothermal catastrophe in stellar systems should develop on a timescale of order t_{relax} (see Inagaki 1980 for confirmation). To investigate the relevance of the gravothermal catastrophe to realistic stellar systems, we must first develop tools that will enable us to follow the evolution of a stellar system over timescales of this order.

---

[5] A version of this argument was given already by Landau (1932). He argued that quantum-mechanical effects led eventually to the existence of a stable equilibrium, and in this way derived the maximum mass limit for a degenerate star (the Chandrasekhar limit).

8.3 The Fokker-Planck Approximation

1 Master Equation

Under the influence of the smooth potential $\Phi(\mathbf{x})$, the distribution function (DF) $f(\mathbf{x}, \mathbf{v}, t)$ obeys the collisionless Boltzmann equation $df/dt = 0$, where the derivative is taken along the path of a star through phase space [eq. (4-13c)]; in words, the phase-space density around a given star always remains the same. When encounters are taken into account, the phase-space density around a star changes with time, and we may write

$$\frac{df}{dt} = \Gamma[f], \tag{8-46}$$

or, from equation (4-13c),

$$\frac{\partial f}{\partial t} + \mathbf{v} \cdot \nabla f - \nabla \Phi \cdot \frac{\partial f}{\partial \mathbf{v}} = \Gamma[f], \tag{8-47}$$

where the **collision term** Γ denotes the rate of change of f due to encounters. The value of the collision term is a function of position $\mathbf{x}$, velocity $\mathbf{v}$ and time t that is determined by the DF $f(\mathbf{x}, \mathbf{v}, t)$ in a manner that we now describe.

Let $\Psi(\mathbf{w}, \Delta\mathbf{w})d^3\Delta\mathbf{w}\Delta t$ be the probability that a star with the six canonical phase-space coordinates $\mathbf{w}$ is scattered to a new volume of phase space $d^3\Delta\mathbf{w}$ around $\mathbf{w} + \Delta\mathbf{w}$ during the time interval Δt. The scattering cross-section Ψ includes only the effects of encounters with other stars but not acceleration by the smooth potential of the stellar system, since the latter is accounted for already in the left side of equation (8-47). To distinguish the star whose trajectory we are following from the stars doing the scattering, we call the former the **test star** and the latter **field stars**.

As a result of encounters, test stars are scattered out of a unit volume of phase space centered on $\mathbf{w}$ at a rate

$$\left.\frac{\partial f(\mathbf{w})}{\partial t}\right|_{-} = -f(\mathbf{w}) \int \Psi(\mathbf{w}, \Delta\mathbf{w})d^3\Delta\mathbf{w}. \tag{8-48}$$

There are also encounters that scatter test stars into this volume, at a rate

$$\left.\frac{\partial f(\mathbf{w})}{\partial t}\right|_{+} = \int \Psi(\mathbf{w} - \Delta\mathbf{w}, \Delta\mathbf{w})f(\mathbf{w} - \Delta\mathbf{w})d^3\Delta\mathbf{w}. \tag{8-49}$$

The sum $(\partial f/\partial t)_{-} + (\partial f/\partial t)_{+}$ is just the collision term $\Gamma[f]$. Hence we arrive at the **master equation** (e.g., Reichl 1980)

$$\frac{df}{dt} = \Gamma[f] = \int [\Psi(\mathbf{w} - \Delta\mathbf{w}, \Delta\mathbf{w})f(\mathbf{w} - \Delta\mathbf{w}) - \Psi(\mathbf{w}, \Delta\mathbf{w})f(\mathbf{w})]d^3\Delta\mathbf{w}. \tag{8-50}$$

2 Fokker-Planck Equation

In the crude estimate of the relaxation time presented in Chapter 4, we found that encounters give rise to a mean-square velocity perturbation per crossing time [eq. (4-8)]

$$\Delta v_\perp^2 \approx \frac{8v^2 \ln(R/b_{\min})}{N}, \tag{8-51}$$

where N is the number of stars in the system and v is the star's velocity. This result arose from integrating over impact parameters between the system's characteristic radius R and the impact parameter $b_{\min} = Gm/v^2$ at which the velocity change in an encounter is comparable to the velocity v. The contribution to $\Delta v_\perp^2$ from impact parameters in any interval (b_1, b_2) can be obtained by simply replacing $\ln(R/b_{\min})$ in equation (8-51) by $\ln(b_2/b_1)$. Thus *equal logarithmic intervals of impact parameter contribute equally to $\Delta v_\perp^2$*; in other words, encounters with impact parameters in the range R to $\frac{1}{2}R$, $\frac{1}{2}R$ to $\frac{1}{4}R$, and so forth, down to the interval $2b_{\min}$ to $b_{\min}$, are all of equal importance for the relaxation process. Since the fractional velocity change in an encounter is $(\delta v_\perp/v) \approx (b_{\min}/b)$ [see eq. (4-3)], it follows that *when $R \gg b_{\min}$, most of the scattering is due to weak encounters, that is, ones with $\delta v \ll v$.*[6]

We can exploit the dominance of weak encounters to derive a simplified form of the collision term. For weak encounters, $|\Delta \mathbf{w}|$ is small, and we can expand the first term of equation (8-50) in a Taylor series

$$\Psi(\mathbf{w} - \Delta\mathbf{w}, \Delta\mathbf{w}) f(\mathbf{w} - \Delta\mathbf{w}) = \Psi(\mathbf{w}, \Delta\mathbf{w}) f(\mathbf{w})$$

$$- \sum_{i=1}^{6} \Delta w_i \frac{\partial}{\partial w_i} [\Psi(\mathbf{w}, \Delta\mathbf{w}) f(\mathbf{w})] \tag{8-52}$$

$$+ \frac{1}{2} \sum_{i,j=1}^{6} \Delta w_i \Delta w_j \frac{\partial^2}{\partial w_i \partial w_j} [\Psi(\mathbf{w}, \Delta\mathbf{w}) f(\mathbf{w})] + O(\Delta \mathbf{v}^3).$$

The **Fokker-Planck approximation** consists of truncating this series after the second-order terms. When we carry out the integral over $\Delta\mathbf{w}$, we then obtain

$$\Gamma[f] = - \sum_{i=1}^{6} \frac{\partial}{\partial w_i} [f(\mathbf{w}) D(\Delta w_i)] + \frac{1}{2} \sum_{i,j=1}^{6} \frac{\partial^2}{\partial w_i \partial w_j} [f(\mathbf{w}) D(\Delta w_i \Delta w_j)], \tag{8-53}$$

---

[6] As a specific example, consider the core of a globular cluster, with typical velocity $v = 10\,\mathrm{km\,s^{-1}}$ and radius $R = 1\,\mathrm{pc}$ (see Table 1-3). For solar mass stars, we have $b_{\min} = 4 \times 10^{-5}\,\mathrm{pc}$. Thus $\ln(R/b_{\min}) = 10.0$, and half of the scattering is due to encounters with impact parameters exceeding b_1, where $\ln(R/b_1) = 5.0$ or $b_1 = 0.007\,\mathrm{pc}$; for these impact parameters the fractional change in velocity is less than 1%.

where $D(\Delta w_i)$ denotes the expectation per unit time of the change in w_i:

$$D(\Delta w_i) \equiv \int \Delta w_i \Psi(\mathbf{w}, \Delta \mathbf{w}) d^3 \Delta \mathbf{w}, \qquad (8\text{-}54)$$

with a similar definition for $D(\Delta w_i \Delta w_j)$. The quantities $D(\Delta w_i)$ and $D(\Delta w_i \Delta w_j)$ are known as **diffusion coefficients** since they measure the rate at which stars diffuse through phase space as a result of encounters. In principle, equation (8-53) can be extended to a higher order of approximation involving diffusion coefficients like $D(\Delta w_i \Delta w_j \Delta w_k)$, but these are generally much smaller.[7] [It might also be thought that $D(\Delta w_i \Delta w_j)$ is much less than $D(\Delta w_i)$, but we shall show that in fact these two diffusion coefficients are usually comparable in magnitude.]

Equations (8-47) and (8-53) together constitute the **Fokker-Planck equation**. The Fokker-Planck equation has the virtue that all of the dependence on the field star DF f_a is collapsed into the diffusion coefficients, which are functions only of the phase-space coordinates of the test star. Once the diffusion coefficients are known, the Fokker-Planck equation is a differential equation, rather than an integro-differential equation like the master equation, and hence is much easier to solve. As a result, it has become the principal tool for the study of the slow evolution of stellar systems that is driven by encounters.

The practical application of the Fokker-Planck equation to the study of stellar systems can be simplified greatly by the use of one or both of the following two approximations.

(a) Orbit-averaged approximation   We argued at the beginning of this chapter that in stellar systems with large N, the relaxation time is much larger than the crossing time. In these systems, the changes in the DF and diffusion coefficients caused by encounters are generally small over timescales of one orbital period. Hence it is useful to separate the slow changes in the phase-space coordinates caused by encounters from the rapid changes associated with orbital motion in the smooth potential; in effect, we **orbit-average** the Fokker-Planck equation.

Orbit averaging is most easily understood by working in action-angle variables (§3.5). Regular orbits are confined to three-dimensional surfaces in the six-dimensional phase space, namely, the orbital tori, which we label by the action three-vector $\mathbf{J}$. There are three angle

---

[7] By of order the Coulomb logarithm $\ln \Lambda$ (eq. 8-60). See Hénon (1960a, 1973a).

variables $\boldsymbol{\theta}$ conjugate to the actions, and the six variables $(\mathbf{J}, \boldsymbol{\theta})$ can be used as canonical coordinates in phase space.

From §4.1 we know that the DF in a stationary collisionless system is a function $f(\mathbf{J})$ of the actions alone. If encounters cause stars to move from one torus to another, f will change. However, if the movement of stars across tori is slow compared with the time required for a star to explore its current torus thoroughly (i.e., if $t_{\text{relax}} \gg t_{\text{cross}}$), the DF will remain at any given time a function of the actions only, although a function that changes slowly in time. Thus $f = f(\mathbf{J}, t)$.

The Fokker-Planck equation in action-angle coordinates can be written as

$$\frac{\partial f}{\partial t} + \dot{J}_i \frac{\partial f}{\partial J_i} + \dot{\theta}_i \frac{\partial f}{\partial \theta_i} = \Gamma[f], \tag{8-55}$$

in which the time derivatives on the left side refer to motion in the smooth potential (that is, neglecting encounters), and the collision term $\Gamma[f]$ is given by equation (8-53), where we have chosen as phase-space coordinates $\mathbf{w} = (\mathbf{J}, \boldsymbol{\theta})$. Since $f = f(\mathbf{J}, t)$ and $\mathbf{J}$ is conserved in the absence of encounters, the left side simplifies to $\partial f / \partial t$.

We now orbit average, that is, we integrate equation (8-55) over $(2\pi)^{-3} \int d^3\boldsymbol{\theta}$. The left side is unchanged. All terms on the right side involving $\partial / \partial \theta_i$ or $\partial / \partial \theta_j$ vanish, since all quantities are periodic in θ_i. Thus equation (8-55) simplifies to the **orbit-averaged Fokker-Planck equation**

$$\frac{\partial f(\mathbf{J}, t)}{\partial t} = -\frac{\partial}{\partial J_i} \left[f \overline{D}(\Delta J_i) \right] + \tfrac{1}{2} \frac{\partial^2}{\partial J_i \partial J_j} \left[f \overline{D}(\Delta J_i \Delta J_j) \right], \tag{8-56a}$$

where the **orbit-averaged diffusion coefficients** are

$$\overline{D}(\Delta J_i) = \frac{1}{(2\pi)^3} \int D(\Delta J_i) d^3\boldsymbol{\theta}. \tag{8-56b}$$

The advantage of orbit averaging is that the Fokker-Planck equation is reduced from an equation involving six phase-space coordinates plus time to one involving only the three actions plus time. Moreover, in the case of spherical symmetry the DF and the diffusion coefficients can depend on energy E and angular momentum L but not the z-component of angular momentum L_z; hence the azimuthal action $J_a = |L_z|$ and the latitudinal action $J_l = L - |L_z|$ can only enter in the combination $J_a + J_l = L$, and the problem is reduced to one involving two phase-space coordinates, say, J_r and L, plus time.

(b) Local approximation   One of the principal obstacles to any exact treatment of the dynamics of stellar encounters is the long range of the gravitational force between two stars. In a sense, every star in a stellar system is influenced by every other star at all times. Thus the distinction between the smooth gravitational potential of the system and scattering by individual stars is somewhat ill-defined. For many purposes, it therefore proves useful to work with an artificially simple model of stellar encounters. All encounters are assumed to be *local*, that is, the impact parameter b is assumed to be much less than the system size R. For local encounters the duration of the encounter is much less than the orbital time, which implies that (i) since the encounter time is short, the encounter affects only the velocity, not the position, of the interacting stars; (ii) during the encounter the stars may be assumed to move on Keplerian hyperbolae, unaffected by the potential of the cluster; (iii) the effects of stellar encounters on a star at $\mathbf{x}$ can be calculated as if the star were embedded in an infinite homogeneous medium in which the DF is everywhere equal to the DF at $\mathbf{x}$.

The local approximation is successful because equal logarithmic intervals of impact parameter contribute equally to the relaxation [see discussion following eq. (8-51)], so that encounters with impact parameters comparable to the system size make only a small relative contribution to the overall relaxation rate.

The Fokker-Planck collision term (8-53) is simplified in the local approximation, because an encounter changes only the velocity of a star and not its position. Hence, if we choose the canonical phase-space coordinates $\mathbf{w}$ to be Cartesian coordinates $(\mathbf{x}, \mathbf{v})$, then $\Psi(\mathbf{w}, \Delta\mathbf{w})$ is zero unless $\Delta\mathbf{x} = 0$, and as a consequence any diffusion coefficient of the form $D(\Delta x_i)$, $D(\Delta x_i \Delta x_j)$ or $D(\Delta x_i \Delta v_j)$ is zero. Thus the collision term simplifies to

$$\Gamma[f] = -\sum_{i=1}^{3} \frac{\partial}{\partial v_i}[f(\mathbf{w})D(\Delta v_i)] + \tfrac{1}{2}\sum_{i,j=1}^{3} \frac{\partial^2}{\partial v_i \partial v_j}[f(\mathbf{w})D(\Delta v_i \Delta v_j)].$$

$$(8\text{-}57)$$

3 Diffusion Coefficients

We next evaluate the diffusion coefficients $D(\Delta v_i)$ and $D(\Delta v_i \Delta v_j)$ in the local approximation. The diffusion coefficients represent mean changes per unit time due to a large number of encounters of the test star with the other stars in the cluster. Each encounter is assumed to be independent of all the others, and to involve only a single pair of stars (i.e., triple and multiple encounters are neglected). Thus the relative orbit of the two stars is a hyperbola [eq. (7-5)], described by the relative velocity V_0 at large separations and the impact parameter b.

In Appendix 8.A we show that the diffusion coefficients can be written as

$$D(\Delta v_i) = 4\pi G^2 m_a (m + m_a) \ln \Lambda \frac{\partial}{\partial v_i} h(\mathbf{v}),$$

$$D(\Delta v_i \Delta v_j) = 4\pi G^2 m_a^2 \ln \Lambda \frac{\partial^2}{\partial v_i \partial v_j} g(\mathbf{v}),$$

(8-58)

where m and m_a are the masses of the test star and field stars, $h(\mathbf{v})$, $g(\mathbf{v})$ are the **Rosenbluth potentials** (Rosenbluth et al. 1957),

$$h(\mathbf{v}) = \int \frac{f_a(\mathbf{v}_a) \, d^3\mathbf{v}_a}{|\mathbf{v} - \mathbf{v}_a|} \quad ; \quad g(\mathbf{v}) = \int f_a(\mathbf{v}_a)|\mathbf{v} - \mathbf{v}_a| \, d^3\mathbf{v}_a, \quad (8\text{-}59)$$

the field star DF is $f_a(\mathbf{v}_a)$, and

$$\Lambda = \frac{b_{\max} v_{\text{typ}}^2}{G(m + m_a)}. \quad (8\text{-}60)$$

Here v_{typ} is a typical velocity of stars in the system, and $b_{\max}$ is the maximum impact parameter considered. Numerical experiments show that the appropriate value for $b_{\max}$ is of order the radius of the system R—even though the local approximation is being abused somewhat by including such distant encounters.[8] The factor $\ln \Lambda$ is the Coulomb logarithm which has appeared already in similar calculations in Chapters 4 and 7. Our expressions for the diffusion coefficients have a fractional accuracy of order $(\ln \Lambda)^{-1}$. Notice that the diffusion coefficient $D(\Delta v_i)$ is identical to the deceleration due to dynamical friction in equation (7-13a), once the same approximations are made in evaluating the Coulomb logarithm. This identity is hardly surprising since the derivation leading to the diffusion coefficients (8-58) is simply a more general version of the derivation leading to the dynamical friction formula.

It is important to remember that this approach does not properly represent the effects of either very close or very distant encounters. Close encounters, those with impact parameter $\lesssim b_{\min} = Gm/v^2$, have $\delta v/v$ of order unity and hence violate the Fokker-Planck approximation. Distant encounters, those with impact parameters of order the system size R, cannot be treated within the context of the local approximation, which is

---

[8] In systems with strong central concentration, R should be taken as the orbital radius of the test star rather than the radius of the system, since the density over most of the system may be much less than the local density.

only strictly valid for $b \ll R$. Nevertheless, the Fokker-Planck plus local approximations yield satisfactory results whenever $\ln \Lambda$ is significantly larger than unity. This is because, as we saw at the beginning of §8.3.2, equal logarithmic intervals of impact parameter contribute equally to the relaxation process. When $\ln \Lambda \approx \ln(R/b_{\min})$ is large, there are many octaves in impact parameter that contribute to the relaxation, and failure of the approximations for an octave or two at either end does not lead to significant error.

The diffusion coefficients are greatly simplified if the field star DF $f_a(\mathbf{v}_a)$ depends only on $|\mathbf{v}_a| \equiv v_a$. (In a spherical system this condition is satisfied if f_a depends only on the energy E and not the angular momentum L.) If f_a is spherically symmetric in velocity space, then the only preferred direction in velocity space is defined by the velocity of the test star $\mathbf{v}$, and therefore it is natural to choose a coordinate system in which $\hat{\mathbf{e}}_z$ is parallel to $\mathbf{v}$, and $\hat{\mathbf{e}}_x$, $\hat{\mathbf{e}}_y$ are perpendicular to $\mathbf{v}$. Then the symmetry of the problem demands that

$$D[(\Delta v_x)^2] = D[(\Delta v_y)^2], \tag{8-61}$$

and that

$$D(\Delta v_x) = D(\Delta v_y) = D(\Delta v_x \Delta v_y) = D(\Delta v_x \Delta v_z) = D(\Delta v_y \Delta v_z) = 0. \tag{8-62}$$

Hence there are only three independent diffusion coefficients, which we may write as

$$\begin{aligned} D(\Delta v_\parallel) &\equiv D(\Delta v_z), \\ D(\Delta v_\parallel^2) &\equiv D[(\Delta v_z)^2], \\ D(\Delta v_\perp^2) &\equiv 2D[(\Delta v_x)^2] = 2D[(\Delta v_y)^2], \end{aligned} \tag{8-63}$$

where the factor 2 has been inserted in the last line so that $D(\Delta v_\perp^2)$ represents the total diffusion rate in the two-dimensional plane perpendicular to $\mathbf{v}$. The evaluation of these diffusion coefficients is described in Appendix 8.A. Their values are:

$$D(\Delta v_\parallel) = -\frac{16\pi^2 G^2 m_a (m + m_a) \ln \Lambda}{v^2} \int_0^v v_a^2 f_a(v_a)\, dv_a,$$

$$D(\Delta v_\parallel^2) = \frac{32\pi^2 G^2 m_a^2 \ln \Lambda}{3v} \left[\int_0^v \frac{v_a^4}{v^2} f_a(v_a)\, dv_a + v \int_v^\infty v_a f_a(v_a)\, dv_a \right],$$

$$D(\Delta v_\perp^2) = \frac{32\pi^2 G^2 m_a^2 \ln \Lambda}{3v}$$
$$\times \left[\int_0^v \left(3v_a^2 - \frac{v_a^4}{v^2} \right) f_a(v_a)\, dv_a + 2v \int_v^\infty v_a f_a(v_a)\, dv_a \right]. \tag{8-64}$$

The relation of these coefficients to the diffusion coefficients with respect to an arbitrarily oriented coordinate system, $D(\Delta v_i)$ and $D(\Delta v_i \Delta v_j)$, is given by equation (8A-20):

$$D(\Delta v_i) = \frac{v_i}{v} D(\Delta v_{\parallel}),$$

$$D(\Delta v_i \Delta v_j) = \frac{v_i v_j}{v^2} [D(\Delta v_{\parallel}^2) - \tfrac{1}{2} D(\Delta v_{\perp}^2)] + \tfrac{1}{2} \delta_{ij} D(\Delta v_{\perp}^2). \tag{8-65}$$

The rate of change of the kinetic energy of the test star is

$$D(\Delta E) = m \sum_{i=1}^{3} [v_i D(\Delta v_i) + \tfrac{1}{2} D(\Delta v_i^2)]$$

$$= m[v D(\Delta v_{\parallel}) + \tfrac{1}{2} D(\Delta v_{\parallel}^2) + \tfrac{1}{2} D(\Delta v_{\perp}^2)]$$

$$= 16\pi^2 G^2 m m_a \ln \Lambda \left[m_a \int_v^{\infty} v_a f_a(v_a) dv_a - m \int_0^v \frac{v_a^2}{v} f_a(v_a) dv_a \right]. \tag{8-66}$$

Both integrals in equation (8-66) are non-negative. The first describes the tendency of the test-star population to be heated by the field-star population. The heating rate is proportional to the field-star mass. The second integral describes the cooling effect of dynamical friction and is proportional to the test-star mass. The energy of the test-star population settles to an equilibrium when the two terms balance. The mean-square speed of the population at which the integrals cancel is proportional to m^{-1}. This is the classical phenomenon of **equipartition of energy**.

The diffusion coefficients can be explicitly evaluated if the field star DF $f_a(v_a)$ is known. The most important special case is the Maxwellian distribution

$$f_a(v_a) = \frac{\rho}{m_a (2\pi\sigma^2)^{3/2}} e^{-v_a^2/2\sigma^2}, \tag{8-67}$$

where ρ and σ are the density and one-dimensional velocity dispersion of the field stars. Evaluating the integrals of equation (8-64) we find

$$D(\Delta v_{\parallel}) = -\frac{4\pi G^2 \rho (m + m_a) \ln \Lambda\, G(X)}{\sigma^2},$$

$$D(\Delta v_{\parallel}^2) = \frac{4\sqrt{2}\pi G^2 \rho m_a \ln \Lambda}{\sigma} \frac{G(X)}{X}, \tag{8-68}$$

$$D(\Delta v_{\perp}^2) = \frac{4\sqrt{2}\pi G^2 \rho m_a \ln \Lambda}{\sigma} \left[\frac{\mathrm{erf}(X) - G(X)}{X} \right],$$

where $X \equiv v/(\sqrt{2}\sigma)$, $\mathrm{erf}(X)$ is the error function (Appendix 1.C.3), and

$$G(X) \equiv \frac{1}{2X^2} \left[\mathrm{erf}(X) - X \frac{d\,\mathrm{erf}(X)}{dX} \right] = \frac{1}{2X^2} \left[\mathrm{erf}(X) - \frac{2X}{\sqrt{\pi}} e^{-X^2} \right]. \tag{8-69}$$

4 Relaxation Time

We can employ the diffusion coefficients to obtain a more precise value for the relaxation time than our crude estimate of equation (4-9). Following Spitzer and Hart (1971), we shall base our estimate on $D(\Delta v_\parallel^2)$ since it plays the most important role in the transfer of energy between particles. Thus we define the relaxation time of a test star to be

$$t_{\mathrm{relax}} \equiv \frac{v^2}{D(\Delta v_\parallel^2)}. \tag{8-70}$$

To measure the mean relaxation time for a population of stars, we set $m_a = m$ and assume that the test stars have the same Maxwellian velocity distribution as the field stars. The RMS value of v is thus $\sqrt{3}\sigma$ and the corresponding value of X is 1.225. Since $G(X)/X$ is not a rapidly varying function of X, we simply set $X = 1.225$ in $D(\Delta v_\parallel^2)$ and replace the factor v^2 in (8-70) by σ^2, which is the mean-square velocity in any one direction. We find

$$\begin{aligned} t_{\mathrm{relax}} &= 0.34 \frac{\sigma^3}{G^2 m \rho \ln \Lambda} \\ &= \frac{1.8 \times 10^{10}\,\mathrm{yr}}{\ln \Lambda} \left(\frac{\sigma}{10\,\mathrm{km\,s^{-1}}} \right)^3 \left(\frac{1\,\mathrm{M_\odot}}{m} \right) \left(\frac{10^3\,\mathrm{M_\odot\,pc^{-3}}}{\rho} \right). \end{aligned} \tag{8-71}$$

This definition is somewhat arbitrary, but the exact value of the numerical coefficient does not matter very much, since the main purpose of our definition is simply to provide a standard time unit for measuring the speed of relaxation processes.

The relaxation time usually varies by several orders of magnitude in different regions of a single system. For reference purposes it is often useful to characterize a system by a single measure of the relaxation time. To this end, we replace the density ρ in equation (8-71) by the mean density inside the system's median radius r_h, which is just $\frac{1}{2}M/(\frac{4}{3}\pi r_h^3)$, and we replace $3\sigma^2$ by the mean-square speed of the system's stars $\langle v^2 \rangle$. By equation (4-80b), $\langle v^2 \rangle \simeq 0.4GM/r_h$. Finally, we set $\Lambda = r_h \langle v^2 \rangle / Gm = 0.4N$, and equation (8-70) then yields the **median relaxation time** (Spitzer & Hart 1971)

$$\begin{aligned} t_{\mathrm{rh}} &= \frac{0.14N}{\ln(0.4N)} \sqrt{\frac{r_h^3}{GM}} \\ &= \frac{6.5 \times 10^8\,\mathrm{yr}}{\ln(0.4N)} \left(\frac{M}{10^5\,\mathrm{M_\odot}} \right)^{\frac{1}{2}} \left(\frac{1\,\mathrm{M_\odot}}{m} \right) \left(\frac{r_h}{1\,\mathrm{pc}} \right)^{\frac{3}{2}}. \end{aligned} \tag{8-72}$$

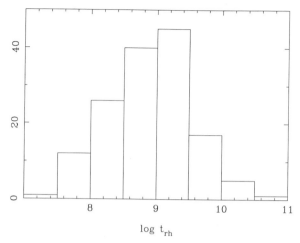

Figure 8-2. The distribution of median relaxation time $t_{\rm rh}$ in years for 147 galactic globular clusters. The results are based on the tabulation of Webbink (1985), assuming a stellar mass $m = 0.7\,{\rm M_\odot}$ and mass-to-light ratio $\Upsilon = 2\Upsilon_\odot$.

Figure 8-2 shows the distribution of $t_{\rm rh}$ for galactic globular clusters; in almost all cases $t_{\rm rh} \lesssim 10^{10}\,{\rm yr}$, confirming that substantial relaxation has occurred in these systems. The relaxation times in the cores are shorter still, by at least an order of magnitude.

5 Numerical Solutions of the Fokker-Planck Equation

(a) Moment equations   In §4.2 we derived the Jeans equations by taking moments of the collisionless Boltzmann equation over velocity space. Since the Fokker-Planck equation is simply the collisionless Boltzmann equation with a collision integral $\Gamma[f]$ on the right side, we can similarly derive collisional analogs of the Jeans equations. These moment equations lead to useful approximate solutions of the Fokker-Planck equation.

We first integrate equations (8-47) and (8-57) over all velocities less than the escape speed, $|\mathbf{v}| < v_e = [-2\Phi(\mathbf{x})]^{1/2}$. The evaluation of the

integrals in equation (8-47) is carried out as described in §4.2, and we
obtain

$$
\frac{\partial \nu}{\partial t} + \frac{\partial (\nu \overline{v}_i)}{\partial x_i} = - \int \frac{\partial}{\partial v_i} \Big[f(\mathbf{v}) D(\Delta v_i) \Big] \, d^3 \mathbf{v}
$$
$$
+ \tfrac{1}{2} \int \frac{\partial^2}{\partial v_i \partial v_j} \Big[f(\mathbf{v}) D(\Delta v_i \Delta v_j) \Big] \, d^3 \mathbf{v},
$$

(8-73)

where the spatial number density of stars $\nu(\mathbf{x})$ and the mean stellar
velocity $\overline{v}_i(\mathbf{x})$ are defined in equation (4-20). The divergence theorem
(1B-42) enables us to rewrite the right side as

$$
\int d^2 v_i \bigg\{ - f(\mathbf{v}) D(\Delta v_i) + \tfrac{1}{2} \frac{\partial}{\partial v_j} \Big[f(\mathbf{v}) D(\Delta v_i \Delta v_j) \Big] \bigg\},
$$

(8-74)

where the integral is over the surface of a sphere of radius v_e in velocity
space, and $d^2\mathbf{v}$ is an outward-pointing vector normal to the surface of
the sphere, whose magnitude is the area of a surface element. We can
assume that $f(\mathbf{v}) = 0$ at $|\mathbf{v}| = v_e$. Thus the first term in equation (8-74)
is zero, and we have

$$
\frac{\partial \nu}{\partial t} + \frac{\partial (\nu \overline{v}_i)}{\partial x_i} = \tfrac{1}{2} \int d^2 v_i \frac{\partial}{\partial v_j} [f(\mathbf{v}) D(\Delta v_i \Delta v_j)].
$$

(8-75)

The left side is just the usual continuity equation (4-21), while the right
side represents losses due to the flux of stars across the escape-velocity
surface in velocity space.

Three more moment equations can be obtained by integrating equations (8-47) and (8-57) over velocity after multiplying by v_k, $k = 1, 2, 3$.
In these equations the moments $\overline{v}_k$ and $\overline{v_k v_i}$ appear. If we multiply by
$v_k v_l$ and integrate over velocity, we obtain six more moment equations
involving $\overline{v_k v_l}$ and $\overline{v_k v_l v_i}$. Thus, as in the case of the Jeans equations
studied in §4.2, we must make some physical assumption about the DF to
terminate this regression to ever higher moments of the velocity distribution. We therefore assume some form for the DF $f(\mathbf{x}, \mathbf{v}, t)$. The strategy
is to choose a rather general form for f, whose behavior depends on,
say, K adjustable parameters, and then determine these parameters by
solving K moment equations.

For example, Larson (1970a,b) investigates the evolution of a spherically symmetric cluster by assuming that the DF can be written as a
Maxwellian times a power series in v_r and $v_\theta^2 + v_\phi^2$. The parameters of the
power series and of the Maxwellian are taken to be functions of radius r
and time t. The diffusion coefficients are computed on the assumption

that the distribution is exactly Maxwellian (although this assumption is not self-consistent, it should not introduce substantial errors). By truncating the power series after only a few terms, he is able to obtain a set of six moment equations that determine the evolution of the parameters in r and t. A closely related procedure is to choose a form of the DF so that the moment equations reduce to the fluid equations for a gas with a particular form for the thermal conductivity (Lynden-Bell & Eggleton 1980).

The value of solutions of the moment equations depends on whether the assumed form of the DF is physically well-motivated. We shall see below that the moment equations lead to results that are often in good agreement with more accurate methods. However, it is probably best to regard the moment equations as describing a model system that shares some properties with self-gravitating stellar systems, rather than as a technique that describes the actual behavior of stellar systems.

(b) Monte Carlo methods   In this approach the Fokker-Planck equation itself is never used. Instead, a representative sample of n test stars is chosen at random from the much larger number of stars comprising the cluster. The orbit of each star is followed numerically, and small random velocity perturbations are applied at frequent intervals along the orbit. The perturbations are chosen by Monte Carlo sampling techniques in such a way that the mean and RMS perturbation in velocity in unit time are consistent with the diffusion coefficients. The number of stars followed must be sufficiently large so that the statistical fluctuations in the results are acceptably small; in practice, typically $n = 1000$. A Monte Carlo program is much faster than an N-body program, both because n is much less than the total number of stars in the systems of interest ($N \approx 10^5$ for a globular cluster), and because the number of operations per crossing time is proportional to n, rather than to $\frac{1}{2}N(N-1)$ as it is in an N-body program using direct summation [cf. §2.8(a)].

The most important case is a spherical cluster. Here it is convenient to think of the ith test star as representing a large number p of stars, each with the same radius r_i, radial velocity v_{ri}, transverse speed v_{ti}, and mass m_i, but with randomly distributed directions of the transverse velocity. The introduction of these shells of stars, or "superstars", has the advantage of forcing the calculation to maintain spherical symmetry, as well as simplifying the potential calculations, since the potential of a superstar is just $\Phi(r) = -Gmp/r$ if $r > r_i$ and $\Phi(r) = -Gmp/r_i$ if $r < r_i$ (see §2.1.1).

Monte Carlo programs of this sort have been developed by Hénon (1972) and by Spitzer and his collaborators at Princeton (see references in Spitzer 1975). The two programs differ in several important respects. In the Princeton program, the orbit $r_i(t)$ of each superstar is followed numerically. After every five timesteps, v_{ri} and v_{ti} are given random perturbations whose average values are determined by the diffusion coefficients at r_i (in calculating the diffusion coefficients, the velocity distribution is assumed to be Maxwellian, and the density and velocity dispersion are determined from means over a region of forty superstars containing the one in question). In contrast, Hénon argues that since encounters only affect the orbit over a time t_{relax} which is much longer than the crossing time t_{cross}, there is little to be gained by integrating the orbits of the superstars in detail. Hence he stores only the energy E and angular momentum L of each superstar and not its radius, and the program is used only to compute energy and angular momentum changes due to encounters, not to integrate orbits. At each timestep (which can now be much longer than t_{cross}) he simply assigns each superstar to a randomly chosen phase of its orbit and perturbs E and L by random amounts, whose average values are determined by the values of the diffusion coefficients at that radius. In contrast to the Princeton program, Hénon does not assume that the field star DF is either Maxwellian or isotropic.

An important advantage of Hénon's method is speed. The timestep in the Princeton code must be a small fraction of t_{cross}, while Hénon's timestep need only be a small fraction of t_{relax}, which is much longer [eq. (8-1)]. On the other hand, in the Fokker-Planck approximation the evolution of a cluster with large N can be determined by scaling the evolution of a cluster with small N,[9] so the Princeton code can be run on models with relatively low N, and then the results can be rescaled to apply to clusters with larger N. In addition, we shall find in §8.4.1 that the Princeton code provides a more accurate treatment of the evaporation process. A hybrid scheme that combines many advantages of both codes is described by Shapiro (1985).

(c) Direct numerical solution   At the present time it is feasible to solve the Fokker-Planck equation only in its orbit-averaged form [§8.3.2(a)]. For spherical systems, the orbit-averaged Fokker-Planck equation is a partial differential equation in three independent variables, time t, radial action J_r, and angular momentum L. The solution is complicated by the fact that the cluster density and potential are constantly

---

[9] If the clusters have the same total mass and radius, the evolution is the same but the timescale is proportional to $N \ln N$.

changing; thus each time the DF $f(J_r, L, t)$ is updated, we must recompute the potential of the self-consistent stellar system that has the new f for its DF.

The details of the numerical methods used to solve the orbit-averaged Fokker-Planck equation (8-56) are described by Cohn (1979).[10]

Notice that Hénon's Monte Carlo method is also an orbit-averaged method, because it keeps track of only the energy and angular momentum of each superstar, and not its orbital phase. Direct orbit-averaged methods share with Hénon's method an inability to handle evaporation satisfactorily.

(d) Checks and comparisons   It is extremely important to compare the results of different methods of solving the Fokker-Planck equation, both with each other and with N-body integrations. Figure 8-3 shows the results of one such comparison. The initial state was a Plummer model with isotropic velocity distribution [§4.4.3(a)]. All of the stars had equal masses. The radii containing 10%, 50%, and 90% of the mass are plotted as a function of time. The time unit is the median relaxation time (8-72). Results from the Larson moment equations, the Princeton and Hénon Monte Carlo codes, Cohn's orbit-averaged Fokker-Planck code, and four different N-body integrations (one with $N = 100$ and three with $N = 250$) are all shown. There is generally good agreement between the different programs. In particular, there is no visible systematic difference between the N-body integrations with $N = 100$ and $N = 250$, despite the fact that their relaxation times differ by a factor of two. The Monte Carlo calculations agree very well among themselves and with the orbit-averaged calculation, which is encouraging because the three approaches are all very different. The agreement with the N-body models is slightly less good; the Monte Carlo codes appear to evolve faster than the N-body models by about a factor of 1.5, but this is only a minor disagreement (see Hénon 1975).

These results confirm the validity of the basic assumptions of the Fokker-Planck and local approximations, that diffusion due to weak, local, two-body encounters is the principal source of relaxation in stellar systems.

---

[10] In fact, Cohn works with a DF $f(E, L)$. This approach has the conceptual complication that any changing potential causes changes in $f(E, L)$, even if there are no encounters. By contrast, since the actions are adiabatic invariants (§3.6), the DF $f(J_r, L, t)$ is invariant under slow changes in the potential.

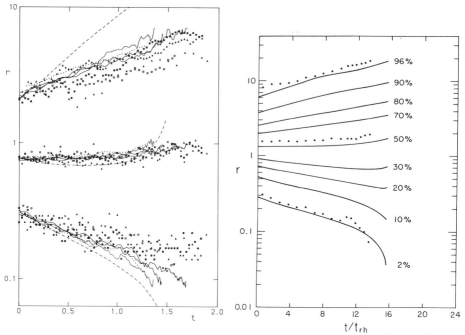

Figure 8-3. The evolution of an isotropic Plummer model according to various programs. Left: radii containing 10%, 50%, and 90% of the mass, plotted versus time, for Hénon's Monte Carlo method (solid lines); Princeton Monte Carlo method (dotted lines); and Larson moment equations (dashed lines). Open triangles and squares denote N-body integrations with $N = 100$ and 250, while filled triangles and squares denote independent N-body integrations with $N = 250$ (Aarseth et al. 1974). Right: comparison of Cohn's (1979) orbit-averaged Fokker-Planck code (solid lines) with the Princeton Monte Carlo code (dots). The curves represent radii containing the indicated percentages of the cluster mass. The time unit in the second diagram is the *initial* median relaxation time. Reprinted by permission from *The Astrophysical Journal*.

8.4 The Evolution of Spherical Stellar Systems

In this section we describe the evolution of a spherical stellar system over timescales of order the relaxation time. The evolution times will be expressed in units of the median relaxation time $t_{\rm rh}$. For concreteness we shall assume that the system in question is a globular cluster, although many of our results can also be applied to other systems such as open clusters or galactic nuclei.

The formation of globular clusters is rather poorly understood, and thus we have only a crude idea of their typical states after they have settled into dynamical equilibrium but before encounters become important (i.e., when their age t satisfies $t_{\rm cross} \ll t \ll t_{\rm relax}$). Fortunately for our

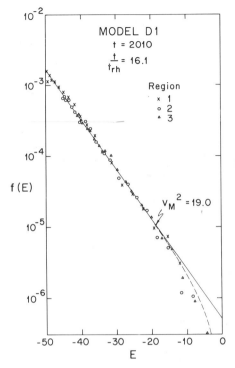

Figure 8-4. The phase-space density of stars $f(E)$ in the inner regions of a model computed by Spitzer and Thuan (1972). Zero energy is defined to be the energy of a star at rest at infinity. The coordinates are chosen so that a Maxwellian distribution appears as a straight line. The best-fitting Maxwellian is plotted as a solid line, while the best-fitting lowered isothermal DF [eq. (4-130)] is shown as a dashed line. From Spitzer (1975), by courtesy of the International Astronomical Union.

purposes, it appears that relaxation tends to erase a cluster's memory of its initial state, and the numerical experiments described below give very similar results for a wide range of initial conditions.

Since the relaxation time is inversely proportional to density, relaxation effects appear first in the central regions of the cluster (often called the cluster "core"). Under the influence of encounters, the DF $f(E, L)$ evolves toward the isothermal or Maxwellian distribution of equation (8-67), but the evolution is rapid only for values of E and L that allow stars to penetrate the central core where the relaxation time is short. This behavior is exhibited in Figure 8-4, which shows $f(E, L \simeq 0)$ in a model constructed with the Princeton Monte Carlo program, plotted in coordinates in which a Maxwellian would appear as a straight line. The agreement is good except near $E = 0$, where the loss of unbound stars causes the points to drop away to zero. The dashed line shows the King DF (4-130), which provides a much better fit to the Monte Carlo model.

In the outer parts of the cluster (the "halo") the relaxation time is long, and encounters have relatively little effect. However, as relaxation proceeds, the halo population is augmented by stars that were originally in the core but that now have reached energies close to escape energy as a result of encounters. Although the apocenters of these orbits lie far out in the halo, their pericenters must still lie in the core—since relaxation is only effective in the core, an orbit that remains bound can never be expelled completely from the core by encounters. These *nouveau riche* halo members, who have risen in the world in consequence of a series of profitable encounters with less fortunate stars, ultimately overwhelm the original halo members, who acquired their wealth at birth. Thus, after a few core relaxation times, the distribution of halo stars is determined by relaxation in the core, rather than by initial conditions.

These arguments suggest that the DF $f(E, L)$ of a relaxed cluster has the following properties: (i) at low energies, f is approximately isothermal, $f(E, L) \propto \exp(-E/\sigma^2)$; (ii) there are few stars with angular momenta greater than some cutoff L_0, corresponding roughly to the angular momentum of a nearly unbound star whose orbit just grazes the core; (iii) $f(E, L)$ tends smoothly to zero as E tends to the escape energy E_e. Detailed analysis shows that in this regime $f \propto (E_e - E)$ (Spitzer & Shapiro 1972). A DF that satisfies all of these qualitative criteria is the Michie DF of §4.4.4(b),

$$
f(E, L) = \begin{cases} Ce^{-L^2/L_0^2}\left[e^{\mathcal{E}/\sigma^2} - 1\right], & \mathcal{E} > 0, \\ 0, & \mathcal{E} \leq 0, \end{cases} \tag{8-76}
$$

where $\mathcal{E} \equiv \Phi_0 - E$ and Φ_0 is the escape energy. Thus, the Michie DF should provide a good empirical model for the DF of globular clusters and other collisionally relaxed stellar systems.

The Michie model—or any other model satisfying our criteria—makes several definite predictions about the radial dependence of observable parameters such as the stellar density $\nu(r)$, the radial velocity dispersion $\overline{v_r^2}$, and the anisotropy parameter $\beta(r) = 1 - \overline{v_\theta^2}/\overline{v_r^2}$. (i) The central density profile will resemble closely the profile of an isothermal sphere; (ii) the asymptotic behavior of $\nu(r)$ as $r \to \infty$ is straightforward to determine; in particular, for an isolated cluster, $\nu \propto r^{-7/2}$ (Problem 4-20). Figure 8-5 shows the density profiles of a Princeton model sequence, along with the profile $\nu \propto r^{-7/2}$. The agreement with the isothermal sphere in the inner parts, and the $r^{-7/2}$ profile in the outer parts, is remarkably good. (iii) The radial and tangential velocity dispersions should be the same in the inner parts of the cluster, and in the outer halo the velocity ellipsoid should become more and more radial, with the tangential dispersion falling off as r^{-1} (since $L = rv_t \lesssim L_0$ so that $v_t < L_0/r$) and the radial dispersion in an isolated system falling off more slowly as $r^{-1/2}$. These predictions are also confirmed by numerical models.

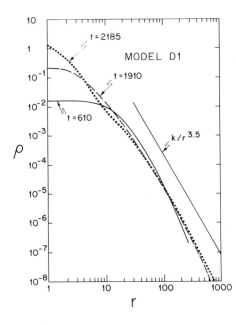

Figure 8-5. Density profiles in a model sequence computed by Spitzer and Thuan (1972). The profile of the central regions resembles that of an isothermal sphere with steadily decreasing core radius, while the profile in the outer parts is close to a power law, $r^{-7/2}$. From Spitzer (1975), by courtesy of the International Astronomical Union.

The collisional evolution of a stellar system involves many different physical processes. In the rest of this section we review the most important known processes individually. However, the reader should bear in mind that in fact all of these processes occur simultaneously, and it may be difficult in practice—and sometimes even in principle—to isolate the effects of any single process on the evolution of a cluster.

1 Evaporation and Ejection

Stars can escape from a cluster by two conceptually different mechanisms: (i) A single close encounter with another star can produce a velocity change comparable with the initial relative velocity of the two stars, thereby leaving one of the stars with a speed exceeding the local escape speed v_e; we shall call this process **ejection**. (ii) A series of weaker, more distant encounters can gradually increase the energy of a star, until a single weak encounter gives the star slightly positive energy and it escapes; we shall call this process **evaporation** to suggest its more gradual nature.

Hénon (1969b) has calculated the ejection rate for an isolated system obeying a Plummer law with isotropic velocity distribution. If all the stars have the same mass and the total number of stars is N, the ejection rate is

$$\frac{dN}{dt} = -8.8 \times 10^{-4} \frac{N}{t_{\rm rh} \ln(0.4N)}, \qquad (8\text{-}77)$$

where we have expressed the result in units of the median relaxation time $t_{\rm rh}$ [eq. (8-72)]. Notice that the Coulomb logarithm $\ln(0.4N)$ is present only to cancel the dependence of $t_{\rm rh}$ on $\ln(0.4N)$; since the Coulomb logarithm arises from the cumulative effect of distant encounters, it does not directly influence the ejection rate, which is due to close encounters.

From equation (8-77) we can define an ejection time

$$t_{\rm ej} = -\left(\frac{1}{N}\frac{dN}{dt}\right)^{-1} = 1.1 \times 10^3 \ln(0.4N) t_{\rm rh}. \qquad (8\text{-}78)$$

For typical values of the Coulomb logarithm $[\ln(0.4N) \approx 10]$, we shall find that $t_{\rm ej}$ is much longer than the evaporation time due to distant encounters. Hence for most purposes we can neglect ejection relative to evaporation.

Evaporation is a more complicated process than ejection. The basic idea is that a myriad of weak encounters causes the star to wander at random through phase space, and some of the most energetic stars wander into the portion of phase space that is associated with unbound orbits. However, many high-energy stars experience very few encounters, because they orbit always within the low-density halo. Thus the evaporation rate is dominated by stars on highly elongated orbits: these stars experience a significant number of encounters each radial period as they pass through the dense cluster core. As the energy of such a star approaches escape energy, the apocenter increases and the orbital period becomes longer, but the pericenter tends to remain at roughly the same distance. Thus the RMS energy change due to encounters is approximately constant, and we denote this constant by ϵ_2. When the energy of the halo star is within ϵ_2 of the escape energy Φ_0, there is a substantial chance that it will escape after its next passage through pericenter.

The behavior described above has important consequences for the orbit-averaged Fokker-Planck equation. Orbit averaging is only valid so long as the fractional changes in the orbital parameters are small in a single orbit. However, we have seen that orbits only escape from an isolated cluster when their binding energy $\Phi_0 - E$ is comparable to the RMS energy change per orbit ϵ_2. Hence *an orbit-averaged calculation cannot accurately predict the rate of escape from a cluster*. The most extreme example of this difficulty with the orbit-averaged equations occurs in an isolated cluster. In this case the period of the stars, and hence the interval between passages through the core, approaches infinity as the stars approach escape energy [see eqs. (3-26) and (3-28)]. Consequently, the rate of diffusion becomes slower and slower as the stars approach escape energy, and in fact the escape energy is never reached (**Hénon's paradox**). A similar problem does not arise in tidally truncated clusters, because the period of an orbit at the escape energy is finite.

The Princeton code was the first Fokker-Planck code to yield reliable evaporation rates for isolated clusters. For clusters of stars with a single mass, Spitzer and Thuan (1972) find that the evaporation rate is given by

$$t_{\text{evap}} = -N(dN/dt)^{-1} \approx 300 t_{\text{rh}}. \tag{8-79}$$

Evaporation has other effects on a cluster besides the loss of stars. Let us consider a simple model of the evolution of a cluster in which the evolution is self-similar, that is, in which the shape of the cluster remains constant while the total mass M and radius r evolve. According to the virial theorem, the total energy of the cluster may be written as

$$E = -k\frac{GM^2}{r}, \tag{8-80}$$

where k is a dimensionless constant of order unity, which is independent of time because the cluster shape is assumed to be fixed. The evaporation is predominantly due to weak encounters, so that stars escape with very nearly zero energy. Hence to a good approximation the total energy E is fixed. Thus, as the cluster loses mass,

$$r = r_0(M/M_0)^2, \tag{8-81}$$

where M_0 and r_0 are the initial mass and radius. The mass loss rate is

$$\frac{dM}{dt} = -\frac{k_e M}{t_{\text{rh}}}, \tag{8-82}$$

where $k_e \approx 0.003$ according to the Princeton models. Finally, because the cluster evolution is self-similar, we may use equation (8-72) to write

$$t_{\text{rh}} = t_{\text{rh}}^0 \sqrt{\frac{Mr^3}{M_0 r_0^3}}, \tag{8-83}$$

where t_{rh}^0 is the initial relaxation time, and we have neglected the slow variation of $\ln(0.4N)$. Combining these results, we have

$$\frac{dM}{dt} = -\frac{k_e M_0^{7/2}}{t_{\text{rh}}^0 M^{5/2}}, \tag{8-84}$$

which is easily solved to yield (Gurevich & Levin 1950; King 1958)

$$M(t) = M_0 \left(1 - \frac{7k_e t}{2t_{\text{rh}}^0} \right)^{2/7}. \tag{8-85}$$

Thus the cluster evaporates completely in a finite time $2t_{\text{rh}}^0/7k_e$, or approximately $100t_{\text{rh}}^0$. Notice that even though the cluster is losing mass, it is becoming more dense, with the density increasing as $\rho \propto M^{-5}$. Evaporation causes a kind of collapse of the cluster, with a shrinkage of the cluster radius and growth of the density. The time remaining before collapse is always $\approx 10^2 t_{rh}$, in gratifying agreement with the crude estimate we made at the start of this chapter.

Tidal forces from the Galaxy can substantially increase the evaporation rate. In one case examined by Spitzer and Chevalier (1973) with a tidal radius $r_t = 9.3r_h$, the escape rate was increased by about a factor of five, yielding $k_e = 0.015$ in equation (8-82).

In the light of this discussion, it is interesting to reexamine Figure 8-2, which shows the distribution of t_{rh} in galactic globular clusters. There are almost no known globulars with t_{rh} less than about $10^{7.5}$ yr, or about 1% of the age of the Galaxy. This may, of course, be a coincidence, but a far more compelling explanation is that there once were many globulars with shorter relaxation times, but that these all evaporated in $\approx 10^2 t_{rh}$. This explanation suggests that the number of globular clusters that existed shortly after the Galaxy formed may have been far larger than the present population of 200 or so.

2 Core Collapse

The evolution of the mass distribution in a simple cluster that began as a Plummer model is shown in Figure 8-3. The radius r_{90} containing 90% of the mass expands, due to the gradual growth of the halo as core stars diffuse toward the escape energy. At the same time, the radius r_{10} containing 10% of the mass decreases, so that the central density is growing; moreover, a modest extrapolation of the curve of r_{10} versus time (particularly in Larson's and Cohn's models) implies that the central density will become infinite in a finite time. Indeed, all of the simulations shown in this figure eventually had to be stopped because of various singular events associated with the growing central density. In Larson's model, the central density became infinite at $t = 16.0t_{\text{rh}}$ (where t_{rh} now denotes the *initial* median relaxation time). In Hénon's three models, the innermost superstar abruptly collapsed under its own self-gravity at times between 14.6 and 18.2 times t_{rh}. In the Princeton models, the computations do not quite reach a singularity, but a small extrapolation indicates that the central density becomes infinite at $t = 14.4t_{rh}$; in Cohn's model the central density becomes infinite at $15.9t_{rh}$. In the N-body integrations, a close binary formed near the center, at times ranging from 15.1 to 17.3 times t_{rh}, and then became more and more tightly bound, soaking up more and more of the binding energy of the cluster (and more and more computing time) until the integration had to be stopped.

This apparent singularity, which invariably appears between 12 and 19 times t_{rh} in systems of stars of a single mass, has come to be known as **core collapse**. Core collapse is a phenomenon of direct observational relevance, since examination of Figure 8-2 shows that many globular clusters have ages that significantly exceed $20t_{rh}$ and hence probably have already undergone core collapse. Core collapse is also probably the single most fascinating theoretical aspect of the collisional evolution of stellar systems.

One of the most illuminating numerical studies of core collapse was carried out by Cohn (1980), who used an orbit-averaged Fokker-Planck equation and assumed that the velocity distribution of the stars was isotropic.[11] The unique feature of Cohn's simulations was their large dynamic range—he was able to follow the collapse over an interval during which the central density increased by a factor of $\approx 10^{20}$, far larger than the range over which the densities of real cluster cores can collapse before other factors intervene. The density profiles of his models are shown in Figure 8-6a. This figure shows that in the late stages of core collapse the density profiles become self-similar, that is, they differ only in normalization and scale. To exhibit the self-similarity more explicitly we use the King radius [eq. (4-124b)]

$$r_0 = \sqrt{\frac{9\sigma^2}{4\pi G\rho_0}}, \tag{8-86}$$

where $3\sigma^2$ is the mean-square velocity at the center and ρ_0 is the central density. The King radius is nearly the same as the core radius, where the projected surface density falls to half its central value. In Figure 8-6b we plot the logarithmic density gradient $(d\ln\rho/d\ln r)$ against r/r_0. Since the King radius shrinks as the cluster evolves, the curves extend to larger values of r/r_0 at later times. By the end of the computation

---

[11] The assumption of isotropy is of course over-restrictive, but Cohn's results are consistent with the results of more general but less accurate codes.

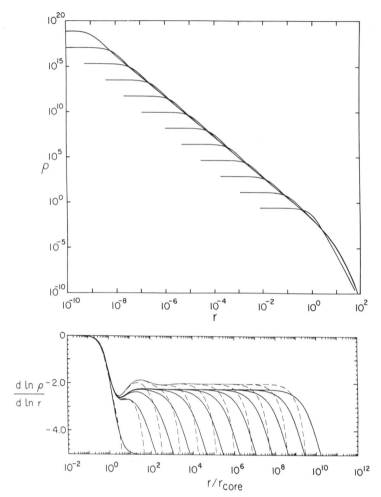

Figure 8-6. An orbit-averaged Fokker-Planck calculation of core collapse by Cohn (1980). Top: evolution of the density profile with time. Bottom: the logarithmic slope of the density profile as a function of time. Reprinted with permission from *The Astrophysical Journal*.

the logarithmic density gradient outside the core is flat out to radii exceeding $10^8 r/r_0$. It is clear that for most practical purposes the curve can be assumed to extend to $r/r_0 \to \infty$ and that this extended curve will describe the density distribution at all subsequent times up to the moment of collapse.

Any self-similar solution of this kind can be written in the form

$$\rho(r,t) = \rho_0(t)\rho_\star(r_\star), \tag{8-87}$$

where

$$r_\star = \frac{r}{r_0(t)}, \tag{8-88}$$

and $\rho_0(t)$ and $r_0(t)$ are the central density and King radius at time t. Figure 8-6a shows that the density is independent of time at radii that lie well outside the core, and hence

$$0 = \frac{\partial \rho(r,t)}{\partial t} = \frac{d\rho_0}{dt}\rho_\star - \rho_0\frac{d\rho_\star}{dr_\star}\frac{r}{r_0^2}\frac{dr_0}{dt} \qquad \text{for } r \gg r_0. \tag{8-89}$$

Thus

$$\frac{r_\star}{\rho_\star}\frac{d\rho_\star}{dr_\star} = \frac{r_0}{\rho_0}\frac{dt}{dr_0}\frac{d\rho_0}{dt} \qquad \text{for } r_\star \gg 1. \tag{8-90}$$

Since the left side is a function only of $r_\star$, and the right side is a function only of t, both must be equal to a constant, which we shall call $-\beta$. Hence we find

$$\begin{aligned} \rho_\star(r_\star) &\propto r_\star^{-\beta} \qquad \text{for } r_\star \gg 1, \\ \rho_0(t) &\propto r_0^{-\beta}(t). \end{aligned} \tag{8-91}$$

The first equation shows that for $r \gg r_0(t)$ the logarithmic density gradient is $(d\ln\rho/d\ln r) = -\beta$. The constancy of the logarithmic density gradient is of course consistent with the fact that the curves in Figure 8-6b are flat for $r/r_0 \gg 1$, and shows that the vertical coordinate of these curves directly yields the value of β to be

$$\beta = 2.23. \tag{8-92}$$

Lynden-Bell and Eggleton (1980) found the same value of β, to within about 1%, by modeling core collapse in a thermally conducting gas.[12]

Let us now define the core mass to be $M_0(t) \equiv \rho_0(t)r_0^3(t)$. From the second of equations (8-91) we find $M_0(t) \propto r_0^{3-\beta}$. According to equation (8-86) the central velocity dispersion is $\sigma \propto r_0^{1-\beta/2}$. From equation (8-71), the relaxation time in the core is $t_{\text{relax}} \propto \sigma^3/\rho_0 \propto r_0^{3-\beta/2}$, where we have neglected variations in $\ln\Lambda$. Since core collapse is associated with relaxation processes, we expect the characteristic timescale for changes in the core to be comparable to the relaxation time in the core. Thus we write

$$\frac{1}{r_0}\frac{dr_0}{dt} \propto \frac{1}{t_{\text{relax}}} \propto r_0^{\beta/2-3}. \tag{8-93}$$

---

[12] Larson (1970a) determined a similar but less accurate value, $\beta \simeq 2.4$, from the self-similar behavior of his solutions to the moment equations.

This equation is easily solved to yield

$$r_0(t) \propto (t_0 - t)^{2/(6-\beta)} \propto \tau^{0.53}, \qquad (8\text{-}94)$$

where t_0 is a constant of integration representing the moment of collapse, $\tau = t_0 - t$ is the time remaining until collapse, and in the second proportionality we have inserted the value $\beta = 2.21$. Similarly,

$$\rho_0(t) \propto \tau^{-2\beta/(6-\beta)} \propto \tau^{-1.17} \quad ; \quad \sigma^2(t) \propto \tau^{(4-2\beta)/(6-\beta)} \propto \tau^{-0.11},$$
$$M_0(t) \propto \tau^{(6-2\beta)/(6-\beta)} \propto \tau^{0.42} \quad ; \quad t_{\text{relax}}(r = 0) \propto \tau.$$
$$(8\text{-}95)$$

The last equation implies that the time to core collapse is always a fixed multiple of the central relaxation time. Cohn finds

$$\tau = 330 t_{\text{relax}}(r = 0). \qquad (8\text{-}96)$$

It is possible to test this proportionality observationally. Suppose that we observe p clusters with central relaxation times between 0 and t_{relax}. These clusters will undergo core collapse within a time $330 t_{\text{relax}}$. However, over any interval that is substantially less than the age of the Galaxy, the number of core collapses should be proportional to the length of the interval. Hence $2p$ clusters should undergo core collapse within the interval $660 t_{\text{relax}}$ and thus should have central relaxation times between 0 and $2 t_{\text{relax}}$. Thus, *the number of globulars with central relaxation times between 0 and t_{relax} should be proportional to t_{relax} so long as t_{relax} is sufficiently small.* This result appears to be consistent with the distribution of central relaxation times compiled by Webbink (1985) for 148 globular clusters: the numbers of clusters with $t_{\text{relax}}(r = 0)$ in the intervals $(0, 10)$ Myr, $(10, 20)$ Myr, and $(20, 30)$ Myr are, respectively, 10, 13, and 10, consistent with a uniform distribution (see Cohn & Hut 1984 for more detailed comparisons and a discussion of whether some of the observed clusters are in a post-collapse phase).

It appears that core collapse (in a cluster with stars of a single mass) is a two-stage process. The initial stage is driven by evaporation. Encounters in the core cause stars to diffuse to higher and higher energy, gradually populating a halo in which, as in a Michie model, the density falls as $r^{-7/2}$. At the same time, the core shrinks, as required by energy conservation. Acting alone, evaporation would cause collapse of the cluster in about $100 t_{\text{rh}}$ [eq. (8-85)]. However, after a time $\simeq 3 t_{\text{rh}}$, a second process intervenes. The rate of core collapse accelerates, and a self-similar evolution satisfying the similarity relations (8-95) develops. This second stage of core collapse, the one seen in Figure 8-6, is almost certainly a manifestation of the gravothermal catastrophe of §8.2. The inner parts of the system have negative heat capacity, and evolve by losing energy and thereby growing hotter. The energy lost is transferred outward by star-star encounters. Hence the temperature (i.e., velocity dispersion) always decreases outward, and the center continually loses energy, shrinks, and heats up.

What is the final outcome of core collapse? We remark first that despite the formal singularity in central density at $t = t_0$, core collapse is a relatively unspectacular process. To see this, let us employ the similarity relations to follow core collapse in a typical globular cluster. Taking parameters from Table 1-3, we estimate the initial parameters as $\rho_0 \simeq 10^4 \, M_\odot \, \mathrm{pc}^{-3}$, $\sigma \simeq 10 \, \mathrm{km \, s}^{-1}$, and $r_0 \simeq 1.5 \, \mathrm{pc}$. The initial core mass is thus $M_0 = \rho_0 r_0^3 = 3.4 \times 10^4 \, M_\odot$. Therefore, in terms of the number of stars N remaining in the core, we have from equation (8-95)

$$r_0 \simeq 3 \times 10^{-6} N^{1.26} \, \mathrm{pc} \quad ; \quad \sigma \simeq 40 N^{-0.13} \, \mathrm{km \, s}^{-1}. \tag{8-97}$$

Thus, when there is only one star in the core, $\sigma \simeq 40 \, \mathrm{km \, s}^{-1}$ and $r_0 \simeq 1 \times 10^{13}$ cm (about the distance of the Earth from the Sun). Clearly, the statistical approximations that form the basis of the theory of core collapse fail long before the density is high enough for stellar coalescence or other exotic phenomena to occur.

The final stages of core collapse, and the point at which the similarity solution (8-97) fails, are largely determined by the physics of binary stars. Hence we defer further discussion of this process until we have examined the formation and evolution of cluster binaries.

3 Equipartition

So far, all of our results have been based on idealized models in which the stars all have the same mass. If several different stellar masses are present, encounters tend to establish equipartition of kinetic energy. The more massive stars lose kinetic energy and sink toward the center, while the lighter stars gain kinetic energy and their orbits expand. The equipartition timescale is comparable to t_{rh}. Equipartition is extremely important in constructing models of globular clusters containing a realistic distribution of stellar masses, since it relates the DFs of stars of different masses (see Gunn & Griffin 1979). However, it is difficult to obtain direct observational evidence of equipartition, since the most luminous stars in globular clusters are giants, which in old stellar systems all have nearly the same mass.

It is instructive to derive a criterion for equipartition between stars of two different masses, m_1 and m_2, where $m_2 > m_1$. In particular, we consider the analytically tractable case where the total mass of the heavy stars, M_2, is much smaller than the core mass of the system of lighter stars, $\rho_{c1} r_{c1}^3$, but the individual heavy stars are much more massive than the light stars, $m_2 \gg m_1$. In this case equipartition will cause the heavy stars to form a small subsystem in the center of the core of the system formed by the light stars. The virial theorem for the heavy stars may be written in the form (Problem 8-2)

$$2K_2 + W_2 - G \int_0^\infty \frac{\rho_2(r) M_1(r)}{r} 4\pi r^2 \, dr = 0. \qquad (8\text{-}98)$$

Here $K_2 = \frac{1}{2} M_2 \langle v_2^2 \rangle$ is the kinetic energy of the heavy stars, W_2 is the potential energy arising from the interactions of the heavy stars with themselves, $\rho_1(r)$ and $\rho_2(r)$ are the densities of the light and heavy stars, and $M_1(r) = 4\pi \int_0^r \rho_1(r) dr$. Since the total mass in heavy stars is small, the heavy system will not strongly perturb the core of light stars, and $\rho_1(r)$ will be approximately constant, $\rho_1(r) \equiv \rho_{c1}$. We may also write $W_2 = -fGM_2^2/r_{h2}$, where r_{h2} is the median radius of the heavy stars and f is a dimensionless constant that is approximately 0.4 for many systems [see eq. (4-80b)]. Thus

$$\langle v_2^2 \rangle = f \frac{GM_2}{r_{h2}} + \frac{4\pi G \rho_{c1}}{3} \langle r_2^2 \rangle, \qquad (8\text{-}99)$$

where $\langle r_2^2 \rangle$, the mean-square radius of the heavy stars, may be written as $g^2 r_{h2}^2$ with g a dimensionless constant of order unity.

In equipartition, $m_2 \langle v_2^2 \rangle = m_1 \langle v_1^2 \rangle = 3 m_1 \sigma^2$, where σ represents the central one-dimensional dispersion of the lights. Using equation (8-86) to express σ in terms of the central density ρ_{c1} and King radius r_{c1} of the lights, we find

$$\frac{4\pi}{3} \frac{m_1}{m_2} G \rho_{c1} r_{c1}^2 = f \frac{GM_2}{r_{h2}} + \frac{4\pi}{3} G \rho_{c1} g^2 r_{h2}^2. \qquad (8\text{-}100)$$

The right side, considered as a function of r_{h2}, has a global minimum value. Hence equipartition cannot be satisfied unless the value of the left side exceeds the value of the right at this minimum. This implies that

$$\frac{M_2}{\rho_{c1} r_{c1}^3} \le \frac{1.61}{fg} \left(\frac{m_1}{m_2} \right)^{3/2}. \qquad (8\text{-}101)$$

There is a simple physical explanation of why equipartition cannot be achieved when this inequality is violated.[13] If the mass in heavy stars is

---

[13] Merritt (1981) has pointed out that formal solutions that violate this inequality can be found; however, they are unlikely to be relevant to the argument presented here.

too large, they form an independent self-gravitating system at the center of the core of light stars. Encounters cause the system of heavy stars to lose energy to the light stars. According to the virial theorem, this energy loss causes the velocity dispersion of the heavies to increase, so that they evolve away from, not toward, equipartition, and the process of energy loss, heating, and contraction of the heavy system must continue indefinitely. This phenomenon is sometimes called the **equipartition instability** (Spitzer 1969).

In realistic systems with a distribution of stellar masses, the chief effect of the equipartition instability is to produce a dense central core of heavy stars, which contracts independently from the rest of the core. However, as this core becomes denser and denser, the gravothermal instability eventually dominates over the equipartition instability, and the core collapses in much the same way as the core in a single-component stellar system.

Equipartition also causes the evaporation rate of the lightest stars to be greatest. Consequently, relaxed globular clusters are expected to have lost most of their low-mass stars (masses less than a few tenths of a solar mass). Since these stars have very high mass-to-light ratios (see Table 3-6 of MB), their loss tends to lower the mass-to-light ratio of the remaining cluster. This process may help to explain why the mass-to-light ratios of globular clusters ($\Upsilon \simeq 2\Upsilon_\odot$, see Table 1-3) are lower than in other Population II systems ($\Upsilon \simeq 10\Upsilon_\odot$ in the centers of ellipticals and the spheroids of spiral galaxies; see Table 4-2).

4 Binary Stars

In this subsection we investigate the formation, evolution, and destruction of binary stars as a result of gravitational encounters with field stars. Our principal interest is in the role played by binaries in the evolution of stellar systems. The discussion is based mainly on the seminal article of Heggie (1975).

Let us consider a homogeneous stellar system consisting of single (or "field" stars) and binary stars. The field stars have mass m_a, density ρ_a, and velocity dispersion σ, with a Maxwellian velocity distribution [eq. (8-67)]. We assume that the velocity distribution of the centers of mass of the binary stars is also Maxwellian, with velocity dispersion σ_b.

Consider a binary composed of two stars of masses m_1 and m_2. Let $\mathbf{x} \equiv \mathbf{x}_1 - \mathbf{x}_2$ be the separation vector between the stars, and let $\mathbf{V} = \dot{\mathbf{x}}$ be their relative velocity. According to equation (1D-32), the internal energy of the binary (i.e., its total energy minus the kinetic energy of the center-of-mass motion) is

$$E = \tfrac{1}{2}\mu V^2 - \frac{Gm_1 m_2}{r}, \tag{8-102}$$

where $\mu \equiv m_1 m_2/(m_1 + m_2)$ is the reduced mass. The separation vector $\mathbf{x}$ satisfies the equation of motion of a fictitious "reduced particle" of mass μ orbiting in the potential $-Gm_1 m_2/r$. Hence it follows a Kepler ellipse with semi-major axis a, where

$$E = -\frac{Gm_1 m_2}{2a}. \tag{8-103}$$

A binary is called **soft** if $|E|/m_a\sigma^2 < 1$ and **hard** if $|E|/m_a\sigma^2 > 1$. The evolution and properties of hard and soft binaries are quite different, and they will be analyzed separately below. We shall often concentrate on the behavior of very soft and very hard binaries, which are simpler to understand, and extrapolate our results to the transition region near $|E| \approx m_a\sigma^2$.

Some readers may wish to skip directly to the subsection on hard binaries, since we shall find that soft binaries play little or no direct role in the evolution of stellar systems.

(a) Soft binaries   The evolution of binaries is easiest to describe in the limit where they are very soft, $|E| \ll m_a\sigma^2$. Consider an encounter of star 1 with a field star at an impact parameter that is much less than the binary separation. Then the encounter changes the velocity of star 2 much less than the velocity of star 1. If we neglect the orbital motion of the binary during the encounter (the impulse approximation of §7.2), we may write the change of internal energy in the encounter as

$$\Delta E = \tfrac{1}{2}\mu\Delta(V^2) = \frac{m_1 m_2}{m_1 + m_2}\left[\Delta\mathbf{v}_1 \cdot (\mathbf{v}_1 - \mathbf{v}_2) + \tfrac{1}{2}\Delta v_1^2\right]. \tag{8-104}$$

The mean value of ΔE per unit time can be obtained by replacing $\Delta\mathbf{v}_1$ and Δv_1^2 by the diffusion coefficients (8-65). Thus we set

$$D(\Delta\mathbf{v}_1) = \frac{\mathbf{v}_1}{v_1}D(\Delta v_\parallel) \quad ; \quad D(\Delta v_1^2) = D(\Delta v_\perp^2)_1 + D(\Delta v_\parallel^2)_1. \tag{8-105}$$

Substituting into (8-104) and using equation (8-68), we find

$$D(\Delta E) = \frac{4\pi G^2 \rho \ln\Lambda}{\sigma}\frac{m_1 m_2}{m_1 + m_2}$$
$$\times \left[-(m + m_a)\frac{\mathbf{v}_1 \cdot (\mathbf{v}_1 - \mathbf{v}_2)}{v_1\sigma}G(X_1) + \frac{m_a}{\sqrt{2}}\frac{\mathrm{erf}(X_1)}{X_1}\right], \tag{8-106}$$

where $X_1 = v_1/(\sqrt{2}\sigma)$, $\ln \Lambda$ is the Coulomb logarithm, and the function $G(X_1)$ is defined by equation (8-69). Now, for a very soft binary $|\mathbf{v}_1 - \mathbf{v}_2| \ll \sigma$, so the first term may be neglected in comparison with the second; moreover, we can replace v_1 by the speed of the binary center of mass v_{cm}. Finally we double the value of $D(\Delta E)$ to account for encounters of field stars with star 2. Thus for very soft binaries

$$D(\Delta E) = \frac{2^{5/2}\pi G^2 \rho \ln \Lambda}{\sigma} \frac{m_a m_1 m_2}{m_1 + m_2} \frac{\mathrm{erf}(X_{cm})}{X_{cm}}, \qquad (8\text{-}107)$$

where $X_{cm} = v_{cm}/\sqrt{2}\sigma$. We now average over v_{cm}, assuming for simplicity that the dispersion σ_b of the binaries is equal to the dispersion σ of the field stars. Using the relation

$$\left\langle \frac{\mathrm{erf}\,X}{X} \right\rangle = \frac{\int_0^\infty X\,\mathrm{erf}(X)e^{-X^2}dX}{\int_0^\infty X^2 e^{-X^2}dX} = \sqrt{\frac{2}{\pi}}, \qquad (8\text{-}108)$$

we obtain

$$D(\Delta E) = \frac{8\sqrt{\pi}G^2 \rho \ln \Lambda}{\sigma} \frac{m_a m_1 m_2}{m_1 + m_2}. \qquad (8\text{-}109)$$

The value of the Coulomb logarithm is given by equation (8-60), $\Lambda = b_{max} v_{typ}^2 / G(m + m_a)$.

We must now verify the condition $\Lambda \gg 1$ for our expressions for the diffusion coefficients to be valid. We set the maximum impact parameter equal to half of the binary semi-major axis, $b_{max} = \frac{1}{2}a$, since our derivation is only valid for encounters that are much closer to one star than the other (in more distant encounters the effect of the field star on one component almost cancels its effect on the other). We replace v_{typ} by σ, and since the masses of the stars involved are usually not very different, we can set $m_1 \approx m_2 \approx m_a$; after these approximations $\Lambda \approx a\sigma^2/4Gm_a$. Using equation (8-103) we can write $\Lambda \approx m_a\sigma^2/|E|$; thus the condition that the binary is very soft guarantees that $\Lambda \gg 1$.

Equation (8-109) implies that on average soft binaries gain energy from encounters with field stars; in other words, *soft binaries become softer*. This result can be interpreted in terms of energy equipartition. Encounters can be regarded as perturbations on the orbit of the reduced particle. Since the binary is very soft, the kinetic energy of the reduced particle is much less than the kinetic energy of the field stars. Because the system tends toward equipartition, encounters tend to increase the energy of the reduced particle and thus to increase the internal energy of the binary.

Encounters lead to disruption of soft binaries by gradually increasing the internal energy until it becomes positive. We shall call this process "evaporation" since it is reminiscent of the evaporation of stars from a cluster. Since $D(\Delta E)$ is independent of E (except for a very slow dependence on the Coulomb logarithm), the lifetime of a very soft binary of energy E before evaporation is simply

$$t_{\text{evap}} = \frac{E}{D(\Delta E)} = \frac{m_1 + m_2}{m_a} \frac{\sigma}{16\sqrt{\pi}G\rho a \ln \Lambda}. \qquad (8\text{-}110)$$

The soft binary evaporation time t_{evap} is generally much shorter than the local relaxation time t_{relax} [eq. (8-71)], since the latter is the average time required for changes in energy of order $m_a\sigma^2$, while disruption of a very soft binary only requires changes of order $|E| \ll m_a\sigma^2$.

The short evaporation time for very soft binaries implies that in relaxed clusters there will be an equilibrium in which the rate of evaporation of soft binaries is balanced by the rate of binary formation due to three-body encounters. It turns out that the equilibrium number of very soft binaries is of order unity, independent of the number of stars in the cluster (Appendix 8.B). Thus, soft binaries play no important role in the evolution of stellar systems.

(b) Hard binaries   The interactions of a hard binary with a field star can be extremely complex. The principal reason for this complexity is that the binding energy of the binary is usually larger in magnitude than the relative kinetic energy of the binary and the field star. Hence the three stars can temporarily form a bound three-body system, a possibility that is not available to soft binaries.

Figure 8-7 shows a typical interaction of a hard binary and a field star. Notice that the encounter is an example of an **exchange**: the original binary containing stars 1 and 2 is dissociated, and stars 1 and 3 join to produce a new binary while star 2 escapes to infinity. The detailed behavior of the stars during the encounter is impossible to describe simply; most of the time the system displays a hierarchy where one star travels in an elongated ellipse with a tightly bound binary at one focus. At each pericenter passage of the outermost star, the three stars interact strongly, one of the three stars is flung into an elongated orbit, and the process repeats. Ultimately one star is flung off with enough energy to escape, and the three-body interaction ends.

It should be stressed that Figure 8-7 does not represent a particularly complicated interaction. Only three temporary binaries were formed before star 2 escaped, while in a typical encounter 10 to 100 temporary binaries are formed before one of the three stars escapes (Hut &

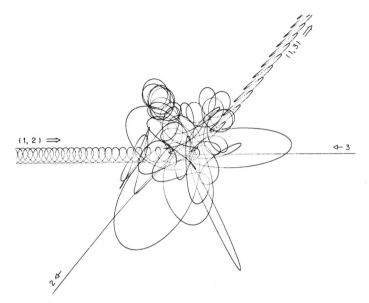

Figure 8-7. A typical interaction between a hard binary and a field star (from Hut & Bahcall 1983). All three stars have equal mass and the orbits are plotted in the center-of-mass frame. The binary, containing stars 1 and 2, enters from the left; the single star (labeled 3) enters from the right. The outcome of the interaction is that star 2 escapes, leaving 1 and 3 behind as a newly formed binary. Reprinted by permission from *The Astrophysical Journal.*

Bahcall 1983). Another measure of the importance of complex encounters is that 45% of the rate of energy exchange between a population of hard binaries and a field star population is due to encounters involving an exchange.

A simple statistical argument can be used to determine the qualitative effect of encounters on the internal energy E of very hard binaries. We first note that the initial relative velocity of the binary and the field star is of order σ, which is much less than the orbital speed of the stars in a very hard binary. We now consider two cases: (i) if $|E|$ is increased by the interaction, then the field star gains energy and escapes back to infinity; (ii) if $|E|$ decreases as the field star approaches the binary, then the initial kinetic energy of the field star is so small that the field star almost certainly becomes bound to the binary, and a temporary triple system is formed. When one of the three stars is finally ejected from the triple system, its escape speed will typically be of the order of the orbital speed. Hence the ejected single star will have a higher speed than the incoming single star. By energy conservation, the internal energy E must therefore decrease, that is, $|E|$ increases. In each of these two cases, $|E|$ increases; in other words, hard binaries become harder. Combining this with our results on soft binaries, we arrive at **Heggie's law**: *hard binaries get harder and soft binaries get softer*. There is a "watershed" energy, near $-m_a\sigma^2$, at which the average rate of energy input from encounters is zero.

Quantitative estimates of the rate of formation, evolution, and destruction of hard binaries have been made analytically by Heggie (1975) and numerically by Hut and Bahcall (1983) and Hut (1983). We shall now list some of their results, restricting ourselves to the case where the masses of the two stars in the binary and the field stars are all equal, $m_1 = m_2 = m_a \equiv m$.

We first consider the disruption rate. (Here "disruption" refers to an encounter from which both the field star and the two original binary members all emerge as single stars. Thus an exchange would not be counted as a disruption.) Hard binaries are much more difficult to disrupt than soft binaries, because, unlike soft binaries, they cannot be disrupted by the gradual growth of their orbital energy through the accumulated effects of many encounters; by Heggie's law, the average effect of encounters is to make them harder (and harder to disrupt). A hard binary can only be disrupted if the total energy of the binary plus the field star in the center-of-mass frame is positive, and for very hard binaries the total energy is only positive for the very few field stars whose speed is many times the velocity dispersion σ.

The probability per unit time that a binary is disrupted by a single close encounter may be written as $1/t_{\rm dis}$, where

$$t_{\rm dis} = \frac{9|E|^2}{16\sqrt{\pi}\nu G^2 m^4 \sigma} \left(1 + \frac{4m\sigma^2}{15|E|}\right) \left[1 + \exp\left(\frac{3|E|}{4m\sigma^2}\right)\right], \qquad (8\text{-}111)$$

and ν is the number density of field stars. This formula was obtained by Hut and Bahcall (1983) from numerical experiments; the functional form is based on formulae derived by Heggie (1975) for very hard and very soft binaries. In principle, $t_{\rm dis}$ depends on the initial eccentricity of the binary, but the numerical experiments show that the formula is accurate to within $\pm 20\%$ at all eccentricities.[14] Equation (8-111) shows that the lifetime of a hard binary against disruption becomes exponentially long as the binary becomes very hard, and thus that primordial hard binaries,

---

[14] For very soft binaries, $|E| \ll m\sigma^2$, the disruption time given by equation (8-111) is similar in form to the evaporation time given by equation (8-110), but longer by a factor of order $\ln \Lambda$; this reflects the fact that the accumulated effects of many weak encounters are more likely to disrupt a soft binary than a single strong encounter.

unlike primordial soft binaries, can survive from the birth of the cluster to the present day.

The rate at which hard binaries become harder was estimated by Heggie to be

$$D(\Delta E) = \langle \dot{E} \rangle = -5.1 \frac{\nu G^2 m^3}{\sigma}, \tag{8-112}$$

Numerical experiments (Hut 1983) show that the coefficient 5.1 is somewhat too large, but by less than a factor of two.[15] The hardening rate (8-112) can be simply expressed in terms of the local relaxation time [eq. (8-71)]. Assuming that $\ln \Lambda \simeq 10$ in the expression for the relaxation time, the average energy change per relaxation time is

$$\langle \Delta E \rangle t_{\text{relax}} \approx -0.2 m \sigma^2. \tag{8-113}$$

Notice that the rate of change of energy is independent of the energy; in other words, the rate of hardening is independent of the hardness. This is the result of the cancelation of two opposing effects: as the binary becomes harder and smaller, the rate of close encounters decreases, but the severity of each encounter increases so that the binary loses energy in larger and larger increments.

5 Inelastic Encounters

We have so far approximated stars as point masses. However, in a region such as the high density core that forms during core collapse, stars may pass so close to one another that they raise tides that dissipate their relative orbital kinetic energy. The dissipation removes kinetic energy from the stellar system and hastens core collapse. In some cases the loss of energy may be so large that the stars form a binary; in other cases, the stars may collide and coalesce into a single star, or a binary star with a common atmosphere.

To investigate these effects we compute the **collision time** t_{coll}, where $1/t_{\text{coll}}$ is the collision rate, that is, the average number of physical collisions that a given star suffers per unit time. For simplicity we restrict ourselves to a cluster in which all stars have the same mass m. Consider an encounter with initial relative velocity $\mathbf{V}_0$ and impact parameter b. The angular momentum per unit mass of the reduced particle is $L = bV_0$ [eq. (7-4)]. At the distance of closest approach, which

---

[15] The accuracy of Heggie's result is quite remarkable in view of the fact that roughly half of the contribution to $D(\Delta E)$ comes from encounters that are at least as complicated as the one shown in Figure 8-7.

we denote by r_{coll}, the radial velocity must be zero, and hence the angular momentum is $L = r_{coll} V_{max}$, where V_{max} is the relative speed at r_{coll}. From the energy equation (1D-32), the energy in the center of mass frame is $E = \frac{1}{2}\mu V^2 - Gm^2/r$, where $\mu = \frac{1}{2}m$ is the reduced mass. Equating the energy at r_{coll} and $r \to \infty$, we have

$$\tfrac{1}{4}mV_0^2 = \tfrac{1}{4}mV_{max}^2 - \frac{Gm^2}{r_{coll}}. \tag{8-114}$$

Since L is conserved, we can eliminate V_{max} to obtain

$$b^2 = r_{coll}^2 + \frac{4Gmr_{coll}}{V_0^2}. \tag{8-115}$$

If we set r_{coll} equal to the sum of the radii of the two stars, then a collision will occur if and only if the impact parameter is less than the value of b determined by equation (8-115).

Let $f(\mathbf{v}_a)d^3\mathbf{v}_a$ be the number of stars per unit volume with velocities in the range $\mathbf{v}_a$ to $\mathbf{v}_a + d^3\mathbf{v}_a$. The number of encounters per unit time with impact parameter less than b that are suffered by a given star is just $f(\mathbf{v}_a)d^3\mathbf{v}_a$ times the volume of an annulus with radius b and length V_0, that is,

$$\int f(\mathbf{v}_a)\pi b^2 V_0 d^3\mathbf{v}_a, \tag{8-116}$$

where $V_0 = |\mathbf{v} - \mathbf{v}_a|$ and $\mathbf{v}$ is the velocity of the star in question. The quantity in equation (8-116) is equal to $1/t_{coll}$ for a star with velocity $\mathbf{v}$; to obtain a mean value of $1/t_{coll}$ we average over $\mathbf{v}$ by multiplying (8-116) by $f(\mathbf{v})/\nu$, where $\nu = \int f(\mathbf{v})d^3\mathbf{v}$ is the number density of stars, and integrating over $d^3\mathbf{v}$. Thus

$$\frac{1}{t_{coll}} = \frac{\pi}{\nu}\int f(\mathbf{v})f(\mathbf{v}_a)b^2|\mathbf{v} - \mathbf{v}_a|\,d^3\mathbf{v}d^3\mathbf{v}_a. \tag{8-117}$$

To evaluate the integrals we assume that the DF is Maxwellian [eq. (8-67)] with dispersion σ. Substituting for b^2 from equation (8-115) we obtain

$$\frac{1}{t_{coll}} = \frac{\nu}{8\pi^2\sigma^6}\int e^{-(v^2+v_a^2)/2\sigma^2}\left(r_{coll}^2|\mathbf{v} - \mathbf{v}_a| + \frac{4Gmr_{coll}}{|\mathbf{v} - \mathbf{v}_a|}\right)d^3\mathbf{v}d^3\mathbf{v}_a. \tag{8-118}$$

We now replace the dummy variable $\mathbf{v}_a$ by $\mathbf{V} = \mathbf{v} - \mathbf{v}_a$. The argument of the exponential is then $-[(\mathbf{v} - \frac{1}{2}\mathbf{V})^2 + \frac{1}{4}V^2]/\sigma^2$, and if we replace the dummy variable $\mathbf{v}$ by $\mathbf{v}_{cm} = \mathbf{v} - \frac{1}{2}\mathbf{V}$ (the center of mass velocity), then we have

$$\frac{1}{t_{coll}} = \frac{\nu}{8\pi^2\sigma^6}\int e^{-(v_{cm}^2+V^2/4)/\sigma^2}\left(r_{coll}^2 V + \frac{4Gmr_{coll}}{V}\right)d^3\mathbf{v}_{cm}d^3\mathbf{V}. \tag{8-119}$$

The integral over $\mathbf{v}_{\text{cm}}$ is given by

$$\int e^{-v_{\text{cm}}^2/\sigma^2} d^3\mathbf{v}_{\text{cm}} = \pi^{3/2}\sigma^3. \qquad (8\text{-}120)$$

Thus

$$\frac{1}{t_{\text{coll}}} = \frac{\pi^{1/2}\nu}{2\sigma^3} \int_0^\infty e^{-V^2/4\sigma^2}\left(V^3 r_{\text{coll}}^2 + 4GmV r_{\text{coll}}\right)dV. \qquad (8\text{-}121)$$

The integrals are easily done, and we find

$$\frac{1}{t_{\text{coll}}} = 4\sqrt{\pi}\nu\sigma r_{\text{coll}}^2 + \frac{4\sqrt{\pi}Gm\nu r_{\text{coll}}}{\sigma}. \qquad (8\text{-}122)$$

The first term of this result can be derived from simple kinetic theory. The rate of a reaction is $\nu\Sigma\langle V\rangle$, where Σ is the cross-section and $\langle V\rangle$ is the mean relative speed. Substituting $\Sigma = \pi r_{\text{coll}}^2$ and $\langle V\rangle = 4\sigma/\sqrt{\pi}$ (which is appropriate for a Maxwellian distribution with dispersion σ) we recover the first term of (8-122). The second term represents the enhancement in the collision rate by gravitational focusing, that is, the deflection of trajectories by the gravitational attraction of the two stars.

If $r_\star$ is the stellar radius, we may set $r_{\text{coll}} = 2r_\star$. It is convenient to introduce the escape speed from the stellar surface, $v_\star = \sqrt{2Gm/r_\star}$, and to write equation (8-122) as

$$\frac{1}{t_{\text{coll}}} = 16\sqrt{\pi}\nu\sigma r_\star^2\left(1 + \frac{v_\star^2}{4\sigma^2}\right) = 16\sqrt{\pi}\nu\sigma r_\star^2(1+\Theta), \qquad (8\text{-}123)$$

where we have introduced the **Safronov number**[16]

$$\Theta = \frac{v_\star^2}{4\sigma^2} = \frac{Gm}{2\sigma^2 r_\star}. \qquad (8\text{-}124)$$

Numerically, we have

$$t_{\text{coll}} = \begin{cases} 6.8 \times 10^{12}\,\text{yr}\left(\frac{10^5\,\text{pc}^{-3}}{\nu}\right)\left(\frac{100\,\text{km s}^{-1}}{\sigma}\right)\left(\frac{R_\odot}{r_\star}\right)^2, & \Theta \ll 1, \\ 7.1 \times 10^{11}\,\text{yr}\left(\frac{10^5\,\text{pc}^{-3}}{\nu}\right)\left(\frac{\sigma}{100\,\text{km s}^{-1}}\right)\left(\frac{R_\odot}{r_\star}\right)\left(\frac{M_\odot}{m}\right), & \Theta \gg 1. \end{cases}$$
$$(8\text{-}125)$$

The ratio of the collision time to the relaxation time [eq. (8-71)] is

$$\frac{t_{\text{coll}}}{t_{\text{relax}}} = 0.8\ln\Lambda\frac{\Theta^2}{1+\Theta}. \qquad (8\text{-}126)$$

---

[16] After V. S. Safronov, who introduced Θ in studies of the collision of planetesimals in the early solar system.

[An approximate version of this equation for $\Theta \ll 1$ was given earlier, as eq. (8-6).] Core collapse occurs in about 300 central relaxation times [see eq. (8-96)], and hence collisions have a substantial influence on core collapse if $t_{\mathrm{coll}}/t_{\mathrm{relax}} \lesssim 300$. Let us now evaluate this ratio in two typical cases. The escape speed from the Sun is $v_\star = 620\,\mathrm{km\,s}^{-1}$. In a typical globular cluster, where $\sigma \simeq 10\,\mathrm{km\,s}^{-1}$, the Safronov number for solar-type stars is $\Theta \approx 10^3$. Hence $t_{\mathrm{coll}} \approx 10^4 t_{\mathrm{relax}}$ for $\ln \Lambda \approx 10$. Thus, in the early stages of core collapse, the effects of collisions of solar-type, or indeed of any dwarf (i.e., main-sequence) stars are negligible compared to the effects of relaxation.[17] However, collisions *can* be important in the late stages of core collapse, since the velocity dispersion grows [cf. eq. (8-97), which predicts that the dispersion grows by a factor of four or so as the core mass falls by a factor 10^4] and hence the Safronov number may be as small as $\Theta \approx 10^2$ (cf. §8.4.5). Collisions also play an important role in a galactic nucleus with $\sigma \approx 150\,\mathrm{km\,s}^{-1}$ and $N \approx 10^8$ (cf. Table 1-5), where the Safronov number is $\Theta \approx 4.3$, and $t_{\mathrm{coll}} \approx 50 t_{\mathrm{relax}}$ for $\ln \Lambda \approx \ln(0.4N) \approx 18$.

The effect of collisions on the evolution of a stellar system are not yet well understood. A nearly head-on collision is very inelastic, and dissipates an energy of order $mv_\star^2$. If $\Theta \gg 1$, the energy dissipated exceeds the relative kinetic energy of the colliding stars, and the stars must coalesce into a single star of mass $2m$. This new star may in turn collide and merge with other stars, thereby becoming very massive. As its mass increases, both the collision time and the nuclear evolution times shorten. If the nuclear evolution time is shorter, the star is likely to explode as a supernova, leaving behind a neutron star or black hole remnant. If, on the other hand, the collision time is shorter, there may be runaway coalescence leading to the formation of a few supermassive objects per cluster. If $\Theta \lesssim 1$, much of the mass in the colliding stars may be liberated, ultimately settling to the cluster center and forming new stars, or perhaps a single supermassive object. One intriguing possibility is that a substantial fraction of the total mass of an evolving galactic nucleus may be swallowed up in a supermassive ($\approx 10^8\,\mathrm{M}_\odot$) black hole. Reviews of some of the complicated effects arising in the evolution of dense galactic nuclei are given by Begelman and Rees (1978) and Lightman and Shapiro (1978).

To close this subsection, we briefly note the effects of near-collisions. An encounter of two stars in which the minimum separation is several stellar radii will raise violent tides on the surface of each star. The

---

[17] Red giants have much larger radii (up to $\approx 30\,\mathrm{R}_\odot$) but are less abundant by about a factor of 100, and so a dwarf star collides with giants even less often than it collides with other dwarfs.

energy that excites the tides comes from the relative kinetic energy of the stars. This effect is especially important in systems where the Safronov number $\Theta \gg 1$, since the loss of a small amount of kinetic energy may leave the two stars with negative total energy, that is, as a bound binary system. Successive pericenter passages will dissipate more energy, until the binary orbit is nearly circular (see Problem 8-5). This mechanism of binary formation, called **tidal capture**, can be the dominant process of forming hard binaries in a globular cluster (Fabian et al. 1975). Press and Teukolsky (1977) and Lee and Ostriker (1986) have computed the **tidal capture time** t_{tid}, where $1/t_{\text{tid}}$ is the tidal capture rate, that is, the probability per unit time that a star will be tidally captured into a bound state. For main-sequence stars in a system with $\Theta \gg 1$ they find

$$
t_{\text{tid}} \simeq 1 \times 10^{12}\,\text{yr} \left(\frac{10^5\,\text{pc}^{-3}}{\nu} \right) \left(\frac{\sigma}{100\,\text{km s}^{-1}} \right)^{1.2} \left(\frac{\text{R}_\odot}{r_\star} \right)^{0.9} \left(\frac{\text{M}_\odot}{m} \right)^{1.1}.
$$
$$(8\text{-}127)$$

This formula is very similar to the formula for t_{coll} when $\Theta \gg 1$; the similarity is not surprising since stars must nearly collide in order to raise strong tides. The formulae do not tell us, however, what fraction of the bound systems form binaries and what fraction coalesce. We argued in the previous paragraph that head-on collisions lead to coalescence, while captures arising from close encounters at several times r_{coll} probably lead to binary formation; in practice it appears that roughly 25% of the captures produce a single coalesced star (Lee & Ostriker 1986).

Tidal capture can produce dozens of close binary systems in a globular cluster over 10^{10} yr.[18] It seems likely that the X-ray sources seen in compact globular clusters are close binary systems formed by this process (see Lightman & Grindlay 1982 for a review).

6 Binaries and Core Collapse

Recent work strongly suggests that hard binary stars are responsible for arresting core collapse (see Ostriker 1985 for a review).[19] According to Heggie's law, hard binaries continually become harder: as a result of interactions with other cluster stars, they lose energy at a rate given by

---

[18] Note that the number of captures per unit volume per unit time is $\frac{1}{2}\nu/t_{\text{tid}}$; the factor $\frac{1}{2}$ arises because each capture has been counted twice.

[19] In a remarkable early paper, Hénon (1961) constructed a model of a globular cluster that contracted self-similarly as it evolved. He found that the boundary conditions at the center required an energy source at $r = 0$ and was the first to argue that the required energy might be produced by the formation of binary and multiple stars.

equation (8-113). Much of the energy released by the hard binary is lost from the cluster, since a close encounter of a hard binary and a single star will often give both enough velocity to escape from the cluster. However, the reaction products that do not escape will eventually share their energy with the other cluster stars through encounters. Furthermore, even the reaction products that do escape give up some of their energy to the cluster: a mass m that escapes to infinity from a cluster potential well of depth $|\Phi|$ does work $m|\Phi|$ on the remaining cluster stars as it travels off to infinity.

Thus, hard binaries in a cluster act as an energy source, somewhat as do nuclear reactions in stars. The energy provided by the binaries can halt core collapse by cooling the central core and thereby reversing the outward flow of energy from core to halo that lies at the root of the gravothermal catastrophe.

Where do the hard binaries come from? We have seen that the number of binaries formed by three-body encounters is about $0.1(N \ln N)^{-1}$ per relaxation time [eq. (8-9)], where N is to be interpreted as the number of stars remaining in the central core. Since the characteristic evolution time during core collapse is of order $300t_{\rm relax}$ [eq. (8-96)], we expect that the first binary will form by a three-body encounter when $N \ln N \approx 30$, corresponding to N of order 10. The number of binaries formed by inelastic encounters can be determined from equation (8-126), assuming that the rate of binary formation is roughly the same as the collision time. When the core has shrunk to, say, $N = 100$ stars, the velocity dispersion is $\sigma \simeq 20\,{\rm km\,s^{-1}}$ [eq. (8-97)], and the Safronov number for solar-type stars is $\Theta \simeq 240$. Thus $t_{\rm coll} \simeq 2000t_{\rm relax}$ (for $\ln \Lambda \simeq 10$), and therefore in the characteristic evolution time of $300t_{\rm relax}$ [eq. (8-96)] we expect about 10% of the stars to be tidally captured into hard binaries.

Thus, binaries formed in a collapsing core by inelastic encounters are far more numerous than binaries formed by three-body encounters. There is an additional, but uncertain, contribution to the binary population from primordial binaries, which tend to concentrate in the center of the cluster because they lose energy to the lighter single stars through equipartition (see Spitzer & Mathieu 1980).

The analysis of encounters of binaries with single stars does not exhaust the possible effects of binaries on a stellar system. During the late stages of collapse of a binary-rich core, binary-binary encounters are likely to predominate over binary-single encounters. Four-body encounters are even more complex than three-body encounters, and investigation of the cross-sections and reaction rates for binary-binary scattering has only just begun.

The details of post-collapse evolution are a subject of active research. At what stage is the collapse halted? How rapidly and for how long does the reexpansion of the core take place? Can post-collapse cores be recognized observationally? Do clusters undergo repeated core collapse? A detailed understanding of core collapse offers exciting possibilities for comparison with observations of the central regions of globular clusters.

7 Influence of a Central Black Hole

It has often been suggested that the power emitted by active galactic nuclei and quasars arises from consumption of gas and stars by a supermassive ($M_{\rm BH} \approx 10^8 \, {\rm M}_\odot$) black hole embedded in a dense stellar system ($\nu \approx 10^6 \, {\rm pc}^{-3}$). Smaller black holes ($M_{\rm BH} \approx 10^3 \, {\rm M}_\odot$) might also be present at the centers of some globular clusters, although there is little direct evidence for their existence.

Let us investigate the effect of a central black hole on a stellar system whose core stars obey a Maxwellian distribution with dispersion σ, central number density ν_0, and King radius r_0 [eq. (8-86)]. We shall assume that the mass of the hole is much less than the core mass, $M_{\rm BH} \ll \rho_0 r_0^3$. The influence of the hole on the distribution of stars in the core depends on whether or not the hole has been present in the core for longer than a relaxation time.

First consider the case where the age of the system is much less than the relaxation time. Here the effect of the hole on the stellar distribution depends on how the hole was formed. The most plausible hypothesis is that the hole grew slowly, over a time much longer than the crossing time r_0/σ. In this case, the potential changes by only a small amount each crossing time, and the Jeans theorem is approximately valid. Hence the DF at any given time is a function only of the integrals of the motion, though a function that changes slowly with time. Thus we may write $f = f(E, L, t)$ or $f = f(\mathbf{J}, t)$. The great advantage of the using the actions $\mathbf{J}$ as independent variables is that they are adiabatic invariants (§3.6), that is, they are conserved as the hole grows slowly. Since each star conserves its actions,

$$\frac{\partial f(\mathbf{J}, t)}{\partial t} = 0, \qquad (8\text{-}128)$$

in other words, $f = f(\mathbf{J})$ is independent of time.

To determine $f(\mathbf{J})$, we note that before the hole is introduced, the initial DF may be assumed to be Maxwellian, and the number of stars in a small volume element of phase space is

$$d^6 N = \frac{\nu_0}{(2\pi\sigma^2)^{3/2}} e^{-E/\sigma^2} d^3\mathbf{x} d^3\mathbf{v} = f(\mathbf{J}) d^3\mathbf{J} d^3\boldsymbol{\theta}, \qquad (8\text{-}129)$$

where $\boldsymbol{\theta}$ represents the three angle variables conjugate to the actions, and the zero point of the potential has been chosen to lie at the cluster center, $\Phi(r = 0) = 0$. Now the volume element of phase space is the same in all canonical coordinate systems [eq. (1D-71)], and thus

$$f(\mathbf{J}) = \frac{\nu_0}{(2\pi\sigma^2)^{3/2}} e^{-E_i(\mathbf{J})/\sigma^2}, \qquad (8\text{-}130)$$

where $E_i(\mathbf{J})$ is the energy as a function of the actions in the initial potential. Similarly, the DF after the hole has been introduced may be written in terms of energy and angular momentum as

$$f_f(E, L) = f[\mathbf{J}_f(E, L)], \qquad (8\text{-}131)$$

where $\mathbf{J}_f(E, L)$ denotes the actions as a function of energy and angular momentum in the final potential.

These transformations generally require numerical evaluation. However, they can be carried out analytically if we restrict ourselves to the stars that eventually become bound to the hole. These stars all had initial energies $E_i \ll \sigma^2$, and in this case equations (8-130) and (8-131) imply $f(\mathbf{J}) = \nu_0/(2\pi\sigma^2)^{3/2} = f_f(E, L)$. Thus the corresponding density distribution of bound stars after the hole has formed is

$$\nu_b(r) = 4\pi \int_0^{\sqrt{2GM_{\text{BH}}/r}} v^2 f_f(E, L) dv = \frac{4\nu_0}{3\sqrt{\pi}} \left(\frac{r_{\text{BH}}}{r}\right)^{\frac{3}{2}}, \qquad (8\text{-}132)$$

where we have introduced a parameter

$$r_{\text{BH}} = \frac{GM_{\text{BH}}}{\sigma^2}, \qquad (8\text{-}133)$$

which is a measure of the radius of the "sphere of influence" of the hole.

The density cusp of stars bound to the hole that is induced by slow growth of the hole is therefore $\nu_b \propto r^{-3/2}$ (at least when the age of the central mass is small compared to the relaxation time). This cusp is strong enough so that it could in principle be detected by number counts of the stars or by surface photometry. The cusp made up of stars not bound to the hole has a much shallower density profile (see Problem 8-11). Numerical calculations of the response of a collisionless isothermal sphere to the adiabatic growth of a central black hole are described by Young (1980).

Now consider the case where the age of the system is much larger than the relaxation time. In such systems an equilibrium distribution of bound stars will be set up. This distribution is independent of the details of the formation process, and reflects a balance between two competing processes: (i) in thermal equilibrium the density should be that of an isothermal sphere $\nu \propto \exp[-\Phi(r)/\sigma^2] \propto \exp(r_{\rm BH}/r)$, which diverges exponentially near the hole; however, (ii) a star whose pericenter is too close to the hole will be tidally shredded or perhaps swallowed entirely by the hole. ·

iiLet us assume that the equilibrium density near the hole is a power law, $\nu(r) \simeq \nu_0(r_{\rm BH}/r)^s$. We can use simple scaling arguments to determine the exponent s. The Jeans equations dictate that the mean-square velocity at any radius should be of order $\langle v^2\rangle \simeq GM_{\rm BH}/r$. Hence the local relaxation time [eq. (8-71)] is $t_{\rm relax} \approx \langle v^2\rangle^{3/2}/(G^2 m^2 \nu) \propto r^{s-3/2}$. A star that is swallowed by the hole from radius r_t has energy of order $E(r_t) = -GM_{\rm BH}m/r_t$, and in a steady state this loss of negative energy implies that a flux of positive energy must flow out through the cusp. Crudely, we can argue that relaxation among the $N(r)$ cusp stars interior to r can carry energy equal to $N(r)E(r)$ through the shell at radius r per relaxation time; since $N(r) \propto r^{3-s}$ the flow of energy through radius r is $N(r)E(r)/t_{\rm relax} \propto r^{\frac{7}{2}-2s}$. In a steady state the flow must be independent of radius and hence $s = \frac{7}{4}$. This scaling law[20] has been verified by numerical solutions of the Fokker-Planck equation (see Cohn & Kulsrud 1978 and references therein). Cohn and Kulsrud estimate that the total rate of consumption of stars by the hole is $dN/dt = 1/t_{\rm eat}$, where

$$
\begin{aligned}
t_{\rm eat} = 3 \times 10^6 \, {\rm yr} & \left(\frac{10^3 \, {\rm M_\odot}}{M_{\rm BH}}\right)^{2.33} \left(\frac{5 \times 10^4 \, {\rm pc}^{-3}}{\nu_0}\right)^{1.6} \\
& \times \left(\frac{\sigma}{10 \, {\rm km \, s^{-1}}}\right)^{5.76} \left(\frac{{\rm M_\odot}}{m}\right)^{1.06} \left(\frac{{\rm R_\odot}}{r_\star}\right)^{0.4},
\end{aligned}
\tag{8-134}
$$

where ν and σ are the number density and dispersion of stars in the core, outside the radius of influence $r_{\rm BH}$. The consumption rate is dominated,

[20] If this argument were applied to the inward flow of stars rather than to the outward flow of energy we would arrive at an erroneous exponent. In fact, the flow of stars is throttled to nearly zero in a steady state, because the accretion rate is limited by the speed at which the cusp can carry off the energy. See Bahcall and Wolf (1976).

not by stars that have diffused inward to the disruption radius r_t, but by stars in much larger orbits that diffuse to near zero angular momentum, and are disrupted at pericenter (see Shapiro 1985 for a review of this subject).

8.5 Summary

Stellar systems evolve in the direction of increasing entropy. There are at least four conceptually distinct processes that can increase the entropy of an isolated self-gravitating system: (i) escape of stars to infinity; (ii) core collapse; (iii) equipartition; and (iv) formation of hard binaries. Both escapes and core collapse occur on a timescale of order the relaxation time t_{rh}, but numerical experiments indicate that core collapse is generally the dominant process (in single component systems, the timescale for core collapse is 10 to $20t_{rh}$ versus $100t_{rh}$ for evaporation).

It appears that core collapse leads to the formation of hard binaries once the number of stars in the core is sufficiently small. These binaries, which can form by inelastic or three-body encounters, act as energy sources, much like the nuclear reactions occurring in the center of a star, and thereby halt the collapse. The later stages of the evolution, after the reversal of core collapse, are not yet fully understood. It appears that many globular clusters have undergone core collapse, but as yet there is no certain way to distinguish post-collapse from pre-collapse clusters.

Stellar collisions are relatively rare in globular clusters because the escape speed from the surface of the stars is much higher than the velocity dispersion in the cluster. In galactic nuclei the velocity dispersion is much higher, and the evolution of the system is therefore complicated by the poorly understood effects of stellar collisions. Many authors have speculated that supermassive black holes may form in dense galactic nuclei, possibly through stellar coalescence or by accretion of gas that has been shredded off stars by close encounters.

Problems

8-1. [1] Consider a system in which the interparticle potential has the form $\Phi_{\alpha\beta} = C|\mathbf{x}_\alpha - \mathbf{x}_\beta|^{-p}$, where C is a constant. Show that the scalar virial theorem has the form

$$2K + pW = 0, \qquad (8P\text{-}1)$$

where K is the kinetic energy and W is the potential energy. For what values of p does the system have negative heat capacity, in the sense of equation (8-36)?

8-2. [1] Consider a stellar system composed of two types of stars, with density distributions $\rho_1(\mathbf{x})$ and $\rho_2(\mathbf{x})$, and corresponding potentials $\Phi_1(\mathbf{x})$ and $\Phi_2(\mathbf{x})$. Show that in a steady state, the scalar virial theorem for component 2 may be written in the form

$$2K_2 + W_2 - \int \rho_2(\mathbf{x})\mathbf{x} \cdot \nabla\Phi_1(\mathbf{x})d^3\mathbf{r} = 0, \qquad (8P\text{-}2)$$

where K_2 is the total kinetic energy of component 2, and W_2 is the potential energy due to the mutual interaction of the stars of component 2.

8-3. [1] Using equation (8-110) for the evaporation time of soft binaries, estimate the maximum semi-major axis of a primordial soft binary that could survive for 10^{10} yr in the solar neighborhood. Assume that the one-dimensional velocity dispersion in the solar neighborhood is $30\,\mathrm{km\,s^{-1}}$, that the binary components and the field stars have mass $1\,\mathrm{M_\odot}$, and that the stellar density is $\rho = 0.044\,\mathrm{M_\odot\,pc^{-3}}$ (from Table 1-1). See Bahcall et al. (1985) for more detail.

8-4. [1] The disruption rate of hard binaries with internal energy $|E|$ contains a factor $\exp[\frac{3}{4}|E|/(m\sigma^2)]$ when the masses of the binary components and field stars are all equal to m [eq. (8-111)]. Give a simple heuristic explanation of the origin of this dependence, including the factor $\frac{3}{4}$.

8-5. [1] A tidal capture binary is formed as a result of a close encounter of two stars of equal mass m. The minimum separation during the encounter is $d_{\min}$, and the orbital energy dissipated in the encounter is $\Delta E \ll Gm^2/d_{\min}$. Once the binary has formed, more energy is dissipated in each successive orbit, until eventually the binary orbit is circularized. If the spin angular momentum of the stars is negligible compared to the orbital angular momentum, show that the radius of the final circular orbit is $2d_{\min}$.

8-6. [2] Consider a finite D-dimensional stellar system containing N identical stars that interact by inverse-square forces ($D = 1$, linear system; $D = 2$, flat disk; $D = 3$, sphere, etc.). Show that the relaxation time and the crossing time in such a system are related by

$$t_{\mathrm{relax}} \approx \begin{cases} N^{D-2}t_{\mathrm{cross}}, & D \leq 3, \\ N t_{\mathrm{cross}}, & D \geq 3, \end{cases} \qquad (8P\text{-}3)$$

where we have neglected factors of order unity and logarithmic factors [for $D = 3$ this is a crude version of eq. (8-1)]. What is the physical reason for the change in behavior at $D = 3$?

8-7. [2] The diffusion coefficients for a Maxwellian field star distribution depend on the functions $\mathrm{G}(X)$ and $[\mathrm{erf}(X) - \mathrm{G}(X)]$, where $\mathrm{G}(X)$ is defined by equation (8-69) and $X = v/(\sqrt{2}\sigma)$. Show that

$$\lim_{X\to 0} \frac{\mathrm{erf}(X)}{X} = \frac{2}{\sqrt{\pi}} \quad ; \quad \lim_{X\to 0} \frac{\mathrm{G}(X)}{X} = \frac{2}{3\sqrt{\pi}}. \qquad (8P\text{-}4)$$

Thus show that as the velocity of the test star $v \to 0$, the diffusion doefficients of equation (8-68) satisfy $D(\Delta v_\parallel^2) = \frac{1}{2}D(\Delta v_\perp^2)$. Explain physically why this must be so.

8-8. [2] In classical thermodynamics it is often stated that the heat capacity of an isolated system must be positive (e.g., Landau & Lifshitz 1980, §21). How can this result be reconciled with our proof that the heat capacity of a self-gravitating system is negative [eq. (8-36)]?

8-9. [2] In the Fokker-Planck approximation, show that the total escape rate from a spherical cluster with a DF $f(r, v, t)$ that is spherically symmetric in velocity space is given by

$$\frac{dN}{dt} = 8\pi^2 \int_0^\infty r^2\, dr \left[v^2 \frac{df}{dv} D(v_\parallel^2) \right]_{v=v_e(r)}. \qquad (8\text{P-}5)$$

8-10. [2] The object of this problem is to determine the behavior of the curve in Figure 8-1 in the limit as the central concentration $\mathcal{R} \to \infty$.

(a) The density of an isothermal sphere satisfies equation (4-125a),

$$\frac{d}{d\tilde{r}}\left(\tilde{r}^2 \frac{d\ln\tilde{\rho}}{d\tilde{r}} \right) = -9\tilde{r}^2\tilde{\rho}. \qquad (8\text{P-}6)$$

As the dimensionless radius $\tilde{r} \to \infty$, the solutions of equation (8P-6) approach the singular isothermal sphere $\tilde{\rho}_S(\tilde{r}) = 2/(9\tilde{r}^2)$. To determine the asymptotic behavior more accurately, define new variables u and $z(u)$ by $u = 1/\tilde{r}$ and $\tilde{\rho} \equiv \tilde{\rho}_S(\tilde{r})e^z$. Show that equation (8P-6) becomes

$$u^2 \frac{d^2 z}{du^2} + 2(e^z - 1) = 0. \qquad (8\text{P-}7)$$

(b) By linearizing equation (8P-7) for small z, show that the asymptotic behavior of the density $\tilde{\rho}$ is described by the equation (see Chandrasekhar 1939)

$$\tilde{\rho}(\tilde{r}) \simeq \tilde{\rho}_S(\tilde{r})\left[1 + \frac{A}{\tilde{r}^{\frac{1}{2}}} \cos\left(\tfrac{1}{2}\sqrt{7}\ln\tilde{r} + \phi \right) \right], \qquad (8\text{P-}8)$$

where A and ϕ are constants determined by the boundary conditions at small radii. Thus, at large $\tilde{r}$, the density of the isothermal sphere *oscillates* around the singular solution, with fractional amplitude decreasing as $\tilde{r}^{-1/2}$.

(c) Now consider an isothermal gas enclosed in a spherical box of radius r_b, with inverse temperature β [cf. eq. (8-41)]. As $\tilde{r}_b \to \infty$, show that the mass M of gas can be written in the form

$$x \equiv \frac{r_b}{GM\beta} \simeq \frac{1}{2} + \frac{A\tilde{r}^{\frac{1}{2}}}{4}\left[\cos\left(\tfrac{1}{2}\sqrt{7}\ln\tilde{r} + \phi \right) + \sqrt{7}\sin\left(\tfrac{1}{2}\sqrt{7}\ln\tilde{r} + \phi \right) \right]. \qquad (8\text{P-}9)$$

Hence, argue that when the central concentration $\mathcal{R} = 1/\tilde{\rho}(\tilde{r}_b)$ is large, the curve in Figure 8-1 becomes vertical at successive values of $\mathcal{R}$ that are in the ratio $\exp(4\pi/\sqrt{7}) = 115.54$. A similar expression may be obtained for the asymptotic form of $y \equiv Er_b/(GM^2)$, using equation (8-40), and these expressions for x and y give a parametric representation of the curve as it spirals toward the point $(x, y) = (\tfrac{1}{2}, -\tfrac{1}{4})$.

8-11. [2] A black hole of mass M is imbedded in the center of an infinite, homogeneous, three-dimensional sea of test particles. Far from the hole, the test particles have a Maxwellian velocity distribution,

$$f_0(\mathbf{v}) = \frac{\nu_0}{(2\pi\sigma^2)^{3/2}} e^{-\frac{1}{2}v^2/\sigma^2}. \tag{8P-10}$$

Show that the density distribution of test particles that are not bound to the hole is

$$\nu(r) = \nu_0 \left\{ 2\sqrt{\frac{r_H}{\pi r}} + e^{r_H/r} \left[1 - \mathrm{erf}\left(\sqrt{\frac{r_H}{r}} \right) \right] \right\}, \tag{8P-11}$$

where the error function $\mathrm{erf}(x)$ is defined in Appendix 1.C.3, and $r_H = GM/\sigma^2$. Show that close to the hole, $r \ll r_H$, $\nu(r) \propto r^{-1/2}$. Thus there is a weak density cusp around the hole.

9 Stellar Evolution in Galaxies

In the foregoing chapters we have modeled galaxies as systems that are made up of fixed numbers of point masses. In reality, however, they are made of stars, which are born, grow old, and die. We now consider ways in which the evolution of stars affects the structure and evolution of galaxies. We shall discuss three topics:

(i) How do the luminosity and color of an isolated galaxy evolve? This question is of direct observational interest because we can observe galaxies at large cosmological redshifts, when they were much younger than they are today. Galaxies evolve as a result of both stellar-dynamical processes, such as those discussed in Chapters 6 and 8, and the evolution of their stars. Only when we have understood both these sources of change will we possess a theory of galactic evolution that can be fruitfully compared with observation. This comparison is extremely important from the point of view of cosmology since it may enable us to determine the relationship between redshift, cosmic time, and distance, and thus to determine whether or not the Universe is spatially and temporally finite.

(ii) What can we learn about the formation and dynamical history of our own Galaxy from the chemical properties of its stars?

(iii) The colors and gas contents of galaxies are correlated with their Hubble types and luminosities (see §5-4 of MB). Does the theory of stellar evolution help us to understand these correlations?

We shall not be able to give definitive answers to any of these questions, but we shall find that many diverse pieces of information fit together into a plausible picture of galaxy evolution.

9.1 Luminosity and Color Evolution of Galaxies

1 Elliptical Galaxies

The Hertzsprung-Russell diagram (the HR diagram) of a typical globular cluster suggests that all the cluster's stars were formed together approximately 15 Gyr ago (see §1.1.1, and §3-10 of MB). That is, the observed colors and luminosities of the cluster stars are compatible with those predicted by stellar evolution theory for an ensemble of stars of the same age and chemical composition, but of different masses. Furthermore, in many respects globular clusters resemble elliptical galaxies (see §5-2 of MB), and this analogy suggests that the stars of elliptical galaxies may also all be of a similar age. If we adopt this as a working hypothesis, we may calculate how the luminosities and colors of elliptical galaxies should have evolved to their present values.

From Table 3-9 of MB we see that stars that are more massive than $1.25 M_\odot$ live less than 5 Gyr. Thus we expect elliptical galaxies to contain only stars of mass $M < 1 M_\odot$. Table 3-9 and Figure 3-17 of MB show that these low-mass stars emit most of their light during the 1 Gyr or so that they are on the giant branches of the HR diagram. Hence we expect most of the integrated light from an elliptical galaxy to come from giant stars. Observations at visual and infrared wavelengths (Frogel et al. 1978) show that this is indeed the case.[1]

Our understanding of how stars evolve after they leave the subgiant branch and become luminous giants is very poor (see §3-7 of MB). The theoretical calculations are unreliable because they are based on uncertain assumptions as to how convection works, how the explosive ignition of helium in a degenerate core affects the envelope of the star, and exactly how and when stars begin to blow off large parts of their envelopes in a stellar wind. The observational situation is equally complex and ill-understood. Since elliptical galaxies derive most of their light from giants, these difficulties will cause evolutionary models of elliptical galaxies to be correspondingly uncertain. However, several important conclusions can be drawn with fair security even in the absence of a detailed understanding of the late stages of stellar evolution.

The first conclusion depends on the fact that the effective temperature of a giant with $M \lesssim 2 M_\odot$ does not depend sensitively on its mass. Consequently, the intrinsic[2] colors of elliptical galaxies should not have evolved strongly since the epoch, about 1.5 Gyr after their formation, when all stars with masses greater than $2 M_\odot$ had died.

By contrast, the luminosity of elliptical galaxies must still be evolving (Tinsley 1972). To see this, let there be $dN(M)$ stars in the galaxy with mass in the range M to $M + dM$, and let the stars that turn off

---

[1] While the bulk of the light comes from K giants, some blue stars are present (Gunn et al. 1981). It is not known, however, whether these are upper main-sequence stars or horizontal branch stars.

[2] The observed colors of a distant elliptical galaxy differ from the intrinsic colors on account of the cosmic redshift—see §9.1.3.

the main sequence at age t be of mass $M_{GB}(t)$. Table 3-9 of MB shows that for $t \gtrsim 1.5\,\mathrm{Gyr}$, $M_{GB} \lesssim 1.5\,M_\odot$, and that the time t_{GB} that the star spends on the giant branch is much less than t. Hence if stars of mass M emit total energy $E_{GB}(M)$ when they are on the giant branch, the luminosity of the galaxy may be estimated as

$$L \approx \left(E_{GB}\frac{dN}{dM} \right)_{M_{GB}} \left| \frac{dM_{GB}}{dt} \right|. \tag{9-1}$$

We may fit the **initial mass function** (IMF; see §4-2 of MB) dN/dM in the neighborhood of M_{GB} by a power law

$$\frac{dN}{dM} \simeq K\left(\frac{M}{M_\odot} \right)^{-(1+x)}. \tag{9-2}$$

In the solar neighborhood $x \simeq 1.5$ for $M < M_\odot$. For the masses of interest, the main-sequence lifetime t_{MS} of a star is related to its mass by

$$\frac{t_{MS}}{10\,\mathrm{Gyr}} \simeq \left(\frac{M}{M_\odot} \right)^{-2.5}, \quad \text{which implies} \quad \frac{M_{GB}(t)}{M_\odot} \simeq \left(\frac{t}{10\,\mathrm{Gyr}} \right)^{-0.4}. \tag{9-3}$$

Hence

$$\frac{dM_{GB}}{dt} \simeq -0.4\left(\frac{M_{GB}}{M_\odot} \right)^{3.5}\left(\frac{M_\odot}{10\,\mathrm{Gyr}} \right). \tag{9-4}$$

Substituting equations (9-2) and (9-4) into equation (9-1), we find

$$L \simeq \frac{K\,M_\odot E_{GB}(M_{GB})}{25\,\mathrm{Gyr}}\left(\frac{M_{GB}}{M_\odot} \right)^{2.5-x}. \tag{9-5}$$

Differentiating this expression yields

$$\begin{aligned}
\frac{d\ln L}{d\ln t} &= \left[\frac{d\ln E_{GB}}{dM_{GB}} + (2.5 - x) \right]\frac{d\ln M_{GB}}{d\ln t} \\
&= 0.4x - \left(1 + 0.4\frac{d\ln E_{GB}}{d\ln M_{GB}} \right).
\end{aligned} \tag{9-6}$$

E_{GB} probably depends only weakly on M_{GB} ($0 < d\ln E_{GB}/d\ln M_{GB} < 1$), so unless $x > 2.5$, the luminosity of an elliptical galaxy is expected to be a decreasing function of time.

Tinsley (1972), Tinsley and Gunn (1976), Bruzual (1980), and others have used theoretical and empirical evolutionary tracks for low-mass stars to model the luminosity evolution of elliptical galaxies in detail. Naturally these calculations are only as reliable as the evolutionary tracks that they employ, but they all confirm the two conclusions that we have just reached: the intrinsic colors of an elliptical galaxy do not evolve rapidly after 1.5 Gyr, but its luminosity can diminish rapidly, especially if the local slope x of the integrated IMF is small. In fact, Tinsley and Gunn's detailed models yield $(d \ln L / d \ln t) \simeq 0.3x - 1.3$, in good agreement with equation (9-6). Thus we can reliably predict the luminosity evolution of elliptical galaxies only if we can accurately determine x.

Observations of nearby globular clusters yield x directly from star-counts, and one might hope that the value of x that is appropriate for giant elliptical galaxies is similar. Unfortunately, x varies from cluster to cluster in an ill-understood way (Freeman 1977), so this approach gives no clear indication of what x might be in an elliptical galaxy. Since no elliptical galaxy is close enough for its main-sequence stars to be counted individually, the possible values of x are restricted only by the colors and line-strengths of galaxies. Two problems make it impossible to determine the slope of the IMF where it matters, namely, just below the turnoff to the giant branch: (i) K dwarfs contribute very little to the overall luminosity at any wavelength, and (ii) no line-strength index is available to distinguish between K giants and dwarfs in a system of unknown metallicity. The best that can be done is to estimate the relative proportions of M dwarfs and the K giants that are responsible for most of the visual luminosity by comparing the visual luminosity with the strength of features in the far red, such as the Wing-Ford band (Whitford 1977) and the CO band at $2.3\,\mu$ (Frogel et al. 1978), to each of which M dwarfs contribute strongly. Tinsley and Gunn (1976) conclude from these features that the *mean* slope $\bar{x}$ of the IMF between K and M stars satisfies $\bar{x} < 1$. If this mean slope is characteristic of the slope just below the turnoff from the main sequence, then $(d \ln L / d \ln t) < -0.6$. Unfortunately, the remaining uncertainty in x makes the predicted rate of luminosity evolution of elliptical galaxies too uncertain for any practical application.

2 Spiral Galaxies

Late-type galaxies such as our own are continuously forming stars of a variety of masses (see §9-3 of MB). As Table 3-9 and Figure 3-17 of MB show, stars more massive than about $2.5\,M_\odot$ emit most of their light before they reach the giant branch, so the simple analytical arguments

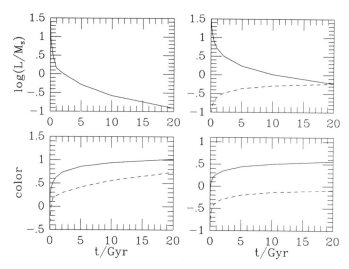

Figure 9-1. The panels on the left show the evolution of the luminosity per unit mass (solar units) and the $U - B$ (dashed curve) and $B - V$ (full curve) colors of a stellar population formed in a single burst of star formation at $t = 0$. This population contains only stars with masses in the range $0.1 < M/M_\odot < 30$, and in this range the IMF has slope $x = 1.3$. The right panels show the same quantities for a system in which the star-formation rate is constant for $t \geq 0$. The dashed curve in the upper right panel shows the evolution of the ratio $L(t)/M_s(20\,\mathrm{Gyr})$. (From data published in Larson & Tinsley 1978.)

that we used in our discussion of color and luminosity evolution of elliptical galaxies cannot be applied to the evolution of systems in which there has been recent star formation. Hence we must rely on numerical models to study the evolution of late-type galaxies.

The use of theoretical or empirical stellar evolution tracks to model the evolution of a galaxy that contains both young and old stars is a straightforward extension of the procedure for modeling the evolution of elliptical galaxies. A good reference to this area is Tinsley's (1980a) paper. Since massive stars are better understood than the K giants that are responsible for most of the light from elliptical galaxies, our present lack of understanding of stellar evolution contributes less to the uncertainty of models of late-type galaxies than to models of early-type galaxies such as ellipticals.

A model for the evolution of a late-type galaxy involves two functions: (i) the IMF that governs the relative numbers of stars that are formed in each mass range; (ii) the star-formation rate $\dot{M}_s$. Two sim-

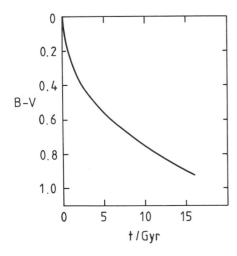

Figure 9-2. The evolution of the $B - V$ color of a galaxy in which the star-formation rate declines exponentially, with time-constant $\tau = 3\,\text{Gyr}$ and $x = 1.3$. After a few billion years, such a galaxy is bluer than one in which star formation has ceased, and redder than a galaxy in which the star-formation rate is constant. (After Larson et al. 1980.)

ple assumptions are that $\dot{M}_s = constant$ and $\dot{M}_s(t) = \dot{M}(0)\exp(-t/\tau)$, where τ is a suitable characteristic time. Formally $\dot{M}_s = constant$ corresponds to $\tau = \infty$.

The right panels of Figure 9-1 show the evolution of the luminosity per unit stellar mass and the $U - B$ and $B - V$ colors of a model in which $\dot{M}_s = constant$ and the IMF has the same slope as that determined observationally for the solar neighborhood, $x \simeq 1.3$ (see §4-2 of MB). Since the number of young blue stars is independent of time, while the population of older stars grows steadily, the system becomes more luminous and redder with the passage of time. After 16 Gyr, the model has $(B-V) \simeq 0.5$, which Figure 5-36 of MB shows to be the $B-V$ color of a typical Sc galaxy. Thus the blueness of late-type spiral galaxies can be understood if their stars form at a roughly constant rate.

Figure 5-36 of MB shows that earlier-type spirals, for example Sa galaxies (Hubble stage $t = 1$), have $(B - V) = 0.7$, which suggests that in these systems either the star-formation rate declines with time, or the IMF is less strongly weighted toward massive and therefore blue stars than is the IMF of later-type galaxies. Figure 9-2 shows the effect of assuming the star-formation rate declines exponentially with time, with time constant $\tau = 3\,\text{Gyr}$. Clearly when $3\,\text{Gyr} < \tau < \infty$, the present $B - V$ color of the galaxy can be placed anywhere in the range $0.5 < (B - V) < 0.7$ in which the $B - V$ colors of disk galaxies lie. The luminosity of the system at most epochs before the present may be either increasing or decreasing according as τ is large or small.

Table 9-1 gives as functions of time the UBV colors of a stellar population that formed in a single burst of star formation with the IMF characteristic of the solar neighborhood. The colors of any stellar

Table 9-1. The UBV colors of a stellar population formed in a single burst (from Larson & Tinsley 1978)

$t/$Gyr	$U-B$	$B-V$	$t/$Gyr	$U-B$	$B-V$	$t/$Gyr	$U-B$	$B-V$
0.01	-1.0	-0.20	0.2	-0.02	0.37	5	0.42	0.86
0.02	-0.79	-0.08	0.5	0.17	0.51	10	0.56	0.94
0.05	-0.37	0.19	1.0	0.22	0.62	20	0.74	1.02
0.1	-0.19	0.26	2.0	0.29	0.73			

population can be predicted by combining a number of such single star-burst models.

3 Observations

When we observe distant galaxies, we see them as they were several billion years ago, so it is in principle possible to test theories of galactic evolution observationally. This is an active and exciting area of research, but as yet it has yielded few firm and well-understood conclusions. Therefore our discussion will be confined to essentials. We start by reviewing some key elements of theoretical cosmology. [For details see Peebles (1971), Weinberg (1972), or Gunn (1978).]

The large-scale structure of the Universe is found to be remarkably well described by the simple **Friedmann cosmological models**. These are the solutions of Einstein's general relativistic field equations that describe a homogeneous and isotropic Universe, that is, a Universe in which it is possible to choose a **cosmic time** t, and to place spatially separated observers, called **fundamental observers**, such that all fundamental observers see the same structure at a given cosmic time, no matter where they are, or in what direction they look.[3] The most important properties of these models are summarized in Appendix 9.A. We shall require three key results:

(i) At cosmic time t the distance between any two nearby fundamental observers is $l = R(t)l_0$, where l_0 is the current separation and the **scale factor** $R(t)$ is unity at the present time t_0. The functional form of $R(t)$ is determined by the current mass density ρ_0 and expansion rate of the Universe. The current rate of change

---

[3] Obviously the structure of the Universe is not homogeneous and isotropic on small scales. For example, there is a large excess of galaxies toward the north galactic pole (the Virgo supercluster). However, inhomogeneities like the Virgo supercluster do not appear to extend over more than a few tens of Mpc, while the size of the visible Universe is several thousand Mpc.

of R, $H_0 \equiv \dot{R}(t_0)$, is the **Hubble constant**, which relates the mutual recession speed v of any two fundamental observers to their separation:

$$v = H_0 l_0. \tag{9-7a}$$

Empirically,

$$H_0 = 100 h \, \text{km s}^{-1} \, \text{Mpc}^{-1} \quad \text{where} \quad 0.5 \lesssim h \lesssim 1. \tag{9-7b}$$

The future of the Universe hangs upon whether the dimensionless **density parameter**

$$\Omega_0 \equiv \frac{8\pi G \rho_0}{3H_0^2} \tag{9-8}$$

is less than, greater than, or equal to unity. If $\Omega_0 < 1$, $R(t)$ is given by equation (9A-22) and the speed with which any two galaxies recede from one another will eventually settle to a constant value, while if $\Omega_0 > 1$, $R(t)$ is given by equation (9A-19) and the Universe will eventually start to recollapse. For illustrative purposes we shall often use the exceptionally simple form taken by $R(t)$ when $\Omega_0 = 1$. Then

$$R(t) = (\tfrac{3}{2} H_0 t)^{\frac{2}{3}}, \tag{9-9}$$

and the cosmic expansion will become arbitrarily slow, while never actually ceasing.

(ii) In Appendix 9.A we show that if light of wavelength λ_e is emitted at time t_e and received at time t_0, it will be observed to have a wavelength λ_0 that is related to $R(t_0)/R(t_e)$ and the **redshift** z by

$$1 + z \equiv \frac{\lambda_0}{\lambda_e} = \frac{R(t_0)}{R(t_e)}. \tag{9-10a}$$

Since $R(t_0) = 1$, we have

$$R(t_e) = \frac{\lambda_e}{\lambda_0}. \tag{9-10b}$$

Thus the redshift of a galaxy tells us directly how much closer together the galaxies were when the light we now observe was emitted. If we know H_0 and Ω_0, the appropriate equations of the set (9A-19) to (9A-23) enable us to find $t_e(z)$. Galaxies can be seen out to $z \approx 1$ and quasars to $z \approx 4$.

(iii) The decrease in the frequency of light that is described by equations (9-10) arises from a time dilation effect; an atomic clock moving

Table 9-2. Distance modulus in Friedmann cosmological models

			Ω_0		
z	0.1	0.25	0.5	0.75	1
0.002	28.88	28.89	28.89	28.89	28.89
0.005	30.90	30.88	30.88	30.88	30.88
0.01	32.39	32.39	32.39	32.39	32.39
0.02	33.91	33.91	33.91	33.90	33.90
0.1	37.48	37.48	37.46	37.45	37.43
0.2	39.08	39.07	39.04	39.01	38.99
0.4	40.76	40.73	40.67	40.62	40.57
1	43.19	43.10	42.96	42.84	42.73
1.5	44.37	44.22	44.02	43.85	43.70
2	45.24	45.05	44.79	44.58	44.40
3	46.52	46.24	45.89	45.62	45.39
4	47.46	47.11	46.67	46.35	46.10

NOTE: Bolometric distance modulus $\Delta = m_{bol} - M_{bol}$ versus redshift z, assuming $h = 1$.

with one fundamental observer appears to run slow when observed by another fundamental observer. Consequently, not only is the energy (hc/λ) associated with each photon diminished by the ratio $R(t_e)/R(t_0)$, but the *rate* at which photons are received is down by the same factor. Hence the energy flux F observed at t_0 from a source that emitted the light at t_e is related to the intrinsic luminosity L of the source by

$$F(t_0) = \frac{L}{4\pi r^2}\left[\frac{R(t_e)}{R(t_0)}\right]^2, \qquad (9\text{-}11)$$

where r is a measure of the distance to the source. In an isotropic Universe, the appropriate distance measure for equation (9-11) must be chosen so that $4\pi r^2$ is the area of the spherical surface illuminated at t_0 by photons emitted at t_e. Equations (9A-9) and (9A-26) show that

$$r = 4c\frac{(1 - \frac{1}{2}\Omega_0)(1 - \sqrt{1 + \Omega_0 z}) + \frac{1}{2}\Omega_0 z}{(1 + z)H_0\Omega_0^2}. \qquad (9\text{-}12)$$

Table 9-2 gives bolometric distance moduli $\Delta_{bol} \equiv m_{bol} - M_{bol}$ calculated from equations (9-11) and (9-12) for different values of z and Ω_0.

Now suppose we have obtained the spectrum of a distant galaxy of known apparent magnitude. If we can identify lines or other features in this spectrum with atomic lines of known rest wavelength, we can obtain from equations (9-10) both the galaxy's redshift z and $R(t_e)$, the scale factor at the time t_e when the light was emitted. For any given pair of values (H_0, Ω_0) of the cosmological parameters, we can then obtain from the appropriate equations of the set (9A-19)–(9A-23) both the present time t_0 and the emission time t_e. Equation (9-9) for the simple case $\Omega_0 = 1$ will give us an idea of the numbers involved. From (9-7b) we have $H_0 = 1.02h \times 10^{-10}\,\mathrm{yr}^{-1}$. Thus if $\Omega_0 = 1$, we have from equation (9-9) that $t_0 = \frac{2}{3}H_0^{-1} \simeq 6.5h^{-1}\,\mathrm{Gyr}$. A very few redshifts have been determined for normal galaxies out to $z \simeq 1.0$. According to equations (9-9) and (9-10), this corresponds to $t_e = 0.35t_0 \simeq 2.3h^{-1}\,\mathrm{Gyr}$. More data are available for smaller redshifts. For example, Dressler and Gunn (1983) have determined several redshifts for galaxies in clusters at $z \simeq 0.5$. By equation (9-9) this corresponds to $t_e \simeq 0.54t_0 \simeq 3.5h^{-1}\,\mathrm{Gyr}$. Thus *we are able to observe galaxies when they were between a third and a half of their present age.* If some galaxies formed relatively recently, or if $\Omega_0 > 1$, the most distant galaxies are being seen at even earlier stages in their lives. Figure 9-1 shows that over this sort of time interval, we expect the colors of both early- and late-type galaxies to be relatively constant, but the luminosities of early-type galaxies are expected to have changed significantly.

(a) Spectra of distant galaxies   A straightforward test of the prediction that the spectra of distant galaxies should resemble those of nearby galaxies is provided by surveys of the broad-band colors of galaxies in distant clusters. The redshifts of these clusters can be determined from the spectra of their brightest galaxies, and modern photometric techniques (see §5-2 of MB) make it possible to measure the broad-band colors of the less luminous cluster galaxies. Unfortunately, surveys of this type have yielded conflicting results. Koo (1981) has shown that the colors of the galaxies in some distant clusters are very similar to the colors of galaxies in nearby clusters of the same type. However, the first survey of galaxy colors in distant rich clusters, which was published by Butcher and Oemler in 1978, suggested that two extremely rich clusters contained a substantial number of very blue galaxies. As we saw in §1.1.5, at the present epoch it is found that bright galaxies in centrally concentrated clusters are mostly early-type galaxies. Since early-type galaxies are now rather red $[(B - V) \simeq 1]$ and, as Figure 9-1 shows, we do not expect them to have been much bluer a mere 5 Gyr ago, the large proportions of blue galaxies in the Butcher-Oemler clusters were quite

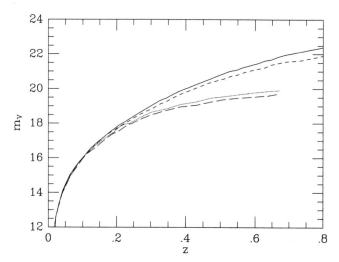

Figure 9-3. A theoretical Hubble diagram for galaxy models in which the star-formation rate declines with time as $e^{-t/\tau}$. Both full and dotted curves correspond to galaxies in which $\tau = 0.83$ Gyr in an $h = 1$ Universe, but $\Omega_0 = 0$ for the full curve while the dotted curve corresponds to $\Omega_0 = 1$. Similarly, both dashed curves are generated by galaxies with $\tau = 1.4$ Gyr in an $h = 1$ Universe, but the short-dashed curve corresponds to $\Omega_0 = 0$ while for the long-dashed curve $\Omega_0 = 1$. At the current epoch all models have absolute magnitude $M_V = -23$. (From data published in Bruzual 1983.)

unexpected. At the present time it is unclear whether these blue objects are normal late-type galaxies, or galaxies that have large contributions to their blue luminosities from nonthermal sources such as an active galactic nucleus; see Dressler and Gunn (1983) for further discussion.

(b) The luminosities of distant galaxies   A fundamental difficulty stands in the way of attempts to verify the theoretically predicted evolutionary paths of the luminosities of different types of galaxies: in order to infer the intrinsic luminosity L of a distant galaxy from the observed flux $F(t_0)$ and redshift z, we need to use equation (9-12), and in order to use this equation we must know Ω_0. Unfortunately, the correct value of Ω_0 is very uncertain. Worse still, the classical means of *determining* Ω_0 is to use equation (9-12) in conjunction with some hypothetical "standard candle" of known redshift, luminosity, and observed flux! Figure 9-3 illustrates this conundrum. Each curve in this figure shows the apparent magnitude of a model galaxy versus the redshift z

at which it is observed—thus these curves are theoretical counterparts of the classical **Hubble diagram** (see Sandage & Tammann 1975). The full and short-dashed curves in Figure 9-3 are for different evolutionary models in an $\Omega_0 = 0$ Universe, while the dotted and long-dashed curves are for the same evolutionary models in an $\Omega_0 = 1$ Universe. Clearly only observations of galaxies at redshifts $z \gtrsim 0.35$ offer any hope of a reasonably secure determination of Ω_0.

(c) Galaxy counts   The majority of objects in the night sky fainter than about $m_V = 21$ are actually distant galaxies (Kron 1980). The most numerous galaxies in any magnitude-limited survey generally have absolute magnitudes $M_V \simeq -21$ (see §5-3 of MB). From Table 9-2 it follows[4] that if $\Omega_0 = 1$, most of the galaxies seen at apparent magnitude $m_V \gtrsim 22.5$ have redshifts $z \gtrsim 1$. Automatic plate-measuring machines can determine with fair reliability whether a photographic image brighter than about $m_V = 24$ is that of a star or of an extended object such as a galaxy. Hence a census of galaxies at $z \gtrsim 1$ can be taken by programming a machine to count extended images in the magnitude range $21 \lesssim m_V \lesssim 24$. The results will clearly depend upon both the true value of Ω_0 and the predicted luminosity and color evolution of galaxies.

The dots in Figure 9-4 show the results of such a census of distant galaxies, together with the results predicted by two models of galaxy evolution. Studies of this kind, which require only a few hours of telescope time rather than the many tens of hours of large-telescope time needed to measure Hubble diagrams like Figure 9-3, strongly suggest that galaxy evolution *is* being detected. Furthermore, diagrams like Figure 9-4 by no means exhaust the cosmological information obtainable from galaxy counts. Bruzual and Kron (1980) argue that counts in many different wavebands may enable us to separate unambiguously the effects of galaxy evolution from those of cosmology. Unfortunately, such a separation will not be easy (Tinsley 1980b) and will require extensive observations of present-day galaxies at ultraviolet wavelengths, and galaxy counts in the near infrared. The Hubble Space Telescope should be able to furnish the required data.

---

[4] The visual distance modulus $\Delta_V \equiv m_V - M_V$ increases more rapidly with z than does the bolometric modulus Δ_{bol} because the proportion of a galaxy's luminosity that falls in the visual band decreases with increasing z. One conventionally writes $\Delta_V = \Delta_{bol} + K_V$, where K_V is the **K-correction** (e.g., C. R. King & Ellis 1985).

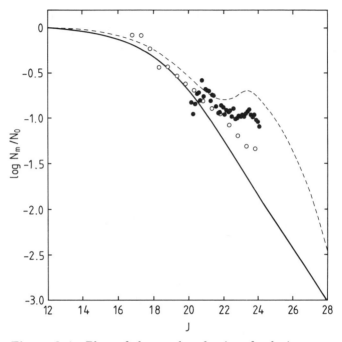

Figure 9-4. Plots of the number density of galaxies per unit magnitude interval versus apparent magnitude for two theoretical models and corresponding observational data. The filled and open circles are from different galaxy samples. The solid curve shows the expected density of unevolving galaxies in an $\Omega_0 = 0.04$, $h = 0.5$ cosmology. The dashed line shows how this prediction is modified when the luminosity evolution of galaxies is modeled by assuming that the star-formation rate in galaxies of each Hubble type declines exponentially with time-constants 3 Gyr (E–Sab), 10 Gyr (Sbc), and ∞ (later than Sbc). Star formation is assumed to have ceased in galaxies of Hubble types E–S0 at $t = 10$ Gyr. (After Tinsley 1980b.)

9.2 Chemical Evolution of Disk Galaxies

The abundances of elements heavier than helium ("metals") in interstellar gas and in stars vary systematically from place to place within galaxies (see Chapters 4, 5, and 7 of MB). Since heavy elements are believed to have been synthesized in massive stars, the metal abundances in present-day galaxies should offer a record through which we may trace the history of star formation within galaxies. Furthermore, if we can infer the metallicity of the interstellar gas at each time in the past, we will be able to date stars by their metallicity much as a dendrochronolgist dates timbers by the patterns of their treerings. So far it has not proved possible fully to attain these goals, but astronomers have gained insight into the formation and evolution of galaxies such as our own. A fuller account of these questions will be found in Tinsley (1980a).

1 The One-Zone Model

We first consider the simplest possible model of the chemical evolution of a portion of a galaxy such as the solar neighborhood. We focus on a narrow annulus of galactocentric radius, and we suppose that in the period under study no material either enters or leaves this region. Initially, the material is entirely gaseous and free of heavy elements. As time goes on, stars are formed from the interstellar gas and massive stars explode, thus returning hydrogen, helium, and heavy elements to the interstellar medium. We assume that turbulent motions keep the gas well stirred and therefore homogeneous. The supply of interstellar gas is gradually consumed, and the remaining gas becomes steadily more polluted with heavy elements.

Suppose that at any time there is a mass M_h of heavy elements in the interstellar gas, which itself has mass M_g. Then the **metallicity** of the interstellar gas is

$$Z \equiv \frac{M_h}{M_g}. \qquad (9\text{-}13)$$

We suppose that at this time the stars have total mass M_s and consider the effect of forming new stars in amount $\delta' M_s$. The very massive stars $(M \gtrsim 9 M_\odot)$ that are responsible for the manufacture of most heavy elements will explode as supernovae in a time that is short compared to the present age of the galaxy, so we may imagine that the heavy elements that are formed by the $\delta' M_s$ of new stars are returned to the interstellar gas immediately. Let the mass of the stars in $\delta' M_s$ that remain after the massive stars have died be δM_s, and let the mass of the heavy elements produced by this stellar generation be $p\delta M_s$, where p is the **yield** of that generation of stars. Then the total change in the heavy element content of the interstellar gas that arises from these new stars is

$$\delta M_h = p\delta M_s - Z\delta M_s = (p - Z)\delta M_s. \qquad (9\text{-}14)$$

The metallicity of the interstellar gas changes by an amount

$$\delta Z = \delta\left(\frac{M_h}{M_g}\right) = \frac{\delta M_h}{M_g} - \frac{M_h}{M_g^2}\delta M_g = \frac{1}{M_g}(\delta M_h - Z\delta M_g). \qquad (9\text{-}15)$$

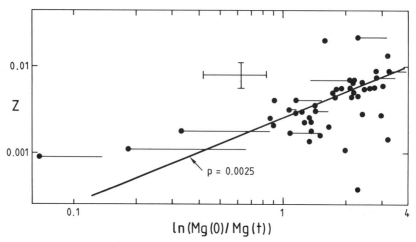

Figure 9-5. Metallicity Z of gas in irregular galaxies versus the galaxies' current gas fraction $M_g(t)/M_{\text{tot}} = M_g(t)/M_g(0)$. The cross indicates a typical uncertainty. (After Pagel 1986.)

By conservation of mass $\delta M_s = -\delta M_g$, so combining equations (9-14) and (9-15), we obtain

$$\delta Z = -p\frac{\delta M_g}{M_g}. \tag{9-16}$$

If the yield p of each generation of stars is the same, we may integrate equation (9-16) to obtain the metallicity at time t as

$$Z(t) = -p\ln\left[\frac{M_g(t)}{M_g(0)}\right]. \tag{9-17}$$

Here we have employed our assumption that the gas is initially free of metals.

Since $M_g(0)$ is by hypothesis the total mass of the annulus, equation (9-17) makes a clear prediction: if we plot the observed metallicity Z of various disk galaxies against the logarithms of the fractions of their total masses that are in gaseous form, we will obtain a straight line. The absolute value of the slope of this straight line will be the yield p. Figure 9-5 shows that a relationship of this sort does indeed hold for Magellanic irregular galaxies. The yield inferred from Figure 9-5 is $p \simeq 0.0025$.

We have not included any spiral galaxies in Figure 9-5 because there are marked metallicity gradients in spiral galaxies (see §5-4 of MB). Thus these galaxies cannot be characterized by a single metallicity. However, we may apply equation (9-17) to different annuli within the same galaxy

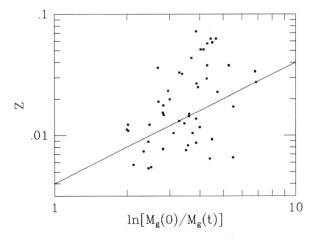

Figure 9-6. Metallicities of HII regions in six Sab–Sbc galaxies, versus the local gas fraction. Much of the scatter in this figure is observational. The line shows the prediction of equation (9-17) for $p = 0.004$. (After Edmunds & Pagel 1984.)

to deduce that the metallicity should be highest where the gas fraction is lowest. In Sa and Sb galaxies, the gas fraction tends to be much lower near the center than at large radii (see §9-1 and §9-2 of MB). Hence equation (9-17) suggests that the metallicity of the interstellar gas in these galaxies should decrease from the center outward. As Figure 5-46 of MB shows, this is exactly what is observed. However, Figure 9-6 shows that the slope in a plot of Z against the logarithm of the local gas fraction in several Sab–Sbc galaxies is rather steeper than is predicted by equation (9-17) with p set equal to the value $p \simeq 0.0025$ deduced from Magellanic irregulars. Peimbert and Serrano (1982) have argued that the mean line in Figure 9-6 has a steeper slope than that of Figure 9-5 because the yield p is not independent of Z, as we have arbitrarily assumed in integrating equation (9-16) to form equation (9-17), but is actually an increasing function of Z. Unfortunately, this modification of our simple theory exacerbates a serious problem with the standard one-zone model that we must now describe.

(a) G-dwarf problem   Equation (9-17) predicts the relative numbers of unevolved stars that have different metallicities at the present epoch. In particular, the mass of the stars that have metallicity less than $Z(t_1)$ is

$$M_s[< Z(t_1)] = M_s(t_1) = M_g(0) - M_g(t_1)$$
$$= M_g(0)[1 - e^{-Z(t_1)/p}]. \tag{9-18}$$

Let us apply this equation to the solar neighborhood. We first calculate the ratio $M_g(t)/M_g(0)$, which we take to be the ratio of the surface density of interstellar gas at the solar radius ($\simeq 5\,M_\odot\,pc^{-2}$, from Table 1-1) to the total surface density of visible stars and gas ($\simeq 50\,M_\odot\,pc^{-2}$ from Table 1-1).[5] The metallicity of this interstellar gas is found to be very similar to that of the Sun,

$$Z_\odot \simeq 0.02. \tag{9-19}$$

Hence from equation (9-17) we obtain $p \simeq 0.43 Z_\odot \simeq 0.009$.

If we use this value of p in equation (9-18) to determine $M_s(< 0.25 Z_\odot)$, we find

$$M_s(< 0.25 Z_\odot) = M_g(0)(1 - e^{-0.25/0.43}) \simeq 0.44 M_g(0). \tag{9-20}$$

Since $M_g(0) \simeq M_s(t_0)$ it follows from this equation that nearly half of all stars in the solar neighborhood, whether luminous stars or cold dwarfs, should have $Z < 0.25 Z_\odot$. Actually only 2% of disk F and G stars in the solar neighborhood have $Z < 0.25 Z_\odot$ (Bond 1970; Clegg & Bell 1973). This contradiction between the standard one-zone model and observation is known as the **G-dwarf problem**.[6] Evidently one of the assumptions of the standard one-zone model is incorrect.

Consider first the possibility that the yield p is not independent of Z, and hence that equation (9-17) is not the correct integral of equation (9-16). One way in which this might happen is if the IMF were metallicity-dependent. For example, if no low-mass stars are born when $Z < 0.25 Z_\odot$, the G-dwarf problem would be resolved. One problem with this proposal is that it implies that the yield p *decreases* with increasing Z, which is the reverse of the trend that is required if the one-zone model is to account satisfactorily for metallicity gradients within galactic disks (Peimbert & Serrano 1982). A second, and perhaps more serious,

---

[5] Table 1-1 gives the total, dynamically determined, surface density of the disk as $\simeq 75\,M_\odot\,pc^{-2}$, and one might argue that this is the relevant surface density. However, it is straightforward to verify that this change would have only a minor effect on our conclusions.

[6] Mould (1978) has argued that a similar result applies to M dwarfs, although the data for M stars are less reliable because the ultraviolet excess $\delta(U-B)$ (see §3-6 of MB) cannot be used as a metallicity indicator for these stars.

problem was pointed out by Thuan et al. (1975), who examined a two-parameter family of one-zone models in which the IMF—and therefore the yield—as well as the overall rate of star formation are continuous functions of time. They found that such one-zone models can reproduce the observed scarcity of low-metallicity dwarf stars only if the IMF at early times is so strongly biased against the formation of low-mass stars that the rate of formation of these stars actually *increases* in time. Unfortunately, an increasing rate of formation of low-mass stars corresponds to an increasing rate of consumption of interstellar gas, so the one-zone models that are compatible with the stellar metallicity distribution predict that the interstellar gas will be exhausted within the next 1 Gyr or so. It seems unlikely that we are privileged to live so close to the end of the epoch during which our Galaxy forms new stars. More plausible explanations of the scarcity of metal-poor stars may be obtained by dropping at least one of the other central assumptions of the standard one-zone model—either that the gas is initially metal-free or that gas neither enters nor leaves any particular annulus of the Galaxy.

If the gas from which the galactic disk formed started from metallicity Z_i, the integral of equation (9-16) becomes

$$Z(t) = Z_i + p \ln \left[\frac{M_g(0)}{M_g(t)} \right], \qquad (9\text{-}21)$$

and there will be no disk stars with $Z < Z_i$. Furthermore, when we insert the observed value of $M_g(0)/M_g(t_0)$ and $Z_i = 0.25 Z_\odot$ into this equation, we deduce $p \simeq 0.006$, in better agreement than our previous solar-neighborhood estimate, $p \simeq 0.009$, with the value that is deduced from the observations of Magellanic irregulars that are shown in Figure 9-5. Therefore, we may retain many of the attractive features of the one-zone model while clearing up the G-dwarf problem, if we can explain why $Z_i \simeq 0.25 Z_\odot$ in the solar neighborhood, and $Z_i \simeq 0$ in Magellanic galaxies.

2 The Disk-Spheroid Model

One difference between our Galaxy and the Magellanic Clouds is that our Galaxy has a prominent spheroid and the Magellanic Clouds do not (see §2.7, and §5-1 of MB). The spheroid of our Galaxy, like a globular cluster or an elliptical galaxy, seems to have been formed in a single great burst of star formation about 10 Gyr ago. Consequently, about 10 Gyr must have elapsed since large numbers of massive spheroid stars formed Type II supernovae and threw off great quantities of freshly synthesized heavy elements. There is evidence that the galactic disk is less than 10 Gyr old [see Demarque & McClure (1977) and §9.2.4(b)]. If this is indeed the case, the gas from which the disk later formed is likely to have been polluted with heavy elements before any disk stars could form. This suggestion has been explored in some detail by Ostriker and Thuan (1975).

Ostriker and Thuan pointed out that the mass of heavy elements that is produced by an early-type system is related in a simple way to the present luminosity L_b of the system and the slope of the IMF x defined by equation (9-2). Most of the present luminosity of an early-type system comes from stars whose original masses lay in a narrow range around the characteristic mass $M_{\rm GB}$ of stars on the present giant branch, while the heavy elements generated in these systems were all contributed by stars more massive than some mass $M_Z \simeq 9\,{\rm M_\odot} \gg M_{\rm GB}$. Suppose these massive stars return a fraction α of their original mass to the interstellar medium in the form of heavy elements. Then the mass of heavy elements produced is

$$M_h = \int_{M_Z}^{\infty} \alpha M \frac{dN}{dM}\, dM. \qquad (9\text{-}22)$$

If we take the IMF, dN/dM, to be of the form (9-2) for $M < M_{\rm max}$ and zero for $M > M_{\rm max}$, we have

$$
\begin{aligned}
M_h &= \alpha K\,{\rm M_\odot}^2 \int_{M_Z/{\rm M_\odot}}^{M_{\rm max}/{\rm M_\odot}} m^{-x}\, dm \\
&= \frac{\alpha K\,{\rm M_\odot}^2}{x-1}\left[\left(\frac{{\rm M_\odot}}{M_Z}\right)^{x-1} - \left(\frac{{\rm M_\odot}}{M_{\rm max}}\right)^{x-1}\right].
\end{aligned}
\qquad (9\text{-}23)
$$

If $x > 1$ and $M_{\rm max} \gg M_Z$, we may neglect the second term in the square bracket of equation (9-23). Then, dividing equation (9-23) by equation (9-5), we obtain the mass of heavy elements produced per unit present luminosity as

$$\frac{M_h}{L_b} = \frac{\alpha\,{\rm M_\odot}}{x-1}\left[\frac{25\,{\rm Gyr}}{E_{\rm GB}(M_{\rm GB})}\right]\left(\frac{M_Z}{{\rm M_\odot}}\right)^{1-x}\left(\frac{M_{\rm GB}}{{\rm M_\odot}}\right)^{x-2.5}. \qquad (9\text{-}24)$$

If we adopt $\alpha = 0.1$, $x = 1.4$, $M_{\rm GB} = 0.85\,{\rm M_\odot}$, $E_{\rm GB} = 2.9 \times 10^{10}\,{\rm L_\odot}$ yr, and $M_Z = 9\,{\rm M_\odot}$, we obtain $(M_h/L_b) = 0.107\,{\rm M_\odot}\,{\rm L_\odot}^{-1}$.

Now consider what will be the metallicity of the proto-disk material if this mass of metals is mixed in with the gas from which the disk is to be formed. If the present disk and spheroid luminosities are in the ratio (D/B) and the mass-to-light ratio of the disk is Υ_d, the disk mass is

$$M_d = \Upsilon_d L_b \left(\frac{D}{B}\right), \qquad (9\text{-}25)$$

from which we obtain the initial metallicity of the disk as

$$Z_i = \frac{M_h}{M_d} \approx \frac{0.11}{\Upsilon_d(D/B)}. \tag{9-26}$$

From Table 1-2, the mass-to-light ratio of the disk at the solar radius is $\Upsilon_d \simeq 5$, and $(D/B) \approx 6$. Thus equation (9-26) gives

$$Z_i = 3.6 \times 10^{-3} = 0.18 Z_\odot, \tag{9-27}$$

in good agreement with the value needed to resolve the G-dwarf problem.

This simple model of pre-enrichment of galactic disks by spheroids has two other attractive features: (i) Since spheroids are more centrally concentrated than disks, Z_i should be higher near the center of a spiral galaxy than at very large radii. Hence we expect the present value of Z to increase toward the centers of disks even more rapidly than equation (9-17) would suggest. As Figure 9-5 shows, this is exactly what is observed. (ii) Most of the luminosity of a rich cluster of galaxies such as that in Coma comes from elliptical galaxies and the spheroids of S0 galaxies. One possible explanation of why these galaxies did not form disks is that encounters between neighboring galaxies tore the proto-disk material out of the galaxies before the gas could fragment into stars (Binney & Silk 1978). Our pre-enrichment picture suggests that this material should be metal-rich (De Young 1978; Binney 1980). X-ray observations show that this material has $Z \simeq 0.3 Z_\odot$, as expected (e.g., Sarazin 1986).

3 The Accretion Model

We have just seen that one way of resolving the G-dwarf problem is to hypothesize that the solar neighborhood had finite metallicity Z_i right from the start of its chemical evolution. However, other resolutions are possible. In particular, a satisfactory model of the chemical evolution of the solar neighborhood can be obtained if we set $Z_i = 0$, but drop the assumption that no material should either enter or leave the solar neighborhood.

Thus suppose that initially the total surface density of gas and dust is very small, and imagine that during some initial period the chemical evolution proceeds according to the standard one-zone model. At the end of an appropriate period, the gas will have solar metallicity, and roughly half of the disk stars will have $Z < 0.25 Z_\odot$. Now feed metal-free gas into the disk at precisely the same rate that the disk turns gas into stars. Imagine a small additional mass δM of primordial gas joining the solar neighborhood. In a steady state, an exactly equal mass is locked up in stellar remnants, and a mass $p\delta M$ of freshly manufactured heavy elements is returned to the interstellar medium. Thus the overall effect on the interstellar gas is to remove mass δM at metallicity Z and to return the same mass at metallicity p. Consequently, if we continue to feed the disk gas at exactly the rate at which the disk is locking material up in stars, the metallicity of the interstellar medium will eventually settle to the value $Z = p$. After a sufficiently long time, most of the stars in the disk will have metallicity $Z \simeq p$, and the fraction of low-metallicity stars will be negligible. This scenario avoids the G-dwarf problem, and will be in agreement with the observed metallicity of interstellar gas provided $p \simeq Z_\odot$.

This qualitative picture can be formulated mathematically. If the total mass of material M_t in some annulus of a galaxy varies, equations (9-14) and (9-15) remain valid. However, we now have

$$0 \neq \delta M_t = \delta M_s + \delta M_g. \tag{9-28}$$

If we eliminate δM_h and δM_s between equations (9-14) and (9-15), we obtain

$$\delta Z = \frac{1}{M_g}[(p - Z)\delta M_t - p\delta M_g]. \tag{9-29}$$

When we divide this equation by δM_t, we obtain a differential equation for the evolution of Z:

$$\frac{dZ}{dM_t} = \frac{1}{M_g}\left[p - Z - p\frac{dM_g}{dM_t}\right]. \tag{9-30}$$

A change of variable enables us to write this equation in a particularly simple form:

$$\frac{dZ}{du} + Z = p\left(1 - \frac{d\ln M_g}{du}\right) \quad \text{where} \quad u \equiv \int \frac{dM_t}{M_g}. \tag{9-31}$$

The general solution is

$$Z = p\left(1 - Ce^{-u} - e^{-u}\int_0^u e^{u'}\frac{d\ln M_g}{du'}du'\right). \tag{9-32}$$

The simplest solution describes the case in which the gas mass M_g is constant and the initial metallicity of the gas is zero. Then

$$Z = p\left[1 - \exp\left(1 - \frac{M_t}{M_g}\right)\right] \quad (M_g = \text{constant}). \tag{9-33}$$

Thus in this model, $Z \simeq p$ once $M_t \gg M_g$. Equation (9-33) predicts that the mass contained in stars that are more metal-poor than Z_1 is

$$M_s(< Z) = M_t(Z) - M_g$$
$$= -M_g \ln\left(1 - \frac{Z}{p}\right) \qquad (M_g = \text{constant}). \tag{9-34}$$

The yield p can be determined from the current gas fraction $M_g/M_t(t_0)$ and the current metallicity $Z(t_0) \simeq Z_\odot$ of the interstellar gas through equation (9-33). For example, in the solar neighborhood, $M_t(t_0) \simeq 10M_g$ and $Z(t_0) \simeq Z_\odot$, so by equation (9-33) we have $p \simeq Z_\odot$. Equation (9-34) then yields

$$M_s(< 0.25Z_\odot) \simeq 0.3M_g \simeq 0.03M_t \qquad (M_g = \text{constant}). \tag{9-35}$$

Thus this simple model predicts that about 3% of all dwarf stars should be more metal poor than $0.25Z_\odot$, which is only slightly in excess of the observed fraction of such stars.

Various modifications of this simple accretion model are possible. Suppose, for example, that only a fraction $1 - q$ of the infalling gas is locked up in star formation, so that $\delta M_g = q\delta M_t$. Then one can show (see Problem 9-1) that the number of low-metallicity stars is reduced below the value given in equation (9-35).

At this point, the natural question to ask is whether there is any evidence that disk galaxies like our own are still accreting gas. There appears to be gas in intergalactic space, and X-ray observations reveal that some galaxies are accreting large quantities of gas (MB §9-2 and Sarazin 1986). However, the only galaxies that are *known* to be accreting gas turn out to be accreting metal-rich gas rather than primordial material. Furthermore, since these accreting galaxies are all elliptical galaxies, they cannot be forming many massive stars at the present epoch. Therefore the accretion model of disk galaxies still rests on circumstantial evidence that spiral galaxies may continue to accrete metal-poor material.

An intriguing line of evidence for accretion involves estimates of the rate at which galaxies turn interstellar gas into stars. In §9-3 of MB it was concluded that there are about $10^{10} M_\odot$ of gas in our Galaxy, while it has been estimated that each year more than $1 M_\odot$ of this gas is turned into stars. Thus, if our Galaxy is not accreting gas, it may run out of gas from which to make stars before the age of the Universe has doubled. Since spiral galaxies like our own are extremely common, it seems unlikely that this breed is teetering on the verge of extinction.

Obviously, spirals would be in no danger of becoming extinct if they were constantly accreting gas. Unfortunately, this argument hangs delicately on very uncertain estimates of the rate at which gas is formed into low mass (and therefore long-lived) stars.[7] These estimates suggest, but do not prove, that spiral galaxies such as our own can sustain their present star-formation rate only because they are accreting on the order of one solar mass per year. Indeed, S0 galaxies may be just those disk galaxies that are not accreting fresh gas and have exhausted their own stocks of material from which to make new stars (Binney & Silk 1978; Larson & Tinsley 1978).

4 Relationship between Metallicity and Age

So far in this chapter we have concentrated on relations between the gas fractions, interstellar metallicities, and stellar metallicity distributions of galaxies. These quantities are of primary interest because they are relatively easy to determine observationally. Something that is harder to determine experimentally, but is at least as interesting, is the relationship between metallicity and *time*. Time can be related to metallicity in several ways: (i) an upper limit on the age of the material of the solar system can be determined by the techniques of **nucleochronology** (Fowler 1978; Tinsley 1980a); (ii) the ages of clusters of known metallicity may be determined by fitting theoretical stellar evolution models to their observed HR diagrams; (iii) the ages of certain field stars may be inferred from their magnitudes and spectra; (iv) the random velocities of field stars are correlated with their ages and metallicities. We first discuss the age of the solar system.

(a) Age of the solar material   The abundances $X_a(t)$ and $X_b(t)$ of two unstable atomic nuclei with decay rates λ_a and λ_b, satisfy

$$\frac{X_a(t_0)}{X_b(t_0)} = \frac{X_a(t_1)}{X_b(t_1)} \exp[-(\lambda_a - \lambda_b)(t_0 - t_1)]. \qquad (9\text{-}36)$$

Fairly reliable calculations of the relative abundances with which certain nuclei are created are possible. These nuclei, for example ^{238}U and ^{232}Th, are believed to form in the envelopes of exploding stars by the

---

[7] In effect, this rate is estimated by first determining the rate of formation of massive, luminous, short-lived stars, and then using other star counts to estimate the IMF that relates the relative numbers of high- and low-mass stars that are formed at any one time. [See Tinsley (1980a), Larson et al. (1980), or §4-2 of MB.]

rapid capture of some of the neutrons that stream out of the collapsing stellar core (Schramm 1974; Tinsley 1980a). The physics of this process is relatively straightforward, and most of the necessary cross-sections and decay times can be measured in the laboratory. Hence, if we assume that the metal enrichment of the solar material occurred suddenly at some time t_1, we may obtain $(t_0 - t_1)$ by substituting the original and present abundances $X_i(t_1)$ and $X_i(t_0)$ into equation (9-36). The value of $(t_0 - t_1)$ that we obtain in this way cannot be smaller than the age of the solar system, and is probably a fair estimate of the mean age of the heavy elements that are now tied up in solar-neighborhood material.

Several different pairs of nuclei yield $(t_0 - t_1) \simeq 4.5\,\text{Gyr}$. Since the metallicity of newly formed stars in the solar neighborhood is still not much greater than that of the Sun, this dating of the solar material implies that the metallicity of the solar neighborhood has been fairly static for the past 5 Gyr, just as the naive accretion model of disk evolution predicts.

(b) Ages of star clusters   The use of stellar evolution models to determine the ages of star clusters is discussed in §3-9 of MB. The key result to emerge from these studies is that while the globular clusters are all older than 8 Gyr (see §3-10 of MB), the oldest galactic cluster, NGC 188, is probably no more than 6 Gyr old (Demarque & McClure 1977). The comparative youth of NGC 188 may imply that the disk as a whole is younger than the spheroid of our Galaxy, but it is equally likely that any clusters that formed before NGC 188 have already been disrupted by the processes we discussed in §7.2.1(a). In any event, since NGC 188 has an essentially solar abundance of heavy elements, its age confirms that the metallicity of the solar neighborhood has not evolved rapidly during the last 5 Gyr.

(c) Ages of field stars   Rough estimates of the ages of field stars can be obtained from photometry (see §3-9 of MB). The best results are obtained for F stars since these stars evolve detectably, but do not die, during the age of the Universe. Some of these stars appear to be considerably older than NGC 188, and may be as old as the globular clusters.

Figure 9-7 shows a recent determination of the age-metallicity relationship for field F stars. This may be compared with three earlier determinations that are shown in Figure 3-23 of MB. These different

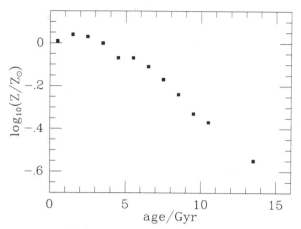

Figure 9-7. Iron abundances of field F stars versus
stellar age. (From data published in Twarog 1980.)

determinations of the age-metallicity relation for disk stars are very uncertain and not in particularly good agreement with one another. However, they all agree that the metallicity of disk stars is several times
larger than when the disk formed, and is still increasing. In terms of the
accretion model, this increase implies that for every $1\,M_\odot$ of gas that
the disk accretes, it locks up 1.5 to $2\,M_\odot$ in stars. If the disk continues
to consume its capital at this rate, it will be bankrupt in about $7\,\mathrm{Gyr}$
(Tinsley 1981).

(d) Metallicity and random velocities   Chapter 7 of MB reviews
the evidence that the random velocities of disk stars increase with age,
and in §7.5 we have discussed how this may have come about. Clearly if
both metallicity and velocity dispersion are correlated with age, metallicity should be correlated with velocity dispersion. Figure 9-8 shows
that this is indeed the case. Vader and de Jong (1981) and Lacey and
Fall (1983) have constructed detailed models of the chemical and kinematic evolution of the solar neighborhood. They find that a satisfactory
model of the solar neighborhood is obtained if one assumes that the disk
accretes metal-free gas at a rate $\dot{M} \propto e^{-t/\tau}$, where $\tau \approx 5\,\mathrm{Gyr}$, and that
the star-formation rate in the disk is proportional to the surface density
of gas in the disk. However, these models do not give satisfactory fits
to either the measured ages of the most metal-poor field stars, or the
radial abundance gradient in the galactic disk.

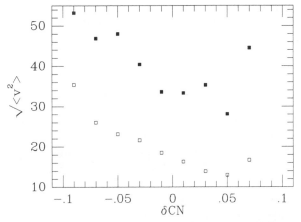

Figure 9-8. Radial (filled squares) and vertical (open squares) velocity dispersions of groups of K giants near the Sun, versus the cyanide line-strength δCN of each group. The metal-poor (low-δCN) stars have large random velocties. (From data published in Janes 1975.)

5 Summary

In the standard one-zone model, each annulus of a galactic disk evolves as a closed system that starts from zero metallicity. This model accounts successfully for (i) the metal-poverty of Magellanic galaxies by comparison with early-type spirals, and (ii) metallicity gradients within disks. However, it predicts more metal-poor stars near the Sun than are observed (G-dwarf problem). The G-dwarf problem may be eliminated in at least two (not necessarily exclusive) ways: (i) early pollution of the protodisk material by spheroid stars (disk-spheroid model), or (ii) steady accretion by the disk of metal-poor material (accretion model). The disk-spheroid model has the attractions that it not only clears up the G-dwarf problem, but also improves the fit to observed metallicity gradients and accounts for the metallicity of hot gas in clusters of galaxies. The accretion picture, on the other hand, is supported by two lines of evidence that the metallicity of the galactic interstellar medium has been roughly constant during the last several Gyr, and by tentative evidence that disk galaxies can be spirals only so long as they have access to fresh supplies of gas. However, a third estimate of the temporal evolution of the interstellar metallicity—that provided by field F stars—shows that the metallicity is still increasing, and hence favors the disk-spheroid model, or perhaps a modified accretion model in which gas is consumed faster than it is accreted. Thus neither the accretion model nor the disk-spheroid model is clearly superior to the other, and the truth almost certainly lies between these two extreme cases (e.g., Tinsley 1981).

9.3 Early Evolution of Spheroidal Components

In this section we seek to extend the analysis of the last section to spheroidal systems such as elliptical galaxies and the spheroids of disk galaxies. In §9.1 we argued that most of the stars of any spheroidal system probably formed in a single violent burst of star formation. Presumably, most of the system's chemical evolution occurred in this burst, so we shall concentrate on this epoch. In §4.7.3 we showed that systems resembling spheroidal components can be formed by allowing large numbers of point masses to fall toward each other from a cold initial configuration. In this section we assume that this picture accurately describes the *dynamics* of spheroid formation, and we focus on the chemical evolution of the protogalactic gas in which and from which the spheroid stars formed. We shall first investigate whether the dynamics of the stars and gas are similar, that is, whether the gas participates in the collapse alongside the stars.

1 Inflows and Outflows

The motion of elements of gas differs from that of compact bodies such as stars, in that elements of gas are subject to a pressure force $-\nabla p$ in addition to the gravitational force $-\rho\nabla\Phi$. Thus stars and gas fall together only if

$$|\nabla p| \ll \rho|\nabla\Phi|. \tag{9-37a}$$

In the protogalactic cloud, p, ρ, and Φ all change by amounts comparable to their central values p_0, ρ_0, and Φ_0, on the same length scale, which we may take to be the median radius r_h of the cloud. Hence the condition (9-37a) is

$$\frac{k_B T_0}{\mu m_p} = \frac{p_0}{\rho_0} \ll |\Phi_0|, \tag{9-37b}$$

where k_B is Boltzmann's constant, T_0 is the cloud's central temperature, m_p is the proton mass, and μ is the mean molecular weight of the gas. Finally, if the cloud has mass M, $|\Phi_0| \approx GM/r_h$, so the condition (9-37a) becomes

$$T_0 \ll \frac{GM\mu m_p}{k_B r_h}. \tag{9-38}$$

We define the cloud's **virial temperature** to be the mean temperature at which the cloud would satisfy the virial theorem. Recalling that in an ideal gas the mean-square velocity is [eq. (1E-24)]

$$\langle v^2 \rangle = \frac{3k_B T}{\mu m_p}, \tag{9-39}$$

and using equation (4-80b), we have

$$T_{\text{vir}} \simeq 0.13 \frac{GM\mu m_p}{k_B r_h} \simeq 6.8 \times 10^5 \mu \left(\frac{M}{10^{11}\,M_\odot}\right)\left(\frac{10\,\text{kpc}}{r_h}\right) \text{K}. \tag{9-40}$$

Thus the condition (9-39) for stars and gas to fall together is $T_0 \ll T_{\text{vir}}$.

What are plausible values of the radius r_h of a protogalactic gas cloud? If galaxies form by collapse, their observed effective radii R_e set lower limits on the characteristic radii r_h of the corresponding protogalactic clouds as the latter started to form large numbers of stars. Let us explore the possibility that $r_h \approx R_e$. In this case, we have from equation (9-40) that $T_{\text{vir}} \approx 4 \times 10^5$ K for $\mu = 0.61$, $R_e = 10\,\text{kpc}$, and $M = 10^{11}\,M_\odot$.[8] We must next decide whether the protogalactic gas could have been maintained at this sort of temperature for at least a dynamical time.

The actual temperature in a protogalactic cloud will have been governed by competition between heating and cooling processes:

(a) Cooling   Since a protogalactic cloud at the virial temperature will have been fully ionized, the only important cooling mechanism will have been the emission of continuum photons by the bremsstrahlung process (see, e.g., Jackson 1975 or Spitzer 1978). The number of times that a continuum photon would have been scattered as it moved from the center of the cloud to the edge, is

$$N_{\text{scat}} \approx r_h n_e \sigma_T \approx \frac{3M\sigma_T}{4\pi r_h^2 m_p} \simeq 0.02 \left(\frac{M}{10^{11}\,M_\odot}\right)\left(\frac{10\,\text{kpc}}{r_h}\right)^2, \tag{9-41}$$

where $\sigma_T = 6.65 \times 10^{-25}$ cm^2 is the Thomson cross-section for electron scattering and $n_e \simeq n_p$ is the electron density. Hence most photons

---

[8] Distributing $10^{11}\,M_\odot$ uniformly through a sphere of radius $10\,\text{kpc}$ corresponds to a mean density of electrons (bound and free) $n_e \simeq 1\,\text{cm}^{-3}$. At densities of this order and $T \gtrsim 10^5$ K, hydrogen and helium are fully ionized. The mass per particle is then $\mu m_p = \dfrac{n_H + 4n_{He}}{2n_H + 3n_{He}} m_p$ since each hydrogen atom contributes one unit of mass and two particles, and each helium atom contributes four units of mass and three particles to the plasma. Cosmic helium synthesis yields $n_{He} \simeq 0.1 n_H$, so $\mu = 0.61$.

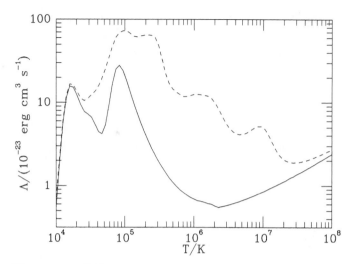

Figure 9-9. Full curve: the cooling function Λ of a hydrogen-helium mixture in collisional equlibrium with 10% helium by number. Dashed curve: the same quantity for gas with solar abundances. (From data kindly supplied by R. J. Edgar.)

generated by the bremsstrahlung process would have escaped from the cloud without scattering. In other words, the cloud would have been **optically thin** to continuum radiation.

An optically thin hydrogen-helium plasma radiates energy per electron E at a rate that is proportional to the electron density n_e:

$$|\dot{E}_{\text{cool}}| = \Lambda(T)n_e. \qquad (9\text{-}42)$$

The full curve in Figure 9-9 shows the variation of Λ with T; we see that for temperatures in the range $5 \times 10^5\,\text{K} \lesssim T \lesssim 2 \times 10^7\,\text{K}$,

$$\Lambda \simeq 10^{-23}\,\text{erg s}^{-1}\,\text{cm}^3. \qquad (9\text{-}43)$$

The energy per electron is $E = \frac{3}{2}k_B T \times 1.92$, where the factor 1.92 is the number of particles per electron when $n_{\text{He}} = 0.1n_{\text{H}}$. Hence, if there were no heat sources, a protogalactic cloud at the virial temperature would have cooled on a time scale

$$t_{\text{cool}} \equiv \frac{E}{|\dot{E}_{\text{cool}}|} \simeq 6.3 \times 10^5 \left(\frac{r_h}{10\,\text{kpc}}\right)^2 \text{yr}. \qquad (9\text{-}44)$$

For comparison, the dynamical time is

$$t_{\text{dyn}} = \pi\sqrt{\frac{r_h^3}{2GM}} = 1.0 \times 10^8 \left(\frac{10^{11}\,\text{M}_\odot}{M}\right)^{\frac{1}{2}} \left(\frac{r_h}{10\,\text{kpc}}\right)^{\frac{3}{2}} \text{yr}, \qquad (9\text{-}45)$$

where we have used equation (2-30) with the density equal to the mean density inside r_h. Thus the dynamical time is more than a factor of 100 longer than the cooling time, so it is clear that pressure could not have impeded the collapse of the cloud, unless there was some additional heat source.

(b) Heating by supernovae A few million years after the onset of widespread star formation, it is likely that a major new heating source came on line: as soon as large numbers of massive stars had had time to burn themselves out, the remaining protogalactic gas would have been violently shaken by a great burst of supernovae (Larson 1974). Could the remaining protogalactic gas have been heated to the virial temperature, and perhaps even expelled from the system, by this burst of supernovae?

When a massive star explodes as a supernova, it ejects of order $5\,M_\odot$ of material at speeds $\simeq 5000\,\mathrm{km\,s^{-1}}$. When the ejected gas first plows into the interstellar medium, it shock-heats it to a temperature of many tens of millions of degrees. In this way most of the kinetic energy of the ejected stellar envelope is rapidly converted to thermal energy in swept-up interstellar gas. The sphere of hot gas that is formed in this way, which is called a **supernova remnant**, continues to expand for tens of thousands of years. As the remnant expands, it radiates some of its energy, and the remainder is continually being shared with freshly entrained interstellar gas. Consequently the temperature, pressure, and kinetic energy of the remnant continually fall. In an infinite interstellar medium, the remnant would eventually merge imperceptibly with the undisturbed medium. However, if the supernova rate in the galaxy is high, the temperature of a swept-up element of interstellar gas may not have returned to its original value before it is entrained in another supernova remnant. In fact, successive supernovae may raise the temperature of the entire interstellar medium until it reaches the virial temperature [eq. (9-40)], at which point gas begins to flow out of the system. We now obtain a naive estimate of the minimum supernova rate that could have heated the gas to the virial temperature (see Chevalier 1977 for more detail).

Since the supernova rate is proportional to the number of young stars, let there be $\alpha\rho_*$ supernovae per unit volume per unit time, where ρ_* is the density of stars. Thus, if each supernova pumped energy E_{SN} into the gas, the rate of energy input per unit mass of gas was

$$\frac{\alpha E_{\mathrm{SN}}\rho_*}{\rho_g}, \qquad (9\text{-}46)$$

where ρ_g is the density of the gas. From equation (9-42) we have that the interstellar gas radiated energy at the rate $\simeq \Lambda n_e/m_p$ per unit mass, and when we equate this to the rate at which the supernovae supplied energy, we obtain a lower limit $\alpha_{\min}$ on the supernova rate that could have heated the gas to $T_{\rm vir}$:

$$\alpha_{\min} = \frac{\Lambda(T_{\rm vir})n_e}{m_p E_{\rm SN}} \left(\frac{\rho_g}{\rho_*} \right). \tag{9-47a}$$

If we adopt $E_{\rm SN} = 10^{51}\,{\rm erg}$ (Spitzer 1978; Blair et al. 1981) and use equation (9-43), this becomes

$$\alpha_{\min} = 38 \left(\frac{n_e}{1\,{\rm cm}^{-3}} \right) \left(\frac{\rho_g}{\rho_*} \right) (10^{11}\,{\rm M_\odot})^{-1}\,{\rm yr}^{-1}. \tag{9-47b}$$

Inserting $n_e = 1\,{\rm cm}^{-3}$ into equation (9-47b), and assuming that soon after star formation got under way $\rho_g \simeq \rho_*$, we conclude that about forty supernovae per year, or 4×10^9 supernovae during the violent collapse phase of a giant protogalaxy, were required to heat a large proportion of the protogalactic gas to the temperature at which pressure forces would have been important.

The number of supernovae expected in a protogalaxy of given mass depends on the form of the IMF according to which stars were formed in that system. However, the IMF of the solar neighborhood gives rise to about one supernova per $200\,{\rm M_\odot}$ of newly formed stars. Furthermore, about this number of supernovae are required to produce the metals seen in and around giant elliptical galaxies.[9] If one supernova explodes per $200\,{\rm M_\odot}$ of new stars, then 5×10^8 supernovae are expected during the collapse, which by equation (9-47b) is insufficient to heat and expel a large proportion of the original gas. Thus, initially stars and gas will have fallen together undisturbed by supernova explosions. Only later, when most of the gas had been turned into stars, will the product $n_e\rho_g/\rho_*$ have fallen to a value $\lesssim 0.1$ at which supernovae were able to heat the gas effectively.

2 Chemical Enrichment of Spheroidal Components

We have seen that during the first stages of the breakup of a protogalactic cloud into stars, before supernova heating became important, stars and gas fell together. Thus we may apply to each region of the cloud the one-zone model of chemical enrichment that we developed in §9.2.

---

[9] Let the abundance by weight of iron in the Sun be $\beta \simeq 0.0016$, and let each Type II supernova (probably the only sources of iron) eject $\gamma\,{\rm M_\odot} \approx 0.5\,{\rm M_\odot}$ (Arnett 1978) of iron. Then $\beta/\gamma \approx 1/300$ supernovae are required to endow each $\rm M_\odot$ with the solar abundance of iron.

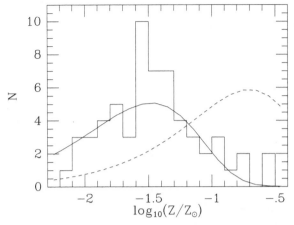

Figure 9-10. The histogram shows the metallicity distribution of sixty globular clusters that are farther than 4 kpc from the galactic plane. The dashed curve shows the fit to this distribution provided by equation (9-48) when $p = 0.004$. The full curve shows the fit provided by equation (9-52) with $p = 0.004$ and $c = 5$. (After Hartwick 1976.)

(a) Standard one-zone model   Any stars that formed in the freely falling gas cloud would have fallen toward the system's center at the same rate as neighboring elements of gas, so each region of the protogalactic cloud would have evolved for a time $t \lesssim t_{\mathrm{dyn}}$ according to the one-zone model of chemical enrichment. Since the very massive $(M \gtrsim 9\,M_\odot)$ stars that are thought to be responsible for the production of heavy elements have lifetimes $t_* \lesssim 10^7\,\mathrm{yr} \ll t_{\mathrm{dyn}}$, the chemical evolution of the protogalaxy was able to proceed through several generations of such stars as the protogalaxy collapsed. Therefore, we may use equation (9-18) to predict the fraction $F(< Z)$ of all stars that formed during this period that have metallicity less than Z:

$$F(< Z) \equiv \frac{M_s(< Z)}{M_s(< Z_1)} = \frac{1 - e^{-Z/p}}{1 - e^{-Z_1/p}}, \qquad (9\text{-}48)$$

where Z_1 is the metallicity of the interstellar gas at the end of the collapse stage. If we assume that *no* stars formed in the galactic spheroid after the initial collapse was complete, we may compare the fractional distribution given by equation (9-48) with that observed for some spheroidal population. Figure 9-10 shows the metallicity distribution of sixty globular clusters. For this sample the maximum metallicity Z_1 is small $(Z_1 \simeq 0.003)$. The dashed curve in Figure 9-10 shows that if the yield

p is of the same order ($p \simeq 0.004$) as that favored by our discussion of the chemical evolution of spiral galaxies, equation (9-48) yields fewer metal-poor clusters than are observed. The observed and predicted distributions agree only if $p \simeq 6 \times 10^{-4}$.

Since physical conditions in the collapsing spheroid and in the present-day galactic disk are likely to have been dissimilar, there is no reason why the yield of stars formed during the collapse of the spheroid should equal that characteristic of star formation in the disk. However, if the yield were as small as 6×10^{-4}, the spheroid could not have provided the heavy elements required in the disk-spheroid model to raise the metallicity of the proto-disk material to $Z \simeq 0.25 Z_\odot$. Therefore it is interesting to see whether our picture of chemical evolution in the collapsing halo can be modified to reproduce the observed stellar metallicity distribution for $p \simeq 0.004$.

(b) Two-phase model   We now show that the difficulty with the standard one-zone model can be eliminated if we suppose that supernovae broke the gas in the collapsing spheroidal system up into a two-phase medium.[10] The cold phase would have consisted of clouds of dense cold gas that fell freely with the stellar component of the spheroid, while the hotter, more rarefied phase expanded, carrying with it many of the heavy elements generated by the supernovae that heated the expanding phase.

A simple mathematical model of this picture will show that it is able to reproduce quantitatively the metallicity distribution of Figure 9-10 (Hartwick 1976).[11] Thus suppose that supernovae drove gas out of the collapsing spheroid at a rate that was proportional to the star-formation rate:

$$\frac{dM_t}{dt} = -c\frac{dM_s}{dt}, \tag{9-49}$$

---

[10] Consideration of Figure 9-9 shows why heated interstellar gas is liable to break up into a two-phase medium; for $9 \times 10^4 \, \mathrm{K} < T < 3 \times 10^6 \, \mathrm{K}$ the cooling function Λ is a *declining* function of T. Consequently equilibrium between heating and cooling processes is unlikely to be established for T in this range, since at such temperatures any small excess of heating over cooling would cause the temperature to rise and the cooling rate to fall, thus exacerbating the original small disequilibrium. Moreover, in the region $10^4 \, \mathrm{K} \lesssim T \lesssim 10^5 \, \mathrm{K}$, cooling is so effective that in the absence of strong heating, gas in this region generally cools to $T \lesssim 10^4 \, \mathrm{K}$. Thus, under equilibrium conditions we expect either $T \lesssim \times 10^4 \, \mathrm{K}$ or $T \gtrsim 3 \times 10^6 \, \mathrm{K}$.

[11] See Arimoto and Yoshii (1987) for a detailed account of wind models of elliptical galaxies.

where c is a constant. From this it follows that $M_t(t) = M_t(0) - cM_s(t)$, and thus by conservation of mass we have for the gas mass at any time,

$$
\begin{aligned}
M_g(t) &= M_t(t) - M_s(t) \\
&= M_t(0) - (1+c)M_s(t).
\end{aligned}
\tag{9-50}
$$

On the other hand, an analysis similar to that which leads to equation (9-16) shows that if the gas expelled had the same metallicity as the rest of the interstellar medium,

$$
\frac{dZ}{dM_s} = \frac{p}{M_g} = \frac{p}{M_t(0) - (1+c)M_s}.
\tag{9-51}
$$

Integrating this differential equation subject to the initial condition $M_s(Z = 0) = 0$ yields

$$
M_s(< Z) = \frac{M_t(0)}{1+c}\left\{1 - \exp\left[-\frac{(1+c)Z}{p}\right]\right\}.
\tag{9-52}
$$

Comparing this expression with equation (9-18), we see that the only effect of the steady outflow of interstellar gas is to reduce the effective yield to $p/(1+c)$. If we set $c = 5$, we obtain the excellent fit to the observations that is shown by the full curve in Figure 9-10.

The ability of this simple two-phase model to account for the data of Figure 9-10 is gratifying but not in itself very persuasive, as this agreement has been obtained at the expense of adding an extra free parameter to the standard one-zone model. Fortunately, the two-phase model is able to account for two other observational facts without further tinkering.

(c) Metallicity-velocity dispersion relationship   Consider the mean stellar metallicity that is predicted by the two-phase model. In this model star formation ceased when the gas was exhausted. By equation (9-50) this occurred when $M_s = M_t(0)/(1+c)$. Thus the present mean stellar metallicity should be

$$
\overline{Z}_s = \frac{1+c}{M_t(0)} \int_0^{M_t(0)/(1+c)} Z\, dM_s.
\tag{9-53}
$$

When we use equation (9-52) to evaluate the integral in this equation, we obtain

$$
\overline{Z}_s = \frac{p}{1+c}.
\tag{9-54}
$$

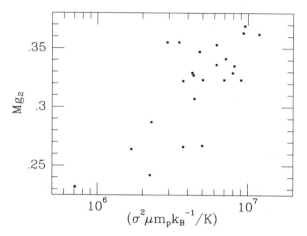

Figure 9-11. A plot of the metallicity index Mg$_2$
versus the square of the central velocity dispersion
σ for twenty-four elliptical galaxies. (From data
published in Terlevich et al. 1981.)

If the system turns all its gas into stars, $c = 0$ and $Z_s = p$, but if
supernovae are able to drive most of the original gas out of the system,
$c \gg 1$ and $\overline{Z}_s \simeq p/c$. Thus the effective yield is expected to decline
in step with the ease with which the hot, metal-rich component of the
interstellar medium could flow out of the galaxy. From the structure
of Figure 9-9 it follows that the temperature of the hot component of
the interstellar medium will have exceeded $\approx 3 \times 10^6$ K. Thus we expect
galaxies with virial temperatures below a few million degrees to have lost
many of the heavy elements synthesized in their first generations of stars
and be less metal-rich than galaxies with higher virial temperatures.
Figure 9-11 shows that the metallicities of elliptical galaxies do indeed
show such a trend. Furthermore, since the central velocity dispersions
of galaxies are related to their luminosities L through the Faber-Jackson
relation, $L \propto \sigma^4$ [see eq. (1-16) and §7-3 of MB], it follows that $\overline{Z}_s$ must
increase with L. It is this correlation of $\overline{Z}_s$ with L that gives rise to the
color-magnitude effect discussed in §5-4 of MB (see Figures 5-39 and
5-41 of MB).

(d) Color gradients   Elliptical and S0 galaxies are redder near
their centers than farther out (see Figure 5-37 of MB). This effect is in
qualitative agreement with the theory of metal enrichment that we have
developed here: far from the center of a galaxy, the electron density
in the gas n_e will, for fixed (ρ_g/ρ_*), be smaller than near the center.

Hence by equation (9-47b), the value of (ρ_g/ρ_*) at which supernovae will initiate mass loss by heating the interstellar medium to $T \gtrsim 3 \times 10^6$ K will be greater at large radii than near the galactic center. Consequently the final mean metallicity should be largest near the center of the galaxy, as is observed to be the case.

(e) Dissipative models All that is required for the two-phase model just described to account for color gradients is that the mass-loss parameter c should decrease inward. Thus, we require that in outlying regions, star formation should cease when a substantial fraction of the local mass is still gaseous, while star formation goes more nearly to completion near the center. We have seen that this goal can be achieved by having supernovae eject material from the outside sooner than from the middle, but this is not the only way to achieve radius-dependent values of c.

Imagine that supernova heating is unable to eject any of the gas from a forming spheroid. Then the gas that remains when the collapse of the protogalaxy is complete will lose its kinetic energy of infall when it collides with oppositely directed gas from the other side of the galaxy. Bereft of its infall energy, the gas will remain at the center of the system as the stars pass by the center and return to large radii. Hence the formation of stars of low binding energy will cease before the gas from which such stars were made was exhausted, and the mean metallicity of these stars will be low. At the center, by contrast, not only will star formation go to completion, but the mean stellar metallicity will eventually be higher than in the standard one-zone model because the central region's own production of metals will have been augmented by metals manufactured far out in the protogalaxy. Larson (1976a,b) has explored this idea in depth, and obtained quantitative agreement between models incorporating this idea and the observed properties of spheroidal systems.

3 Summary

In our discussions of the chemical evolution of disks and spheroids, we have contrasted several simple models that offer alternative explanations of the observed properties of galaxies. Such contrasts serve to clarify the principles involved in theories of galaxy formation. But it would be a mistake to assume that reality can be adequately described by any of these models in isolation—galaxy formation was surely a complex process, involving aspects of *all* the simple models discussed here, and perhaps additional physical processes on which we have not touched at all.

Problems

9-1. [2] In this problem we investigate a more elaborate version of the accretion model presented in §9.2.3. We assume that initially all of the disk mass is in metal-free gas, and that a constant fraction $1 - q$ of each infalling gas parcel δM_t is locked up by star formation. Thus the corresponding change in gas mass is $\delta M_g = q\delta M_t$.

(a) Using equation (9-32), show that a parametric solution for $Z(M_g)$ is

$$Z = p(1 - q)(1 - e^{-u}) \quad ; \quad M_g = M_{g0}e^{qu}, \tag{9P-1}$$

where M_{g0} is the initial gas mass.

(b) Show that the ratio of the stellar mass at t_1 to the mass in gas at the present time t_0 is

$$\frac{M_s(u_1)}{M_g(u_0)} = \frac{1 - q}{q}(e^{qu_1} - 1)e^{-qu_0}, \tag{9P-2}$$

where u_i is the value of the parameter u at time t_i. Explain why this relation together with observations of our galactic disk imply $q \ll 1$.

(c) Now consider the case $u_0 \gg 1$, and let u_1 be an epoch at which the metallicity Z_1 was substantially lower than Z_0. Show (i) that the present metallicity $Z_0 \simeq p(1 - q)$; (ii) that $u_1 \simeq -\ln(1 - Z_1/Z_0) \ll 1$; (iii) that

$$\frac{M_s(u_1)}{M_g(u_0)} \simeq -\ln\left(1 - \frac{Z_1}{Z_0}\right)e^{-qu_0}. \tag{9P-3}$$

This formula differs from that [eq. (9-34)] obtained for the simple accretion model of §9.2.3 only by the presence of the factor e^{-qu_0}. Since this factor can take any value from 0 to 1 depending on q and u_0, it follows that in this modified accretion model the fraction of low-metallicity stars can be made arbitrarily small.

9-2. [2] From time $t = 0$ to time $t = t_i$, a galaxy accretes gas at exactly the rate required to hold the gas mass M_g constant, after which the galaxy ceases to accrete material. At $t = 0$ the system consists of only metal-free gas, and at time t_i the total mass $M_t(t_i) \gg M_g(t_i)$. Show that for any $Z_1 \ll Z(t_i)$ the mass of metal-poor stars is

$$M_s(< Z_1) \simeq M_g(t_i)\frac{Z_1}{Z_0}\left\{1 + \ln\left[\frac{M_g(t_i)}{M_g(t_0)}\right]\right\}, \tag{9P-4}$$

where Z_0 and $M_g(t_0)$ are the present values of Z and M_g. This model suggests that any model in which the mass of gas never increases with time will always predict a greater mass in metal-poor stars than the value $M_s(< Z_1) \simeq M_g(t_0)[Z_1/Z_0]$ predicted by the model in which the gas mass is strictly constant. Since the latter model already slightly overproduces metal-poor stars, we infer that in our Galaxy the mass of the interstellar gas must have increased for a period after disk stars had started to form.

10 Dark Matter

Virtually all of the information we have about the Universe has come to us through photons: optical photons from stars, radio photons from neutral hydrogen gas, X-ray photons from ionized gas, and so forth. Yet there is no reason to suppose that every type of matter in the Universe emits a train of readily detected photons. For example, we know of no reason why the star-formation process should not produce stars with masses below the lower limit for hydrogen burning on the main sequence $(0.08\,M_\odot)$, and any such stars would have to be very nearby $(\ll 1\,\mathrm{pc})$ to be detectable with our present instruments. Neutral hydrogen gas in interstellar space would be far more difficult to detect if there were no 21-cm hyperfine transition. Dust in galaxies was discovered only because dust grains happen to be small compared to optical wavelengths, with the result that stars are not only dimmed by the dust, but also reddened.

Even within a given type of astronomical object, there is no *a priori* reason why mass and luminosity ought to be well correlated. This point is illustrated by the main-sequence luminosity function in the solar neighborhood. Stars brighter than the Sun contribute 95% of the luminosity while stars fainter than the Sun contain at least 75% of the mass (from Table 4-5 of MB). Hence even modest variations in the relative numbers of low-mass and high-mass stars can produce substantial changes in the overall mass-to-light ratio Υ. Nevertheless, the mass-to-light ratios in the well-studied central parts of many different spiral and elliptical galaxies somehow contrive to be roughly the same (see §10.1.2). Presumably, some aspect of the star-formation process is very similar in all these systems. This similarity leads naturally to the conjecture that mass-to-light ratios of this order also apply to regions of smaller luminosity density, such as the outer parts of individual galaxies, or clusters of galaxies. We shall see that this extrapolation has proved to be very wrong. It turns out that on large scales, most astronomical systems have much larger mass-to-light ratios than the central parts of galaxies, in many cases by more than an order of magnitude. Moreover, the volume of these regions of high mass-to-light ratio is so great that they contain over 90% of the mass of the Universe.

What is the reason for these unexpectedly high mass-to-light ratios? Could it be that in low-density regions the star-formation process produces predominantly low-mass objects, such as faint M dwarfs or even planetary-sized objects? Could there be a vast population of black holes and neutron stars that are remnants of a generation of primordial stars? Or is most of the mass of the Universe tied up in some exotic elementary particle?

The first step toward answering these questions is to examine critically the *minimum* mass-to-light ratios that are compatible with our current understanding of stars and galaxies. We begin in §10.1.1 with the environment for which we have the fullest information, the solar neighborhood. The known proportions of main-sequence and giant stars, stellar remnants, gas, and dust yield a ratio of mass density to luminosity density Υ_{min}(local) that is the minimum possible in any model of the composition of the solar neighborhood that is consistent with observations and the basic principles of stellar and galactic evolution. The concept of a minimum mass-to-light ratio can be extended to other Population I and II systems (see §10.1.2). As our understanding improves, we may discover additional components, and these can only raise Υ_{min}. Since our understanding is by no means complete at the present time, we expect that in general the dynamically measured mass-to-light ratio of any system Υ will exceed Υ_{min}. In this case the material of unknown nature that is responsible for the discrepancy is called **dark matter**.

Thus we use the term dark matter to denote *any form of matter whose existence is inferred solely from its gravitational effects.* For example, the white dwarfs in the solar neighborhood are *not* considered to be dark matter, even though most of them have cooled to invisibility, since their existence is inferred directly from the present density of visible white dwarfs and main-sequence stars, the theory of stellar evolution, and an estimate of the history of the star-formation rate in the solar neighborhood.

The first evidence of dark matter was found by the Swiss astronomer Fritz Zwicky in 1933. By that time, Hubble's law relating distance and velocity had been used to establish an extragalactic distance scale (although the normalization of the distance scale, i.e., Hubble's constant, was incorrect). In addition, rotation curves were available for several spiral galaxies, and these had been used to estimate mass-to-light ratios for the luminous central parts of these galaxies.

Zwicky's work was based on measurements of the radial velocities of seven galaxies belonging to the Coma cluster of galaxies. He pointed out that the individual galaxies had radial velocities that differed from the mean velocity of the cluster, with an RMS dispersion of about $700\,\mathrm{km\,s^{-1}}$. He interpreted this dispersion as a measure of the kinetic energy per unit mass of the galaxies in the cluster, and by making a crude estimate of the cluster radius he was able to measure the total mass of the cluster using the virial theorem.

Zwicky's next step was to compare the cluster mass-to-light ratio measured in this way with the mass-to-light ratio as measured from the rotation curves of nearby spirals. He found that the cluster mass-to-light ratio exceeded the rotation curve mass-to-light ratio by a factor of at least 400.[1] He concluded that virtually all of the cluster mass is in the form of some invisible or dark matter that is undetectable except through its gravitational force (Zwicky 1933; see also Smith 1936 for a similar conclusion based on measurements of the Virgo cluster).

Zwicky's analysis was of course extremely crude, based on minimal statistics, an uncertain cluster radius, and a distance scale that was at least a factor of five too small (see Problem 10-1 for a discussion of the effect of the error in distance scale on his conclusions). Nevertheless, his results have held up remarkably well: using the best cluster and galaxy models, we still find (§10.2.4) that the mass-to-light ratio of the Coma cluster as a whole exceeds the mass-to-light ratio of the luminous parts of typical galaxies in the cluster by more than a factor of thirty.

More recently, Ostriker et al. (1974) and Einasto et al. (1974) have proposed that there are large amounts of dark matter around even isolated galaxies: they argue that the dark matter in spiral galaxies is located in giant "halos" extending to several times the radius of the luminous matter and containing most of the total galaxy mass. A remarkably broad range of observational data supports this picture (§10.2), which represents the most radical revision in the theory of galactic structure in the last fifty years. It is possible that the dark matter in halos is the same as the dark matter in clusters like the Coma cluster, except that in clusters the galaxies are so closely packed that the dark matter resides in a common background rather than in separate halos.

Excellent reviews of the observational evidence for dark matter are given in articles by Faber and Gallagher (1979), Peebles (1980a), and Trimble (1987), and in the proceedings of a recent symposium on dark matter (Kormendy & Knapp 1987).

---

[1] This discrepancy is often referred to as the "missing mass" problem. This terminology seems inappropriate since the problem is that there is too much mass, not too little.

10.1 Dark Matter in Individual Galaxies

1 The Solar Neighborhood

An estimate of the minimum mass-to-light ratio in the solar neighborhood can easily be assembled from the data in Table 1-1 and §4.2.1(b). The local mass density in main-sequence and giant stars, stellar remnants (both directly observed and inferred from models of galactic evolution), gas, and dust yields a lower limit to the total density $\rho_{min} \simeq 0.11 \, M_\odot \, pc^{-3}$, while the luminosity density in the V band is $j_V \simeq 0.067 \, L_\odot \, pc^{-3}$. Thus in the galactic midplane

$$\Upsilon_{min}(local, z = 0) = \frac{\rho_{min}}{j_V} \simeq 1.7 \Upsilon_\odot. \tag{10-1}$$

Surface density and surface brightness in a column perpendicular to the plane are more fundamental quantities than volume density or luminosity density, since different stellar populations have different thicknesses and the thickness of a given population changes with time (§7.5). According to Table 1-1, the surface density in known components is $50 \, M_\odot \, pc^{-2}$ and the surface brightness (neglecting interstellar absorption) is $15 \, L_\odot \, pc^{-2}$ in V. Almost all of the light comes from heights $< 700 \, pc$ above the midplane. Thus

$$\Upsilon_{min}(local, |z| < 700 \, pc) = 3.3 \Upsilon_\odot. \tag{10-2}$$

This is probably the best available estimate of the minimum mass-to-light ratio in a typical Population I system.

The measurement of the actual mass-to-light ratio in the solar neighborhood is described in §4.2.1(b). The local mass density is determined from carefully selected star samples by analyzing the velocity dispersion and density profile in the direction normal to the galactic plane. These studies yield a total density $\rho_0 = 0.18 \pm 0.03 \, M_\odot \, pc^{-3}$ (the Oort limit). This yields a mass-to-light ratio in the visible band

$$\Upsilon_V(local, z = 0) = \frac{\rho_0}{j_V} = 2.7 \Upsilon_\odot. \tag{10-3}$$

By a similar analysis, the mean mass-to-light ratio within $700 \, pc$ of the plane is found to be [see eq. (4-43)]

$$\Upsilon_V(local, |z| < 700 \, pc) = 5 \Upsilon_\odot. \tag{10-4}$$

Hence both in the plane and in a column through the plane, the observed mass-to-light ratio Υ exceeds Υ_{min} by about 50%. Consequently, roughly a third of material in the solar neighborhood must be considered to be dark matter.

2 The Central Regions of Elliptical Galaxies

The mass-to-light ratios in the central cores of elliptical galaxies can be measured by comparing the surface-brightness profiles and velocity dispersions to an isothermal sphere [see §4.4.3(b)]. The median core mass-to-light ratio of the sample of ellipticals in Lauer (1985) (see Table 4-2) is

$$\Upsilon_V = 12h\Upsilon_\odot, \tag{10-5}$$

where h is the Hubble constant in units of $100 \, \mathrm{km \, s^{-1} \, Mpc^{-1}}$.

The mass-to-light ratios in elliptical cores are not very different from the solar neighborhood once two simple corrections are taken into account. First, there are fewer young stars in the ellipticals, since star formation almost ceased 10^{10} yr ago. Removing these stars (main-sequence stars more massive than about $1 \, \mathrm{M_\odot}$) from the solar neighborhood would decrease the luminosity density by about a factor of two. Removing the interstellar gas would decrease the mass density by about 7%. These two changes would increase the mass-to-light ratio in a column through the Sun from $5\Upsilon_\odot$ to $9\Upsilon_\odot$, which is consistent with equation (10-5). This result suggests that the composition of the solar neighborhood is similar to the composition of the centers of ellipticals. Hence we may apply the same two corrections to $\Upsilon_{\min}(\mathrm{local}, |z| < 700 \, \mathrm{pc})$ to estimate the minimum V-band mass-to-light ratio for elliptical galaxy cores to be about $6\Upsilon_\odot$. Even smaller mass-to-light ratios, $\simeq 3\Upsilon_\odot$, are seen in very low-luminosity ellipticals like M32 (Nolthenius & Ford 1986).

3 Rotation Curve and Escape Speed in the Solar Neighborhood

In §2.7 we constructed models of the galactic potential based on the two major visible components, the disk and spheroid. These models assumed that the mass-to-light ratio of each component was independent of radius, that the local surface density was determined from the Oort limit, and that the disk surface-brightness profile was exponential, with parameters similar to those of other spiral galaxies. The spheroid parameters were determined by comparison with other galaxies and by matching star counts in our Galaxy. With these models it proved impossible to produce a rotation curve that—like the observed one—is flat or rising near the Sun.

Thus the shape of the rotation curve outside the solar radius R_0 strongly suggests that the Galaxy contains a great deal of dark matter. Unfortunately, due to the absence of suitable tracers it is difficult to determine the rotation curve at galactocentric distances exceeding about $2R_0$, and hence rotation curve observations cannot constrain the distribution of dark matter at larger distances.

A more powerful constraint on the extent of the dark halo is that the escape speed from the solar neighborhood must exceed the maximum velocity observed among local stars. In §2.7 we analyzed a simple spherical mass model in which the rotation curve is flat out to a maximum radius $r_\star$ [eq. (2-191)]. We showed that the presence of local stars with velocities exceeding $500\,\mathrm{km\,s^{-1}}$ implies that the Galaxy must extend to at least $r_\star = 41\,\mathrm{kpc}$ and have total mass at least $M_\star = 4.6 \times 10^{11}\,\mathrm{M_\odot}$. Thus there is at least six times as much dark mass as there is in the visible components of the Galaxy. Based on a total luminosity $L_V = 1.4 \times 10^{10}\,\mathrm{L_\odot}$ the mass-to-light ratio is $\Upsilon_V \gtrsim 30\Upsilon_\odot$, almost ten times the minimum value derived for the solar neighborhood [eq. (10-2)].

A similar result is obtained from the requirement that the globular cluster NGC 5694 remain bound to the Galaxy (Problem 10-3).

Does the dark matter at large radii form a distinct population from the dark matter that we found in the solar neighborhood? The answer depends on the scale height of the dark matter in the solar neighborhood. We showed in §4.2.1(b) that to generate a flat rotation curve with $v_c = 220\,\mathrm{km\,s^{-1}}$ with a thin disk, we require a local surface density of dark matter $\simeq 160\,\mathrm{M_\odot\,pc^{-2}}$ [eq. (4-44) minus the contribution of stars and gas given in Table 1-1]. From Table 1-1 the local density of dark matter is about $0.07\,\mathrm{M_\odot\,pc^{-3}}$. Hence, if there is a single component of dark matter with vertical distribution of the form $\rho(z) \propto \exp(-|z|/z_0)$, the scale height would have to be $z_0 \simeq 1.1\,\mathrm{kpc}$. If the scale height of the local dark matter is smaller than this, then there must be two dark components: a disk that contributes to the Oort limit but not the rotation curve, and a spheroidal component that helps produce a flat rotation curve while making a negligible contribution to the Oort limit. This hypothetical spheroidal component is usually referred to as the **dark halo**.[2] For simplicity we shall generally refer to the component of dark matter that is needed to flatten the rotation curve and to bind the high-velocity stars as the dark halo, although there is little direct evidence that the required material is spheroidal rather than disklike.

4 The Dynamics of Population II Tracers

We can investigate the force field of the Galaxy by examining the dynamics of Population II objects such as stars, globular clusters, or satellite

---

[2] The nomenclature is confusing, since the visible spherical component, which we have called the spheroid, is sometimes also called the "halo", while the dark halo is occasionally called the "corona" or the "envelope".

galaxies. Since these objects have large random velocities, we must use statistical methods to compare the predictions of a given model of the galactic potential with the distribution of observed radial velocities.

Because globular clusters and satellite galaxies are found out to distances $\gg R_0$, they can be used to constrain the force field at large distances, where the rotation curve cannot be measured directly. Let us approximate the galactic potential at large distances as that of a point mass, $\Phi(r) = -GM/r$, and assume that the velocity ellipsoid is spherical. Then it can be shown that the radial velocities v_r and galactocentric distances r of the satellite objects are related to the mass M by (Problems 10-5, 10-6)

$$\langle v_r^2 r \rangle = \frac{GM}{4}, \tag{10-6}$$

where $\langle \cdot \rangle$ denotes the average over a large steady-state ensemble of satellites. If we have N satellites with measured radial velocities v_{ri} and galactocentric distances r_i, then according to equation (10-6) the observable quantity

$$M_{\text{est}} = \frac{4}{GN} \sum_{i=1}^{N} v_{ri}^2 r_i \tag{10-7}$$

should approach M, if N is sufficiently large.

This analysis has been carried out by Lynden-Bell et al. (1983) using nine satellite galaxies listed in Table 10-1 (the galaxies labeled by "G" in column 3). Column 8 gives the heliocentric line-of-sight velocity v_p (negative velocity means the galaxy is approaching, positive means the galaxy is receding), which we convert to the line-of-sight velocity relative to the galactic center, v_G, by the following procedure. We first add the velocity of the Sun relative to the local standard of rest [eq. (1-10)]. Next we add the line-of-sight component of the $220\,\mathrm{km\,s^{-1}}$ circular speed of the local standard of rest. Thus for a galaxy at (ℓ, b) the line-of-sight velocity in a frame at rest with respect to the galactic center is

$$v_G = v_p + (16.5\,\mathrm{km\,s^{-1}})[\cos b \cos 25° \cos(\ell - 53°) + \sin b \sin 25°] \\ + (220\,\mathrm{km\,s^{-1}}) \sin \ell \cos b. \tag{10-8}$$

Since the satellites are all at galactocentric distances $r_i \gg R_0$, the line of sight is nearly radially outward from the galactic center, and so the line-of-sight velocity v_G is nearly equal to the radial velocity v_r that is needed for equation (10-7).

Using Lynden-Bell et al.'s procedure, we find a total mass $M_{\text{est}} = 3.8 \times 10^{11}\,M_\odot$, with an uncertainty of about 40%. The corresponding

Table 10-1. The Local Group

Name	Type		$-M_B$	ℓ	b	d	v_p	v_G
M31=NGC 224	SbI-II		21.6	121.2	−21.6	730	−297	−119
The Galaxy	SbcI-II		20.0 :	−	−	−	−	−
M33=NGC 598	ScII-III	A?	19.1	133.6	−31.3	900	−180	−45
Large Magellanic Cloud	SBm	G	18.4	280.5	−32.9	50	270	76
Small Magellanic Cloud	Im	G	17.0	302.8	−44.3	63	163	22
IC 10	Im		16.2	119.0	−3.3	1300	−343	−145
NGC 205	E5p	A	15.7	120.7	−21.1	730	−239	−60
M32=NGC 221	E2	A	15.5	121.2	−22.0	730	−200	−23
NGC 6822	Im		15.3	25.3	−18.4	680	−49	51
WLM=DDO 221	IBm		15.3	75.9	−73.6	1600	−116	−59
IC 1613	Im		14.8	129.7	−60.6	850	−235	−156
NGC 185	E3p	A	14.6	120.8	−14.5	730	−227	−40
NGC 147	E5	A	14.4	119.8	−14.3	730	−187	2
IC 5152	Sdm		14.4	343.9	−50.2	1500	121	80
Peg=DDO 216	Im		13.7	94.8	−43.6	1600:	−183	−21
Leo A=DDO 69	IBm		13.1	196.9	52.4	1600:	26	−15
Fornax	dE0	G	12.0	237.3	−65.7	138	55	−34
DDO 210	Im		11.3	34.1	−31.4	1500:	−131	−17
Sculptor	dE3	G	10.6	287.7	−83.1	79	107	74
And I	dE3	A	10.6	121.7	−24.9	730	−	−
And II	dE	A	10.6	128.9	−29.2	730	−	−
And III	dE	A	10.6	119.3	−26.3	730	−	−
Sag	Im		10.5	21.1	−16.2	1100	−79	7
Leo I=DDO 74	dE3	G	9.6	226.0	49.1	220	185	77
Leo II=DDO 93	dE0	G	8.5	220.1	67.2	220	95	41
Ursa Minor=DDO 199	dE4	G	8.2	105.0	44.8	63	−249	−87
Draco=DDO 208	dE0	G	8.0	86.4	34.7	75	−289	−94
Carina	dE	G	−	260.1	−22.2	91	230	14

NOTES: "G" or "A" means the galaxy is probably a bound satellite of our Galaxy or the Andromeda galaxy (M31), respectively. Distance d is measured from the Sun, in kpc. M_B is the absolute magnitude in the B band, which is related to the blue luminosity in solar units by $-M_B = -5.48 + 2.5\log(L_B / L_\odot)$. v_p is the heliocentric line-of-sight velocity in $\mathrm{km\,s^{-1}}$; v_G is the galactocentric relative velocity as determined by equation (10-8). A colon after the absolute magnitude or distance denotes an especially uncertain value. Data are mostly from Kraan-Korteweg and Tammann (1979), except that the absolute magnitude of the Galaxy has been taken from de Vaucouleurs and Pence (1978) and Bahcall and Soneira (1980); also, some velocities and distances of galactic satellites are taken from Aaronson (1983), Aaronson et al. (1983), Lynden-Bell et al. (1983), and Armandroff and Da Costa (1986).

mass-to-light ratio is $27\Upsilon_\odot$. This value confirms the large mass indicated by the high-velocity stars, and suggests that if the mass model of equation (2-191) is correct, the dark halo extends to about $r_* = 34\,\mathrm{kpc}$. However, there are a number of uncertainties that beset the method: (i) We have assumed that the velocity ellipsoid is isotropic, but if it were assumed to be radially elongated the estimated mass would be smaller, while if the satellite velocities were assumed to be primarily tangential, the estimated mass would be larger. The unknown shape of the velocity ellipsoid leads to uncertainties of a factor of two or three in the final mass. (ii) The number of objects in the sample is relatively small, so the statistical error is large. This can be remedied by using large samples of distant halo stars instead of satellite galaxies, but only at a large expense in telescope time. (iii) The velocities of satellites are difficult to measure, and a few bad velocities could render the results meaningless.

Thus the Population II tracers appear to confirm that the Galaxy contains substantial amounts of dark mass, but the results are not yet reliable enough to constrain the distribution of dark mass or accurately measure its extent. The results imply that the halo extends to about $r_\star \simeq 30\,\mathrm{kpc}$, although other techniques, which we discuss below, suggest that even this large halo radius may be too conservative by a factor of two or more.

5 The Magellanic Stream

The Magellanic Stream is a long trail of neutral hydrogen that stretches across the sky in a great circle, starting from the Magellanic Clouds (§7.1.1(b) and MB, §9-2). It is believed to consist of debris torn off the Clouds by tidal forces during a previous close passage near the Galaxy. Dynamical models of the stream constructed by Murai and Fujimoto (1980) and by D. N. C. Lin and Lynden-Bell (1982) were discussed in §7.1.1(b). These models show that it is very difficult to fit the data if the galactic potential is that of a point mass; a much better fit is obtained with an extended mass distribution. The principal reason is that the material at the tip of the stream, some 100° from the Clouds, is falling toward the Galaxy at high speed ($v_G \simeq -220\,\mathrm{km\,s^{-1}}$). If there were no massive halo, this large infall velocity could arise only if the material at the tip had fallen deep into the Galaxy's potential well, to a galactocentric distance $\lesssim 15\,\mathrm{kpc}$. However, at this distance parallax effects due to the 8.5 kpc offset of the Sun from the galactic center would spoil the great circle shape of the Stream. The stronger force field due to a halo permits the same large velocities to be reached at larger galactocentric distances.

A detailed fit of the Stream to a spherical mass model with constant circular speed yields a circular speed $v_c = 244 \pm 20\,\mathrm{km\,s^{-1}}$, consistent with the circular speed at the solar radius of $220\,\mathrm{km\,s^{-1}}$ (D. N. C. Lin & Lynden-Bell 1982). Since the Stream extends from a galactocentric distance of about 50 kpc out to $\gtrsim 100\,\mathrm{kpc}$, this result suggests that the outer radius of the dark halo $r_\star$ is at least 100 kpc and possibly much more. If so, then more than 90% of the mass of the Galaxy consists of dark matter, and the mass-to-light ratio exceeds $80\Upsilon_\odot$.

6 Rotation Curves of Galaxies

The rotation curve provides the most direct method of measuring the mass distribution of a galaxy. The rotation curve of our own Galaxy was discussed in §2.7 and in §8-3 of MB. However, so far the most accurate information on the behavior of rotation curves at large radii has been obtained from other galaxies.

Rotation curves can be measured optically from emission lines in HII regions, or at radio wavelengths using the 21-cm emission line of neutral hydrogen (see MB, §8-4). Neutral hydrogen observations generally extend to larger radii.

The early optical measurements (in the 1950s and 1960s) were usually restricted to the inner parts of galaxies. These curves typically showed a steep rise in the rotation speed near the center, then a short level section before the last data point was reached. This behavior is characteristic of the rotation curve of the exponential disk (Fig. 2-17), which can be divided into three regions: (i) an inner region in which the speed rises linearly with distance from the center; (ii) a region where the speed reaches a maximum and then begins to decline (at the so-called **turnover radius**); and (iii) a **Keplerian region** in which the potential of the disk resembles a point mass potential, so that the rotation speed falls as $R^{-1/2}$. These three regions are found in most rotation curves arising from a smooth, azimuthally symmetric density distribution with finite mass, whether spherical or flattened. Hence it was natural for observers to identify the location of the level section in their rotation curves with the turnover radius and to assume a Keplerian falloff in rotation speed past the last measured point, especially since the light of the galaxy was already mostly contained within the radius of the last measured point. This extrapolation allowed the observers to estimate the total mass of the galaxy.

By 1970 some thirty spiral galaxy rotation curves and masses had been published (Burbidge & Burbidge 1975), all based on the assumption that the unobserved region was Keplerian. In retrospect, it is remarkable that this drastic extrapolation was almost universally accepted, without even any discussion of its dangers.[3] The situation changed around 1970, as improved sensitivity in both optical and 21-cm observations permitted rotation curves to be extended to larger radii. These observations began to show that the flat portion of the rotation curve extended farther than an exponential disk model would predict, and that there was no sign of a Keplerian falloff. In an optical study of the rotation curve of the spiral galaxy M31, Rubin and Ford (1970) found that the mass rose steadily

---

[3] With the notable exception of Schwarzschild (1954).

out to their last measured point at 24 kpc, and carefully stated that "extrapolation beyond that point is clearly a matter of taste". Shortly after, in a 21-cm study of five spiral galaxies, Rogstad and Shostak (1972) stated that all their rotation curves had a similar shape, with the usual steep rise in rotation speed near the center and then a flat outer region extending a factor of two or three in radius out to their last measured point.

By now there are over seventy spiral galaxies with reliable rotation curves out to large radii[4] (see Fig. 10-1, Faber & Gallagher 1979, and Rubin et al. 1980, 1982, 1985). In almost all of them the rotation curve is flat or slowly rising out to the last measured point. Very few galaxies show falling rotation curves, and the ones that do either (i) fall less rapidly than Keplerian; (ii) have nearby companions that may perturb the velocity field; or (iii) have large spheroids that may increase the rotation speed near the center. There is no well-established example of a Keplerian region in any galaxy rotation curve, even those that extend to radii large enough to contain essentially all of the galaxy's light.[5] Consequently, *there is no spiral galaxy with a well-determined total mass.*

The simplest interpretation of these results is that other spiral galaxies, like our own, possess massive dark halos that extend to larger radii than the optical disks, a conclusion first stated by Freeman (1970).[6] If we approximate the dark halo as spherical, and are at sufficiently large radii that the gravitational force from the disk can be neglected, then a rotation curve with constant velocity v_c implies that the halo mass increases linearly with radius [eq. (2-191)] out to a radius beyond the last measured point. We have already seen that this model for the mass distribution in our own Galaxy fits observations of the Magellanic Stream; the rotation curve observations imply that a similar mass distribution is common to most spiral galaxies. The corresponding dark halo density

---

[4] The size of a galaxy is usually measured by the Holmberg radius [§1.1.3(a)]. The Homberg radius of our Galaxy is probably about 17 kpc (de Vaucouleurs & Pence 1978). Modern optical rotation curves typically extend to ≈ 0.5 Holmberg radii, while 21-cm rotation curves extend to about 1.5 Holmberg radii.

[5] It has sometimes been argued that apparently flat rotation curves can arise in a Keplerian region because of warps in the outer disk (§6.6) that change the projection angle of the velocity vector relative to the line of sight. However, in a number of cases warps have been fully modeled using the two-dimensional rotation field on the plane of the sky, and a flat rotation curve persists. In addition, it is hard to see why warps would always conspire to produce constant rotation speed rather than a wide range of shapes of the rotation curve.

[6] "For NGC 300 and M33 21-cm data give turnover points near the photometric outer edges of these systems...there must be in these galaxies additional matter which is undetected...its mass must be at least as large as the mass of the detected galaxy..."

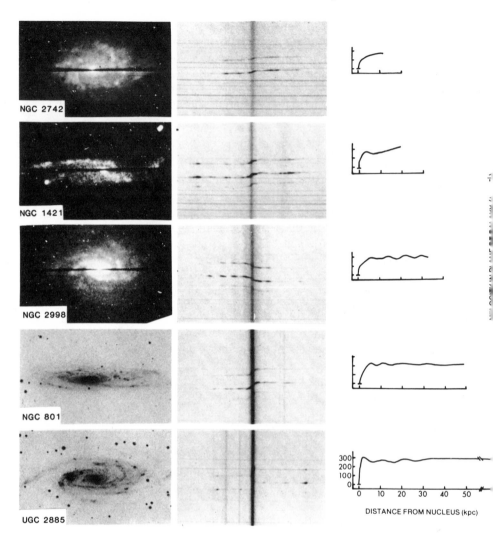

Figure 10-1. Photographs, spectra, and rotation curves for five Sc galaxies, arranged in order of increasing luminosity from top to bottom. The top three images are television pictures, in which the spectrograph slit appears as a dark line crossing the center of the galaxy. The vertical line in each spectrum is continuum emission from the nucleus. The distance scales are based on a Hubble constant $h = 0.5$. Reproduced from Rubin (1983), by permission of *Science*.

is[7]

$$\rho(r) = \frac{1}{4\pi r^2} \frac{dM(r)}{dr} = \frac{v_c^2}{4\pi G r^2}. \tag{10-9}$$

The dark halo density must fall below the value given by equation (10-9) at small radii, since the observed rotation curves remain flat even when the disk mass contributes a substantial fraction of the rotation speed. A better parametrization of the halo density is therefore given by the fitting formula

$$\rho(r) = \frac{\rho_0}{1 + (r/a)^\gamma}. \tag{10-10}$$

If both a rotation curve and accurate photometry are available, then the mass-to-light ratio of the disk and the halo parameters ρ_0, a, and γ can all be fitted simultaneously. The best examples are late-type galaxies, in particular the Sc galaxy NGC 3198 (van Albada et al. 1985; see also Carignan & Freeman 1985 for other examples). The luminosity profile is fitted very well by an exponential disk with scale length $R_d = 2h^{-1}\,\text{kpc}$ and total luminosity $L_V = 4 \times 10^9 h^{-2} L_\odot$. The luminosity of the spheroid is negligible. Its rotation curve has been measured with the Westerbork Synthesis Radio Telescope and is shown in Figure 10-2, along with a theoretical curve based on an exponential disk with constant mass-to-light ratio and a spherical dark halo whose density distribution is given by equation (10-10). The principal uncertainty is in the mass or mass-to-light ratio of the disk; the figure shows the model with the maximum disk mass, which yields a mass-to-light ratio $\Upsilon_V = 5.8h\Upsilon_\odot$, $\rho_0 = 0.013h^2\,\text{M}_\odot\,\text{pc}^{-3}$, $a = 6.4h^{-1}\,\text{kpc}$, and $\gamma = 2.1$. At the other extreme, reasonable fits can be achieved even in a model in which the disk has zero mass; in this case the halo parameters are $\rho_0 = 0.58h^2\,\text{M}_\odot\,\text{pc}^{-3}$, $a = 1.1h^{-1}\,\text{kpc}$, $\gamma = 2.25$. It is reassuring that the disk mass-to-light ratios in these two models bracket the mass-to-light ratio for the solar neighborhood determined in equation (10-4), $\Upsilon_V = 5\Upsilon_\odot$. In either model the total mass inside the last measured point of the rotation curve, $r = 22h^{-1}\,\text{kpc}$, is $1.1 \times 10^{11}h^{-1}\,\text{M}_\odot$, which yields a total mass-to-light ratio $\Upsilon_V = 28h\Upsilon_\odot$. The dark halo is at least four times as massive as the disk.

Rotation curves provide the most direct evidence for dark mass in other spiral galaxies. However, the rotation curves give no indication of the maximum extent of the dark halo, since they are still flat at the point where the surface brightness falls below detectable limits. Thus it

---

[7] This is also the density distribution for the isothermal sphere at large radii [eq. (4-123)]. However, there is no compelling theoretical argument to suggest why the dark halo should resemble an isothermal sphere.

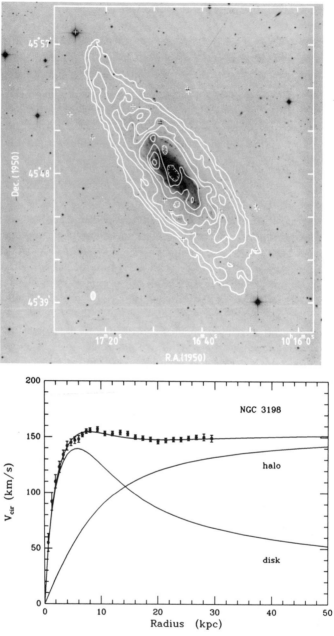

Figure 10-2. The Sc galaxy NGC 3198. Top: neutral hydrogen column density contours superimposed on an optical photograph. Bottom: circular-speed curve plus model fits using an exponential disk with constant mass-to-light ratio and the halo density profile (10-10). The model curve is for the maximum possible disk mass-to-light ratio. The horizontal scale assumes $h = 0.75$. Reprinted from van Albada et al. (1985), by permission of *The Astrophysical Journal.*

is possible that the dark halo ends near the edge of the visible disk; on the other hand, it may equally well extend many times farther, perhaps even out to several hundred kiloparsecs.

7 Other Probes

(a) Tidal radii   In principle, the galactic mass distribution can be determined from the tidal radii of distant globular clusters and dwarf elliptical galaxies. Equation (7-84) relates the mass, distance, and limiting radius of the satellite to the galactic mass. Unfortunately, there appear to be major obstacles to the application of this method: (i) at present, the mass-to-light ratios of dwarf elliptical galaxies are only poorly known so the results are based on an uncertain mass for the satellite; (ii) the tidal radii are based on uncertain extrapolations of the luminosity profiles of the dwarf galaxies—and according to equation (7-84) the estimated Galaxy mass varies as the cube of the tidal radius; (iii) as we discussed in §7.3, the applicability of the tidal radius to systems in eccentric orbits is uncertain, and even if an accurate theory were available, the eccentricities of the orbits of the dwarf satellites are unknown; (iv) it is likely that some dwarf ellipticals have only passed through pericenter once or twice, and they may therefore still be surrounded by a cloud of escaped stars that have not yet dispersed (Barnes & White 1984). In view of all these difficulties, tidal radii are presently of little use as probes of the galactic force field.

(b) Disk stability   In §6.3 we showed that models of galactic disks are often violently unstable to a large-scale bar mode. The instability can be cured if the galaxy contains a dark halo and the halo mass *inside* the disk outer radius is comparable to or greater than the disk mass.[8] One of the first proposals that spiral galaxies had massive dark halos was based on this argument (Ostriker & Peebles 1973). However, as we have seen in §6.3.2, the bar instability in disk galaxies can be suppressed in other ways, for example by a rotation curve that rises sharply near the galactic center. Hence, disk stability arguments do not provide a compelling reason to believe that disk galaxies have massive halos, even though in the end it may well be that a massive halo is an important stabilizing influence on most galactic disks (see Sellwood 1985 for an investigation of the stability of the Bahcall-Soneira model of our Galaxy).

---

[8] Of course, the mass and extent of the halo beyond the disk outer radius are irrelevant because of Newton's first and third theorems.

(c) X-ray halos of elliptical galaxies   Many—and possibly all—
luminous elliptical galaxies contain up to $10^{10}\,M_\odot$ of hot, X-ray-emitting
gas out to radii of order $50\,\mathrm{kpc}$ (Forman et al. 1985). The gas is probably produced by normal stellar mass loss. The X-rays are continuum
photons emitted by gas at temperature $\approx 10^7\,\mathrm{K}$ as it cools through the
bremsstrahlung process [§9.3.1(a)]. The principal heat source for the
gas is supernovae [§9.3.1(c)]. However, the heating and cooling rates for
the gas do not balance at every radius. Since the cooling rate per ion
[eq. (9-42)] is proportional to the local electron density, cooling will generally dominate over heating at small radii, where the density is high.
Thus the gas in the central regions will steadily cool and flow in to the
center of the galaxy, where it presumably forms stars. However, the flow
velocity is usually much less than the sound speed, so that approximate
hydrostatic equilibrium is maintained in the gas.

In a spherically symmetric galaxy, hydrostatic equilibrium implies
[eq. (1E-8) with $\mathbf{v} = 0$]

$$\frac{dp}{dr} = -\frac{GM(r)\rho}{r^2}, \tag{10-11}$$

where ρ and p are the density and pressure, and $M(r)$ is the mass interior
to radius r. Using the ideal gas law, (1E-22), this can be rewritten as

$$M(r) = \frac{k_B T r}{G \mu m_p} \left[-\frac{d\ln\rho}{d\ln r} - \frac{d\ln T}{d\ln r} \right], \tag{10-12}$$

where T is the gas temperature, μ is the mean molecular weight, and
m_p is the proton mass. Thus if we can measure the temperature profile
$T(r)$ and the density profile $\rho(r)$, we can use equation (10-12) to find
the mass distribution $M(r)$. This method has an important advantage
over analogous methods of mass determination based on stellar velocity
dispersions [e.g., eq. (4-55)] in that the gas pressure is isotropic so that
no uncertainty arises from the shape of the velocity ellipsoid.

High-resolution maps of the X-ray surface brightness for about a
dozen galaxies were provided by the *Einstein* satellite. Since the gas is
optically thin [cf. §9.3.1(a)], and the luminosity density at any radius is
proportional to the square of the density in a fully ionized gas [eq. (9-
42)], it is straightforward to determine the density profile $(d\ln\rho/d\ln r)$
in equation (10-12), if the temperature is known. The mean temperature
can be determined by fitting the X-ray spectrum to a bremsstrahlung
spectrum, or by modeling the strengths of X-ray spectral lines. Unfortunately, spatially resolved temperature measurements are usually not
available from *Einstein* observations so that the temperature profile is
difficult to determine.

The best data are for the giant elliptical galaxy M87 (Figure 1-6), where at least a crude temperature profile is available. Separate analyses by Fabricant and Gorenstein (1983) and Stewart et al. (1984) show that in this galaxy $M(r)$ rises roughly linearly with radius out to more than 300 kpc, with $M(< 300\,\text{kpc}) \simeq 3 \times 10^{13}\,\text{M}_\odot$. The corresponding mass-to-light ratio is $\Upsilon \simeq 750\Upsilon_\odot$, far larger than the minimum mass-to-light ratio for Population II systems of $6\Upsilon_\odot$. Apparently over 99% of the mass of M87 is composed of dark matter.

With better data, X-ray observations can be expected to provide detailed maps of the mass distribution around elliptical galaxies; however, at the moment only M87 provides a clear case for a massive dark halo. Most other galaxies are too faint in X-rays for a good temperature determination. M87 is a most atypical galaxy by virtue of its position at the center of the Virgo cluster, and the presence of a dark halo around M87 need not imply that similar halos surround most other galaxies.

10.2 Dark Matter in Systems of Galaxies

1 Timing the Local Group

The nearest giant spiral galaxy is the Sb galaxy M31, at a distance of about 700 kpc. Like our own Galaxy, M31 is surrounded by a number of companion galaxies, most of which are probably gravitationally bound to it. Beyond these, the next nearest prominent galaxies are in the Sculptor and M81 groups (de Vaucouleurs 1975), at a distance of 3 Mpc. Thus our Galaxy, M31, and their companions form a relatively isolated system known as the **Local Group**, whose members are listed in Table 10-1.

The heliocentric line-of-sight velocity of the center of mass of M31 is $v_p = -297\,\text{km s}^{-1}$. Using equation (10-8) we find that the center of mass of M31 is approaching the center of mass of the Galaxy at $v_G = -119\,\text{km s}^{-1}$. This is an unexpected result, since most galaxies are moving apart with the general Hubble expansion. One possible explanation is that M31 and the Galaxy have random velocities relative to the Hubble flow such that their paths happen to cross, like ships that pass in the night. In this case, the Local Group is no more than a temporary and accidental superposition of the two subgroups associated with the Galaxy and M31. However, the probability of such a superposition is small—perhaps 5% or 10%, based on the average density of giant spirals like M31 (van den Bergh 1971). An equally fundamental difficulty with this hypothesis is that in the absence of mutual gravitational interactions between galaxies, random velocities relative to the Hubble flow decay as $R(t)^{-1}$ where $R(t)$ is the scale factor of the Universe [eq. (9A-4)]. Hence the present relative velocity would have required a much higher random velocity in the past, which cannot be generated by any plausible means.

A much more natural explanation is that the relative Hubble expansion of M31 and the Galaxy has been halted and reversed by their mutual gravitational attraction. Kahn and Woltjer (1959) pointed out that this hypothesis leads directly to an estimate of the total mass of the Local Group (see also Gunn 1975; Lynden-Bell 1982; and Mishra 1985). Since M31 and the Galaxy are by far the most luminous members of the group, we can treat the two galaxies as an isolated system of two point masses. Their separation r as a function of time t is therefore governed by Kepler's equations (3-27),

$$r = a(1 - e\cos\eta) \quad ; \quad t = \sqrt{\frac{a^3}{GM}}(\eta - e\sin\eta) + C, \qquad (10\text{-}13)$$

where C is a constant, a is the semi-major axis, e is the eccentricity, η is the eccentric anomaly, and M is the mass of the Local Group (that is, the Galaxy plus M31). At the initial singularity $r = 0$ and $t = 0$, so we must choose $e = 1$ and $C = 0$, corresponding to a radial orbit.[9] The radial velocity may be written

$$\frac{dr}{dt} = \frac{dr/d\eta}{dt/d\eta} = \sqrt{\frac{GM}{a}}\frac{\sin\eta}{1 - \cos\eta} = \frac{r}{t}\frac{\sin\eta(\eta - \sin\eta)}{(1 - \cos\eta)^2}. \qquad (10\text{-}14)$$

At the present time t_0 we have $dr/dt = v_G = -119\,\mathrm{km\,s^{-1}}$ and $r = 0.73$ Mpc. The age of the Universe t_0 is roughly between 1×10^{10} yr and 2×10^{10} yr (§10.3.5). Solving equation (10-14) numerically we find that η must lie between 4.11 and 4.46 radians modulo 2π. From the first line of equation (10-13) the semi-major axis is then found to be between 0.47 and 0.58 Mpc, and if M31 is on its first approach to the Galaxy ($\pi < \eta < 2\pi$), then the second line of equation (10-13) yields the mass of the Local Group to be between $5.5 \times 10^{12}\,\mathrm{M_\odot}$ and $3.2 \times 10^{12}\,\mathrm{M_\odot}$. The larger age leads to the smaller mass.

Since the luminosity of the Galaxy in the V band is $1.4 \times 10^{10} L_\odot$ (Table 1-2) and M31 is about twice as luminous, the corresponding mass-to-light ratio for the Local Group is between 130 and $76\Upsilon_\odot$. Furthermore, the estimated mass of the Local Group is increased if either (i)

---

[9] In fact, the orbit is probably not precisely radial. The galaxies are not point masses and hence tidal torques can lead to angular momentum exchange between the orbit and the galaxy spins. This is likely to be the origin of the spins. However, even if the total angular momentum now in the spins were transferred to the orbit, the eccentricity of the orbit would still be nearly unity. See Gott and Thuan (1978) and Lynden-Bell (1982) for more general discussions of the angular momentum of the Local Group.

the orbit is not exactly radial, or (ii) M31 and the Galaxy have already had one or more pericenter passages since the origin of the Universe.

The large mass-to-light ratio we have derived for the Local Group strongly corroborates the evidence we have already presented that most of the galactic mass is dark. If the relative masses of the Galaxy and M31 are proportional to their relative luminosities, and if the galactic mass is distributed in a spherical dark halo with a constant rotation speed $v_c = 220\,\mathrm{km\,s^{-1}}$, then the halo must extend out to at least $r_\star \simeq 95\,\mathrm{kpc}$.

2 Binary Galaxies

All of our data on the masses of stars come from measurements of binary stars. Thus it is natural to attempt to measure galaxy masses by studying binary galaxies. Unfortunately, the orbital periods of binary galaxies are so long that we cannot even hope to measure their relative proper motion, much less follow them around a complete orbit as we do with binary stars. Consequently, any investigation of binary galaxies must be based on statistical studies of the relative line-of-sight velocities in a large sample of galaxy pairs.

The statistical measurement of masses of binary galaxies has proven to be a very thorny problem. Although the first paper on the subject dates back to Holmberg (1937), the subject is still plagued with difficulties. The mass-to-light ratio of binary galaxies is probably large, but not so large as the ratio of the mass of papers on this subject to the light they have shed on it.

The first difficult step is to choose a sample of binaries. Since the galaxy distribution is strongly clustered on all scales, it is very difficult to isolate an uncontaminated sample of binary galaxies. If two unrelated galaxies are treated as a real binary, then their mass may be wildly overestimated, since the difference in their Hubble velocities will be treated as orbital velocity.

One of the best binary samples was constructed by Turner (1976) from Zwicky et al.'s (1961–1968) *Catalog of Galaxies and Clusters of Galaxies* (MB, Table 2-6), containing over 30,000 objects. After rejecting galaxy pairs whose angular separation exceeded 8′ (to minimize contamination by accidental pairs), those that had a third galaxy nearby (to help ensure that the galaxy pairs were isolated), and those that were so faint that the accuracy of the catalog was doubtful or that were heavily obscured by dust in our own Galaxy, Turner was left with 156 pairs. However, even this rigorous selection procedure proved insufficient to isolate an uncontaminated sample. Later measurements (both visual inspection and velocity determinations by White et al. 1983) show that, in fact, only about half of these pairs are relatively "pure". The rest appear to be members of groups or clusters or accidental superpositions. Clearly, it is not feasible to find a binary sample that is entirely free from contamination, and extreme care is needed to minimize the contamination and its effects on the mass determination.[10]

Once a sample has been chosen, the relative line-of-sight velocity of the members of each pair must be measured. The relative velocities observed are usually very small: for example, the median velocity difference for the 76 "pure" pairs from Turner's sample is $89\,\mathrm{km\,s^{-1}}$ as determined optically (White et al. 1983). The RMS error in the relative velocity was estimated to be $42\,\mathrm{km\,s^{-1}}$. Clearly, if there has been any substantial underestimate of the velocity errors, the relative velocity (and hence the mass) will be overestimated. Fortunately, the error estimate can be confirmed by comparing the optical velocities to accurate radio velocities measured from the 21-cm line of HI (Rood 1982).

Finally, the measurements must be fit to a dynamical model. Let the probability that a pair has separation in the volume element $d^3\mathbf{r}$ around $\mathbf{r}$ be $\nu(r)d^3\mathbf{r}$. We first attempt to fit the velocity measurements to a simple model in which each galaxy is represented as a point mass. As shown by the discussion in Appendix 1.D, the relative motion of the two point masses is the same as the motion of a single test particle around a fixed body of mass equal to the total mass M. Thus the velocity dispersion of an ensemble of binary galaxies may be described by the Jeans equation (4-54) in the form

$$\frac{d(\nu\overline{v_r^2})}{dr} + \frac{2\nu\beta\overline{v_r^2}}{r} = -\frac{GM}{r^2}\nu(r), \qquad (10\text{-}15)$$

where $\overline{v_r^2}$ is the radial relative velocity dispersion at separation r, and $\beta(r)$ is a measure of the anisotropy of the velocity ellipsoid [eq. (4-53b)]. The line-of-sight dispersion at projected separation R is denoted

---

[10] Above all, it is essential that the selection criteria for the sample be well defined and objective, since selection criteria that reduce contamination always introduce a bias that must be modeled in the analysis. The following example illustrates this point. Assume that all binary galaxies are on circular orbits with radius r and relative velocity v, and that in order to minimize contamination by unrelated galaxies we examine only pairs with projected separation $R < R_{\max}$, where $R_{\max} \ll r$. This selection criterion implies that the pairs in our sample will have velocities nearly perpendicular to the line of sight, and hence that the relative line-of-sight velocity v_p will be very small. Thus the apparent binary mass, which is always based on $v_p^2 R/G$ times some factor of order unity, will be severely underestimated.

by $\sigma_p(R)$ and is given by equation (4-57).[11] Analyses of the two-point correlation function of galaxies show that the probability density in separation is well approximated by a power law, $\nu(r) \propto r^{-\gamma}$, where $\gamma = 1.8$ (Peebles 1974). The normalization of $\nu(r)$ does not concern us here since it cancels out of equation (10-15).

We shall consider two specific orbit distributions: isotropic orbits ($\beta = 0$) and radial orbits ($\beta = 1$). Solving the differential equation (10-15) subject to the boundary condition $\overline{v_r^2} = 0$ as $r \to \infty$, we find

$$\overline{v_r^2} = \begin{cases} \dfrac{GM}{(\gamma+1)r}, & \text{isotropic orbits,} \\[2mm] \dfrac{GM}{(\gamma-1)r}, & \text{radial orbits.} \end{cases} \tag{10-16}$$

These results can then be substituted into equation (4-57) to find the line-of-sight dispersion $\sigma_p(R)$. The resulting values of $\sigma_p(R)$ can be directly compared with the RMS velocity differences in the binary galaxy sample. (Note that by comparing the predicted and observed velocity differences at a given projected separation, we are unaffected by any bias due to selection criteria based on projected separations.) However, it is already clear from the form of equation (10-16) that $\sigma_p^2(R)$ will be equal to some numerical constant times GM/R. Hence we can test the validity of the point mass model by searching for the predicted correlation of the line-of-sight velocity differences and the projected separation. This search has been carried out by White et al. (1983) and no such correlation is found. They conclude that point mass models for binary galaxies must be rejected.

The simplest interpretation of this result is that the galaxies have dark halos that extend well beyond their optical boundaries. By analogy with our results for single galaxies, we are therefore led to models in which we replace the point-mass gravitational force $-GM/r^2$ by a force $-v_c^2/r$, which would produce a flat rotation curve with constant circular speed v_c. Making this substitution in the Jeans equation (10-15), and taking $\nu(r) \propto r^{-\gamma}$ as before, we find

$$\overline{v_r^2} = \begin{cases} \dfrac{v_c^2}{\gamma} & \text{isotropic orbits,} \\[2mm] \dfrac{v_c^2}{\gamma-2} & \text{radial orbits.} \end{cases} \tag{10-17}$$

---

[11] An analysis of the type used in §10.1.4 for Population II tracers is inappropriate for binary galaxies, since it is based on the assumption that the objects are visible at all phases of their orbit, whereas the selection criteria for a binary galaxy sample, which are usually based on apparent separation, discriminate against pairs that happen to be near apocenter.

First consider the case of isotropic orbits. The integration to find the line-of-sight dispersion is trivial and yields $\sigma_p = v_c/\sqrt{\gamma}$. For comparison with the data, we restrict ourselves to 55 "pure" pairs from the Turner sample in which both members are spiral galaxies (we eliminate elliptical galaxies since we want to compare the value of v_c to rotation curve measurements of spiral galaxies). For these pairs the RMS velocity difference is $127\,\mathrm{km\,s^{-1}}$, which should be equal to σ_p. Thus for $\gamma = 1.8$ we find $v_c = 170\,\mathrm{km\,s^{-1}}$. This result is somewhat lower than the typical speeds seen in the flat rotation curves of bright spiral galaxies (200 to $300\,\mathrm{km\,s^{-1}}$), but the discrepancy is not serious.

The case of radial orbits is very different. Equation (10-17) shows that radial orbit models with $\gamma = 1.8$ cannot be constructed (since either v_c^2 or $\overline{v_r^2}$ must be negative). The reason for this is easy to understand. For radial orbits, the density rises with decreasing radius $\propto r^{-2}$ even if the gravitational force is negligible, simply from the crowding of the stars as they approach the origin. The attractive gravitational force leads to an even steeper rise, and hence a density distribution characterized by $\gamma < 2$ cannot arise from radial orbits. Clearly, by adjusting the anisotropy parameter β we can derive a circular speed v_c that is as small as we want. Hence the binary galaxy data cannot be used to determine v_c until we have an independent estimate of the anisotropy parameter for these systems.

To summarize, the binary galaxy data appear to be inconsistent with point mass models, and consistent with models having extended massive halos so long as the velocity ellipsoid is roughly isotropic. Unfortunately, the results are extremely sensitive to the assumed shape of the velocity ellipsoid, and may still be biased by errors in the velocity measurements and contamination. An additional problem is that the dynamical behavior of binary galaxies with extended massive halos is not fully understood. How severely are the halos truncated by tidal forces? Can the binaries survive against dynamical friction, which could lead many binaries to merge after one or two orbits [§7.1.1(c)]? If the velocity ellipsoid is isotropic rather than nearly radial, then how did the galaxies acquire their orbital angular momentum? So far, the study of binaries has confirmed that the picture of galaxies as point masses obeying Kepler's laws is seriously flawed, but it has not yet provided strong evidence in favor of any alternative model.

3 Groups of Galaxies

Binary galaxies are only a special example of bound galaxy systems. Groups of galaxies are collections of three or more galaxies whose separations are much smaller than the typical intergalactic separation. The high density of galaxies in groups suggests that they are gravitationally bound, so that an analysis of the positions and velocities of the group members can yield an estimate of the mass and mass-to-light ratio of the group.

Consider a group of N point mass galaxies with masses M_i, positions r_i, and velocities v_i. Both positions and velocities are measured relative to the center of mass of the group. If the group is bound and in stationary equilibrium, then according to the virial theorem (8-16)

$$\sum_{i=1}^{N} M_i \langle v_i^2 \rangle_t = \sum_{i=1}^{N} \sum_{j<i} G M_i M_j \left\langle \frac{1}{|r_i - r_j|} \right\rangle_t, \tag{10-18}$$

where the angle brackets $\langle \cdot \rangle_t$ denote a time average. We can measure only the line-of-sight velocity component $v_{p,i}$, but if we take an average over all possible orientations of the group (denoted by $\langle \cdot \rangle_\Omega$) we expect that $\langle v_{p,i}^2 \rangle_\Omega = \frac{1}{3} v_i^2$, since the line-of-sight component and the two orthogonal transverse components should all have the same mean-square value. Similarly, if θ_{ij} is the angle between the vector $r_i - r_j$ and the line of sight, and R_i is the projection of r_i onto the plane of the sky, then

$$\left\langle \frac{1}{|R_i - R_j|} \right\rangle_\Omega = \frac{1}{|r_i - r_j|} \left\langle \frac{1}{\sin \theta_{ij}} \right\rangle_\Omega$$
$$= \frac{1}{|r_i - r_j|} \frac{\int d\Omega / \sin \theta}{\int d\Omega} = \frac{1}{|r_i - r_j|} \frac{\int_0^\pi d\theta}{\int_0^\pi \sin \theta d\theta} = \frac{\pi}{2|r_i - r_j|}. \tag{10-19}$$

Thus, if we assume that the mass-to-light ratio Υ of each member galaxy is the same, then

$$\Upsilon = \frac{3\pi}{2G} \frac{\sum_{i=1}^{N} L_i \langle v_{p,i}^2 \rangle_{t,\Omega}}{\sum_{i=1}^{N} \sum_{j<i} L_i L_j \langle |R_i - R_j|^{-1} \rangle_{t,\Omega}}. \tag{10-20}$$

Naturally, we cannot measure the temporal and angular averages in equation (10-20); however, if the galaxy orbits have random phases and orientations, then the observable quantity

$$\Upsilon_{\text{est}} = \frac{3\pi}{2G} \frac{\sum_{i=1}^{N} L_i v_{p,i}^2}{\sum_{i=1}^{N} \sum_{j<i} L_i L_j |R_i - R_j|^{-1}} \tag{10-21}$$

should approach Υ if N is large or if we average over a large number of groups (Limber & Mathews 1960).

Many groups contain a single dominant galaxy surrounded by a number of small satellites, and for these groups it is natural to assume that most of the mass resides in the dominant galaxy. In this case, equation (10-21) makes inefficient use of the data: in the limit where the luminosity of a satellite approaches zero it makes no contribution to $\Upsilon_{\rm est}$, yet the line-of-sight velocity and projected separation obviously still contain important information about the mass of the central galaxy. For these groups it is more appropriate to regard the satellites as N test particles moving in the gravitational field of a single body of mass M and luminosity L; it is not hard to show that in this type of system the appropriate estimator is

$$\Upsilon_{\rm est} = \frac{3\pi}{2GL} \frac{\sum_{i=1}^{N} v_{p,i}^2}{\sum_{i=1}^{N} R_i^{-1}}, \qquad (10\text{-}22)$$

where $v_{p,i}$ and R_i are the line-of-sight velocity and projected distance relative to the dominant galaxy.

A related question is whether (10-21) is the correct form to use when there is dark matter in the groups. In this case it may be more appropriate to regard the galaxies as test particles swimming in a diffuse sea of dark matter. If the spatial distribution of the galaxies is similar to the distribution of dark matter, the proper form of the virial theorem is

$$\Upsilon_{\rm est} = \frac{3\pi N}{2G} \frac{\sum_{i=1}^{N} v_{p,i}^2}{\left(\sum_{i=1}^{N} L_i\right) \sum_{i=1}^{N} \sum_{j<i} |\mathbf{R}_i - \mathbf{R}_j|^{-1}}. \qquad (10\text{-}23)$$

If, on the other hand, the dark matter distribution is more extended than the galaxy distribution, equation (10-23) tends to underestimate the mass-to-light ratio.

Alternative estimators to equations (10-21) to (10-23) are discussed by Bahcall and Tremaine (1981) and Heisler et al. (1985).

Mass estimates of groups of galaxies, like those of binary galaxies, can be biased by the selection procedure. If an unrelated galaxy is treated as a member of the group, the estimated mass-to-light ratio can be much too large. On the other hand, the mass-to-light ratio can be seriously underestimated if galaxies with relatively large velocity differences or separations are trimmed off the group in an attempt to remove interlopers. In many groups, the final mass-to-light ratio depends strongly on whether or not one or two outlying galaxies are included in the group.

Catalogs of groups can be compiled by several different approaches. **Subjective** catalogs (e.g., de Vaucouleurs 1975) assign galaxies to groups

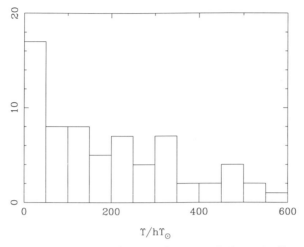

Figure 10-3. Distribution of mass-to-light ratio Υ for groups of galaxies from the catalog of Huchra and Geller (1982). Twenty-five of the 92 groups in the catalog have $\Upsilon > 600h\Upsilon_\odot$ and are not shown.

using subjective judgments, based on the proximity of the galaxies in position and velocity, similarity in distance as judged from appearance, and evidence of interactions such as tidal tails and bridges. **Objective** catalogs (e.g., Turner & Gott 1976, Huchra & Geller 1982) use automatic algorithms to search for clumps of galaxies that exceed a fixed density enhancement relative to the background (usually a factor of 10 or 20). Objective catalogs have two principal advantages: first, selection effects can be explicitly modeled and corrected for, and second, the results can be compared directly with numerical simulations of the Universe by applying the same objective criteria to choose groups from the simulations.[12] It is reassuring that, in general, both subjective and objective techniques yield similar catalogs (Huchra & Geller 1982).

Huchra and Geller determined Υ_{est} for their groups using the virial theorem in the form (10-23). They find a median value in the V band of $260h\Upsilon_\odot$, but with a wide spread (see Fig. 10-3).[13] Much of the spread appears to be statistical, and could be reduced if more galaxy velocities were available (half of the groups have only three measured velocities).

[12] A third technique is to use a statistical approach, which abandons the assignment of galaxies to individual groups but instead describes the clustering by statistics such as the n-point correlation functions in position and velocity (see Davis et al. 1978 or Peebles 1980b).
[13] Their luminosities are quoted in the system used by Zwicky et al. (1961–1968), which we have converted to visual luminosity by multiplying by 1.3.

Despite the spread, it is clear that the mass-to-light ratios in groups are much larger than the values seen in the luminous parts of galaxies, and once again we are forced to invoke the presence of large amounts of dark matter to explain the observations.

4 Clusters of Galaxies

The Coma cluster of galaxies was the system in which Zwicky (1933) first found evidence for large amounts of dark matter, and rich clusters of galaxies still provide some of the best available sites for studying the nature and distribution of dark matter.

Clusters have the advantage of containing many more galaxies than typical groups. In some nearby clusters, several hundred galaxy velocities have been measured, thereby almost eliminating the statistical uncertainties that plague measurements of galaxy groups. Furthermore, the number of galaxies is so large that we can use the solutions of the collisionless Boltzmann equation discussed in Chapter 4 to model the cluster. Contamination by interloper galaxies remains a problem, especially in the outer parts of the cluster where the surface density of cluster galaxies is low, so that the most reliable mass-to-light ratio determinations are for the central regions of the clusters.

As was the case in Zwicky's time, the best available data are for the Coma cluster. Figure 10-4 shows the line-of-sight velocities of galaxies within $350'$ of the center of this cluster (Kent & Gunn 1982). Within about $200'$ (corresponding to $4h^{-1}$ Mpc) there is a fairly clear separation between the portion of velocity space dominated by cluster members and the portion dominated by presumably unrelated interlopers. This separation is marked by two curves on the diagram. Beyond $200'$ the cluster has faded into the background.

Kent and Gunn bin the velocities from the region between the two curves to generate a velocity dispersion profile. They also derive a profile of the surface density of galaxies obtained from galaxy counts. Assuming that the mass density is proportional to the number density of galaxies, these profiles can be fit to Michie models [§4.4.4(b)] to derive an estimate for the mass-to-light ratio of the cluster, in much the same way that velocity-dispersion measurements and the surface-brightness profile in the cores of elliptical galaxies are used to determine galaxy mass-to-light ratios.

Kent and Gunn's analysis yields a mass-to-light ratio $\Upsilon_V \simeq 360h\Upsilon_\odot$ (as reported by Kent & Sargent 1983). The uncertainty in this result is

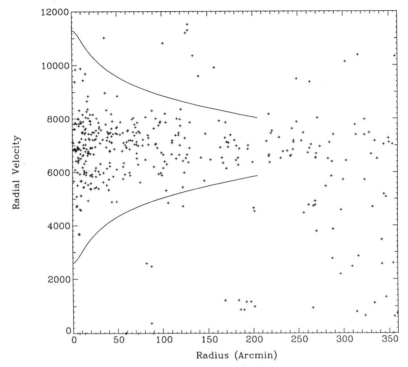

Figure 10-4. Line-of-sight velocities of galaxies in the Coma cluster (in $km\,s^{-1}$) as a function of distance from the cluster center in minutes of arc (Kent & Gunn 1982). The curves mark the authors' estimate of the boundary between cluster members and interlopers. At the distance of Coma 1 arcmin $= 20h^{-1}$ kpc. Reprinted by permission from *The Astronomical Journal*.

difficult to estimate, but is probably no more than a factor of two in the downward direction, due mainly to the uncertain anisotropy parameter $\beta(r)$ [eq. (4-53b)] (Merritt 1987a). A similar analysis of the Perseus cluster of galaxies yields $\Upsilon_V \simeq 600h\Upsilon_\odot$ with a similar uncertainty. These values are some 30 to 50 times larger than the mass-to-light ratios in the cores of elliptical galaxies, and 60 to 100 times larger than the minimum mass-to-light ratio for the stellar population in elliptical cores (§10.1.2).

Like the individual galaxies discussed in §10.1.7(c), many clusters of galaxies contain hot gas that emits X-rays (Figure 10-5; Jones & Forman 1984; Sarazin 1986). Since the gas is in hydrostatic equilibrium, it can be used to trace the gravitational potential and mass distribution in clusters using equation (10-12). The principal obstacle in applying this method is the absence of spatially resolved spectral information,

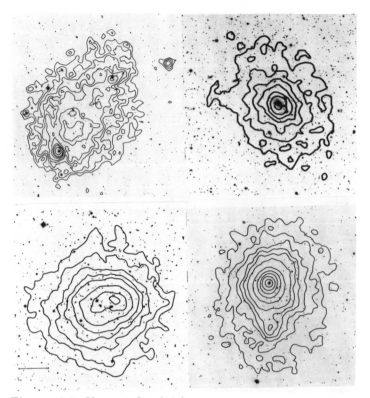

Figure 10-5. X-ray surface-brightness contours superimposed on photographs of several clusters of galaxies. Clockwise from top left, the clusters are A1367, A262, A85, and A2256 (see Jones & Forman 1984).

so that the temperature gradient term $d\ln T/d\ln r$ in equation (10-12) must be estimated by indirect methods. A recent analysis by Cowie et al. (1987) suggests $\Upsilon_V \simeq 180h\Upsilon_\odot$ for Coma, a factor of two less than the optical observations indicate, but still large enough that most of the cluster mass must be dark.

It is likely that the composition of the dark matter in groups and clusters is the same as that of the matter in dark halos of individual galaxies. In the absence of evidence to the contrary, it is natural to assume that the dark matter that would normally form the halo of a galaxy has been stripped off by tidal interactions with the other galaxies in the cluster and now forms a diffuse background that comprises most of the mass of the cluster.

10.3 Dark Matter in Cosmology

1 The Mean Luminosity Density of the Universe

A central parameter in the Friedmann cosmological models described in Appendix 9.A is the mean mass density of the Universe, ρ_0, which is related to the luminosity density j_0 and the mass-to-light ratio Υ by $\rho_0 = j_0 \Upsilon$. In terms of the critical density $\rho_c = 3H_0^2/(8\pi G)$ [eq. (9A-18)] and the density parameter $\Omega_0 = \rho_0/\rho_c$ we may write $\Omega_0 = j_0 \Upsilon/\rho_c$. Thus, by determining the mean luminosity density j_0, we can convert the mass-to-light ratios already determined for galaxies and systems of galaxies into an estimate of the density parameter Ω_0.

The local luminosity density is biased by the nearby Virgo cluster of galaxies, so j_0 must be measured on scales that are large enough so that Virgo has negligible effect. Furthermore, any flux-limited sample is dominated by bright galaxies, so we must extrapolate to include the contribution of faint galaxies using an assumed form of the luminosity function [usually Schechter's law, eq. (1-15)]. Fortunately, the result is relatively insensitive to the details of the extrapolation since most of the luminosity density is in bright galaxies. Two recent independent measurements (Davis & Huchra 1982; Kirshner et al. 1983) agree within a factor of two; the average is

$$j_0 = 1.7 \pm 0.6 \times 10^8 h \, L_\odot \, \text{Mpc}^{-3} \quad \text{in } V. \qquad (10\text{-}24)$$

Thus

$$\Omega_0 = 6.1 \times 10^{-4} h^{-1} \frac{\Upsilon_V}{\Upsilon_\odot}. \qquad (10\text{-}25)$$

The critical mass-to-light ratio required to make $\Omega_0 = 1$ is therefore

$$\Upsilon_c = 1600 h \Upsilon_\odot, \qquad (10\text{-}26)$$

in the V band. The uncertainty in this value is probably about 50%.

Table 10-2 contains a summary of the mass-to-light ratio estimates we obtain in this chapter and the corresponding values of Ω_0. Note the happy circumstance that in many of these cases, the dependence on the poorly known Hubble constant has canceled out so that Ω_0 is independent of h. Notice also that none of the dynamical estimates yields a mass-to-light ratio large enough to imply a closed Universe ($\Omega_0 > 1$).

Table 10-2.   Estimates of the density parameter

Method	$\Upsilon_V/\Upsilon_\odot$	Ω_0
Solar neighborhood	5	$0.003h^{-1}$
Elliptical galaxy cores	$12h$	0.007
Local escape speed	30	$0.018h^{-1}$
Satellite galaxies	30	$0.018h^{-1}$
Magellanic Stream	> 80	$> 0.05h^{-1}$
Rotation curve of NGC 3198	$> 28h$	> 0.017
X-ray halo of M87	> 750	$> 0.46h^{-1}$
Local Group timing	100	$0.06h^{-1}$
Groups of galaxies	$260h$	0.16
Clusters of galaxies	$400h$	0.25
Virgocentric flow	–	0.25
Nucleosynthesis	–	$(0.01 - 0.05)h^{-2}$
Inflation	–	1

NOTES: All lines except the last three are based on the luminosity density (10-24). Nucleosynthesis estimate omits density in non-baryonic matter. Several methods, such as Local Group timing and X-ray halo of M87, depend on h in complicated ways, and this dependence has been suppressed. See text for further detail.

2 Virgocentric Flow

The nearest large cluster of galaxies is the Virgo cluster, which is receding from the Local Group at about $1000\,\mathrm{km\,s^{-1}}$. The gravitational acceleration of this mass concentration should perturb the velocity field of galaxies around Virgo—including our own Local Group—away from a pure Hubble flow. In other words, our recession velocity from Virgo should be smaller than $H_0 r_{v0}$, where r_{v0} is the distance to Virgo and H_0 is the Hubble constant as determined from the velocity field well beyond Virgo. By measuring this difference we can estimate the total mass and mass-to-light ratio associated with the Virgo cluster.[14] The basic idea behind this test dates back to the 1950s (see Davis & Peebles 1983 for a historical review) but it was only in the late 1970s that the theory and observations were sufficiently well developed that significant results began to emerge.

Suppose that the mass distribution in the Virgo supercluster is spherically symmetric, that our distance from the center of Virgo at

---

[14] Usually the term **Virgo supercluster**, rather than Virgo cluster, is used in this context. A supercluster is a very large concentration of galaxies that has not yet collapsed, so that its mean density is not much larger than the background density. A supercluster may or may not have a high density cluster in virial equilibrium at its core.

time t is $r_v(t)$, and that the total mass contained inside the radius $r_v(t)$ is M. Assuming that the velocity field is smooth and single-valued, and that the random velocities of galaxies are small compared to the spherically symmetric expansion velocity, then the mass M is independent of time. The equations of motion governing $r_v(t)$ are precisely the same as the equations governing the evolution of the scale factor in a Friedmann cosmology (Appendix 9.A). In both cases the evolution is completely independent of the mass outside the radius in question, because of Birkhoff's theorem. Thus, the sphere centered on Virgo with the Local Group on its surface can be treated as its own Universe, with its own mean interior density $\rho_v(t) = 3M/[4\pi r_v(t)^3]$, its own Hubble parameter $H_{v0} = (d\ln r_v/dt)_{t_0}$ (for the rest of this section we reserve the term "Hubble constant" for the present velocity/distance ratio of galaxies much farther than Virgo), and its own density parameter

$$\Omega_{v0} = \frac{8\pi G\rho_{v0}}{3H_{v0}^2}, \qquad (10\text{-}27)$$

where t_0 is the present time and $\rho_{v0} = \rho_v(t_0)$, $r_{v0} = r_v(t_0)$.

By analogy with equations (9A-22), for $\Omega_{v0} < 1$ the evolution of $r_v(t)$ is given parametrically by the equations

$$\frac{r_v(\eta_v)}{r_{v0}} = A_v(\cosh\eta_v - 1) \quad ; \quad t(\eta_v) = B_v(\sinh\eta_v - \eta_v), \qquad (10\text{-}28)$$

where A_v and B_v are constants determined by equations (9A-23) and the condition that $r_v = r_{v0}$ when $t = t_0$. A similar set of equations holds for the case $\Omega_{v0} > 1$. The present value of the parameter η_v is related to the density parameter Ω_{v0} by the condition

$$\Omega_{v0} = \frac{2}{1 + \cosh\eta_{v0}}. \qquad (10\text{-}29)$$

We can rearrange these equations to relate η_{v0} to t_0 and ρ_{v0}:

$$H_{v0}t_0 = \frac{\sinh\eta_{v0}(\sinh\eta_{v0} - \eta_{v0})}{(\cosh\eta_{v0} - 1)^2} \quad ; \quad \rho_{v0} = \frac{3}{4\pi G t_0^2}\frac{(\sinh\eta_{v0} - \eta_{v0})^2}{(\cosh\eta_{v0} - 1)^3}. \qquad (10\text{-}30)$$

At large distances, where the gravitational influence of Virgo is negligible, we return to the usual Friedmann solution, which can be obtained from equation (10-30) by replacing Ω_{v0} by Ω_0, H_{v0} by H_0, ρ_{v0} by ρ_0, and η_{v0} by η_0. Since the time t_0 from the initial singularity is the same in both solutions, it can be eliminated to yield

$$\begin{aligned} \frac{H_{v0}}{H_0} &= \frac{h(\eta_{v0})}{h(\eta_0)} \\ \frac{\rho_{v0}}{\rho_0} &= \frac{g(\eta_{v0})}{g(\eta_0)} \end{aligned} \quad \text{where} \quad \begin{aligned} h(\eta) &= \frac{\sinh\eta(\sinh\eta - \eta)}{(\cosh\eta - 1)^2} \\ g(\eta) &= \frac{(\sinh\eta - \eta)^2}{(\cosh\eta - 1)^3} \end{aligned} \qquad (10\text{-}31)$$

and Ω_0 and $\Omega_{v0} < 1$, with similar expressions if Ω_0 or $\Omega_{v0} > 1$.

Equations (10-31) give two relations between the parameters η_{v0} and η_0 and the ratios H_{v0}/H_0 and ρ_{v0}/ρ_0. Thus, if ρ_{v0}/ρ_0 and H_{v0}/H_0 are known, we can determine η_0 and hence the density parameter Ω_0 through equation (10-29).

Note that we do not need the actual value of the Hubble constant H_0 but only the ratio H_{v0}/H_0. Thus we need only measure (i) the mean velocity of the galaxies in the core of the Virgo cluster, v_1; (ii) the velocity of some much more distant cluster, v_2; (iii) the distance ratio between the galaxies in the core of Virgo and the galaxies in the more distant cluster, $\Delta = \sqrt{f_1/f_2}$, where f_1 and f_2 are the fluxes from galaxies of the same type in the two clusters. Then $H_{v0}/H_0 = v_1 \Delta / v_2$ can be measured quite accurately, even though H_0 itself is uncertain by a factor of two or so.[15]

Similarly, to measure ρ_{v0}/ρ_0 we assume that the mean mass-to-light ratio in the sphere of radius r_{v0} centered on Virgo is the same as the mass-to-light ratio of the Universe as a whole, so that the ratio of mass densities can be determined from the ratio of luminosity densities. Here a subtle difficulty arises. In order to determine the luminosity density ratio we must know the distances of nearby galaxies to within a scale factor. The usual procedure for determining distances is to use a parameter such as the central velocity dispersion (for ellipticals) or the velocity width in HI (for spirals) that correlates with luminosity and has been calibrated using a large sample of nearby galaxies. However, the distances to the calibrating galaxies are usually determined from Hubble's law, and if the perturbation by Virgo is substantial, the usual Hubble law for determining distances is inaccurate for nearby galaxies. Instead, we must fit the luminosity indicators, fluxes and velocities of nearby galaxies to a complete model of the flow pattern of the Virgo supercluster (e.g., Schechter 1980; Aaronson et al. 1982). An example of such a flow model is shown in Figure 10-6.

There are several additional uncertainties that must be considered: (i) The Local Group may have a substantial peculiar velocity relative to

---

[15] A completely different approach to measuring H_{v0}/H_0 is based on the peculiar velocity of the Local Group relative to the cosmic microwave background radiation, which we may call $\mathbf{v}_3$. This velocity can be measured directly from the dipole anisotropy of the microwave background (Smoot & Lubin 1979, Boughn et al. 1981). Assuming that Virgo is at rest with respect to the microwave background and that the Virgo supercluster is spherically symmetric, the peculiar velocity $\mathbf{v}_3$ should point toward the center of Virgo and we would have $H_{v0}/H_0 = v_1/(v_1 + v_3)$. However, the peculiar velocity vector determined in this way points about $45°$ from Virgo, possibly as a result of the gravitational acceleration of more distant mass concentrations, so the use of this velocity vector in a spherically symmetric model is not really satisfactory.

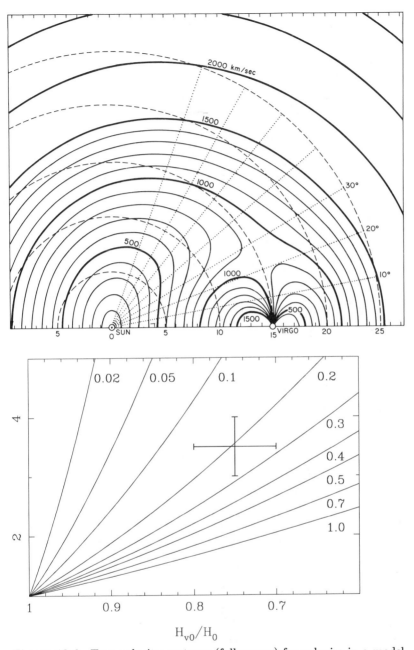

Figure 10-6. Top: velocity contours (full curves) for galaxies in a model of the Virgo supercluster. Dashed curves represent unperturbed Hubble flow. The Sun and the center of the Virgo cluster are on the horizontal axis. The distance scale is in units of $100\,\mathrm{km\,s^{-1}}$ unperturbed Hubble flow. The ratio $H_{v0}/H_0 = 0.733$ and the excess density around Virgo is assumed to vary with radius as r^{-2}. Reprinted from Tonry and Davis (1981) by permission of *The Astrophysical Journal*. Bottom: the density parameter Ω_0 as a function of H_{v0}/H_0 and ρ_{v0}/ρ_0, computed from equations (10-31). The cross shows our best estimates of the parameters. (After Davis et al. 1980.)

the mean velocity of the galaxies around it, and this should be fitted in any complete model. (ii) The core of the Virgo cluster has a complex, clumpy structure in velocity space, and thus the mean velocity of the galaxies in Virgo cannot be defined precisely. Different authors have used Virgo velocities that differ by up to 20%. (iii) Several models of the flow pattern in the Virgo supercluster that have allowed tangential motions have found substantial rotation around the center of Virgo. This effect may be related to the fact that the peculiar velocity with respect to the microwave background points at about 45° from Virgo, and suggests that spherically symmetric models are oversimplified.

Estimates of ρ_{v0}/ρ_0 in the literature range between about 3 and 4 (Davis & Huchra 1982; Yahil et al. 1980), and we shall adopt 3.5 ± 0.5. Most of the range probably arises from differences in the volume used to define ρ_0. Estimates of H_{v0}/H_0 range from 0.8 to 0.7 (see Aaronson et al. 1982 or Davis and Peebles 1983 for a review) with the range arising from differences in the assigned Virgo velocity and differences in the volume used to define H_0; we adopt $H_{v0}/H_0 = 0.75 \pm 0.05$. These values are marked on Figure 10-6 and lead us to adopt $\Omega_0 = 0.25 \pm 0.15$, which corresponds to a mass-to-light ratio of $400h\Upsilon_\odot$ by equation (10-25).

The Virgo distance represents the largest scale on which we have a direct dynamical measurement of Ω_0. The measurement confirms that large quantities of dark matter are present on this scale, but that there is not enough dark matter to close the Universe (since the upper limit to Ω_0 is substantially less than unity).

3 Gravitational Lenses

One of the classic tests of general relativity is the deflection of light by the Sun (cf. Weinberg 1972). In fact, any mass concentration bends the paths of photons traveling nearby and hence acts as a **gravitational lens** (see Problem 10-9). It has long been recognized that galaxies act as gravitational lenses, and that we might therefore expect that occasionally two or more images of a single distant object such as a quasar might be visible to observers on Earth.

The first gravitationally lensed quasar was discovered in 1979 by Walsh et al. (1979). These authors argued that the pair of quasars Q0957+561 A, B, which appeared separated on the sky by 6.2 arcsec, were really two images of a single quasar. The principal evidence for this claim was that the spectra of the two objects were nearly identical: in particular their redshifts ($z = 1.41$) were the same to better than one part in 10^4. Subsequent observations showed that there was indeed a galaxy (redshift $z = 0.39$) located between the two images in approximately the required location for a gravitational lens. The galaxy is located in a rich cluster of galaxies, and it seems likely that the cluster also contributes to the lensing action.

By now there are a half-dozen or so probable lenses, with up to four images produced by a single lens and image separations of up to 7 arcsec (see Turner 1987 for a review). Since the image characteristics are determined by the mass distribution in the lens, gravitational lenses provide a promising probe of the mass distribution in galaxies; moreover, they offer the exciting possibility of detecting mass concentrations that are *completely* dark, i.e., that have no gas, stars, or galaxies nearby.

Unfortunately, this promise has not yet been realized, in part because the images are technically difficult to observe, being rare, faint, and generally on angular scales that are close to the resolution limit of an optical telescope. Moreover, the complexities of the mass distributions of the lenses, cosmological effects, and poorly known quasar properties mean that simple models often yield misleading or ambiguous results. About the only firm conclusion so far is that models in which the mass density in the lens is everywhere proportional to the luminosity density do *not* fit the data, thus reinforcing the evidence that substantial quantities of dark matter are present in galaxies.

4 Primordial Nucleosynthesis

In Friedmann models of the Universe, the temperatures and densities in the first few seconds after the Big Bang were high enough that all nuclear species were in thermal equilibrium. As the Universe expanded and cooled, the nuclear reaction rates slowed, and the nuclides dropped out of equilibrium within a few minutes of the Big Bang. By tracing the reaction chains, we can predict the mass fraction that should be present in various nuclear species at this time. Further nucleosynthesis in stellar interiors modifies these mass fractions, but in predictable ways. Thus, comparison of the expected primordial element abundances (where "primordial" means after the first few minutes but before stellar nucleosynthesis) with observations provides us with a direct observational window on the early evolution of the Universe. The relevant calculations are described by many authors, so we will only give a brief qualitative description of the principal processes.[16]

[16] See Peebles (1971), Weinberg (1972), and Yang et al. (1984) for more detail. Note that some of the details in the earlier two references are incorrect, in particular the relation between time and temperature, because the contribution of the τ-neutrino to the energy density was not included.

We begin 10^{-4} seconds after the Big Bang, when the temperature of the black-body radiation was about $T = 10^{12}$ K, corresponding to a characteristic energy $k_B T = 86.2$ MeV. At this point the density is so high that even weakly interacting particles are in thermal equilibrium. In particular, the reactions

$$p + e^- \rightleftharpoons n + \nu_e \quad \text{and} \quad n + p \rightleftharpoons d + \gamma \qquad (10\text{-}32)$$

are in equilibrium, where n, p, and d denote the neutron, proton, and deuteron. As a consequence of the first of these reactions, the number densities of neutrons and protons are in the ratio

$$\frac{n_n}{n_p} = \exp\left(-\frac{\Delta mc^2}{k_B T}\right) = \exp\left(-\frac{1.50 \times 10^{10}\text{K}}{T}\right), \qquad (10\text{-}33)$$

where $\Delta mc^2 = 1.29$ MeV is the mass difference between neutron and proton (see Landau & Lifshitz 1980, §105, for a discussion of statistical mechanics in relativistic gases). Thus at $T = 10^{12}$K, the number density of neutrons is nearly equal to the number density of protons; the number density of deuterons, however, is much lower, basically because there are far more photons than neutrons or protons, so it is much more probable that a given deuteron will be photodissociated than that a proton will collide with a neutron to form a deuteron. The low deuteron density inhibits the formation of heavier species by reactions like $d + d \rightleftharpoons \text{He}^3 + n$.

As the Universe expands and cools, the rates of reactions such as (10-32) decrease, both because the cross-sections are lower at lower energies and because the number density of all the reagents is decreasing. By one second after the Big Bang, when the temperature has cooled to $T \simeq 1 \times 10^{10}$K, the reaction rates are not high enough to maintain the neutron-proton number density ratio in equilibrium with the cooling black body radiation. The ratio n_n/n_p, which has been decreasing rapidly according to equation (10-33), now "freezes out" at a value

$$n_n/n_p \simeq 0.20. \qquad (10\text{-}34)$$

From now on, no new neutrons are formed, and the existing neutrons decay with a half-life of 10.6 minutes.

The binding energy of the deuteron is 2.225 MeV, which equals $k_B T$ at a temperature of 2.6×10^{10}K. Thus, once the temperature has dropped well below 10^{10}K, the black-body photons are no longer able to dissociate the deuterons. However, the deuterons that do form are rapidly burned into heavier nuclei, primarily H^3, He^3, and He^4. It is difficult to build up heavier elements since there is no stable nucleus with $A = 5$, although detailed calculations show that a small amount of Li^7 is produced. Because He^4 has by far the largest binding energy of any nucleus with $A < 5$, and because of the rapid rates of the reactions that incorporate free neutrons into deuterons and then He^4, almost all of the neutrons will end up inside He^4 nuclei, with only trace amounts of deuterium, He^3, and Li^7 left over.

Since almost all of the neutrons are used in making He^4, the mass fraction in He^4 at the end of nucleosynthesis should be simply $X_{He^4} = 2n_n/(n_n + n_p)$, where the neutron and proton densities are those prevailing at the onset of He^4 production. This occurs about 10^2 seconds after the Big Bang, so the decay of the free neutrons is not yet important, and the neutron/proton ratio should be nearly equal to the value in equation (10-34). Thus we expect $X_{He^4} \simeq 0.3$. More detailed calculations show a slightly lower abundance and a weak dependence on the number density of protons and neutrons, generally giving values in the range $X_{He^4} = 0.21 - 0.28$. This result is in excellent agreement with observed values of the He^4 abundance (principally from emission lines in HII regions), $X_{He^4} = 0.26 \pm 0.04$ (Yang et al. 1984). Most of this helium must arise from primordial nucleosynthesis, since helium production in stellar nucleosynthesis would be accompanied by the synthesis of more heavy elements than are observed. The ability of the standard Big Bang model to explain the primordial helium abundance by these arguments is one of its greatest triumphs.

Since deuterons are burned so rapidly, the present deuterium abundance depends strongly (exponentially) on the competition between the rates of the nuclear reactions in the early Universe and the rate of cosmic expansion when $T \approx 10^{10}$ K. A higher density of baryons (protons plus neutrons) means the reactions proceed faster, so fewer deuterons survive. Thus the primordial deuterium abundance provides a measure of the baryon density in the early Universe, and hence—since baryons are conserved—of the present density.

The present deuterium abundance can be measured from molecular spectra in Jupiter's atmosphere, from ultraviolet line strengths in the interstellar medium, and from lunar samples. In terms of the ratio of the number densities of deuterons and protons, these measurements yield $n_d/n_p = (1 - 2) \times 10^{-5}$. This should be regarded as a lower limit to the primordial density, since some of the deuterium may have already been burned in stellar interiors. Combining this result with data on He^3 and Li^7 yields the present mass density in baryons, $\rho_B = (2 - 9) \times 10^{-31}$ g cm^{-3} (Yang et al. 1984). Comparison to the critical density $\rho_c = 3H_0^2/8\pi G$ yields

$$\Omega_B \equiv \frac{\rho_B}{\rho_c} = (0.011 - 0.048)h^{-2}, \tag{10-35}$$

where Ω_B is the fraction of the critical density supplied by baryons alone. Since $0.5 \lesssim h \lesssim 1$ [eq. (1-12)], we have $0.011 \lesssim \Omega_B \lesssim 0.19$.

Thus, baryons alone fail to close the Universe by at least a factor of five. However, it is worth noting that the range of Ω_B allowed by nucleosynthesis arguments is roughly consistent with the measurements of Ω_0 from the Virgocentric flow, and from the mass-to-light ratios of groups and clusters of galaxies. In other words, essentially all the mass in structures up to the size of the Virgo supercluster may be in baryons.

5 The Age of the Universe

A lower limit to the age of the Universe is provided by the ages of the oldest stars. Globular clusters are thought to be the oldest stellar systems in the Galaxy, and they are ideal candidates for age measurements since they provide many stars that are known to have the same distance and metallicity. Fitting color-magnitude diagrams for stars in globular clusters (MB, §§3-6 and 3-10) to theoretical models of stellar evolution yields ages of $(15 - 18) \times 10^9$ yr (Harris et al. 1983, VandenBerg 1983). Radioactive decay measurements yield similar values but with a larger spread, about $(7 - 20) \times 10^9$ yr (Hainebach & Schramm 1977).

On the other hand, the age predicted by a Friedmann model can be written $t_0 = f(\Omega_0)H_0^{-1} = 9.78 \times 10^9 h^{-1} f(\Omega_0)$ yr, where the function $f(\Omega_0)$ is 1 for $\Omega_0 = 0$ and $\frac{2}{3}$ for $\Omega_0 = 1$ (Appendix 9.A). Thus, if $\Omega_0 = 1$ and the stellar evolution chronometer is accurate, then we must have $h < 0.44$; alternatively, if we accept the consensus among most cosmologists that $0.5 \lesssim h \lesssim 1$, then we can rule out models with $\Omega_0 = 1$.

At the present time it is unclear how seriously these arguments should be taken. In particular, the same arguments imply that *any* Friedmann model is ruled out unless $h \lesssim 0.65$, and yet there is no obvious error in the estimates of the Hubble constant that yield $h \simeq 1$. Perhaps we must wait until there is general agreement on the value of the Hubble constant before using stellar evolution chronometers to constrain cosmological models.

6 Is $\Omega_0 = 1$?

There are at least two attractive arguments that imply that the density parameter Ω_0 should be almost *exactly* 1. The first of these is sometimes termed the "Copernican" or "coincidence" argument, since it is based on the Copernican principle that we should not occupy a special place or time in the Universe (see Dicke 1970 for an early version of this argument). To take a concrete example, suppose that $\Omega_0 = 0.3$ at the present time t_0. Then, according to the Friedmann equations (9A-22), an observer at $t = 0.1t_0$ would measure a density parameter $\Omega = 0.709$, at $t = 0.01t_0$, $\Omega = 0.9227$, and so on, with $\Omega \to 1$ as $t \to 0$. Thus, on a logarithmic time scale, Ω was nearly unity over most of the past history of the Universe. Similarly, $\Omega \to 0$ as $t \to \infty$: in our example, at $t = 10t_0$, $\Omega = 0.054$ and at $t = 100t_0$, $\Omega = 0.006$. Thus, over most of the future of the Universe Ω will be nearly zero. Similarly, in a closed Universe Ω is usually near unity or much greater than unity. In summary, if $\Omega(t)$ in any Friedmann model is plotted on a logarithmic time scale, then it is almost always nearly 1, nearly 0, or very large. There is no obvious reason why the process of evolution of intelligent life should bring us to the point where we ask these questions during the relatively short time when Ω_0 is neither near zero, near unity, nor very large, and we know that Ω_0 is neither near zero nor very large. Hence, in the absence of an inexplicable coincidence, Ω_0 must be near unity.

The second argument that suggests that Ω_0 is almost exactly 1 is based on the concept of an "inflationary" stage in the early history of the Universe (Guth 1981; Linde 1982; Albrecht & Steinhardt 1982; see Guth 1986 for a review). This idea arose from investigations of grand unified theories of particle physics, which attempt to unite theories of the strong, electromagnetic, and weak forces. In some of these theories there is an equilibrium state of matter at very high temperatures ($T \gtrsim 10^{27}\text{K}$) called the **false vacuum**. The false vacuum has the curious property that its mass density is fixed at a constant value $\rho_{\text{vac}} \approx 10^{74}$ g cm^{-3}. The equation of state of the false vacuum can be determined by the following simple argument. Consider a cylinder containing a piston that is filled with false vacuum and surrounded by ordinary vacuum. Let the piston be pulled out so that the volume of the chamber increases by V. The mass of false vacuum has increased by $\rho_{\text{vac}}V$, and its energy has therefore increased by $\rho_{\text{vac}}c^2V$, which must equal the work done on the piston, $-pV$. Thus the pressure of the false vacuum is $p = -\rho_{\text{vac}}c^2$. If the Universe is filled with false vacuum, then the Friedmann equation for the evolution of the scale factor $R(t)$ [eq. (9A-11)] becomes

$$\ddot{R}(t) = -\frac{4\pi}{3}G\left(\rho + \frac{3p}{c^2}\right)R(t) = \frac{8\pi G\rho_{\text{vac}}R(t)}{3}, \tag{10-36}$$

which has the solution

$$R(t) \propto \exp(\chi t), \tag{10-37}$$

where $\chi = \sqrt{\frac{8}{3}\pi G \rho_{vac}}$. The scale factor expands exponentially with a time constant $\chi^{-1} \approx 10^{-34}$ seconds, and the interval described by equation (10-37) is known as the **inflationary phase**. Inflation ends when the false vacuum undergoes a phase transition into normal matter. The effect of inflation on the density parameter can be seen by rewriting the Friedmann equation (9A-14) in the form

$$\Omega(t) = 1 + \frac{kc^2}{r_u^2 \dot{R}(t)^2}, \qquad (10\text{-}38)$$

where $\Omega(t) = \frac{8}{3}\pi G \rho(t)/H(t)^2$ and $H(t) = \dot{R}/R$ [at the present time t_0 equation (10-38) reduces to (9A-16)]. During inflation the curvature k is constant but $\dot{R}$ increases exponentially. After a few e-foldings, the density parameter $\Omega(t)$ differs from unity by an extremely small factor. Hence Ω is almost exactly unity at the end of the inflationary phase. Even after inflation is over, Ω will remain near unity since $\dot{R}^2(1 - \Omega)$ is constant in any Friedmann model [by eq. (9A-14)] and $1 - \Omega$ is exponentially small as a result of the inflation.

The great beauty of the inflationary scenario is that it explains at least three aspects of the observed Universe that most other models have to put in as initial conditions: (i) as we have shown, inflationary models produce $\Omega_0 = 1$ without any fine-tuning of the initial conditions; (ii) the density fluctuations that produce galaxies and clusters of galaxies arise naturally and inevitably from zero-point quantum fluctuations; (iii) the homogeneity of the microwave background is a consequence of the fact that all parts of the present Universe were causally connected before the inflationary phase. Thus the basic idea of inflation is extremely attractive, although many details may be expected to change as our understanding of fundamental physics develops.

Since both inflation and the "Copernican" argument given above suggest that $\Omega_0 = 1$, we must ask why our dynamical measurements—in particular the Virgocentric flow—all yield $\Omega_0 < 1$. One possible explanation is that galaxy formation is **biased**, that is, galaxies preferentially form in high density regions. If this were so, then the excess of luminous matter around Virgo would not reflect an equivalent excess of mass. Consequently our estimate of the density excess ρ_{v0}/ρ_0 would have to be revised downward, which would increase the implied value of Ω_0.

We must also ask how the nucleosynthesis arguments, which give $\Omega_B \lesssim 0.2$, are consistent with $\Omega_0 = 1$. One possible solution, which we discuss further below, is that most of the mass in the Universe is in some species of non-baryonic particle that has not yet been detected.[17]

10.4 The Composition of the Dark Matter

There is a simple model of the mass distribution in the Universe that is consistent with most of the available data. In this model, the Universe is open, with $\Omega_0 \simeq 0.2$, as suggested by dynamical measurements of the mass in groups and clusters and by the Virgocentric flow. Furthermore, this mass density is consistent with the upper limit on the mass density in baryons for $h \simeq 0.5$, so that virtually all of the mass in the Universe may be in the form of baryons. With $\Omega_0 = 0.2$ and $h = 0.5$, the age of the Universe is 16.6 Gyr, consistent with the ages of globular clusters.

However, this model has some serious flaws. It is incompatible with inflation and the coincidence argument of §10.3.6, both of which suggest that $\Omega_0 = 1$. Moreover, it appears to be inconsistent with observational constraints on fluctuations in the microwave background (see Bond & Efstathiou 1984).

Hence we are led to consider a second model, in which $\Omega_0 = 1$, as suggested by inflation. In this case, the lower values of Ω_0 measured by dynamical tests arise because galaxies preferentially form in high-density regions. Since the total mass density in baryons can contribute at most 20% of the critical density, the majority of the mass in the Universe must be in some non-baryonic form.

In either of these two models, the central question is: what is the nature of the dark matter? This is probably the single most important unresolved question in extragalactic astronomy, and we are far from a definitive answer. In this section we will sketch the present constraints on the composition of the dark matter, and explore whether there are any alternatives to the conclusion that large amounts of dark matter must be present.

1 Dark Matter in Baryons

Whether or not the majority of the mass in the Universe is non-baryonic, baryonic dark matter must be a major constituent of galaxies and the Universe. Recall that even in the solar neighborhood, roughly half of

---

[17] Of course, in this case our attempt to avoid one improbable coincidence—that Ω_0 is neither zero nor unity—will have landed us in another—that the densities of nucleons and non-baryonic matter are approximately the same.

the mass is dark, and that this dark mass was almost certainly collisional at some time in the past since it now has a disklike configuration.
This implies that it was once gaseous and hence must be composed of
baryons. In addition, there must be other forms of baryonic dark matter containing still more mass, for the mass-to-light ratio in the solar
neighborhood of $5\Upsilon_\odot$ [eq. (10-4)] would yield only $\Omega_0 = 0.003h^{-1}$ by
equation (10-25), a factor of almost four less than the lower limit on the
baryonic contribution to Ω_0 implied by the nucleosynthesis arguments
in §10.3.4. What are the candidates for baryonic dark matter?

(a) Low-luminosity stars and stellar remnants   One possible
explanation is that the star formation process happens to convert most
of a given mass of interstellar gas into **brown dwarfs**, that is, stars with
masses too low to burn hydrogen ($\lesssim 0.08\,\mathrm{M}_\odot$). These stars are very difficult to detect unless they are in binary systems (McCarthy et al. 1985),
and even if all of the local dark mass were in brown dwarfs, the nearest
one to the Sun would still be too faint to be easily detectable. Alternatively, star formation may produce mainly high mass stars, which have
by now completed their evolution and turned into white dwarfs, neutron
stars, or black holes. In the latter case the principal observational constraint is that the density in such remnants cannot exceed $\Omega_0 \approx 0.03$
or else their integrated light output would contribute too much to the
background radiation density (Carr et al. 1984). Thus stellar remnants
could supply all the dark mass in the solar neighborhood but not all the
dark mass in clusters of galaxies or the Virgo supercluster.

If the dark mass consists of brown dwarfs, there must be an unex-
pected upturn in the initial mass function (IMF) below $0.08\,\mathrm{M}_\odot$, since an
extrapolation of the observed IMF (MB, §4-2) predicts very little mass
in stars with $M < 0.08\,\mathrm{M}_\odot$. Similarly, if the dark mass consists of remnants of high-mass stars, the IMF must be bimodal (two-peaked) and
the formation rate of massive stars must decline with time more rapidly
than the formation rate of low-mass stars; Larson (1986) has argued
that models of just this kind are consistent with all of the available constraints on the evolution and stellar content of the solar neighborhood.
Unfortunately, at the present time there is no known way to detect the
required population of either brown dwarfs or massive remnants directly.

(b) Small solid bodies   Objects such as comets, asteroids or dust
grains[18] are unlikely sources of the dark mass because they are mainly

---

[18] Small solid bodies are distinct from very low mass stars in that they are held
together primarily by molecular forces rather than by gravity.

composed of elements such as silicon, carbon, and oxygen, which are always much less abundant than hydrogen and helium. These elements are never found in abundances exceeding about 1% of the hydrogen abundance, and should not be present at all unless the dark matter has somehow been processed through stars. Thus it seems unreasonable that there could be appreciable amounts of nonvolatile dark matter without much larger amounts of hydrogen. One might argue that the hydrogen is hidden in the form of solid snowballs, but these would evaporate (Hegyi & Olive 1983).

(c) Neutral and ionized gas   Most of the mass in any primordial gas must be in the form of hydrogen. We have already seen [§10.1.7(c)] that many elliptical galaxies contain hot gas that is detectable in the X-ray band. However, the mass in this gas is too small to account for the dark mass. For example, in M87 the total mass within 100 kpc, as determined from the density and temperature profiles of the X-ray-emitting gas, is over $10^{13} M_\odot$, but the total gas mass is only $3 \times 10^{11} M_\odot$ (Fabricant & Gorenstein 1983).

An alternative hypothesis is that the gas is neutral and resides in an extended disk distribution. However, radio observations at 21 cm show that the mass of neutral hydrogen gas in the outer parts of galaxies is far too small to provide the mass required for flat rotation curves.

It is also possible to place stringent limits on the density of inter-galactic hydrogen gas. Shortly after quasars were discovered, Gunn and Peterson (1965) searched for neutral intergalactic gas by looking for attenuation of quasar radiation due to absorption by the Lyman-α ($n = 1$ to $n = 2$) transition of hydrogen at $\lambda = 1215$Å. By using quasars that are sufficiently distant, the Lyman-α line is redshifted to visible wavelengths, so that the measurement can be made with ordinary optical telescopes rather than space-based ultraviolet instruments. This test is extremely sensitive because of the large cross-section of this transition, and yields the remarkably strong limit $\Omega_{HI} \lesssim 4 \times 10^{-7} h^{-1}$, where Ω_{HI} is the fraction of the critical density in neutral hydrogen (see Peebles 1971 or Field 1972 for details). A similar test can be used to rule out large quantities of molecular hydrogen.

Recently, more sensitive measurements of Lyman-α absorption in quasar spectra have actually detected some intergalactic neutral hydrogen in the form of discrete clouds (Sargent et al. 1980). The neutral hydrogen content of these clouds is consistent with the limit set by the Gunn-Peterson test. However, only a small fraction of the mass in these clouds is neutral, since most of the atoms have been photoionized by ultraviolet radiation from quasars. Hence their contribution to the mean density in ionized hydrogen corresponds to the much larger—but still negligible—value $\Omega_{HII} \approx 10^{-3}$. Moreover, the condition that the clouds have not been overheated by conduction sets a limit on the density of a homogeneous ionized intergalactic medium of $\Omega_{HII} \approx 0.02h^{-2}$.

Thus, there do not appear to be cosmologically significant quantities of either ionized or neutral gas, whether in galactic halos, discrete intergalactic clouds, or a uniform intergalactic medium.

(d) Massive black holes   One interesting possibility is that an early epoch of star formation, which possibly occurred even before galaxy formation, produced stars of very high mass that by now have collapsed into black holes (Carr et al. 1984). If the stars are sufficiently massive, $M > M_c$, where M_c is probably a few hundred solar masses, then they collapse directly to a black hole, without expelling mass in an explosion as less massive stars do. The difference between these two cases is crucial, since strong limits can be set on the number density of exploding stars by requiring that they do not pollute the interstellar or intergalactic medium with too many heavy elements.[19] Thus the maximum allowed density in black-hole remnants of stars with $M < M_c$ is only $\Omega_0 \lesssim 10^{-4}$ (Carr et al. 1984), while for $M > M_c$ the observational constraints are much less severe. For example, all of the dark mass in halos could be in black holes with masses satisfying $M_c < M < 10^6 \, M_\odot$; the upper limit at $10^6 \, M_\odot$ arises from the requirement that gravitational perturbations from the holes do not heat up the disk stars so much that the velocity dispersion-age relation for stars in the solar neighborhood is violated (see §7.5 and Lacey & Ostriker 1985). More stringent limits may be set from the gravitational effects of the black holes on photons arriving from distant quasars: Gorenstein et al. (1984) argue that the absence of substructure on scales $\gtrsim 10^{-3}$ arcsec in their very high resolution maps of the gravitationally lensed quasar Q0957+561 implies that $M \lesssim 10^5 \, M_\odot$, while Canizares (1982) argues that the absence of visible lensing effects in most quasars implies that the density in a cosmological population of black holes with masses $\gtrsim 0.01 \, M_\odot$ is negligible.

2 Other Kinds of Dark Matter

If the inflationary scenario is correct and our understanding of primordial nucleosynthesis is correct, then at least 80% of the matter in the

---

[19] However, a modest amount of pollution is desirable, to explain why no stars have been observed with zero or near-zero metal abundance (Carney 1984).

Universe cannot be composed of baryons. A large zoo of exotic particles has been proposed at one time or another as candidates for this dark mass, including massive neutrinos, gravitinos, axions, monopoles, and photinos. Although exotic particles are regarded by many cosmologists as the most attractive candidates for dark matter in galactic halos and on larger scales (see Primack 1987 for a review), they cannot provide the dark matter in galactic disks since they are not dissipative.

In many respects the discussion of these particles is the province of high-energy physics rather than cosmology, and we might hope that when the correct grand unified theory of the strong, electromagnetic, and weak interactions is developed, it will predict not only the properties of the particles that make up the dark mass, but also the initial baryon asymmetry that produced the present density of baryonic matter relative to dark matter. At the present time, however, cosmology appears to be better understood than particle physics, and hence the requirement of consistency with cosmological observations has become one of the strongest available constraints on possible grand unified theories.

To provide a concrete example, we discuss the possibility that the dark matter is composed of massive neutrinos. First, let us consider the role of zero-mass neutrinos in cosmology. There are three known species of neutrino, electron, muon, and tau neutrinos, and each species has its own antineutrino. Shortly after the Big Bang, all of these species are in thermal equilibrium, so the number density of a given species of neutrinos in phase space is given by the Fermi-Dirac distribution,

$$f(\mathbf{p}) = \frac{2g}{h_P^3} \frac{1}{\exp(E/kT) + 1}, \tag{10-39}$$

where E is the neutrino energy, h_P is Planck's constant, k is Boltzmann's constant, g is the number of spin states ($g = 1$), and the factor 2 accounts for particles and antiparticles. As the Universe expands and cools, the typical neutrino collision rate decreases, both because the weak interaction cross-section is lower at lower energy and because the number density is falling. Eventually, at a temperature $T_d \approx 2 \times 10^{10}$K, the time between collisions exceeds the expansion time, and the neutrinos drop out of thermal equilibrium. As the Universe continues to expand and cool, the neutrinos are redshifted to lower energies but they do not interact any further; thus the Universe today is filled with a uniform background of neutrinos left over from the Big Bang in much the same way that it is filled with a background of microwave photons. The number density of neutrinos can therefore be calculated in terms of the number density of photons; for each of the three species we have (e.g., Weinberg 1972),

$$n_\nu = \tfrac{3}{11} n_\gamma. \tag{10-40}$$

Now consider what happens if one or more species of neutrino have a non-zero rest mass m_ν such that $m_\nu c^2 \ll kT_d \approx 2$ MeV. Since the neutrinos are still relativistic when they drop out of thermal equilibrium, and since they do not interact afterward, the number density of neutrinos is unchanged and is still given by equation (10-40). For interesting values of m_ν the neutrinos are now non-relativistic, so the mass density is

$$\rho_\nu = \tfrac{3}{11} n_\gamma \sum_{i=1}^{3} m_{\nu i}, \qquad (10\text{-}41)$$

where the sum is over the three species. From the black-body radiation formula, we have $n_\gamma = 16\pi\varsigma(3)(kT_\gamma/h_P c)^3$, where $\varsigma(3) = 1.202$ is the Riemann zeta function and $T_\gamma = 2.7$ K is the present temperature of the cosmic background radiation; thus $n_\gamma = 403\,\mathrm{cm}^{-3}$. In terms of the critical density $\rho_c = 3H_0^2/8\pi G$, we may therefore write

$$\Omega_\nu = \frac{\rho_\nu}{\rho_c} = 0.010 h^{-2} \sum_i \frac{m_{\nu i} c^2}{1\ \mathrm{eV}}. \qquad (10\text{-}42)$$

A remarkable feature of this equation is that the present density of the neutrinos is directly related to the mass of the neutrino, with no free parameters. Thus, if we assume that one species of the neutrino is substantially more massive than the others, and $\Omega_\nu = 1$, then the mass of this neutrino must be $m_\nu = 100 h^2 \mathrm{eV}/c^2$. It is also remarkable that the cosmologically interesting neutrino masses are close to the experimental upper limits to the electron neutrino rest mass, $\lesssim 50$ eV/c^2 (see Primack 1987 for references). If future experiments detect a non-zero neutrino mass in the range of tens of eV, then it is inevitable that massive neutrinos make a substantial contribution to the dark mass.

Despite these agreeable features, there are serious problems with massive neutrino dark matter. While the neutrinos are still relativistic they travel in random directions at close to the speed of light, thereby washing out any small-scale primordial density fluctuations by phase mixing (§4.7.2). For example, with a neutrino mass of 30 eV/c^2 any density fluctuations with a mass $\lesssim 3 \times 10^{15}$ M$_\odot$ are erased by the time the neutrinos become non-relativistic and their velocities drop below the speed of light (Bond et al. 1980). Without any small-scale fluctuations it proves to be difficult to construct a model of galaxy formation that is consistent with the observations; in particular, galaxy formation must occur at redshifts $z \lesssim 2$, which is inconsistent with the presence of quasars with $z > 3$, since quasars appear to be associated with galaxies. (See Problem 10-10 for another possible difficulty.)

The process of galaxy formation clearly depends strongly on the degree of phase mixing experienced by the dark matter. This dependence has led cosmologists to classify hypothetical dark matter candidates into three broad categories (Bond et al. 1984): (i) **hot dark matter**, for which phase mixing occurs on all scales $\lesssim 10^{15}\,M_{\odot}$; hot dark matter candidates are usually low-mass particles like neutrinos that have large random velocities; (ii) **warm dark matter**, which phase mixes on scales $\lesssim 10^{11}\,M_{\odot}$, similar to the mass of a galaxy; and (iii) **cold dark matter**, for which phase mixing is negligible, usually because the particles are so massive that their random velocities are small. The only category that can produce models that are consistent with data on the spatial distribution of galaxies is cold dark matter (Davis et al. 1985). Cold dark matter models with $\Omega_0 = 1$ and biased galaxy formation can explain the observed masses of galaxies and many of the features of galaxy clustering, and are consistent with limits on microwave background fluctuations. Many exotic particles are candidates for cold dark matter, including the axion and the photino; however, at present there is no direct evidence that any of them exist. Note that massive black holes formed in an early generation of stars (as discussed above) would behave like cold dark matter insofar as galaxy formation models are concerned.

3 New Physics

It is worth remembering that all of the discussion so far has been based on the premise that Newtonian gravity and general relativity are correct on large scales. In fact, there is little or no direct evidence that conventional theories of gravity are correct on scales much larger than a parsec or so. Newtonian gravity works extremely well on scales of $\sim 10^{14}$ cm (the solar system) and there is some evidence that it works on scales of $10^{17} - 10^{19}$ cm (the conventional models of the Oort comet cloud and of the cores of globular and open clusters seem to work well), but on larger scales there are no direct quantitative tests of its validity. It is principally the elegance of general relativity theory and its success in solar system tests that lead us to the bold extrapolation that the gravitational acceleration has the form GM/r^2 on the scales $10^{21} - 10^{26}$ cm that are relevant for the solar neighborhood, galaxies, clusters of galaxies, and superclusters.

(a) Modified gravity   A number of authors, starting with Finzi (1963), have proposed that the gravitational potential might deviate from the usual $-GM/r$ form at large distances. For example, Sanders (1984) has suggested that the potential has the form

$$U = -\frac{GM}{(1-\beta)r}\left(1 - \beta e^{-r/r_0}\right). \qquad (10\text{-}43)$$

Thus the gravitational force has the usual inverse square form at both very small and very large distances, but the gravitational constant at very large distances is $G' = G/(1 - \beta)$. It is possible to choose values for the adjustable parameters r_0 and β so that observations of mass-to-light ratios in the outer parts of galaxies and in groups and clusters of galaxies can be fitted without any dark mass. However, the dark mass in the solar neighborhood cannot be explained in this way, since the distances involved are small enough that Newtonian gravity should still apply.

An alternative proposal (Milgrom 1983; Bekenstein & Milgrom 1984) is that Newton's law of gravitation fails when the potential gradient is small rather than when the distance is large. Thus, Poisson's equation (2-10) is modified to

$$\nabla\left(\frac{|\nabla\Phi|}{a_0}\nabla\Phi\right) = 4\pi G\rho, \qquad \text{when } |\nabla\Phi| \ll a_0, \qquad (10\text{-}44)$$

where a_0 is a universal constant. This modification can be shown to arise in a fairly natural way from a non-relativistic Lagrangian formulation. Milgrom finds that with an appropriate value of a_0 ($\approx 8 \times 10^{-8}h^{-2}\,\mathrm{cm\ s^{-2}}$) he does not require dark matter in any systems, either the solar neighborhood, galactic halos, clusters and groups of galaxies, or the Virgo supercluster. At the present time there is no satisfactory cosmological theory consistent with the Milgrom-Bekenstein modification.

(b) The cosmological constant   There is one nonstandard theory of gravitation that can be easily incorporated into cosmological models. This is a modification to Einstein's equations that is equivalent to the addition of a fictitious fluid with density and pressure[20]

$$\rho_\lambda = \frac{\lambda}{8\pi G}, \qquad p_\lambda = -\frac{\lambda c^2}{8\pi G}, \qquad (10\text{-}45)$$

---

[20] Note that the equation of state is that of the false vacuum, as described in §10.3.6.

where λ is a universal constant with units $(\text{time})^{-2}$, the **cosmological constant** (see Weinberg 1972, Gunn & Tinsley 1975, and Peebles 1984 for more detailed discussion).

When Einstein introduced the general theory of relativity, the Universe was believed to be static. There is no static Friedmann model with zero cosmological constant and non-zero density. Hence Einstein was led to consider the possibility that λ was non-zero. Of course, a static model of the Universe is now known to be incorrect because distant galaxies are receding from us, according to Hubble's law. Nevertheless, the existence of a non-zero cosmological constant remains an interesting possibility that cannot be excluded.

An attractive feature of the cosmological constant is that it can reconcile the density determined from cosmological tests, $\Omega_0 \simeq 0.2$, with the predictions of inflation. The enormous expansion in the scale factor $R(t)$ during inflation [eq. (10-37)] guarantees that the integration constant E in equation (9A-14) is presently negligible, and hence an inflationary Universe satisfies the relation

$$\Omega_0 = 1 - \frac{\lambda}{3H_0^2}. \tag{10-46}$$

Thus if $\Omega_0 = 0.2$ we have $\lambda = 2.5 \times 10^{-35} h^2 \text{sec}^{-2}$. It appears that a model Universe with this value of λ is consistent with all available observations (Peebles 1984) as well as with inflation. All of the mass can be in baryons and there is no need for any exotic particles to comprise most of the mass of the Universe. However, these models are subject to the same "coincidence" objection that was made to models with $\Omega_0 < 1$ in §10.3.6: during most of the evolution of this model, Ω is either near unity or near zero, and why should we be living at the special epoch when Ω_0 is neither 1 nor 0? In addition, there appears to be no natural combination of fundamental constants that would produce a numerical value close to the one required for λ.

At present, all of the suggestions that the need for dark matter can be eliminated by new physics seem somewhat ad hoc and contrived. It may well be that "dark matter" is really a manifestation of the breakdown of Newtonian gravity and general relativity, but overall the dark matter hypothesis provides a more economical explanation. If a new theory of gravity is required, it will ultimately be accepted because of its beauty and unifying properties rather than because it eliminates the need for dark matter.

10.5 Summary

The largest astronomical systems whose composition and dynamics we think we understand are the globular and open clusters, which are a few parsecs in size. On all larger scales, there is strong evidence that a substantial fraction of the total mass is in some form that we cannot detect, except by its gravitational effects. There is dark mass in the solar neighborhood, in the outer parts of individual galaxies, in groups and clusters of galaxies, and in the Virgo supercluster. The ratio of dark to luminous matter increases in systems of larger size, but the mean density of dark matter implied by dynamical measurements does not appear to be sufficient to close the Universe if the galaxy density is proportional to the mass density.

The most attractive model at the present time is that the Universe is at the critical density, $\Omega_0 = 1$, as a result of an early inflationary phase. This model requires that most of the mass should reside in some unknown non-baryonic form, and the galaxies should be more clumped than the mass.

Dark matter is likely to prove to be one of the most important discoveries of modern astronomy. The questions of the nature, origin, and distribution of dark matter are probably the most important issues in contemporary extragalactic astronomy; and if the dark matter is composed of non-baryonic elementary particles, then these issues are of major relevance to particle physics as well. For this reason alone the study of galactic dynamics and evolution is likely to remain a central and exciting area of astrophysics for many decades to come.

The continuing search for dark matter is an exciting detective story, and as the search continues it will be appropriate to bear in mind a famous maxim of the great fictional detective Sherlock Holmes: "When you have eliminated the impossible, whatever remains, however improbable, must be the truth."

Problems

10-1. [1] This problem explores the effect of errors in the distance scale (as determined by Hubble's law with Hubble constant H_0) on mass determinations.

(a) Suppose that the mass of a disk galaxy is determined by measuring the rotation curve out to some given angular radius from the center of the galaxy. Show that the mass contained within this radius is proportional to H_0^{-1}.

(b) Show that the mass-to-light ratio as determined from the mass measured in part (a) is proportional to H_0.

(c) Show that the mass-to-light ratio of a group or cluster of galaxies, as determined using the virial theorem, is proportional to H_0.

(d) Zwicky (1933) compared the mass-to-light ratio of the Coma cluster as measured using the virial theorem with the mass-to-light ratio of the luminous parts of individual galaxies as measured by rotation curves, and concluded that there was 400 times as much dark matter as luminous matter in the Coma cluster. However, Zwicky's conclusion was based on a Hubble constant $H_0 = 558 \, \text{km s}^{-1} \text{Mpc}^{-1}$. How is his conclusion about the ratio of dark to luminous matter affected, now that we believe that the Hubble constant is smaller by a factor of 5 to 10?

10-2. [1] Suppose that the dark matter is composed of iron asteroids of density $\rho = 8 \, \text{g cm}^{-3}$ and radius r that are uniformly distributed throughout intergalactic space (ignore for the moment the question of how to produce so much iron from primordial helium and hydrogen). For a Universe with $\Omega_0 = 1$, find an approximate lower limit on r from the condition that the Universe is not opaque, i.e., that we can see distant quasars. Your answer need only be correct to within an order of magnitude.

10-3. [1] The globular cluster NGC 5694 is found at position $\ell = 332°$, $b = 30°$, $26 \pm 5 \, \text{kpc}$ from the galactic center. Its radial velocity relative to the local standard of rest is $-180 \pm 20 \, \text{km s}^{-1}$. Assume that the Galaxy is spherical and has a flat rotation curve with circular speed $v_c = 220 \, \text{km s}^{-1}$ out to some maximum radius r_*, after which the rotation curve is Keplerian [cf. eq. (2-192)]. Find a lower limit to r_* from the requirement that NGC 5694 is bound (see Harris & Hesser 1976).

10-4. [1] Suppose that most of the mass interior to the solar radius R_0 were in a spherical dark halo. If the rotation curve is flat, $v_c = 220 \, \text{km s}^{-1}$, and $R_0 = 8.5 \, \text{kpc}$, what is the density of the dark halo in the solar neighborhood? Show that the contribution of the halo density to the Oort limit is negligible.

10-5. [2] This problem and the next describe derivations of equation (10-6) relating GM to the average value of $v_r^2 r$ for an ensemble of particles on Kepler orbits with an isotropic velocity ellipsoid.

(a) We begin with the Jeans equation for spherical systems with isotropic velocity ellipsoid [eq. (4-54) with $\beta = 0$]. Let the potential be that of a point mass, $\Phi(r) = -GM/r$. Multiply the Jeans equation by $\nu(r)r^4$ and integrate over r to show that

$$4 \int_0^\infty \nu \overline{v_r^2} r^3 \, dr = GM \int_0^\infty \nu r^2 \, dr, \qquad (10P\text{-}1)$$

where $\nu(r)$ is the number density of particles and $\overline{v_r^2}$ is the mean-square radial velocity at r.

(b) Show that equation (10P-1) is equivalent to equation (10-6).

10-6. [2] This problem describes an alternative derivation of equation (10-6).

(a) Using the properties of Kepler orbits from §3.1(b), prove that for an orbit of eccentricity e

$$\langle v_r^2 r \rangle_t = \tfrac{1}{2} GM e^2, \qquad (10\text{P-}2)$$

where v_r is the radial velocity, r is the distance, and the angle brackets $\langle \cdot \rangle_t$ denote a time average over one orbit. [Hint: Use eq. (3-26)].

(b) For isotropic orbits the mean-square eccentricity is $\tfrac{1}{2}$ (see Problem 4-22). Use this result together with equation (10P-2) to derive equation (10-6).

10-7. [2] In principle, the density of matter in the solar neighborhood can be measured from its effects on planetary orbits. Assume that the solar system is permeated by a uniform medium of density $0.1\,M_\odot\,\mathrm{pc}^{-3}$. Estimate the rate of precession of the perihelion of Neptune (radius 30 astronomical units) due to the perturbing force from this medium, and compare your result to the minimum measurable precession ≈ 0.01 arcsec yr^{-1}.

10-8. [2] Is the Local Group bound to the Virgo cluster? You may base your answer on the best values derived in §10.3.2, $\rho_{v0}/\rho_0 = 3.5 \pm 0.5$ and $H_{v0}/H_0 = 0.75 \pm 0.05$, and you may assume a spherically symmetric Virgo potential.

10-9. [2] In this problem we investigate the simple theory of gravitational lenses. For further detail, see Young et al. (1980).

(a) Consider a non-relativistic particle that travels with speed v past a spherically symmetric galaxy, in which $M(r)$ is the mass contained within radius r. Assume that the velocity is sufficiently large that the deflection of the trajectory due to the gravitational force from the galaxy is small, and let the distance of closest approach to the center of the galaxy be b. If $\alpha_{\mathrm{nr}}(v) \ll 1$ is the angle through which the trajectory is deflected, show that

$$\alpha_{\mathrm{nr}}(v) = \frac{2Gb}{v^2} \int_b^\infty \frac{dr}{\sqrt{r^2 - b^2}} \frac{M(r)}{r^2}. \qquad (10\text{P-}3)$$

[This is a generalization of equation (4-3).] This result holds only for non-relativistic particles, but in general relativity it can be shown that the deflection angle for photons is simply

$$\alpha = 2\alpha_{\mathrm{nr}}(c). \qquad (10\text{P-}4)$$

(b) Assume that the galaxy has a flat rotation curve with circular speed v_c [eq. (10-9)]. Show that the deflection angle for photons is

$$\alpha = \frac{2\pi v_c^2}{c^2}. \qquad (10\text{P-}5)$$

(c) If $\Omega_0 = 1$, so that space is Euclidean, show that the angular separation of two images formed by a singular isothermal sphere is

$$\Delta\theta = \frac{2\alpha D_{sl}}{D_s} = \frac{8\pi D_{sl}\sigma^2}{D_s c^2}, \qquad (10\text{P-6})$$

where D_{sl} is the distance between the source and the lens (the galaxy), and D_s is the distance between the source and the observer.

(d) In the case of Q0957+561, $D_{sl}/D_s = 0.573$ for $\Omega_0 = 1$ (see Young et al. 1980). Given that the separation of the images is $\Delta\theta = 6.2$ arcsec, show that the rotation speed v_c of the lensing galaxy is $610\,\text{km}\,\text{s}^{-1}$, assuming that its rotation curve is flat and that the contribution of the cluster to the lensing is negligible.

10-10. [2] This problem describes a possible argument against massive neutrinos as the source of the dark matter. The velocity dispersions of bright stars in dwarf elliptical galaxies imply that most of the mass in these systems may be dark. The central parts of the Draco dwarf galaxy (Table 10-1) may be modeled as an isothermal sphere [eq. (4-124b)] with King radius $r_0 = 150\,\text{pc}$ and velocity dispersion $\sigma = 9\,\text{km}\,\text{s}^{-1}$ (Kormendy 1987). If most of the mass in Draco resides in one species of massive neutrino and its antineutrino, find a lower limit on the neutrino mass in eV/c^2 from the condition that the Pauli exclusion principle is not violated. Show that this limit is inconsistent with the neutrino mass derived in §10.4.2 for $\Omega_\nu \leq 1$ (Tremaine & Gunn 1979).

Appendixes

Appendix 1. A Useful Numbers[1]

Physical Constants

Gravitational constant	$G = 6.672(4) \times 10^{-8}$ cm^3 g^{-1} sec^{-2}
Speed of light	$c = 2.99792458 \times 10^{10}$ cm s^{-1} (by definition)
Planck's constant	$h_P = 6.62618(4) \times 10^{-27}$ erg s
	$\hbar_P = 1.054589(6) \times 10^{-27}$ erg s
Boltzmann's constant	$k_B = 1.38066(4) \times 10^{-16}$ erg K^{-1}
Electron charge	$e = 4.80324(1) \times 10^{-10}$ esu
	$= 1.602189(5) \times 10^{-19}$ coulomb
Proton mass	$m_p = 1.672649(9) \times 10^{-24}$ g
Electron mass	$m_e = 9.10953(5) \times 10^{-28}$ g
Stefan-Boltzmann constant	$\sigma = \pi^2 k_B^4/(60\hbar_P^3 c^2)$
	$= 5.6703(7) \times 10^{-5}$ erg s^{-1} cm^{-2} K^{-4}
Thomson cross-section	$\sigma_T = 8\pi e^4/(3m_e^2 c^4)$
	$= 6.65245(6) \times 10^{-25}$ cm^2

Astronomical Constants

Astronomical unit	1 AU $= 1.49597892(1) \times 10^{13}$ cm
Parsec	1 pc $= 648,000/\pi$ AU
	$= 3.08567802(2) \times 10^{18}$ cm
1 sidereal year (1900.0)	1 yr $= 3.1558149984 \times 10^7$ sec
Hubble constant	$H_0 = 100h$ km s^{-1} Mpc^{-1} where $0.5 \lesssim h \lesssim 1$
Hubble time	$H_0^{-1} = 9.78h^{-1} \times 10^9$ yr
Solar mass	$\mathrm{M}_\odot = 1.989(2) \times 10^{33}$ cm
Gaussian constant	$G\,\mathrm{M}_\odot = 1.32712497(1) \times 10^{26}$ cm^3 s^{-2}
Solar radius	$\mathrm{R}_\odot = 6.9599(7) \times 10^{10}$ cm
Solar luminosity (bolometric)	$\mathrm{L}_\odot = 3.826(8) \times 10^{33}$ erg s^{-1}
Escape speed from Sun	$v_\star = \sqrt{2G\,\mathrm{M}_\odot/\mathrm{R}_\odot}$
	$= 617.5$ km s^{-1}
Solar absolute magnitude	$M_V = +4.83$
	$M_B = +5.48$
Earth mass	$\mathrm{M}_\oplus = 5.976(4) \times 10^{27}$ g

---

[1] Taken from Allen (1973), Lang (1980), and Particle Data Group (1984). Numbers given in parentheses indicate one standard deviation uncertainty in the last digits of the preceding number.

Useful Relations

$$1\,\mathrm{km\,s^{-1}} \simeq 1\,\mathrm{pc} \text{ per million years (actually 1.023)}$$
$$\mathrm{M_\odot}/\mathrm{L_\odot} \simeq 0.5 \text{ in cgs units (actually 0.520)}$$
$$1 \text{ radian} = 206{,}265"$$

Appendix 1.B Mathematical Background

The text presupposes an acquaintance with mathematical physics at the level of Arfken (1970), Margenau and Murphy (1956), or Mathews and Walker (1970). This Appendix contains a summary and review of some of the material and formulae that will be needed.

1 Vectors

The position of the point with Cartesian coordinates (x, y, z) may be described by a **position vector**,

$$\mathbf{x} = x\hat{\mathbf{e}}_x + y\hat{\mathbf{e}}_y + z\hat{\mathbf{e}}_z, \tag{1B-1}$$

where $\hat{\mathbf{e}}_x$, $\hat{\mathbf{e}}_y$, and $\hat{\mathbf{e}}_z$ are fixed unit vectors that point along the x, y and z axes. The distance of the point from the origin is written r or $|\mathbf{x}|$ and is equal to $(x^2 + y^2 + z^2)^{1/2}$.

Similarly, we may represent an arbitrary vector $\mathbf{A}$ in component form as

$$\mathbf{A} = A_x\hat{\mathbf{e}}_x + A_y\hat{\mathbf{e}}_y + A_z\hat{\mathbf{e}}_z. \tag{1B-2}$$

The magnitude of a vector $\mathbf{A}$ is $A \equiv |\mathbf{A}| \equiv (A_x^2 + A_y^2 + A_z^2)^{1/2}$.

The **scalar** or **dot product** of two vectors $\mathbf{A}$ and $\mathbf{B}$ is

$$\mathbf{A} \cdot \mathbf{B} \equiv |\mathbf{A}||\mathbf{B}|\cos\psi, \tag{1B-3}$$

where ψ is the (smaller) angle between the two vectors. Note that $\mathbf{A}\cdot\mathbf{B} = \mathbf{B}\cdot\mathbf{A}$ and $\mathbf{A} \cdot \mathbf{A} = |\mathbf{A}|^2$. Since $\hat{\mathbf{e}}_x \cdot \hat{\mathbf{e}}_x = \hat{\mathbf{e}}_y \cdot \hat{\mathbf{e}}_y = \hat{\mathbf{e}}_z \cdot \hat{\mathbf{e}}_z = 1$, and $\hat{\mathbf{e}}_x \cdot \hat{\mathbf{e}}_y = \hat{\mathbf{e}}_x \cdot \hat{\mathbf{e}}_z = \hat{\mathbf{e}}_y \cdot \hat{\mathbf{e}}_z = 0$, we may write the dot product in component form as

$$\mathbf{A} \cdot \mathbf{B} = \sum_{i=1}^{3} A_i B_i, \tag{1B-4}$$

where the subscripts 1, 2, and 3 stand for x, y, and z, respectively. For simplicity we generally adopt the **summation convention**, i.e., we automatically sum from 1 to 3 over any dummy subscript that appears repeatedly in one term of an equation. Thus equation (1B-4) may be written

$$\mathbf{A} \cdot \mathbf{B} = A_i B_i. \tag{1B-5}$$

The **vector** or **cross product** of two vectors is

$$\mathbf{A} \times \mathbf{B} \equiv AB\sin\psi\,\hat{\mathbf{p}}, \tag{1B-6}$$

where $\hat{\mathbf{p}}$ is a unit vector that is perpendicular to the plane containing $\mathbf{A}$ and $\mathbf{B}$ and points in the direction of movement of a right-hand screw when $\mathbf{A}$ is rotated about the origin into $\mathbf{B}$. Note that $\mathbf{A} \times \mathbf{B} = -\mathbf{B} \times \mathbf{A}$, that $\mathbf{A} \times \mathbf{A} = 0$, and that $\hat{\mathbf{e}}_x \times \hat{\mathbf{e}}_y = \hat{\mathbf{e}}_z$, $\hat{\mathbf{e}}_y \times \hat{\mathbf{e}}_z = \hat{\mathbf{e}}_x$, $\hat{\mathbf{e}}_z \times \hat{\mathbf{e}}_x = \hat{\mathbf{e}}_y$. In component form the cross product may be written

$$\mathbf{A} \times \mathbf{B} = \epsilon_{ijk}\hat{\mathbf{e}}_i A_j B_k, \tag{1B-7}$$

where a sum over i, j, and k is implied by the summation convention. Here ϵ_{ijk} is the **Levi-Civita tensor** which is defined to be zero if any two or more of the indices i, j, and k are equal, $+1$ if (i,j,k) is an even permutation of $(1,2,3)$ [the even permutations are $(2,3,1)$ and $(3,1,2)$] and -1 if (i,j,k) is an odd permutation of $(1,2,3)$.

Some identities that involve the dot and cross product include:

$$\mathbf{A} \cdot (\mathbf{B} \times \mathbf{C}) = \mathbf{C} \cdot (\mathbf{A} \times \mathbf{B}) = \mathbf{B} \cdot (\mathbf{C} \times \mathbf{A}), \tag{1B-8}$$

$$\mathbf{A} \times (\mathbf{B} \times \mathbf{C}) = (\mathbf{A} \cdot \mathbf{C})\mathbf{B} - (\mathbf{A} \cdot \mathbf{B})\mathbf{C}, \tag{1B-9}$$

$$(\mathbf{A} \times \mathbf{B}) \cdot (\mathbf{C} \times \mathbf{D}) = (\mathbf{A} \cdot \mathbf{C})(\mathbf{B} \cdot \mathbf{D}) - (\mathbf{A} \cdot \mathbf{D})(\mathbf{B} \cdot \mathbf{C}). \tag{1B-10}$$

In proving these identities is it useful to remember the relation

$$\epsilon_{ijk}\epsilon_{klm} = \delta_{il}\delta_{jm} - \delta_{im}\delta_{jl}, \tag{1B-11}$$

where the summation convention has been used and the **Kronecker delta** δ_{pq} is defined to be 1 if $p = q$ and zero otherwise.

The velocity and acceleration of a particle may be written in Cartesian components as

$$\mathbf{v} \equiv \dot{\mathbf{x}} \equiv \frac{d\mathbf{x}}{dt} = \dot{x}\hat{\mathbf{e}}_x + \dot{y}\hat{\mathbf{e}}_y + \dot{z}\hat{\mathbf{e}}_z \quad ; \quad \mathbf{a} \equiv \ddot{\mathbf{x}} \equiv \frac{d^2\mathbf{x}}{dt^2} = \ddot{x}\hat{\mathbf{e}}_x + \ddot{y}\hat{\mathbf{e}}_y + \ddot{z}\hat{\mathbf{e}}_z. \tag{1B-12}$$

2 Curvilinear Coordinate Systems

Let (q_1, q_2, q_3) denote the coordinates of a point in an arbitrary coordinate system. A fundamental quantity of any coordinate system is the **metric tensor** h_{ij}, defined such that the distance ds between the points (q_1, q_2, q_3) and $(q_1 + dq_1, q_2 + dq_2, q_3 + dq_3)$ is given by

$$ds^2 = h_{ij}dq_i dq_j. \tag{1B-13}$$

The coordinate systems used in this book are all **orthogonal**, that is, $h_{ij} = 0$ if $i \neq j$. In this case we write $h_{ii} \equiv h_i^2$, so that

$$ds^2 = h_i^2 dq_i^2. \tag{1B-14}$$

The velocity is

$$\dot{\mathbf{x}} = \sum_i h_i \frac{dq_i}{dt}\hat{\mathbf{e}}_i, \tag{1B-15}$$

where $\hat{\mathbf{e}}_1$ is a unit vector pointing in the direction from (q_1, q_2, q_3) to $(q_1 + dq_1, q_2, q_3)$. The volume element in orthogonal coordinates is

$$d^3\mathbf{x} = h_1 h_2 h_3 dq_1 dq_2 dq_3. \tag{1B-16}$$

For Cartesian coordinates, $(q_1, q_2, q_3) = (x, y, z)$ and $h_1 = h_2 = h_3 = 1$.

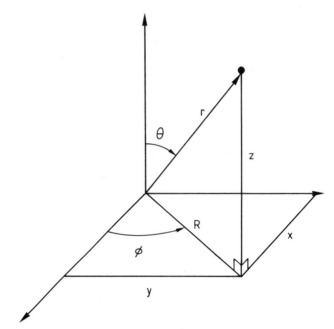

Figure 1B-1. The three main coordinate systems: Cartesian (x, y, z), cylindrical (R, ϕ, z), and spherical (r, θ, ϕ).

Cylindrical coordinate system   In this system the location of a particle is denoted by the triple (R, ϕ, z), where R is the perpendicular distance from the z-axis to the particle, and ϕ is the azimuthal angle between the x-axis and the projection of the position vector onto the (x, y) plane (Figure 1B-1). Thus the relation to Cartesian coordinates is

$$x = R \cos \phi \quad ; \quad y = R \sin \phi \quad ; \quad z = z. \tag{1B-17}$$

In cylindrical coordinates the position vector is

$$\mathbf{x} = R \hat{\mathbf{e}}_R + z \hat{\mathbf{e}}_z. \tag{1B-18}$$

An arbitrary vector may be written $\mathbf{A} = A_R \hat{\mathbf{e}}_R + A_\phi \hat{\mathbf{e}}_\phi + A_z \hat{\mathbf{e}}_z$, where $\hat{\mathbf{e}}_\phi = \hat{\mathbf{e}}_z \times \hat{\mathbf{e}}_R$ and

$$A_x = A_R \cos \phi - A_\phi \sin \phi \quad ; \quad A_y = A_R \sin \phi + A_\phi \cos \phi \quad ; \quad A_z = A_z. \tag{1B-19}$$

The expressions for dot and cross products in cylindrical coordinates are simply equations (1B-5) and (1B-7), with the subscripts $(1, 2, 3)$ denoting (R, ϕ, z) instead of (x, y, z). Note that the decomposition into components must be carried out at the same position for both vectors in the product, since the directions of $\hat{\mathbf{e}}_R$ and $\hat{\mathbf{e}}_\phi$ depend on position.

The velocity in cylindrical coordinates is

$$\mathbf{v} = \frac{d\mathbf{x}}{dt} = \dot{R}\hat{\mathbf{e}}_R + R\dot{\hat{\mathbf{e}}}_R + \dot{z}\hat{\mathbf{e}}_z. \tag{1B-20}$$

To compute $\dot{\hat{\mathbf{e}}}_R$ we use equation (1B-19) with $A_R = 1$, $A_\phi = 0$, $A_z = 0$. Thus $\hat{\mathbf{e}}_R = \cos\phi\hat{\mathbf{e}}_x + \sin\phi\hat{\mathbf{e}}_y$, and $d\hat{\mathbf{e}}_R = (-\sin\phi\hat{\mathbf{e}}_x + \cos\phi\hat{\mathbf{e}}_y)d\phi$. The expression in parentheses is just $\hat{\mathbf{e}}_\phi$. After carrying out a similar analysis for $\dot{\hat{\mathbf{e}}}_\phi$ we have

$$\frac{d\hat{\mathbf{e}}_R}{d\phi} = \hat{\mathbf{e}}_\phi \quad ; \quad \frac{d\hat{\mathbf{e}}_\phi}{d\phi} = -\hat{\mathbf{e}}_R, \tag{1B-21}$$

and

$$\dot{\hat{\mathbf{e}}}_R = +\dot{\phi}\hat{\mathbf{e}}_\phi \quad ; \quad \dot{\hat{\mathbf{e}}}_\phi = -\dot{\phi}\hat{\mathbf{e}}_R. \tag{1B-22}$$

Thus the velocity is

$$\mathbf{v} = \dot{R}\hat{\mathbf{e}}_R + R\dot{\phi}\hat{\mathbf{e}}_\phi + \dot{z}\hat{\mathbf{e}}_z. \tag{1B-23}$$

The acceleration is

$$\mathbf{a} = \frac{d\mathbf{v}}{dt} = (\ddot{R} - R\dot{\phi}^2)\hat{\mathbf{e}}_R + (2\dot{R}\dot{\phi} + R\ddot{\phi})\hat{\mathbf{e}}_\phi + \ddot{z}\hat{\mathbf{e}}_z. \tag{1B-24}$$

For cylindrical coordinates the metric tensor is given by equation (1B-14) with

$$h_R = 1 \quad ; \quad h_\phi = R \quad ; \quad h_z = 1. \tag{1B-25}$$

The volume element is $d^3\mathbf{x} = R\,dR\,d\phi\,dz$.

Spherical coordinate system   The position of a particle is denoted by (r, θ, ϕ) (Figure 1B-1). The coordinate r is the radial distance from the origin to the particle; θ is the angle between the position vector and the z-axis; and ϕ is the same azimuthal angle used in cylindrical coordinates. The relation to Cartesian coordinates is

$$x = r\sin\theta\cos\phi \quad ; \quad y = r\sin\theta\sin\phi \quad ; \quad z = r\cos\theta. \tag{1B-26}$$

In spherical coordinates the position vector is simply

$$\mathbf{x} = r\hat{\mathbf{e}}_r. \tag{1B-27}$$

An arbitrary vector may be written $\mathbf{A} = A_r\hat{\mathbf{e}}_r + A_\theta\hat{\mathbf{e}}_\theta + A_\phi\hat{\mathbf{e}}_\phi$, where $\hat{\mathbf{e}}_\theta = \hat{\mathbf{e}}_\phi \times \hat{\mathbf{e}}_r$ and

$$\begin{aligned} A_x &= A_r\sin\theta\cos\phi + A_\theta\cos\theta\cos\phi - A_\phi\sin\phi, \\ A_y &= A_r\sin\theta\sin\phi + A_\theta\cos\theta\sin\phi + A_\phi\cos\phi, \\ A_z &= A_r\cos\theta - A_\theta\sin\theta. \end{aligned} \tag{1B-28}$$

Once again, the expressions for dot and cross products in spherical coordinates are simply equations (1B-5) and (1B-7), with the subscripts $(1, 2, 3)$ denoting

(r, θ, ϕ) instead of (x, y, z), and with the understanding that the decomposition into components must be carried out at the same position for both vectors in the product.

The rate of change of the unit vectors in spherical coordinates is

$$\dot{\hat{\mathbf{e}}}_r = \dot{\theta}\hat{\mathbf{e}}_\theta + \dot{\phi}\sin\theta\hat{\mathbf{e}}_\phi \; ; \; \dot{\hat{\mathbf{e}}}_\theta = -\dot{\theta}\hat{\mathbf{e}}_r + \dot{\phi}\cos\theta\hat{\mathbf{e}}_\phi \; ; \; \dot{\hat{\mathbf{e}}}_\phi = -\dot{\phi}\sin\theta\hat{\mathbf{e}}_r - \dot{\phi}\cos\theta\hat{\mathbf{e}}_\theta. \tag{1B-29}$$

Thus the velocity is

$$\mathbf{v} = \dot{r}\hat{\mathbf{e}}_r + r\dot{\theta}\hat{\mathbf{e}}_\theta + r\sin\theta\dot{\phi}\hat{\mathbf{e}}_\phi. \tag{1B-30}$$

The acceleration is

$$\mathbf{a} = \frac{d\mathbf{v}}{dt} = (\ddot{r} - r\dot{\theta}^2 - r\sin^2\theta\dot{\phi}^2)\hat{\mathbf{e}}_r + (2\dot{r}\dot{\theta} + r\ddot{\theta} - r\sin\theta\cos\theta\dot{\phi}^2)\hat{\mathbf{e}}_\theta$$
$$+ (r\sin\theta\ddot{\phi} + 2\dot{r}\sin\theta\dot{\phi} + 2r\cos\theta\dot{\theta}\dot{\phi})\hat{\mathbf{e}}_\phi. \tag{1B-31}$$

For spherical coordinates the metric tensor is given by

$$h_r = 1 \quad ; \quad h_\theta = r \quad ; \quad h_\phi = r\sin\theta. \tag{1B-32}$$

The volume element is $d^3\mathbf{x} = r^2\sin\theta\, dr d\theta d\phi$.

For brevity the two angular coordinates are sometimes written as $\Omega \equiv (\theta, \phi)$, and the element of area on the unit sphere is written $d^2\Omega \equiv \sin\theta d\theta d\phi$.

3 Vector Calculus

Gradient   Consider a scalar function of position $f(\mathbf{x})$. Using Cartesian coordinates we define the **gradient** of f to be the vector function

$$\boldsymbol{\nabla} f = \hat{\mathbf{e}}_x \frac{\partial f}{\partial x} + \hat{\mathbf{e}}_y \frac{\partial f}{\partial y} + \hat{\mathbf{e}}_z \frac{\partial f}{\partial z} = \hat{\mathbf{e}}_i \frac{\partial f}{\partial x_i}, \tag{1B-33}$$

where $(x_1, x_2, x_3) = (x, y, z)$ and we have used the summation convention. The symbol $\boldsymbol{\nabla}$ is called grad, del, or nabla and is considered to be a vector operator defined by

$$\boldsymbol{\nabla} = \hat{\mathbf{e}}_i \frac{\partial}{\partial x_i}. \tag{1B-34}$$

In cases where it is necessary to distinguish the variable used in the gradient, we write

$$\boldsymbol{\nabla}_{\mathbf{x}} \quad \text{or} \quad \frac{\partial}{\partial \mathbf{x}}. \tag{1B-35}$$

The change in the value of f between the points $\mathbf{x}$ and $\mathbf{x} + d\mathbf{x}$ is given by

$$df = \frac{\partial f}{\partial x_i} dx_i = \boldsymbol{\nabla} f \cdot d\mathbf{x}. \tag{1B-36}$$

If $\mathbf{x}$ and $\mathbf{x} + d\mathbf{x}$ lie on a surface S on which f is constant, that is, if $f(\mathbf{x}) = f(\mathbf{x} + d\mathbf{x})$, then $df = 0$ and $\nabla f \cdot d\mathbf{x} = 0$. Hence ∇f is orthogonal to $d\mathbf{x}$. Since $d\mathbf{x}$ can be chosen to be any infinitesimal vector lying in S, ∇f must be normal to S itself. Hence the gradient of f is normal to the surfaces of constant f.

In cylindrical coordinates the expression for the gradient must be modified, since the path length along each axis is not dR, $d\phi$, dz but rather dR, $R\,d\phi$, dz. Hence for consistency with equation (1B-36) we must have

$$\nabla = \hat{\mathbf{e}}_R \frac{\partial}{\partial R} + \frac{\hat{\mathbf{e}}_\phi}{R} \frac{\partial}{\partial \phi} + \hat{\mathbf{e}}_z \frac{\partial}{\partial z}. \tag{1B-37}$$

In spherical coordinates

$$\nabla = \hat{\mathbf{e}}_r \frac{\partial}{\partial r} + \frac{\hat{\mathbf{e}}_\theta}{r} \frac{\partial}{\partial \theta} + \frac{\hat{\mathbf{e}}_\phi}{r \sin \theta} \frac{\partial}{\partial \phi}. \tag{1B-38}$$

These are special cases of the general form valid in any orthogonal coordinate system,

$$\nabla = \frac{\hat{\mathbf{e}}_i}{h_i} \frac{\partial}{\partial q_i}. \tag{1B-39}$$

Divergence   Next consider a vector function $\mathbf{F}(\mathbf{x})$. Using Cartesian coordinates we define the divergence of $\mathbf{F}$ to be

$$\nabla \cdot \mathbf{F} = \frac{\partial F_x}{\partial x} + \frac{\partial F_y}{\partial y} + \frac{\partial F_z}{\partial z} = \frac{\partial F_i}{\partial x_i}, \tag{1B-40}$$

where $(x, y, z) = (x_1, x_2, x_3)$.

To clarify the physical meaning of the divergence, consider a volume V, enclosed by a surface S. For simplicity, assume initially that the volume is cubical, occupying the region $x_{ia} \leq x_i \leq x_{ib}$, $i = 1, 2, 3$. Then

$$\int_V \nabla \cdot \mathbf{F} \, d^3x = \int_{x_{1a}}^{x_{1b}} dx_1 \int_{x_{2a}}^{x_{2b}} dx_2 \int_{x_{3a}}^{x_{3b}} dx_3 \left(\frac{\partial F_1}{\partial x_1} + \frac{\partial F_2}{\partial x_2} + \frac{\partial F_3}{\partial x_3} \right)$$

$$= \int_{x_{2a}}^{x_{2b}} dx_2 \int_{x_{3a}}^{x_{3b}} dx_3 [F_1(x_{1b}, x_2, x_3) - F_1(x_{1a}, x_2, x_3)] + \text{two similar terms.}$$

$$\tag{1B-41}$$

This expression may be written more concisely as the area integral $\int_S \mathbf{F} \cdot d^2\mathbf{S} = \int_S \mathbf{F} \cdot \hat{\mathbf{n}} \, d^2S$, where d^2S is a small element of the surface of the cube, and $\hat{\mathbf{n}}$ is the normal to the surface pointing outward from the volume V. (Note that in this notation $d^3\mathbf{x}$ is a scalar but $d^2\mathbf{S}$ is a vector.) It is easy to generalize this result to an arbitrary volume by dividing the volume into many small cubes and noting that the surface integrals from the inside faces of the cubes cancel. Hence for an arbitrary volume V,

$$\int_V \nabla \cdot \mathbf{F} \, d^3x = \int_S \mathbf{F} \cdot d^2\mathbf{S}. \tag{1B-42}$$

This result is known as the **divergence theorem**.

One simple consequence of the divergence theorem is that for arbitrary scalar and vector functions g and $\mathbf{F}$,

$$\int g \boldsymbol{\nabla} \cdot \mathbf{F} \, d^3 \mathbf{x} = \int_S g \mathbf{F} \cdot d^2 \mathbf{S} - \int (\mathbf{F} \cdot \boldsymbol{\nabla}) g \, d^3 \mathbf{x}, \qquad (1\text{B-}43)$$

which is a three-dimensional analog of integration by parts. Also, if $\mathbf{F} = f \hat{\mathbf{e}}_i$, then we have $\int_V (\partial f / \partial x_i) d^3 \mathbf{x} = \int_S f \, d^2 S_i$, which can be written more compactly after multiplication by $\hat{\mathbf{e}}_i$ and summing over indices as the vector equation

$$\int_V \boldsymbol{\nabla} f \, d^3 \mathbf{x} = \int_S f \, d^2 \mathbf{S}. \qquad (1\text{B-}44)$$

In cylindrical coordinates

$$\boldsymbol{\nabla} \cdot \mathbf{F} = \frac{1}{R} \frac{\partial}{\partial R} (R F_R) + \frac{1}{R} \frac{\partial F_\phi}{\partial \phi} + \frac{\partial F_z}{\partial z}. \qquad (1\text{B-}45)$$

This result can be derived from the divergence theorem, or—at least formally— by writing $\boldsymbol{\nabla} \cdot \mathbf{F} = (\hat{\mathbf{e}}_R \partial/\partial R + \hat{\mathbf{e}}_\phi \partial/\partial \phi + \hat{\mathbf{e}}_z \partial/\partial z) \cdot (F_R \hat{\mathbf{e}}_R + F_\phi \hat{\mathbf{e}}_\phi + F_z \hat{\mathbf{e}}_z)$ and expanding the expression using equation (1B-21).

In spherical coordinates

$$\boldsymbol{\nabla} \cdot \mathbf{F} = \frac{1}{r^2} \frac{\partial}{\partial r} (r^2 F_r) + \frac{1}{r \sin \theta} \frac{\partial}{\partial \theta} (\sin \theta F_\theta) + \frac{1}{r \sin \theta} \frac{\partial F_\phi}{\partial \phi}. \qquad (1\text{B-}46)$$

In arbitrary orthogonal coordinates

$$\boldsymbol{\nabla} \cdot \mathbf{F} = \frac{1}{h_1 h_2 h_3} \left[\frac{\partial}{\partial q_1} (h_2 h_3 F_1) + \frac{\partial}{\partial q_2} (h_3 h_1 F_2) + \frac{\partial}{\partial q_3} (h_1 h_2 F_3) \right]. \qquad (1\text{B-}47)$$

Laplacian   The divergence of the gradient of a scalar function is called the **Laplacian** of that function. Thus the Laplacian of $F(\mathbf{x})$ is

$$\boldsymbol{\nabla}^2 F \equiv \boldsymbol{\nabla} \cdot (\boldsymbol{\nabla} F). \qquad (1\text{B-}48)$$

In different coordinate systems we have

$$\boldsymbol{\nabla}^2 F = \frac{\partial^2 F}{\partial x^2} + \frac{\partial^2 F}{\partial y^2} + \frac{\partial^2 F}{\partial z^2}; \qquad (1\text{B-}49)$$

$$\boldsymbol{\nabla}^2 F = \frac{1}{R} \frac{\partial}{\partial R} \left(R \frac{\partial F}{\partial R} \right) + \frac{1}{R^2} \frac{\partial^2 F}{\partial \phi^2} + \frac{\partial^2 F}{\partial z^2}; \qquad (1\text{B-}50)$$

$$\boldsymbol{\nabla}^2 F = \frac{1}{r^2} \frac{\partial}{\partial r} \left(r^2 \frac{\partial F}{\partial r} \right) + \frac{1}{r^2 \sin \theta} \frac{\partial}{\partial \theta} \left(\sin \theta \frac{\partial F}{\partial \theta} \right) + \frac{1}{r^2 \sin^2 \theta} \frac{\partial^2 F}{\partial \phi^2}; \qquad (1\text{B-}51)$$

$$\boldsymbol{\nabla}^2 F = \frac{1}{h_1 h_2 h_3} \left[\frac{\partial}{\partial q_1} \left(\frac{h_2 h_3}{h_1} \frac{\partial F}{\partial q_1} \right) + \frac{\partial}{\partial q_2} \left(\frac{h_3 h_1}{h_2} \frac{\partial F}{\partial q_2} \right) + \frac{\partial}{\partial q_3} \left(\frac{h_1 h_2}{h_3} \frac{\partial F}{\partial q_3} \right) \right].$$
$$(1\text{B-}52)$$

Convective operator   In many operations we need the convective operator $(\mathbf{A} \cdot \boldsymbol{\nabla})\mathbf{B}$. In Cartesian coordinates we have

$$[(\mathbf{A} \cdot \boldsymbol{\nabla})\mathbf{B}]_x = A_x \frac{\partial B_x}{\partial x} + A_y \frac{\partial B_x}{\partial y} + A_z \frac{\partial B_x}{\partial z}, \tag{1B-53}$$

with similar expressions for the y and z components. In cylindrical coordinates

$$(\mathbf{A} \cdot \boldsymbol{\nabla})\mathbf{B} = \left(A_R \frac{\partial B_R}{\partial R} + \frac{A_\phi}{R} \frac{\partial B_R}{\partial \phi} + A_z \frac{\partial B_R}{\partial z} - \frac{A_\phi B_\phi}{R} \right) \hat{\mathbf{e}}_R$$

$$+ \left(A_R \frac{\partial B_\phi}{\partial R} + \frac{A_\phi}{R} \frac{\partial B_\phi}{\partial \phi} + A_z \frac{\partial B_\phi}{\partial z} + \frac{A_\phi B_R}{R} \right) \hat{\mathbf{e}}_\phi \tag{1B-54}$$

$$+ \left(A_R \frac{\partial B_z}{\partial R} + \frac{A_\phi}{R} \frac{\partial B_z}{\partial \phi} + A_z \frac{\partial B_z}{\partial z} \right) \hat{\mathbf{e}}_z.$$

In spherical coordinates

$$(\mathbf{A} \cdot \boldsymbol{\nabla})\mathbf{B} = \left(A_r \frac{\partial B_r}{\partial r} + \frac{A_\theta}{r} \frac{\partial B_r}{\partial \theta} + \frac{A_\phi}{r \sin \theta} \frac{\partial B_r}{\partial \phi} - \frac{A_\theta B_\theta + A_\phi B_\phi}{r} \right) \hat{\mathbf{e}}_r$$

$$+ \left(A_r \frac{\partial B_\theta}{\partial r} + \frac{A_\theta}{r} \frac{\partial B_\theta}{\partial \theta} + \frac{A_\phi}{r \sin \theta} \frac{\partial B_\theta}{\partial \phi} + \frac{A_\theta B_r}{r} - \frac{A_\phi B_\phi \cot \theta}{r} \right) \hat{\mathbf{e}}_\theta$$

$$+ \left(A_r \frac{\partial B_\phi}{\partial r} + \frac{A_\theta}{r} \frac{\partial B_\phi}{\partial \theta} + \frac{A_\phi}{r \sin \theta} \frac{\partial B_\phi}{\partial \phi} + \frac{A_\phi B_r}{r} + \frac{A_\phi B_\theta \cot \theta}{r} \right) \hat{\mathbf{e}}_\phi. \tag{1B-55}$$

In arbitrary orthogonal coordinates

$$[(\mathbf{A} \cdot \boldsymbol{\nabla})\mathbf{B}]_j = \sum_{i=1}^{3} \left[\frac{A_i}{h_i} \frac{\partial B_j}{\partial q_i} + \frac{B_i}{h_i h_j} \left(A_j \frac{\partial h_j}{\partial q_i} - A_i \frac{\partial h_i}{\partial q_j} \right) \right]. \tag{1B-56}$$

4 Abel's Integral Equation

Let

$$f(x) = \int_0^x \frac{g(t) dt}{(x-t)^\alpha}, \qquad 0 < \alpha < 1. \tag{1B-57a}$$

Then

$$g(t) = \frac{\sin \pi \alpha}{\pi} \frac{d}{dt} \int_0^t \frac{f(x) dx}{(t-x)^{1-\alpha}} = \frac{\sin \pi \alpha}{\pi} \left[\int_0^t \frac{df}{dx} \frac{dx}{(t-x)^{1-\alpha}} + \frac{f(0)}{t^{1-\alpha}} \right]. \tag{1B-57b}$$

This can be proved by substituting equation (1B-57a) into the first part of (1B-57b), interchanging the order of integration, and using the integral

$$\int_0^1 \frac{du}{u^\alpha (1-u)^{1-\alpha}} = \frac{\pi}{\sin \pi \alpha}. \tag{1B-58}$$

Another useful result, which can also be proved using the integral (1B-58), is

$$f(x) = \int_x^\infty \frac{g(t) dt}{(t-x)^\alpha}, \qquad 0 < \alpha < 1, \tag{1B-59a}$$

$$g(t) = -\frac{\sin \pi \alpha}{\pi} \frac{d}{dt} \int_t^\infty \frac{f(x) dx}{(x-t)^{1-\alpha}}$$

$$= -\frac{\sin \pi \alpha}{\pi} \left[\int_t^\infty \frac{df}{dx} \frac{dx}{(x-t)^{1-\alpha}} - \lim_{x \to \infty} \frac{f(x)}{(x-t)^{1-\alpha}} \right]. \tag{1B-59b}$$

5 Calculus of Variations

Consider a curve $\mathbf{x} = \mathbf{x}(t)$, $t_0 \leq t \leq t_1$, which we label by γ. We define a functional

$$I(\gamma) = \int_{t_0}^{t_1} L[\mathbf{x}(t), \dot{\mathbf{x}}(t), t] dt, \tag{1B-60}$$

where the notation indicates that I depends on the curve γ (mathematically, the functional is a mapping from the space of curves to the space of real numbers).

Now consider a nearby curve γ' defined by $\mathbf{x} = \mathbf{x}(t) + \epsilon \mathbf{h}(t)$, where $\mathbf{h}(t_0) = \mathbf{h}(t_1) = 0$. As $\epsilon \to 0$ we have

$$
\begin{aligned}
I(\gamma') - I(\gamma) &= \int_{t_0}^{t_1} [L(\mathbf{x} + \epsilon \mathbf{h}, \dot{\mathbf{x}} + \epsilon \dot{\mathbf{h}}, t) - L(\mathbf{x}, \dot{\mathbf{x}}, t)] dt \\
&= \epsilon \int_{t_0}^{t_1} \left[\mathbf{h} \cdot \frac{\partial L}{\partial \mathbf{x}} + \dot{\mathbf{h}} \cdot \frac{\partial L}{\partial \dot{\mathbf{x}}} \right] dt + \mathrm{O}(\epsilon^2),
\end{aligned}
\tag{1B-61}
$$

where the second integral is evaluated along the unperturbed curve γ. On this curve L can be considered to be a function of time only, $L(t) = L[\mathbf{x}(t), \dot{\mathbf{x}}(t), t]$. Hence we may integrate by parts to obtain

$$I(\gamma') - I(\gamma) = -\epsilon \int_{t_0}^{t_1} \mathbf{h} \cdot \left[\frac{d}{dt} \left(\frac{\partial L}{\partial \dot{\mathbf{x}}} \right) - \frac{\partial L}{\partial \mathbf{x}} \right] dt + \left(\mathbf{h} \cdot \frac{\partial L}{\partial \dot{\mathbf{x}}} \right)_{t_0}^{t_1} + \mathrm{O}(\epsilon^2). \tag{1B-62}$$

The boundary term following the integral is zero since $\mathbf{h}(t_0) = \mathbf{h}(t_1) = 0$.

The curve γ is an **extremal** curve if $I(\gamma) - I(\gamma') = \mathrm{O}(\epsilon^2)$ for all variations $\mathbf{h}$. The extremal curves have the largest or smallest values of the functional I of any continuous curves connecting the fixed endpoints $\mathbf{x}(t_0)$ and $\mathbf{x}(t_1)$. Since the integral in equation (1B-62) must vanish for all variations $\mathbf{h}(t)$ for which $\mathbf{h}(t_0) = \mathbf{h}(t_1) = 0$, the condition for an extremal curve is

$$\frac{d}{dt} \left(\frac{\partial L}{\partial \dot{\mathbf{x}}} \right) - \frac{\partial L}{\partial \mathbf{x}} = 0. \tag{1B-63}$$

This is the **Euler-Lagrange equation** for the functional I. The solution of this differential equation yields the curves with the extreme (maximum or minimum) values of I.

Appendix 1.C Special Functions

More complete lists of formulae are given in Abramowitz and Stegun (1965).

1 Dirac Delta Function

This is a singular function defined by the properties

$$\delta(x) = 0, \quad x \neq 0 \quad ; \quad \int_{-\infty}^{\infty} f(x) \delta(x)\, dx = f(0), \tag{1C-1}$$

where $f(x)$ is an arbitrary function continuous at $x = 0$. The Dirac delta function can be written as

$$\delta(x) = \frac{1}{\pi} \lim_{\epsilon \to 0} \frac{\epsilon}{\epsilon^2 + x^2} = \frac{1}{\pi} \lim_{a \to \infty} \frac{\sin ax}{x} = \frac{1}{2\pi} \int_{-\infty}^{\infty} e^{ixt} dx. \tag{1C-2}$$

If $p(x_1, \ldots, x_N) dx_1 \ldots dx_N$ is the number of stars or other objects with parameters in the range $x_1 \to x_1 + dx_1, \ldots, x_N \to x_N + dx_N$, and g is some function of $x_1, \ldots, x_N$, then the number of objects with g in the range $g_0 \to g_0 + dg_0$ is $n(g_0)dg_0$, where

$$n(g_0) = \int dx_1 \ldots dx_N p(x_1, \ldots, x_N) \delta[g_0 - g(x_1, \ldots, x_N)]. \tag{1C-3}$$

2 Factorial or Gamma Function

$$\Gamma(z + 1) \equiv z! \equiv \int_0^\infty t^z e^{-t} dt, \qquad \mathrm{Re}(z) > -1.$$
$$= \lim_{n \to \infty} \frac{1 \cdot 2 \cdot 3 \cdots n}{z(z+1)(z+2)\ldots(z+n+1)} n^{z+1}. \tag{1C-4}$$

The function $z!$ is defined for all complex numbers z except $-1, -2, \ldots$

$$z! = z(z-1)!, \tag{1C-5}$$
$$(z-1)!(-z)! = \pi \csc \pi z, \tag{1C-6}$$
$$(2z)! = \pi^{-1/2} 2^{2z} z! (z - \tfrac{1}{2})!, \tag{1C-7}$$
$$z!^* = (z^*)! \tag{1C-8}$$
$$x! = \sqrt{2\pi} x^{x+1/2} \exp\left(-x + \frac{\theta}{12x}\right), \qquad x > 0, \ 0 < \theta < 1. \tag{1C-9}$$

Stirling's approximation to $x!$ for large x is obtained from the last formula by neglecting the second term in parentheses.

Special values of $x!$ include

$$n! = 1 \cdot 2 \cdot 3 \cdots n,$$
$$0! = 1,$$
$$(-\tfrac{1}{2})! = \sqrt{\pi} = 1.77245,$$
$$(\tfrac{1}{2})! = \tfrac{1}{2}\sqrt{\pi} = 0.88623. \tag{1C-10}$$

3 Error Function

$$\mathrm{erf}\, z \equiv \frac{2}{\sqrt{\pi}} \int_0^z e^{-t^2} dt = \frac{2}{\sqrt{\pi}} \sum_{n=0}^{\infty} \frac{(-1)^n z^{2n+1}}{n!(2n+1)}. \tag{1C-11}$$

$$\mathrm{erf}(0) = 0, \qquad \mathrm{erf}(\infty) = 1, \qquad \mathrm{erf}(-z) = -\mathrm{erf}(z). \tag{1C-12}$$

$$\lim_{x \to \infty} (1 - \mathrm{erf}\, x) = \frac{e^{-x^2}}{\sqrt{\pi} x}. \tag{1C-13}$$

4 Elliptic Integrals

The incomplete elliptic integrals of the first and second kinds are

$$F(\theta, k) \equiv \int_0^\theta \frac{d\phi}{\sqrt{1 - k^2 \sin^2 \phi}} \quad ; \quad E(\theta, k) \equiv \int_0^\theta \sqrt{1 - k^2 \sin^2 \phi}\, d\phi. \quad \text{(1C-14)}$$

The complete elliptic integrals of the first and second kinds are

$$K(k) \equiv F(\tfrac{1}{2}\pi, k) = \int_0^{\pi/2} \frac{d\phi}{\sqrt{1 - k^2 \sin^2 \phi}} = \int_0^1 \frac{dt}{\sqrt{(1 - t^2)(1 - k^2 t^2)}},$$

$$E(k) \equiv E(\tfrac{1}{2}\pi, k) = \int_0^{\pi/2} \sqrt{1 - k^2 \sin^2 \phi}\, d\phi = \int_0^1 \sqrt{\frac{1 - k^2 t^2}{1 - t^2}}\, dt.$$

$$\text{(1C-15)}$$

The derivatives of the complete elliptic integrals are related by

$$\frac{d[kK(k)]}{dk} = \frac{E(k)}{1 - k^2}, \qquad \frac{dE(k)}{dk} = \frac{E(k) - K(k)}{k}. \quad \text{(1C-16)}$$

5 Associated Legendre Functions

The associated Legendre functions of the first and second kinds, $P_\lambda^\mu(z)$ and $Q_\lambda^\mu(z)$, are linearly independent solutions of the differential equation

$$\frac{d}{dz}\left[(1 - z^2)\frac{dw}{dz}\right] - \frac{\mu^2}{1 - z^2}w + \lambda(\lambda + 1)w = 0. \quad \text{(1C-17)}$$

For $\text{Re}(\nu) > 0$, the associated Legendre functions of the first kind diverge ($\propto z^\nu$) as $|z| \to \infty$, while the functions of the second kind vanish [$\propto z^{-(\nu+1)}$] as $|z| \to \infty$. As $z \to 0$

$$\left[\frac{d\ln P_\lambda^\mu(z)}{dz}\right]_{z=0} = 2\tan[\tfrac{1}{2}\pi(\lambda + \mu)]\frac{[\tfrac{1}{2}(\lambda + \mu)]![\tfrac{1}{2}(\lambda - \mu)]!}{[\tfrac{1}{2}(\lambda + \mu - 1)]![\tfrac{1}{2}(\lambda - \mu - 1)]!},$$

$$\left[\frac{d\ln Q_\lambda^\mu(z)}{dz}\right]_{z=0} = 2\exp\{\tfrac{1}{2}\pi i\, \text{sign}[\text{Im}(z)]\}\frac{[\tfrac{1}{2}(\lambda + \mu)]![\tfrac{1}{2}(\lambda - \mu)]!}{[\tfrac{1}{2}(\lambda + \mu - 1)]![\tfrac{1}{2}(\lambda - \mu - 1)]!}.$$

$$\text{(1C-18)}$$

For many purposes we need only consider real arguments, $z = x$, in the interval $-1 \leq x \leq 1$; furthermore, we often only use functions of the first kind $P_\lambda^\mu(x)$ (functions of the second kind diverge at $x = 1$ or -1). We also restrict ourselves to the case where μ is equal to an integer, m, and since only the square of m appears in equation (1C-17) we may assume without loss of generality that $m \geq 0$.[2] We have

$$P_{-\lambda-1}^m(x) = P_\lambda^m(x), \quad \text{(1C-19)}$$

---

[2] Our convention follows Arfken (1970) and Mathews and Walker (1970) but differs from Abramowitz and Stegun (1965) by a factor $(-1)^m$.

$$P_\lambda^m(x) = (1 - x^2)^{m/2} \frac{d^m P_\lambda^0(x)}{dx^m}. \tag{1C-20}$$

For most purposes it is sufficient to consider non-negative integer values of λ, $\lambda = l = 0, 1, \ldots$, since for other values of λ, $P_\lambda^m(z)$ diverges at $z = 1$ or -1. Then the Legendre functions with $m = 0$ are polynomials given by the formula

$$P_l(x) \equiv P_l^0(x) = \frac{1}{2^l l!} \frac{d^l}{dx^l} (x^2 - 1)^l. \tag{1C-21}$$

These **Legendre polynomials** are also generated by the relation

$$\frac{1}{\sqrt{1 - 2xt + t^2}} = \sum_{l=0}^{\infty} P_l(x) t^l, \qquad |t| < 1, \ |x| \le 1, \tag{1C-22}$$

which leads to an expression for the inverse distance between the points $\mathbf{x}$ and $\mathbf{x}'$,

$$\frac{1}{|\mathbf{x} - \mathbf{x}'|} = \sum_{l=0}^{\infty} \frac{r_<^l}{r_>^{l+1}} P_l(\cos\gamma), \tag{1C-23}$$

where $r_< = \min(|\mathbf{x}|, |\mathbf{x}'|)$, $r_> = \max(|\mathbf{x}|, |\mathbf{x}'|)$, and γ is the angle between the two vectors.

The associated Legendre functions are orthogonal in the sense that

$$\int_{-1}^{1} P_l^m(x) P_n^m(x) dx = \frac{2}{2l + 1} \frac{(l + m)!}{(l - m)!} \delta_{ln}. \tag{1C-24}$$

The associated Legendre functions can be written most compactly using the substitution $x = \cos\theta$; since $-1 \le x \le 1$ we take $0 \le \theta \le \pi$ and let $c = \cos\theta$, $s = \sin\theta$:

$$\begin{array}{llll}
P_0(c) = 1 & & & \\
P_1(c) = c & P_1^1(c) = s & & \\
P_2(c) = \frac{1}{2}(3c^2 - 1) & P_2^1(c) = 3cs & P_2^2(c) = 3s^2 & \\
P_3(c) = \frac{1}{2}(5c^3 - 3c) & P_3^1(c) = \frac{3}{2}s(5c^2 - 1) & P_3^2(c) = 15cs^2 & P_3^3(c) = 15s^3.
\end{array} \tag{1C-25}$$

Note that $P_l^m(x)$ vanishes for $m > l$, and that $P_l^m(x)$ is even in x if $l - m$ is even, and odd if $l - m$ is odd.

For $l \gg 1$ and $l \gg m$

$$P_l^m(\cos\theta) \simeq l^{m-\frac{1}{2}} \sqrt{\frac{2}{\pi \sin\theta}} \cos[(l + \tfrac{1}{2})\theta + (\tfrac{1}{2}m - \tfrac{1}{4})\pi]. \tag{1C-26}$$

6 Spherical Harmonics

A spherical harmonic is defined by the expression[3]

$$Y_l^m(\theta, \phi) = \sqrt{\frac{2l + 1}{4\pi} \frac{(l - |m|)!}{(l + |m|)!}} P_l^{|m|}(\cos\theta) e^{im\phi} \times \begin{cases} (-1)^m & \text{for } m \ge 0 \\ 1 & \text{for } m < 0, \end{cases} \tag{1C-27}$$

---

[3] We adopt the so-called Condon-Shortley convention for defining spherical harmonics. Other conventions may differ by factors of $(-1)^m$ for positive m, negative m, or both.

where $l = 0, 1, 2, \ldots$ and $m = -l, -l + 1, \ldots, l - 1, l$. The variables lie in the range $0 \leq \theta \leq \pi$ and $0 \leq \phi \leq 2\pi$ and usually represent the two angular coordinates in a spherical coordinate system (see Figure 1B-1). Note that

$$Y_l^{-m}(\theta, \phi) = (-1)^m Y_l^{m*}(\theta, \phi), \tag{1C-28}$$

where the asterisk denotes complex conjugation.

The most important feature of the spherical harmonics, which is easily proved using equation (1C-24), is that they are orthonormal in the sense that

$$\int Y_l^{m*}(\Omega) Y_{l'}^{m'}(\Omega) d^2\Omega = \int_0^{2\pi} d\phi \int_0^\pi \sin\theta \, d\theta \, Y_l^{m*}(\theta, \phi) Y_{l'}^{m'}(\theta, \phi) = \delta_{ll'} \delta_{mm'}. \tag{1C-29}$$

The addition theorem for spherical harmonics states that if the directions (θ, ϕ) and (θ', ϕ') are separated by an angle γ, then

$$P_l(\cos\gamma) = \frac{4\pi}{2l + 1} \sum_{m=-l}^{l} Y_l^{m*}(\theta', \phi') Y_l^m(\theta, \phi). \tag{1C-30}$$

Together with equation (1C-23), this leads to an expression for the inverse distance between the points $\mathbf{x} = (r, \theta, \phi)$ and $\mathbf{x}' = (r', \theta', \phi')$:

$$\frac{1}{|\mathbf{x} - \mathbf{x}'|} = \sum_{l=0}^{\infty} \sum_{m=-l}^{l} \frac{4\pi}{2l+1} \frac{r_<^l}{r_>^{l+1}} Y_l^{m*}(\theta', \phi') Y_l^m(\theta, \phi), \tag{1C-31}$$

where $r_< = \min(r, r')$, and $r_> = \max(r, r')$.

Using equations (1B-51) and (1C-17) we can show that

$$\nabla^2 [f(r) Y_l^m(\theta, \phi)] = \left[\frac{1}{r^2} \frac{d}{dr} \left(r^2 \frac{df}{dr} \right) - l(l+1) \frac{f(r)}{r^2} \right] Y_l^m(\theta, \phi). \tag{1C-32}$$

The first few spherical harmonics are:

$$Y_0^0(\theta, \phi) = \frac{1}{\sqrt{4\pi}}$$

$$Y_1^0(\theta, \phi) = \sqrt{\frac{3}{4\pi}} \cos\theta \qquad Y_1^{\pm 1}(\theta, \phi) = \mp\sqrt{\frac{3}{8\pi}} \sin\theta \, e^{\pm i\phi}$$

$$Y_2^0(\theta, \phi) = \sqrt{\frac{5}{16\pi}} (3\cos^2\theta - 1) \quad Y_2^{\pm 1}(\theta, \phi) = \mp\sqrt{\frac{15}{8\pi}} \sin\theta \cos\theta \, e^{\pm i\phi}$$

$$Y_2^{\pm 2}(\theta, \phi) = \sqrt{\frac{15}{32\pi}} \sin^2\theta \, e^{\pm 2i\phi}. \tag{1C-33}$$

7 Bessel Functions

The Bessel functions of the first and second kind, $J_\nu(z)$ and $Y_\nu(z)$, are linearly independent solutions of the differential equation

$$\frac{1}{z} \frac{d}{dz} \left(z \frac{dw}{dz} \right) + \left(1 - \frac{\nu^2}{z^2} \right) w = 0. \tag{1C-34}$$

In series form,

$$J_\nu(z) = \sum_{k=0}^{\infty} \frac{(-1)^k}{k!(\nu+k)!}(\tfrac{1}{2}z)^{\nu+2k}. \tag{1C-35}$$

The function $Y_\nu(z)$ is defined by the relation

$$Y_\nu(z) = \frac{\cos\nu\pi\, J_\nu(z) - J_{-\nu}(z)}{\sin\nu\pi}, \tag{1C-36}$$

or by its limiting value if ν is an integer. The function $Y_\nu(z)$ diverges as $z \to 0$. As $|z| \to \infty$

$$J_\nu(z) \simeq \sqrt{\frac{2}{\pi z}} \cos(z - \tfrac{1}{2}\nu\pi - \tfrac{1}{4}\pi) + O(|z|^{-1}). \tag{1C-37}$$

If $\nu \equiv n$ is an integer

$$J_{-n}(z) = (-1)^n J_n(z) \quad ; \quad Y_{-n}(z) = (-1)^n Y_n(z), \tag{1C-38}$$

$$J_n(z) = \frac{1}{\pi} \int_0^\pi \cos(z\sin\theta - n\theta)\, d\theta. \tag{1C-39}$$

If C_ν denotes either J_ν or Y_ν,

$$C_{\nu-1}(z) + C_{\nu+1}(z) = \frac{2\nu}{z}C_\nu(z) \quad ; \quad C_{\nu-1}(z) - C_{\nu+1}(z) = 2\frac{dC_\nu(z)}{dz}, \tag{1C-40}$$

which implies

$$\frac{dC_\nu(z)}{dz} = C_{\nu-1}(z) - \frac{\nu}{z}C_\nu(z) = -C_{\nu+1}(z) + \frac{\nu}{z}C_\nu(z); \tag{1C-41}$$

in particular,

$$J_0'(z) = -J_1(z). \tag{1C-42}$$

Note that

$$\int_0^\infty t^\mu J_\nu(t)\, dt = 2^\mu \frac{[\tfrac{1}{2}(\nu+\mu-1)]!}{[\tfrac{1}{2}(\nu-\mu-1)]!}, \quad \mathrm{Re}(\mu+\nu) > -1, \quad \mathrm{Re}(\mu) < \tfrac{1}{2}, \tag{1C-43}$$

and

$$\int_0^\infty e^{-\alpha t} J_\nu(t) t^{\nu+1}\, dt = \frac{2^{\nu+1}\alpha(\nu+\tfrac{1}{2})!}{\sqrt{\pi}(1+\alpha^2)^{\nu+3/2}}, \quad \mathrm{Re}(\nu) > -1, \quad \mathrm{Re}(\alpha) > 0. \tag{1C-44}$$

We shall also use the result (Watson 1944, §13.22)

$$\int_0^\infty J_0(kr) J_0(kr')\, dk = \frac{2}{\pi r_>} K\left(\frac{r_<}{r_>}\right), \tag{1C-45}$$

where $r_< = \min(r, r')$, $r_> = \max(r, r')$, and K denotes an elliptic integral. A useful integral identity is (Watson 1944, §14.3)

$$\int_0^\infty k\, dk \int_0^\infty F(R) J_\nu(kR) J_\nu(kr) R\, dR = F(r), \quad \text{for } \nu \geq -\tfrac{1}{2}, \qquad (1C\text{-}46)$$

where $F(R)$ is an arbitrary function. This identity is the basis of Hankel transforms: if

$$g(k) = \int_0^\infty f(r) J_\nu(kr) r\, dr, \qquad (1C\text{-}47a)$$

then g is called the **Hankel transform** of f, and equation (1C-46) yields

$$f(r) = \int_0^\infty g(k) J_\nu(kr) k\, dk. \qquad (1C\text{-}47b)$$

The **modified Bessel functions** are

$$I_\nu(z) = e^{-i\pi\nu/2} J_\nu(z e^{i\pi/2}) \quad ; \quad K_\nu(z) = K_{-\nu}(z) = \frac{\pi}{2} \frac{I_{-\nu}(z) - I_\nu(z)}{\sin \nu\pi}. \qquad (1C\text{-}48)$$

As $z \to 0$,

$$I_\nu(z) \to \frac{1}{\nu!} \left(\tfrac{1}{2}z\right)^\nu \quad ; \quad K_\nu(z) \to \frac{(\nu-1)!}{2} \left(\tfrac{1}{2}z\right)^{-\nu}. \qquad (1C\text{-}49)$$

At large $|z|$,

$$I_\nu(z) \to \frac{e^z}{\sqrt{2\pi z}} \quad ; \quad K_\nu(z) \to \sqrt{\frac{\pi}{2z}} e^{-z}. \qquad (1C\text{-}50)$$

If $\nu \equiv n$ is an integer,

$$I_{-n}(z) = I_n(z) = \frac{1}{\pi} \int_0^\infty e^{z\cos\theta} \cos(n\theta)\, d\theta. \qquad (1C\text{-}51)$$

If Z_ν denotes either I_ν or $e^{i\pi\nu} K_\nu$,

$$Z_{\nu-1}(z) + Z_{\nu+1}(z) = 2\frac{dZ_\nu(z)}{dz} \quad ; \quad Z_{\nu-1}(z) - Z_{\nu+1}(z) = \frac{2\nu}{z} Z_\nu(z), \qquad (1C\text{-}52)$$

which implies

$$\frac{dZ_\nu(z)}{dz} = Z_{\nu-1}(z) - \frac{\nu}{z} Z_\nu(z) = Z_{\nu+1}(z) + \frac{\nu}{z} Z_\nu(z). \qquad (1C\text{-}53)$$

In particular,

$$I_0'(z) = I_1(z), \qquad\qquad K_0'(z) = -K_1(z). \qquad (1C\text{-}54)$$

We shall use the identity

$$e^{z\cos\theta} = I_0(z) + 2\sum_{n=1}^\infty I_n(z) \cos n\theta = \sum_{n=-\infty}^\infty I_n(z) \cos n\theta. \qquad (1C\text{-}55)$$

Appendix 1.D Mechanics

We assume a background in elementary mechanics. Useful texts are Goldstein (1980) and Landau and Lifshitz (1976). The most elegant and concise treatment of the subject is found in Arnold (1978). This Appendix contains a brief summary of the main concepts employed in this book.

1 Single Particles

The motion of a particle is described by **Newton's second law**,

$$\mathbf{F} = \frac{d\mathbf{p}}{dt}, \tag{1D-1}$$

where $\mathbf{F}$ is the total force acting on the particle, and $\mathbf{p}$ is its momentum. The momentum $\mathbf{p} = m\mathbf{v}$, where m is the mass of the particle and $\mathbf{v}$ is its velocity. Thus, if the mass of the particle is constant,

$$\frac{d\mathbf{v}}{dt} = \frac{d^2\mathbf{x}}{dt^2} = \frac{\mathbf{F}}{m}. \tag{1D-2}$$

The **work** done by the force $\mathbf{F}$ on a particle traveling from $\mathbf{x}_1$ to $\mathbf{x}_2$ is

$$W_{12} = \int_1^2 \mathbf{F} \cdot d\mathbf{x}, \tag{1D-3}$$

where the limits indicate that the integral is a line integral that is to be taken along the particle's trajectory from $\mathbf{x}_1$ to $\mathbf{x}_2$. For a particle whose mass is constant,

$$W_{12} = m \int_1^2 \frac{d^2\mathbf{x}}{dt^2} \cdot d\mathbf{x} = m \int_1^2 \frac{d^2\mathbf{x}}{dt^2} \cdot \frac{d\mathbf{x}}{dt} dt = m \int_1^2 \mathbf{v} \cdot \frac{d\mathbf{v}}{dt} dt$$
$$= \tfrac{1}{2} m[v^2(\mathbf{x}_2) - v^2(\mathbf{x}_1)]. \tag{1D-4}$$

The **kinetic energy** of a particle is $K \equiv \tfrac{1}{2} mv^2$ and hence

$$W_{12} = K_2 - K_1. \tag{1D-5}$$

Many of the forces encountered in nature are **conservative**, that is, the work W_{12} is independent of the path taken between the endpoints $\mathbf{x}_1$ and $\mathbf{x}_2$. In this case, we may choose $\mathbf{x}_1$ to be some fixed point and define the **potential energy** $W(\mathbf{x})$ by

$$W(\mathbf{x}) \equiv \int_{\mathbf{x}_1}^{\mathbf{x}} \mathbf{F} \cdot d\mathbf{x}. \tag{1D-6}$$

A frequently employed convention that fixes the otherwise arbitrary zero point of W is to place the point $\mathbf{x}_1$ at "infinity," that is, far from all interacting bodies, where $\mathbf{F}$ is negligibly small. All such points yield the same zero point, since the work done in moving from one point "at infinity" to another is negligible. There are two immediate consequences of equation (1D-6). Taking the gradient of W, we obtain

$$\mathbf{F} = -\nabla W(\mathbf{x}) = -\frac{\partial W}{\partial \mathbf{x}}, \tag{1D-7}$$

and substituting (1D-6) into (1D-5) yields

$$K_1 + W(\mathbf{x}_1) = K_2 + W(\mathbf{x}_2). \tag{1D-8}$$

Thus if the **energy** of a particle is defined to be $E = K + W = \frac{1}{2}mv^2 + W(\mathbf{x})$, we find that *if the forces acting on a particle are conservative, then its energy is conserved.* We shall also encounter cases where the forces are conservative if the trajectory is traversed instantaneously, but the force at a given position is time-dependent; in these cases we can still write $\mathbf{F} = -\nabla W(\mathbf{x}, t)$, but the energy E will no longer be conserved.

Forces due to gravity are conservative (the proof is given at the beginning of Chapter 2); for gravity the potential energy of a particle is proportional to its mass and we may write $W(\mathbf{x}) = m\Phi(\mathbf{x})$, where Φ is the **gravitational potential**. Thus the energy per unit mass $\frac{1}{2}v^2 + \Phi(\mathbf{x})$ is conserved. Despite the resulting minor ambiguity we shall often shorten the term "energy per unit mass" to simply "energy," and refer to $\frac{1}{2}v^2$ as the kinetic energy and Φ as the potential energy.

The **angular momentum** of a particle relative to some origin O is defined as

$$\mathbf{L} = \mathbf{x} \times \mathbf{p}, \tag{1D-9}$$

where the position vector $\mathbf{x}$ is measured from O. The torque is

$$\mathbf{N} = \mathbf{x} \times \mathbf{F}. \tag{1D-10}$$

We have

$$\frac{d\mathbf{L}}{dt} = \frac{d\mathbf{x}}{dt} \times \mathbf{p} + \mathbf{x} \times \frac{d\mathbf{p}}{dt} = \mathbf{v} \times \mathbf{p} + \mathbf{x} \times \mathbf{F}. \tag{1D-11}$$

The first term is proportional to $\mathbf{p} \times \mathbf{p} = 0$, and thus

$$\mathbf{N} = \frac{d\mathbf{L}}{dt}; \tag{1D-12}$$

in other words, *the torque is equal to the rate of change of angular momentum.*

2 Systems of Particles

Consider an isolated system of N particles of masses m_α and positions $\mathbf{x}_\alpha$. The total mass of the system is $M = \sum_{\alpha=1}^{N} m_\alpha$, and the total force on particle α is

$$\mathbf{F}_\alpha = \sum_{\beta \neq \alpha} \mathbf{F}_{\alpha\beta}, \tag{1D-13}$$

where $\mathbf{F}_{\alpha\beta}$ is the force exerted on particle α by particle β, and the sum is taken over $\beta = 1, \ldots, N$ excluding $\beta = \alpha$. According to **Newton's third law**

$$\mathbf{F}_{\alpha\beta} = -\mathbf{F}_{\beta\alpha}. \tag{1D-14}$$

The **center of mass** is located at

$$\mathbf{x}_{\text{cm}} = \frac{\sum_{\alpha=1}^{N} m_\alpha \mathbf{x}_\alpha}{M}. \tag{1D-15}$$

Thus

$$\frac{d^2 x_{cm}}{dt^2} = \frac{1}{M} \sum_{\alpha=1}^{N} m_\alpha \frac{d^2 x_\alpha}{dt^2} = \frac{1}{M} \sum_{\alpha=1}^{N} \sum_{\beta \neq \alpha} \mathbf{F}_{\alpha\beta}. \tag{1D-16}$$

The sum over $\mathbf{F}_{\alpha\beta}$ vanishes since each pair $\mathbf{F}_{\alpha\beta} + \mathbf{F}_{\beta\alpha}$ sums to zero. Thus

$$M \frac{d^2 x_{cm}}{dt^2} = 0; \tag{1D-17}$$

and we conclude that *the center of mass moves at uniform velocity.*

Similarly, the total angular momentum is

$$\mathbf{L} = \sum_{\alpha=1}^{N} x_\alpha \times \mathbf{p}_\alpha, \tag{1D-18}$$

and

$$\frac{d\mathbf{L}}{dt} = \sum_{\alpha=1}^{N} \sum_{\beta \neq \alpha} x_\alpha \times \mathbf{F}_{\alpha\beta}. \tag{1D-19}$$

The first term is a sum of pairs of the form $x_\alpha \times \mathbf{F}_{\alpha\beta} + x_\beta \times \mathbf{F}_{\beta\alpha} = (x_\alpha - x_\beta) \times \mathbf{F}_{\alpha\beta}$. In most cases the interparticle force acts along the line joining the two particles, and hence this term also vanishes. In other words, *the total angular momentum is conserved if the interparticle force acts along the line joining the particles.*

As in the case of a single particle, we compute the work done in moving the system from configuration 1 to configuration 2,

$$W_{12} = \sum_{\alpha=1}^{N} \int_1^2 \mathbf{F}_\alpha \cdot dx_\alpha, \tag{1D-20}$$

and as in equation (1D-5) we may write

$$W_{12} = K_{tot,2} - K_{tot,1}, \tag{1D-21}$$

where $K_{tot} = \frac{1}{2} \sum_{\alpha=1}^{N} m_\alpha v_\alpha^2$ is the total kinetic energy of the system. Note that

$$K_{tot} = \frac{1}{2} M v_{cm}^2 + \frac{1}{2} \sum_{\alpha=1}^{N} m_\alpha v_\alpha'^2, \tag{1D-22}$$

where $v_\alpha' = v_\alpha - v_{cm}$ and $v_{cm} = \dot{x}_{cm}$ is the velocity of the center of mass. Thus the total kinetic energy is the sum of the kinetic energy of motion about the center of mass, and the kinetic energy of a single particle of mass M moving at the velocity of the center of mass.

In many cases the interparticle forces are conservative, and can be written in the form

$$\mathbf{F}_{\alpha\beta} = -\frac{\partial}{\partial x_\alpha} W(|x_\alpha - x_\beta|). \tag{1D-23}$$

(Note that this form automatically guarantees the validity of Newton's third law.) Thus

$$W_{12} = -\sum_{\alpha=1}^{N}\sum_{\beta\neq\alpha}\int_{1}^{2}\frac{\partial}{\partial\mathbf{x}_\alpha}W(|\mathbf{x}_\alpha - \mathbf{x}_\beta|)\cdot d\mathbf{x}_\alpha. \tag{1D-24}$$

We define the **potential energy** of the system to be

$$W_{\text{tot}} = \tfrac{1}{2}\sum_{\alpha=1}^{N}\sum_{\beta\neq\alpha} W(|\mathbf{x}_\alpha - \mathbf{x}_\beta|). \tag{1D-25}$$

Then, as a result of any small change $\mathbf{x}_\alpha \to \mathbf{x}_\alpha + d\mathbf{x}_\alpha$, $\alpha = 1,\ldots,N$,

$$dW_{\text{tot}} = \sum_{\alpha=1}^{N}\sum_{\beta\neq\alpha}\frac{\partial W(|\mathbf{x}_\alpha - \mathbf{x}_\beta|)}{\partial\mathbf{x}_\alpha}\cdot d\mathbf{x}_\alpha, \tag{1D-26}$$

where the factor $\tfrac{1}{2}$ has disappeared because the term involving a given pair of particles [say, $W(|\mathbf{x}_1 - \mathbf{x}_2|)$] appears twice in equation (1D-25). Thus equations (1D-26) and (1D-24) yield $W_{12} = W_{\text{tot},1} - W_{\text{tot},2}$, and equation (1D-21) yields

$$K_{\text{tot},1} + W_{\text{tot},1} = K_{\text{tot},2} + W_{\text{tot},2}; \tag{1D-27}$$

in other words, *the total energy of the system $K_{\text{tot}} + W_{\text{tot}}$ is conserved.*

The behavior of an isolated two-body system is particularly simple. We have

$$m_1\frac{d^2\mathbf{x}_1}{dt^2} = \mathbf{F}_{12} = -m_2\frac{d^2\mathbf{x}_2}{dt^2}. \tag{1D-28}$$

The center of mass is at $\mathbf{x}_{\text{cm}} = (m_1\mathbf{x}_1 + m_2\mathbf{x}_2)/M$ and moves at uniform velocity. The relative separation vector $\mathbf{r} = \mathbf{x}_2 - \mathbf{x}_1$ obeys the equation

$$\frac{d^2\mathbf{r}}{dt^2} = -\frac{\mathbf{F}_{12}}{\mu} = \frac{1}{\mu}\frac{\partial W(|\mathbf{r}|)}{\partial\mathbf{x}_1}, \tag{1D-29}$$

where the **reduced mass** is

$$\mu = \frac{m_1 m_2}{m_1 + m_2}. \tag{1D-30}$$

Since $\partial W/\partial\mathbf{x}_1 = -\partial W/\partial\mathbf{r}$, we have

$$\frac{d^2\mathbf{r}}{dt^2} = -\frac{1}{\mu}\frac{\partial W(|\mathbf{r}|)}{\partial\mathbf{r}}, \tag{1D-31}$$

which is the equation of motion of a fictitious single particle (the **reduced particle**) of mass μ in the fixed potential $W(|\mathbf{r}|)$. Thus the two-body problem has been reduced to a one-body problem. The total energy is

$$E = \tfrac{1}{2}Mv_{\text{cm}}^2 + \tfrac{1}{2}(m_1v_1'^2 + m_2v_2'^2) + W = \tfrac{1}{2}Mv_{\text{cm}}^2 + \tfrac{1}{2}\mu\dot{r}^2 + W(|\mathbf{r}|). \tag{1D-32}$$

3 Lagrange's Equations

Consider a particle moving in a conservative force field defined by the potential energy $W(\mathbf{x}, t)$. We define the **Lagrangian**

$$\mathcal{L}(\mathbf{x}, \dot{\mathbf{x}}, t) = K - W = \tfrac{1}{2} m \dot{\mathbf{x}}^2 - W(\mathbf{x}, t). \tag{1D-33}$$

The **principle of least action** or **Hamilton's principle** states that *the motion of the particle from time t_0 to t_1 is along a curve $\mathbf{x}(t)$ that is an extremal of the* **action**

$$I = \int_{t_0}^{t_1} \mathcal{L} \, dt. \tag{1D-34}$$

The proof is simple. According to the Euler-Lagrange equation (1B-63), the trajectory is an extremal of I if and only if

$$0 = \frac{d}{dt} \left(\frac{\partial \mathcal{L}}{\partial \dot{\mathbf{x}}} \right) - \frac{\partial \mathcal{L}}{\partial \mathbf{x}} = m\ddot{\mathbf{x}} + \frac{\partial W}{\partial \mathbf{x}}, \tag{1D-35}$$

which is simply a restatement of Newton's second law.◁

The great advantage of this approach is that the Lagrangian $\mathcal{L}$ is a *scalar* function. Hence it is straightforward to compute $\mathcal{L}$ as a function $\mathcal{L}(\mathbf{q}, \dot{\mathbf{q}}, t)$ of arbitrary—sometimes called **generalized**—coordinates $\mathbf{q}$ and their time derivatives $\dot{\mathbf{q}}$. Extremizing the action with $\mathcal{L}$ expressed in this form, we obtain **Lagrange's equations**

$$\frac{d}{dt} \left(\frac{\partial \mathcal{L}}{\partial \dot{\mathbf{q}}} \right) - \frac{\partial \mathcal{L}}{\partial \mathbf{q}} = 0, \tag{1D-36}$$

which are the equations of motion in the generalized coordinates. This approach avoids the heavy algebra required to express vector equations in curvilinear coordinates. In equation (1D-36), $\mathcal{L} = K - W$, where $W = W(\mathbf{q}, t)$ is the potential energy expressed in terms of the generalized coordinates. The kinetic energy K is

$$K = \tfrac{1}{2} m \dot{\mathbf{x}}^2 = \tfrac{1}{2} m \sum_{i=1}^{3} \left(\sum_{j=1}^{3} \dot{q}_j \frac{\partial x_i}{\partial q_j} + \frac{\partial x_i}{\partial t} \right)^2, \tag{1D-37}$$

where the Cartesian coordinates have been written in terms of the generalized coordinates and time as $x_i(\mathbf{q}, t)$.

As an example, we derive the motion of a particle in a Cartesian coordinate system $\mathbf{x}'$ that rotates at an instantaneous angular speed Ω with respect to inertial space. The inertial velocity $\mathbf{v}_{\text{in}}$ is related to the rate of change of the rotating-frame coordinates $d\mathbf{x}'/dt$ by

$$\mathbf{v}_{\text{in}} = \frac{d\mathbf{x}'}{dt} + \mathbf{\Omega} \times \mathbf{x}'. \tag{1D-38}$$

Hence the Lagrangian of a particle of mass m is

$$\mathcal{L} = \tfrac{1}{2} m (\dot{\mathbf{x}}' + \mathbf{\Omega} \times \mathbf{x}')^2 - W(\mathbf{x}') = \tfrac{1}{2} m \sum_{i=1}^{3} (\dot{x}'_i + \epsilon_{ijk} \Omega_j x'_k)^2 - W(x'_i). \tag{1D-39}$$

Thus

$$\frac{\partial \mathcal{L}}{\partial \dot{x}'_l} = m(\dot{x}'_i + \epsilon_{ijk}\Omega_j x'_k)\delta_{il} = m(\dot{\mathbf{x}}' + \mathbf{\Omega} \times \mathbf{x}')_l, \tag{1D-40}$$

and

$$\frac{\partial \mathcal{L}}{\partial x'_l} = m(\dot{x}'_i + \epsilon_{ijk}\Omega_j x'_k)\epsilon_{ipq}\Omega_p \delta_{ql} - \frac{\partial W}{\partial x'_l} \tag{1D-41}$$

$$= m[(\dot{\mathbf{x}}' + \mathbf{\Omega} \times \mathbf{x}') \times \mathbf{\Omega}]_l - (\mathbf{\nabla}'W)_l.$$

If we now drop the primes for simplicity, Lagrange's equations yield

$$m[\ddot{\mathbf{x}} + \dot{\mathbf{\Omega}} \times \mathbf{x} + 2\mathbf{\Omega} \times \dot{\mathbf{x}} + \mathbf{\Omega} \times (\mathbf{\Omega} \times \mathbf{x})] + \mathbf{\nabla}W = 0. \tag{1D-42}$$

This result can be interpreted as follows: in a rotating coordinate system, particles move as if they were subject to three **fictitious forces**: the **inertial force of rotation**, $-m\dot{\mathbf{\Omega}} \times \mathbf{x}$; the **Coriolis force** $-2m\mathbf{\Omega} \times \dot{\mathbf{x}}$; and the **centrifugal force** $-m\mathbf{\Omega} \times (\mathbf{\Omega} \times \mathbf{x})$. In particular, in a frame rotating at constant angular speed $\mathbf{\Omega} = \Omega\hat{\mathbf{e}}_z$ the fictitious force is

$$\mathbf{F}_{\text{fict}} = -2m\mathbf{\Omega} \times \dot{\mathbf{x}} + \Omega^2 R\,\hat{\mathbf{e}}_R. \tag{1D-43}$$

4 Hamiltonian Dynamics

(a) Hamilton's equations   For a given set of generalized coordinates $\mathbf{q}$ we define the **generalized momentum p** to be

$$\mathbf{p} \equiv \left(\frac{\partial \mathcal{L}}{\partial \dot{\mathbf{q}}}\right)_{\mathbf{q},t}. \tag{1D-44}$$

The **Hamiltonian** H is defined by

$$H(\mathbf{q}, \mathbf{p}, t) \equiv \mathbf{p} \cdot \dot{\mathbf{q}} - \mathcal{L}(\mathbf{q}, \dot{\mathbf{q}}, t), \tag{1D-45}$$

where it is understood that $\dot{\mathbf{q}}$ is to be eliminated in favor of $\mathbf{q}$, $\mathbf{p}$, and t using equation (1D-44).

The total derivative of the Hamiltonian is

$$dH = \left(\frac{\partial H}{\partial \mathbf{q}}\right)_{\mathbf{p},t} \cdot d\mathbf{q} + \left(\frac{\partial H}{\partial \mathbf{p}}\right)_{\mathbf{q},t} \cdot d\mathbf{p} + \left(\frac{\partial H}{\partial t}\right)_{\mathbf{q},\mathbf{p}} dt. \tag{1D-46}$$

We may also write

$$dH = \mathbf{p} \cdot d\dot{\mathbf{q}} + \dot{\mathbf{q}} \cdot d\mathbf{p} - \left(\frac{\partial \mathcal{L}}{\partial \mathbf{q}}\right)_{\dot{\mathbf{q}},t} \cdot d\mathbf{q} - \left(\frac{\partial \mathcal{L}}{\partial \dot{\mathbf{q}}}\right)_{\mathbf{q},t} \cdot d\dot{\mathbf{q}} - \left(\frac{\partial \mathcal{L}}{\partial t}\right)_{\mathbf{q},\dot{\mathbf{q}}} dt$$

$$= \dot{\mathbf{q}} \cdot d\mathbf{p} - \left(\frac{\partial \mathcal{L}}{\partial \mathbf{q}}\right)_{\dot{\mathbf{q}},t} \cdot d\mathbf{q} - \left(\frac{\partial \mathcal{L}}{\partial t}\right)_{\mathbf{q},\dot{\mathbf{q}}} dt,$$

$$\tag{1D-47}$$

where the first and fourth terms have canceled because of equation (1D-44). Since equations (1D-46) and (1D-47) must be the same, we have

$$\dot{\mathbf{q}} = \left(\frac{\partial H}{\partial \mathbf{p}}\right)_{\mathbf{q},t} ; \quad \left(\frac{\partial H}{\partial \mathbf{q}}\right)_{\mathbf{p},t} = -\left(\frac{\partial \mathcal{L}}{\partial \mathbf{q}}\right)_{\dot{\mathbf{q}},t} ; \quad \left(\frac{\partial H}{\partial t}\right)_{\mathbf{q},\mathbf{p}} = -\left(\frac{\partial \mathcal{L}}{\partial t}\right)_{\mathbf{q},\dot{\mathbf{q}}}.$$
(1D-48)

Using Lagrange's equations (1D-36) and simplifying the notation, the first two of these equations lead us to **Hamilton's equations**

$$\dot{\mathbf{q}} = \frac{\partial H}{\partial \mathbf{p}} ; \quad \dot{\mathbf{p}} = -\frac{\partial H}{\partial \mathbf{q}}.$$
(1D-49)

The **phase space** of a system with n generalized coordinates $(q_1, \ldots, q_n)$ is the $2n$-dimensional space with coordinates $(q_1, \ldots, q_n, p_1, \ldots, p_n) \equiv (\mathbf{q}, \mathbf{p})$. Since Hamilton's equations (1D-49) are first-order differential equations, if we are given a particle's phase-space coordinates $(\mathbf{q}_0, \mathbf{p}_0)$ at time $t = 0$, we can solve Hamilton's equations for the coordinates $(\mathbf{q}_t, \mathbf{p}_t)$ at any later time t. Thus through each point $(\mathbf{q}_0, \mathbf{p}_0)$ in phase space there passes a unique phase-space trajectory $[\mathbf{q}(\mathbf{q}_0, \mathbf{p}_0, t), \mathbf{p}(\mathbf{q}_0, \mathbf{p}_0, t)]$, which gives the future and past phase-space coordinates of the particle that at $t = 0$ has coordinates $(\mathbf{q}_0, \mathbf{p}_0)$. Note that no two of these trajectories can ever intersect, since if they did, the future trajectory of a particle that started from the intersection point would not be unique.

We define the **time-evolution operator** $\mathbf{H}_t$ by

$$\mathbf{H}_t(\mathbf{q}_0, \mathbf{p}_0) \equiv [\mathbf{q}(\mathbf{q}_0, \mathbf{p}_0, t), \mathbf{p}(\mathbf{q}_0, \mathbf{p}_0, t)].$$
(1D-50)

We say that the *operator* $\mathbf{H}_t$ is generated by the *function* $H(\mathbf{q}, \mathbf{p})$.

Along a trajectory $\{\mathbf{q}(t), \mathbf{p}(t)\}$, the Hamiltonian $H[\mathbf{q}(t), \mathbf{p}(t), t]$ changes at a rate

$$\frac{dH}{dt} = \frac{\partial H}{\partial \mathbf{q}} \cdot \dot{\mathbf{q}} + \frac{\partial H}{\partial \mathbf{p}} \cdot \dot{\mathbf{p}} + \frac{\partial H}{\partial t} = \frac{\partial H}{\partial t}.$$
(1D-51)

Hence, if $\partial \mathcal{L}/\partial t = 0$, it follows from equation (1D-48) that the Hamiltonian is conserved along all dynamical trajectories.

Thus, for example, consider motion in the time-independent potential $W(\mathbf{x})$. If we work in Cartesian coordinates, the Lagrangian $\mathcal{L} = \frac{1}{2}m\dot{x}^2 - W(\mathbf{x})$ depends only on $\mathbf{x}$ and $\dot{\mathbf{x}}$, so $\partial \mathcal{L}/\partial t = 0$. Hence the Hamiltonian H is conserved. The physical quantity to which H corresponds is easily found. We have $\mathbf{p} = \partial \mathcal{L}/\partial \dot{\mathbf{x}} = m\dot{\mathbf{x}}$ and

$$H(\mathbf{x}, \mathbf{p}) = \mathbf{p} \cdot \dot{\mathbf{q}} - \mathcal{L}$$
$$= \frac{p^2}{2m} + W(\mathbf{x}),$$
(1D-52)

which is simply the total energy $E = K + W$. Thus *for motion in a fixed potential the Hamiltonian is equal to the total energy.*

The velocity $\mathbf{v}$ of a particle of mass m is just $1/m$ times the particle's Cartesian momentum $\mathbf{p}$, so by a slight abuse of language, we shall also use the term phase space to denote the six-dimensional space with coordinates $(\mathbf{x}, \mathbf{v})$.

(b) Poincaré invariants   Let S_0 be any two-dimensional surface in phase space, and (u, v) any pair of coordinates that may be used to specify points of S_0. The time-evolution operator H_t maps each point of S_0 into a new surface S_t and we denote by (u, v) the point of S_t into which H_t maps the point (u, v) of S_0. With these definitions, all six phase-space cordinates q_i and p_i are functions of the three variables u, v, and t.

We define

$$A(t) \equiv \iint_{S_t} d\mathbf{p} \cdot d\mathbf{q} \equiv \sum_{i=1}^{3} \iint_{S_t} dp_i dq_i$$
$$= \sum_{i=1}^{3} \iint_{S_t} \frac{\partial(p_i, q_i)}{\partial(u, v)} \, du dv,$$

(1D-53)

and calculate dA/dt. We set $t' \equiv t + \delta t$, where δt is small, and introduce the conventions $q(u, v) \equiv q(u, v, t)$, $q'(u, v) \equiv q(u, v, t')$ and similar notations for p and p'. To first order in the small time interval δt, Hamilton's equations (1D-49) yield

$$(\mathbf{q}', \mathbf{p}') = H_{\delta t}(\mathbf{q}, \mathbf{p}) = \left(\mathbf{q} + \frac{\partial H}{\partial \mathbf{p}} \delta t, \ \mathbf{p} - \frac{\partial H}{\partial \mathbf{q}} \delta t \right).$$

(1D-54)

Differentiating these equations with respect to u and v, we find

$$\frac{\partial(p_i', q_i')}{\partial(u, v)} = \frac{\partial(p_i, q_i)}{\partial(u, v)} - \left[\frac{\partial q_i}{\partial v} \frac{\partial^2 H}{\partial u \partial q_i} - \frac{\partial p_i}{\partial u} \frac{\partial^2 H}{\partial v \partial p_i} + \frac{\partial p_i}{\partial v} \frac{\partial^2 H}{\partial u \partial p_i} - \frac{\partial q_i}{\partial u} \frac{\partial^2 H}{\partial v \partial q_i} \right] \delta t$$
$$+ O(\delta t)^2.$$

(1D-55)

Thus

$$\frac{dA}{dt} = \lim_{\delta t \to 0} \left\{ \frac{1}{\delta t} \iint du dv \sum_i \left[\frac{\partial(p_i', q_i')}{\partial(u, v)} - \frac{\partial(p_i, q_i)}{\partial(u, v)} \right] \right\}$$
$$= -\sum_i \iint du dv \left[\frac{\partial q_i}{\partial v} \frac{\partial^2 H}{\partial u \partial q_i} - \frac{\partial p_i}{\partial u} \frac{\partial^2 H}{\partial v \partial p_i} + \frac{\partial p_i}{\partial v} \frac{\partial^2 H}{\partial u \partial p_i} - \frac{\partial q_i}{\partial u} \frac{\partial^2 H}{\partial v \partial q_i} \right].$$

(1D-56)

One may show that the sum of the square brackets in equation (1D-56) vanishes by replacing every occurrence of $\dfrac{\partial}{\partial u}$ in the second derivatives by $\sum_k \left(\dfrac{\partial q_k}{\partial u} \dfrac{\partial}{\partial q_k} + \dfrac{\partial p_k}{\partial u} \dfrac{\partial}{\partial p_k} \right)$ and similarly for $\dfrac{\partial}{\partial v}$. Hence $\dot{A} = 0$ and we have:

Poincaré invariant theorem *If $S(0)$ is any two-surface in phase space, and $S(t)$ is the surface into which $S(0)$ is mapped by the time-evolution operator H_t, then*

$$\iint_{S(0)} d\mathbf{p} \cdot d\mathbf{q} = \iint_{S(t)} d\mathbf{p} \cdot d\mathbf{q}.$$

(1D-57)

Corollary If $\gamma(0)$ is any closed path through phase space, and $\gamma(t)$ is the path to which $\gamma(0)$ is mapped by the time-evolution operator, then

$$\oint_{\gamma(0)} \mathbf{p} \cdot d\mathbf{q} = \oint_{\gamma(t)} \mathbf{p} \cdot d\mathbf{q}. \tag{1D-58}$$

Proof. By Green's theorem,

$$\oint_{\gamma(t)} \mathbf{p} \cdot d\mathbf{q} = \sum_i \oint_{\gamma(t)} p_i dq_i = \sum_i \iint_{S(t)} dp_i dq_i, \tag{1D-59}$$

where $S(t)$ is any surface that has $\gamma(t)$ as its boundary. The result now follows from the Poincaré invariant theorem.◁

(c) Canonical maps[4]   Any mapping of phase space onto itself which, like $\mathbf{H}_t$, conserves line integrals of the form $\oint \mathbf{p} \cdot d\mathbf{q}$, is called a **canonical map**.

Let $A(\mathbf{q}, \mathbf{p})$ and $B(\mathbf{q}, \mathbf{p})$ be any two functions of the phase-space coordinates. Then the **Poisson bracket** $[A, B]$ is defined by

$$[A, B] \equiv \frac{\partial A}{\partial \mathbf{q}} \cdot \frac{\partial B}{\partial \mathbf{p}} - \frac{\partial A}{\partial \mathbf{p}} \cdot \frac{\partial B}{\partial \mathbf{q}}. \tag{1D-60}$$

It is straightforward to verify the following properties of Poisson brackets:

(i) $[A, B] = -[B, A]$ and $[A + B, C] = [A, C] + [B, C]$,

(ii) $[[A, B], C] + [[B, C], A] + [[C, A], B] = 0$ (**Jacobi identity**),

(iii) Hamilton's equations may be written

$$\dot{q}_i = [q_i, H] \quad ; \quad \dot{p}_i = [p_i, H]. \tag{1D-61}$$

(iv) The coordinates $(\mathbf{q}, \mathbf{p})$ satisfy the **canonical commutation relations** $[p_i, p_j] = [q_i, q_j] = 0$ and $[q_i, p_j] = \delta_{ij}$.

If we write $(w_i \equiv q_i, w_{3+i} \equiv p_i \quad i = 1, 2, 3)$, and define the **symplectic matrix c** by

$$c_{\alpha\beta} \equiv [w_\alpha, w_\beta] = \begin{cases} \pm 1 & \text{for } \alpha = \beta \mp 3, \ 1 \leq \alpha, \beta \leq 6; \\ 0 & \text{otherwise}, \end{cases} \tag{1D-62a}$$

we have

$$[A, B] = \sum_{\alpha, \beta = 1}^{6} c_{\alpha\beta} \frac{\partial A}{\partial w_\alpha} \frac{\partial B}{\partial w_\beta}. \tag{1D-62b}$$

---

[4] We adopt Arnold's (1978) approach, which stresses the parallels between canonical maps and orthogonal and Lorentz transformations, rather than the traditional approach of Goldstein (1980), which is based on generating functions.

Any set of six phase-space coordinates $\{W_\alpha, \ \alpha = 1, \ldots, 6\}$ is called **canonical** if $[W_\alpha, W_\beta] = c_{\alpha\beta}$. Let $\{W_\alpha\}$ be such a set; then with equation (1D-62b) and the chain rule we have

$$[A, B] = \sum_{\alpha,\beta=1}^{6} c_{\alpha\beta} \frac{\partial A}{\partial w_\alpha} \frac{\partial B}{\partial w_\beta} = \sum_{\kappa\lambda} \left(\sum_{\alpha\beta} c_{\alpha\beta} \frac{\partial W_\kappa}{\partial w_\alpha} \frac{\partial W_\lambda}{\partial w_\beta} \right) \frac{\partial A}{\partial W_\kappa} \frac{\partial B}{\partial W_\lambda}$$

$$= \sum_{\kappa\lambda} [W_\kappa, W_\lambda] \frac{\partial A}{\partial W_\kappa} \frac{\partial B}{\partial W_\lambda} = \sum_{\kappa\lambda} c_{\kappa\lambda} \frac{\partial A}{\partial W_\kappa} \frac{\partial B}{\partial W_\lambda}.$$

$$\text{(1D-63)}$$

Thus the derivatives involved in the definition (1D-60) of the Poisson bracket can be taken with respect to any set of canonical coordinates, just as the vector formula $\nabla \cdot \mathbf{a} = \sum_i (\partial a_i / \partial x_i)$ is valid in any Cartesian coordinate system.

The rate of change of an arbitrary canonical coordinate W_α along an orbit is

$$\dot{W}_\alpha = \sum_{\beta=1}^{6} \frac{\partial W_\alpha}{\partial w_\beta} \dot{w}_\beta, \tag{1D-64}$$

where, as usual, $\mathbf{w} \equiv (\mathbf{q}, \mathbf{p})$. With Hamilton's equations (1D-61) and equation (1D-63) this becomes

$$\dot{W}_\alpha = \sum_{\beta=1}^{6} \frac{\partial W_\alpha}{\partial w_\beta} [w_\beta, H] = \sum_{\beta\gamma\delta} \frac{\partial W_\alpha}{\partial w_\beta} c_{\gamma\delta} \frac{\partial w_\beta}{\partial w_\gamma} \frac{\partial H}{\partial w_\delta} = \sum_{\gamma\delta} c_{\gamma\delta} \frac{\partial W_\alpha}{\partial w_\gamma} \frac{\partial H}{\partial w_\delta}$$

$$= [W_\alpha, H].$$

$$\text{(1D-65)}$$

Thus Hamilton's equations (1D-61) are valid in *any* canonical coordinate system.

Finally, we show that equation (1D-53) for the Poincaré invariant of a two-surface is valid in any canonical coordinate system. We first note that if we are given any function $B(\mathbf{q}, \mathbf{p})$ we may obtain a one-parameter family of maps $\mathbf{B}_a$ of phase space onto itself by the following procedure. From each point $(\mathbf{q}_0, \mathbf{p}_0)$ of some five-dimensional surface in phase space we integrate the coupled ordinary differential equations

$$\frac{d\mathbf{q}}{da} = [\mathbf{q}, B] \quad , \quad \frac{d\mathbf{p}}{da} = [\mathbf{p}, B] \tag{1D-66}$$

from the initial conditions $\mathbf{q}(0) = \mathbf{q}_0$, $\mathbf{p}(0) = \mathbf{p}_0$. If the initial five-surface is large enough, the **integral curves** $\{\mathbf{q}(a), \mathbf{p}(a)\}$ of B reach every point of phase space. Then the map $\mathbf{B}_a$ is defined by

$$\mathbf{B}_a(\mathbf{q}(a'), \mathbf{p}(a')) = (\mathbf{q}(a + a'), \mathbf{p}(a + a')). \tag{1D-67}$$

The function $B(\mathbf{q}, \mathbf{p})$ is indistinguishable from a Hamiltonian, since it satisfies Hamilton's equations (1D-66), with a playing the role of the time t. Thus the Poincaré invariant theorem shows that $\mathbf{B}_a$ is a canonical map.

Now let $\mathbf{S}$ be any two-surface in phase space, and U and V two functions on phase space whose integral curves lie within $\mathbf{S}$. Then we may use the maps

$\mathbf{U}_u$ and $\mathbf{V}_v$ to lay out a coordinate grid on $\mathcal{S}$: we pick a point $(\mathbf{q}_0, \mathbf{p}_0)$ of $\mathcal{S}$ and define $(\mathbf{q}, \mathbf{p})_{(u,v)} \equiv \mathbf{V}_v \mathbf{U}_u(\mathbf{q}_0, \mathbf{p}_0)$. The Poincaré invariant (1D-53) of $\mathcal{S}$ may now be written

$$A = \iint_{\mathcal{S}} d\mathbf{p} \cdot d\mathbf{q} = \iint \sum_i \frac{\partial(p_i, q_i)}{\partial(u, v)} \, du \, dv$$

$$= \iint \sum_i ([p_i, U][q_i, V] - [p_i, V][q_i, U]) \, du \, dv, \tag{1D-68}$$

where the last equality follows by expanding the Jacobian and using equation (1D-66). But by equation (1D-60), $[p_i, U] = -(\partial U/\partial q_i)$ and $[q_i, V] = (\partial V/\partial p_i)$, so we may rewrite equation (1D-68) as

$$A = -\iint \sum_i \left(\frac{\partial U}{\partial q_i} \frac{\partial V}{\partial p_i} - \frac{\partial V}{\partial q_i} \frac{\partial U}{\partial p_i} \right) du \, dv = -\iint [U, V] \, du \, dv. \tag{1D-69}$$

Now notice that the Poisson bracket $[U, V]$ is the same when evaluated in terms of derivatives with respect to any set of canonical coordinates. For the same reason, the map between (u, v) and the surface $\mathcal{S}$ is independent of the canonical coordinates used in equation (1D-66). Hence we may change to a new set of canonical coordinates $(\mathbf{Q}, \mathbf{P})$, retrace each step in the derivation of equation (1D-69), and thus show that

$$\iint_{\mathcal{S}} d\mathbf{p} \cdot d\mathbf{q} = \iint_{\mathcal{S}} d\mathbf{P} \cdot d\mathbf{Q} \qquad \text{for any } \mathcal{S}. \tag{1D-70}$$

(d) Phase-space volumes   A result closely related to equation (1D-70) concerns phase-space volumes. We define the volume associated with a region $\mathcal{V}$ of phase space to be

$$V = \iint_{\mathcal{V}} d^3p \, d^3q. \tag{1D-71a}$$

It can be shown that if $(\mathbf{Q}, \mathbf{P})$ are any canonical coordinates, then

$$V = \iint_{\mathcal{V}} d^3P \, d^3Q. \tag{1D-71b}$$

To prove this, we consider a small six-dimensional parallelepiped. We introduce six independent functions $U_i(\mathbf{q}, \mathbf{p})$, $(i = 1, \ldots, 6)$ whose integral curves run parallel to the edges of the parallelepiped. We denote by u_i the parameter a that measures displacement along the integral curve generated by the function U_i through equations (1D-66). Let us define the canonical volume of the parallelepiped to be

$$dV \equiv \frac{1}{3!} \sum (-1)^\nu [U_{i_1}, U_{i_2}][U_{i_3}, U_{i_4}][U_{i_5}, U_{i_6}] \, du_1 du_2 du_3 du_4 du_5 du_6, \tag{1D-72}$$

where the sum is over all permutations $(i_1, \ldots, i_6)$ of the numbers $(1, \ldots, 6)$ such that $i_1 < i_2$, $i_3 < i_4$, and $i_5 < i_6$ and ν is 0 or 1 if the permutation is even or odd. The motivation for this equation can be seen by applying equation (1D-72) to a phase-space element whose sides are parallel to the axes of the coordinate system $(\mathbf{q}, \mathbf{p})$. In this case we may take $U_1 = q_1$, $U_2 = p_1$, $U_3 = q_2, \ldots, U_6 = p_3$, and a straightforward calculation yields $dV = dp_1 dq_1 dp_2 dq_2 dp_3 dq_3$, which is consistent with the definition of volume given in equation (1D-71).

Now notice that the Poisson bracket $[U_i, U_j]$ is the same when evaluated in terms of derivatives with respect to any set of canonical coordinates. For the same reason, the map between $\{u_i\}$ and the body of the parallelepiped is independent of the canonical coordinates used in equation (1D-66). Hence we may change to a new set of canonical coordinates $(\mathbf{Q}, \mathbf{P})$, retrace each step in the derivation of equation (1D-72), and thus show that equations (1D-71) and (1D-72) provide equivalent expressions for the volume of an arbitrary region of phase space $\mathcal{V}$.

(e) Point transformations   If $(Q_i(\mathbf{q}), \ i = 1, 2, 3)$ are any three independent functions of the generalized coordinates $\mathbf{q}$, then by equation (1D-44) we obtain the new momenta $P_i = (\partial \mathcal{L}/\partial \dot{Q}_i)$ by expressing the Lagrangian as a function $\mathcal{L}(\mathbf{Q}, \dot{\mathbf{Q}})$ of the Q_i and their time derivatives. The coordinate change $(\mathbf{q}, \mathbf{p}) \rightarrow (\mathbf{Q}, \mathbf{P})$ is called a **point transformation**, because the new coordinates are functions only of the old. It is straightforward to show that the new coordinates are canonical, by evaluating their Poisson brackets. In fact, if we interpret $(\mathbf{Q}, \mathbf{P})$ not as new coordinates for a single point $(\mathbf{q}, \mathbf{p})$, but as the coordinates in the $(\mathbf{q}, \mathbf{p})$ system of a new point in phase space, then one may show that the map $(\mathbf{q}, \mathbf{p}) \rightarrow (\mathbf{Q}, \mathbf{P})$ is canonical.

The importance of these results for our purposes is that we will often work in curvilinear coordinates $\mathbf{Q}$ and derive the corresponding momenta $\mathbf{P} = (\partial \mathcal{L}/\partial \dot{\mathbf{Q}})$. We have shown that $(\mathbf{Q}, \mathbf{P})$ is canonical, and hence the Poisson bracket (1D-60) can be equally well evaluated by taking derivatives with respect to $\mathbf{Q}$ and $\mathbf{P}$ as with respect to $\mathbf{q}$ and $\mathbf{p}$. Hence all curvilinear coordinates have equal status in Hamiltonian mechanics.

Appendix 1.E Fluid Mechanics

The basic principles of fluid mechanics are important in studying galaxy dynamics, both because fluid and stellar systems often behave in similar ways and because gas dynamics are important for the formation and evolution of galaxies. We review here some of the concepts of fluid mechanics that will be used in Chapters 4, 5, and 6.

An excellent introduction to fluid mechanics is given in Landau and Lifshitz (1959).

1 Basic Equations

The state of a fluid is specified by its density $\rho(\mathbf{x}, t)$, pressure $p(\mathbf{x}, t)$, and velocity field $\mathbf{v}(\mathbf{x}, t)$, and possibly by other thermodynamic functions such as the temperature $T(\mathbf{x}, t)$ or specific entropy $s(\mathbf{x}, t)$.

(a) **Continuity equation**   Consider an arbitrary closed volume V that is fixed in position and shape and bounded by a surface S. The mass of fluid in this volume is $M(t) = \int_V \rho(\mathbf{x}, t) d^3\mathbf{x}$; $M(t)$ changes with time at a rate $dM/dt = \int_V (\partial \rho / \partial t) d^3\mathbf{x}$. The mass flowing out through the area element d^2S per unit time is $\rho \mathbf{v} \cdot d^2\mathbf{S}$, where $d^2\mathbf{S}$ is an outward-pointing vector, normal to the surface, with magnitude d^2S. Thus $dM/dt = -\int_S \rho \mathbf{v} \cdot d^2\mathbf{S}$ and hence

$$\int_V \frac{\partial \rho}{\partial t} d^3\mathbf{x} + \int_S \rho \mathbf{v} \cdot d^2\mathbf{S} = 0. \tag{1E-1}$$

Using the divergence theorem (1B-42),

$$\int_V \left[\frac{\partial \rho}{\partial t} + \boldsymbol{\nabla} \cdot (\rho \mathbf{v}) \right] d^3\mathbf{x} = 0, \tag{1E-2}$$

and since this result must hold for any volume, we arrive at the **continuity equation**

$$\frac{\partial \rho}{\partial t} + \boldsymbol{\nabla} \cdot (\rho \mathbf{v}) = 0. \tag{1E-3}$$

In perturbation theory, we sometimes need to know the change in density resulting from a small displacement of the fluid elements. Let the fluid element at position $\mathbf{x}$ be displaced to $\mathbf{x} + \epsilon \boldsymbol{\xi}(\mathbf{x})$, where $\epsilon \ll 1$. If the displacement occurs at $t = t_0$, we may integrate equation (1E-3) from $t_0 - \delta$ to $t_0 + \delta$ and let δ shrink to zero, to obtain

$$\rho_1 = -\boldsymbol{\nabla} \cdot (\rho_0 \boldsymbol{\xi}), \tag{1E-4}$$

where $\epsilon \rho_1(\mathbf{x})$ is the change in density at $\mathbf{x}$. Note that we have replaced the density by its unperturbed value in the divergence, since it is multiplied by the small quantity $\boldsymbol{\xi}$ (for more details see Chandrasekhar 1969, §13).

(b) **Euler's equation**   In an inviscid fluid, the total pressure force acting on the volume is $-\int_S p \, d^2\mathbf{S}$. In addition, there may be some external force, in particular the force from a gravitational potential $\Phi(\mathbf{x}, t)$. Thus Newton's second law reads

$$M \frac{d\mathbf{v}}{dt} = -\int_S p \, d^2\mathbf{S} - M \boldsymbol{\nabla} \Phi. \tag{1E-5}$$

According to the divergence theorem (1B-44), $\int_S p \, d^2\mathbf{S} = \int_V \boldsymbol{\nabla} p \, d^3\mathbf{x}$, and since equation (1E-5) must hold for every small volume V,

$$\rho \frac{d\mathbf{v}}{dt} = -\boldsymbol{\nabla} p - \rho \boldsymbol{\nabla} \Phi. \tag{1E-6}$$

We must now relate $d\mathbf{v}/dt$ to the velocity field $\mathbf{v}(\mathbf{x}, t)$. The change $d\mathbf{v}$ in velocity of a given particle during the interval dt is the sum of the change in velocity at a given point in space, $(\partial\mathbf{v}/\partial t)dt$, and the difference in velocities between two points separated by $d\mathbf{x} = \mathbf{v}dt$ at the same instant. The latter change is $(\partial\mathbf{v}/\partial x_i)dx_i = (d\mathbf{x} \cdot \nabla)\mathbf{v}$. Thus

$$\frac{d\mathbf{v}}{dt} = \frac{\partial\mathbf{v}}{\partial t} + (\mathbf{v} \cdot \nabla)\mathbf{v}; \tag{1E-7}$$

the quantity $d\mathbf{v}/dt$ is sometimes referred to as the **Lagrangian** or **convective** derivative of $\mathbf{v}$ as opposed to the **Eulerian** derivative $\partial\mathbf{v}/\partial t$. Expressions for $(\mathbf{v} \cdot \nabla)\mathbf{v}$ are given in equations (1B-53) to (1B-56).

Combining equations (1E-6) and (1E-7) we arrive at **Euler's equation**

$$\frac{\partial\mathbf{v}}{\partial t} + (\mathbf{v} \cdot \nabla)\mathbf{v} = -\frac{1}{\rho}\nabla p - \nabla\Phi. \tag{1E-8}$$

(c) Equation of state   To relate the pressure and density, we generally need an **equation of state** $p = p(\rho, s)$ or $p = p(\rho, T)$ together with an auxiliary equation determining the specific entropy or temperature. Since we will only use fluid systems as analogs of stellar systems, it is sufficient to consider the much simpler case of a **barotropic** equation of state, where the pressure is uniquely determined by the density,

$$p = p(\rho). \tag{1E-9}$$

In a barotropic fluid the Euler equation can be simplified by defining the **specific enthalpy** (the relation of this quantity to the thermodynamic enthalpy is discussed below),

$$h(\rho) = \int_0^\rho \frac{dp}{\rho} = \int_0^\rho \frac{dp(\rho)}{d\rho}\frac{d\rho}{\rho}. \tag{1E-10}$$

Equation (1E-8) can then be written

$$\frac{\partial\mathbf{v}}{\partial t} + (\mathbf{v} \cdot \nabla)\mathbf{v} = -\nabla(h + \Phi). \tag{1E-11}$$

The two equations (1E-3) and (1E-9), together with the three components of the vector equation (1E-8), provide a complete description of the evolution of the variables p, ρ and the three components of $\mathbf{v}$.

The most important examples of barotropic fluids are those that are **isentropic** or **adiabatic**, that is, those whose specific entropy is constant. In this case it is instructive to examine the relation between the entropy S, the internal energy U, and the enthalpy $H \equiv U + pV$. From the second law of thermodynamics,

$$dU = TdS - pdV \quad ; \quad dH = TdS + Vdp, \tag{1E-12}$$

where $V = M/\rho$ is the volume occupied by the fluid. If we now consider a unit mass of fluid, then $V = 1/\rho$ and S, H, and U are replaced by their specific values s, h, and u. For isentropic fluids $ds = 0$, and we have

$$du = \frac{p}{\rho^2}d\rho \quad ; \quad dh = \frac{dp}{\rho}. \tag{1E-13}$$

The second of these relations confirms that the quantity defined in equation (1E-10) is the usual thermodynamic enthalpy. The first shows that the internal energy per unit mass is

$$u(\rho) = \int_0^\rho \frac{p(\rho')}{\rho'^2}d\rho'. \tag{1E-14}$$

2 Sound Waves

Consider a stationary barotropic fluid of constant density and pressure, and assume that the gravitational field $\nabla\Phi = 0$. If the fluid is subject to a small perturbation, we may write

$$\rho(\mathbf{x},t) = \rho_0 + \epsilon\rho_1(\mathbf{x},t), \quad h(\mathbf{x},t) = h_0 + \epsilon h_1(\mathbf{x},t), \quad \mathbf{v}(\mathbf{x},t) = \epsilon\mathbf{v}_1(\mathbf{x},t), \tag{1E-15}$$

where $\epsilon \ll 1$. Substituting (1E-15) into equations (1E-3), (1E-10), and (1E-11), we find that the terms that are independent of ϵ vanish, and discarding terms proportional to ϵ^2 we obtain

$$\frac{\partial\rho_1}{\partial t} + \rho_0\nabla\cdot\mathbf{v}_1 = 0, \quad \frac{\partial\mathbf{v}_1}{\partial t} = -\nabla h_1, \quad h_1 = \left(\frac{dp}{d\rho}\right)_{\rho_0}\frac{\rho_1}{\rho_0}. \tag{1E-16}$$

We differentiate the first of equations (1E-16) with respect to time and eliminate $\mathbf{v}_1$ and h_1 to obtain the **wave equation**

$$\frac{\partial^2\rho_1}{\partial t^2} - v_s^2\nabla^2\rho_1 = 0, \tag{1E-17a}$$

where

$$v_s = \sqrt{\left.\frac{dp}{d\rho}\right|_{\rho_0}}. \tag{1E-17b}$$

The solution of this equation is simplest to understand in the case where ρ_1 depends only on one coordinate, say, x. Then

$$\frac{\partial^2\rho_1}{\partial t^2} - v_s^2\frac{\partial^2\rho_1}{\partial x^2} = 0. \tag{1E-18}$$

The general solution is

$$\rho_1 = f_+(x - v_st) + f_-(x + v_st), \tag{1E-19}$$

where f_+ and f_- are arbitrary functions. This solution consists of two super-imposed traveling waves, one (f_+) traveling to the right at v_s and one (f_-) traveling to the left at v_s. The disturbances may be regarded as sound waves that propagate through the fluid at a speed v_s, and hence v_s is known as the **sound speed**.

The simplest examples of sound waves are uniform wave trains of the form

$$\rho_1 = A\cos(kx - \omega t), \tag{1E-20}$$

which satisfy equation (1E-18) if

$$\omega^2 = v_s^2 k^2. \tag{1E-21}$$

This is the **dispersion relation** for sound waves.

3 The Ideal Gas

One of the simplest fluids is the ideal gas, whose equation of state is

$$p = \frac{\rho k_B T}{m}, \tag{1E-22}$$

where k_B is Boltzmann's constant and m is the molecular mass. The pressure in an ideal gas is related to the velocity dispersion of the molecules in each direction by

$$p = \rho\overline{v_x^2} = \rho\overline{v_y^2} = \rho\overline{v_z^2}, \tag{1E-23}$$

and thus

$$\overline{v_x^2} = \overline{v_y^2} = \overline{v_z^2} = \tfrac{1}{3}\overline{v^2} = \frac{k_B T}{m}. \tag{1E-24}$$

There are two important cases in which the ideal gas is barotropic. (i) Isothermal gas. If the temperature of the gas is kept fixed everywhere at T_0, then the fluid is said to be **isothermal**. In this case the equation of state is barotropic, with

$$p = K'\rho, \tag{1E-25}$$

and $K' = k_B T_0/m$. (ii) Isentropic gas. Consider the thermodynamic relation (Landau & Lifshitz 1980, §16)

$$\left(\frac{\partial p}{\partial \rho}\right)_s = \gamma\left(\frac{\partial p}{\partial \rho}\right)_T, \tag{1E-26}$$

where the subscripts s and T denote derivatives at constant entropy and temperature, and γ is the ratio of the heat capacity at constant pressure to the heat capacity at constant volume, which is usually constant over a wide range of conditions. For an ideal gas $(\partial p/\partial \rho)_T = k_B T/m = p/\rho$, and hence equation (1E-26) can be integrated to yield

$$p = K\rho^\gamma, \tag{1E-27}$$

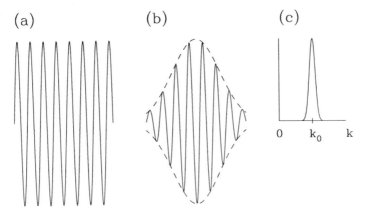

Figure 1E-1. (a) A wave train with wavevector k_0. (b) A typical wave packet. The dashed curve is the envelope or amplitude of the wave packet. (c) The Fourier amplitude of the packet.

where K is a constant. Thus an isentropic ideal gas is barotropic; a barotropic equation of state with the particular form (1E-27) is known as a **polytropic** equation of state.

In general the frequency of sound waves is so high that the perturbations are adiabatic, and hence equation (1E-18) yields for the sound speed in an ideal gas

$$v_s = \sqrt{\frac{\gamma k_B T}{m}}. \tag{1E-28}$$

4 Group Velocity

The dispersion relation for sound waves (1E-21) has the simplifying feature that $\omega(k) = v_s k$ is linear in the wavenumber k. This is not a general property of all types of waves; for example, the dispersion relation for sound waves in a medium with self-gravity, equation (5-20), has the form $\omega(k) = v_s \sqrt{k^2 - 4\pi G \rho_0 / v_s^2}$. If $\omega(k)$ is not linear in k, the medium is said to be **dispersive**.

One example of a wave in a dispersive medium is a uniform wave train of the form (1E-20), as shown in Figure 1E-1a. However, any wave train of this sort is somewhat unphysical, since it extends throughout all space. To describe a spatially localized wave is more complicated, since simple solutions of the form (1E-19) do not exist in dispersive media. Instead, we must construct a **wave packet**, a wave whose amplitude is non-zero only over a region of finite extent $\Delta x \gg |k|^{-1}$. A typical wave packet is shown in Figure 1E-1b. The wave packet can be represented mathematically by superimposing various monochromatic waves:

$$f(x,t) = \int_{-\infty}^{\infty} F(k) e^{i[kx - \omega(k)t]} dk. \tag{1E-29}$$

In any small region, the wave packet looks like a wave train of fixed wavenumber, say, k_0. Hence $F(k)$ will only be non-zero for values of k near k_0 [see Figure 1E-1c], and we may therefore expand $\omega(k)$ in a Taylor series, $\omega(k) \simeq \omega(k_0) + (k - k_0)v_g$, where

$$v_g \equiv \left[\frac{d\omega(k)}{dk}\right]_{k_0}. \tag{1E-30}$$

Equation (1E-29) becomes

$$f(x,t) \simeq e^{i[k_0 x - \omega(k_0)t]} \int_{-\infty}^{\infty} F(k_0 + u)e^{iu(x - v_g t)}\, du. \tag{1E-31}$$

We can rewrite this expression as

$$f(x,t) \simeq e^{i[k_0 x - \omega(k_0)t]} A(x - v_g t), \tag{1E-32}$$

where

$$A(x) = \int_{-\infty}^{\infty} F(k_0 + u)e^{iux}\, du. \tag{1E-33}$$

The function $A(x)$ is simply the amplitude of the packet at $t = 0$, $|f(x, t = 0)|$. Equation (1E-32) represents a plane wave with amplitude $A(x - v_g t)$, and shows that the envelope of the wave propagates with a velocity v_g, which is known as the **group velocity**. The group velocity represents the velocity of any true physical disturbance, since a disturbance arising from any physical source is always localized. Thus, for example, it can be shown that the energy of the wave is propagated at the group velocity (see Whitham 1974 for details).

The group velocity is distinct from the **phase velocity**, the velocity of a given crest of the wave. On a given crest $kx - \omega t = \text{constant}$; thus

$$v_p = \frac{\omega}{k}. \tag{1E-34}$$

The phase velocity and group velocity are only equal in a non-dispersive medium, where $\omega(k)$ is linear in k.

Appendix 4.A Proof of the Strong Jeans Theorem

This theorem states that the DF of a steady-state galaxy in which almost all orbits are regular with incommensurable frequencies may be presumed to be a function only of three independent isolating integrals. We define a galaxy to be in a steady state if all of its observable properties are time-independent. Any observational property involves averaging the distribution function over some non-zero region of phase space; we may formalize this by stating that all observations are based on moments of the form

$$\langle Q \rangle = \int Q(\mathbf{x}, \mathbf{v}) f(\mathbf{x}, \mathbf{v}, t) d^3\mathbf{x}\, d^3\mathbf{v}, \tag{4A-1}$$

where $f(\mathbf{x}, \mathbf{v}, t)$ is the DF of the galaxy at time t, and $Q(\mathbf{x}, \mathbf{v})$ is some smooth phase-space function.

Since the galaxy is in a steady state, $\langle Q \rangle$ is equal to its time average $\overline{\langle Q \rangle}$;

$$\langle Q \rangle = \overline{\langle Q \rangle} = \int Q(\mathbf{x}, \mathbf{v}) \overline{f}(\mathbf{x}, \mathbf{v}, t) d^3\mathbf{x} d^3\mathbf{v}, \qquad (4A\text{-}2a)$$

where

$$\overline{f} \equiv \lim_{T \to \infty} \frac{1}{T} \int_0^T f(\mathbf{x}, \mathbf{v}, t) dt. \qquad (4A\text{-}2b)$$

In this integral $\mathbf{x}$ and $\mathbf{v}$ are independent of t. But since f satisfies the collisionless Boltzmann equation, $(df/dt) = 0$, and we have

$$f(\mathbf{x}, \mathbf{v}, t) = f(\mathbf{x}_t, \mathbf{v}_t, 0), \qquad (4A\text{-}3a)$$

where $[\mathbf{x}_t(\mathbf{x}, \mathbf{v}), \mathbf{v}_t(\mathbf{x}, \mathbf{v})]$ are the coordinates at time $t = 0$ of the star that at time t is at $(\mathbf{x}, \mathbf{v})$. Thus

$$\overline{f}(\mathbf{x}, \mathbf{v}) = \lim_{T \to \infty} \frac{1}{T} \int_0^T f(\mathbf{x}_t, \mathbf{v}_t, 0) dt. \qquad (4A\text{-}3b)$$

Since almost all orbits are regular, action-angle coordinates $(\mathbf{J}, \boldsymbol{\theta})$ exist everywhere[5] in phase space. Transforming equation (4A-3b) to these coordinates and using equation (3-150b), we obtain

$$\overline{f}(\mathbf{J}, \boldsymbol{\theta}) = \lim_{T \to \infty} \frac{1}{T} \int_0^T f(\mathbf{J}, \boldsymbol{\theta} - \boldsymbol{\omega} t) dt. \qquad (4A\text{-}4)$$

In computing observable properties, $\overline{f}$ is always convolved with a smooth function Q. Hence it is sufficient to know the value of $\overline{f}$ averaged over small but non-zero volumes of phase space. By the time averages theorem of §3.5.2, when the frequencies ω_i are incommensurable, time averages over non-zero volumes are equal to angle averages. Thus the time average (4A-4) can be written

$$\overline{f} = (2\pi)^{-3} \int f(\mathbf{J}, \boldsymbol{\theta}) d^3\boldsymbol{\theta}, \qquad (4A\text{-}5)$$

and hence $\overline{f}$ is a function of the actions only. Thus equation (4A-2a) is of the form

$$\langle Q \rangle = \int Q(\mathbf{x}, \mathbf{v}) \overline{f}[\mathbf{J}(\mathbf{x}, \mathbf{v})] d^3\mathbf{x} d^3\mathbf{v}. \qquad (4A\text{-}6)$$

The DF $\overline{f}$, which is a function only of the three actions, therefore generates the same moments and hence the same observables as the true DF f. Hence we may adopt $\overline{f}$ as the galaxy's DF. Since the actions may (in principle) be expressed as functions of any three independent isolating integrals, the theorem now follows.

---

[5] Except at a set of points that has zero phase-space volume.

Appendix 4.B N–Body Program

We reproduce here the FORTRAN listing for a program that computes the time evolution of N point masses under the influence of their mutual self-gravity. The acceleration of each mass is computed by the direct summation of the forces due to the other $N - 1$ bodies [cf. §2.8(a)], so the computation time per crossing time grows roughly as N^2. The program was written by S. J. Aarseth, whose programs have set the industry standard in this area for many years. A key strategy of the program is that each particle is followed with its own timestep—an essential feature in view of the wide range of orbital times in a typical stellar system. A complete description of the numerical method is given by Aarseth (1985).

The code contains both input and output through READ and WRITE statements and hence is completely self-contained. The gravitational constant G is taken to be unity. The input variables are the number of bodies N; an accuracy parameter η; the output interval Δt; the length of the integration t_{crit}; the square of the softening parameter, ϵ^2 [eq. (2-194)]; and the masses, initial positions, and velocities m_i, $\mathbf{x}_i$, $\dot{\mathbf{x}}_i$, $i = 1, \ldots, N$. The timestep chosen for a given particle, δt, is related to the force F and its time derivatives by $\delta t = (\eta F/\ddot{F})^{1/2}$ (this is a slight simplification of the criterion given in Aarseth 1985). A typical value $\eta = 0.02$ usually conserves total energy to better than one part in 10^4 over one crossing time, in the absence of close encounters. After each interval Δt, the program outputs the mass, timestep, position, and velocity of each body, as well as the cumulative number of steps and the energy.

Some warnings: For $N > 50$ the DIMENSION statements must be modified. The code is not designed to work efficiently with large N or zero softening parameter. Notice that although certain critical variables are REAL*8, the force calculations are done in single precision to save time.

```
C          S. J. Aarseth's Standard N-Body Program.
C
      REAL*8   X,XO,XODOT,TO,TIME
      DIMENSION  X(3,50),XO(3,50),XODOT(3,50),TO(50),BODY(50),STEP(50),
     1  F(3,50),FDOT(3,50),D1(3,50),D2(3,50),D3(3,50),T1(50),T2(50),
     2  T3(50),A(17),F1(3),F1DOT(3),F2DOT(3),F3DOT(3)
      DATA  TIME,TNEXT,NSTEPS  /0.0,0.0,0/
      READ (5,*)  N,ETA,DELTAT,TCRIT,EPS2
      DO 1 I = 1,N
    1 READ (5,*)  BODY(I),(XO(K,I),K=1,3),(XODOT(K,I),K=1,3)
C          Obtain total force and first derivative for each body.
      DO 20 I = 1,N
      DO 2 K = 1,3
      F(K,I) = 0.0
      FDOT(K,I) = 0.0
      D2(K,I) = 0.0
    2 D3(K,I) = 0.0
      DO 10 J = 1,N
      IF (J.EQ.I)  GO TO 10
      DO 5 K = 1,3
      A(K) = XO(K,J) - XO(K,I)
```

```
    5 A(K+3) = XODOT(K,J) - XODOT(K,I)
      A(7) = 1.0/(A(1)**2 + A(2)**2 + A(3)**2 + EPS2)
      A(8) = BODY(J)*A(7)*SQRT (A(7))
      A(9) = 3.0*(A(1)*A(4) + A(2)*A(5) + A(3)*A(6))*A(7)
      DO 8 K = 1,3
      F(K,I) = F(K,I) + A(K)*A(8)
    8 FDOT(K,I) = FDOT(K,I) + (A(K+3) - A(K)*A(9))*A(8)
   10 CONTINUE
   20 CONTINUE
C          Form second and third force derivative.
      DO 40 I = 1,N
      DO 30 J = 1,N
      IF (J.EQ.I) GO TO 30
      DO 25 K = 1,3
      A(K) = XO(K,J) - XO(K,I)
      A(K+3) = XODOT(K,J) - XODOT(K,I)
      A(K+6) = F(K,J) - F(K,I)
   25 A(K+9) = FDOT(K,J) - FDOT(K,I)
      A(13) = 1.0/(A(1)**2 + A(2)**2 + A(3)**2 + EPS2)
      A(14) = BODY(J)*A(13)*SQRT (A(13))
      A(15) = (A(1)*A(4) + A(2)*A(5) + A(3)*A(6))*A(13)
      A(16) = (A(4)**2 + A(5)**2 + A(6)**2 + A(1)*A(7) + A(2)*A(8)
      1                            + A(3)*A(9))*A(13) + A(15)**2
      A(17) = (9.0*(A(4)*A(7) + A(5)*A(8) + A(6)*A(9)) + 3.0*(A(1)*A(10)
      1+ A(2)*A(11) + A(3)*A(12)))*A(13) + A(15)*(9.*A(16) -12.*A(15)**2)
      DO 28 K = 1,3
      F1DOT(K) = A(K+3) - 3.0*A(15)*A(K)
      F2DOT(K) = (A(K+6) - 6.0*A(15)*F1DOT(K) - 3.0*A(16)*A(K))*A(14)
      F3DOT(K) = (A(K+9) - 9.0*A(16)*F1DOT(K) - A(17)*A(K))*A(14)
      D2(K,I) = D2(K,I) + F2DOT(K)
   28 D3(K,I) = D3(K,I) + F3DOT(K) - 9.0*A(15)*F2DOT(K)
   30 CONTINUE
   40 CONTINUE
C          Initialize integration steps and convert to force differences.
      DO 50 I = 1,N
      STEP(I) = SQRT (ETA*SQRT ((F(1,I)**2 + F(2,I)**2 + F(3,I)**2)/
      1                    (D2(1,I)**2 + D2(2,I)**2 + D2(3,I)**2)))
      TO(I) = TIME
      T1(I) = TIME - STEP(I)
      T2(I) = TIME - 2.0*STEP(I)
      T3(I) = TIME - 3.0*STEP(I)
      DO 45 K = 1,3
      D1(K,I) = (D3(K,I)*STEP(I)/6.0 - 0.5*D2(K,I))*STEP(I) + FDOT(K,I)
      D2(K,I) = 0.5*D2(K,I) - 0.5*D3(K,I)*STEP(I)
      D3(K,I) = D3(K,I)/6.0
      F(K,I) = 0.5*F(K,I)
   45 FDOT(K,I) = FDOT(K,I)/6.0
   50 CONTINUE
C          Energy check and output.
  100 E = 0.0
      DO 110 I = 1,N
```

```
      DT = TNEXT - TO(I)
      DO 101 K = 1,3
      F2DOT(K) = D3(K,I)*((TO(I) - T1(I)) + (TO(I) - T2(I))) + D2(K,I)
      X(K,I) = ((((0.05*D3(K,I)*DT + F2DOT(K)/12.0)*DT + FDOT(K,I))*DT
     1                           + F(K,I))*DT + XODOT(K,I))*DT + XO(K,I)
  101 A(K) = (((0.25*D3(K,I)*DT + F2DOT(K)/3.0)*DT + 3.0*FDOT(K,I))*DT +
     1                                  2.0*F(K,I))*DT + XODOT(K,I)
      WRITE (6,105)  I,BODY(I),STEP(I),(X(K,I),K=1,3),(A(K),K=1,3)
  105 FORMAT (1H ,I10,F10.2,F12.4,3X,3F10.2,3X,3F10.2)
  110 E = E + 0.5*BODY(I)*(A(1)**2 + A(2)**2 + A(3)**2)
      DO 130 I = 1,N
      DO 120 J = 1,N
      IF (J.EQ.I)  GO TO 120
      E = E - 0.5*BODY(I)*BODY(J)/SQRT ((X(1,I) - X(1,J))**2 +
     1            (X(2,I) - X(2,J))**2 + (X(3,I) - X(3,J))**2 + EPS2)
  120 CONTINUE
  130 CONTINUE
      WRITE (6,140)  TNEXT,NSTEPS,E
  140 FORMAT (1HO,5X,'TIME =',F7.2,'  STEPS =',I6,'  ENERGY =',F10.4,/)
      IF (TIME.GT.TCRIT)  STOP
      TNEXT = TNEXT + DELTAT
C           Find next body to be advanced and set new time.
  200 TIME = 1.0E+10
      DO 210 J = 1,N
      IF (TIME.GT.TO(J) + STEP(J))  I = J
      IF (TIME.GT.TO(J) + STEP(J))  TIME = TO(J) + STEP(J)
  210 CONTINUE
C           Predict all coordinates to first order in force derivative.
      DO 220 J = 1,N
      S = TIME - TO(J)
      X(1,J) = ((FDOT(1,J)*S + F(1,J))*S + XODOT(1,J))*S + XO(1,J)
      X(2,J) = ((FDOT(2,J)*S + F(2,J))*S + XODOT(2,J))*S + XO(2,J)
  220 X(3,J) = ((FDOT(3,J)*S + F(3,J))*S + XODOT(3,J))*S + XO(3,J)
C           Include second and third order and obtain the velocity.
      DT = TIME - TO(I)
      DO 230 K = 1,3
      F2DOT(K) = D3(K,I)*((TO(I) - T1(I)) + (TO(I) - T2(I))) + D2(K,I)
      X(K,I) = (0.05*D3(K,I)*DT + F2DOT(K)/12.0)*DT**4 + X(K,I)
      XODOT(K,I) = (((0.25*D3(K,I)*DT + F2DOT(K)/3.0)*DT +
     1                   3.0*FDOT(K,I))*DT + 2.0*F(K,I))*DT + XODOT(K,I)
  230 F1(K) = 0.0
C           Obtain the current force on i th body.
      DO 240 J = 1,N
      IF (J.EQ.I)  GO TO 240
      A(1) = X(1,J) - X(1,I)
      A(2) = X(2,J) - X(2,I)
      A(3) = X(3,J) - X(3,I)
      A(4) = 1.0/(A(1)**2 + A(2)**2 + A(3)**2 + EPS2)
      A(5) = BODY(J)*A(4)*SQRT (A(4))
      F1(1) = F1(1) + A(1)*A(5)
      F1(2) = F1(2) + A(2)*A(5)
```

```
      F1(3) = F1(3) + A(3)*A(5)
  240 CONTINUE
C            Set time intervals for new differences and update the times.
      DT1 = TIME - T1(I)
      DT2 = TIME - T2(I)
      DT3 = TIME - T3(I)
      T1PR = TO(I) - T1(I)
      T2PR = TO(I) - T2(I)
      T3PR = TO(I) - T3(I)
      T3(I) = T2(I)
      T2(I) = T1(I)
      T1(I) = TO(I)
      TO(I) = TIME
C            Form new differences and include fourth-order semi-iteration.
      DO 250 K = 1,3
      A(K) = (F1(K) - 2.0*F(K,I))/DT
      A(K+3) = (A(K) - D1(K,I))/DT1
      A(K+6) = (A(K+3) - D2(K,I))/DT2
      A(K+9) = (A(K+6) - D3(K,I))/DT3
      D1(K,I) = A(K)
      D2(K,I) = A(K+3)
      D3(K,I) = A(K+6)
      F1DOT(K) = T1PR*T2PR*T3PR*A(K+9)
      F2DOT(K) = (T1PR*T2PR + T3PR*(T1PR + T2PR))*A(K+9)
      F3DOT(K) = (T1PR + T2PR + T3PR)*A(K+9)
      XO(K,I) = (((A(K+9)*DT/30.0 + 0.05*F3DOT(K))*DT +
     1                 F2DOT(K)/12.0)*DT + F1DOT(K)/6.0)*DT**3 + X(K,I)
  250 XODOT(K,I) = (((0.2*A(K+9)*DT + 0.25*F3DOT(K))*DT +
     1                 F2DOT(K)/3.0)*DT + 0.5*F1DOT(K))*DT**2 + XODOT(K,I)
C            Scale F and FDOT by factorials and set new integration step.
      DO 260 K = 1,3
      F(K,I) = 0.5*F1(K)
      FDOT(K,I) = ((D3(K,I)*DT1 + D2(K,I))*DT + D1(K,I))/6.0
  260 F2DOT(K) = 2.0*(D3(K,I)*(DT + DT1) + D2(K,I))
      STEP(I) = SQRT (ETA*SQRT ((F1(1)**2 + F1(2)**2 + F1(3)**2)/
     1                 (F2DOT(1)**2 + F2DOT(2)**2 + F2DOT(3)**2)))
      NSTEPS = NSTEPS + 1
      IF (TIME - TNEXT) 200,100,100
      END
```

Appendix 5.A Modes of an Infinite Stellar System

Here we analyze the linearized perturbations of an infinite homogeneous stellar system whose unperturbed DF is Maxwellian,

$$f_0(\mathbf{v}) = \frac{\rho_0}{(2\pi\sigma^2)^{3/2}} e^{-v^2/2\sigma^2}. \tag{5A-1}$$

The linear response of the system is described by the dispersion relation (5-29), which reads

$$1 - \frac{2\sqrt{2\pi}G\rho_0}{k\sigma^3} \int_{-\infty}^{\infty} \frac{v_x e^{-v_x^2/2\sigma^2}}{kv_x - \omega}\, dv_x = 0. \qquad (5A\text{-}2)$$

The neutral ($\omega = 0$) solution to (5A-2) occurs at the Jeans wavenumber k_J, where

$$k^2 = k_J^2 \equiv \frac{4\pi G\rho_0}{\sigma^2}. \qquad (5A\text{-}3)$$

To evaluate the integral in the dispersion relation (5A-2) for real, non-zero values of ω, we must decide how to integrate around the singularity at $\omega = kv_x$. One natural choice is to take the Cauchy principal value of the integral, which we denote by the symbol $\wp$ (see Mathews & Walker 1970). In this case we may write the dispersion relation as

$$1 - \frac{k_J^2}{k^2} \mathcal{W}_V\left(\frac{\omega}{k\sigma}\right) = 0, \qquad (5A\text{-}4)$$

where the function $\mathcal{W}_V$ is defined by

$$\mathcal{W}_V(Z) = \frac{1}{\sqrt{2\pi}} \wp \int_{-\infty}^{\infty} \frac{x}{x - Z} e^{-x^2/2}\, dx. \qquad (5A\text{-}5)$$

This relation is similar to the dispersion relation derived by Vlasov for longitudinal oscillations in a collisionless Maxwellian plasma,

$$1 + \frac{k_P^2}{k^2} \mathcal{W}_V\left(\frac{\omega}{k\sigma}\right) = 0, \qquad (5A\text{-}6)$$

where $k_P = 4\pi n e^2/m\sigma^2$, and n, e, and m are the electron number density, charge, and mass. Unfortunately, the use of the principal value can produce misleading results; the correct procedure for handling the singularity is more subtle, and was first described by Landau (Landau 1946; see Lifshitz & Pitaevskii 1981 or Krall & Trivelpiece 1973 for plasmas, and Lynden-Bell 1967b or Sweet 1963 for stellar systems).

The method of analysis that we have used so far leads us to the **modes** of the stellar system, that is, the solutions whose spatial and temporal dependence is $\propto \exp[i(\mathbf{k}\cdot\mathbf{x} - \omega t)]$. The problem with modes is that they are somewhat unphysical, because they always vary as $\exp(-i\omega t)$, both as $t \to -\infty$ and as $t \to \infty$, whereas a real system is set up at a fixed time $t = t_0$ (we usually take $t_0 = 0$). Physically it is more instructive to consider the response of the system to a particular perturbation that is imposed at $t = 0$. Thus we study the *initial value problem*.

We start from the linearized collisionless Boltzmann and Poisson equations (5-25) and consider solutions of the form

$$f_1(\mathbf{x}, \mathbf{v}, t) = f_b(\mathbf{v}, t)e^{i\mathbf{k}\cdot\mathbf{x}} \quad ; \quad \Phi_1(\mathbf{x}, t) = \Phi_b(t)e^{i\mathbf{k}\cdot\mathbf{x}}. \qquad (5A\text{-}7)$$

The equations for f_b and Φ_b read

$$\frac{\partial f_b}{\partial t} + i\mathbf{k}\cdot\mathbf{v}f_b - i\Phi_b\mathbf{k}\cdot\frac{\partial f_0}{\partial \mathbf{v}} = 0 \quad ; \quad -k^2\Phi_b = 4\pi G \int f_b\, d^3\mathbf{v}. \qquad (5A\text{-}8)$$

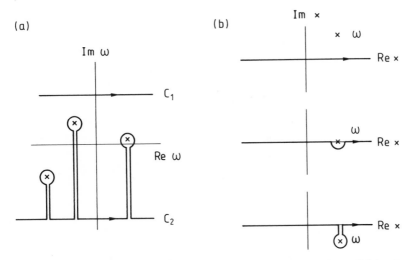

Figure 5A-1. Integration contours (a) in the complex ω plane; (b) in the complex x plane.

Now multiply both equations by $\exp(i\omega t)$ and integrate with respect to t from $t = 0$ to $t = \infty$. We define the transforms

$$\tilde{f}(\omega, \mathbf{v}) = \int_0^\infty e^{i\omega t} f_b(\mathbf{v}, t)\, dt, \qquad \mathrm{Im}(\omega) > \mu,$$

$$\tilde{\Phi}(\omega) = \int_0^\infty e^{i\omega t} \Phi_b(t)\, dt, \qquad \mathrm{Im}(\omega) > \mu. \tag{5A-9}$$

$\tilde{\Phi}(\omega)$ and $\tilde{f}(\omega, \mathbf{v})$ are the Laplace transforms of $\Phi_b(t)$ and $f_b(\mathbf{v}, t)$ (although the conventional Laplace transform uses $-s$ in place of $i\omega$ in the integrand), and μ is some positive number such that $f_b \exp(-\mu t)$ and $\Phi_b \exp(-\mu t)$ approach zero as $t \to \infty$. In terms of these transforms, equations (5A-8) become

$$(\mathbf{k} \cdot \mathbf{v} - \omega)\tilde{f} - \tilde{\Phi}\mathbf{k} \cdot \frac{\partial f_0}{\partial \mathbf{v}} = -i f_b(\mathbf{v}, t = 0) \quad ; \quad -k^2\tilde{\Phi} = 4\pi G \int \tilde{f}\, d^3\mathbf{v}, \tag{5A-10}$$

which have the solution

$$\tilde{\Phi}(\omega) = \frac{\dfrac{4\pi G i}{k^2} \displaystyle\int \frac{f_b(\mathbf{v}, t = 0)\, d^3\mathbf{v}}{\mathbf{k} \cdot \mathbf{v} - \omega}}{1 + \dfrac{4\pi G}{k^2} \displaystyle\int \frac{(\mathbf{k} \cdot \partial f_0/\partial \mathbf{v})\, d^3\mathbf{v}}{\mathbf{k} \cdot \mathbf{v} - \omega}}. \tag{5A-11}$$

We can now obtain $\Phi_b(t)$ by the Laplace inversion integral (see, e.g., Mathews & Walker 1970, p. 108)

$$\Phi_b(t) = \frac{1}{2\pi} \int_{-\infty + iw}^{\infty + iw} \tilde{\Phi}(\omega) e^{-i\omega t}\, d\omega, \qquad w > \mu. \tag{5A-12}$$

The contour of integration in (5A-12) is shown as C_1 in Figure 5A-1a.

Equations (5A-11) and (5A-12) represent a well-defined solution to the problem because the contour lies above the real axis where there are no singularities in the integrands of (5A-11) [we assume that the initial perturbation $f_b(\mathbf{v}, t = 0)$ is nonsingular]. If the function $\tilde{\Phi}(\omega)$ is analytically continued over the whole ω-plane, it will have a number of poles, which in Figure 5A-1a are marked schematically by $\times$'s. We deform the contour C_1 into C_2, and note that as $t \to \infty$ the contribution from the horizontal part of C_2 becomes negligible. Thus, by the residue theorem (see Mathews & Walker 1970, p. 65) $\Phi_b(t) = -i \sum_p R_p \exp(-i\omega_p t)$, where R_p is the residue of $\tilde{\Phi}$ at the pole $\omega = \omega_p$. The system is stable if the maximum value of $\mathrm{Im}(\omega_p)$ is negative, and thus it is clear that the stability of the system is determined by the location of the poles of $\tilde{\Phi}(\omega)$ in equation (5A-11).

To find these poles, let us first rewrite the numerator of (5A-11). Put $\mathbf{v} = (v_x, v_y, v_z)$, where the x-axis is in the direction of $\mathbf{k}$. Then

$$\int \frac{f_b(\mathbf{v}, t = 0)\, d^3\mathbf{v}}{\mathbf{k} \cdot \mathbf{v} - \omega} = \frac{1}{k} \int_{-\infty}^{\infty} \frac{F_b(x/k, t = 0)\, dx}{x - \omega}, \qquad (5A\text{-}13)$$

where $x = kv_x$ and $F_b(v_x, t = 0) = \int f_b(\mathbf{v}, t = 0)\, dv_y dv_z$. Since $\tilde{\Phi}(\omega)$ for $\mathrm{Im}(\omega) < \mu$ is defined to be the analytic continuation of $\tilde{\Phi}(\omega)$ for $\mathrm{Im}(\omega) > \mu$, we must choose the x-contour in (5A-13) so that the integral is always an analytic continuation of the integral defined for $\mathrm{Im}(\omega) > \mu$. This can be done by deforming the contour in the x-plane as $\mathrm{Im}(\omega)$ decreases, as shown in Figure 5A-1b. This choice of contour is the keystone of Landau's analysis.

The integral in the denominator of (5A-11) must also use Landau's contour. In particular, for a Maxwellian distribution (5A-1), the denominator can be written as

$$1 - \frac{k_J^2}{k^2} \mathcal{W}_L\left(\frac{\omega}{k\sigma}\right),$$

where

$$\mathcal{W}_L(Z) = \frac{1}{\sqrt{2\pi}} \int_C \frac{x}{x - Z} e^{-x^2/2}\, dx, \qquad (5A\text{-}14)$$

and the contour C is shown in Figure 5A-1b. From equation (5A-5), $\mathcal{W}_L(Z) = \mathcal{W}_V(Z)$ for $\mathrm{Im}(Z) > 0$, but for $\mathrm{Im}(Z) = 0$, $\mathcal{W}_L(Z)$ and $\mathcal{W}_V(Z)$ differ by a term $\frac{1}{2}\sqrt{2\pi} i Z \exp(-\frac{1}{2}Z^2)$, and for $\mathrm{Im}(Z) < 0$, they differ by $\sqrt{2\pi} i Z \exp(-\frac{1}{2}Z^2)$.

Now let us go back to searching for the poles of $\tilde{\Phi}(\omega)$. Since $f_b(\mathbf{v}, t = 0)$ has been assumed to be nonsingular, the integral in the numerator (5A-13) has no poles. Hence, any poles must arise from zeros in the denominator, i.e., from solutions of the dispersion relation

$$1 - \frac{k_J^2}{k^2} \mathcal{W}_L\left(\frac{\omega}{k\sigma}\right) = 0. \qquad (5A\text{-}15)$$

There are three cases to consider:

(i) $\mathrm{Im}(\omega) > 0$ (unstable solutions). Let $Z = \omega/k\sigma$. The imaginary part of $\mathcal{W}_L(Z)$, which must vanish, is equal to

$$\mathrm{Im}(\mathcal{W}_L) = \frac{\mathrm{Im}(Z)}{\sqrt{2\pi}} \int_{-\infty}^{\infty} \frac{x e^{-x^2/2}\, dx}{[x - \mathrm{Re}(Z)]^2 + [\mathrm{Im}(Z)]^2}. \qquad (5A\text{-}16)$$

But unless $\mathrm{Re}(\omega) = 0$, this integral cannot be zero: if, for example, $\mathrm{Re}(\omega) > 0$, the absolute value of the integrand at $+x$ will always be smaller than its value at $-x$, and the integral will be negative, while if $\mathrm{Re}(\omega) < 0$ the integral will be positive. Hence the only unstable solutions have ω imaginary (i.e., there are no **overstable** modes with $\mathrm{Im}(\omega) > 0$ and $\mathrm{Re}(\omega) \neq 0$). The evaluation of the integrals in (5A-16) yields the dispersion relation (5-31) for unstable modes.

(ii) $\mathrm{Im}(\omega) = 0$, $\mathrm{Re}(\omega) \neq 0$ (undamped solutions). $\mathcal{W}_L$ is now given by

$$\mathcal{W}_L(Z) = \frac{1}{\sqrt{2\pi}} \wp \int_{-\infty}^{\infty} \frac{x e^{-x^2/2}\, dx}{x - Z} + i\sqrt{\tfrac{1}{2}\pi} Z e^{-Z^2/2}, \qquad (5A\text{-}17)$$

where $\wp$ denotes the Cauchy principal value of the integral. The principal value is real, and so $\mathrm{Im}(\mathcal{W}_L) = 0$ only when $\omega = 0$; hence we conclude that there are no undamped traveling waves. In this respect the stellar system is quite different from a gaseous system, whose dispersion relation [eq. (5-20)] allows undamped waves with all wavelengths less than the Jeans length.

(iii) $\mathrm{Im}(\omega) < 0$ (damped solutions). Since only waves having $k < k_J$ are unstable, and there are no undamped waves, it follows that all waves with $k > k_J$ must be damped. Numerical evaluation of $\mathcal{W}_L$ (see Fried & Conte 1961 and Ikeuchi et al. 1974) shows that all these solutions are strongly damped, that is, $|\mathrm{Im}(\omega)| \gtrsim |\mathrm{Re}(\omega)|$. Thus, they are of relatively little interest. This damping is called **Landau damping**. Landau damping is different from phase mixing (§4.7.2), although both damp out perturbations in a stellar system. The difference is twofold: (a) the rate of phase mixing depends on the initial perturbation $f_b(\mathbf{v}, t = 0)$, while the Landau damping rate depends only on the unperturbed DF; (b) Landau damping is due to collective effects arising because of self-gravity, while phase mixing occurs even in systems with no self-gravity.

Landau-damped waves are more important in plasmas where the plasma dispersion relation permits long wavelength ($k \ll k_P$) modes corresponding to traveling waves that decay only slowly due to Landau damping.

It is important to recognize that Landau-damped waves are not true modes. Landau's analysis only shows that as $t \to \infty$, $\Phi_b \to \exp(-i\omega_x t)$, where ω_x is determined from the dispersion relation (5A-15) using the contour of Figure 5A-1b. By contrast, a mode is a solution of the collisionless Boltzmann and Poisson equations (5-26) that behaves like $\exp(-i\omega t)$ at all times. In other words, the Landau-damped waves are not true modes because they satisfy equation (5A-15) instead of (5A-4). The distinction does not arise for exponentially growing solutions because $\mathcal{W}_L(Z) = \mathcal{W}_V(Z)$ for $\mathrm{Im}(Z) > 0$.

If the Landau-damped waves are not modes, then do modes exist with wavenumbers $k > k_J$? Van Kampen (1955) has shown that such modes do exist, and that their oscillation frequencies ω are real; however, the modes are singular. Their DF takes the form

$$f_a(\mathbf{v}) = \frac{k_J^2}{(2\pi\sigma^2)^{3/2} k^2} \wp\left(\frac{k v_x e^{-v^2/2\sigma^2}}{k v_x - \omega} \right)$$
$$+ \left[1 - \frac{k_J^2}{k^2} \mathcal{W}_V\left(\frac{\omega}{k\sigma} \right) \right] \delta(v_x - \omega/k)\delta(v_y)\delta(v_z). \qquad (5A\text{-}18)$$

As usual δ denotes the Dirac delta function and the x-axis is in the direction of $\mathbf{k}$. The symbol $\wp$ denotes that the Cauchy principal value should be taken when integrating over $\mathbf{v}$. It is straightforward to verify that (5A-18) satisfies the collisionless Boltzmann and Poisson equations (5-26). Moreover, it can be shown (van Kampen 1955; Case 1959) that the van Kampen modes are complete and that Landau-damped waves can be regarded as a superposition of van Kampen modes. Note that the modes do not satisfy a dispersion relation since they exist for all real ω and $k > k_J$. We shall not use these results in future sections; we include them here mainly to illustrate that a stellar system has a much richer, more complicated and more pathological set of modes than a fluid.

Appendix 5.B Perturbation Energy in Fluid Systems

In this Appendix we show that Chandrasekhar's variational principle, equation (5-49), is really a stability criterion based on energy.

The total energy of a gaseous body may be written $E = T + W + U$, where the three terms are the kinetic energy due to bulk motion

$$T = \tfrac{1}{2} \int \rho(\mathbf{r}, t)[\mathbf{v}(\mathbf{r}, t)]^2 \, d^3\mathbf{r}, \tag{5B-1}$$

the gravitational potential energy [eq. (2-19)]

$$W = -\frac{G}{2} \iint \frac{\rho(\mathbf{r}, t)\rho(\mathbf{r}', t)}{|\mathbf{r} - \mathbf{r}'|} d^3\mathbf{r} \, d^3\mathbf{r}', \tag{5B-2}$$

and the internal energy

$$U = \int \rho(\mathbf{r}, t) u(\rho) \, d^3\mathbf{r} = \int \rho(\mathbf{r}, t) \left[\int_0^\rho \frac{p(\rho')}{\rho'^2} \, d\rho' \right] d^3\mathbf{r}, \tag{5B-3}$$

where u is the internal energy per unit mass for a barotropic gas [eq. (1E-14)]. We now write $\rho(\mathbf{r}, t) = \rho_0(\mathbf{r}) + \epsilon \rho_1(\mathbf{r}, t)$ and expand $W + U$ in ascending powers of the small parameter ϵ. The Taylor series for the internal energy per unit mass is

$$u = \int_0^{\rho_0} \frac{p(\rho')}{\rho'^2} \, d\rho' + \epsilon \frac{p_0}{\rho_0^2} \rho_1 + \tfrac{1}{2} \epsilon^2 \rho_1^2 \frac{d}{d\rho} \left[\frac{p(\rho)}{\rho^2} \right]_{\rho_0} + O(\epsilon^3). \tag{5B-4}$$

If we write the expansion of $W + U$ as

$$W + U = (W + U)_0 + \epsilon(W + U)_1 + \epsilon^2(W + U)_2 + O(\epsilon^3), \tag{5B-5}$$

then we find

$$(W + U)_1 = \int \rho_1 \, d^3\mathbf{r} \left[\int_0^{\rho_0} \frac{p(\rho')}{\rho'^2} \, d\rho' + \frac{p_0}{\rho_0} - G \int \frac{\rho_0(\mathbf{r}') \, d^3\mathbf{r}'}{|\mathbf{r} - \mathbf{r}'|} \right], \tag{5B-6}$$

$$(W + U)_2 = \frac{1}{2} \int \left| \frac{d\Phi}{d\rho} \right|_0 \rho_1^2 \, d^3\mathbf{r} - \frac{G}{2} \iint \frac{d^3\mathbf{r} \, d^3\mathbf{r}'}{|\mathbf{r} - \mathbf{r}'|} \rho_1(\mathbf{r})\rho_1(\mathbf{r}'), \qquad (5\text{B-}7)$$

where equation (5-36) has been used in deriving the last line. Now consider the gradient of the quantity in square brackets. We can replace the last term by $\Phi_0(\mathbf{r})$, and we have

$$\nabla \left[\int_0^{\rho_0} \frac{p(\rho')}{\rho'^2} \, d\rho' + \frac{p_0}{\rho_0} + \Phi_0 \right] = \frac{\nabla p_0}{\rho_0} + \nabla \Phi_0,$$

which vanishes because of the condition of hydrostatic equilibrium (5-35). Hence the quantity in square brackets must be some constant C, and thus $(W + U)_1 = C \int \rho_1 \, d^3\mathbf{r}$, which also vanishes because mass conservation ensures that $\int \rho_1 \, d^3\mathbf{r} = 0$. Thus the change in the potential plus internal energy is second order in the small quantity ϵ.

Next consider the kinetic energy T in equation (5B-1). Since the unperturbed star is static, the velocity $\mathbf{v}(\mathbf{r}, t)$ is of order ϵ. Hence T is also of order ϵ^2, and we can write $T = \epsilon^2 T_2 + O(\epsilon^3)$. The total energy of the star is $E = E_0 + \epsilon^2(T + W + U)_2 + O(\epsilon^3)$. Now consider a growing mode. As the mode grows, the magnitudes of T_2, W_2, and U_2 all increase, but the total energy is conserved; thus we must have $(T + W + U)_2 = 0$. In other words, *the total energy associated with a growing mode must be zero.* From equation (5B-1) it is clear that T_2 is positive definite. Hence $(W + U)_2$ must be negative in a growing mode, and if $(W + U)_2$ is non-negative for all possible perturbations ρ_1 that conserve mass, then the star must be stable. Comparison of equation (5B-7) with equation (5-49) shows that we have once again proved Chandrasekhar's variational principle, and that $\mathcal{E}[\rho_1]$ is just twice the sum of the internal and potential energies associated with the perturbation ρ_1.

Appendix 5.C The Antonov-Lebovitz Theorem

This theorem states that a barotropic star with $d\rho_0/dr < 0$ and $\rho_0(r_m) = 0$ is stable to all nonradial perturbations.

To prove this, we must show that $\mathcal{E}[\rho_1]$ in equation (5-49) is never negative for nonradial perturbations. To do this we maximize the quantity

$$J[\rho_1] = G \iint \frac{d^3\mathbf{r} \, d^3\mathbf{r}'}{|\mathbf{r} - \mathbf{r}'|} \rho_1(\mathbf{r})\rho_1(\mathbf{r}'), \qquad (5\text{C-}1)$$

subject to the constraint

$$K[\rho_1] = \int \left| \frac{d\Phi}{d\rho} \right|_0 \rho_1^2 \, d^3\mathbf{r} = 1, \qquad (5\text{C-}2)$$

and show that $\max J[\rho_1] \leq 1$. Using a Lagrange multiplier λ, the maximization condition can be written $\delta J - \lambda \delta K = 0$, or

$$\int \delta\rho_1(\mathbf{r}) \, d^3\mathbf{r} \left[G \int \frac{\rho_1(\mathbf{r}') \, d^3\mathbf{r}'}{|\mathbf{r} - \mathbf{r}'|} - \lambda \rho_1 \left| \frac{d\Phi}{d\rho} \right|_0 \right] = 0. \qquad (5\text{C-}3)$$

Since (5C-3) must be true for all variations $\delta\rho_1$, we have

$$G \int \frac{\rho_1(\mathbf{r}')\, d^3\mathbf{r}'}{|\mathbf{r}-\mathbf{r}'|} - \lambda\rho_1 \left.\frac{d\Phi}{d\rho}\right|_0 = 0. \tag{5C-4}$$

Substituting back in equation (5-49) we find that the system is stable if $\lambda \le 1$ for all functions $\rho_1(\mathbf{r})$ satisfying the integral equation (5C-4).

It is more convenient to work with the potential Φ_1 that is generated by ρ_1. In terms of Φ_1, (5C-4) reads

$$\nabla^2\Phi_1 + \frac{4\pi G}{\lambda}\left.\left|\frac{d\rho}{d\Phi}\right|\right._0 \Phi_1 = 0. \tag{5C-5}$$

We now decompose Φ_1 into spherical harmonics,

$$\Phi_1(\mathbf{r}) = \frac{d\Phi_0}{dr} \sum_{l=l_{\min}}^{\infty} \sum_{m=-l}^{l} s_{lm}(r)\, Y_l^m(\theta,\phi), \tag{5C-6}$$

where $l_{\min} = 1$ since only nonradial ($l \neq 0$) perturbations are considered. The common factor $d\Phi_0/dr$ has been introduced to simplify later equations. We substitute (5C-6) into (5C-5) and use equation (1C-32) to evaluate $\nabla^2 Y_l^m(\theta,\phi)$. Because the spherical harmonics are linearly independent, the coefficient of each $Y_l^m(\theta,\phi)$ must vanish. Simplifying the result using Poisson's equation in the form

$$\frac{d^2\Phi_0}{dr^2} + \frac{2}{r}\frac{d\Phi_0}{dr} = 4\pi G\rho_0, \tag{5C-7}$$

we find[6]

$$\left(r^2\frac{d\Phi_0}{dr}\right)\frac{d^2 s_{lm}}{dr^2} + \left(8\pi Gr^2\rho_0 - 2r\frac{d\Phi_0}{dr}\right)\frac{ds_{lm}}{dr}$$
$$+ \left[(2-l-l^2)\frac{d\Phi_0}{dr} + 4\pi Gr^2\frac{d\rho_0}{dr}\left(1-\frac{1}{\lambda}\right)\right]s_{lm} = 0. \tag{5C-8}$$

We multiply (5C-8) by $s_{lm}d\Phi_0/dr$ and integrate from $r = 0$ to $r = r_m$, the surface of the system. After integrating the first term by parts and using equation (5C-7) to simplify the result, we obtain

$$\left[\left(r\frac{d\Phi_0}{dr}\right)^2 s_{lm}\frac{ds_{lm}}{dr}\right]_0^{r_m} + \int_0^{r_m} dr\left[-\left(r\frac{d\Phi_0}{dr}\frac{ds_{lm}}{dr}\right)^2\right.$$
$$\left. + (2-l-l^2)\left(\frac{d\Phi_0}{dr}s_{lm}\right)^2 + 4\pi Gr^2\left(1-\frac{1}{\lambda}\right)\frac{d\rho_0}{dr}\frac{d\Phi_0}{dr}s_{lm}^2\right] = 0. \tag{5C-9}$$

At $r = r_m$, Φ_1 must match onto a decaying solution of Laplace's equation. It is easy to show from equation (1C-32) that for $r > r_m$ the coefficient

---

[6] Note that a trivial solution of (5C-8) is $s_{lm} = $ constant, $l = 1$, $\lambda = 1$. This corresponds to a displacement of the entire system.

of each spherical harmonic in the decaying solution of Laplace's equation is proportional to $r^{-l-1} Y_l^m(\theta, \phi)$; hence for the potential and its derivative to be continuous at r_m we must have

$$\frac{d}{dr}\left[\ln\left(\frac{d\Phi_0}{dr} s_{lm}\right)\right]_{r=r_m} = -\frac{l+1}{r_m}. \tag{5C-10}$$

Using Poisson's equation (5C-7) for the unperturbed star plus the condition $\rho_0(r_m) = 0$, we find

$$\left(\frac{ds_{lm}}{dr}\right)_{r=r_m} = \frac{(1-l)s_{lm}(r_m)}{r_m}. \tag{5C-11}$$

This relation permits us to evaluate the boundary term in (5C-9) at $r = r_m$. At $r = 0$ the boundary term vanishes. Hence we can write

$$4\pi G\left(1 - \frac{1}{\lambda}\right)\int_0^{r_m} \frac{d\rho_0}{dr}\frac{d\Phi_0}{dr}s_{lm}^2 r^2\,dr = (l-1)r_m\left(\frac{d\Phi_0}{dr}s_{lm}\right)^2_{r=r_m}$$
$$+ \int_0^{r_m}\left(r\frac{d\Phi_0}{dr}\frac{ds_{lm}}{dr}\right)^2 dr + (l^2 + l - 2)\int_0^{r_m}\left(\frac{d\Phi_0}{dr}s_{lm}\right)^2 dr. \tag{5C-12}$$

Every term on the right side is non-negative for $l \geq 1$; moreover, $d\Phi_0/dr > 0$ and $d\rho_0/dr < 0$. Thus $1 - 1/\lambda \leq 0$ or $0 \leq \lambda \leq 1$, and since $\lambda \leq 1$ was the condition for stability, the system must be stable. This completes the proof.

Appendix 6.A Derivation of the Reduction Factor

Our aim in this Appendix is to derive the reduction factor $\mathcal{F}$, the factor by which the response of a stellar disk to an imposed potential is reduced below the response of a cold disk. We assume that the potential perturbation is tightly wound, that the stellar orbits in the disk can be described by the epicycle approximation, and that the disk is two-dimensional with DF given by equation (6-44).

As usual we assume that the perturbations are small, so that we can linearize the equations of motion. The linearized collisionless Boltzmann equation (5-5) reads

$$\frac{\partial f_1}{\partial t} + \mathbf{v}\cdot\frac{\partial f_1}{\partial \mathbf{x}} - \nabla\Phi_0\cdot\frac{\partial f_1}{\partial \mathbf{v}} - \nabla\Phi_1\cdot\frac{\partial f_0}{\partial \mathbf{v}} = 0. \tag{6A-1}$$

Here $f_0(\mathbf{x}, \mathbf{v})$ and $f_1(\mathbf{x}, \mathbf{v}, t)$ are the equilibrium and perturbed DFs, and $\Phi_0(\mathbf{x})$ and $\Phi_1(\mathbf{x}, t)$ are the equilibrium and perturbed gravitational potentials.

Recall that the convective derivative $d/dt = \partial/\partial t + \mathbf{v}\cdot\partial/\partial\mathbf{x} - \nabla\Phi\cdot\partial/\partial\mathbf{v}$ measures the rate of change of a quantity as viewed from a particle moving on an orbit in phase space [see eq. (4-14)]. By analogy we introduce the operator

$$\left(\frac{d}{dt}\right)_0 = \frac{\partial}{\partial t} + \mathbf{v}\cdot\nabla - \nabla\Phi_0\cdot\frac{\partial}{\partial\mathbf{v}}, \tag{6A-2}$$

which measures the rate of change as viewed from a particle orbiting in the unperturbed potential. Hence equation (6A-1) can be written

$$\left(\frac{df_1}{dt}\right)_0 = \boldsymbol{\nabla}\Phi_1 \cdot \frac{\partial f_0}{\partial \mathbf{v}}. \tag{6A-3}$$

Integrating along the unperturbed orbit, we find

$$f_1(\mathbf{x}, \mathbf{v}, t) = \int_{-\infty}^t \boldsymbol{\nabla}'\Phi_1(\mathbf{x}', \mathbf{v}', t) \cdot \frac{\partial f_0}{\partial \mathbf{v}'}(\mathbf{x}', \mathbf{v}') \, dt', \tag{6A-4}$$

where $(\mathbf{x}', \mathbf{v}')$ is the phase-space position occupied at t' by the particle whose unperturbed orbit passes through $(\mathbf{x}, \mathbf{v})$ at time t. In writing equation (6A-4) we have assumed that $f_1 \to 0$ as $t \to -\infty$, i.e., the perturbation was small in the distant past.

We write the potential perturbation Φ_1 in the disk plane in the form [see eqs. (6-29) and (6-36)][7]

$$\Phi_1(\mathbf{x}, t) = F(R)e^{i(\int^R k\, dR + m\phi - \omega t)}. \tag{6A-5}$$

The component of $\boldsymbol{\nabla}\Phi_1$ in the disk plane is

$$\boldsymbol{\nabla}\Phi_1 = \left[\hat{\mathbf{e}}_R\left(\frac{dF}{dR} + ikF\right) + \hat{\mathbf{e}}_\phi \frac{imF}{R}\right] e^{i(\int k\, dR + m\phi - \omega t)}. \tag{6A-6}$$

Since the potential Φ_1 is tightly wound ($|kR| \gg 1$) we keep only the term proportional to k:

$$\boldsymbol{\nabla}\Phi_1 \simeq \hat{\mathbf{e}}_R ikFe^{i(\int k\, dR + m\phi - \omega t)}. \tag{6A-7}$$

Thus equation (6-44) for the unperturbed DF, together with (6A-4), yields

$$f_1(R, \phi, v_R, v_\phi, t) = -\frac{i}{2\pi}\int_{-\infty}^t \left(\frac{kF\Sigma}{\sigma_R^3\sigma_\phi}\right)_{R'} e^{i(\int^{R'} k''\, dR'' + m\phi' - \omega t')} e^{-w'} v_R'\, dt', \tag{6A-8}$$

where the unperturbed trajectory at time t' is $(\mathbf{x}', \mathbf{v}') = (R', \phi', v_R', v_\phi')$ and

$$w' = \frac{v_R'^2}{2\sigma_R^2} + \frac{[v_\phi' - v_c(R')]^2}{2\sigma_\phi^2}. \tag{6A-9}$$

Since the epicycle amplitude is assumed to be small, $|R' - R|$ is always much less than R. Hence to a good approximation $k(R')$, $F(R')$, $\Sigma(R')$,

---

[7] Since we have assumed in equation (6A-4) that the perturbation vanishes as $t \to -\infty$, we must have $\text{Im}(\omega) > 0$. To investigate density waves with real ω, we replace ω by $\omega + i\epsilon$, where ϵ is positive but very small.

$\sigma_R(R')$, and $\sigma_\phi(R')$ can all be replaced by their values at R and taken outside the integral. At the same level of approximation we can write

$$e^{i\int^{R'} k'' dR''} \simeq e^{i[\int^R k'' dR''+k(R)(R'-R)]}. \qquad (6A\text{-}10)$$

Thus

$$f_1(R, \phi, v_R, v_\phi, t) = -\frac{i}{2\pi} \left(\frac{kF\Sigma}{\sigma_R^3 \sigma_\phi}\right)_R e^{i\int^R k'' dR''} \qquad (6A\text{-}11)$$
$$\times \int_{-\infty}^t e^{i[k(R)(R'-R)+m\phi'-\omega t']} e^{-w'} v_R' \, dt'.$$

The epicyclic orbit equations (3-67) and (3-74) yield the relations

$$R' = R_g + X\cos(\kappa t' + \psi), \qquad (6A\text{-}12a)$$
$$v_R' = -\kappa X \sin(\kappa t' + \psi), \qquad (6A\text{-}12b)$$
$$v_\phi' - v_c(R') = 2BX \cos(\kappa t' + \psi), \qquad (6A\text{-}12c)$$

where the epicycle amplitude X and phase ψ are determined by the boundary condition that $R' = R$, $v_R' = v_R$, and $v_\phi' = v_\phi$ at $t' = t$. Equations (6A-12), together with the relation between σ_R and σ_ϕ implied by the epicycle approximation [eq. (4-52)], imply that $w' = \kappa^2 X^2/(2\sigma_R^2)$. Since X is an integral of the motion, w' must therefore be conserved along the trajectory.[8] Thus $w' = w$ and the factor $\exp(-w)$ can be taken out of the integral. Straightforward manipulation of equations (6A-12) and the boundary conditions at $t = t'$ leads to the results:

$$R' - R = \frac{v_\phi - v_c}{2B}(\cos\tau - 1) + \frac{v_R}{\kappa}\sin\tau \quad ; \quad v_R' = v_R\cos\tau - \frac{\kappa(v_\phi - v_c)}{2B}\sin\tau,$$
$$(6A\text{-}13)$$

where $\tau = \kappa(t' - t)$ and v_ϕ, v_R and $v_c(R)$ are all evaluated at t. Since the azimuthal angle ϕ' only enters equation (6A-11) through the factor $\exp(im\phi')$, and the epicycle amplitude is small, we may neglect the epicyclic motion in ϕ and set $\phi' = \phi + \Omega(t' - t)$ (we cannot use the same argument to neglect the epicyclic motion in R since the factor $(R' - R)$ is multiplied by the large factor k). Inserting these approximations into equation (6A-11), we find

$$f_1(R, \phi, v_R, v_\phi, t) = -\frac{ikF\Sigma}{2\pi\kappa\sigma_R^2\sigma_\phi} e^{i(\int^R k'' dR''+m\phi-\omega t)} e^{-\frac{1}{2}(u^2+v^2)}$$
$$\times \int_{-\infty}^0 e^{i(k\sigma_R/\kappa)[u\sin\tau - v(\cos\tau - 1)]-is\tau}(u\cos\tau + v\sin\tau)d\tau.$$
$$(6A\text{-}14)$$

Here we have introduced the notations

$$s = \frac{\omega - m\Omega}{\kappa}, \qquad u = \frac{v_R}{\sigma_R}, \qquad v = \frac{v_\phi - v_c}{\sigma_\phi} = -\frac{\kappa(v_\phi - v_c)}{2B\sigma_R}, \qquad (6A\text{-}15)$$

---

[8] This is a direct consequence of the Jeans theorem: since the unperturbed DF depends on velocity only through w, w must be an integral.

where the last equality follows from equation (4-52).

The mean radial velocity at a given point is

$$\overline{v_{R1}} = \frac{\int (f_0 + f_1) v_R d^2 \mathbf{v}}{\int (f_0 + f_1) d^2 \mathbf{v}}. \tag{6A-16}$$

Since $\int f_0 v_R d^2 \mathbf{v} = 0$, to first order in the perturbation we have

$$\overline{v_{R1}} = \frac{\int f_1 v_R d^2 \mathbf{v}}{\int f_0 d^2 \mathbf{v}} = \frac{\sigma_R^2 \sigma_\phi \int f_1 u \, du \, dv}{\Sigma}. \tag{6A-17}$$

Writing $\overline{v_{R1}} = \overline{v_{Ra}} \exp[i(m\phi - \omega t)]$ [eq. (6-29)] and using equation (6A-5) and the definition of the reduction factor $\mathcal{F}$ [eq. (6-42)], we find

$$\mathcal{F} = i \frac{(1 - s^2)}{2\pi s} \int_{-\infty}^{\infty} du \int_{-\infty}^{\infty} dv \, u e^{-\frac{1}{2}(u^2 + v^2)}$$

$$\times \int_{-\infty}^{0} e^{i(k\sigma_R/\kappa)[u \sin\tau - v(\cos\tau - 1)] - is\tau}(u \cos\tau + v \sin\tau) d\tau. \tag{6A-18}$$

We now use the following integrals:

$$\int_{-\infty}^{\infty} e^{-\frac{1}{2}x^2} e^{i\mu x} dx = \sqrt{2\pi} e^{-\frac{1}{2}\mu^2}, \quad \int_{-\infty}^{\infty} e^{-\frac{1}{2}x^2} x e^{i\mu x} dx = i\sqrt{2\pi}\mu e^{-\frac{1}{2}\mu^2},$$

$$\int_{-\infty}^{\infty} e^{-\frac{1}{2}x^2} x^2 e^{i\mu x} dx = \sqrt{2\pi}(1 - \mu^2) e^{-\frac{1}{2}\mu^2}. \tag{6A-19}$$

Evaluating the u and v integrals and replacing τ by $-\tau$, we obtain

$$\mathcal{F}(s, \chi) = i \frac{(1 - s^2)}{s} \int_0^{\infty} e^{is\tau - \chi(1 - \cos\tau)}(\cos\tau - \chi \sin^2 \tau) d\tau, \tag{6A-20}$$

where $\chi = (k\sigma_R/\kappa)^2$. Now write the integral as a sum of integrals from $\tau = 0$ to 2π, 2π to 4π, etc.

$$\mathcal{F}(s, \chi) = i \frac{(1 - s^2)}{s} \sum_{n=0}^{\infty} e^{2\pi i n s} \int_0^{2\pi} e^{is\tau - \chi(1 - \cos\tau)}(\cos\tau - \chi \sin^2 \tau) d\tau. \tag{6A-21}$$

The geometric series in (6A-21) is of the form $\sum_{n=0}^{\infty} p^n$, where $|p| = |e^{2\pi i s}| = e^{-2\pi \text{Im}(s)} < 1$, since $\text{Im}(s) > 0$ (see footnote at beginning of this Appendix). Hence the series is convergent, with sum $1/(1 - p)$. Thus

$$\mathcal{F}(s, \chi) = i \frac{(1 - s^2)}{s} \frac{1}{1 - e^{2\pi i s}} \int_0^{2\pi} e^{is\tau - \chi(1 - \cos\tau)}(\cos\tau - \chi \sin^2 \tau) d\tau. \tag{6A-22}$$

Replacing the variable τ by $\tau' = \tau - \pi$, we have

$$\mathcal{F}(s, \chi) = i \frac{(1 - s^2)}{s} \frac{e^{\pi i s}}{1 - 2\pi i s} \int_{-\pi}^{\pi} e^{is\tau' - \chi(1 + \cos\tau')}(-\cos\tau' - \chi \sin^2 \tau') d\tau'. \tag{6A-23}$$

We can write $(1 - e^{2\pi i s})/e^{\pi i s} = -2i \sin \pi s$. Also, if we write $e^{i s \tau'} = \cos s \tau' + i \sin s \tau'$, only the cosine term will contribute to the integral since the sine produces an integrand that is odd in τ'. Dropping the prime on τ we find

$$\mathcal{F}(s, \chi) = \frac{1 - s^2}{s \sin \pi s} \int_0^\pi e^{-\chi(1 + \cos \tau)} \cos s\tau (\cos \tau + \chi \sin^2 \tau) d\tau. \tag{6A-24}$$

The integral can be simplified by writing the contribution from the second term of the integrand as $\chi \cos s\tau e^{-\chi(1 + \cos \tau)} \sin^2 \tau = \cos s\tau \sin \tau d[e^{-\chi(1 + \cos \tau)}]/d\tau$ and integrating by parts:

$$\mathcal{F}(s, \chi) = \frac{1 - s^2}{\sin \pi s} \int_0^\pi e^{-\chi(1 + \cos \tau)} \sin s\tau \sin \tau \, d\tau. \tag{6A-25}$$

Yet another integration by parts yields a second form:

$$\mathcal{F}(s, \chi) = \frac{1 - s^2}{\chi} \left[1 - \frac{s}{\sin \pi s} \int_0^\pi e^{-\chi(1 + \cos \tau)} \cos s\tau \, d\tau \right]. \tag{6A-26}$$

We obtain a third form, involving the modified Bessel function I_n, by using the identity (1C-55) and the relation $I_n(-\chi) = (-1)^n I_n(\chi)$:

$$\mathcal{F}(s, \chi) = \frac{1 - s^2}{\chi} \left[1 - \frac{s}{\sin \pi s} e^{-\chi} \sum_{n=-\infty}^\infty (-1)^n I_n(\chi) \int_0^\pi \cos n\tau \cos s\tau \, d\tau \right]. \tag{6A-27}$$

It is simple to show that $\int_0^\pi \cos n\tau \cos s\tau d\tau = (-1)^{n+1} s \sin \pi s / (n^2 - s^2)$; hence

$$\mathcal{F}(s, \chi) = \frac{1 - s^2}{\chi} \left[1 + s^2 e^{-\chi} \sum_{n=-\infty}^\infty \frac{I_n(\chi)}{n^2 - s^2} \right]. \tag{6A-28}$$

This equation can be simplified by rewriting the unit term in the square brackets as $1 = e^{-\chi} \sum_{n=-\infty}^\infty I_n(\chi)$, an identity that follows from equation (1C-55) by setting $\theta = 0$. After some rearrangement, we finally obtain

$$\mathcal{F}(s, \chi) = \frac{2}{\chi} (1 - s^2) e^{-\chi} \sum_{n=1}^\infty \frac{I_n(\chi)}{1 - s^2/n^2}. \tag{6A-29}$$

Appendix 8.A The Diffusion Coefficients

We consider a test star of mass m and velocity $\mathbf{v}$ moving through a homogeneous sea of field stars of mass m_a and DF $f_a(\mathbf{v}_a)$. Due to encounters with the field stars, the test star suffers a gradually accumulating random velocity change $\Delta \mathbf{v}$. Our goal is to calculate the average changes per unit time $D(\Delta v_i)$ and $D(\Delta v_i \Delta v_j)$. Our derivation follows Rosenbluth et al. (1957).

Suppose that in a particular encounter the test star has velocity $\mathbf{v}$ and the field star has a velocity that lies in the velocity-space volume $d^3 \Delta \mathbf{v}_a$ around

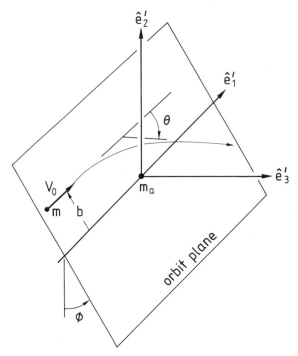

Figure 8A-1. Geometry of an encounter. The angle
θ is the deflection angle θ_{defl} used in §7.1.

$\mathbf{v}_a$. The relative velocity is $\mathbf{V} = \mathbf{v} - \mathbf{v}_a$. Its initial value is $\mathbf{V}_0$ and the change
in velocity caused by the encounter is $\Delta\mathbf{V}$. We introduce a coordinate system
$\hat{\mathbf{e}}'_1, \hat{\mathbf{e}}'_2, \hat{\mathbf{e}}'_3$, such that $\hat{\mathbf{e}}'_1$ is parallel to $\mathbf{V}_0$ (see Figure 8A-1). Thus

$$\mathbf{V}_0 \cdot \hat{\mathbf{e}}'_1 = |\mathbf{V}_0| \equiv V_0 \quad ; \quad \mathbf{V}_0 \cdot \hat{\mathbf{e}}'_2 = \mathbf{V}_0 \cdot \hat{\mathbf{e}}'_3 = 0. \tag{8A-1}$$

We denote the angle between the plane of the orbit and $\hat{\mathbf{e}}'_2$ by ϕ (there is an
ambiguity of 180° in the definition, but this does not affect our results). The
change in $\mathbf{v}$ during the encounter may be written

$$\Delta\mathbf{v} = -\Delta v_\parallel \hat{\mathbf{e}}'_1 + \Delta v_\perp (\cos\phi\,\hat{\mathbf{e}}'_2 + \sin\phi\,\hat{\mathbf{e}}'_3), \tag{8A-2}$$

where $\Delta v_\parallel$ and $\Delta v_\perp$ are the magnitudes of the components of $\Delta\mathbf{v}$ that
are parallel and perpendicular to $\mathbf{V}_0$. (The minus sign preceding $\Delta v_\parallel$
arises because the figure shows that $\Delta\mathbf{V} \cdot \hat{\mathbf{e}}'_1 < 0$, and we also have that
$\Delta\mathbf{v} = m_a \Delta\mathbf{V}/(m + m_a)$; thus $\Delta\mathbf{v} \cdot \hat{\mathbf{e}}'_1 < 0$.)

Writing $\Delta\mathbf{v} = \sum_{k=1}^3 (\Delta\mathbf{v} \cdot \hat{\mathbf{e}}'_k)\hat{\mathbf{e}}'_k$ we have

$$\Delta v_i = \sum_{k=1}^3 (\Delta\mathbf{v} \cdot \hat{\mathbf{e}}'_k)(\hat{\mathbf{e}}_i \cdot \hat{\mathbf{e}}'_k),$$

$$\Delta v_i \Delta v_j = \sum_{k,l=1}^3 (\Delta\mathbf{v} \cdot \hat{\mathbf{e}}'_k)(\Delta\mathbf{v} \cdot \hat{\mathbf{e}}'_l)(\hat{\mathbf{e}}_i \cdot \hat{\mathbf{e}}'_k)(\hat{\mathbf{e}}_j \cdot \hat{\mathbf{e}}'_l). \tag{8A-3}$$

Since all angles $0 < \phi < 2\pi$ are equally probable, we can take averages of equations (8A-2) and (8A-3) over ϕ. Denoting these averages by $\langle \cdot \rangle_\phi$, we may write $\langle \cos \phi \rangle_\phi = \langle \sin \phi \rangle_\phi = 0$, $\langle \cos^2 \phi \rangle_\phi = \langle \sin^2 \phi \rangle_\phi = \frac{1}{2}$. Thus equation (8A-3) yields

$$\langle \Delta v_i \rangle_\phi = -\Delta v_\| (\hat{\mathbf{e}}_i \cdot \hat{\mathbf{e}}_1'),$$
$$\langle \Delta v_i \Delta v_j \rangle_\phi = (\Delta v_\|)^2 (\hat{\mathbf{e}}_i \cdot \hat{\mathbf{e}}_1')(\hat{\mathbf{e}}_j \cdot \hat{\mathbf{e}}_1') \qquad (8A\text{-}4)$$
$$+ \tfrac{1}{2}(\Delta v_\perp)^2 [(\hat{\mathbf{e}}_i \cdot \hat{\mathbf{e}}_2')(\hat{\mathbf{e}}_j \cdot \hat{\mathbf{e}}_2') + (\hat{\mathbf{e}}_i \cdot \hat{\mathbf{e}}_3')(\hat{\mathbf{e}}_j \cdot \hat{\mathbf{e}}_3')],$$

where from equation (7-10) we have

$$\Delta v_\perp = \frac{2m_a b V_0^3}{G(m + m_a)^2} \left[1 + \frac{b^2 V_0^4}{G^2 (m + m_a)^2} \right]^{-1},$$
$$\Delta v_\| = \frac{2m_a V_0}{m + m_a} \left[1 + \frac{b^2 V_0^4}{G^2 (m + m_a)^2} \right]^{-1}. \qquad (8A\text{-}5)$$

We now sum the effects of all the encounters. The number density of field stars in the velocity range that we are considering is $f_a(\mathbf{v}_a) d^3 \mathbf{v}_a$. The number of encounters per unit time with impact parameters between b and $b + db$ is just this density times the volume of an annulus with inner radius b, outer radius $b + db$, and length V_0, that is,

$$2\pi b\, db V_0 f_a(\mathbf{v}_a) d^3 \mathbf{v}_a. \qquad (8A\text{-}6)$$

Thus

$$D(\Delta v_i) = \int f_a(\mathbf{v}_a) V_0 \langle \Delta v_i \rangle_\phi 2\pi b\, db\, d^3 \mathbf{v}_a, \qquad (8A\text{-}7)$$

with a similar equation for $D(\Delta v_i \Delta v_j)$. We can carry out the integration over the impact parameter b using equations (8A-5). The range of integration is from 0 to $b_{\max}$. The integrals involved are:

$$\int_0^{b_{\max}} \Delta v_\| b\, db = \frac{G^2 m_a (m + m_a)}{V_0^3} \ln(1 + \Lambda^2),$$
$$\int_0^{b_{\max}} (\Delta v_\|)^2 b\, db = 2 \frac{G^2 m_a^2}{V_0^2} \left(1 - \frac{1}{1 + \Lambda^2} \right), \qquad (8A\text{-}8)$$
$$\int_0^{b_{\max}} (\Delta v_\perp)^2 b\, db = 2 \frac{G^2 m_a^2}{V_0^2} \left[\ln(1 + \Lambda^2) - 1 + \frac{1}{1 + \Lambda^2} \right],$$

where $\Lambda = b_{\max} V_0^2 / G(m + m_a)$. The appropriate value for $b_{\max}$ is approximately R, where R is the radius of the system, since this is the maximum possible impact parameter. In most applications, Λ is very large, and therefore without loss of accuracy we can discard terms involving $(1 + \Lambda^2)^{-1}$ and replace $\ln(1 + \Lambda^2)$ by $2 \ln \Lambda$ in equation (8A-8). Furthermore, although this is a less accurate approximation, we can discard terms of order unity compared to those of order $\ln \Lambda$. In this manner we arrive at a simplified version of equations (8A-8), valid to a fractional accuracy of order $(\ln \Lambda)^{-1}$,

$$\int_0^{b_{max}} \Delta v_{\parallel} b \, db = 2 \frac{G^2 m_a (m + m_a)}{V_0^3} \ln \Lambda,$$

$$\int_0^{b_{max}} (\Delta v_{\parallel})^2 b \, db = 0 \quad ; \quad \int_0^{b_{max}} (\Delta v_{\perp})^2 b \, db = 4 \frac{G^2 m_a^2}{V_0^2} \ln \Lambda. \tag{8A-9}$$

Using equations (8A-4), (8A-7), and (8A-9) we can write the diffusion coefficients as

$$D(\Delta v_i) = -4\pi G^2 m_a (m + m_a) \int \frac{f_a(\mathbf{v}_a) \ln \Lambda}{V_0^2} (\hat{\mathbf{e}}_i \cdot \hat{\mathbf{e}}_1') \, d^3 \mathbf{v}_a,$$

$$D(\Delta v_i \Delta v_j) = 4\pi G^2 m_a^2 \int \frac{f_a(\mathbf{v}_a) \ln \Lambda}{V_0} [(\hat{\mathbf{e}}_i \cdot \hat{\mathbf{e}}_2')(\hat{\mathbf{e}}_j \cdot \hat{\mathbf{e}}_2') + (\hat{\mathbf{e}}_i \cdot \hat{\mathbf{e}}_3')(\hat{\mathbf{e}}_j \cdot \hat{\mathbf{e}}_3')] d^3 \mathbf{v}_a. \tag{8A-10}$$

Since we assume that $\ln \Lambda$ is large, we do not make much additional error by replacing V_0 by some typical stellar speed v_{typ} in the expression for Λ so that $\ln \Lambda$ is independent of $\mathbf{v}_a$. Thus

$$\Lambda = \frac{b_{max} v_{typ}^2}{G(m + m_a)}, \tag{8A-11}$$

and $\ln \Lambda$ may be taken outside the integral. Also, the square bracket in the second line can be simplified, since $\sum_{p=1}^3 (\hat{\mathbf{e}}_i \cdot \hat{\mathbf{e}}_p')(\hat{\mathbf{e}}_j \cdot \hat{\mathbf{e}}_p') = \delta_{ij}$. If we write $\hat{\mathbf{e}}_1' = \mathbf{V}_0 / V_0$, so that $(\hat{\mathbf{e}}_i \cdot \hat{\mathbf{e}}_1') = V_{0i}/V_0$, then we obtain

$$D(\Delta v_i) = -4\pi G^2 m_a (m + m_a) \ln \Lambda \int \frac{f_a(\mathbf{v}_a)}{V_0^3} V_{0i} \, d^3 \mathbf{v}_a,$$

$$D(\Delta v_i \Delta v_j) = 4\pi G^2 m_a^2 \ln \Lambda \int \frac{f_a(\mathbf{v}_a)}{V_0} \left[\delta_{ij} - \frac{V_{0i} V_{0j}}{V_0^2} \right] d^3 \mathbf{v}_a. \tag{8A-12}$$

These equations can be simplified by noting that $V_0 = [\sum_{i=1}^3 (v_i - v_{ai})^2]^{1/2}$, so that

$$\frac{\partial}{\partial v_i} \frac{1}{V_0} = -\frac{V_{0i}}{V_0^3} \quad ; \quad \frac{\partial^2}{\partial v_i \partial v_j} V_0 = \frac{\delta_{ij}}{V_0} - \frac{V_{0i} V_{0j}}{V_0^3}; \tag{8A-13}$$

thus we can write

$$D(\Delta v_i) = 4\pi G^2 m_a (m + m_a) \ln \Lambda \frac{\partial}{\partial v_i} h(\mathbf{v}),$$

$$D(\Delta v_i \Delta v_j) = 4\pi G^2 m_a^2 \ln \Lambda \frac{\partial^2}{\partial v_i \partial v_j} g(\mathbf{v}), \tag{8A-14}$$

where the **Rosenbluth potentials** are

$$h(\mathbf{v}) \equiv \int \frac{f_a(\mathbf{v}_a) \, d^3 \mathbf{v}_a}{|\mathbf{v} - \mathbf{v}_a|} \quad ; \quad g(\mathbf{v}) = \int f_a(\mathbf{v}_a) |\mathbf{v} - \mathbf{v}_a| \, d^3 \mathbf{v}_a. \tag{8A-15}$$

To within a proportionality constant, $h(\mathbf{v})$ is simply the gravitational potential at $\mathbf{v}$ generated by a fictitious body in velocity space that has density $f_a(\mathbf{v}_a)$ [eq. (2-3)].

When the field star DF $f_a(\mathbf{v}_a)$ depends only on $v_a = |\mathbf{v}_a|$, more explicit expressions can be obtained. In this case, symmetry dictates that the Rosenbluth potentials can depend only on $v = |\mathbf{v}|$. To evaluate $g(v)$ and $h(v)$, we use equation (1C-23) to write

$$\frac{1}{|\mathbf{v} - \mathbf{v}_a|} = \sum_{l=0}^{\infty} \frac{v_<^l}{v_>^{l+1}} P_l(\cos\gamma), \qquad (8A\text{-}16)$$

where $v_<$ and $v_>$ are the smaller and larger of v and v_a, $P_l(x)$ is a Legendre polynomial, and γ is the angle between $\mathbf{v}$ and $\mathbf{v}_a$. Thus

$$h(v) = 2\pi \sum_{l=0}^{\infty} \int_0^\infty \frac{v_a^2 v_<^l}{v_>^{l+1}} f_a(v_a)\, dv_a \int_0^\pi P_l(\cos\gamma) \sin\gamma\, d\gamma. \qquad (8A\text{-}17)$$

Using the relation $\int_{-1}^1 P_l(x)\, dx = 2\delta_{l0}$ [eq. (1C-24) with $m = n = 0$] we obtain

$$h(v) = 4\pi \left[\frac{1}{v} \int_0^v v_a^2 f_a(v_a)\, dv_a + \int_v^\infty v_a f_a(v_a)\, dv_a \right]. \qquad (8A\text{-}18)$$

Similarly, to evaluate $g(v)$ we write $|\mathbf{v} - \mathbf{v}_a| = (v^2 - 2vv_a \cos\gamma + v_a^2)/|\mathbf{v} - \mathbf{v}_a|$ and use equation (8A-16) to expand $1/|\mathbf{v} - \mathbf{v}_a|$. Using the identity $\int_{-1}^1 x P_l(x)\, dx = \frac{2}{3}\delta_{l1}$ [eq. (1C-24) with $m = 0$, $n = 1$] we find after some algebra that

$$g(v) = \frac{4\pi v}{3} \left[\int_0^v \left(3v_a^2 + \frac{v_a^4}{v^2} \right) f_a(v_a)\, dv_a + \int_v^\infty \left(3\frac{v_a^3}{v} + vv_a \right) f_a(v_a)\, dv_a \right]. \qquad (8A\text{-}19)$$

Using the relation $\partial v/\partial v_i = v_i/v$, equations (8A-14) can be written in the form

$$D(\Delta v_i) = \frac{v_i}{v} D(\Delta v_\parallel),$$
$$D(\Delta v_i \Delta v_j) = \frac{v_i v_j}{v^2} [D(\Delta v_\parallel^2) - \tfrac{1}{2} D(\Delta v_\perp^2)] + \tfrac{1}{2}\delta_{ij} D(\Delta v_\perp^2), \qquad (8A\text{-}20)$$

where

$$D(\Delta v_\parallel) = 4\pi G^2 m_a(m + m_a) \ln \Lambda \frac{dh(v)}{dv},$$

$$D(\Delta v_\parallel^2) = 4\pi G^2 m_a^2 \ln \Lambda \frac{d^2 g(v)}{dv^2} \quad ; \quad D(\Delta v_\perp^2) = \frac{8\pi G^2 m_a^2 \ln \Lambda}{v} \frac{dg(v)}{dv}. \qquad (8A\text{-}21)$$

The reason for this notation can be seen by considering the case in which $\mathbf{v}$ lies along one of the coordinate axes, say $\hat{\mathbf{e}}_1$, so that $v_1 = v$ and $v_2 = v_3 = 0$. Then $D(\Delta v_\parallel^2) = D(\Delta v_1^2)$ is the total diffusion rate parallel to the velocity vector, and $D(\Delta v_\perp^2) = 2D(\Delta v_2^2) = 2D(\Delta v_3^2)$ is the total diffusion rate in the two-dimensional plane perpendicular to the velocity vector.

Substituting from equations (8A-18) and (8A-19) into equation (8A-21), we have

$$D(\Delta v_\parallel) = -\frac{16\pi^2 G^2 m_a \ln \Lambda}{v^2}(m + m_a) \int_0^v v_a^2 f_a(v_a)\, dv_a,$$

$$D(\Delta v_\parallel^2) = \frac{32\pi^2 G^2 m_a^2 \ln \Lambda}{3v}\left[\int_0^v \frac{v_a^4}{v^2}f_a(v_a)\, dv_a + v\int_v^\infty v_a f_a(v_a)\, dv_a\right],$$

$$D(\Delta v_\perp^2) = \frac{32\pi^2 G^2 m_a^2 \ln \Lambda}{3v}\left[\int_0^v \left(3v_a^2 - \frac{v_a^4}{v^2}\right)f_a(v_a)\, dv_a + 2v\int_v^\infty v_a f_a(v_a)\, dv_a\right].$$

$$(8A\text{-}22)$$

Appendix 8.B The Equilibrium Soft Binary Distribution

Consider a system of N stars of mass m contained in a volume $\mathcal{V}$. As in §8.1.3, we define the two-particle DF $f^{(2)}$ such that there are $N^2 f^{(2)}(\mathbf{x}_1, \mathbf{v}_1, \mathbf{x}_2, \mathbf{v}_2) \times d^3\mathbf{x}_1 d^3\mathbf{v}_1 d^3\mathbf{x}_2 d^3\mathbf{v}_2$ pairs of stars in the volume element $d^3\mathbf{x}_1 d^3\mathbf{v}_1 d^3\mathbf{x}_2 d^3\mathbf{v}_2$.

By analogy with other statistical reaction theories, it is natural to assume that in equilibrium $f^{(2)}$ follows a Maxwell-Boltzmann distribution,

$$N^2 f^{(2)}(\mathbf{x}_1, \mathbf{v}_1, \mathbf{x}_2, \mathbf{v}_2) = Ce^{-\beta H}, \tag{8B-1}$$

where C and β are constants and the Hamiltonian H is given by

$$H = \tfrac{1}{2}m(v_1^2 + v_2^2) - \frac{Gm^2}{|\mathbf{x}_1 - \mathbf{x}_2|}. \tag{8B-2}$$

We shall defer a discussion of the validity of this assumption until we have worked out its consequences.

If the separation of stars 1 and 2 is large, then their distributions should be independent, and $f^{(2)}$ should be simply the product of the one-particle DFs. If these are Maxwellian, with dispersion σ and density $\nu = N/\mathcal{V}$, then we have

$$N^2 f^{(2)}(\mathbf{x}_1, \mathbf{v}_1, \mathbf{x}_2, \mathbf{v}_2) = \frac{\nu^2}{(2\pi\sigma^2)^3}e^{-\frac{1}{2}v_1^2/\sigma^2}e^{-\frac{1}{2}v_2^2/\sigma^2} \quad \text{as } |\mathbf{x}_1 - \mathbf{x}_2| \to \infty.$$
$$(8B\text{-}3)$$

Comparison of equations (8B-3) and (8B-1) permits us to evaluate the constants C and β and replace them in equation (8B-1). We find

$$N^2 f^{(2)}(\mathbf{x}_1, \mathbf{v}_1, \mathbf{x}_2, \mathbf{v}_2) = \frac{\nu^2}{(2\pi\sigma^2)^3}\exp\left[-\frac{1}{2\sigma^2}\left(v_1^2 + v_2^2 - \frac{2Gm}{|\mathbf{x}_1 - \mathbf{x}_2|}\right)\right]. \tag{8B-4}$$

From equation (1D-32), the internal energy of the binary can be written

$$E = \tfrac{1}{2}\mu V^2 - \frac{Gm^2}{|\mathbf{x}_1 - \mathbf{x}_2|}, \tag{8B-5}$$

where $\mathbf{V} = \mathbf{v}_1 - \mathbf{v}_2$ is the relative velocity and $\mu = \tfrac{1}{2}m$ is the reduced mass. Thus the number density of binaries with internal energies in the range $E \to E + dE$ is $\nu_b(E)dE$, where

$$\nu_b(E) = \frac{N^2}{2\mathcal{V}} \int d^3\mathbf{x}_1 d^3\mathbf{v}_1 d^3\mathbf{x}_2 d^3\mathbf{v}_2 f^{(2)}(\mathbf{x}_1, \mathbf{v}_1, \mathbf{x}_2, \mathbf{v}_2) \delta\left(\tfrac{1}{4}mV^2 - \frac{Gm^2}{|\mathbf{x}_1 - \mathbf{x}_2|} - E\right).$$
$$\tag{8B-6}$$

In this expression δ denotes the Dirac delta function, we have used the identity (1C-3), and we have multiplied by a factor of $\tfrac{1}{2}$ because each binary pair is counted twice in the integration. Since we are interested in bound pairs, we shall assume from now on that $E < 0$.

At this point it is convenient to convert from $\mathbf{v}_1$ and $\mathbf{v}_2$ to the relative velocity $\mathbf{V}$ and the center of mass velocity $\mathbf{v}_{cm} = \tfrac{1}{2}(\mathbf{v}_1 + \mathbf{v}_2)$. We have

$$\mathbf{v}_1 = \mathbf{v}_{cm} + \tfrac{1}{2}\mathbf{V} \quad ; \quad \mathbf{v}_2 = \mathbf{v}_{cm} - \tfrac{1}{2}\mathbf{V}. \tag{8B-7}$$

The Jacobian satisfies $|\partial(\mathbf{v}_1, \mathbf{v}_2)/\partial(\mathbf{v}_{cm}, \mathbf{V})| = 1$, and hence the velocity-space volume element $d^3\mathbf{v}_1 d^3\mathbf{v}_2$ can be replaced by $d^3\mathbf{v}_{cm} d^3\mathbf{V}$. In addition $v_1^2 + v_2^2 = 2v_{cm}^2 + \tfrac{1}{2}V^2$. Equations (8B-4) and (8B-6) now yield

$$\nu_b(E) = \frac{\nu^2}{16\pi^3\sigma^6\mathcal{V}} \int d^3\mathbf{x}_1 d^3\mathbf{x}_2 d^3\mathbf{v}_{cm} d^3\mathbf{V}$$
$$\times \exp\left[-\frac{1}{\sigma^2}\left(v_{cm}^2 + \tfrac{1}{4}V^2 - \frac{Gm}{|\mathbf{x}_1 - \mathbf{x}_2|}\right)\right] \delta\left(\tfrac{1}{4}mV^2 - \frac{Gm^2}{|\mathbf{x}_1 - \mathbf{x}_2|} - E\right).$$
$$\tag{8B-8}$$

The integral over $\mathbf{v}_{cm}$ is given in equation (8-120). Because of the δ-function, we can set $\tfrac{1}{4}V^2 - Gm/|\mathbf{x}_1 - \mathbf{x}_2| = E$ in the argument of the exponential. We can replace the dummy variable $\mathbf{x}_1$ by $\mathbf{x} = \mathbf{x}_1 - \mathbf{x}_2$; moreover, since we are primarily interested in binaries whose separation $|\mathbf{x}|$ is much less than the system size, we can just as well extend the integral over $\mathbf{x}$ to cover all space (i.e., we neglect edge effects). In this approximation, the integral over $\mathbf{x}_2$ is independent of $\mathbf{x}$ and yields $\int d^3\mathbf{x}_2 = \mathcal{V}$. Thus

$$\nu_b(E) = \frac{\nu^2}{16\pi^{3/2}\sigma^3} e^{-E/m\sigma^2} \int d^3\mathbf{x} d^3\mathbf{V} \delta\left(\tfrac{1}{4}mV^2 - \frac{Gm^2}{|\mathbf{x}|} - E\right). \tag{8B-9}$$

Since the integrand depends only on $|\mathbf{x}|$ and $|\mathbf{v}|$, we can write

$$\nu_b(E) = \frac{\pi^{1/2}\nu^2 e^{-E/m\sigma^2}}{\sigma^3} \int_0^\infty V^2 dV \int_0^\infty x^2 dx \delta\left(\tfrac{1}{4}mV^2 - \frac{Gm^2}{x} - E\right). \tag{8B-10}$$

Performing the integration over x, replacing the number density ν by the mass density $\rho = m\nu$, and recalling that $E < 0$, we obtain

$$\nu_b(E) = \frac{\pi^{1/2}\rho^2 G^3 m^4 e^{|E|/m\sigma^2}}{\sigma^3} \int_0^\infty \frac{V^2 dV}{(\frac{1}{4}mV^2 + |E|)^4}. \tag{8B-11}$$

The integral over v can be performed using the result $\int_0^\infty x^2 dx/(x^2+1)^4 = \pi/32$. We find

$$\nu_b(E) = \frac{\pi^{3/2}\rho^2 G^3 m^{5/2} e^{|E|/m\sigma^2}}{4\sigma^3 |E|^{5/2}}. \tag{8B-12}$$

There are several limitations on the validity of this result. First of all, the use of the equilibrium Maxwell-Boltzmann distribution in equation (8B-1) is suspect, since we argued in §8.1 that stellar systems have no maximum-entropy state and therefore no state of thermodynamic equilibrium. A heuristic justification is possible in the case of soft binaries, since their disruption time is much shorter than the local relaxation time, so that they are always in instantaneous equilibrium with the field star distribution. A second difficulty is that the number of binaries diverges as $|E| \to \infty$; however, this is not a serious problem because the hard binaries are not in equilibrium (see §8.4.4). Therefore, equation (8B-12) ought to apply to soft binaries but not hard ones, and so we may as well drop the factor $\exp[|E|/(m\sigma^2)]$. We thereby obtain our final expression for the equilibrium distribution of soft binaries,

$$\nu_b(E) = \frac{\pi^{3/2}\rho^2 G^3 m^{5/2}}{4\sigma^3 |E|^{5/2}}, \tag{8B-13}$$

which is valid for $|E| \ll m\sigma^2 \equiv |E|_{\max}$. Notice that the total number density of soft binaries with $|E| > \epsilon$ diverges as $\epsilon \to 0$. The divergence arises because of the great volume of phase space available to a loosely bound binary. However, binaries with a very large separation are disrupted by tidal forces from the cluster; the maximum semi-major axis is $a_t \approx (m/M)^{1/3} R$ [eq. (7-84)], where $M = Nm$ and R are the mass and radius of the system. Hence the minimum $|E|$ for a binary is $|E|_{\min} \approx Gm^2/a_t \approx GmM/(N^{2/3}R)$, and using the relation $\sigma^2 \approx GM/R$ we can write $|E|_{\min} \approx m\sigma^2/N^{2/3}$.

We may use these results to estimate the equilibrium total number of soft binaries in a cluster. Although equation (8B-13) is strictly valid only for $|E|_{\min} \ll |E| \ll |E|_{\max}$, we should not make any serious error by assuming that it is valid for all energies between $|E|_{\max}$ and $|E|_{\min}$. We can then integrate $\nu_b(E)$ over E from $|E|_{\min}$ to $|E|_{\max}$; since ν_b diverges strongly as $|E| \to 0$, the result depends only on $|E|_{\min}$. We find the total number density of soft binaries to be

$$\nu_b = \frac{\pi^{3/2}\rho^2 G^3 m^{5/2}}{6\sigma^3 |E|_{\min}^{3/2}}. \tag{8B-14}$$

The total number of soft binaries is $N_b \approx \nu_b R^3$, and if we set $\rho \approx Nm/R^3$ and use our approximate expressions for $|E|_{\min}$ and σ^2, we can make a crude estimate of N_b. We find $N_b \approx 1$, that is, there is only of order one soft binary in the whole system! Obviously, soft binaries play no role of any consequence in the evolution of stellar systems; even a large primordial population of soft binaries will rapidly be destroyed and virtually no new soft binaries will be created.

Appendix 9.A Cosmology

This Appendix provides a brief introduction to the aspects of cosmology that we use in this book. For more detail see Peebles (1971), Weinberg (1972), or Gunn (1978).

To a very good approximation, the Universe is observed to be homogeneous and isotropic on large scales (say, greater than a few tens of Mpc). Therefore, it is often useful to average over the small-scale structure and approximate the Universe as *exactly* homogeneous and isotropic. Of course, the Universe does not appear isotropic to all observers: an observer traveling rapidly with respect to the local matter will see galaxies moving toward him in one direction and away from him in another. Furthermore, the Universe is evolving, so that the Universe will only appear to be homogeneous if the clocks of spatially separated observers are properly synchronized. Thus, we must formalize the assumption of homogeneity and isotropy as follows: there exists a set of **fundamental observers** and a **cosmic time**, such that the Universe appears homogeneous and isotropic in all its properties to all fundamental observers at a given cosmic time (Gunn 1978).

The random velocities of galaxies, and the velocities of stars within galaxies, are small compared to the relative velocities of galaxies separated by tens of Mpc. Thus, to a good approximation, any observer—like ourselves—who is associated with a star in a galaxy is a fundamental observer.

1 Kinematics

Consider the triangle defined by any three fundamental observers. As the Universe evolves, the triangle may change in size, but cannot change in shape or orientation (or else it would define a preferred direction, thus violating the isotropy assumption). Thus, if $l_{ij}(t)$ is the length of the side joining observers i and j at cosmic time t, we must have $l_{ij}(t) = l_{ij}(t_0)R(t)$, where $R(t)$ is independent of i, j and equals unity at the present time t_0. This argument can be extended to any two fundamental observers, so the distance between any two fundamental observers must have the form

$$l(t) = l(t_0)R(t), \qquad (9A\text{-}1)$$

where the **scale factor** $R(t)$ is a universal function and $R(t_0) = 1$. The relative velocity of the fundamental observers is

$$v(t) = \frac{dl}{dt} = l(t_0)\dot{R}(t) = l(t)\frac{\dot{R}(t)}{R(t)} \equiv l(t)H(t), \qquad (9A\text{-}2)$$

where $H(t)$ is the **expansion rate**. At the present time, $H(t_0) = \dot{R}(t_0) \equiv H_0$ is the Hubble constant, and equation (9A-2) is a statement of Hubble's law [eq. (1-11)].

Next we consider a free particle that travels past a fundamental observer at cosmic time t, moving with relative or **peculiar** speed $v_p \ll c$. After a time interval dt, it has moved a distance $dl = v_p dt$ and hence is overtaking a fundamental observer moving away at speed $H(t)dl = Hv_p dt$. Thus its speed relative to a local fundamental observer steadily decreases, at a rate

$$\frac{dv_p}{dt} = -H(t)v_p = -\frac{v_p}{R}\frac{dR}{dt}, \tag{9A-3}$$

which integrates to yield

$$v_p(t) \propto \frac{1}{R(t)}. \tag{9A-4}$$

Similarly, the frequency ν of propagating photons as measured in the rest frame of a local fundamental observer decreases at a rate

$$\frac{d\nu}{dt} = -H(t)\nu; \quad \text{thus} \quad \nu(t) \propto \frac{1}{R(t)}. \tag{9A-5}$$

In other words, a photon of frequency ν_e or wavelength $\lambda_e = c/\nu_e$ emitted by a fundamental observer at time t_e, and received by a second fundamental observer at the present time t_0, will be observed to have frequency ν_0 and wavelength λ_0 given by

$$\frac{\nu_e}{\nu_0} = \frac{\lambda_0}{\lambda_e} = \frac{R(t_0)}{R(t_e)} = \frac{1}{R(t_e)} \equiv 1 + z. \tag{9A-6}$$

Here z is the **redshift**.

2 Geometry

Let the position of any fundamental observer be labeled by coordinates (q_1, q_2, q_3). At a given cosmic time t, the distance ds between the observers at (q_1, q_2, q_3) and $(q_1 + dq_1, q_2 + dq_2, q_3 + dq_3)$ can be written in the form

$$ds^2 = R^2(t)h_{ij}dq_i dq_j, \tag{9A-7}$$

where we have used the summation convention, and h_{ij} is the metric tensor [cf. eq. (1B-13)]. It can be shown that homogeneity and isotropy demand that the metric tensor can always be written in the form (the **Robertson-Walker metric**)

$$ds^2 = R^2(t)\left[\frac{dr^2}{1 - kr^2/r_u^2} + r^2(d\theta^2 + \sin^2\theta d\phi^2)\right]. \tag{9A-8}$$

Here θ and ϕ are the usual angles from spherical coordinates (Appendix 1.B.2), r is a radial coordinate, r_u is a constant, and the **curvature constant** k is $+1$, 0 or -1. The case $k = 0$ corresponds to ordinary flat space [cf. eq. (1B-32)], while the case $k = +1$ is the three-dimensional generalization of the metric on the surface of a sphere of radius r_u, $ds^2 = dr^2/(1 - r^2/r_u^2) + r^2 d\phi^2$, where r is the perpendicular distance from the polar axis to the point in question. Unfortunately, the case $k = -1$ has no analogous 2-surface imbedded in Euclidean 3-space.

For determining the relation between flux and luminosity of a distant source, it is important to know the area $A(r,t)$ of the spherical surface with coordinate distance r. It is clear from equation (9A-8) that

$$A(r,t) = 4\pi R^2(t)r^2. \tag{9A-9}$$

3 Dynamics

To determine the temporal evolution of the scale factor $R(t)$, we need a theory of gravity, which we take to be general relativity. We shall use two results from general relativity:

(i) In a spherically symmetric system, the gravitational acceleration at any radius is determined by the matter distribution inside that radius. This is **Birkhoff's theorem**, the relativistic analog of Newton's first theorem (§2.1.1).

(ii) A fluid with density ρ and pressure p has an active mass density $\rho' = \rho + 3p/c^2$.

The isotropy assumption assures us that the Universe is spherically symmetric as viewed by any fundamental observer. Now draw a sphere of radius l around any such observer, where l is sufficiently small that Newtonian physics applies within the sphere (i.e., $lH \ll c$). Because of Birkhoff's theorem, we can ignore the effects of material outside the sphere. Then Newton's law of gravity tells us that a fundamental observer on the surface of the sphere is accelerated toward its center at a rate

$$\frac{d^2 l}{dt^2} = -\frac{GM}{l^2}, \tag{9A-10}$$

where M is the mass inside the sphere. Note that there are no pressure forces since $\nabla p = 0$ by homogeneity. Since the mass $M = \rho'V$, where $V = \frac{4}{3}\pi l^3$, and $l = l_0 R(t)$, we may rewrite equation (9A-10) as

$$\frac{\ddot{R}}{R} = -\frac{4\pi G\rho'}{3} = -\frac{4\pi G}{3}\left(\rho + \frac{3p}{c^2}\right). \tag{9A-11}$$

To integrate this equation, we must know how p and ρ vary with the scale factor $R(t)$. The internal energy U of the material in the sphere is $U = \rho c^2 V$, where we have included the rest mass energy in the internal energy. As the Universe evolves, the material satisfies $dU + p\,dV = 0$ [eq. (1E-12) with $dS = 0$, since there can be no heat flow in an isotropic Universe], and hence

$$d(\rho c^2 V) + p\,dV = 0 \quad \text{or} \quad d\rho + \left(\rho + \frac{p}{c^2}\right)\frac{dV}{V} = 0. \tag{9A-12}$$

Since $V \propto R(t)^3$, we have $dV/V = 3dR/R$, and equations (9A-11) and (9A-12) can be combined to eliminate p:

$$\frac{\ddot{R}}{R} = \frac{4\pi G}{3}\left(2\rho + R\frac{d\rho}{dR}\right). \tag{9A-13}$$

This is a linear differential equation for ρ, which can be integrated to yield

$$\dot{R}^2 - \frac{8\pi G\rho}{3} R^2 = 2E, \tag{9A-14}$$

where E is a constant of integration, closely analogous to the Newtonian energy. Equations (9A-11), (9A-12), and (9A-14) specify the **Friedmann models** of the Universe.

These equations can also be derived directly from general relativity. The relativistic derivation also relates the parameters of the Robertson-Walker metric (9A-8) to the integration constant E:

$$2E = -\frac{kc^2}{r_u^2}. \tag{9A-15}$$

The constant E can be evaluated from observations of the density at the present epoch, $\rho_0 \equiv \rho(t_0)$, and the Hubble constant $H_0 = \dot{R}(t_0)$. We may rewrite equation (9A-14) as

$$\frac{2E}{H_0^2} = 1 - \Omega_0, \tag{9A-16}$$

where

$$\Omega_0 = \frac{8\pi G\rho_0}{3H_0^2} \tag{9A-17}$$

is the **density parameter**. One sometimes writes

$$\Omega_0 = \frac{\rho_0}{\rho_c}, \quad \text{where} \quad \rho_c \equiv \frac{3H_0^2}{8\pi G} = 1.88 \times 10^{-29} h^2 \, \text{g cm}^{-3} \tag{9A-18}$$

is the **critical density**. The sign of E is positive if $\Omega_0 < 1$, and negative if $\Omega_0 > 1$. We may now divide the Friedmann models into three classes:

(i) **Open** or **unbound models** ($E > 0$, $\Omega_0 < 1$, $k = -1$). Here equation (9A-14) implies that $\dot{R}^2$ is never zero. Since the Universe is expanding at the present epoch ($H_0 > 0$) it must expand forever. As $t \to \infty$, the density becomes negligibly small, so the expansion rate $\dot{R} \to \sqrt{2E}$.

(ii) **Critical** or **Einstein-de Sitter model** ($E = 0$, $\Omega_0 = 1$, $k = 0$). Once again the Universe expands forever, but at a rate that becomes slower and slower as the density falls.

(iii) **Closed** or **bound models** ($E < 0$, $\Omega_0 > 1$, $k = +1$). As the Universe expands, the second term on the left side of equation (9A-14) becomes smaller, and eventually the expansion halts ($\dot{R} = 0$) and the Universe begins to contract.

The temporal behavior of the Friedman models can be determined explicitly in the case where the density is dominated by non-relativistic matter (it is generally believed that this has been the case over most of the history of the Universe). Then the active mass density ρ' is simply the rest-mass density ρ, and thus the mass M in equation (9A-10) is time-independent. The behavior of $l(t)$ or $R(t)$ is then given by the motion of a test particle on a radial orbit in the gravitational field of a point mass. After minor changes in the notation of equation (3-27) with eccentricity $e = 1$, we have for a bound model ($E < 0$),

$$R = A(1 - \cos \eta) \quad ; \quad t = B(\eta - \sin \eta), \tag{9A-19}$$

where η is the eccentric anomaly. The boundary conditions are

$$\frac{A^3}{B^2} = \frac{4\pi G \rho_0}{3} \quad ; \quad A = \frac{1}{1 - \cos \eta_0} \quad ; \quad B = \frac{\sin \eta_0}{H_0(1 - \cos \eta_0)^2}, \tag{9A-20}$$

which yield $\Omega_0 = \sec^2 \frac{1}{2}\eta_0$. Similarly, for $E = 0$,

$$R(t) = \left(\frac{t}{t_0}\right)^{\frac{2}{3}}, \quad \text{where} \quad H_0 = \frac{2}{3t_0}. \tag{9A-21}$$

Finally, for an unbound model ($E > 0$),

$$R = A(\cosh \eta - 1) \quad ; \quad t = B(\sinh \eta - \eta), \tag{9A-22}$$

where

$$\frac{A^3}{B^2} = \frac{4\pi G \rho_0}{3} \quad ; \quad A = \frac{1}{\cosh \eta_0 - 1} \quad ; \quad B = \frac{\sinh \eta_0}{H_0(\cosh \eta_0 - 1)^2}, \tag{9A-23}$$

which yield $\Omega_0 = \operatorname{sech}^2 \frac{1}{2}\eta_0$. In all three cases there is an initial singularity or **Big Bang** at $t = 0$.

It is useful to know our coordinate distance r as measured by a fundamental observer at redshift z or cosmic time t. Since photons travel a distance ds in time $dt = ds/c$, the Robertson-Walker metric (9A-8) yields

$$cdt = \frac{R(t)dr}{\sqrt{1 - kr^2/r_u^2}} \quad \text{or} \quad c\int_t^{t_0} \frac{dt}{R(t)} = \int_0^r \frac{dr}{\sqrt{1 - kr^2/r_u^2}}. \tag{9A-24}$$

For simplicity, let us specialize to the case of a bound model. Using equations (9A-19) to eliminate R and t in favor of η, the time integral is found to be simply $(B/A)(\eta_0 - \eta)$. Since $k = +1$ for bound models, the integral over r is simply $r_u \arcsin(r/r_u)$. Using equations (9A-15), (9A-16), and (9A-20), it can be shown that $(B/A) = r_u/c$, and hence our coordinate distance as measured by an observer at η is

$$r = r_u \sin(\eta_0 - \eta). \tag{9A-25}$$

After some algebra, we find that this can be recast in the form

$$r = 4c \frac{(1 - \frac{1}{2}\Omega_0)(1 - \sqrt{1 + \Omega_0 z}) + \frac{1}{2}\Omega_0 z}{(1 + z)H_0\Omega_0^2}. \tag{9A-26}$$

It is not hard to show that this expression is also valid for critical and unbound models.

References

Aaronson, M. 1983. *Astrophys. J. Lett.*, **266**, L11.

Aaronson, M., Huchra, J., Mould, J., Schechter, P. L., & Tully, R. B. 1982. *Astrophys. J.*, **258**, 64.

Aaronson, M., Olszewski, E. W., & Hodge, P. W. 1983. *Astrophys. J.*, **267**, 271.

Aarseth, S. J. 1972. In *Gravitational N-Body Problem*, IAU Colloquium No. 10, ed. M. Lecar, p. 373. Dordrecht: Reidel.

Aarseth, S. J. 1985. In *Multiple Time Scales*, ed. J. U. Brackbill & B. I. Cohen, p. 377. Orlando: Academic Press.

Aarseth, S. J., & Binney, J. J. 1978. *Mon. Not. Roy. Astron. Soc.*, **185**, 227.

Aarseth, S. J., & Fall, S. M. 1980. *Astrophys. J.*, **236**, 43.

Aarseth, S. J., Hénon, M., & Wielen, R. 1974. *Astron. Astrophys.*, **37**, 183.

Abramowitz, M., & Stegun, I. A. 1965. *Handbook of Mathematical Functions.* New York: Dover.

Aguilar, L. A., & White, S.D.M. 1985. *Astrophys. J.*, **295**, 374.

Ahmad, A., & Cohen, L. 1973. *J. Comp. Phys.*, **12**, 389.

Albrecht, A., & Steinhardt, P. J. 1982. *Phys. Rev. Letters*, **48**, 1220.

Allen, C. W. 1973. *Astrophysical Quantities*, 3rd ed. London: Athlone Press.

Ambarzumian, V. A. 1938. *Ann. Leningrad State U.*, No. 22 [translated in Goodman & Hut (1985)].

Antonov, V. A. 1960. *Astr. Zh.*, **37**, 918 (translated in *Sov. Astron.* **4**, 859).

Antonov, V. A. 1962a. *Vestnik Leningrad Univ.*, **7**, 135 [translated in Goodman & Hut (1985)].

Antonov, V. A. 1962b. *Vestnik Leningrad Univ.*, **19**, 96 [translated in de Zeeuw (1987)].

Antonov, V. A. 1973. In *The Dynamics of Galaxies and Star Clusters*, ed. G. B. Omarov, p. 139. Alma Ata: Nauka [translated in de Zeeuw (1987)].

Aoki, S., Noguchi, M., & Iye, M. 1979. *Publ. Astron. Soc. Japan*, **31**, 737.

Arfken, G. 1970. *Mathematical Methods for Physicists*, 2nd ed. New York: Academic Press.

Arimoto, N., & Yoshii, Y. 1987. *Astron. Astrophys.*, **173**, 23.

Armandroff, T. E., & Da Costa, G. S. 1986. *Astron. J.*, **92**, 777.

Arnett, W. D. 1978. *Astrophys. J.*, **219**, 1008.

Arnold, V. I. 1978. *Mathematical Methods of Classical Mechanics.* New York: Springer.

Arp, H. 1966. *Atlas of Peculiar Galaxies.* Washington, D.C.: Carnegie Institution. Also *Astrophys. J. Suppl.*, **14**, 1.

Arp, H. C., & Madore, B. S. 1987. *A Catalogue of Southern Peculiar Galaxies and Associations.* 2 vols. New York: Cambridge University Press.

Athanassoula, E. 1984. *Phys. Reports*, **114**, 319.

Athanassoula, E., Bienaymé, O., Martinet, L., & Pfenniger, D. 1983. *Astron. Astrophys.*, **127**, 349.

Baade, W. 1963. *Evolution of Stars and Galaxies*, p. 63. Cambridge, Mass.: Harvard University Press.

Bahcall, J. N. 1984a. *Astrophys. J.*, **276**, 169.

Bahcall, J. N. 1984b. *Astrophys. J.*, **287**, 926.

Bahcall, J. N. 1986. *Ann. Rev. Astron. Astrophys.*, **24**, 577.

Bahcall. J. N., & Soneira, R. M. 1980. *Astrophys. J. Suppl.*, **44**, 73.

Bahcall, J. N., & Tremaine, S. 1981. *Astrophys. J.*, **244**, 805.

Bahcall, J. N., & Wolf, R. A. 1976. *Astrophys. J.*, **209**, 214.

Bahcall, J. N., Hut, P., & Tremaine, S. 1985. *Astrophys. J.*, **290**, 15.

Bahcall, J. N., Schmidt, M., & Soneira, R. M. 1982. *Astrophys. J. Lett.*, **258**, L23.

Bahcall, J. N., Schmidt, M., & Soneira, R. M. 1983. *Astrophys. J.*, **265**, 730.

Bailey, M. E., & MacDonald, J. 1981. *Mon. Not. Roy. Astron. Soc.*, **194**, 195.

Balbus, S. 1984. *Astrophys. J.*, **277**, 550.

Barbanis, B., & Woltjer, L. 1967. *Astrophys. J.*, **150**, 461.

Bardeen, J. M. 1975. In *Dynamics of Stellar Systems*, IAU Symposium No. 69, ed. A. Hayli, p. 297. Dordrecht: Reidel.

Barnes, J. 1986. In *Dynamics of Star Clusters*, IAU Symposium No. 113, ed. J. Goodman & P. Hut, p. 297. Dordrecht: Reidel.

Barnes, J., & Hut, P. 1986. *Nature*, **324**, 446.

Barnes, J., & White, S.D.M. 1984. *Mon. Not. Roy. Astron. Soc.*, **211**, 753.

Barnes, J., Goodman, J., & Hut, P. 1986. *Astrophys. J.*, **300**, 112.

Bean, A. J., Efstathiou, G., Ellis, R. S., Peterson, B. A., & Shanks, T. 1983. *Mon. Not. Roy. Astron. Soc.*, **205**, 605.

Becker, W., & Contopoulos, G., ed. 1970. *The Spiral Structure of Our Galaxy*, IAU Symposium No. 38. Dordrecht: Reidel.

Becker, W., & Fenkart, R. 1971. *Astron. Astrophys. Suppl.*, **4**, 241.

Begelman, M. C., & Rees, M. J. 1978. *Mon. Not. Roy. Astron. Soc.*, **185**, 847.

Bekenstein, J., & Milgrom, M. 1984. *Astrophys. J.*, **286**, 7.

Berkhuijsen, E. M., & Wielebinski, R., ed. 1978. *Structure and Properties of Nearby Galaxies*, IAU Symposium No. 77. Dordrecht: Reidel.

Berry, M. V. 1978. *Topics in Nonlinear Mechanics*, ed. S. Jorna, p. 16. New York: American Institute of Physics.

Binney, J. J. 1976. *Mon. Not. Roy. Astron. Soc.*, **177**, 19.

Binney, J. J. 1977. *Mon. Not. Roy. Astron. Soc.*, **181**, 735.

Binney, J. J. 1978. *Mon. Not. Roy. Astron. Soc.*, **183**, 501.

Binney, J. J. 1980. In *X-ray Astronomy*, ed. R. Giacconi & G. Setti, p. 245. Dordrecht: Reidel.

Binney, J. J. 1981. *Mon. Not. Roy. Astron. Soc.*, **196**, 455.

Binney, J. J. 1982. In *Morphology and Dynamics of Galaxies*, Twelfth Advanced Course of the Swiss Society of Astronomy and Astrophysics, ed. L. Martinet & M. Mayor, p. 1. Sauverny: Geneva Observatory.

Binney, J. J. 1987. In *The Galaxy*, ed. G. Gilmore & R. Carswell, p. 399. Dordrecht: Reidel.

Binney, J. J., & Mamon, G. A. 1982. *Mon. Not. Roy. Astron. Soc.*, **200**, 361.

Binney, J. J., & May, A. 1986. *Mon. Not. Roy. Astron. Soc.*, **218**, 743.

Binney, J. J., & Silk, J. 1978. *Comments Astrophys.*, **7**, 139.

Binney, J. J., & Spergel, D. N. 1983. In *The Nearby Stars and the Stellar Luminosity Function*, IAU Colloquium No. 76, ed. A.G.D. Philip & A. R. Upgren, p. 259. Schenectady, N.Y.: Davis Press.

Binney, J. J., & Spergel, D. N. 1984. *Mon. Not. Roy. Astron. Soc.*, **206**, 159.

Binney, J. J., Gerhard, O. E., & Hut, P. 1985. *Mon. Not. Roy. Astron. Soc.*, **215**, 59.

Blair, W. P., Kirshner, R. P., & Chevalier, R. A. 1981. *Astrophys. J.*, **247**, 879.

Bond, H. E. 1970. *Astrophys. J. Suppl.*, **22**, 117.

Bond, J. R., & Efstathiou, G. 1984. *Astrophys. J. Lett.*, **285**, L45.

Bond, J. R., Centrella, J., Szalay, A. S., & Wilson, J. R. 1984. In *Formation and Evolution of Galaxies and Large Structures in the Universe*, ed. J. Audouze & J. Tran Thanh Van, p. 87. Dordrecht: Reidel.

Bond, J. R., Efstathiou, G., & Silk, J. 1980. *Phys. Rev. Lett.*, **45**, 1980.

Bontekoe, Tj. R., & van Albada, T. S. 1987. *Mon. Not. Roy. Astron. Soc.*, **224**, 349.

Boughn, S. P., Cheng, E. S., & Wilkinson, D. T. 1981. *Astrophys. J.*, **243**, L113.

Brandt, J. C. 1960. *Astrophys. J.*, **131**, 293.

Brandt, J. C., & Belton, M. J. 1962. *Astrophys. J.*, **136**, 352.

Bruzual, G. A. 1980. Unpublished Ph.D. thesis, University of California, Berkeley.

Bruzual, G. A. 1983. *Rev. Mexicana Astron. Astrof.*, **8**, 63.

Bruzual, G. A., & Kron, R. G. 1980. *Astrophys. J.*, **241**, 25.

Bryan, G. H. 1888. *Phil. Trans. Roy. Soc. London A*, **180**, 187.

Burbidge, E. M., & Burbidge, G. R. 1975. In *Galaxies and the Universe*, ed. A. Sandage, M. Sandage, & J. Kristian, p. 81. Chicago: University of Chicago Press.

Burton, W. B., & Gordon, M. A. 1978. *Astron. Astrophys.*, **63**, 7.

Butcher, H., & Oemler, A. 1978. *Astrophys. J.*, **219**, 18.

Caldwell, J.A.R., & Ostriker, J. P. 1981. *Astrophys. J.*, **251**, 61.

Canizares, C. 1982. *Astrophys. J.*, **263**, 508.

Carignan, C., & Freeman, K. C. 1985. *Astrophys. J.*, **294**, 494.

Carlberg, R. G. 1984. In *Formation and Evolution of Galaxies and Large Structures in the Universe*, ed. J. Audouze & J. Tran Thanh Van, p. 343. Dordrecht: Reidel.

Carlberg, R. G., & Sellwood, J. A. 1985. *Astrophys. J.*, **292**, 79.

Carney, B. 1984. *Publ. Astron. Soc. Pac.*, **96**, 841.

Carney, B., & Latham, D. W. 1987. In *Dark Matter in the Universe*, IAU Symposium No. 117, ed. J. Kormendy & G. R. Knapp, p. 39. Dordrecht: Reidel.

Carr, B. J., Bond, J. R., & Arnett, W. D. 1984. *Astrophys. J.*, **277**, 445.

Case, K. M. 1959. *Ann. Phys.*, **7**, 349.

Chandrasekhar, S. 1939. *An Introduction to the Theory of Stellar Structure.* Chicago: University of Chicago Press. Reissued by Dover 1958.

Chandrasekhar, S. 1942. *Principles of Stellar Dynamics.* Chicago: University of Chicago Press. Reissued by Dover 1960.

Chandrasekhar, S. 1943. *Astrophys. J.*, **97**, 255.

Chandrasekhar, S. 1961. *Hydrodynamic and Hydromagnetic Stability.* Oxford: Oxford University Press. Reissued by Dover 1981.

Chandrasekhar, S. 1963. *Astrophys. J.*, **138**, 896.

Chandrasekhar, S. 1964. *Astrophys. J.*, **139**, 664.

Chandrasekhar, S. 1969. *Ellipsoidal Figures of Equilibrium.* New Haven, Conn.: Yale University Press.

Chevalier, R. A. 1977. *Ann. Rev. Astron. Astrophys.*, **15**, 175.

Clausius, R. 1870 *Sitz. Niederrheinischen Gesellschaft, Bonn*, p. 114 [translated in *Phil. Mag.* **40**, 112 (1870)].

Clegg, R.E.S., & Bell, R.A.M. 1973. *Mon. Not. Roy. Astron. Soc.*, **163**, 13.

Clutton-Brock, M. 1972. *Astrophys. Space Sci.*, **16**, 101.

Cohn, H. 1979. *Astrophys. J.*, **234**, 1036.

Cohn, H. 1980. *Astrophys. J.*, **242**, 765.

Cohn, H., & Hut, P. 1984. *Astrophys. J. Lett.*, **277**, L45.

Cohn, H., & Kulsrud, R. M. 1978. *Astrophys. J.*, **226**, 1087.

Contopoulos, G. 1954. *Zeitschrift f. Astrophysik*, **35**, 67.

Contopoulos, G. 1986. *Astron. Astrophys.*, **161**, 244.

Contopoulos, G., & Papayannopoulos, Th. 1980. *Astron. Astrophys.*, **92**, 33.

Cooley, J. W., & Tukey, J. W. 1965. *Math. Comp.*, **19**, 297.

Cowie, L. L., & Rybicki, G. B. 1982. *Astrophys. J.*, **260**, 504.

Cowie, L. L., Henriksen, M., & Mushotzsky, R. 1987. *Astrophys. J.*, in press.

Cox, J. P. 1980. *Theory of Stellar Pulsation.* Princeton, N.J.: Princeton University Press.

Davies, R. L., Efstathiou, G., Fall, S. M., Illingworth, G., & Schechter, P. L. 1983. *Astrophys. J.*, **266**, 41.

Davis, M., & Huchra, J. 1982. *Astrophys. J.*, **254**, 437.

Davis, M., & Peebles, P.J.E. 1983. *Ann. Rev. Astron. Astrophys.*, **21**, 109.

Davis, M., Efstathiou, G., Frenk, C. S., & White, S.D.M. 1985. *Astrophys. J.*, **292**, 371.

Davis, M., Geller, M. J., & Huchra, J. 1978. *Astrophys. J.*, **221**, 1.

Davis, M., Tonry, J., Huchra, J., & Latham, D. W. 1980. *Astrophys. J.*, **238**, L113.

Dejonghe, H. 1986. *Physics Reports*, **133**, 217.

Dekel, A., Lecar, M., & Shaham, J. 1980. *Astrophys. J.*, **241**, 946.

Demarque, P., & McClure, R. D. 1977. In *The Evolution of Galaxies and*

Stellar Populations, ed. B. M. Tinsley & R. B. Larson, p. 199. New Haven, Conn.: Yale Universty Observatory.

de Vaucouleurs, G. 1948. *Ann. d'Astrophys.*, **11**, 247.

de Vaucouleurs, G. 1959. In *Handbuch der Physik*, **53**, ed. S. Flügge, p. 275. Berlin: Springer-Verlag.

de Vaucouleurs, G. 1975. In *Galaxies and the Universe*, ed. A. Sandage, M. Sandage, & J. Kristian, p. 557. Chicago: University of Chicago Press.

de Vaucouleurs, G., & Freeman, K. C. 1972. *Vistas in Astronomy*, **14**, 163.

de Vaucouleurs, G., & Pence, W. D. 1978. *Astron. J.*, **83**, 1163.

De Young, D. S. 1978. *Astrophys. J.*, **223**, 47.

de Zeeuw, T. 1985. *Mon. Not. Roy. Astron. Soc.*, **216**, 273, 599.

de Zeeuw, T., ed. 1987. *Structure and Dynamics of Elliptical Galaxies*, IAU Symposium No. 127. Dordrecht: Reidel.

Dicke, R. H. 1970 *Gravitation and the Universe*. Philadelphia: Am. Phil. Soc.

Doremus, J. P., Feix, M. R., & Baumann, G. 1971. *Phys. Rev. Lett.*, **26**, 725.

Dressler, A. 1980. *Astrophys. J.*, **236**, 351.

Dressler, A., & Gunn, J. E. 1983. *Astrophys. J.*, **270**, 7.

Dreyer, J. L. E. 1890. *Mem. Roy. Astron. Soc.*, **49**, 1.

Eddington, A. S. 1916. *Mon. Not. Roy. Astron. Soc.*, **76**, 572.

Eddington, A. S. 1926. *The Internal Constitution of the Stars*, p. 188. Cambridge, Eng.: Cambridge University Press.

Edmunds, M. G., & Pagel, B. E. J. 1984. *Mon. Not. Roy. Astron. Soc.*, **211**, 507.

Efstathiou, G., & Eastwood, J. W. 1981. *Mon. Not. Roy. Astron. Soc.*, **194**, 503.

Efstathiou, G., Davis, M., Frenk, C. S., & White, S.D.M. 1985. *Astrophys. J. Suppl.*, **57**, 241.

Efstathiou, G., Lake, G., & Negroponte, J. 1982. *Mon. Not. Roy. Astron. Soc.*, **199**, 1069.

Einasto, J., Kaasik, A., & Saar, E. 1974. *Nature*, **250**, 309.

Elmegreen, B. G., & Lada, C. 1977. *Astrophys. J.*, **214**, 725.

Elmegreen, D. M. 1981. *Astrophys. J. Suppl.*, **47**, 229.

Faber, S. M., & Gallagher, J. S. 1979. *Ann. Rev. Astron. Astrophys.*, **17**, 135.

Fabian, A. C., Pringle, J. E., & Rees, M. J. 1975. *Mon. Not. Roy. Astron. Soc.*, **172**, 15P.

Fabricant, D., & Gorenstein, P. 1983. *Astrophys. J.*, **267**, 535.

Farouki, R. T., & Shapiro, S. L. 1982. *Astrophys. J.*, **259**, 103.

Ferrers, N. M. 1877. *Quart. J. Pure and Appl. Math.*, **14**, 1.

Field, G. B. 1972. *Ann. Rev. Astron. Astrophys.*, **10**, 227.

Finzi, A. 1963. *Mon. Not. Roy. Astron. Soc.*, **127**, 21.

Forman, W., Jones, C., & Tucker, W. 1985. *Astrophys. J.*, **293**, 535.

Fort, B. P., Prieur, J.-L., Carter, D., Meatheringham, S. J., & Vigroux, L. 1986. *Astrophys. J.*, **306**, 110.

Fowler, W. A. 1978. In *Cosmochemistry*, ed. W. O. Mulligan, p. 61. Houston: Robert A. Welch Foundation.

Freeman, K. C. 1966a. *Mon. Not. Roy. Astron. Soc.*, **133**, 47.

Freeman, K. C. 1966b. *Mon. Not. Roy. Astron. Soc.*, **134**, 1, 15.

Freeman, K. C. 1970. *Astrophys. J.*, **160**, 811.

Freeman, K. C. 1977. In *The Evolution of Galaxies and Stellar Populations*, ed. B. M. Tinsley & R. B. Larson, p. 133. New Haven, Conn.: Yale University Observatory.

Frenk, C. S., & White, S.D.M. 1980. *Mon. Not. Roy. Astron. Soc.*, **193**, 295.

Fridman, A. M., & Polyachenko, V. L. 1984. *Physics of Gravitating Systems*. 2 vols. New York: Springer.

Fried, B. D., & Conte, S. D. 1961. *The Plasma Dispersion Function*. New York: Academic Press.

Frogel, J. A., Persson, S. E., Aaronson, M., & Matthews, K. 1978. *Astrophys. J.*, **220**, 75.

Fujimoto, M. 1968. In *Non-Stable Phenomena in Galaxies*, IAU Symposium No. 29, p. 453. Yerevan: Armenian Academy of Sciences.

Gallagher, J. S., & Hunter, D. A. 1984. *Ann. Rev. Astron. Astrophys.*, **22**, 37.

Geller, M. J. 1984. In *Clusters and Groups of Galaxies*, ed. F. Mardirossian, G. Giuricin, & M. Mezzetti, p. 353. Dordrecht: Reidel.

Gerhard, O. E. 1981. *Mon. Not. Roy. Astron. Soc.*, **197**, 179.

Gerhard, O. E. 1985. *Astron. Astrophys.*, **151**, 279.

Gibbs, J. W. 1884. *Proc. Am. Assoc. Adv. Sci.*, **33**, 57.

Gillon, D., Doremus, J. P., & Baumann, G. 1976. *Astron. Astrophys.*, **48**, 467.

Gliese, W., Jahreiss, H., & Upgren, A. R. 1986. In *The Galaxy and the Solar System*, ed. R. Smoluchowski, J. N. Bahcall, & M. S. Matthews, p. 13. Tucson: University of Arizona Press.

Goldreich, P., & Lynden-Bell, D. 1965. *Mon. Not. Roy. Astron. Soc.*, **130**, 125.

Goldreich, P., & Tremaine, S. 1978. *Astrophys. J.*, **222**, 850.

Goldreich, P., & Tremaine, S. 1979. *Astrophys. J.*, **233**, 857.

Goldreich, P., & Tremaine, S. 1981. *Astrophys. J.*, **243**, 1062.

Goldstein, H. 1980. *Classical Mechanics*, 2nd ed. Reading, Penn.: Addison-Wesley.

Goodman, J., & Binney, J. J. 1984. *Mon. Not. Roy. Astron. Soc.*, **207**, 511.

Goodman, J., & Hut, P., ed. 1985. *Dynamics of Star Clusters*, IAU Symposium No. 113. Dordrecht: Reidel.

Gorenstein, M., Shapiro, I. I., Rogers, A.E.E., Cohen, N. L., Corey, B. E., Porcas, R. W., Falco, E. E., Bonometti, R. J., Preston, R. A., Rius, A., & Whitney, A. R. 1984. *Astrophys. J.*, **287**, 538.

Gott, J. R., & Thuan, T. X. 1978. *Astrophys. J.*, **223**, 426.

Gradshteyn, I. S., & Ryzhik, I. M. 1965. *Tables of Integrals, Series and Products*. New York: Academic.

Gunn, J. E. 1975. *Comm. Astrophys. Sp. Phys.*, **6**, 7.

Gunn, J. E. 1978. In *Observational Cosmology*, Eighth Advanced Course of the Swiss Society of Astronomy and Astrophysics, ed. A. Maeder, L.

Martinet, & G. Tammann, p. 1. Sauverny: Geneva Observatory.

Gunn, J. E., & Griffin, R. F. 1979. *Astron. J.*, **84**, 752.

Gunn, J. E., & Peterson, B. A. 1965. *Astrophys. J.*, **142**, 1633.

Gunn, J. E., & Tinsley, B. M. 1975. *Nature*, **257**, 454.

Gunn, J. E., Knapp, G. R., & Tremaine, S. 1979. *Astron. J.*, **84**, 1181.

Gunn, J. E., Stryker, L. L., & Tinsley, B. M. 1981. *Astrophys. J.*, **249**, 48.

Gurevich, L. E., & Levin, B. Ya. 1950. *Doklady Akad. Nauk. USSR*, **70**, 781.

Guth, A. 1981. *Phys. Rev.*, **D23**, 347.

Guth, A. 1986. In *Inner Space/Outer Space*, ed. E. W. Kolb, M. S. Turner, D. Lindley, K. Olive, & D. Seckel, p. 287. Chicago: University of Chicago Press.

Hainebach, K. L., & Schramm, D. N. 1977. *Astrophys. J.*, **212**, 347.

Harris, W. E., & Hesser, J. E. 1976. *Publ. Astr. Soc. Pacific*, **88**, 377.

Harris, W. E., & Hesser, J. E. 1987. In preparation.

Harris, W. E., Hesser, J. E., & Atwood, B. 1983. *Astrophys. J. Lett.*, **268**, L111.

Hartwick, F.D.A. 1976. *Astrophys. J.*, **209**, 418.

Hartwick, F.D.A., & Sargent, W.L.W. 1978. *Astrophys. J.*, **221**, 512.

Heggie, D. C. 1975. *Mon. Not. Roy. Astron. Soc.*, **173**, 729.

Hegyi, D. J., & Olive, K. A. 1983. *Phys. Lett.*, **126B**, 28.

Heisler, J., Merritt, D., & Schwarzschild, M. 1982. *Astrophys. J.*, **258**, 490.

Heisler, J., Tremaine, S., & Bahcall, J. N. 1985. *Astrophys. J.*, **298**, 8.

Hénon, M. 1959. *Ann. d'Astrophys*, **22**, 126.

Hénon, M. 1960a. *Ann. d'Astrophys*, **23**, 467, 474.

Hénon, M. 1960b. *Ann. d'Astrophys*, **23**, 668.

Hénon, M. 1961. *Ann. d'Astrophys*, **24**, 369.

Hénon, M. 1969a. *Astron. Astrophys.*, **1**, 223.

Hénon, M. 1969b. *Astron. Astrophys.*, **2**, 151.

Hénon, M. 1970. *Astron. Astrophys.*, **9**, 24.

Hénon, M. 1972. In *Gravitational N-Body Problem*, IAU Colloquium No. 10, ed. M. Lecar, pp. 44, 406. Dordrecht: Reidel. Also in *Astrophys. Sp. Sci.*, **13**, 284 & **14**, 151.

Hénon, M. 1973a. In *Dynamical Structure and Evolution of Stellar Systems*, Third Advanced Course of the Swiss Society of Astronomy and Astrophysics, ed. L. Martinet & M. Mayor, p. 183. Sauverny: Geneva Observatory.

Hénon, M. 1973b. *Astron. Astrophys.*, **24**, 229.

Hénon, M. 1975. In *Dynamics of Stellar Systems*, IAU Symposium No. 69, ed. A. Hayli, p. 133. Dordrecht: Reidel.

Hénon, M. 1983. *Astron. Astrophys.*, **114**, 211.

Hernquist, L., & Quinn, P. J. 1987. *Astrophys. J.*, **312**, 1.

Hockney, R. W., & Brownrigg, D.R.K. 1974. *Mon. Not. Roy. Astron. Soc.*, **167**, 351.

Hohl, F. 1971. *Astrophys. J.*, **168**, 343.

Hohl, F., & Zang, T. A. 1979. *Astron. J.*, **84**, 585.

Holmberg, E. 1937. *Ann. Lund Obs.*, **6**.

Horwitz, G., & Katz, J. 1978. *Astrophys. J.*, **222**, 941.

Hubble, E. P. 1930. *Astrophys. J.*, **71**, 231.

Hubble, E. P. 1936. *The Realm of the Nebulae.* New Haven, Conn.: Yale University Press.

Huchra, J. P., & Geller, M. J. 1982. *Astrophys. J.*, **257**, 423. See correction in Geller (1984).

Huchtmeier, W. K., & Bohnenstengel, U.-D. 1975. *Astron. Astrophys.*, **41**, 477.

Hunter, C. 1963. *Mon. Not. Roy. Astron. Soc.*, **126**, 299.

Hunter, C. 1970. *Astrophys. J.*, **162**, 97.

Hunter, C. 1975. *Astron. J.*, **80**, 783.

Hunter, C. 1977a. *Astron. J.*, **82**, 271.

Hunter, C. 1977b. *Astrophys. J.*, **213**, 497.

Hunter, C., & Toomre, A. 1969. *Astrophys. J.*, **155**, 747.

Hut, P. 1983. *Astrophys. J. Lett.*, **272**, L29.

Hut, P., & Bahcall, J. N. 1983. *Astrophys. J.*, **268**, 319.

Hut, P., & Tremaine S. 1985. *Astron. J.*, **90**, 1548.

Icke, V. 1982. *Astrophys. J.*, **254**, 517.

Ikeuchi, S., Nakamura, T., & Takahara, F. 1974. *Prog. Theor. Phys.*, **52**, 1807.

Illingworth, G. 1981. In *The Structure and Evolution of Normal Galaxies*, ed. S. M. Fall & D. Lynden-Bell, p. 27. Cambridge, Eng.: Cambridge University Press.

Inagaki, S. 1980. *Publ. Astr. Soc. Japan*, **32**, 213.

Innanen, K. A., Harris, W. E., & Webbink, R. F. 1983. *Astron. J.*, **88**, 338.

Innanen, K. A., Kamper, K. W., Papp, K. A., & van den Bergh, S. 1982. *Astrophys. J.*, **254**, 515.

Ipser, J. R., & Kandrup, H. E. 1980. *Astrophys. J.*, **241**, 1141.

Ipser, J. R., & Managan, R. A. 1981. *Astrophys. J.*, **250**, 362.

Jackson, J. D. 1975. *Classical Electrodynamics.* New York: Wiley.

Jacobi, C.G.J. 1834. *Poggendorff Annalen der Physik u. Chemie*, **33**, 229.

Jaffe, W. 1983. *Mon. Not. Roy. Astron. Soc.*, **202**, 995.

James, R. A. 1964. *Astrophys. J.*, **140**, 552.

Janes, K. A. 1975. *Astrophys. J. Suppl.*, **29**, 161.

Jeans, J. H. 1915. *Mon. Not. Roy. Astron. Soc.*, **76**, 70.

Jeans, J. H. 1919. *Phil. Trans. Roy. Soc. London A*, **218**, 157.

Jeans, J. H. 1929. *Astronomy and Cosmogony*, 2nd ed. Cambridge, Eng.: Cambridge University Press.

Jeans, J. H. 1940. *Kinetic Theory of Gases.* Cambridge, Eng.: Cambridge University Press.

Jefferys, W. H. 1974. *Astron. J.*, **79**, 710.

Jefferys, W. H. 1976. *Astron. J.*, **81**, 983.

Jones, C., & Forman, W. 1984. *Astrophys. J.*, **276**, 38.

Julian, W. H., & Toomre, A. 1966. *Astrophys. J.*, **146**, 810.

Kahn, F. D., & Woltjer, L. 1959. *Astrophys. J.*, **130**, 705.

Kalnajs, A. J. 1965. Unpublished Ph.D. thesis, Harvard University.

Kalnajs, A. J. 1971. *Astrophys. J.*, **166**, 275.

Kalnajs, A. J. 1972a. *Astrophys. Lett.*, **11**, 41.

Kalnajs, A. J. 1972b. *Astrophys. J.*, **175**, 63.

Kalnajs, A. J. 1972c. In *Gravitational N-body Problem*, IAU Colloquium No. 10, ed. M. Lecar, p. 13. Dordrecht: Reidel.

Kalnajs, A. J. 1973. *Proc. Astron. Soc. Australia*, **2**, 174.

Kalnajs, A. J. 1976. *Astrophys. J.*, **205**, 745 & 751.

Kalnajs, A. J. 1978. In *Structure and Properties of Nearby Galaxies*, IAU Symposium No. 77, ed. E. M. Berkhuijsen & R. Wielebinski, p. 113. Dordrecht: Reidel.

Kalnajs, A. J. 1983. In *Internal Kinematics and Dynamics of Galaxies*, IAU Symposium No. 100, ed. E. Athanassoula, pp. 87, 109. Dordrecht: Reidel.

Kalnajs, A. J., & Athanassoula-Georgala, E. 1974. *Mon. Not. Roy. Astron. Soc.*, **168**, 287.

Kandrup, H. E., & Sygnet, J. F. 1985. *Astrophys. J.*, **298**, 27.

Katz, J. 1978. *Mon. Not. Roy. Astron. Soc.*, **183**, 765.

Keenan, D. W. 1980. *Astron. Astrophys.*, **95**, 334.

Kellogg, O. D. 1953. *Foundations of Potential Theory*. New York: Dover.

Kennicutt, R. C. 1981. *Astron. J.*, **86**, 1847.

Kent, S. M., & Gunn, J. E. 1982. *Astron. J.*, **87**, 945.

Kent, S. M., & Sargent, W.L.W. 1983. *Astron. J.*, **88**, 692.

King, C. R., & Ellis, R. S. 1985. *Astrophys. J.*, **288**, 456.

King, I. R. 1958. *Astron. J.*, **63**, 114.

King, I. R. 1962. *Astron. J.*, **67**, 471.

King, I. R. 1966. *Astron. J.*, **71**, 64.

King, I. R. 1981. *Quart. J. Roy. Astron. Soc.*, **22**, 227.

Kirshner, R. P., Oemler, A., Schechter, P. L., & Shectman, S. A. 1983. *Astron. J.*, **88**, 1285.

Koo, D. C. 1981. *Astrophys. J. Lett.*, **251**, L75.

Kormendy, J. 1977. *Astrophys. J.*, **218**, 333.

Kormendy, J. 1981. In *Structure and Evolution of Normal Galaxies*, ed. S. M. Fall & D. Lynden-Bell, p. 85. Cambridge, Eng.: Cambridge University Press.

Kormendy, J. 1982. In *Morphology and Dynamics of Galaxies*, Twelfth Advanced Course of the Swiss Society of Astronomy and Astrophysics, ed. L. Martinet & M. Mayor, p. 113. Sauverny: Geneva Observatory.

Kormendy, J. 1987. In *Dark Matter in the Universe*, IAU Symposium No. 117, ed. J. Kormendy & G. R. Knapp, p. 139. Dordrecht: Reidel.

Kormendy, J., & Illingworth, G. 1982. *Astrophys. J.*, **256**, 460.

Kormendy, J., & Knapp, G. R., ed. 1987. *Dark Matter in the Universe*, IAU Symposium No. 117. Dordrecht: Reidel.

Kormendy, J., & Norman, C. 1979. *Astrophys. J.*, **233**, 539.

Kraan-Korteweg, R. C., & Tammann, G. A. 1979. *Astron. Nachr.*, **300**, 181.

Krall, N. A., & Trivelpiece, A. W. 1973. *Principles of Plasma Physics*. New York: McGraw-Hill.

Kron, R. G. 1980. *Astrophys. J. Suppl.*, **43**, 305.

Kulsrud, R. M., & Mark, J.W.-K. 1970. *Astrophys. J.*, **160**, 471.

Kuzmin, G. 1952. *Publ. Astr. Obs. Tartu*, **32**, 211.

Kuzmin, G. 1956. *Astron. Zh.*, **33**, 27.

Kwee, K. K., Muller, C. A., & Westerhout, G. 1954. *Bull. Astron. Inst. Netherlands*, **12**, 211.

Lacey, C. G. 1984. *Mon. Not. Roy. Astron. Soc.*, **208**, 687.

Lacey, C. G., & Fall, S. M. 1983. *Mon. Not. Roy. Astron. Soc.*, **204**, 791.

Lacey, C. G., & Ostriker, J. P. 1985. *Astrophys. J.*, **299**, 633.

Lake, G. 1981. *Astrophys. J.*, **243**, 111, 121.

Landau, L. D. 1932. *Phys. Z. Sowjet.*, **1**, 285. [translated in D. ter Haar, ed. (1965), *Collected Papers of L. D. Landau*, p. 60 (New York: Gordon & Breach)].

Landau, L. D. 1946. *J. Phys. (USSR)*, **10**, 25 [translated in D. ter Haar (1969), *Men of Physics: L. D. Landau*, vol. 2 (Oxford: Pergamon)].

Landau, L. D., & Lifshitz, E. M. 1959. *Fluid Mechanics*. Oxford: Pergamon.

Landau, L. D., & Lifshitz, E. M. 1976. *Mechanics*, 3rd ed. Oxford: Pergamon.

Landau, L. D., & Lifshitz, E. M. 1980. *Statistical Physics*, 3rd ed. Oxford: Pergamon.

Lang, K. R. 1980. *Astrophysical Formulae*, 2nd ed. New York: Springer.

Laplace, P. S. 1802. *Celestial Mechanics*, trans. N. Bowditch, vol. 2, chap. 6. New York: Chelsea.

Larson, R. B. 1970a. *Mon. Not. Roy. Astron. Soc.*, **147**, 323.

Larson, R. B. 1970b. *Mon. Not. Roy. Astron. Soc.*, **150**, 93.

Larson, R. B. 1974. *Mon. Not. Roy. Astron. Soc.*, **169**, 229.

Larson, R. B. 1976a. *Mon. Not. Roy. Astron. Soc.*, **176**, 31.

Larson, R. B. 1976b. In *Galaxies*, Sixth Advanced Course of the Swiss Society of Astronomy and Astrophysics, ed. L. Martinet & M. Mayor, p. 67. Sauverny: Geneva Observatory.

Larson, R. B. 1986. *Mon. Not. Roy. Astron. Soc.*, **218**, 409.

Larson, R. B., & Tinsley, B. M. 1978. *Astrophys. J.*, **219**, 46.

Larson, R. B., Tinsley, B. M., & Caldwell, C. N. 1980. *Astrophys. J.*, **237**, 692.

Latham, D. W., Tonry, J., Bahcall, J. N., Soneira, R. M., & Schechter, P. L. 1984. *Astrophys. J. Lett.*, **281**, L41.

Lauer, T. R. 1985. *Astrophys. J.*, **292**, 104.

Laval, G., Mercier, C., & Pellat, R. 1965. *Nuclear Fusion*, **5**, 156.

Lebovitz, N. R. 1965. *Astrophys. J.*, **142**, 229.

Ledoux, P., & Walraven, T. 1958. In *Handbuch der Physik*, **51**, ed. S. Flugge, p. 353. Berlin: Springer-Verlag.

Lee, H.-M., & Ostriker, J. P. 1986. *Astrophys. J.*, **310**, 176.

Lichtenberg, A. J., & Lieberman, M. A. 1983. *Regular and Stochastic Motion.* New York: Springer.

Lifshitz, E. M., & Pitaevskii, L. P. 1981. *Physical Kinetics.* Oxford: Pergamon.

Lightman, A. P., & Grindlay, J. E. 1982. *Astrophys. J.,* **267**, 145.

Lightman, A. P., & Shapiro, S. L. 1978. *Rev. Mod. Phys.,* **50**, 437.

Limber, D. N., & Mathews, W. G. 1960. *Astrophys. J.,* **132**, 286.

Lin, C. C., & Shu, F. H. 1964. *Astrophys. J.,* **140**, 646.

Lin, C. C., & Shu, F. H. 1966. *Proc. Nat. Acad. Sci.,* **55**, 229.

Lin, C. C., Yuan, C., & Shu, F. H. 1969. *Astrophys. J.,* **155**, 721.

Lin, D.N.C., & Lynden-Bell, D. 1982. *Mon. Not. Roy. Astron. Soc.,* **198**, 707.

Lin, D.N.C., & Tremaine, S. 1983. *Astrophys. J.,* **264**, 364.

Linde, A. D. 1982. *Phys. Letters,* **114B**, 431.

Lindoff, U. 1968. *Ark. Astron.,* **5**, 1.

Liszt, H. S., & Burton, W. B. 1980. *Astrophys. J.,* **236**, 779.

Lucey, J. R., Dickens, R. J., Mitchell, R. J., & Dawe, J. A. 1983. *Mon. Not. Roy. Astron. Soc.,* **203**, 545.

Lucy, L. B. 1974. *Astron. J.,* **79**, 745.

Lynden-Bell, D. 1960. *Mon. Not. Roy. Astron. Soc.,* **120**, 204.

Lynden-Bell, D. 1962a. *Mon. Not. Roy. Astron. Soc.,* **123**, 447.

Lynden-Bell, D. 1962b. *Mon. Not. Roy. Astron. Soc.,* **124**, 1.

Lynden-Bell, D. 1962c. *Mon. Not. Roy. Astron. Soc.,* **124**, 279.

Lynden-Bell, D. 1967a. *Mon. Not. Roy. Astron. Soc.,* **136**, 101.

Lynden-Bell, D. 1967b. In *Relativity Theory and Astrophysics,* vol. 2, *Galactic Structure,* ed. J. Ehlers, p. 131. Providence, R.I.: Am. Math. Soc.

Lynden-Bell, D. 1982. In *Astrophysical Cosmology,* ed. H. A. Brück, G. V. Coyne, & M. S. Longair, p. 85. Vatican City: Pont. Acad. Scient.

Lynden-Bell, D., & Eggleton, P. P. 1980. *Mon. Not. Roy. Astron. Soc.,* **191**, 483.

Lynden-Bell, D., & Kalnajs, A. J. 1972. *Mon. Not. Roy. Astron. Soc.,* **157**, 1.

Lynden-Bell, D., & Ostriker, J. P. 1967. *Mon. Not. Roy. Astron. Soc.,* **136**, 293.

Lynden-Bell, D., & Wood, R. 1968. *Mon. Not. Roy. Astron. Soc.,* **138**, 495.

Lynden-Bell, D., Cannon, R. D., & Godwin, P. J. 1983. *Mon. Not. Roy. Astron. Soc.,* **204**, 87P.

Lynds, R., & Toomre, A. 1976. *Astrophys. J.,* **209**, 382.

Lyttleton, R. A. 1953. *Theory of Rotating Fluid Masses.* Cambridge, Eng.: Cambridge University Press.

McCarthy, D. W., Probst, R. G., & Low, F. J. 1985. *Astrophys. J.,* **290**, L9.

McGlynn, T. 1982. Unpublished Ph.D. thesis, Princeton University.

McGlynn, T. 1984. *Astrophys. J.,* **281**, 13.

Margenau, H., & Murphy, G. M. 1956. *Mathematics of Physics and Chemistry.* New York: Van Nostrand.

Mark, J.W.-K. 1974. *Astrophys. J.,* **193**, 539.

Mark, J.W.-K. 1976. *Astrophys. J.,* **205**, 363.

Marochnik, L. S. 1968. *Sov. Astron., A. J.*, **11**, 873.

Mathews, J., & Walker, R. 1970. *Mathematical Methods of Physics*, 2nd ed. Menlo Park, Calif.: W. A. Benjamin.

Mathewson, D. S., van der Kruit, P. C., & Brouw, W. N. 1972. *Astron. Astrophys.*, **17**, 468.

Merritt, D. 1980. *Astrophys. J. Suppl.*, **43**, 435.

Merritt, D. 1981. *Astron. J.*, **86**, 318.

Merritt, D. 1985a. *Astron. J.*, **90**, 1027.

Merritt, D. 1985b. *Mon. Not. Roy. Astron. Soc.*, **214**, 25P.

Merritt, D. 1987a. *Astrophys. J.*, **313**, 121.

Merritt, D. 1987b. In *Structure and Dynamics of Elliptical Galaxies*, IAU Symposium No. 127, ed. T. de Zeeuw, p. 315. Dordrecht: Reidel.

Merritt, D., & Aguilar, L. A. 1985. *Mon. Not. Roy. Astron. Soc.*, **217**, 787.

Mestel, L. 1963. *Mon. Not. Roy. Astron. Soc.*, **126**, 553.

Michard, R. 1980. *Astron. Astrophys.*, **91**, 122.

Michie, R. W. 1963. *Mon. Not. Roy. Astron. Soc.*, **125**, 127.

Michie, R. W., & Bodenheimer, P. H. 1963. *Mon. Not. Roy. Astron. Soc.*, **126**, 269.

Mihalas, D., & Binney, J. J. 1981. *Galactic Astronomy*, 2nd ed. San Francisco: Freeman.

Milgrom, M. 1983. *Astrophys. J.*, **270**, 365.

Miller, R. H. 1971. *Astrophys. Space Sci.*, **14**, 73.

Miller, R. H. 1978. *Astrophys. J.*, **223**, 122.

Miller, R. H., & Smith, B. F. 1979. *Astrophys. J.*, **227**, 785.

Mishra, R. 1985. *Mon. Not. Roy. Astron. Soc.*, **212**, 163.

Miyamoto, M., & Nagai, R. 1975. *Publ. Astron. Soc. Japan*, **27**, 533.

Moser, J. 1973. *Stable and Random Motions in Dynamical Systems*. Princeton, N.J.: Princeton University Press.

Mould, J. R. 1978. *Astrophys. J.*, **226**, 923.

Mueller, M. W., & Arnett, W. D. 1976. *Astrophys. J.*, **210**, 670.

Mulder, W. A. 1983. *Astron. Astrophys.*, **117**, 9.

Murai, T., & Fujimoto, M. 1980. *Publ. Astron. Soc. Japan*, **32**, 581.

Negroponte, J., & White, S.D.M. 1983. *Mon. Not. Roy. Astron. Soc.*, **205**, 1009.

Newton, A. J. 1986. Unpublished D. Phil. thesis, Oxord University.

Newton, A. J., & Binney, J. J. 1984. *Mon. Not. Roy. Astron. Soc.*, **210**, 711.

Nolthenius, R., & Ford, H. 1986. *Astrophys. J.*, **305**, 600.

Ogorodnikov, K. F. 1965. *Dynamics of Stellar Systems*. Oxford: Pergamon.

Oort, J. H. 1932. *Bull. Astron. Inst. Netherlands*, **6**, 349.

Oort, J. H. 1958. In *Stellar Populations*. ed. D.J.K. O'Connell. Amsterdam: North Holland.

Oort, J. H. 1965. In *Galactic Structure*, ed. A. Blaauw & M. Schmidt, p. 455. Chicago: University of Chicago Press.

Oort, J. H. 1977. *Ann. Rev. Astron. Astrophys.*, **15**, 295.

Osipkov, L. P. 1979. *Pis'ma Astr. Zh.*, **5**, 77.

Ostriker, J. P. 1980. *Comments on Astrophys*, **8**, 177.

Ostriker, J. P. 1985. In *Dynamics of Star Clusters*, IAU Symposium No. 113, ed. J. Goodman & P. Hut, p. 347. Dordrecht: Reidel.

Ostriker, J. P., & Peebles, P.J.E. 1973. *Astrophys. J.*, **186**, 467.

Ostriker, J. P., Peebles, P.J.E., & Yahil, A. 1974. *Astrophys. J. Lett.*, **193**, L1.

Ostriker, J. P., Spitzer, L., & Chevalier, R. A. 1972. *Astrophys. J. Lett.*, **176**, L51.

Ostriker, J. P., & Thuan, T. X. 1975. *Astrophys. J.*, **202**, 353.

Ostriker, J. P., & Turner, E. L. 1979. *Astrophys. J.*, **234**, 785.

Pagel, B.E.J. 1986. In *Highlights of Astronomy*, ed. J.-P. Swings, p. 51. Dordrecht: Reidel.

Palmer, P. L., & Papaloizou, J. 1987. *Mon. Not. Roy. Astron. Soc.*, **224**, 1043.

Particle Data Group. 1984. *Rev. Mod. Phys.*, **56**, S1.

Peebles, P.J.E. 1971. *Physical Cosmology.* Princeton, N.J.: Princeton University Press.

Peebles, P.J.E. 1974. *Astrophys. J.*, **189**, L54.

Peebles, P.J.E. 1980a. In *Physical Cosmology*, ed. R. Balian, J. Audouze, & D. N. Schramm, p. 213. Amsterdam: North-Holland.

Peebles, P.J.E. 1980b. *The Large-Scale Structure of the Universe.* Princeton, N.J.: Princeton University Press.

Peebles, P.J.E. 1984. *Astrophys. J.*, **284**, 439.

Peimbert, M., & Serrano, A. 1982. *Mon. Not. Roy. Astron. Soc.*, **198**, 563.

Pence, W. D. 1986. *Astrophys. J.*, **310**, 597.

Peterson, C. J., & King, I. R. 1975. *Astron. J.*, **80**, 427.

Pfenniger, D. 1984a. *Astron. Astrophys.*, **134**, 373.

Pfenniger, D. 1984b. *Astron. Astrophys.*, **141**, 171.

Piran, T., & Villumsen, J. V. 1987. In *Structure and Dynamics of Elliptical Galaxies*, IAU Symposium No. 127, ed. T. de Zeeuw, p. 473. Dordrecht: Reidel.

Plummer, H. C. 1911. *Mon. Not. Roy. Astron. Soc.*, **71**, 460.

Prendergast, K. H., & Tomer, E. 1970. *Astron. J.*, **75**, 674.

Press, W. H., & Teukolsky, S. A. 1973. *Astrophys. J.*, **181**, 513.

Press, W. H., & Teukolsky, S. A. 1977. *Astrophys. J.*, **213**, 183.

Press, W. H., Flannery, B. P., Teukolsky, S. A., & Vetterling, W. T. 1986. *Numerical Recipes.* Cambridge, Eng.: Cambridge University Press.

Primack, J. R. 1987. In *Proceedings of the International School of Physics "Enrico Fermi"*, vol 92, ed. N. Cabibbo. Bologna: Italian Physical Society.

Quinn, P. J. 1984. *Astrophys. J.*, **279**, 596.

Reichl, L. E. 1980. *A Modern Course in Statistical Physics.* Austin: University of Texas Press.

Reynolds, J. H. 1913. *Mon. Not. Roy. Astron. Soc.*, **74**, 132.

Richstone, D. O. 1975. *Astrophys. J.*, **200**, 535.

Richstone, D. O. 1980. *Astrophys. J.*, **238**, 103.

Richstone, D. O. 1982. *Astrophys. J.*, **252**, 496.

Richstone, D. O. 1987. In *Structure and Dynamics of Elliptical Galaxies*, IAU Symposium No. 127, ed. T. de Zeeuw, p. 261. Dordrecht: Reidel.

Richstone, D. O., & Potter, M. D. 1982. *Astrophys. J.*, **254**, 451.

Richstone, D. O., & Tremaine, S. 1984. *Astrophys. J.*, **286**, 27.

Richstone, D. O., & Tremaine, S. 1986. *Astron. J.*, **92**, 72.

Roberts, P. H. 1962. *Astrophys. J.*, **136**, 1108.

Roberts, W. W. 1969. *Astrophys. J.*, **158**, 123.

Roberts, W. W. 1979. In *The Large-Scale Characteristics of the Galaxy*, IAU Symposium No. 84, ed. W. B. Burton, p. 175. Dordrecht: Reidel.

Roberts, W. W., Huntley, J. M., & van Albada, G. D. 1979. *Astrophys. J.*, **233**, 67.

Rogstad, D. H., & Shostak, G. S. 1972. *Astrophys. J.*, **176**, 315.

Rood, H. J. 1982. *Astrophys. J. Suppl.*, **49**, 111.

Rood, H. J., Page, T. L., Kintner, E. C., & King, I. R. 1972. *Astrophys. J.*, **175**, 627.

Rosenbluth, M. N., MacDonald, W. M., & Judd, D. L. 1957. *Phy. Rev.*, **107**, 1.

Rubin, V. C. 1983. *Science*, **220**, 1339.

Rubin, V. C., & Ford, W. K. 1970. *Astrophys. J.*, **159**, 379.

Rubin, V. C., Burstein, D., Ford, W. K., & Thonnard, N. 1985. *Astrophys. J.*, **289**, 81.

Rubin, V. C., Ford, W. K., & Thonnard, N. 1980. *Astrophys. J.*, **238**, 471.

Rubin, V. C., Ford, W. K., Thonnard, N., & Burstein, D. 1982. *Astrophys. J.*, **261**, 439.

Safronov, V. S. 1960. *Ann. d'Astrophysique*, **23**, 979.

Sancisi, R. 1976. *Astron. Astrophys.*, **53**, 159.

Sandage, A. 1961. *The Hubble Atlas of Galaxies*. Washington, D.C.: Carnegie Institution.

Sandage, A., & Brucato, R. 1979. *Astron. J.*, **84**, 472.

Sandage, A., & Tammann, G. A. 1975. *Astrophys. J.*, **197**, 265.

Sandage, A., & Tammann, G. A. 1981. *A Revised Shapley-Ames Catalog of Bright Galaxies*. Washington, D.C.: Carnegie Institution.

Sanders, R. H. 1977. *Astrophys. J.*, **216**, 916.

Sanders, R. H. 1984. *Astron. Astrophys.*, **136**, L21.

Sanders, R. H., & Huntley, J. M. 1976. *Astrophys. J.*, **209**, 53.

Sarazin, C. 1986. *Rev. Mod. Phys.*, **58**, 1.

Sargent, W.L.W., Young, P. J., Boksenberg, A., Shortridge, K., Lynds, C. R., & Hartwick, F.D.A. 1978. *Astrophys. J.*, **221**, 731.

Sargent, W.L.W., Young, P. J., Boksenberg, A., & Tytler, D. 1980. *Astrophys. J. Suppl.*, **42**, 41.

Satoh, C. 1980. *Publ. Astron. Soc. Japan*, **32**, 41.

Schechter, P. L. 1980. *Astron. J.*, **85**, 801.

Schmidt, M. 1956. *Bull. Astron. Inst. Netherlands*, **13**, 15.

Schmidt, M. 1975. *Astrophys. J.*, **202**, 22.

Schramm, D. N. 1974. *Ann. Rev. Astron. Astrophys.*, **12**, 383.

Schuster, A., 1883. *British Assoc. Report*, p. 427.

Schwarz, M. P. 1981. *Astrophys. J.*, **247**, 77.

Schwarzschild, K. 1907. *Göttingen Nachr.*, 614.

Schwarzschild, M. 1954. *Astron. J.*, **59**, 273.

Schwarzschild, M. 1979. *Astrophys. J.*, **232**, 236.

Schwarzschild, M. 1982. *Astrophys. J.*, **263**, 599.

Schwarzschild, M., & Bernstein, S. 1955. *Astrophys. J.*, **122**, 200.

Schweizer, F. 1975. In *La Dynamique des Galaxies Spirales*, ed. L. Weliachew, p. 337. Paris: CNRS.

Schweizer, F. 1982. *Astrophys. J.*, **252**, 455.

Seiden, P. E., & Gerola, H. 1982. *Fund. Cosmic Phys.*, **7**, 241.

Sellwood, J. A. 1980. *Astron. Astrophys.*, **89**, 296.

Sellwood, J. A. 1981. *Astron. Astrophys.*, **99**, 362.

Sellwood, J. A. 1985. *Mon. Not. Roy. Astron. Soc.*, **217**, 127.

Sellwood, J. A., & Carlberg, R. G. 1984. *Astrophys. J.*, **282**, 61.

Shapiro, S. L. 1985. In *Dynamics of Star Clusters*, IAU Symposium No. 113, ed. J. Goodman & P. Hut, p. 373. Dordrecht: Reidel.

Shiveshwarkar, S. W. 1936. *Mon. Not. Roy. Astron. Soc.*, **96**, 749.

Shu, F. H. 1969. *Astrophys. J.*, **158**, 505.

Shu, F. H. 1970. *Astrophys. J.*, **160**, 99.

Shu, F. H. 1978. In *Structure and Properties of Nearby Galaxies*, IAU Symposium No. 77, ed. E. Berkhuisen & R. Wielebinski, p. 139. Dordrecht: Reidel.

Shu, F. H. 1982. *The Physical Universe.* Mill Valley, Calif.: University Science Books.

Smith, S. 1936. *Astrophys. J.*, **83**, 23.

Smith, W. M. 1979. *Astron. J.*, **84**, 979.

Smoot, G. F., & Lubin, P. M. 1979. *Astrophys. J.*, **234**, L83.

Sparke, L. S. 1984. *Astrophys. J.*, **280**, 117.

Spitzer, L. 1940. *Mon. Not. Roy. Astron. Soc.*, **100**, 396.

Spitzer, L. 1942. *Astrophys. J.*, **95**, 329.

Spitzer, L. 1958. *Astrophys. J.*, **127**, 17.

Spitzer, L. 1969. *Astrophys. J. Lett.*, **158**, L139.

Spitzer, L. 1975. In *Dynamics of Stellar Systems*, IAU Symposium No. 69, ed. A. Hayli, p. 3. Dordrecht: Reidel.

Spitzer, L. 1978. *Physical Processes in the Interstellar Medium.* New York: Wiley.

Spitzer, L. 1987. *Dynamical Evolution of Globular Clusters.* Princeton, N.J.: Princeton University Press.

Spitzer, L., & Chevalier, R. A. 1973. *Astrophys. J.*, **183**, 565.

Spitzer, L., & Hart, M. H. 1971. *Astrophys. J.*, **164**, 399.

Spitzer, L., & Mathieu, R. D. 1980. *Astrophys. J.*, **241**, 618.

Spitzer, L., & Schwarzschild, M. 1953. *Astrophys. J.*, **118**, 106.

Spitzer, L., & Shapiro, S. L. 1972. *Astrophys. J.*, **173**, 529.

Spitzer, L., & Thuan, T. X. 1972. *Astrophys. J.*, **175**, 31.

Stark, A. A. 1977. *Astrophys. J.*, **213**, 368.

Statler, T. 1986. Unpublished Ph.D. thesis, Princeton University.

Stewart, G. C., Canizares, C. R., Fabian, A. C., & Nulsen, P.E.J. 1984. *Astrophys. J.*, **278**, 536.

Strom, S. E., Jensen, E. B., & Strom, K. M. 1976. *Astrophys. J. Lett.*, **206**, L11.

Sweet, P. A. 1963. *Mon. Not. Roy. Astron. Soc.*, **125**, 285.

Sygnet, J. F., Des Forets, G., Lachieze-Rey, M., & Pellat, R. 1984. *Astrophys. J.*, **276**, 737.

Szebehely, V. G. 1967. *Theory of Orbits.* New York: Academic Press.

Takahara, F. 1976. *Prog. Theor. Phys.*, **56**, 1665.

Tassoul, J. L. 1978. *Theory of Rotating Stars.* Princeton, N.J.: Princeton University Press.

Terlevich, R., Davies, R. L., Faber, S. M., & Burstein, D. 1981. *Mon. Not. Roy. Astron. Soc.*, **196**, 381.

Thuan, T. X., Hart, M. H., & Ostriker, J. P. 1975. *Astrophys. J.*, **201**, 756.

Tinsley, B. M. 1972. *Astrophys. J. Lett.*, **173**, L93.

Tinsley, B. M. 1980a. *Fund. Cosmic Physics*, **5**, 287.

Tinsley, B. M. 1980b. *Astrophys. J.*, **241**, 41.

Tinsley, B. M. 1981. *Astrophys. J.*, **250**, 758.

Tinsley, B. M., & Gunn, J. E. 1976. *Astrophys. J.*, **203**, 52.

Tonry, J. L., & Davis, M. 1981. *Astrophys. J.*, **246**, 680.

Toomre, A. 1962. *Astrophys. J.*, **138**, 385.

Toomre, A. 1964. *Astrophys. J.*, **139**, 1217.

Toomre, A. 1969. *Astrophys. J.*, **158**, 899.

Toomre, A. 1974. In *Highlights of Astronomy*, ed. G. Contopoulos, p. 457. Dordrecht: Reidel.

Toomre, A. 1977a. *Ann. Rev. Astron. Astrophys.*, **15**, 437.

Toomre, A. 1977b. In *The Evolution of Galaxies and Stellar Populations*, ed. B. M. Tinsley & R. B. Larson, p. 401. New Haven, Conn.: Yale University Observatory.

Toomre, A. 1978. In *The Large-Scale Structure of the Universe*, IAU Symposium No. 79, ed. M. S. Longair & J. Einasto, p. 109. Dordrecht: Reidel.

Toomre, A. 1981.   In *The Structure and Evolution of Normal Galaxies*, ed. S. M. Fall & D. Lynden-Bell, p. 111. Cambridge, Eng.: Cambridge University Press.

Toomre, A. 1982. *Astrophys. J.*, **259**, 535.

Toomre, A. 1983. In *Internal Kinematics and Dynamics of Galaxies*, IAU Symposium No. 100, ed. E. Athanassoula, p. 177. Dordrecht: Reidel.

Toomre, A., & Toomre, J. 1972. *Astrophys. J.*, **178**, 623.

Tremaine, S. 1976. *Astrophys. J.*, **203**, 72.

Tremaine, S., & Gunn, J. E. 1979. *Phys. Rev. Lett.*, **42**, 407.

Tremaine, S., & Weinberg, M. 1984a. *Mon. Not. Roy. Astron. Soc.*, **209**, 729.

Tremaine, S., & Weinberg, M. 1984b. *Astrophys. J. Lett.*, **282**, L5.

Tremaine, S., Hénon, M., & Lynden-Bell, D. 1986. *Mon. Not. Roy. Astron. Soc.*, **219**, 285.

Tremaine, S., Ostriker, J. P., & Spitzer, L. 1975. *Astrophys. J.*, **196**, 407.

Trimble, V. 1987. *Ann. Rev. Astron. Astrophys.*, in press.

Turner, E. L. 1976. *Astrophys. J.*, **208**, 20.

Turner, E. L. 1987. In *Dark Matter in the Universe*, IAU Symposium No. 117, ed. J. Kormendy & G. R. Knapp, p. 227. Dordrecht: Reidel.

Turner, E. L., & Gott, J. R. 1976. *Astrophys. J. Suppl.*, **32**, 409.

Twarog, B. A. 1980. *Astrophys. J.*, **242**, 242.

Unsöld, A., & Baschek, B. 1983. *The New Cosmos*, 3rd ed. New York: Springer.

Vader, J. P., & de Jong T. 1981. *Astron. Astrophys.*, **100**, 124.

van Albada, T. S. 1982. *Mon. Not. Roy. Astron. Soc.*, **201**, 939.

van Albada, T. S., Bahcall, J. N., Begeman, K., & Sancisi, R. 1985. *Astrophys. J.*, **295**, 305.

van Albada, T. S., Kotanyi, C. G., & Schwarzschild, M. 1981. *Mon. Not. Roy. Astron. Soc.*, **198**, 303.

VandenBerg, D. A. 1983. *Astrophys. J. Suppl.*, **51**, 29.

van den Bergh, S. 1971. *Astron. Astrophys.*, **11**, 154.

van der Kruit, P. C., & Freeman, K. C. 1984. *Astrophys. J.*, **278**, 81.

van der Kruit, P. C., & Searle, L. 1981. *Astron. Astrophys.*, **95**, 105.

van der Kruit, P. C., & Searle, L. 1982. *Astron. Astrophys.*, **110**, 61.

Vandervoort, P. O. 1980. *Astrophys. J.*, **240**, 478.

van Kampen, N. G. 1955. *Physica*, **21**, 949.

van Woerden, H. 1979. In *The Large-Scale Characteristics of the Galaxy*, IAU Symposium No. 84, ed. W. B. Burton, p. 510. Dordrecht: Reidel.

Villumsen, J. V. 1982. *Mon. Not. Roy. Astron. Soc.*, **199**, 493.

Villumsen, J. V. 1983. *Astrophys. J.*, **274**, 632.

Villumsen, J. V. 1985. *Astrophys. J.*, **290**, 75.

Villumsen, J. V., & Binney, J. J. 1985. *Astrophys. J.*, **295**, 388.

Visser, H.C.D. 1980. *Astron. Astrophys.*, **88**, 159.

von Hörner, S. 1958. *Z. Astrophys.*, **44**, 221.

Vorontsov-Velyaminov, B. A. 1968. *Morphological Catalog of Galaxies*. Moscow: Moscow University Press.

Walsh, D., Carswell, R. F., & Weymann, R. J. 1979. *Nature*, **279**, 381.

Wasserman, I., & Weinberg, M. D. 1987. *Astrophys. J.*, **312**, 390.

Watson, G. N. 1944. *Theory of Bessel Functions*. Cambridge, Eng.: Cambridge University Press.

Weaver, H., & Williams, D.R.W. 1974. *Astron. Astrophys. Suppl.*, **17**, 251.

Webbink, R. F. 1985. In *Dynamics of Star Clusters*, IAU Symposium No. 113, ed. J. Goodman & P. Hut, p. 541. Dordrecht: Reidel.

Weinberg, S. 1972. *Gravitation and Cosmology*. New York: Wiley.

Weliachew, L., ed. 1975. *La Dynamique des Galaxies Spirales*. Paris: CNRS.

White, S.D.M. 1976. *Mon. Not. Roy. Astron. Soc.*, **177**, 717.

White, S.D.M. 1978. *Mon. Not. Roy. Astron. Soc.*, **184**, 185.

White, S.D.M. 1979. *Mon. Not. Roy. Astron. Soc.*, **189**, 831.

White, S.D.M., & Valdes, F. 1980. *Mon. Not. Roy. Astron. Soc.*, **190**, 55.

White, S.D.M., Huchra, J., Latham, D., & Davis, M. 1983. *Mon. Not. Roy. Astron. Soc.*, **203**, 701.

Whitford, A.E. 1977. *Astrophys. J.*, **211**, 527.

Whitham, G. B. 1974. *Linear and Nonlinear Waves*, p. 382. New York: Wiley.

Wielen, R. 1971. *Astron. Astrophys.*, **13**, 309.

Wielen, R. 1977. *Astron. Astrophys.*, **60**, 263.

Wielen, R. 1985. In *Dynamics of Star Clusters*, IAU Symposium No. 113, ed. J. Goodman & P. Hut, p. 449. Dordrecht: Reidel.

Wilczynski, E. J. 1896. *Astrophys. J.*, **4**, 97.

Wilkinson, A., & James, R. A. 1982. *Mon. Not. Roy. Astron. Soc.*, **199**, 171.

Wilson, C. P. 1975. *Astron. J.*, **80**, 175.

Woltjer, L., ed. 1962. *The Distribution and Motion of Interstellar Matter in Galaxies*. New York: Benjamin.

Woolley, R., & Candy, M. P. 1968. *Mon. Not. Roy. Astron. Soc.*, **139**, 231 and **141**, 277.

Woolley, R., & Dickens, R. J. 1961. *Roy. Greenwich Obs. Bulletin*, No. 42.

Yahil, A., Sandage, A., & Tammann, G. A. 1980. *Astrophys. J.*, **242**, 448.

Yang, J., Turner, M. S., Steigman, G., Schramm, D. N., & Olive, K. A. 1984. *Astrophys. J.*, **281**, 493.

Young, P. 1980. *Astrophys. J.*, **242**, 1232.

Young, P., Gunn, J. E., Kristian, J., Oke, J. B., & Westphal, J. A. 1980. *Astrophys. J.*, **241**, 507.

Zang, T. A. 1976. Unpublished Ph.D. thesis, Massachusetts Institute of Technology.

Zang, T. A., & Hohl, F. 1978. *Astrophys. J.*, **226**, 521.

Zweibel, E. 1978. *Astrophys. J.*, **222**, 103 & 110.

Zwicky, F. 1933. *Helvetica Physica Acta*, **6**, 110.

Zwicky, F., Herzog, E., Wild, P., Karpowicz, M., & Kowal, C. T. 1961–1968. *Catalog of Galaxies and Clusters of Galaxies*. Pasadena: California Institute of Technology.

Index

Abel's integral equation, 651

accretion, 571–574, 588; *see also* mergers

action, 163–165, 168, 179–186, 508–509, 663; adiabatic invariance of, 178–183, 545; principle of least, 663; space, 166–168, 172, 255, 265–266

adiabatic: changes, 178; fluid, 672; growth of central mass, 545; invariance, 165, 178–183, 185, 436, 441; mass loss, 282

age: of elliptical galaxies, 553; of field stars, 340, 575–576; and metallicity, 574–576; solar, 574–575; of star clusters, 24–25, 443, 575; of the Universe, 626, 705

Andromeda galaxy. *See* M31

angle variables, 165–166

angular momentum, 103–104, 660–661; as integral, 111, 221–222, 246; of Local Group, 606

anisotropic velocity dispersion: anisotropy parameter, 204, 216; in binary galaxies, 609–610; in Coma cluster, 614–615; in ellipticals, 218; in galactic halo, 596–597; in globular clusters, 522; in M87, 206–209; spherical models with, 239–242; spherical models without, 222–235

anomaly: eccentric, 108, 705; true, 108

anti-spiral theorem, 384, 397

Antonov's laws, 305–308

Antonov's variational principle, 304

Antonov-Lebovitz theorem, 300; proof, Appendix 5.C

apocenter, 105

Arnold diffusion, 177

astronomical unit (AU), 643

asymmetric drift, 14, 198–199

Atlas of Peculiar Galaxies, 463–466

Bahcall-Soneira Galaxy model, 85–90, 348–349; stability, 603

barotrope, 672

barred galaxies, 24, 340, §6.5, 416; bar instability, 321, 372–375, 380–383, *see also* halo; weak bars, 146–153, 401–402

baryons, density in, 623–626, 629–632

BBGKY hierarchy, 497–500

bending waves, 333–334, 407–411, 414

Bessel functions, 656–658

Big Bang, 705

binary galaxies, 607–610

binary stars: and core collapse, 543–545; disruption of wide, 443–444, 534–536, 549; formation, 492–493, 543–544; in globular clusters, 533–539, 542–544; hard, 534, 536–539, 549; lifetimes, 536, 538, 549; primordial, 493; soft, 534–536, 549, Appendix 8.B; tidal capture, 542–544, 549

Birkhoff's theorem, 703

black holes: in centers of stellar systems, 545–548, 551; as dark matter, 630, 632

bolometric: distance modulus, 560; luminosity, 8

bremsstrahlung, 579–580, 604

brown dwarfs, 630

calculus of variations, 652

canonical: commutation relations, 667; coordinates, 668; map, 667

catalogs of galaxies, 607, 612–613

center of mass, 660

centrifugal: barrier, 115, 448; force, 135, 664

Chandrasekhar's variational principle, 300; relation to perturbation energy, 686–687

Library of Congress Cataloging-in-Publication Data

Binney, James, 1950–
    Galactic dynamics.

    (Princeton series in astrophysics)
    Bibliography: p.
    Includes index.
    1. Galaxies. 2. Stars—Clusters. I. Tremaine, Scott, 1950–   .
II. Title. II. Series. QB857.B524 1988   523.1'12   86-43129
ISBN 0-691-08444-0 (alk. paper)
ISBN 0-691-08445-9 (pbk.)

DEMCO

SIR EDWARD GREY
WITCHFINDER

CREATED BY MIKE MIGNOLA

SIR EDWARD GREY

WITCHFINDER™

City *of the* Dead

———————◆———————

story
MIKE MIGNOLA and CHRIS ROBERSON

art
BEN STENBECK

colors
MICHELLE MADSEN

letters
CLEM ROBINS

cover and chapter break art by
JULIÁN TOTINO TEDESCO

———————◆———————

president and publisher MIKE RICHARDSON

editor SCOTT ALLIE

associate editor SHANTEL LAROCQUE

assistant editor KATII O'BRIEN

collection designer RICK DeLUCCO

digital art technician CHRISTINA McKENZIE

DARK HORSE BOOKS

Published by Dark Horse Books
A division of Dark Horse Comics, Inc.
10956 SE Main Street
Milwaukie, OR 97222

Advertising Sales: 503-905-2237
International Licensing: 503-905-2377
Comic Shop Locator Service: 888-266-4226

DarkHorse.com
Facebook.com/DarkHorseComics
Twitter.com/DarkHorseComics

First edition: April 2017
ISBN 978-1-50670-166-0

Library of Congress Cataloging-in-Publication Data

Names: Mignola, Michael, author. | Roberson, Chris, author. | Stenbeck, Ben,
  artist. | Madsen, Michelle (Illustrator), colourist. | Robins, Clem, 1955-
  letterer.
Title: Witchfinder. Volume 4, City of the dead / story by Mike Mignola and
  Chris Roberson ; art by Ben Stenbeck ; colors by Michelle Madsen ; letters
  by Clem Robins.
Other titles: City of the dead
Description: First edition. | Milwaukie, OR : Dark Horse Books, 2017. | "This
  volume collects Witchfinder: City of the Dead #1-#5"
Identifiers: LCCN 2016043702 | ISBN 9781506701660 (paperback)
Subjects: LCSH: Comic books, strips, etc. | BISAC: COMICS & GRAPHIC NOVELS /
  Horror. | COMICS & GRAPHIC NOVELS / Crime & Mystery. | FICTION / Occult &
  Supernatural.
Classification: LCC PN6728.W5888 M54 2017 | DDC 741.5/973--dc23
LC record available at https://lccn.loc.gov/2016043702

CHAPTER ONE

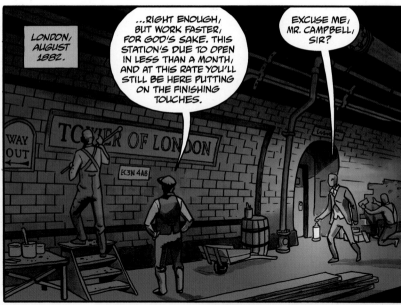

LONDON, AUGUST 1882.

...RIGHT ENOUGH, BUT WORK FASTER, FOR GOD'S SAKE. THIS STATION'S DUE TO OPEN IN LESS THAN A MONTH, AND AT THIS RATE YOU'LL STILL BE HERE PUTTING ON THE FINISHING TOUCHES.

EXCUSE ME, MR. CAMPBELL, SIR?

TOWER OF LONDON

EC3N 4AB

WAY OUT

WHAT IS IT, HAWKINS?

WELL, IT'S... THERE'S...

SOMETHING BACK THERE YOU SHOULD SEE, IS ALL.

WE WAS CLEARING OUT THE EQUIPMENT FROM THE SOUTHERN TUNNEL, JUST LIKE YOU SAID, SIR. ME, DAVIES, AND WILKES.

WE SHIFTED SOME OF THE UNUSED PIPING TO ONE SIDE, RESTED IT AGAINST THE TUNNEL WALL, AND... WELL...

THIS HAPPENED.

BLIMEY!

DAVIES, WILKES, AND ME CAME DOWN HERE, TO SEE HOW FAR IT WENT.

MY OLD DAD WAS A MOLEMAN, AND HIS DAD BEFORE HIM, DIGGING SEWERS AND SUCH. I'VE HEARD ABOUT FINDS LIKE THIS, DOWN IN THE SOFT CLAY.

CHAMBERS DATING BACK TO ELIZABETH'S TIME, MEDIEVAL WARRENS, SAXON FIRE PITS... THE HOMES OF THE DEAD, UNDER THE FEET OF THE LIVING.

WELL, I DON'T KNOW NAUGHT ABOUT THAT, BUT WHOEVER BUILT IT, IT'S BEEN A GOOD LONG WHILE.

AND WHAT WE FOUND AT THE END OF THE TUNNEL? IT'S LIKE NO HOME I'VE EVER SEEN BEFORE.

SOME KIND OF CHURCH, MORE LIKE, YOU ASK ME.

BLOODY HELL.

...AT WHICH POINT I WAS FINALLY ABLE TO LOCATE THE DUKE'S MISSING NEPHEW, DEEP IN THE WILD WOOD, ALONG WITH THE LUNATIC WHO HAD ABDUCTED HIM.

FROM THE SECRET JOURNALS OF SIR EDWARD GREY.

THE DUCHESS'S INSISTENCE THAT "FAIRIES" HAD TAKEN THE BOY PROVED AS DUBIOUS AS I HAD ANTICIPATED, WITH THE CULPRIT INSTEAD BEING AN ALL-TOO-MORTAL MAN WITH GROTESQUE APPETITES.

THE OFFICIAL STORY SUPPLIED TO THE PAPERS WAS THE BOY HAD BEEN KIDNAPPED FOR RANSOM BY FENIANS. IT WAS FELT THE PUBLIC WAS ILL PREPARED FOR THE TRUTH. I WAS ILL PREPARED FOR IT MYSELF.

BUT WITH ALL THAT I HAVE SEEN OVER THE YEARS, THE UNCANNY AND THE INEXPLICABLE, HOW CAN I STILL BE SURPRISED BY THE MUNDANE DEPRAVITIES OF MY FELLOW MAN? THE SUPERNATURAL CREATURES FROM FAIRY TALES AND PENNY DREADFULS MAY SOMETIMES PROVE TO BE REAL, BUT THEY ARE NOTHING COMPARED WITH THE NIGHTMARES THAT LURK IN THE HUMAN MIND.

WITCHFINDER RECOVERS MISSING CHILD FROM IRISH RADICALS

KING OF ZULULAND VISITS LONDON

SPEAKING OF WHICH, I HAVE HAD THE SAME TROUBLING DREAM, NIGHT AFTER NIGHT, BUT UPON WAKING CAN NEVER RECALL THE DETAILS. ONLY THE SENSATION OF

WHAT WAS IT AGAIN...?

REMEMBER.

SAINT JOHN OF THE CROSS POLICE HOSPITAL.

I'VE JUST REACHED THE END OF A LONG AND DIFFICULT ASSIGNMENT, AND I'M IN NO MOOD FOR JOKES.

AND I MUST CONFESS, MR. SILK, THAT I CANNOT SEE THIS AS ANYTHING OTHER THAN AN ILL-ADVISED ATTEMPT AT HUMOR.

I ASSURE YOU, I'M AS SERIOUS AS THE GRAVE. AH, SEE WHAT I DID THERE? "GRAVE." AN UNINTENTIONAL PUN, BUT FITTING. DEADLY SERIOUS, YOU MIGHT EVEN SAY.

YES, WELL, THERE IS NO COMPELLING REASON TO DOUBT THE MAN'S TESTIMONY, AFTER ALL.

BUT MISS GOAD, THE MAN WAS AN ADMITTED **GRAVE ROBBER.** WHY SHOULD WE TAKE HIM AT HIS WORD ABOUT **ANYTHING?**

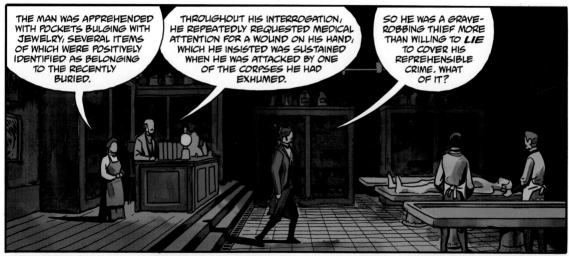

THE MAN WAS APPREHENDED WITH POCKETS BULGING WITH JEWELRY, SEVERAL ITEMS OF WHICH WERE POSITIVELY IDENTIFIED AS BELONGING TO THE RECENTLY BURIED.

THROUGHOUT HIS INTERROGATION, HE REPEATEDLY REQUESTED MEDICAL ATTENTION FOR A WOUND ON HIS HAND, WHICH HE INSISTED WAS SUSTAINED WHEN HE WAS ATTACKED BY ONE OF THE CORPSES HE HAD EXHUMED.

SO HE WAS A GRAVE-ROBBING THIEF MORE THAN WILLING TO *LIE* TO COVER HIS REPREHENSIBLE CRIME. WHAT OF IT?

I WAS DISPATCHED TO SEE TO HIS INJURIES, BUT THE SUSPECT PERISHED BEFORE I ARRIVED.

DR. MANLEY AND I HAVE YET TO ESTABLISH THE CAUSE OF DEATH, BUT THE INJURY HE REPORTED IS CLEARLY EVIDENT.

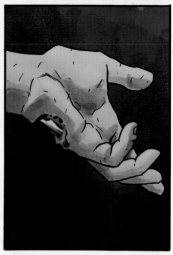

IT IS A HUMAN BITE MARK, DR. LEWIS, WHAT IS SO BLESSED MYSTERIOUS ABOUT THAT? HE COULD EASILY HAVE BITTEN HIMSELF, JUST TO ADD CREDENCE TO HIS STORY.

SEE HERE, GREY, JUST WHOM DO YOU THINK YOU'RE ADDRESSING? OF *COURSE* THE FIRST THING WE DID WAS TO MEASURE THE BITE'S CIRCUMFERENCE AGAINST THE MAN'S OWN INDENTATION.

WAS MY SUGGESTION, TO BE FAIR.

I FAIL TO SEE WHAT ANY OF THIS HAS TO DO WITH ME.

MY REMIT FROM HER MAJESTY IS VERY CLEAR. BUT THIS IS A SIMPLE CASE OF GRAVE ROBBING, AND I FIND NO EVIDENCE HERE OF "HOODOO" OF ANY SORT--

THAT'S QUITE ENOUGH OF *THAT*.

STHUNK

THUD

I ≷GASP≷ I DON'T... DON'T...

FOR PITY'S SAKE, LEWIS, PULL YOURSELF TOGETHER.

CLATTER

NOW, WHERE EXACTLY DID THIS GRAVE ROBBER SAY HE WAS ATTACKED?

LAMB STREET CEMETERY.

THE TIMID FEAR THE GRAVEYARD INSTINCTUALLY, HARBORING ATAVISTIC, GROUNDLESS ANXIETIES.

BUT HAVING SEEN WHAT PRACTITIONERS OF THE DARK ARTS CAN DO WITH A LIFELESS CORPSE, I HAVE **EARNED** THE DISQUIET I FEEL WHENEVER I TRESPASS AMONG THE DEAD.

DO YOU HAVE A MOMENT?

I'M HERE ON OFFICIAL BUSINESS, AND HAVE A FEW QUESTIONS FOR YOU.

SURE. SO LONG AS IT DOESN'T TAKE LONG.

JACOB COHEN
1786-1829

...AND I AM CURIOUS TO KNOW WHETHER YOU HAD ANY DETAILS TO SHARE ABOUT THE RECENT THEFTS.

DON'T KNOW ANYTHING ABOUT ANY *THEFTS*.

WHERE'S ALL THIS COMING FROM, ANYWAY?

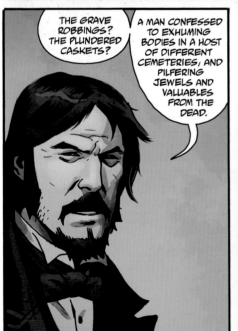

THE GRAVE ROBBINGS? THE PLUNDERED CASKETS?

A MAN CONFESSED TO EXHUMING BODIES IN A HOST OF DIFFERENT CEMETERIES, AND PILFERING JEWELS AND VALUABLES FROM THE DEAD.

THIS CEMETERY FEATURED PROMINENTLY IN THAT LIST.

NOTHING LIKE THAT HERE. COURSE, THIS ISN'T LIKE A LOT OF THOSE *OTHER* BURIAL GROUNDS.

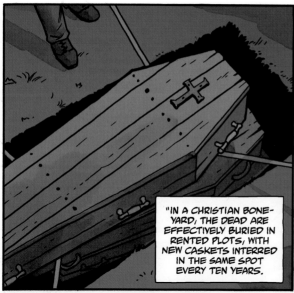

"THIS IS A JEWISH CEMETERY, AND ALWAYS HAS BEEN. JUDAIC LAW SAYS THE DEAD HAVE TO BE BURIED WITHIN TWENTY-FOUR HOURS, ONLY ONE BODY TO A GRAVE."

"IN A CHRISTIAN BONE-YARD, THE DEAD ARE EFFECTIVELY BURIED IN RENTED PLOTS, WITH NEW CASKETS INTERRED IN THE SAME SPOT EVERY TEN YEARS."

IT'S INDECENT, HOW SOME OF THOSE OTHERS TREAT THEIR DEAD. BUT HERE, WE TREAT THEM WITH **RESPECT.**

I'M NOT QUESTIONING THAT, I ASSURE YOU.

BUT YOU **INSIST** THAT YOU KNOW NOTHING ABOUT THE GRAVE ROBBINGS?

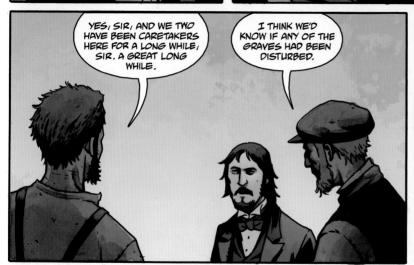

YES, SIR, AND WE TWO HAVE BEEN CARETAKERS HERE FOR A LONG WHILE, SIR. A GREAT LONG WHILE.

I THINK WE'D KNOW IF ANY OF THE GRAVES HAD BEEN DISTURBED.

YES, I'M CERTAIN THAT YOU **WOULD,** AT THAT.

I HAVE ENCOUNTERED MANY VILE, HORRID CREATURES OVER THE YEARS, AND KNOW ALL TOO WELL THE DANGERS POSED BY THE SUPERNATURAL.

BUT NONE SO DANGEROUS AS MY FELLOW MAN. THE CARETAKERS ARE LYING, THAT MUCH IS CERTAIN. WHAT DO THEY GAIN FROM THE FALSEHOOD? AND WHO ELSE BENEFITS?

SIR EDWARD GREY?

PRECISELY THE MAN I HAVE COME TO SEE. I TRUST YOU RECALL OUR PREVIOUS MEETING?

ON BEHALF OF THE HELIOPIC BROTHERHOOD OF RA, I MUST INFORM YOU THAT THERE IS A MATTER OF SOME URGENCY--

AUGUST SWAIN.

YOU DARE SPEAK TO ME? WHEN YOU STILL HAVE THE BLOOD OF MY FRIENDS ON YOUR HANDS?!

"MARY WOLF.

"THE CAPTAIN.

"AND THE BLOODY-HANDED BASTARDS WHO KILLED THEM."

YOU AND YOUR WHOLE BAND OF **OCCULT FANATICS** SHOULD BE **HANGED** FOR WHAT YOU'VE DONE.

I WOULD **KILL** YOU WHERE YOU STAND WITH MY OWN BARE HANDS IF--

NOW, NOW, SIR EDWARD. CALM YOUR-SELF.

I HAVE NOT COME TO QUARREL WITH YOU.

AND BESIDES, IT'D HARDLY BE PROPER FOR AN AGENT OF THE CROWN TO MAKE A SCENE IN THE BROAD DAYLIGHT, WOULDN'T YOU SAY?

WHAT DO YOU WANT, SWAIN?

I COME WITH A REQUEST FOR ASSISTANCE, AND AN OFFER OF THE SAME.

WE HAVE INTELLIGENCE DATING BACK YEARS THAT WE BELIEVE WOULD AID YOU GREATLY IN YOUR PRESENT INVESTIGATIONS, AND WE THINK YOU ARE IN A POSITION TO ASSIST US, AS WELL.

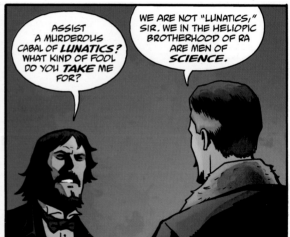

ASSIST A MURDEROUS CABAL OF *LUNATICS?* WHAT KIND OF FOOL DO YOU *TAKE* ME FOR?

WE ARE NOT "LUNATICS," SIR. WE IN THE HELIOPIC BROTHERHOOD OF RA ARE MEN OF *SCIENCE.*

"AND WITH THE ELECTRIC EYE OF SCIENCE, WE WILL LAY BARE THE *GREATER MYSTERIES.*"

"PERHAPS EVEN THE MYSTERY OF *LIFE* ITSELF."

I HAVE *HEARD* THIS LINE FROM YOU LOT *BEFORE,* I *DON'T* NEED TO HEAR IT AGAIN.

YOU WILL COME TO A BETTER UNDERSTANDING, IN TIME. YOU SEE, AT PRESENT, WE SHARE A COMMON CAUSE.

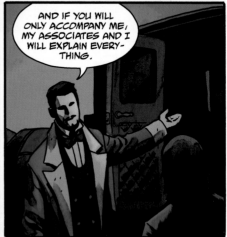

AND IF YOU WILL ONLY ACCOMPANY ME, MY ASSOCIATES AND I WILL EXPLAIN EVERY-THING.

YOU MUST BE JOKING.

I DID NOT BELIEVE FOR A MOMENT THE TESTIMONY OF THE CEMETERY CARE-TAKERS, AS THEY'RE CLEARLY HIDING SOMETHING. BUT IS THERE SOME TIE TO THE HELIOPIC BROTHERHOOD...?

I AM ELECTING TO TAKE A CLOSER LOOK AT THOSE GRAVES THE CARETAKERS WERE TAMPING DOWN, WHEN THEY ARE NOT AROUND TO INTERFERE.

THOUGH I WILL REQUIRE ASSISTANCE...

...QUIETLY, YOU TWO.

MY WARRANT ALLOWS FOR THIS INSPECTION TO BE CARRIED OUT LEGALLY, BUT I WOULD PREFER NOT TO ALERT THE PROPRIETORS OF OUR VISIT JUST YET, IF POSSIBLE.

I DON'T LIKE THIS, GREY. MUCKING ABOUT WITH CORPSES AFTER A DEAD MAN HOPPED UP OFF THE SLAB ONLY THIS MORNING.

WHOLE MESS OF DEAD MEN AROUND HERE, I'D SAY.

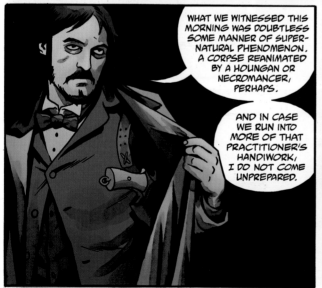

WHAT WE WITNESSED THIS MORNING WAS DOUBTLESS SOME MANNER OF SUPER-NATURAL PHENOMENON. A CORPSE REANIMATED BY A HOUNGAN OR NECROMANCER, PERHAPS.

AND IN CASE WE RUN INTO MORE OF THAT PRACTITIONER'S HANDIWORK, I DO NOT COME UNPREPARED.

"SILVER BULLETS LOADED IN MY WEBLEY-PRYSE, CHARMS AND AMULETS, SACRED WARDS OF ALL VARIETIES."

NOW, THE CARETAKERS INSISTED THAT THERE HAD BEEN NO GRAVE ROBBERS HERE.

BUT THEY WERE IN THE PROCESS OF FILLING A GRAVE, THE HEADSTONE OF WHICH DATED BACK HALF A CENTURY. WHICH SUGGESTS THAT THEY WERE REBURYING A RECENTLY EXHUMED CASKET.

IT WAS RIGHT ABOUT... *THERE.*

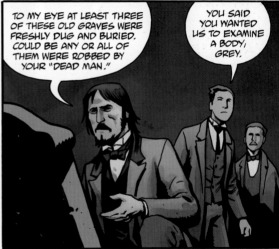

TO MY EYE AT LEAST THREE OF THESE OLD GRAVES WERE FRESHLY DUG AND BURIED. COULD BE ANY OR ALL OF THEM WERE ROBBED BY YOUR "DEAD MAN."

YOU SAID YOU WANTED US TO EXAMINE A BODY, GREY.

I MIGHT HAVE IMPLIED IT, YES. BUT WHAT I NEED MORE AT THE MOMENT IS A COUPLE OF STRONG BACKS TO HELP WITH THE DIGGING.

SO GET TO WORK, WON'T YOU? THERE'S A GOOD MAN.

NO SIGN OF DECAY. NO DECOMPOSITION.

SAME HERE.

I'VE SEEN LIVING MEN WHO DON'T LOOK AS HALE AND HEARTY AS THE ONE DOWN THERE.

AND YET ALL THREE WERE BURIED IN THE GROUND BETWEEN FIFTY AND EIGHTY YEARS AGO.

JCCB COHEN
1756-1571

COULD THE CASKETS JUST BE PARTICULARLY WELL MADE? AIRTIGHT, WATERTIGHT? THE BODIES NATURALLY MUMMIFIED, AS SOMETIMES HAPPENS IN DESERT CAVES?

EVEN THEN THEY'D BE DESICCATED, DRIED OUT. THESE ARE MORE LIKE THOSE BODIES PRESERVED IN PEAT BOGS, THE BODILY FLUIDS TRAPPED WITHIN, THE FLESH TANNED TO LEATHER.

JACOB COHEN
- 1786-1829 -

BUT THESE ARE NOT DESERT CAVES OR PEAT BOGS, GENTLEMEN. THESE ARE SIMPLE WOODEN COFFINS BURIED IN THE DAMP EARTH OF LONDON.

I KNOW YOU'D BOTH PREFER THAT THESE WERE SOMEHOW NATURALLY OCCURRING PHENOMENA. IT SEEMS, HOWEVER, THAT SOME FORM OF NECROMANCY SURELY WAS INVOLVED, AND--

SKRITCH

THE DEVIL...?

CHAPTER TWO

STAY BEHIND ME, DOCTORS.

URRHHHNNN...

NOW, WE'LL SEE HOW THEY LIKE THE TASTE OF SILVER.

BLAM BLAM BLAM

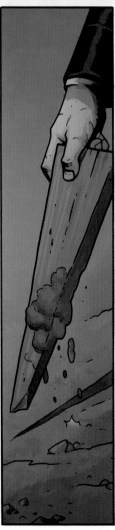

AAARRR!

FSSSSSS

≥URK≤ THANK YOU ≥GRN≤ GREY--

JUST GET A HOLD OF YOURSELF, LEWIS, FOR GOD'S SAKE!

MANLEY! IMPALE THE CREATURE! DO IT NOW!

NOW THERE CAN BE NO DOUBT THAT SOMETHING SUPERNATURAL IS AFOOT.

NOT ONE BUT THREE BODIES EXHUMED FROM GRAVES AT THE LAMB STREET CEMETERY, THOUGH DEAD, PROVED TO POSSESS SOME SEMBLANCE OF LIFE.

I HAD THOUGHT THAT A PRACTITIONER OF THE DARK ARTS MIGHT BE RESPONSIBLE FOR ANIMATING THE DEAD, BUT THAT EXPLANATION NO LONGER SUFFICES.

THESE "UNDEAD" ARE THEMSELVES SUPERNATURAL CREATURES, PERHAPS SOME SPECIES OF REVENANT, OR GHOUL, OR EVEN VAMPIRE.

FOR CENTURIES ENGLISH FOLKLORE HAS BEEN FILLED WITH STORIES OF THE UNQUIET DEAD RISING FROM THEIR GRAVES TO PLAGUE THE LIVING. BUT I NEVER CREDITED SUCH

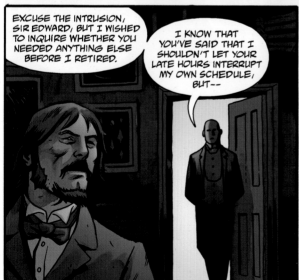

EXCUSE THE INTRUSION, SIR EDWARD, BUT I WISHED TO INQUIRE WHETHER YOU NEEDED ANYTHING ELSE BEFORE I RETIRED.

I KNOW THAT YOU'VE SAID THAT I SHOULDN'T LET YOUR LATE HOURS INTERRUPT MY OWN SCHEDULE, BUT--

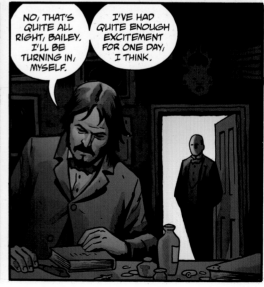

NO, THAT'S QUITE ALL RIGHT, BAILEY. I'LL BE TURNING IN, MYSELF.

I'VE HAD QUITE ENOUGH EXCITEMENT FOR ONE DAY, I THINK.

DING DING

BE WARY, ENGLISHMAN, AND BE WARNED.

DANGERS UNTOLD LIE HIDDEN BENEATH THE EARTH, BUT NOTHING STAYS BURIED FOREVER.

THOUGH THE GODDESS OF THE DARK MAY SOMETIMES SLUMBER, HER CHILDREN NEVER REST, AND SHADOWS WILL SPREAD AS DAY DRAWS INTO NIGHT.

THIS MORNING I VISITED A NUMBER OF GOVERNMENT OFFICES, TO SEE WHAT I COULD DISCOVER ABOUT THE LAMB STREET CEMETERY.

I KNEW THAT IT DATED TO THE EIGHTEENTH CENTURY AND, BECAUSE IT CATERED TO A JEWISH CLIENTELE, WAS EXEMPT FROM MUCH OF THE TRADITION THAT GOVERNED OTHER LONDON GRAVEYARDS.

BUT I CAN FIND NO HISTORY OF UNUSUAL ACTIVITY ASSOCIATED WITH THE CEMETERY, NOR ANYTHING SPECIFIC ABOUT THE BODIES WE DISINTERRED, AND WHICH SUBSEQUENTLY ATTACKED US.

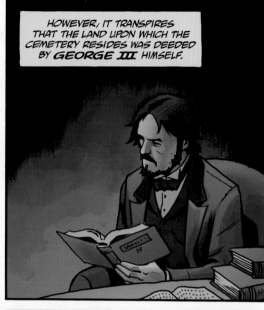

HOWEVER, IT TRANSPIRES THAT THE LAND UPON WHICH THE CEMETERY RESIDES WAS DEEDED BY **GEORGE III** HIMSELF.

A EUROPEAN FAMILY HAS HELD THE LANDS AND TITLE IN TRUST EVER SINCE, WITH CONTROL PASSING THROUGH SUCCESSIVE GENERATIONS.

THE CURRENT DEED HOLDER RESIDES IN LONDON, AND I WAS ABLE TO OBTAIN HIS NAME AND ADDRESS WITHOUT TOO MUCH DIFFICULTY.

AS THE CARETAKERS APPEAR TO HAVE VANISHED WITHOUT A TRACE, I BELIEVE I SHALL PAY THE DEED HOLDER A VISIT AND SEE IF HE CAN SHED ANY LIGHT ON MATTERS.

MY SLEEP LAST NIGHT WAS RESTLESS, FILLED WITH NIGHTMARES.

DISQUIETING VISIONS OF RAVENS AND SERPENTS, OF TOLLING BELLS AND CAVERNS BENEATH THE EARTH.

BUT NOW STRANGE AFRICAN MEDICINE MEN APPEAR IN MY DREAMS AS WELL? WHAT INSPIRED THAT FANCY?

NOCK NOCK

I BEG FORGIVENESS, MASTER, BUT--

I TOLD YOU I WAS NOT TO BE DISTURBED. WHAT IS THE MEANING OF THIS?

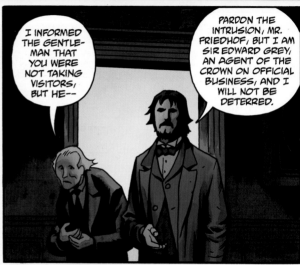

I INFORMED THE GENTLE-MAN THAT YOU WERE NOT TAKING VISITORS, BUT HE--

PARDON THE INTRUSION, MR. FRIEDHOF, BUT I AM SIR EDWARD GREY, AN AGENT OF THE CROWN ON OFFICIAL BUSINESS, AND I WILL NOT BE DETERRED.

IT IS MY UNDER-STANDING THAT YOU ARE THE DEED HOLDER FOR THE LAND UPON WHICH THE LAMB STREET CEMETERY RESIDES, AND THE SOLE OWNER OF THE CEMETERY ITSELF.

IS THAT CORRECT?

YES. WHAT OF IT?

THERE HAVE BEEN CERTAIN...*UNUSUAL* OCCURRENCES CONNECTED WITH THE CEMETERY GROUNDS IN RECENT DAYS.

NOTHING THAT HAS BEEN MADE KNOWN TO THE PUBLIC, OF COURSE.

EVIDENCE SUGGESTS THAT THE CEMETERY'S CARETAKERS MIGHT HAVE *DISQUIETING* ALLEGIANCES. AND ONCE THEY WERE AWARE OF MY INVESTIGATION, THEY APPEAR TO HAVE GONE TO GROUND.

IF I MAY ASK, ARE YOU AWARE OF ANY STRANGE BEHAVIOR ON THEIR PART, OR ON THE PROPERTY?

YOU MAY ASK.

BUT IT DOES NOT NECESSARILY FOLLOW THAT I MUST ANSWER.

INTERESTING.

IT IS A SIMPLE QUESTION, MR. FRIEDHOF. YOUR EVASION SUGGESTS THAT YOU KNOW MORE ABOUT THIS BUSINESS THAN YOU LET ON.

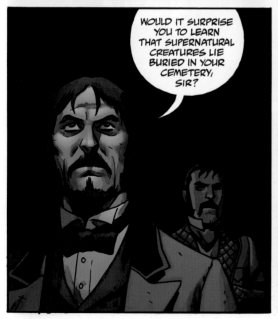

WOULD IT SURPRISE YOU TO LEARN THAT SUPERNATURAL CREATURES LIE BURIED IN YOUR CEMETERY, SIR?

OR ARE YOU ALREADY ALL TOO AWARE OF THAT FACT?

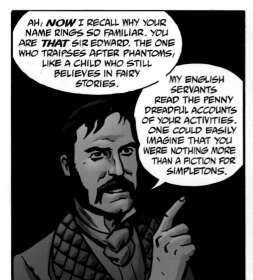

AH, **NOW** I RECALL WHY YOUR NAME RINGS SO FAMILIAR. YOU ARE **THAT** SIR EDWARD. THE ONE WHO TRAIPSES AFTER PHANTOMS, LIKE A CHILD WHO STILL BELIEVES IN FAIRY STORIES.

MY ENGLISH SERVANTS READ THE PENNY DREADFUL ACCOUNTS OF YOUR ACTIVITIES. ONE COULD EASILY IMAGINE THAT YOU WERE NOTHING MORE THAN A FICTION FOR SIMPLETONS.

WHAT IS IT THAT THE RABBLE CALLS YOU AGAIN...?

WITCHFINDER.

I SAY AGAIN, **SIR**, THAT I AM AN AGENT OF THE CROWN ON OFFICIAL BUSINESS.

I WILL NOT BE DETERRED.

NO, I DON'T IMAGINE THAT YOU WILL, AT THAT.

WHAT A BORE YOU ARE.

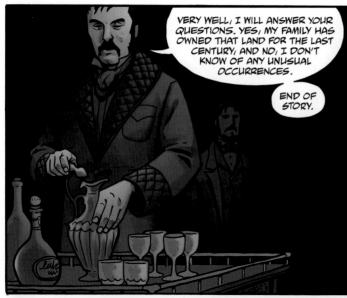

VERY WELL, I WILL ANSWER YOUR QUESTIONS. YES, MY FAMILY HAS OWNED THAT LAND FOR THE LAST CENTURY, AND NO, I DON'T KNOW OF ANY UNUSUAL OCCURRENCES.

END OF STORY.

SO YOU MAINTAIN THAT YOU KNOW NOTHING ABOUT THOSE BURIED THERE?

ASIDE FROM THE FACT THAT THEY ARE DEAD? NO, I'VE NOTHING TO ADD.

THERE ARE NO GHOSTS OR GHOULS TO BE EXORCISED HERE, SIR EDWARD. ONLY BUSINESSMEN WITH IMPORTANT MATTERS TO WHICH TO ATTEND.

I TRUST YOU CAN FIND YOUR OWN WAY OUT?

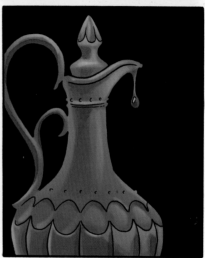

MY MASTER BID ME INFORM YOU THAT THIS MATTER IS AT AN END.

I RATHER THINK THAT'S NOT HIS DECISION TO MAKE.

INFORM YOUR "MASTER" THAT I WILL BE--

SLAM

THE WINDOWS DRAPED, EVERY ONE OF THEM. DARK AND SILENT AS A TOMB WITHIN. I WONDER IF--

≤COUGH≥

WHERE--?

And Zarod said.
Behold.
and light sprang forth.

OH. OF
COURSE.

YOU MUST EXCUSE THE BRUSQUE NATURE OF YOUR ARRIVAL, GREY.

BUT AFTER ALL, YOU *DID* REFUSE OUR FIRST INVITATION, AND AFTER THAT MEETING YOU JUST LEFT WE SIMPLY WERE NO LONGER IN A POSITION TO TAKE "NO" FOR AN ANSWER.

DAMN YOUR EYES, SWAIN. I *KNEW* THAT YOU WERE MIXED UP IN THIS BUSINESS *SOMEHOW.*

BEFORE THE, SHALL WE SAY, *UNTIMELY* DEATH OF THE MAN KNOWN AS "THE CAPTAIN," THE HELIOPIC BROTHERHOOD TOOK POSSESSION OF MANY OF HIS FILES ON LONDON'S SECRET HISTORY.

THEY MAKE FOR FASCINATING READING, I ASSURE YOU. THE CAPTAIN WAS A MAN OF CONSIDERABLE RESOURCES, AND HE KNEW WHERE THE BODIES WERE BURIED. SO TO SPEAK.

BUT ULTIMATELY THEY PROVIDED US WITH VERY LITTLE IN THE WAY OF *USEFUL* INFORMATION.

IN THE FINAL ANALYSIS, THE WORK OF THE CAPTAIN'S LONG LIFE AMOUNTED TO LITTLE MORE THAN MILD DIVERSION.

BUT THAT CHANGED WHEN THE HEAD OF OUR ORDER RECENTLY ARRIVED FROM FRANCE.

ALLOW ME TO PRESENT TEFNUT TRIONUS, QUEEN OF HELIOPOLIS, SECOND INCARNATION OF EUGENE REMY, AND GRAND MISTRESS OF THE HELIOPIC BROTHERHOOD OF RA.

EVER SINCE I WAS A CHILD, I HAVE BEEN ABLE TO SEE BEYOND REALITY'S VEIL. TO PERCEIVE THAT WHICH IS HIDDEN, OR LOST, OR YET TO COME.

IT IS BECAUSE OF WHAT I HAVE SEEN THAT I AM COME TO ENGLAND.

YES, WELL, I KNOW SOMETHING OF MEDIUMS AND PSYCHICS. DON'T EXPECT ME TO BE *TOO* IMPRESSED.

I HAVE SEEN *YOU* IN MY VISIONS, SIR EDWARD, AT THE SHORES OF A RIVER OF DARKNESS SOMEWHERE FAR UNDERGROUND.

ACHERON...

I HAVE SEEN A *BLACK GODDESS* RISING FROM THE DARK WATERS OF THAT RIVER, COME TO RECLAIM HER LOST KINGDOM.

I HAVE SEEN A MAN LEADING AN ARMY OF THE DEAD IN HER NAME, LAYING WASTE TO THE LIVING WORLD.

I CAME HERE BECAUSE I SAW THAT *THIS* IS WHERE THIS MAN WOULD RAISE HIS ARMY.

NOT ONLY THAT, BUT SHE WAS ABLE TO IDENTIFY A PORTRAIT OF THE MAN, ON LOAN TO THE BRITISH MUSEUM FROM A PRIVATE COLLECTION IN STUTTGART, GERMANY.

I SPOKE TO THIS MAN ONLY A SHORT WHILE AGO. HIS NAME IS FRIEDHOF. BUT WHY COMMISSION A PORTRAIT IN NAPOLEONIC-ERA COSTUME? SOME SORT OF FANCY DRESS BALL?

HA!

"FRIEDHOF" IS THE GERMAN WORD FOR "CEMETERY," SIR EDWARD.

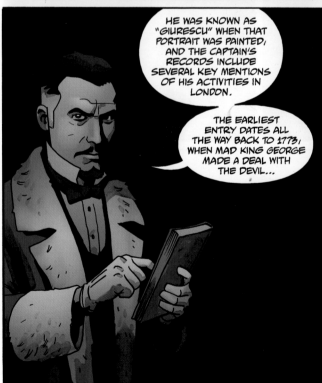

HE WAS KNOWN AS "GIURESCU" WHEN THAT PORTRAIT WAS PAINTED, AND THE CAPTAIN'S RECORDS INCLUDE SEVERAL KEY MENTIONS OF HIS ACTIVITIES IN LONDON.

THE EARLIEST ENTRY DATES ALL THE WAY BACK TO 1773, WHEN MAD KING GEORGE MADE A DEAL WITH THE DEVIL...

YOU HAVE MY INTEREST. CONTINUE.

THE PLAN IS IN MOTION, MY QUEEN.

THE TIME FAST APPROACHES WHEN WE WILL RING DOWN THE CURTAIN ON THE AGE OF MAN.

YOUR ARMIES WILL DELIVER THE WORLD BACK INTO CHAOS, AND YOU SHALL REIGN SUPREME IN ETERNAL DARKNESS.

SOON.

LONDON,
OCTOBER 1773.

...AND THIS IS THE ONLY PRICE THAT YOU ASK?

YES, YOUR MAJESTY. THE MADELYN ROSE SETS SAIL TOMORROW, AND MY PRUSSIAN MERCENARIES ARE PREPARED TO BOARD.

THEY WILL SQUASH YOUR NOISOME COLONIAL REVOLT, AND ALL I ASK IN RETURN IS THE DEED TO A SMALL AMOUNT OF REAL ESTATE.

PAID, AND GLADLY. FOR THIS GOOD SERVICE, THAT SMALL PATCH OF LAND SHALL BE YOURS FROM NOW TILL DOOMSDAY.

THAT SHOULD SUFFICE.

LONDON,
AUGUST 1882.

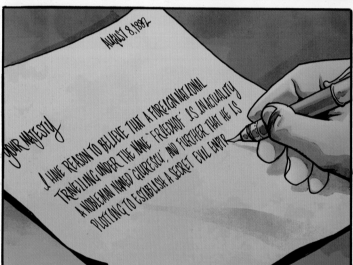

August 8, 1882.

YOUR MAJESTY

I HAVE REASON TO BELIEVE THAT A FOREIGN NATIONAL TRAVELLING UNDER THE NAME "FRIEDHOF" IS IN ACTUALITY A NOBLEMAN NAMED "QUIRESCU", AND FURTHER THAT HE IS PLOTTING TO ESTABLISH A SECRET EVIL EMPIR

BAILEY, I KNOW IT IS OUTSIDE YOUR USUAL REMIT, BUT I MUST ASK YOU TO DELIVER THIS MISSIVE PERSONALLY.

I'M AFRAID THAT HER MAJESTY WOULD APPEAR TO BE SUMMERING AT OSBORNE HOUSE ON THE ISLE OF WIGHT, BUT IF YOU CATCH THE MORNING TRAIN YOU SHOULD BE BACK BY NIGHTFALL.

THE COOK WILL HAVE TO ATTEND TO YOUR DUTIES IN YOUR ABSENCE.

REMEMBER, ONLY A MEMBER OF THE QUEEN'S PERSONAL STAFF CAN TAKE RECEIPT OF THIS LETTER. I KNOW I CAN TRUST YOUR DISCRETION IN THIS MATTER.

THANK YOU, SIR. OF COURSE, SIR.

THE INTELLIGENCE I HAVE RECEIVED FROM THE HELIOPIC BROTHERHOOD OF RA IS TROUBLING, TO SAY THE LEAST.

UNDER NORMAL CIRCUMSTANCES I WOULD BE DISINCLINED TO PUT MUCH FAITH IN ANYTHING THAT SWAIN AND HIS CRONIES HAD TO SAY, BUT CIRCUMSTANCES ARE FAR FROM NORMAL.

I HAD ASSUMED THAT WE WERE DEALING WITH A GARDEN-VARIETY NECROMANCER OR OCCULTIST, SOMEONE ON THE LEVEL OF A GUSTAV STROBL OR MARY AND ELIZABETH WASHBROOK AND SARA WEBB.

BUT I'M FORCED TO CONSIDER THE POSSIBILITY THAT WHAT WE ARE DEALING WITH IS NOT A MORTAL MAN WHO EMPLOYS SUPERNATURAL FORCES, BUT A SUPERNATURAL BEING HIMSELF.

LILLIPUT

WE WILL HAVE TO ATTEND TO THIS "GIURESCU," AND DECISIVELY.

BEFORE IT'S TOO LATE...

OH!

...AND THAT'S ALL YOU CAN TELL US ABOUT THIS COVE, IS IT?

I'M SORRY, SERGEANT, BUT THE DETAILS OF THE MATTER ARE CLASSIFIED. SUFFICE IT TO SAY THAT THE MAN IS A THREAT TO QUEEN AND COUNTRY, AND MUST BE APPREHENDED AT ONCE.

THAT ROYAL WARRANT OF YOURS IS GOOD ENOUGH FOR ME, SIR EDWARD.

ALL RIGHT, YOU LOT! GET ON WITH IT!

UNF!

SMASH

OF COURSE YOU DID, ENGLISHMAN. YOU SAW ME BECAUSE I *WISHED* YOU TO SEE ME.

MY NAME IS MOHLOMI.

MOHLOMI? YOU'LL FORGIVE ME IF I'M SOMEWHAT WARY, BUT I'M NOT ACCUSTOMED TO MEETING PEOPLE FIRST IN MY *DREAMS* AND ONLY AFTERWARD IN THE FLESH.

OR DO YOU DENY THAT WE HAVE ENCOUNTERED ONE ANOTHER BEFORE?

I HAVE TRAVELED A GREAT DISTANCE TO MEET YOU. FARTHER THAN YOU CAN IMAGINE.

AN EVIL THAT HAS SLUMBERED BENEATH THE EARTH FOR A GREAT MANY YEARS IS THREATENING TO AWAKEN, AND I BELIEVE THAT *YOU* ARE THE ONE WHO MUST PREVENT IT.

YES, YES, ALL VERY PORTENTOUS.

I DON'T SUPPOSE YOU HAVE MORE THAN RIDDLES AND CRYPTIC WARNINGS TO SHARE?

I SPEAK AS PLAINLY AS I CAN.

THERE ARE SOME MATTERS THAT ARE NOT EASILY ENCOMPASSED IN THE SPEECH OF MEN.

"THERE ARE MORE THINGS IN HEAVEN AND EARTH..."

YOU CLEARLY ARE MORE THAN YOU SEEM. WHAT IS YOUR CONNECTION WITH THIS UNDEAD BUSINESS? WHAT DO YOU KNOW ABOUT GIURESCU?

I KNOW THAT THE DEAD DO NOT ALWAYS REMAIN IN THEIR RESTING PLACES, AND THAT SOME MEN WALK THE EARTH BEYOND THEIR APPOINTED TIME.

AS FOR THE ONE YOU SPEAK OF, I KNOW ONLY THE DANGER THAT HE POSES, FAR GREATER THAN HIMSELF.

IF YOU DESIRE ANSWERS, YOU MUST GO TO THE PLACE WHERE THE RAVENS GATHER.

RAVENS HAVE ALWAYS FAVORED THE *MISTRESS* OF THE ONE WHOM YOU SEEK.

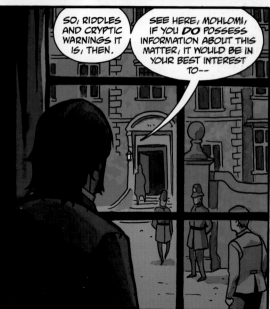

SO, RIDDLES AND CRYPTIC WARNINGS IT IS, THEN.

SEE HERE, MOHLOMI, IF YOU *DO* POSSESS INFORMATION ABOUT THIS MATTER, IT WOULD BE IN YOUR BEST INTEREST TO--

MOHLOMI?

THE POLICE SEARCH THE HOUSE NOW, BUT THEY WILL FIND NOTHING.

WE WERE *MOST* THOROUGH, MASTER.

IN THAT CASE, YOUR SERVICES ARE NO LONGER REQUIRED.

AND AS TO THE MATTER OF OUR *PAYMENT...?*

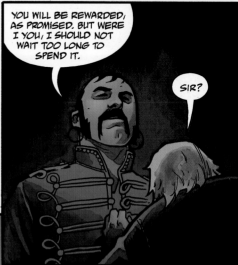

YOU WILL BE REWARDED, AS PROMISED. BUT WERE I YOU, I SHOULD NOT WAIT TOO LONG TO SPEND IT.

SIR?

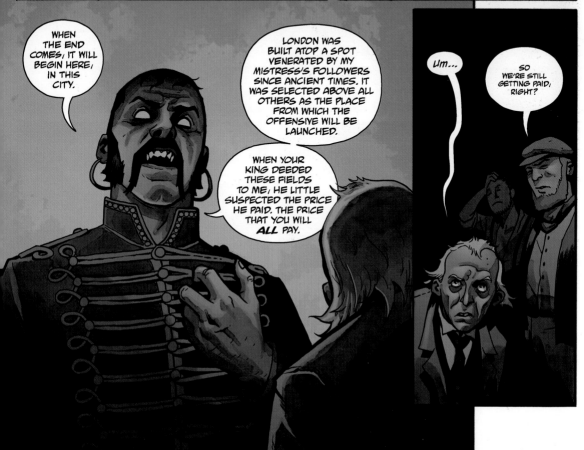

WHEN THE END COMES, IT WILL BEGIN HERE, IN THIS CITY.

LONDON WAS BUILT ATOP A SPOT VENERATED BY MY MISTRESS'S FOLLOWERS SINCE ANCIENT TIMES. IT WAS SELECTED ABOVE ALL OTHERS AS THE PLACE FROM WHICH THE OFFENSIVE WILL BE LAUNCHED.

WHEN YOUR KING DEEDED THESE FIELDS TO ME, HE LITTLE SUSPECTED THE PRICE HE PAID. THE PRICE THAT YOU WILL *ALL* PAY.

UM...

SO WE'RE STILL GETTING PAID, RIGHT?

SAINT JOHN OF THE CROSS POLICE HOSPITAL

SWEET

...AND THE SEARCH OF THE PREMISES FOUND NO USEFUL EVIDENCE, THE SERGEANT INFORMS ME.

THIS FRIEDHOF, OR *GIURESCU*, OR WHATEVER NAME HE CHOOSES, WOULD APPEAR TO HAVE GONE TO GROUND.

SO WE MUST FIND HIM BEFORE HE MAKES HIS NEXT MOVE.

I SAY AGAIN, THAT WE SEE BEFORE US PROOF OF THE EASTERN EUROPEAN LEGEND OF THE UPYR OR VAMPIR.

LIKE "VARNEY THE VAMPIRE" MADE FLESH!

ALL TOO REAL FLESH, GIVEN OUR ENCOUNTER IN THE LAMB STREET CEMETERY.

THAT'S PUTTING IT MILDLY...

AND WHAT OF THE AFRICAN WITCH DOCTOR'S WARNINGS?

THE MENTION OF RAVENS GATHERING *MUST* MEAN THE TOWER OF LONDON.

MY THOUGHTS EXACTLY, MISS GOAD.

HAVING EXHAUSTED OUR LEADS RELATING TO THE CEMETERY GROUNDS AND TO THE DEED HOLDER HIMSELF, THIS CRYPTIC HINT SEEMS OUR NEXT BEST OPTION.

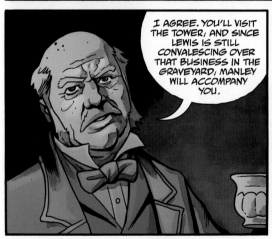

I AGREE. YOU'LL VISIT THE TOWER, AND SINCE LEWIS IS STILL CONVALESCING OVER THAT BUSINESS IN THE GRAVEYARD, MANLEY WILL ACCOMPANY YOU.

NOT ON YOUR *LIFE.* I'M A *SURGEON*, FOR PITY'S SAKE, NOT A PUPPET IN A PUNCH AND JUDY SHOW.

CALM YOURSELF, MANLEY. THERE IS NOTHING TO FEAR AT THE WHITE TOWER. WHY ELSE WOULD HER MAJESTY ENTRUST SO MANY ARCANE OBJECTS OF POWER TO ITS KEEPING?

OH, NO, THAT DOESN'T SOUND DANGEROUS AT *ALL.*

OH, IF HE WON'T GO, *I* WILL! I'VE BEEN COOPED UP INDOORS FOR *FAR* TOO LONG.

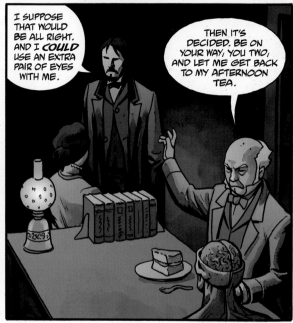

I SUPPOSE THAT WOULD BE ALL RIGHT. AND I *COULD* USE AN EXTRA PAIR OF EYES WITH ME.

THEN IT'S DECIDED. BE ON YOUR WAY, YOU TWO, AND LET ME GET BACK TO MY AFTERNOON TEA.

MYTHOLOGY IS SOMETHING OF A PASSION OF MINE, YOU KNOW.

THERE ARE LEGENDS WHICH LINK RAVENS WITH THE WHITE HILL BACK TO THE TIME OF THE ANCIENT BRITONS, EVEN BEFORE THE ROMANS, LONG BEFORE THE TOWER WAS BUILT.

THE *MABINOGION* RECOUNTS THAT THE SEVERED HEAD OF BRAN THE BLESSED WAS BURIED BENEATH THIS SPOT, AND AS I'M SURE YOU KNOW "BRAN" MEANS BOTH "KING" *AND* "RAVEN."

AND WHILE SOME INSIST THAT THE IDEA IS A MODERN INVENTION, I'VE READ FAR TOO MANY ACCOUNTS DATING BACK CENTURIES THAT MENTION A FLOCK OF RAVENS RESIDING HERE.

NOT "FLOCK," MISS GOAD. "UNKINDNESS."

SO HAS THERE BEEN ANYTHING OUT OF THE ORDINARY THAT YOU HAVE ENCOUNTERED?

ANY... *UNUSUAL* OCCURRENCES THAT MIGHT HAVE CAUGHT YOUR ATTENTION?

NO, SIR, I CAN'T SAY THAT THERE'VE BEEN ANY SUCH.

WE TAKE OUR DUTIES SERIOUSLY, SIR EDWARD. IF THERE HAD BEEN ANYTHING UNUSUAL AFOOT, WE'D HAVE SEEN IT.

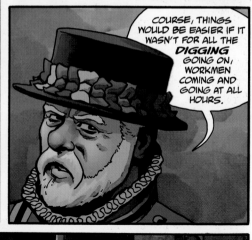

COURSE, THINGS WOULD BE EASIER IF IT WASN'T FOR ALL THE *DIGGING* GOING ON, WORKMEN COMING AND GOING AT ALL HOURS.

"DIGGING"?

THAT SOUNDS PROMISING.

...AND YOU SAY THE TWO WORKMEN SIMPLY *VANISHED?*

NO ONE HAS SEEN OR HEARD FROM THEM SINCE?

THAT'S RIGHT, SIR. HAD TO SEND HAWKINS HOME, EVEN, WHEN HIS NERVES GOT THE BEST OF 'IM.

SAID HE COULDN'T STAY UNDERGROUND A MOMENT LONGER, WHAT WITH THE FEAR THAT WHATEVER DONE FOR DAVIES AND WILKES MIGHT BE BACK TO FINISH THE JOB.

SAID HE WAS SURE SOMETHING WAS *BURIED* DOWN HERE. SOMETHING *BAD.*

THIS IS IT. WE'RE PLANNING ON BRICKING IT UP BEFORE THE LINE OPENS; BUT HAVEN'T GOTTEN 'ROUND TO IT YET.

I'LL TELL YOU, I'M NOT TOO KEEN TO COME BACK DOWN HERE, MYSELF. NOT THAT I'M SUPERSTITIOUS, MIND YOU.

SO *OLD*. TO THINK THAT ALL OF THIS LAY BENEATH OUR FEET, ALL THE WHILE, AS WE WALKED ABOVE, UNAWARE.

LONDON HAS ALWAYS BEEN BUILT ATOP THE BONES OF THE PAST, MISS GOAD. THE CITY IS CONSTRUCTED ON CLAY AND SHIFTING SOIL THAT IS CONTINUALLY SUBSIDING INTO THE EARTH.

TODAY IS CARRIED ON THE BACKS OF COUNTLESS YESTERDAYS.

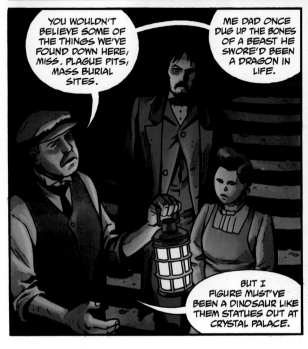

YOU WOULDN'T BELIEVE SOME OF THE THINGS WE'VE FOUND DOWN HERE, MISS. PLAGUE PITS, MASS BURIAL SITES.

ME DAD ONCE DUG UP THE BONES OF A BEAST HE SWORE'D BEEN A DRAGON IN LIFE.

BUT I FIGURE MUST'VE BEEN A DINOSAUR LIKE THEM STATUES OUT AT CRYSTAL PALACE.

ANYWAY, HERE IT IS. THE LAST PLACE ANYONE SAW WILKES AND DAVIES.

I'VE SEEN ETRUSCAN DESIGNS SIMILAR TO THIS. THERE ARE ECHOES OF GREEK AND EGYPTIAN MOTIFS, AS WELL.

AND THROUGHOUT, THE RECURRENT IMAGERY OF THE SERPENT, AND THE WOMAN. CERTAINLY ENOUGH MYTHOLOGICAL PRECEDENT FOR *THAT*.

A ROMAN MYSTERY CULT, PERHAPS, DATING BACK TO THE TIME OF LONDINIUM? OR A CELTIC TEMPLE WITH COINCIDENTALLY SIMILAR DESIGNS?

WOULD THAT I COULD COMPARE THE IMAGES TO THE AVAILABLE REFERENCE IN BETTER LIGHTING CONDITIONS.

AH, I THINK I HAVE JUST THE THING!

WAX PAPER LEFT OVER FROM A TRIP TO THE BUTCHER'S, AND A MARKING PENCIL.

NOW, MR. CAMPBELL, WAS IT?

IN THE TIME THAT ELAPSED BETWEEN THE LAST TIME YOUR WORKERS WERE SEEN AND THE MOMENT THEIR DISAPPEARANCE WAS DISCOVERED, WAS ANYTHING *HEARD* IN THE TUNNELS ABOVE?

BLESSED HARD TO HEAR MUCH OF ANYTHING UP THERE WHEN THE DIGGING WORKS ARE IN OPERATION.

BUT NO, NOTHING THAT I KNOW OF.

AN "EXTRA PAIR OF EYES," HE SAYS. MORE LIKE "AN EXTRA PAIR OF HANDS TO DO THE WORK HE DEIGNS NOT DO."

THEY CANNOT SIMPLY HAVE *VANISHED.*

EVEN IF SOMETHING SUPERNATURAL BEFELL THEM, THERE WOULD SURELY BE SOME EVIDENCE LEFT BEHIND.

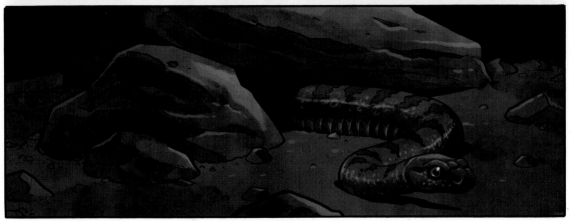

TOO OFTEN IN RECENT DAYS HAVE I HEARD PORTENTS AND OMENS ABOUT THINGS BURIED BENEATH THE EARTH.

I CANNOT BUT THINK THAT THE DISAPPEARANCE OF THESE TWO MEN IS SOMEHOW CONNECTED TO THE MATTER AT HAND.

THAT SHOULD JUST ABOUT DO IT. NOW TO--

OOOH!

SOME-THING THE MATTER, MISS GOAD?

BRING THAT LIGHT HERE, MR. CAMPBELL. QUICKLY, MAN.

OH, I WAS JUST STARTLED FOR A MOMENT.

BUT SEE? IT'S NOTHING TO BE SCARED OF.

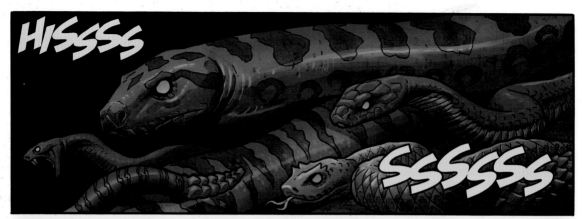

NOW WE'RE GETTING SOME-WHERE...

WHAT WAS THAT YOU SAID ABOUT THERE BEING "NOTHING TO FEAR," SIR EDWARD?

I AM MISTAKEN ON RARE OCCASION, MISS GOAD.

LET'S JUST HOPE IT'S NOT FOR THE *LAST* TIME.

BLAM

WE NEED TO GET OUT OF HERE, AND *QUICKLY.*

NEEDN'T TELL *ME* TWICE.

IT'S...IT'S *IMPOSSIBLE.* IT'S *UNNATURAL.*

THERE'S NO POINT GAWPING, MR. CAMPBELL. *RUN!*

BLAM

BLAM

SIR EDWARD, WHAT JUST *HAPPENED?*

A *HANDFUL* OF SNAKES I MIGHT ACCEPT, BUT THAT WAS A COORDINATED *ATTACK.*

OF COURSE. I CAN'T SEE HOW YOU COULD DESCRIBE IT ANY OTHER WAY.

BUT ON THE CREDIT SIDE OF THE EQUATION, IT MEANS THAT OUR INVESTIGATIONS ARE GETTING SOMEWHERE, TO PROVOKE THAT KIND OF RESPONSE.

NEVER... NEVER SEEN THE LIKE...

SO WHAT NOW?

FOR NOW, I WISH TO GET BACK TO THE SURFACE.

I'VE HAD QUITE ENOUGH OF BEING UNDER-GROUND FOR ONE DAY.

GIURESCU IS FAR MORE THAN HE SEEMS.

NOT SIMPLY A HUMAN SORCERER, NOR MERELY AN INHUMAN VAMPIRE. HE IS IN LEAGUE WITH FORCES BOTH POWERFUL AND ANCIENT.

I AM CONVINCED HE HAS GONE TO GROUND, TO CONTINUE FROM THE SHADOWS HIS PLANS TO CLAIM ENGLAND FOR HIS OWN.

THE TRICK WILL BE LOCATING HIM, SINCE THERE ARE COUNTLESS MILES OF TUNNELS THAT CRISSCROSS BENEATH LONDON, SOME DATING BACK CENTURIES OR MORE.

FORTUNATELY, WE CAN CALL UPON POWERFUL FORCES OURSELVES...

NEW SCOTLAND YARD.

...AND I SPEAK WITH THE FULL AUTHORITY OF HER MAJESTY'S GOVERNMENT.

I'M AFRAID I CANNOT BE SPECIFIC ABOUT THE CIRCUMSTANCES OF THE CASE, SIMPLY THAT IT IS A MATTER OF SUPREME IMPORTANCE.

THIS INVESTIGATION MUST TAKE PRECEDENCE OVER **ALL** OTHER POLICE MATTERS.

YOUR ROYAL WARRANT IS OFFICIAL ENOUGH, SIR--YOU WON'T HEAR **ME** ARGUING AGAINST YOUR AUTHORITY.

BUT I'M BLOODY WELL BAFFLED BY THE **CONDITIONS** OF YOUR REQUEST. YOU CANNOT SERIOUSLY MEAN FOR MY MEN TO--

MY INSTRUCTIONS MIGHT SOUND UNUSUAL, I GRANT YOU, BUT THEY **MUST** BE FOLLOWED TO THE LETTER.

I KNOW HER MAJESTY APPRECIATES YOUR COOPERATION IN THIS MATTER.

ON MY ORDERS, THE METROPOLITAN POLICE ARE MOUNTING A FULL-SCALE MANHUNT IN THE CITY'S TUNNELS.

AS I WRITE, THE POLICE SEARCH THE CITY'S SEWERS AND UNDERGROUND RAILROAD TUNNELS, EXPLORING ITS CRYPTS, BASEMENTS, AND VAULTS.

OF COURSE, THERE IS AN ENTIRE **POPULATION** OF VAGRANTS WHO MAKE THOSE SUBTERRANEAN PASSAGES THEIR HOME...

...AND CRIMINALS AND MALCONTENTS OF THE MUNDANE VARIETY WHO HAVE EVADED CAPTURE BEFORE NOW BY DESCENDING UNDERGROUND.

I UNDERSTAND THAT THE CITY'S JAIL CELLS ARE CROWDED WITH PICKPOCKETS AND CUTPURSES THAT HAVE BEEN ROUSTED IN THE COURSE OF THE SEARCH.

BUT THERE HAS BEEN NO SIGN OF GIURESCU, AFTER A WEEK OF SEARCHING.

HE **MUST** BE LOCATED, THOUGH. THE ALTERNATIVE IS UNTHINKABLE.

AND WHILE THE POLICE SEARCH BELOW GROUND, I SEEK ANSWERS ABOVE.

THE BRITISH MUSEUM.

THE SUBTERRANEAN TEMPLE IS LOST TO US NOW, BURIED UNDER A TON OF RUBBLE AND ROCK.

BUT THANKS TO THE QUICK THINKING OF MISS GOAD, I WAS ABLE TO INVESTIGATE THE STRANGE MARKINGS AND CARVINGS WE FOUND THERE.

I WAS SURE THAT I HAD SEEN THEM BEFORE...

AH, *THERE* IT IS.

SIMILAR DESIGNS ARE FOUND IN A MOSAIC RECOVERED FROM A TEMPLE TO SELENE, PRESERVED BENEATH THE ASH IN POMPEII IN THE FIRST CENTURY A.D.

SELENE WAS THE GODDESS OF THE MOON, CONFLATED IN SACRED MYSTERIES WITH THE ROMAN GODDESS **TRIVIA,** GODDESS OF THE CROSSROADS.

THERE CAN BE NO DOUBT.

HECATE.

WITH THE POLICE ALL **BELOW** GROUND, THERE IS NO ONE LEFT TO MAINTAIN ORDER **ABOVE.**

OI! THAT'S MY PURSE!

WE MUST FIND GIURESCU AND SOON, OR WE WILL HAVE MORE TO WORRY ABOUT THAN--

Oh.

SSSSSSSSILVER SSSSSSTINGS, WITCHFINDER...

...BUT IT WILL ONLY... DELAY ME... NOT **STOP** ME.

GIURESCU.

THE SILVER DOES MORE THAN **STING**, I THINK. LET US END THIS, HERE AND NOW.

YOU ≷GRN≷ HAVE BEEN A THORN IN MY SIDE LONG ENOUGH.

A THORN IN YOUR SIDE, OR A **STAKE** IN YOUR **HEART**?

LET US SEE HOW YOU FARE WITH SIX INCHES OF **STOUT ENGLISH OAK** THROUGH YOUR CHEST!

NO.

THERE ARE FORCES HERE BEYOND YOUR COMPREHENSION, LITTLE MAN.

WHATEVER YOU AND THAT WITCH GODDESS HAVE PLANNED, I **WILL** STOP YOU.

HAVE A CARE, WITCHFINDER. YOU HAVE A POWERFUL ENEMY IN ME, BUT YOU COURT DAMNATION BY TORMENTING MY MISTRESS. I WOULD SIMPLY **KILL** YOU...

...BUT **SHE** WOULD SHOW YOU THAT THERE ARE FATES **FAR** WORSE THAN DEATH--

...A **VAMPIRE** ABROAD IN THE STREETS OF LONDON, POSING AN **IMMINENT** THREAT TO THE DOMAIN. I SAW HIM WITH MY OWN **EYES** LAST NIGHT, AND **WOUNDED** HIM IN THE ENCOUNTER.

MY **SIMPLE** REQUEST WAS THAT THE METROPOLITAN POLICE SEARCH THE AREA IN WHICH I ENCOUNTERED HIM, AS HE **CANNOT** HAVE GONE FAR.

NOW YOU TELL ME THE COMMISSIONER HAS CALLED **OFF** THE HUNT AFTER **JUST ONE DAY**?!

AFRAID SO, OLD MAN. HE HAD OFFICERS BANGING ON DOORS AND SEARCHING CARRIAGE HOUSES ALL MORNING AND AFTERNOON. WHEN THEY CAME UP EMPTY, THAT WAS THE LAST STRAW.

HE INSISTS THAT TOO MANY RESOURCES HAVE BEEN DEPLETED IN YOUR "WILD-GOOSE CHASE," AS HE CALLS IT.

BEFORE TODAY, OFFICERS HAD COMPLAINED ABOUT THE BAD AIR BELOW GROUND, MANY OF THEM TAKEN ILL AFTER THE EXPOSURE TO THE MUCK AND THE GRIME.

TWO OF THEM DIDN'T EVEN BOTHER REPORTING IN AFTER THEIR SHIFT, IT SEEMS.

DIDN'T REPORT IN AT ALL? NOT EVEN TO BE ADDED TO THE SICK ROLLS?

THE COMMISSIONER THINKS THEY JUST WANDERED OFF TO A PUB ALL DAY.

WHERE HAD THE OFFICERS BEEN SEARCHING WHEN LAST THEY REPORTED IN...?

IT APPEARS THAT I NO LONGER HAVE THE RESOURCES OF THE METROPOLITAN POLICE AT MY DISPOSAL.

BUT THE DANGER IS ALL TOO REAL, AND ALL TOO IMMINENT-- THAT MUCH I KNOW FOR CERTAIN.

AND SO I MUST TAKE MATTERS INTO MY OWN HANDS.

I JUST PRAY THAT IT WILL SUFFICE.

SIR EDWARD? THE COOK WAS ASKING ABOUT YOUR PLANS FOR DINNER.

INFORM HER THAT SHE IS AT HER LIBERTY TONIGHT.

I HAVE OTHER PLANS. AND IF THEY DO NOT SUCCEED, SHE WON'T NEED TO WORRY ABOUT DINNER EVER AGAIN.

OH.

THAT SOUND AHEAD...?

RUSHING WATER?

AH.

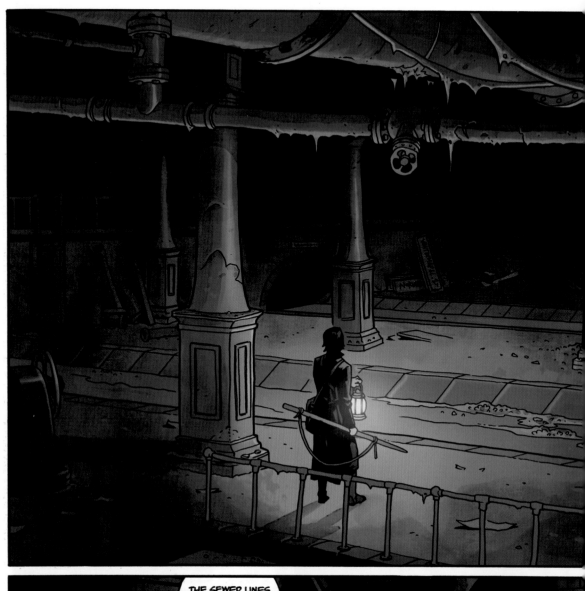

THE SEWER LINES CONVERGE, WHERE A BURIED RIVER STILL COURSES TOWARD THE SEA.

THIS SHOULD DO NICELY...

I DON'T KNOW WHAT YOU'VE BEEN SCHEMING IN THE SHADOWS--

--BUT I *PROMISE* YOU THAT I WILL STOP YOU.

HA! I ADMIRE BLUSTER, *WITCH-FINDER*--HOWEVER MISPLACED. BUT MY PLANS HAVE BEEN IN MOTION SINCE LONG BEFORE YOU WERE BORN.

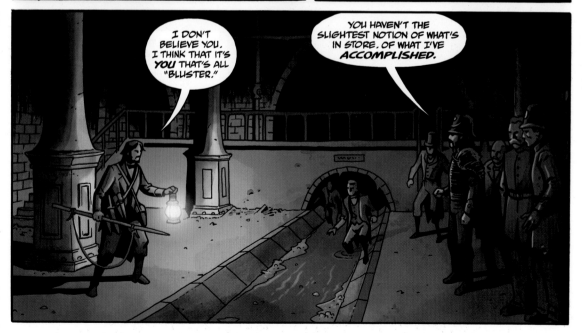

I DON'T BELIEVE YOU. I THINK THAT IT'S *YOU* THAT'S ALL "BLUSTER."

YOU HAVEN'T THE SLIGHTEST NOTION OF WHAT'S IN STORE, OF WHAT I'VE *ACCOMPLISHED.*

"MEN AND WOMEN WERE CHOSEN FROM THE RABBLE TO BE *RAISED* UP, MADE *WOLVES* AMONG THE *SHEEP.*

"THEY WERE NOT LOOSENED THEN, AND WERE INSTEAD HIDDEN AWAY, SLEEPING DREAMLESSLY BENEATH THE EARTH. BUT THE DAY APPROACHES WHEN THEY WILL BE *WOKEN* FROM THEIR SLUMBER.

"AND THEN THE AGE OF MEN WILL COME TO AN *END* AS THE WORLD IS SWEPT CLEAN WITH *FIRE AND BLOOD,* AND *DARKNESS* WILL FOREVER AFTER REIGN."

MAD KING GEORGE DID A DEAL WITH THE DEVIL A CENTURY PAST, AND ALMOST DOOMED ALL OF ENGLAND FOR IT.

SO I WAS FORCED TO DEAL WITH OTHER DEVILS TO MAKE THINGS RIGHT.

KZZZZT

FULL CHARGE, STANDING BY!

FWASSSHH

THE HELIOPIC BROTHERHOOD OF RA ARE NO FRIENDS OF MINE, BUT THEY HAVE NO MORE DESIRE TO SEE ENGLAND FALL TO A **REIGN OF BLOOD** THAN I DO.

ONE MAN OR ONE HUNDRED--IT MAKES NO DIFFERENCE! WE WILL NOT BE DEFEATED!

ATTACK, MY BROTHERS!

KZZZZZZZZZTTT AARRRRG!

DON'T LET UP! FINISH IT!

GRN!

STHUNK

THAT'S ONE DOWN! BUT THERE'S STILL MORE TO--

AIIIE!

CRUNCH

URK!

BITE THIS!

SPLOOT

MEDDLING *FOOL!* YOU ARE MERELY A *MAN.*

NOTHING YOU CAN DO WILL MATTER IN THE END! YOU HAVE ALREADY *LOST!*

I WOULDN'T BE SO SURE ABOUT THAT, IF I WERE YOU.

MY ARMY LIES BURIED AND READY!

I HAVE ONLY TO GIVE THE WORD AND--

AH, YES, BUT YOU SEE--

--YOUR ARMY ISN'T *QUITE* SO BIG AS YOU THINK, ANYMORE.

AAARR!

KILL YOU!

YOU LOSE WHAT LITTLE HUMANITY REMAINS TO YOU.

YOU'RE NO BETTER THAN AN *ANIMAL*. YOU'RE *LESS* THAN AN ANIMAL.

AND YOUR FOLLOWERS ARE *FINISHED*.

RRRAAAARRRI

BAM

UNGH!

RRRRRRRRRR!

GRN!

SHUN NKKK

URK!

UUUU GRN RRRRK

BUT...I....

FROM ALL REPORTS, IT SOUNDS AS THOUGH THE OPERATION WAS A COMPLETE SUCCESS.

SEVERAL GOOD MEN WERE LOST, BUT THEY ACCEPTED THE RISKS WILLINGLY.

IF YOU SAY SO, SWAIN.

THOUGH I'M NOT SURE THAT *ANY* OF US KNOWS QUITE WHAT WE WERE RISKING DOWN THERE IN THE DARKNESS.

WELL, THAT'S WHY THE BROTHERHOOD EXISTS, GREY. TO SHINE A LIGHT INTO DARK PLACES.

PERHAPS NOW THAT WE'RE ON BETTER FOOTING, YOU MIGHT CONSIDER JOINING OUR NUMBER? WE COULD USE A MAN OF YOUR EXPERIENCE.

DO NOT MISTAKE PRAGMATIC NECESSITY FOR A CHANGE OF HEART. THE ENEMY OF MY ENEMY IS **NOT** MY FRIEND, THOUGH WE MIGHT MOMENTARILY FIND COMMON CAUSE.

TODAY DOES NOT ABSOLVE ANY OF THE SINS YOU HAVE COMMITTED IN THE PAST, OR ANY YOU MIGHT COMMIT IN THE **FUTURE.** IN THE END, I **WILL** HOLD YOU TO ACCOUNT.

IN THE FUTURE, THE END WILL BE MERELY THE BEGINNING.

ANCIENT MONSTERS THAT YET SLUMBER BENEATH THE EARTH WILL WAKEN AND THE SUN WILL SET ON THE DAY OF MAN.

THERE WILL BE DEATH, AND FIRE, AND BLOOD.

AND THE WORKS OF MAN WILL BE WIPED AWAY. LARZOD HAS SHOWN ME.

BUT MAN'S FUTURE LIES BENEATH THE EARTH, AS WELL.

WE MUST GO **UNDER-GROUND** IF WE ARE TO SURVIVE...

ANYTHING ELSE, SIR?

NO, THANK YOU, BAILEY. THAT WILL BE ALL. THOUGH IN THE MORNING I'LL HAVE ANOTHER LETTER FOR YOU TO DELIVER TO THE QUEEN.

THERE ARE DARK DAYS AHEAD.

OH. IT'S YOU.

DARKER DAYS THAN YOU CAN IMAGINE, ENGLISHMAN.

OH, SPLENDID.

TODAY YOU PUSHED BACK THE DARKNESS, FOR A TIME.

BUT THE END CAN ONLY BE POSTPONED, NOT AVOIDED.

LISTEN, OLD MAN, YOUR COUNSEL WAS USEFUL, I'LL BE THE FIRST TO ADMIT.

BUT MUST YOUR EVERY UTTERANCE BE SO CRYPTIC? SPEAK PLAINLY FOR ONCE.

WHY DID YOU INSIST THAT I WAS THE ONLY ONE CAPABLE OF STOPPING GIURESCU?

IT ISN'T A MATTER OF BEING CAPABLE.

I SAID THAT *YOU* MUST BE THE ONE TO PREVENT THE GREAT EVIL FROM AWAKENING.

THERE WAS A LESSON FOR YOU TO LEARN. ONE THAT YOU WILL *NEED* IN THE COMING DAYS.

TOK

UHN?

THE GODDESS SLEEPS, BUT STIRS IN HER SLUMBER.

WHEN SHE WAKES, DARKNESS REIGNS.

THE END

WITCHFINDER™

SKETCHBOOK

Notes by Julián Totino-Tedesco and Ben Stenbeck

Julián Totino-Tedesco: One of the things that I love the most about doing covers for *Witchfinder* is that I get the chance to work with more somber moods than what I'm used to when doing superhero covers and such. If it were just for me, I'd go crazy dark, but it's still a comic book cover, and it's got to be catchy (And I say this in a good way, of course!).

The cover sketch on the next page was my favorite. It has a kind of Victorian vibe that I really liked, but I had to admit that the first sketch was more suitable as a cover. I would later take a more minimalistic approach for the rest of the covers.

Second page following: Pencils for the issue #1 cover.

I was sad that the second sketch for issue #2 (*facing, bottom*) wasn't chosen. I really liked that one. As I said, I opted for very simple and direct compositions for the covers. (I have a soft spot for minimalism, but I'm always concerned that people might think I'm lazy.) The cover for issue #4 (*sketches above*) turned out to be one of my favorite covers ever. I'm rarely happy with a cover, but I loved that one.

Sketches for the issue #5 cover.

Ben Stenbeck: This was the cover (*facing*) for the short story "Beware the Ape" (collected in *Witchfinder* Vol. 3). This is one of my favorite covers I've done. And of course Dave Stewart's colors really make it work.

When I first started out I was going to be a "grey-tone comic-art guy," but people told me I'd never get work because it takes too long. So I gave up on that. Then when I was asked to draw this story, I thought, Well, if they let Tyler use grey tone for the previous volume, I'm just gonna go ahead and do it too. I was really out of practice working with grey tone—I think I only got it to start working somewhere around issue #4. But it was fun to do. I did experiments with watercolor and special paper, but in the end opted for a digital watercolor brush. That was very fast, only took about an hour or two for each page. (*See following pages.*)

HELLBOY

by MIKE MIGNOLA